Music and Space

natur, wissenschaft und die künste
nature, science et les arts
nature, science and the arts

Volume 7

PETER LANG
Bern · Berlin · Bruxelles · Frankfurt am Main · New York · Oxford · Wien

Dorothea Baumann

Music and Space

A systematic and historical investigation
into the impact of architectural acoustics
on performance practice
followed by a study of Handel's Messiah

PETER LANG
Bern · Berlin · Bruxelles · Frankfurt am Main · New York · Oxford · Wien

Bibliographic information published by Die Deutsche Nationalbibliothek
Die Deutsche Nationalbibliothek lists this publication in the Deutsche Nationalbibliografie; detailed bibliographic data is available on the Internet at ‹http://dnb.d-nb.de›.

British Library and Library of Congress Cataloguing-in-Publication Data:
A catalogue record for this book is available from *The British Library,* Great Britain

Library of Congress Cataloging-in-Publication Data

Baumann, Dorothea.
Music and space : a systematic and historical investigation into the impact of architectural acoustics on performance practice followed by a study of Handel's Messiah / Dorothea Baumann.
p. cm. -- (Natur, Wissenschaft und die Künste Nature = Science et les arts nature = Science and the arts ; v. 7)
Includes bibliographical references.
ISBN 978-3-03-430615-7
1. Music--Acoustics and physics. 2. Music-halls. 3. Music--Performance--History. 4. Music and architecture. 5. Handel, George Frideric, 1685-1759. Messiah. I. Title.
ML3805.B315 2011
781.2--dc23
2011041542

The editors and publisher gratefully acknowledge the permission granted to reproduce the copyright material in this book. Every effort has been made to trace copyright holders and to obtain their permission for the use of copyright material. The publisher apologizes for any errors or omissions and would be grateful if notified of any corrections that should be incorporated in future reprints or editions of this book.

This study was accepted in spring 2000 by the Faculty of Arts of the University Zurich for the conferral of the *venia legendi* in musicology.

Cover illustration: Oxford, Sheldonian Theatre, University Commemoration (1781), ink wash on paper by Samuel Hieronymous Grimm, © The British Library Board

Cover design: Thomas Jaberg, Peter Lang AG

ISBN 978-3-0343-0615-7

Contents

PART 2: HANDEL'S *MESSIAH*

PART 3: APPENDIX

List of figures

List of music examples

List of tables

Tables in the appendix

Abbreviations

a	absorption
A. D.	Anno domini
AIM	American Institute of Musicology
Aml	Acta Musicologica
AMZ	Allgemeine Musikalische Zeitung
B. C.	before Christ
bc	basso continuo
bn	basson
cf.	cited from
cl	clarinet
dB	decibel
db	double bass
ed.	edited, edition, editor(s)
EDT	Early Decay Time
enl.	enlarged
EM	Early Music
EMH	Early Music History
et al.	*et alii,* and others
f.	folio
f.	following
facs.	facsimile
fl	flute
Fs.	Festschrift
GdM	Gesellschaft der Musikfreunde Wien
GS	Gesammelte Schriften
GW	Gesammelte Werke
HG	Händel Gesellschaft; Ausgabe der Deutschen Händel Gesellschaft
HHA	Hallische Händel-Ausgabe
HHdb.	Händel Handbuch
HJb.	Händel-Jahrbuch
hn	horn
HWV	Händel Werkverzeichnis
Hz	Herz
ibid.	*ibidem,* cited work, same page
instr.	instruments
IRASM	International Review of the Aesthetics and Sociology of Music
IRT	Initial Reverberation Time
ISO	International Standard Organisation
JASA	Journal of the Acoustical Society of America
JAES	Journal of the Audio Engineering Society
Jb.	Jahrbuch
JAMS	Journal of the American Musicological Society
NMA	Neue Mozart Ausgabe
m	meter
m3	cubic meter
MGG	Die Musik in Geschichte und Gegenwart
MJb.	Mozart Jahrbuch
ML	Music and Letters
mm	millimeter
ms	millisecond
ms.	manuscript
mss.	manuscripts
MQ	Musical Quarterly
MT	Musical Times
n.d.	no date
n.p.	no place
NG	New Grove Dictionary of Music and Musicians
ob	oboe
ÖMZ	Österreichische Musikzeitschrift
op. cit.	cited work
or.	orchestra
orig.	original
P.	Palais, Palazzo
perf.	performance
PRMA	Proceedings of the Royal Music Association
publ.	publication, published
r. p. m.	rotations per minute
repr.	reprint
rev.	revised
RAA	Répertoire d'Art et Archéologie

RIBA	Royal Institute of British Architects
RILM	Répertoire International de Littérature Musicale
RIM	Rivista Italiana di Musicologia
RIPM	Répertoire International de Presse Musicale
RISM	Répertoire International des Sources Musicales
RMA	Royal Music Association
RTM	Rundfunktechnische Mitteilungen
s	second(s)
s.a.	sine anno
SIMG	Sammelbände der Internationalen Musikgesellschaft
s.l.	sine loco
SMG	Schweizerische Musikforschende Gesellschaft
SS	Sämtliche Schriften
suppl.	supplement
T	Reverberation Time
tpt	trumpet
transl.	translated, translation
trbn	trombone
UP	University Press
v	violin
v.	voices
va	viola
vc	violoncello
vn	violin
vol.	volume
Zs.	Zeitschrift

Preface

The idea for this long term study on architectural acoustics in relation to music and music history arose from the author's experience as a performing musician, during sound recording and while teaching music acoustics at the University of Zurich, other universities and on the Swiss Radio and Television's education program.

The revival of Early Music since the 1950s, and even more so since the 1980s, paid special attention to the reconstruction of musical instruments, playing technique, and scores. Astonishingly, this interest very rarely included investigations of the original space of performance and even of subsequent acoustical conditions, although the impact of room acoustics on performance practice is generally acknowledged. Among others, this study explores methods that describe and qualify the acoustics of rooms, whether extant or not, in relation to the music performed.

The book consists of two parts. Following the methodology, the first part introduces the basic theoretical concepts of the related historical and exact disciplines: theoretical and applied acoustics, the history of acoustics, music history and music perception, sound recording technique and its historical development, architectural acoustics and the definition of quality factors, and, finally, the relation between room acoustics and the different types of music and performance situations.

The methods developed to describe and qualify the acoustics of historical rooms (including those no longer extant) in relation to the performed music rely on an extensive database of collected pictures, plans, dimensions of halls, theaters and churches of historical importance for music performance, as well as literary documentation related to these spaces. Where not yet published, dimensions of spaces were extracted from plans and, where necessary and possible, measured on site. Information on all elements relevant to room acoustics was added whenever possible. From this database several groups of spaces were selected and classified by certain criteria in order to present the whole spectrum of architectural space used for musical performance and to define characteristic types of architectural space in relation to music history, music genre, and performance situation.

The second part traces the performance history of one musical work from its first performance to the early 20th century. Handel's oratorio *Messiah* was selected with its uninterrupted performance tradition in Great Britain, which brought this famous work from the chamber music hall in Dublin where it was first performed in 1742, small baroque theaters, and the chapel of London's *Foundling Hospital,* after Handel's death from the choir of large cathedrals to "amphitheatrical" stages in the nave of these cathedrals, to the new and large 19th-century concert halls, and finally to the immense *Crystal Palace* in Sydenham and the *Royal Albert Hall* in London. Within this period the number of performers increased from roughly two dozen to nearly four thousand, while the audience grew from a few hundred to a maximum of nearly ninety thousand.

This book, furthermore, explores a paradox: how can a musical work that was written specifically for a certain architectural space "survive" such dramatic changes in performance conditions? Are there boundaries determining an adequate performance? How can we define the quality of room acoustics and how does this quality affect the performance as well as the presentation of a musical work? In short, how do different acoustical conditions affect basic aesthetic premises? There are no simple answers to these complex questions, which elicit different responses according to varying points of view. This aspect of cultural history necessarily calls for an investigation based on systematic, historical, and psychological methods.

Acknowledgments

The German version of this book was accepted by the Philosophical Faculty of the University of Zurich as Habilitation in spring 2000. The author owes gratitude to many, especially to the Musicology Department of the University of Zurich and their directors Ernst Lichtenhahn and Max Lütolf, the Swiss Foundation Pro Helvetia und Barry S. Brook †, director of the Musicology Department at the Graduate Center of the City University of New York for the invitation to the Swiss Lectureship in 1987, Tilman Seebass for the invitation to a guest lectureship at the University of Innsbruck in 1998, the Department of Architecture at the Swiss Federal Institute of Technology ETH Zurich for the invitation to seminars on room acoustics in 1992 and 2000 and 2007. Thanks go to many colleagues, students, listeners and friends for their questions, exchange of personal room acoustic experiences, observations and hints to sound recordings, documents and publications, as well as to many libraries and collections, mainly of the Institutes of Musicology and Art History of the University of Zurich, the Music Department of the Zurich University of the Arts, the Zentralbibliothek Zürich, the ETH libraries, the Theatersammlung Berne, the Eidgenössisches Archiv für Denkmalpflege EAD Berne (legacy Ernst Schiess), the Avery Architectural Library (Columbia University), New York, the Burghauptmannschaft, Vienna, the Fondazione Cini, Venice, the Prefettura di Roma, Ufficio Economato (Ceriello Cino). My personal thanks go to Fritz

Winckel † (Berlin), Jürg Jecklin (Universität für Musik und darstellende Kunst, Vienna, earlier Swiss Radio, Studio Basel), Kurt Eggenschwiler (Swiss Federal Laboratories for Materials, Science and Technology EMPA, Dübendorf), Russell Johnson † (ARTEC, New York), Jürgen Meyer (Technische Bundesanstalt, Braunschweig), Deborah Howard and Laura Moretti (Cambridge and Venice), Isabel Rucki (Zürich), Beate Schnitter (Küsnacht), and Ellen Taller (Küsnacht). Special thanks go to Thomas Baer-Loy † (Dübendorf) and Christina Niederstätter (architect Bolzano, Italy) for geometrical analyses, Beatrice Smedley (Tel Aviv) for the control of the English text, Elena Abramov-van Rijk (Jerusalem) for the layout and her help with the index, and Rosmarie Niggli (Männedorf) for her help with corrections.

Part 1

Music performance and architectural space: Theoretical and historical fundamentals

1.1 Introduction

1.1.1 Room acoustics and related disciplines

With the establishment of academic disciplines in the nineteenth century room acoustics became part of several scientific and humanistic fields, such as physical acoustics, medicine, psychology, sociology, history of architecture, musicology and others. This is one of the reasons why no comprehensive history of acoustics nor a history of room acoustics has been written thus far. Furthermore, the difficulty of measuring acoustic phenomena delayed the development of acoustics as part of physics and related exact disciplines. Despite the progress during the past hundred and mainly during the past thirty years, important questions in room acoustics still remain open. Nevertheless, the scientific knowledge available in all the involved disciplines provides a solid ground for a basic survey of the history of room acoustics and current theoretical concepts. The first part of this study will present such a survey from the different points of view of the main disciplines relevant to this research.[1]

Information on the history of room acoustics had to be compiled mostly from special studies, which are listed in the bibliography at the end of this book. Only very few larger historical surveys have been published so far, such as Frederick H. Hunt's *Origins in Acoustics,* printed posthumously in 1978 from an unfinished manuscript, and Robert B. Lindsay's history of acoustics, written as an introduction to the 1945 reprint of the 1877 edition of Lord Rayleigh's *Theory of Sound.*

The following publications were important to theoretical and applied room acoustics: *Planning for Good Acoustics* (1931) by the architect Hope Bagenal and the physicist Alexander Wood, which features plans of historical buildings and geometrical analyses that address also the point of view of musicians and listeners; Leo L. Beranek's *Music, Acoustics and Architecture* (1962) and *Concert and Opera Halls: How They Sound* (1997), both with an invaluable collection of plans, documents and acoustic analyses, though later research has rendered some assessments in the first book obsolete; Lothar Cremer's *Principles and Applications of Room Acoustics: Geometrical, Statistical and Psychological Room Acoustics,* published with Helmut A. Müller and Theodore J. Schultz (1982), based on Cremer's former German edition *Die wissenschaftlichen*

1 For more on the role of the various disciplines and the structure of this study, see 1.1.4.

Grundlagen der Raumakustik (1976–1978); *Bau- und Raumakustik* (1987) by Fasold, Sonntag and Winkler, a handbook on applied acoustics and engineering based on experience gained during the reconstruction of such famous historical buildings as Semper's Opera of Dresden destroyed during the Second World War.

The following publications pay special attention to the musicians' point of view: Fritz Winckel's *Music, Sound and Sensation* (1967), first published in German as *Phänomene des musikalischen Hörens* (1960);[2] Juan G. Roederer's *Introduction to the Physics and Psychophysics of Music* (1973); the two volumes on *Acoustique musicale* (1977, last reprint 1984) and *La machine à écouter* (1977) by the Paris physicist Emile Leipp; several publications by the German engineer Jürgen Meyer, mainly his *Acoustics and Performance of Music* (1972, 5th edition 2009) and *Kirchenakustik* (2003), which apply current theoretical knowledge and practice to music performance.

Basic introductions to psychoacoustics are provided by Jens Blauert in *Spatial Hearing, the Psychophysics of Human Sound Localization* (revised edition 1996), Eberhard Zwicker *Psychoacoustics, Facts and Models* (1999) and Stephen Handel *Listening, an Introduction to the Perception of Auditory Events* (1989).

In the cognitive neurosciences, which have developed rapidly in the last 30 years, the following publications were used, as they offer clear explanations of complex facts: *The Self and Its Brain* (1977) by John Eccles and Karl R. Popper; Richard F. Thompson's *The Brain* (1993); Manfred Spitzer's *Musik im Kopf: Hören, Musizieren, Verstehen und Erleben im neuronalen Netzwerk* (2002) and Lutz Jäncke's *Macht Musik schlau? Neue Erkenntnisse aus den Neurowissenschaften und der kognitiven Psychologie* (2008), a survey and bibliography on recent research on music in neurology.

Several publications on the history of rooms for music present the view of various disciplines. Michael Forsyth's *Buildings for Music* (1995, translated into German and French, unfortunately all out of print) is one of the best in the field of architectural history with a plethora of illustrations and a chronological survey of churches, opera houses and concert halls. Recent research has widely confirmed Forsyth's often independent acoustic assessments of spaces. Further important information is contained in publications on the history of theater construction, such as Manfred Semper's volume on theater building in the series *Handbuch der Architektur* (IV/6/5, 1904), Hammitzsch's dissertation *Der moderne Theaterbau* (1906) and Richard and Helen Leacroft's *Theatre and Playhouse: an Illustrated Survey of Theatre Building from Ancient Greece to*

2 A professional singer and engineer, Fritz Winckel (1907–2000) started to develop criteria for the assessment of room acoustics after the Second World War, during the reconstruction of so many famous destroyed halls and theaters.

the Present Day (1984). Information on the history of concert halls is featured in Hans-Ulrich Glogau's *Der Konzertsaal: Zur Struktur alter und neuer Konzerthäuser* (1989) and in several special studies, such as Heinrich Habel's *Das Odeon in München* (1967), with an important second part on early concert hall building; the excellent documentation on the *Gewandhaus Leipzig* by Rudolf Skoda (1984), which includes a historical survey on famous concert halls of the eighteenth and nineteenth century; and the two publications by Deborah Howard and Laura Moretti on architecture and music in Renaissance Venice published 2006 and 2009, with detailed analyses of several churches based on extensive tests and sound recordings in different churches, from small parish and *ospedali* churches to *San Marco*, *San Giorgio Maggiore* and *Il Redentore*.

The following bibliographic tools of architectural history were helpful: *Répertoire International de la Littérature d'Art (RILA), Répertoire d'Art et d'Archéologie (RAA), Bibliography of the History of Art (BHA,* since 1991), Edward H. Teague's *World Architecture Index* (1991, with an index of plans and pictures) and Terence Russel's *The Built Environment* (1989). Nikolaus Pevsner's *History of Building Types* (1976) and *The Penguin Dictionary of Architecture and Landscape Architecture* edited with Hugh Honour and John Fleming (2000) were consulted in order to define building types and to clarify terminological questions.

Many historical documents and pictures of rooms are published in Heinrich W. Schwab's *Das Konzert* in the series *Musikgeschichte in Bildern* (1971) and in Walter Salmen's book with the same title (1988). Howard M. Brown and Stanley Sadie present a wealth of information on the history of performance practice in the two volumes *Performance Practice* (1989) published in the series of the *New Grove Handbooks of Music.* Daniel J. Koury's dissertation *Orchestral Performance Practices in the Nineteenth Century: Size, Proportions, and Seating* (1986) offers a basic survey on this special field, as does Ottmar Schreiber's dissertation *Orchester und Orchesterpraxis in Deutschland zwischen 1780 und 1850* (1938), whose systematic evaluation of 30 music periodicals of the time underpins a survey on the development of the orchestra and orchestra seating in German-speaking countries. In *Musik und Raum: Gesellschaftliche und ästhetische Perspektiven zur Situation um 1800* (1989) Ernst Lichtenhahn presents important, thus far unknown, comments on room acoustics of the eighteenth and nineteenth century culled from writings on the sociology of music and aesthetics. More documents were found thanks to the indexes of nineteenth-century music periodicals indexed in *Répertoire International de la Presse Musicale (RIPM)* edited by Robert Cohen since 1980 (recent title: *Retrospective Music to Music Periodicals).* As a preliminary stage of this study, the author herself published several papers on acoustics and performance practice and their relation to room acoustics, sound recording and the building of musical instruments, theaters, concert halls and churches. These papers are listed in the bibliography at the end of this book.[3]

1.1.2 Architectural space, room acoustics and music performance

An analysis of the relation between architectural space, room acoustics and music performance requires, first of all, a survey on the history of buildings and their architectural change over time. This information must be broad yet detailed enough to allow general assessments of the acoustic situation of a given performance at a given time. The descriptions and comments presented in this study draw on various sources, such as plans, designs, paintings, reports on buildings, halls and events, as well as material for musical performance, lists of musicians' payments, public reports, private letters from and diaries of listeners, performers, composers and critics. The available documents and publications contain more or less detailed information of variable reliability and significance for the event.

Since historical sources are often lacking or have been lost, it is not always possible to gather complete architectural documentation on a specific hall, theater or church. For example, it is especially problematic to obtain full documentation on an opera house covering a longer period of time, as these buildings were often renovated, destroyed by fire, rebuilt or adapted to changing conditions.

Regarding music performance, information on musical instruments, orchestra size or such details as seating plans of musicians has rarely been preserved and, if extant, is often unreliable. Nevertheless, certain conditions permit a comparison between one performance and other, better documented performances. Such comparisons warrant conclusions about the expected or presumable number of instruments, the expected seating plan or the probable number of listeners.

Architectural spaces can be grouped by building type as theaters, halls and churches, and then classified according to shape, volume and other relevant criteria that enable comparisons between spaces of the same class or type used at the same time or even at different times. The information for this study had to

3 D. Baumann, "Performance Practice and Architectural Acoustics: Bibliographic Sources in Related Disciplines" (1991).

be collected systematically, with such comparisons in mind, and its amount and quality had to be defined.[4]

The documents were classified according to the following main categories:

1. Documents on architectural space: shape, dimension, interior decoration,
   a) chronologically, by building type (church, theater, hall);
   b) by musical centers, grouped by building type.
2. Documents on music performance:
   size of orchestra, orchestra seating, number of attendants,
   a) by musical genre;
   b) by city / musical centre;
   c) by building type.
3. Documents on musical composition in relation to architectural space,
   a) by social function of the event;
   b) by musical genre;
   c) by composer.
4. Documents on music performance practice:
   performance material (scores, parts and annotations), instrumentation, instrument construction and playing techniques, etc.
5. Documents on the reception of musical works:
   reports by composers, musicians, listeners and critics.

Each group is related to respective theoretical writings:

1. Theory of architecture;
2. Theory of orchestration and conducting;
3. Theory of composition;
4. Theory of instrumentation;
5. Music aesthetics, sociology of music, psychology of music.

An initial overview of documents shows multiple overlapping among data from different groups and partial completion of content among the groups and subgroups. Furthermore, this classification permits generalization of information only under certain conditions. For example, data from group 2 on performance practice provide a general chronological survey on the orchestration and specific number of instruments used in musical performances from the seventeenth to the twentieth century. This information grid can serve as a background for the interpretation of a certain event or its comparison with a similar event, such as another performance in the same hall or a performance in a hall of similar shape and size, but also for a comparison between performances in the same

4 See 1.1.4: Methodology.

category of architectural space. Across this grid different subgroups, such as documents on a specific musical work related to a certain building type, can be regrouped by the event's social function in order to gain a broader view on the social history of a music genre (subgroups 3a and b). Only if information is embedded in this complex network of relations can one avert the risk of incomplete reconstructions or speculations that would inevitably lead to a distorted view of the historical situation.

1.1.3 Music for rooms and rooms for music: two points of view and three levels of approach

Given such complex subject matter, how can one select and group the musical and music-related architectural aspects of room acoustics to develop clear definitions and allow for significant comparisons between related elements? Two central points of view command attention: one relies on the musical composition, the other is underpinned by the architectural space used for the performance of certain musical works. A brief examination is already sufficient to show that these two different approaches affect the interpretation of facts.

Let us take a well-known example: polychoral compositions, a genre that emerged in the fifteenth century with music for two small choirs and by the seventeenth century had developed into a complex musical event with several ensembles placed on different balconies, filling churches in Northern Italy and elsewhere with impressive, splendid sound. As Iain Fenlon said on the adoption of the *cori spezzati* style at *San Marco* in Venice, "any reconstruction of music and liturgical practice inside *San Marco* during the sixteenth and seventeenth centuries must take account of the three musical elements: chant, improvised polyphony, and composed polyphony. Throughout the period these three co-existed. […] While chant and improvised polyphony were sung by the canons of the Basilica, polyphony was performed by the professional singers of the Cappella Marciana. – Those charged with performing these different strands of the musical component of the liturgical rituals inside the Basilica were located in different places at different times."[5] Laura Moretti's research and the documents presented by Iain Fenlon reveal the following main positions for musicians in *San Marco* (figure 1 and 2):[6] a) the *pulpitum magnum cantorum* or *bigonzo* (dating from the first half of the thirteenth century), an octagonal

5 I. Fenlon, "The Performance of *cori spezzati* in San Marco" (2006), 94–5.

6 L. Moretti, "Architectural Spaces for Music: Jacopo Sansovino and Adrian Willaert at St Mark's" (2004).

structure located in the crossing in front of the iconostasis on the right-hand side; b) the two *pergoli* or balconies just behind the iconostasis within the main pier on either side of the choir, at about 2.10 meters above the floor level;[7] c) different positions on the upper level, mainly the two organ lofts on either side of the choir at 5.60 meters above floor level; d) the usual position for liturgy held at the floor level, with singers standing or sitting on benches in the choir.[8]

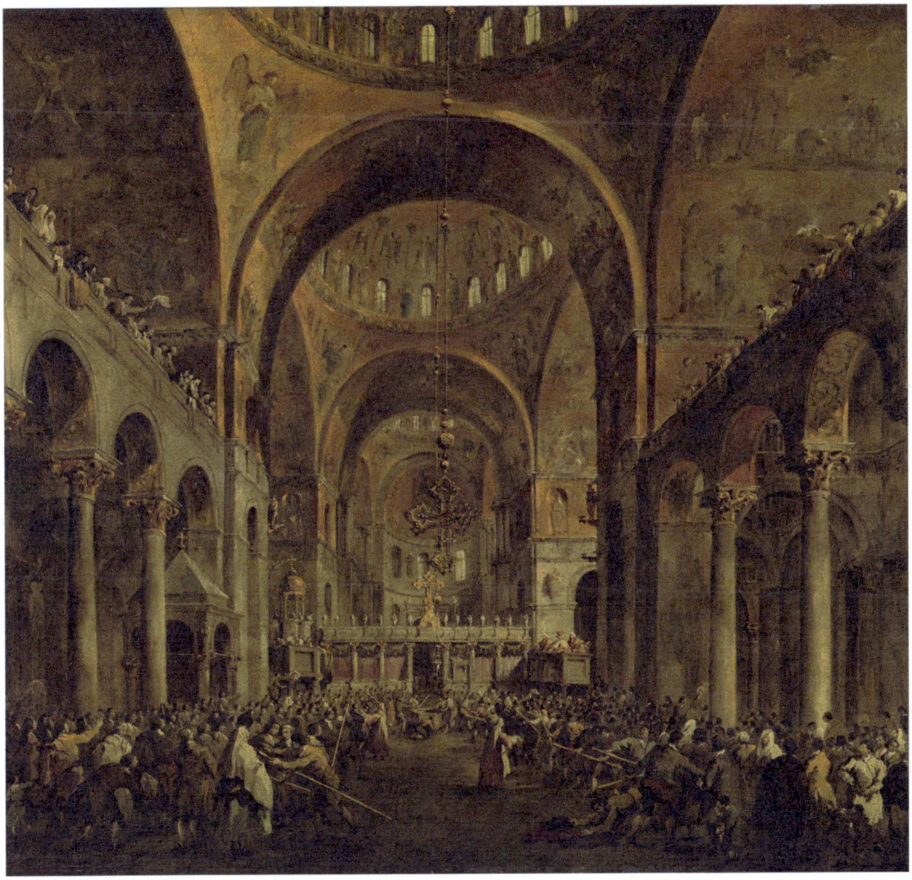

Figure 1: Venice, *San Marco,* view from the main nave toward the iconostasis with the presbyterium behind and the ocatagonal *bigonzo* on the right-hand side: the presentation of doge Alvise IV Mocenigo (Francesco Guardi, ca. 1763).

7 Measures taken by the author, who would like to thank Laura Moretti and Patrizia Lerco for their help and Monsignor Antonio Meneguolo for permission to enter the *presbyterium* of the basilica. For a detailed acoustic analysis, see D. Baumann: "Geometrical Analysis of Acoustical Conditions in San Marco and San Giorgio Maggiore in Venice" (2006).

8 See Moretti (2004), figures 2–7; Fenlon (2006), figures 4–8.

Singers and musicians in *San Marco,* the private chapel of the doge and the chapel of the Republic of Venice had to respect liturgical and ceremonial functions. Acoustic considerations were but one further element. As Iain Fenlon wrote, "the principal 'audience' for the civic and religious rituals which took place in *San Marco* consisted of the doge and senate who occupied [...] the space in front of the High Altar and within the central sacral area. [...] This was to create a private chapel within the context of the broader uses of the building as a whole. [...] Members of the general public [...] were confined to the areas west of the iconostasis."

Musicians knew that the elevated positions in the crossing and in the choir helped cope with acoustic difficulties in this complex space with its five domes and large pillars separating the main nave, side naves and transept. This was certainly one of the reasons why the medieval singer balcony on the south side behind the iconostasis was raised higher in 1536–37 and why in 1541–44, just at the time Adrian Willaert wrote his first double-choir compositions, a similar balcony had been built on the north side.[9] The two *pergoli* stand at a distance of about 10 meters, directly under the large arch connecting the two piers, which is about 17.30 meters above the singers' heads. The *pergoli* provide excellent acoustic conditions for the interplay of two four-voice choirs with one or two singers in each voice and for listeners within the central sacred area, especially for the doge, who sat in the choir, his back to the iconostasis, on a new throne made by Jacopo Sansovino.

Listeners outside the choir could hear the music as though it were coming from another space, not only because the iconostasis was closed by carpets during ceremonies, but also because sound coming from the *pergoli* reached the areas west of the iconostasis only through wave reflections directed upwards to the side walls and into the cupola (see figure 2). Astonishingly we read that around the 1560s psalms during vespers and most masses in *coro spezzato* technique were sung by the two choirs crowded into the octagonal *bigonzo* or *pulpitum magnum cantorum* (position a) located in the crossing in front of the iconostasis on the right-hand side, or, if it was occupied by the doge, in the *pulpitum novum lectionum* on the left hand side, "even if they had very little space there."[10] If singers faced the main nave, the sound projected directly to the public. If they faced towards the pier's west front, as shown in a

9 Moretti (2004); Fenlon (2006); D. Howard and L. Moretti, *Sound and Space in Renaissance Venice: Architecture, Music, Acoustics* (2009), 17–42, and 243–5.

10 In the *Ceremoniale* of 1564 we read: "In Vigilia vero Ascensionis cantore [...] cantant divisi in duobus choris alternatim. Sua Serenitas ascendit pulpitum magnum et ibi audit vesperas [...] cantores cantant in pulpito novo lectionum, licet anguste maneant in eo." Cf. D. Bryant, "The 'cori spezzati' of St Mark's: Myth and Reality" (1981), 172, footnote 27.

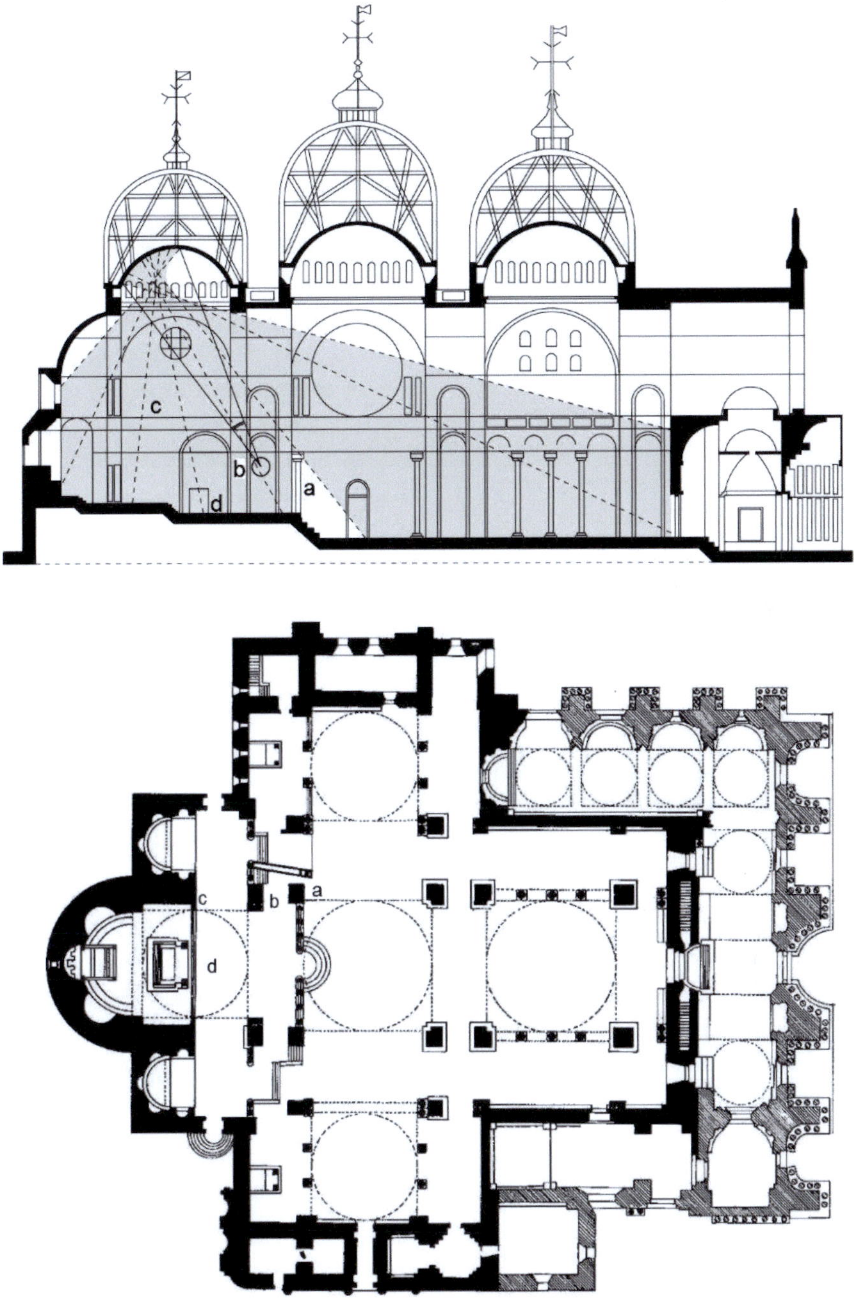

Figure 2: Venice, San Marco, section showing sound distribution from position b and ground plan.

drawing by Canaletto from 1766,[11] sound was reflected backwards to the main nave but also reached the choir, passing through the uppermost, open part of the iconostasis.

Larger-scale polychoral pieces, such as those by Andrea and Giovanni Gabrieli, many of which are not strictly liturgical and were probably written for special occasions, required special performance arrangements.[12] For such solemn feast days singers and musicians could be placed in the two organ lofts on either side of the *presbyterium* at 5.60 meters above floor level with a distance of about 12.50 meters between them (position c). This emplacement was probably also less frequent, as no stairs led directly from the church to the organ lofts.[13] Once up in the lofts, musicians and singers could not serve other functions down in the church. In case of two or more organs and groups of wind players, musicians were also placed in special balconies aside and below the organ lofts on both sides of the *presbyterium.* Sometimes musicians and singers were placed also on special *pulpiti* in the *presbyterium.* Canons always remained on floor level for liturgical and ceremonial tasks. The choir space functioned as a separate acoustic space connected to the rest of the church only via the dome and the vaulted ceiling above the apse (position d).

In all positions higher than the iconostasis musicians had to contend with a fuller acoustic reaction from the main volume of the church, especially at higher dynamic levels and with low bass instruments.

These examples from *San Marco* in Venice, which have been analyzed more deeply in other publications,[14] show that each point of view, the musical and the architectural, runs on three significant levels.

From the musical perspective, the relation between music and performance practice operates on the following levels:

1. the performance of a *specific piece of music* (for instance, a polychoral *symphonia sacra* by Giovanni Gabrieli);
2. the performance practice of a *specific musical genre* that demands certain acoustic conditions, as does polychoral church music in general which, for acoustic reasons, needs a full choir with bass for each group, if the choirs are placed apart;[15]

11 Baumann (2006), figure 8.

12 Fenlon (2006), 95.

13 Bryant (1981), 165–86.

14 See Moretti (2004); Fenlon (2006); Howard and Moretti (2009).

15 G. Zarlino, *Le istitutioni harmoniche,* Venezia 1558, III, cap. 66; see V. Ravizza, “Musikalischer Satz und räumliche Disposition – Zur frühen venezianischen Doppelchörigkeit” (1994), 177 and footnote 3; Bryant (1981), 167, footnote 7; Fenlon (2006), 89, footnote 23.

3. the most general level, that is, *church music in general* with its peculiarities related to the liturgical function, which in itself may have developed partly as a result of acoustic conditions of performance.

In acoustics-related architectural elements the following levels must be distinguished:

1. the analysis of a *specific space,* its specific shape, special decoration, arrangement of platforms and seats for a specific event and number of attendants at the moment of a certain performance, as, in this case, the church of *San Marco* in Venice with its organ lofts and special *palchetti;*
2. the *type of architectural space* with its typical construction properties according to style and function, as, in this case, a Byzantine-style church in the form of a Greek cross with side naves and several domes;
3. the *category of space,* that is, the *church in general,* which elicits in the listener the expectation of a long reverberation, regardless of its dimensions, construction and acoustics. The church then becomes an acoustic *topos* closely related to the symbol of the dome as *domus dei* (house of God),[16] even though churches of different shapes and dimensions, with various ceilings (flat and vaulted), ground plans and, therefore, very different acoustic properties, have been built throughout history.

1.1.4 Methodology: the study's two parts

The main subject of this musicological research is music as a sonorous art, the substance of music in its actual sounding presentation and perception. Its primary aim is to elucidate the spatial-acoustic properties of music and their relevance to performance practice.

Since the Middle Ages the volume of documents related to music, space and room acoustics has steadily grown in all fields of research. The relations among the disciplines involved are complex, as each discipline has its own terminology, often not easily accessible to non-specialists, but also because methods used in history, science and cultural studies must be applied. Whereas the exact or natural sciences (as in the present study physical room acoustics and the physiology of hearing) seek laws and principles, historical disciplines (such as the history of architecture, of music, of its performance and reception) ad-

16 E. Baldwin Smith, *The Dome* (1971), 4.

dress phenomena subject to change and investigate them using conventions of style and assumptions that shift over time.[17]

According to the Neo-Kantian Wilhelm Windelband (1848–1915) historical disciplines use the so-called *idiographic* procedure, which describes phenomena created by human thought and consciousness, whereas the exact or natural sciences proceed *nomothetically,* that is, by defining laws and principles. Natural sciences, which teach "what has always been", seek laws, whereas human sciences, which describe "what happened in former times", seek forms or "Gestalten". Still, as Windelband remarked in *History and Science,*[18] the same topics can be studied both nomothetically and idiographically.

Another Neo-Kantian, Wilhelm Dilthey (1833–1911), mainly in *Introduction to the Human Sciences* (1883), *Ideas for a descriptive and analytic psychology* (1894), *The Rise of Hermeneutics* (1900) and *The Formation of the Historical World in the Human Sciences* (1895 and 1907),[19] defined the humanities as an understanding informed by the "re-experience" of a historical or foreign existence expressed in writing, language, gestures, mimics, art and other human phenomena. These active processes are not based on rational thinking alone, as in these fields cognition cannot be proven in the sense of a definitive truth or falsehood as in natural science but is always a hermeneutic interpretation underpinned by a certain point of view within a certain context. Meaning cannot be constructed from single elements removed from their context. Rather, both the element and the whole can be understood only in reference to each other. This procedure, which Dilthey termed "hermeneutic circle", sets psychology as the foundation of humanities – though not the branch of psy-

17 D. Baumann, "Systematische Musikwissenschaft – eine Disziplin zwischen Kulturgeschichte und Naturwissenschaften" (2009), 40–51.

18 W. Windelband, *Geschichte und Naturwissenschaft* (1894): "Die einen sind Gesetzeswissenschaften, die anderen Ereigniswissenschaften; jene lehren, was immer ist, diese, was einmal war. Das wissenschaftliche Denken ist – wenn man neue Kunstausdrücke bilden darf – in dem einen Falle nomothetisch, in dem andern idiographisch. [...] die eine sucht Gesetze, die andere Gestalten. [...] Es bleibt möglich und zeigt sich in der Tat, dass dieselben Gegenstände zum Object einer nomothetischen und daneben auch einer idiographischen Untersuchung gemacht werden können." See also C. Dahlhaus, "Musikwissenschaft und Systematische Musikwissenschaft" (1982), 28 ff.

19 W. Dilthey, "Einleitung in die Geisteswissenschaften: Versuch einer Grundlegung für das Studium der Gesellschaft und der Geschichte" (1883), *Gesammelte Schriften* vol. 1 (1973), xv–xx; "Ideen über eine beschreibende und zergliedernde Psychologie" (1894); "Beiträge zum Studium der Individualität" (1895/96), "Die Entstehung der Hermeneutik" (1900); "Die geistige Welt: Einleitung in die Philosophie des Lebens", *GS* vol. 5 (1968); "Der Aufbau der geschichtlichen Welt in den Geisteswissenschaften" (1895, 1907), *GS* vol. 7 (1973). English editions: W. Dilthey, *Selected Works* vol. 1 (1989), vol. 3 (2002), vol. 4 (1996); see also R. A. Makkreel, "Wilhelm Dilthey" (2008), and B. Ramber, G. Gjesdal, "Hermeneutics", *Stanford Encyclopedia of Philosophy* (2005).

chology informed by the natural sciences, but a primarily descriptive psychology that seeks to understand human expression within its context.

The philosopher Ernst Cassirer (1874–1945) explored these questions extensively in the 1940s in his five studies on cultural history and in *An Essay on Man* (1944).[20] According to him a discipline that investigates processes of human culture calls for a special methodology based on physics, history and psychology, as the concepts of these three fields are indispensable to the description of cultural objects. Yet we cannot understand these objects on the basis of their discrete elements but only through the latter's mutual, interpenetrating relations.[21] Such understanding relies on the formal and stylistic concepts of cultural studies, which differ from both scientific and historical concepts. Here the topic is not "dead" matter but human action.

Following this approach to cultural history, a study of the relationship between room acoustics and music performance requires a finely tuned combination of methods. Normative nomothetic models alone cannot point out why certain rooms are better or worse for certain kinds of music, nor would it be appropriate to use only historical idiographic descriptions to explain why music has been performed in certain rooms at certain times and why musicians and attendants qualified these performances as better or worse. A strictly idiographic procedure would also require a full survey of the history of performance practice of Western music and its social background, which would exceed by far the scope of this study.[22] This research aims, rather, to *combine* normative, idiographic and psychological methods in order to develop clear ideas about the relationship between music and room acoustics. As a result, criteria can be defined for a better understanding of the subjective process of qualification of the acoustic background of music performance then and now. However, the combination of history, exact sciences and cultural studies does not imply a fusion of methods. The different symbolic realms remain distinct, and their symbols preserve specific meanings. But we may follow Cassirer, who said in *Structuralism in modern linguistics,* his last lecture shortly before his death in 1945, about the relationship among various symbolic realms: "The dissimilarity of the objects of natural science [or acoustics] and linguistics [or music and musical performance practice] does not exclude a correspondence in the structure of

20 E. Cassirer: "Naturbegriffe und Kulturbegriffe" (1980; orig. publ. 1942), 56–86; *An Essay on Man: an Introduction to a Philosophy of Human Culture* (2006; orig. publ. 1944); see also R. A. Makkreel: "Cassirer zwischen Kant und Dilthey" (1997), 150–1.

21 Makkreel (1997), 57.

22 A wealth of new documents and studies has been published in the context of the European Research Project: *Musical Life in Europe 1600–1900. Circulation, Institutions, Representation* (2003 ff.). Important in the context of this study is vol. 9: *Espaces et lieux de concert en Europe, 1700–1920: architecture, musique, société* (2008).

the judgments that we find in both sciences."[23] Not only objects of natural sciences but also functions of the mind and sensorial perception have a clear logical structure. The concept of "Gestalt", a form of both objects and processes (according to Cassirer),[24] enables us to connect apparently very disparate things if their Gestalts show similarities. While physical processes and mental functions that generate human culture may share formal or logical analogies, such sharing does not prove that matter and meaning are identical for human beings. In the analysis of a process of change, physics investigates the cause of change, but cultural studies address the human meaning of that change. Still, with its detailed scientific description, the knowledge of physical processes can become a basis for the understanding of these processes of human existence.

Hence the two-part structure of this study: the first part presents a survey on the history of room acoustics and charts the current knowledge about room acoustics as viewed by the different disciplines involved and introduces, explains and compares their terminology. Using the general historical background and the basic systematic rules presented in the first part, the second part describes and assesses the performance history of one specific musical work from its first performance up to the early twentieth century.

The first part opens with a brief historical outline of basic acoustic knowledge applied to building construction from antiquity through the time acoustics developed as a science and up to the late nineteenth century. Follows a discussion on the changing relations between room acoustics and music since the beginning of public concerts up to the First World War. The various ways of perception are then addressed from the viewpoints of musicians, composers and listeners. The basic theories of acoustics, hearing, sound perception and room acoustics in physics and the disciplines dealing with auditory phenomena (physiology of hearing, cognitive neurology, psychology of perception) constitute the next section. Finally, a historical survey of sound recording techniques and their spatial aspects serves as a basis for the analyses of sound recordings used in the second part as both historical sources and a tool for the auditory analyses of room acoustics.

Based on the current knowledge on physical room acoustics and the disciplines related to auditory perception, architecture-related quality factors are defined to be used for extant and no longer extant historical rooms. Music-related quality factors of room acoustics are assessed next. The material from the five fields defined in 1.1.2 (documents on architectural space, music performance, musical composition in relation to architectural space, performance practice, and the reception of musical works) is then regrouped differently in order to elucidate the relationship between, on the one hand, the period a specific space

23 E. Cassirer, "Structuralism in Modern Linguistics" (2007; orig. publ. 1945), 313.
24 see 1.4.3

was used and, on the other hand, the three types of space (church, theater and music room or concert hall) and performance-related aspects. Musical genres are then examined in relation to the typical size of the space and the musical ensemble.

Although a room's acoustic quality must always be evaluated in relation to a specific musical work, these general definitions produce a grid that allows to describe the general background of acoustic experience for a certain group of listeners at a certain period in a certain room category. This approach permits, among others, a better understanding of the new definition of chamber music in relation to hall size, which emerged around 1800, well before the widespread use of large concert halls for symphony orchestras and the even later preference for small halls for chamber music.[25]

The second part of this study called for a musical composition with an uninterrupted performance tradition from its first performance to the twentieth century, well documented in all five fields and performed already during the composer's lifetime in all three room categories. Of the main music genres – opera, oratorio and symphony – only the oratorio was performed in all three room categories, namely, in theaters, churches and small halls called "oratorios" or music rooms. The history of the oratorio is well documented from the seventeenth to the twentieth century.[26] A bibliographic survey of documents indicates that Handel, an oratorio composer with a vast contemporary documentation, is an appropriate object for such an investigation. Handel discovered the Italian *oratorio volgare* around 1708 in Rome and consequently developed the genre of the English oratorio which, from 1732 onward, played an important role in the development of the public concert in England. From 1770 a selection of his oratorios was performed in various music centers on the continent in translated versions.[27] Among Handel's oratorios we chose *Messiah*, as many contemporary documents on music performance and reports on the reception of musical works are extant. Furthermore, this oratorio has enjoyed a nearly uninterrupted performance tradition from 1742 to our time in many different spaces of all three categories, some of which are still extant. The inclusion of other selected compositions (some by other composers) enables documentation of the main aspects of the research matter. The choice of Handel's *Messiah* set down the main research lines of the second part, presented in detail at its beginning.[28]

In the first, mainly systematic, part as well as in the second, mainly historical, part on Handel's *Messiah* questions emerged that could not be answered, as they need further special research or broader research in music history, the

25 See 1.3.3: New definition of chamber music.
26 See H. Smither, *A history of the Oratorio* (1977–1987), and many special studies.
27 See appendix 3.5.2, table B.
28 See 2.1: Handel's *Messiah,* the change of performance practice and room acoustics.

history of composition and reception and its sociological and aesthetic aspects, but also because further interdisciplinary research must involve scholars from several related disciplines in order to apply their methodologies. Further research on the history of architecture and art would fill lacunae in the history of rooms used for music. Psychological tests would answer questions on the psychological process of perception, and more exact acoustic measurements and calculations carried out by engineers in conjunction with musicians and instrument makers would hone our understanding about the significance of room acoustics for both musical production and perception. Nevertheless, the systematic and historical overview presented in the first part provides a solid ground for the analysis of the performance tradition and the interpretation of the music examples in the second part.

The first part lists examples of halls, theaters and churches that are important for music history. Some readers may know many of them well from personal experience. Whenever possible, the main room dimensions are indicated. Many of those partial or not yet published were completed from scaled plans or measured on site. With the exception of some famous twentieth-century concert halls and opera houses, the study is limited to spaces erected before the end of the First World War, mainly because new construction techniques have introduced important changes in architectural design, and modern scientific acoustics have been applied on the basis of measurements of absorption factors and the calculation of reverberation time according to the formula defined by Wallace C. Sabine around 1900.[29] New shapes and interior design styles of halls and theaters were developed, and large-scale spaces were built, mainly in the United States. Moreover, the invention of sound recording and reproduction and, later, of sound amplification affected the perception of music and its spatial aspects.[30] The reconstruction of old and the construction of new opera houses and concert halls after the devastation of the Second World War has accelerated the disappearance of old spaces. Some famous old buildings have been reconstructed, many of them with slightly or considerably different acoustics. Yet all these changes in the exterior aspects of music performance and perception in the twentieth century require further research that is beyond the scope of this study.[31]

29 For the development of acoustics as a science, see 1.2.5.

30 1.3.5: Electro acoustic reproduction of space; 1.7: Spatial impression in sound recording.

31 See also D. Baumann, "Konzertsäle und Opernhäuser des 20. Jahrhunderts im Spannungsfeld zwischen Umbruch und Tradition" (2002); M. Forsyth, *Buildings for Music* (1985): "Science and the Auditorium", "The Hi-Fi Concert Hall" and "Toward the Future: A New Context for Music"; L. L. Beranek, *Music, Acoustics, and Architecture* (1979) and *Concert and Opera Halls, How They Sound* (1996).

1.2 Acoustic knowledge applied to the construction and use of rooms

Understanding the relation between the history of architecture and music performance requires, first of all, an examination of the acoustic knowledge available to those who built or chose rooms for music performance.

1.2.1 Acoustic knowledge in ancient Greece and Rome

One of the few physical laws known in the ancient world was the law of reflection, according to which light rays reflect off a mirror at the incident angle.[1] Since Aristotle (fourth century B.C.) this law was also known to be valid for sound. Aristotle himself refers several times to the analogy between light and sound, mainly in his explanation of the echo heard when a voice hits a hard surface without being shattered: "The whole remains intact or the two parts are separated because reflection occurs at the same angle. That is why the voice of the echo is similar to the original voice."[2] This law was also important for the understanding of sound in ancient theaters and closed spaces. *De architectura libri decem* by the Roman architect and engineer Marcus Vitruvius Pollio (active around 30 B.C.) contains important knowledge in room acoustics. From the Middle Ages to the seventeenth century this treatise was widely transmitted to the Arab and Christian cultures.[3] Vitruvius not only describes the ancient theater but also gives a detailed account of sound propagation and distinguishes between helpful reflections (*reflexiones consonantes* – sounding together and reinforcing sound) and disturbing reflections (*reflexiones resonantes* – resounding like an echo; *reflexiones dissonantes* – coming from above, impeding the free distribution of the following sound waves; *reflexiones circumsonantes* –

1 C. B. Boyer, "Aristotelian References to the Law of Reflection" (1945), 92–5.

2 Aristotle, *Problemata,* XI, 23, 901/b, cf. Boyer (1945), 93.

3 Vitruvius, *De architectura libri X,* ed. and translated by M. H. Morgan (1960); on acoustics see book V, chapter 3–9; on the transmission of Vitruvius, *ibid.,* introduction, 10 ff.; on the transmission of general acoustic knowledge of the ancient Greeks in Arabic and Greek translations, see D. Baumann, "Musical Acoustics in the Middle Ages" (1990), 199–210. An adequate critical edition of Vitruvius' treatise is still lacking.

moving in a circle, decreasing the clarity of the voice).[4] These remarkable definitions must have been based on a long tradition of exact observation. Excavations show that the ancient Greeks developed naturally shell-shaped sites with good acoustics into half-circle open-air theaters with several thousand seats in steeply raking rows where all citizens of a town could enjoy not only an excellent view but also astonishingly good speaking conditions (figure 3).[5]

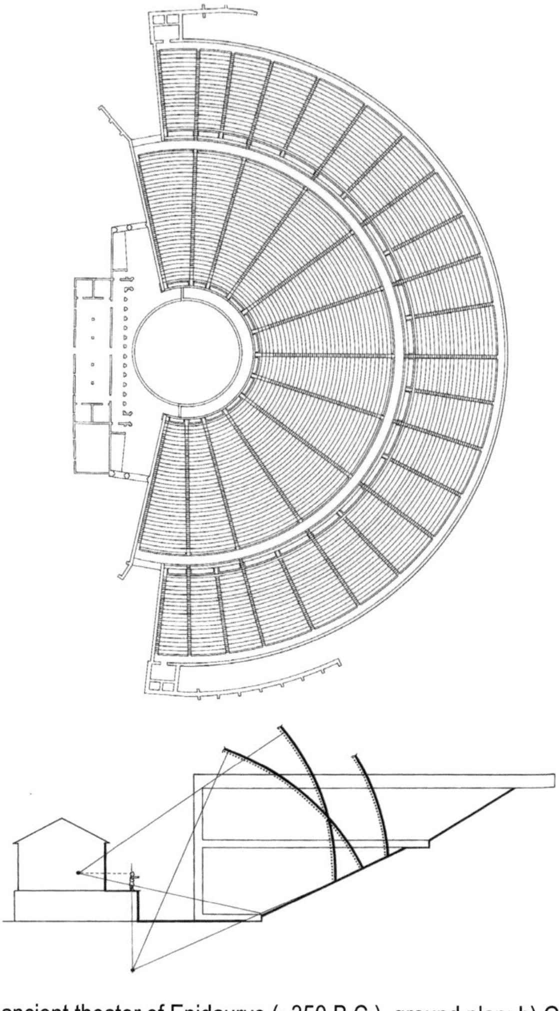

Figure 3: a) The ancient theater of Epidaurus (~350 B.C.), ground plan; b) Greek theater showing direct sound and reflected wavefronts from the orchestra floor and the back wall.

4 See Hunt, *op. cit.,* 34, footnote 52; 167.

5 M. Bieber, *The History of the Greek and Roman Theatre* (2nd rev. ed. 1961), 54 f.; see also M. Semper, *Handbuch der Architektur,* vol. IV, 6, 5 (1904), 7–13.

Sound coming from a raised platform was supported by reflections off the free surface, the so-called *orchestra,* and off the back wall behind the stage.[6]

The famous *echea* described by Vitruvius – the vessels or vases under the seats of these theaters – were already obsolete at his time, but their usefulness has been discussed ever since.[7] Modern measuring could confirm however that these vessels could produce a resonance of up to two seconds if excited by sound exactly at their resonance frequency.[8] Based on this evidence, proven by modern physics, a geometrical analysis of the site and its acoustic properties supports a new hypothesis on the use of this sophisticated installation: in the very dry acoustics of the theater the resonance may not have been heard close to the vases only, which were placed under the seats, but also exactly at the center of the orchestra, as all sound reflections were thrown back to this point from the vertical fronts of the half-circle steps. Since resonance vases were tuned according to the Greek scales, they probably allowed actors and choir singers standing exactly at the center to control intonation, loudness, timbre and voice carrying because each vase reproduced only its own resonance frequency and only if this specific frequency of sound entering the vase was loud enough.[9]

The Greeks knew that the clear but dry acoustics of the open-air *theatron* (in Greek "a place to see") were excellent for speaking but inadequate for music. Therefore, the *odeion* (Greek "a place to hear") was developed for song performances during musical competitions.[10] A hall of limited size with excellent acoustics erected on a rectangular ground plan, the *odeion* featured steeply raking rows of seats and a flat horizontal ceiling. The walls and the ceiling with structured surfaces produced well supporting *reflexiones consonantes,* the open windows absorption.[11] At the time of the Roman emperors, a mixed type between *odeion* and open *theatron* evolved – a semicircular theater covered not only by a *velarium* as protection against the sun but partly or completely by a fixed wooden roof that may have been added for climatic reasons. This kind of space was also called *odeion,* in Latin *odeon* or *theatrum tectum* as the odeion of Agrippa in Athens (figure 4) or in Pompeii close to the large theater (figure 5).

6 H. Bagenal and A. Wood, *Planning for Good Acoustics* (1931), 344–46.

7 R. Lewcock et al.: "Acoustics", § I, 7: "Room Acoustics: Classical times", *NG2,* 1, 79–82; D. Baumann et al., "Exkurs über die akustischen Gefässe im Jöchelsthurn" (1992), 213–16; Baumann (1990), 201–202.

8 L. Cremer, *Die wissenschaftlichen Grundlagen der Raumakustik,* vol. II (1976), 179–80.

9 Bagenal and Wood, *ibid.*

10 R. Meinel, *Das Odeion* (1980), 427, footnotes 8–12: "Odeion" as described by Heschios (5th to 6th century A.D.), by Photios, in *Bibliotheké* (ca. 820–891) and in the encyclopedia *Sudias* (around 1000 B.C.); Bieber (1961), 9, 3, 176 and 220.

11 Meinel (1980); G. Izenour, "The ancient Roman roofed theater" (1990). The acoustics of these types of space are explained in 1.6; for dimensions see table C in the appendix.

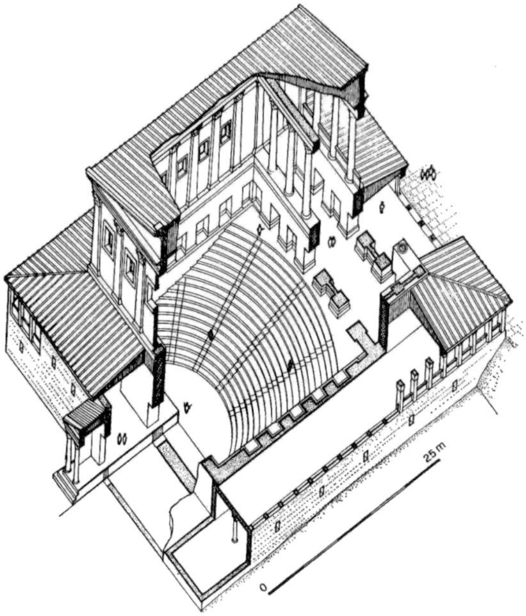

Figure 4: Athens, Odeion of Agrippa (~ 15 B.C.), axonometry restored.

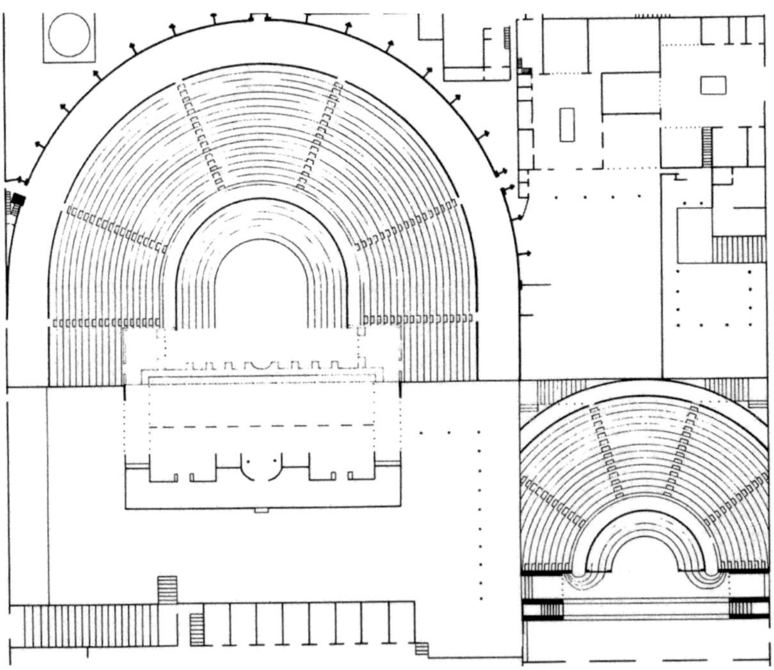

Figure 5: Pompeii, Forum: large theater (200 B.C.) and small *theatrum tectum* (80 B.C.).

In Western architecture, these two types of spaces – the semicircular theater and the rectangular *odeion* – and the concept of steeply raking rows of seats have strongly influenced the design of many public halls, such as law courts, University lecture halls, assembly halls, theaters, music rooms and concert halls from the Renaissance up to our time.[12]

1.2.2 Pragmatic room acoustics in the Middle Ages and the Renaissance

The few surviving medieval sources indicate that the construction of buildings in the Middle Ages was based on measurement and geometrical rules. The designs preserved in a small portfolio with *exempla* by Villard de Honnecourt, a draughtsman who worked for the Cistercian order between 1225 and 1250, reveal how important geometry was in the stonemason's craft.[13] Medieval chronicles refer to "celestial inspiration" as a source of building design that follows the rules of harmonic proportions inherent in the eternal "harmonia mundi".[14]

No evidence could be found for the practical use of ancient acoustic knowledge in the formal conception of medieval buildings, especially the large cathedrals. However, the presence of ancient knowledge is obvious in acoustic vases still extant in churches and halls, mainly under vaulted ceilings, either to amplify certain frequencies by resonance or to absorb these same frequencies by means of absorbing material placed inside the throat of these vases.[15] Once the rooms were ready for use, priests, church musicians and organ builders devised pragmatic rules on how to reach the best sound effect under the given circumstances.[16] Exact observation and acoustic tests, possibly accompanied by ideas on geometrical sound distribution, were behind the positioning of singers in churches on steps or special balconies. Since the twelfth century organs were placed on technically difficult, expensive constructions of lofts and rood galle-

12 See for instance N. Pevsner, *A history of building types* (1976).

13 Experts do not agree on whether Villard de Honnecourt's *exempla,* Paris, Bibliothèque Nationale, Ms. fr. 19093 (written between ca. 1220 and 1240) is a pattern book – see *Architectural Theory from the Renaissance to the Present* (2003), 196–201 – or a personal collection of drawings made for a hitherto unknown reason – see C. F. Barnes jr.: "Villard de Honnecourt, a new critical edition and facsimile" (2009) – or a builders' association manual – see facsimile edition ed. by H. R. Hahnloser: *Das Bauhüttenbuch des Villard de Honnecourt* (2nd ed. 1972); A. Reinle, "Architekturlehre" (1981), 214.

14 Reinle (1981), 212.

15 R. Lewcock et al., "Acoustics", § I, 8: "Room acoustics: Medieval times", *NG2,* 1, 82–7; Baumann (1990), 199–210; Baumann (1992), 213–16.

16 Baumann (1990), 199–210; D. Baumann: "Acustica", *Enciclopedia dell'Arte Medievale,* vol. 1 (1991), 21–3.

ries and from the thirteenth century onward in "swallow nests" hanging on upper walls of the nave or choir.[17]

During that period some scholars developed an interest in ancient geometrical theories of optics transmitted and partly improved by Arab scholars.[18] Robert de Grosseteste (ca. 1175–1253), a professor in Oxford and later bishop of Lincoln, was the first to fully explain the law of reflection on the basis of experiments and mathematical induction. As he had access to Greek and Arab sources in Oxford, he was able to rely on Aristotles' *lex parsimoniae* (law of the shortest path), proven by Heron of Alexander in his *Catroptrics,* and the observations by the Arab mathematician Alhazen (Ibn al-Haitam, 965–ca. 1040) that the angles of light incidence and reflection are equal and lie in a tangential plane vertical to the mirror.[19]

With its new conception of man and his needs, the Renaissance bequeathed to us documents that provided rules not only for the users of musical spaces but also for architects on how to design them with good acoustics. Frater Francesco Giorgi gave this remarkable advice for the construction of the new church of *S. Francesco della Vigna* in Venice in 1535:

> I recommend to have all the chapels and the choir vaulted, because the word or song of the minister echoes better from the vault than it would from rafters. But in the nave of the church, where there will be sermons, I recommend a ceiling, (so that the voice of the preacher may not escape nor re-echo from the vaults). I should like to have it coffered with as many squares as possible, with their appropriate measurements and proportions. (...) And these coffers, I recommend, amongst other reasons, because they will be very convenient for preaching: this the experts know and experience will prove it.[20]

17 B. Owen and P. Williams, "Organ", § IV, 6: "The church organ, 1100–1450", *NG2,* 18, 588; F. Jakob, "Die Orgel des westlichen Kulturkreises im Mittelalter" (2001), 56; St. Bicknell, *The History of the English Organ* (1998), 17; D. Baumann, "Acoustics in Gothic Cathedrals, Theory and Practical Experience in the Middle Ages" (2001), 37–48; E. Leipp, "A propos du Grand Orgue de la cathédrale de Strasbourg: Pourquoi des orgues en nids d'hirondelles" (1986), 17–30. Theophilus' organ could be mounted within a recess in an interior wall, presumably at gallery level, although this is not explicitly said. See W. Stromer, comment on the revised edition with partial German translation of Theophilus presbyter, *Diversarum artium schedula,* Book 3, chapter LXXXIII (written in the first 3rd of the 12th century; ed. 1984), 396, footnote 16 and 17. For church acoustics, see 1.8.9.

18 A. C. Crombie, *Science in the Middle Ages* (1959), 188.

19 Heron of Alexander, *Katoptrika,* §4, 324, German translation by W. Schmidt (1900), 16 ff.; see Hunt (1978), 39.

20 Francesco Giorgi, *Memorandum di S. Francesco della Vigna,* Engl. translation by R. Wittkower, *Architectural Principles in the Age of Humanism* (1949), appendix I, 137, from G. Moschini, *Guida per la città di Venezia,* vol. I, Venezia 1815, 55–61; Italian ed.: *Principi architettonici nell'età dell'umanesimo* (1964), 149–50: "Il volto luodo, che si facci in tutte le capelle et in choro: Imperochè il dir è cantar dell'Ufficio meglio rimbomba nel volto, che nella travadura. Ma nel corpo della Chiesa, ove si ha a predicare (conciossa che le prediche non

The application of pragmatic rules derived from the priests' daily experience is also evident in a category of halls developed during the second half of the sixteenth century, the so-called "oratorios" built for the *Congregazione dell'Oratorio* of S. Filippo Neri and other groups of that order in Rome and other Italian cities. These halls were meant to provide good acoustic conditions for sacred *exercitia* in lectures, the singing of *laudae* and the performance of sacred madrigals and *sacre rappresentazioni*.

We have no information on the size and shape of the *Piccolo oratorio* on the right side of the transept of the *Chiesa S. Maria in Vallicella,* Rome, where Cavalieri's *Rappresentatione di anima e di corpo* was performed in 1600,[21] nor do we have any architectural documents on the *Oratorio del S. Girolamo della Carità* of 1554, the *Oratorio del SS. Crocifisso* of 1568 or the *Oratorio della Pietà* of 1574. But one of the early halls, the *Oratorio del Gonfalone* in Rome, which was built from 1544 to 1547, is still extant and regularly used for concerts. A narrow, long rectangular hall with excellent acoustics, it features a high wooden ceiling with deep coffers and wooden paneling covering the lower part of the side walls (figure 6).[22]

In 1640 Borromini built the oratorio for the *Chiesa nuova* on the request of the *Congregazione dell'Oratorio* as a narrow, relatively long rectangular room with a high flat plaster ceiling with vaulted edges which offers excellent acoustics.[23] This design, based on harmonic proportions of 3:2:3 (length to width to height), is all the more remarkable, as Borromini, following the contemporary trend, erected the *refectorium* (dining room) in the same building complex on an elliptical ground plan.[24]

reescino, ne sintendino nelli volti) luodo la contiguatione. Ben la vorrei in quadri sfondratti, al più che si può, con le suoe misure et proporzioni perhò. [...] E questi sfondri i luodo, fra le altre ragioni per esser molto convenevoli al prdicare: il che sanno li periti dell'arte et l'esperienza il comprobarà." See also R. Lewcock et al., "Acoustics", § I, 9: "Renaissance and Baroque periods", *NG2,* 1, 86 (incomplete citation). We cannot enter here into a discussion of the influence of harmonic proportions on room acoustics because systematic studies are lacking. Evidence for an influence on church acoustics going beyond the rather mystic believe is given by Bagenal and Wood (1931), 216, in a paragraph on "sympathetic notes" of certain churches with a marked reverberation. For the acoustic relevance of harmonic proportions in rectangular concert halls, see 1.8.2. Further evidence for spaces with vaulted ceilings acting as resonance chambers is presented by Athanasius Kirchner in his description of the Vitruvian theater; see 1.2.4.

21 Emilio de' Cavalieri, *Rappresentatione di anima e di corpo,* ed. M. C. Bradshaw, AIM, *Miscellanea,* vol. 5 (2007), preface.

22 F. Onorati, "La confraternità del Gonfalone" (1993), 24–5.

23 J. Connors, *Borromini and the Roman Oratory* (1980), 29, footnote 19 and figure 35.

24 P. Portoghesi, *Borromini* (German ed. 1977), 49 f. For the acoustic properties of elliptical rooms, see 1.2.4, 1.2.5 and 1.2.6.

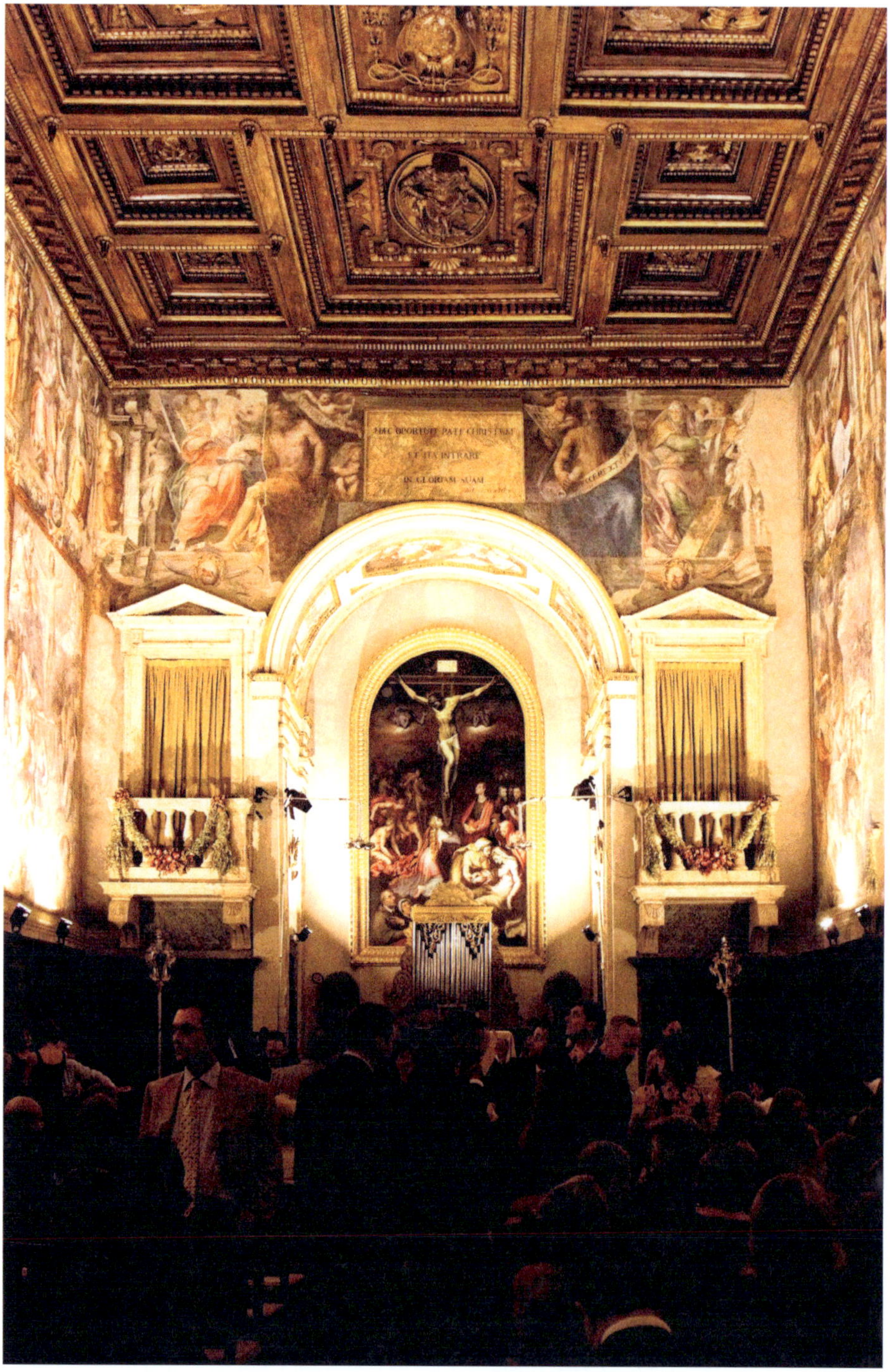

Figure 6: Rome, *Oratorio del Gonfalone* (1544).

Many small rectangular *oratorios* with excellent acoustics are extant in the cities of Italy and other Catholic countries. In Protestant cities the same shape was used for church assembly halls and early music rooms. In theaters and opera houses it was the typical shape for ball rooms and foyers also used for concerts. Thus the early history of the public concert halls can be traced back to this type of rectangular hall built for para-liturgical functions.[25]

1.2.3 Rediscovery of the ancient theater

The rediscovery of ancient buildings in the fifteenth and sixteenth century sparked a new interest in Vitruvius' *De architectura libri decem* and his description of the ancient theater.[26] Several theoretical treatises of architecture of that time contain examples of theater construction based on the Vitruvian model.[27] Leonardo da Vinci (1452–1519), who was familiar with Vitruvius' treatise, left sketches for sacred rooms with steeply raked rows of seats indicated as "teatri per uldire [sic!] messa" and "loco dove si predica" (figure 7).[28]

In the first case the choir is placed in the center, in the second case the lecture pulpit is on a column in the center of a circular ground plan. The center pulpit, familiar from synagogues of the Ashkenazy tradition,[29] was used also in seventeenth-century Protestant churches by architects in the Netherlands and Scandinavia.[30]

25 For further information, see 1.9.2.1.
26 Vitruvius' treatise was printed for the first time by Giovanni Sulpicio da Veroli (1487(?), no date, no place).
27 For instance Leon Battista Alberti's *De re aedificatoria libri X* of about 1450, printed in 1485. Many treatises on theater building that address perspective, change of scenery, lighting, etc., have been published since; see I. Mackintosh, *Architecture, Actor, and Audience* (1993), bibliography.
28 Leonardo da Vinci, *Literary Works,* ed. P. Richter (1939), Vol. 2, 42–52; M. Forsyth, *Buildings for music* (1985), figure 1.9, facsimiles of the drawings from Paris, Institut de France, ms. B (2173), f. 55a; the same drawings and a further sketch with the title *Teatro da predicare* from ms. B, f. 52 recto also published in *Leonardo da Vinci, das Lebensbild eines Genies* (6th ed. 1972). The sketch f. 55a is close to a ground plan for the church of *Santa Maria in Pertica* in Pavia and other studies on centralized churches. The drawings are from 1486–88, see C. Vecco and C. Pedretti, *Leonardo* (1998); see also Wittkower (1949), 16, footnote 5 and plate 5.
29 See H. A. Meek, *The Synagogue* (1995).
30 G. Germann, *Der protestantische Kirchenbau in der Schweiz von der Reformation bis zur Romantik* (1963), 159 f.

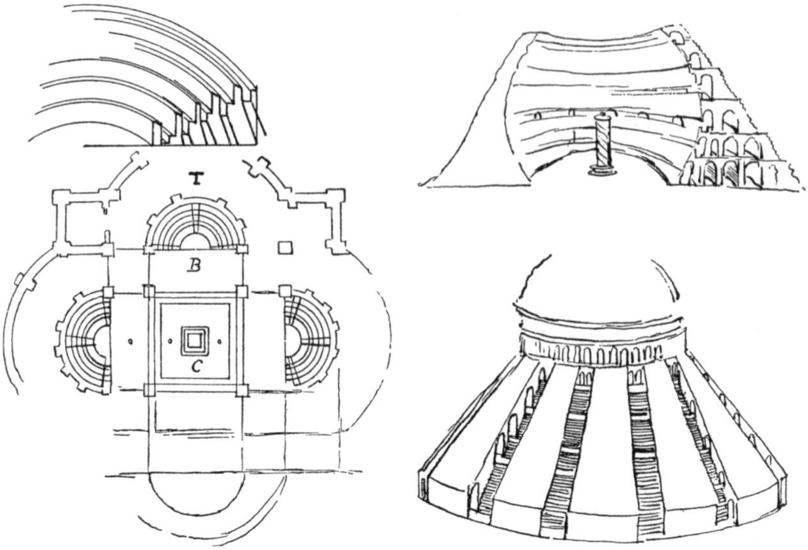

Figure 7: Leonardo da Vinci: *Loco dove si predica e teatri per uldire messe,* 1478.

Leonardo left three drawings of sound rays produced by hammer blows transmitted through a wall, passing through a hole in the wall and reflected off the back wall.[31] These important sketches, which were certainly based on experiments, belong to a group of studies on the propagation of water waves, sound waves and light rays that are directly related to the already mentioned observations by Aristotle and Alhazen on equal incident and reflected angles. These ideas had to wait until the seventeenth century to be developed into a scientific theory of optics and acoustics.[32]

As mentioned, the interest in ancient theoretical knowledge around 1500 led to the construction of open-air theaters in Italy, England, France and other countries. Serlio's 1545 plan for an open-air theater is an important early document on the subject.[33] Covered theaters with semicircular raked rows of seats were also built following the model of the ancient *Odeion.* One of the best known examples still extant is the *Teatro Olimpico* in Vicenza (1500 seats, fi-

31 E. Winternitz, *Leonardo da Vinci and Music* (1982), 97 f.: Milan, Bibl. Ambros. Codex Atlantico C.A. f. 126 ra. The drawings are from 1478 and 1519, see Vecco and Pedretti, *Leonardo* (1998).

32 D. Argentieri, "Die Optik bei Leonardo" (1972), 405–36; on further studies of mirrors by Leonardo, see London, British Library, Codex Arundel 263, folio 84 verso–87; on optical and acoustic experiments in the seventeenth century, see 1.2.4.

33 Sebstiano Serlio: *Della Architettura,* Venice 1545, vol. II, appendix; ground plan and section, see M. Hammitzsch, *Theaterbau* (1906), 12.

gure 8a),[34] built by Palladio in 1580 for the *Accademia Olimpica.* For the inaugural representation of *Oedipus rex* in 1585, Vincenzo Scamozzi added behind the proscenium foreshortened scenery showing "The Seven roads of Thebes", which thereafter remained an integral part of the building. The limited size of the construction ground imposed a semi-elliptic rather than semicircular ground plan for this flat-ceiling theater. The principle of raked seats was applied also to rectangular theater halls with several straight rows of seats along the side walls and the back wall, such as the famous *Teatro Mediceo* in Florence (about 2000 seats), built at the Uffici in 1589 by Buontalenti with Giulio and Alfonso Parigi for the marriage festivities of Ferdinando de' Medici and Christine of Lothringen, specifically for the performance of the comedy *La Pellegrina* with musical *intermedii.*[35]

Raked seats were also featured in U-shaped ground plans, as in the *Gran Teatro Farnese,* built in 1617–1618 by Giovanni Battista Aleotti on the first floor of the *Palazzo della Pilotta* in Parma with an auditorium nearly 29 meters wide, 44 meters long, 22.65 meters high and accommodating about 2500 seats (figure 8b).[36] The theater, still one of the largest auditoria in Europe, was severely damaged in 1944 and reconstructed in 1956 following the original plans, though unfortunately without the horizontal wooden ceiling and the rich stucco decoration. Its current acoustic properties are therefore considerably different. The theater's enormous size prompted the composer Claudio Monteverdi, who was commissioned to write the music for the inauguration, to visit the theater "in order to write music adapted to the dimensions of the hall as much as possible" (Monteverdi in a letter to Bentivoglio on 25 September 1627).[37] As no attention was paid to the orchestra placement for the music he composed to Claudio Acchillini's *Torneo di Mercurio e Marte,* he wrote several letters to the architect between November 1627 and February 1628, requesting the erection of a special platform half-way down the stairs that would connect the stage with the auditorium floor (see figure 8b with exact shape and position of the orchestra podium in front of the stage).

34 Vincenzo Scamozzi: *Dell'Idea della Archittetura Universale,* Venice 1615, introduction to book VIII; ground plan and section, see Hammitzsch (1906), 15 f.; on the significance of room width for acoustics, see 1.6, especially 1.6.6 and 1.8.5: Critical room width.

35 S. Bertelli, *The Courts of the Italian Renaissance* (1986), 138 ff.; P. Roselli, *I teatri di Firenze* (1978), 43.

36 For figures, see also P. Patte, *Essai sur l'architecture théâtrale,* Paris 1782, 65 f., Hammitzsch (1906), 36, fig. 15–18; R. Leacroft, *Theatre and Playhouse* (1984), fig. 103 (perspective view). Several contemporary reports indicated the number of seats was "several thousand", Furttenbach 5300. Such a capacity would need the whole surface of the ground floor and standees. For the calculation of seating capacity, see 1.8.10.

37 "Sarà cosa cauta l'andar a vedere il [*sic!*] Theatro in Parma per poterli applicare più che sia possibile le proprie armonie decenti al gran sito" (letter to Bentivoglio from 25 September 1627), cf. G. F. Malipiero, *Claudio Monteverdi* (1929), 275.

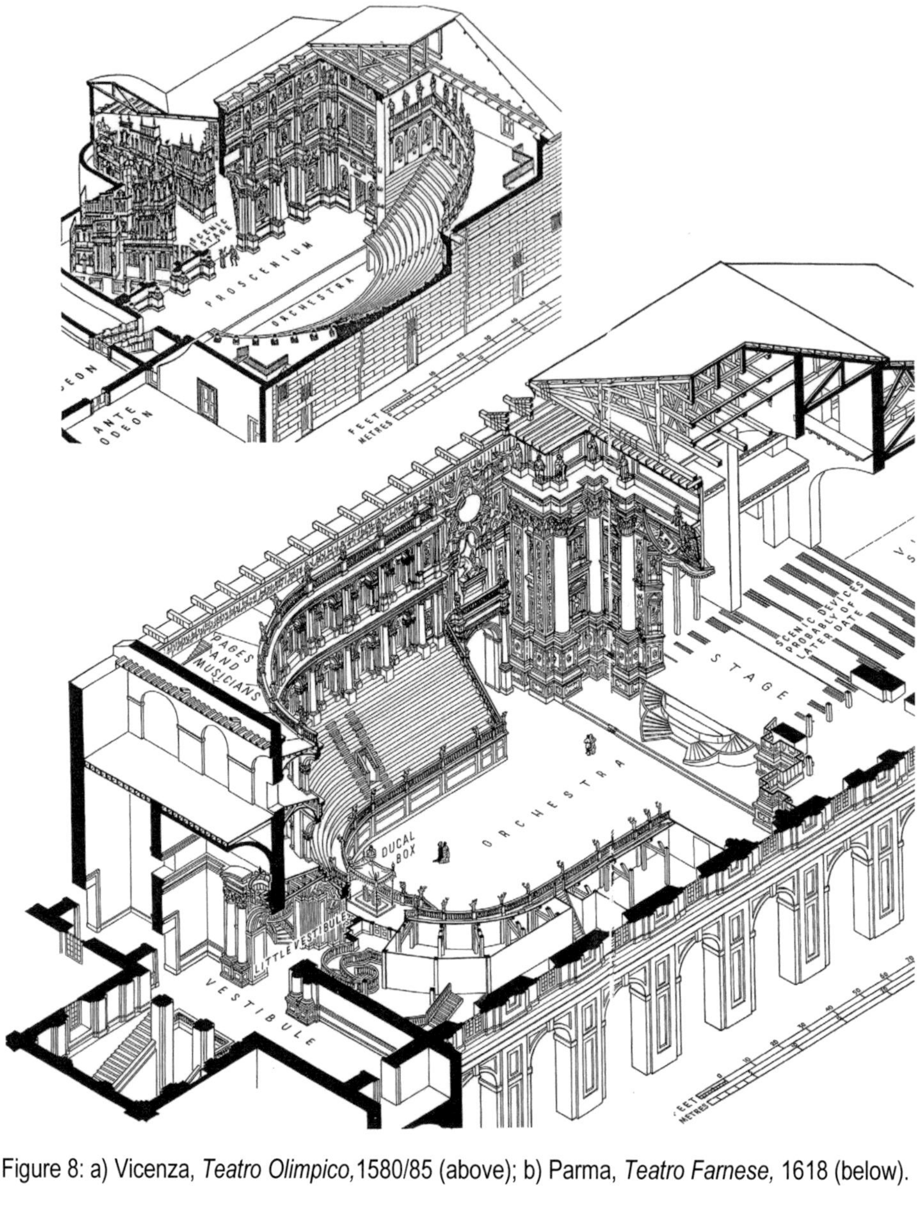

Figure 8: a) Vicenza, *Teatro Olimpico,*1580/85 (above); b) Parma, *Teatro Farnese,* 1618 (below).

The raised orchestra platform for *Mercurio e Marte* was also necessary to protect the musicians from the water flooding the parterre during the scene of the sea battle *(naumachia).*[38] The orchestra's position in front of the stage, which

38 I. Lavin, "Lettres de Parme (1618, 1627-28) et débuts du Théâtre Baroque" (1986), 110, 158: Pl. II, Fig. 3. This information was used to correct the exact shape and position of the orchestra podium in front of the stage in Leacroft's perspective view used for figure 9b.

later became the standard for opera and other staged works, was probably first used in 1622 for an oratorio performance in a Florence theater built by the theater architect Giulio Parigi.[39] Previously, musicians played on stage, behind or above it.

Already in these early theaters it became evident that a good view of the stage did not necessarily entail good acoustics. To cite just one, though prominent, criticism voiced by the English King James I, reported by the Earl of Suffolk and others before Council: during a festival held in a theater built in Serlio's style in *Christ Church Hall* at Oxford in 1605:

> in the end the place [of the King] [...] was sett in the midst of the hall, but too far from the stage (vizt) xxviij. feete, soe that there were manye longe speeches delivered, which neyther the kinge nor anye neere him could here or vnderstand.[40]

1.2.4 Debates on acoustics in the seventeenth century

In the seventeenth century the reflections off elliptic surfaces, much used in architecture at that time, were discussed amid the learned academies. The Franciscan friar of the Minimite order Marin Mersenne (1588–1648), a mathematician, philosopher and music theorist, invited prominent scientists to his room in the Parisian monastery every week, among them Pierre de Fermat (1601–1665),[41] Gilles de Personne Roberval (1602–1672),[42] Pierre Gassendi (1592–1655) and the young Blaise Pascal (1623–1662). In the first volume of *Harmonie universelle,* printed in 1627, he presented ray diagrams of sound reflections in an elliptic ground plan. To calculate the velocity of sound and the delay necessary to produce echoes, he first used pulse beats, a totally inadequate means of time measuring, but later attained a better value based on pendulum movements.[43]

39 C. Monteverdi, letters from 25 September, 2, 16 and 23 November 1627, and 18 February 1628; see D. Stevens, *The letters of Claudio Monteverdi* (1980); see also *Quellentexte zur Konzeption der europäischen Oper im 17. Jahrhundert* (1981), 37 and the literature cited there.

40 J. Orrell, *The Theatres of Inigo Jones and John Webb* (1985), 14 and 30.

41 The principle of Fermat, named after him, is the mathematical basis for the proof of the law of reflection based on the shortest path.

42 In 1666 he was among the founders of *Académie des Sciences.*

43 F. V. Hunt, *Origins in acoustics : the science of sound from antiquity to the age of Newton* (1978), 82 f. and 95 f. Pierre Gassendi in 1635 gave a value of 478 m/s, Mersenne based on pulse beats 450 m/s and in 1636 based on pendulum movements 319 m/s (see Marin Mersenne, vol. 3 of *Harmonie universelle,* Paris 1636).

In *Musurgia universalis,* printed in Rome in 1650, the German-born Jesuit Athanasius Kircher (1602–1680),[44] in *Musurgia universalis* (Rome 1650), used geometric ray diagrams to show sound reflection and distribution in architectural contexts. A universal scholar whose erudition was enhanced by a world-wide correspondence, he had used already in 1646 the same method in his treatise *Ars magna lucis et umbrae* for an optical analysis of light ray distribution.[45] At least since 1623 he conducted acoustic experiments. His writings feature instructions on how to build a wall that would produce a sequence of several echoes and a plan for a very effective whispering gallery with an elliptical dome that would concentrate all sound emanating from focus C at focus D on the other side due to the reflections off the ceiling (figure 9).

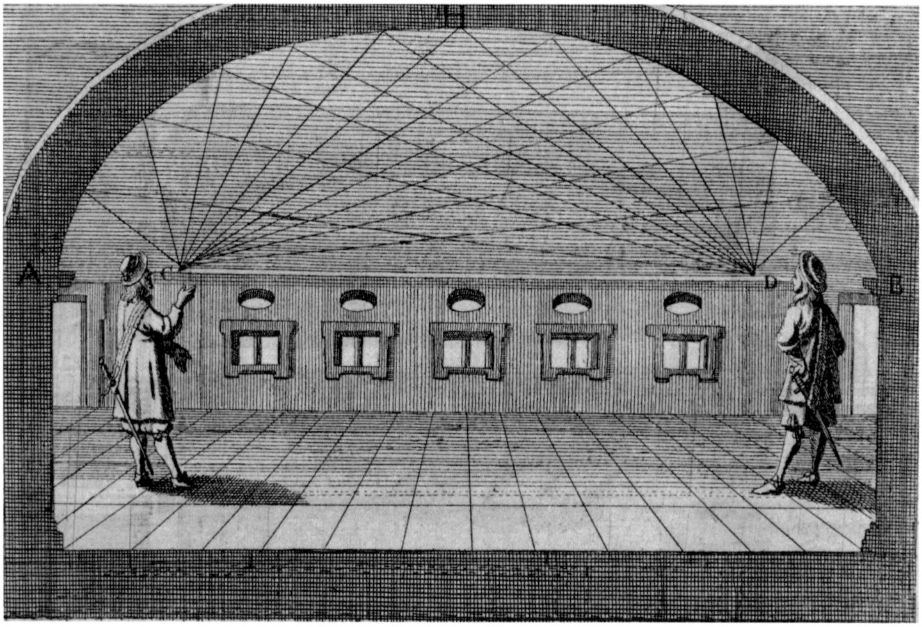

Figure 9: Athanasius Kircher: Whispering gallery (Phonurgia nova, 1673).

Between these two points very soft whispering can be exchanged without being perceived in the rest of the hall. The increase in sound intensity is astonishing, as the path length of all reflections is identical under an elliptical vault, and,

44 Kircher was a professor of mathematics, Hebrew and Syrian at the *Collegium Romanum.* He was also a surveyor, moral theologist, orientalist and music theoretist with special interests in the relation between music and medicine.

45 The material on these acoustic topics was extracted and republished in Kempten in 1673 under the title *Phonurgia nova* and in 1684 in German translation by Agathus Carion (Christoph Fischer) as *Neue Hall- und Tonkunst* (partly carelessly copied designs).

therefore, they all arrive at the same time. In his comment Kircher also explains how these rooms were used: "Rooms and halls for the use and pleasure of noble men by means of which they can reveal to each other their consultations and secret attacks." [46] A famous still extant whispering gallery with an elliptic ceiling constructed probably for such a purpose is hidden below the octagonal chapel of the palace of Charles V built after 1526 adjacent to the Alhambra palace in Granada.[47]

Offering partly correct, partly untenable theoretical explanations, Kircher also describes such curiosities as megaphone-like acoustic horns and loudspeaking trumpets used as hearing aids and sound transmitters. He describes the Vitruvian theater incorrectly: without the raking rows of seats but with a semicylindrical exterior wall and several ranks of boxes with vaulted ceilings.[48] In this design many boxes are closed, with only small round openings thought to serve as resonance chambers or "vases". Kircher was, however, disappointed to find out that exactly tuned bells, when put inside such chambers, did not produce any resonance.[49]

1.2.5 The development of acoustics as a science from the seventeenth to the twentieth century

Acoustics as a science made but slow progress.[50] Since the seventeenth century many scholars and interested laymen studied the vibrations of strings, tubes, plates and harmonic partials. The wave theory of sound propagation was developed mainly by Isaac Newton (1687), Christiaan Huygens (1690), Leonhard Euler (1727, 1748, 1759, 1765), Joseph Louis Lagrange (1759, 1760) and Pierre-Simon de Laplace (1816, 1825). In 1738, several scholars under the supervision of the *Académie des Sciences* in Paris reached the nearly exact value

46 "Zimmer und Gemächer / zum Gebrauch und Lust hoher Personen / durch deren Mittel und Hülff sie auch heimlich einander Ihre *Consilia,* und gehaime Anschläge offenbaren können." A. Kircher, *Neue Hall- und Tonkunst* (1684), 72.

47 See D. Baumann, "Whispering galleries and special places for music: Arab and Iberian acoustic sources and practices" (2001), 481–91.

48 Similar galleries and an early form of boxes with vaulted ceilings possibly thought to produce resonance still exist at the *Teatro Farnese* in Parma, see 1.2.3, figure 9b

49 Kircher, *Neue Hall- und Tonkunst* (1684), 72 with figure: "Eigentlich Gestalt deß Vitruvianischen Spihlplatzes"; see also Hunt (1978), 123 f.

50 R. B. Lindsay, *The Nature of Physics* (1968), 161 ff.; A. B. Wood, *A Textbook of Sound* (1944); E. Skudrzyk, *Die Grundlagen der Akustik* (1954), 2–9, 936–38; S. Dostrovsky, J. F. Bell, C. Truesdall, "Physics of Music", *NG2,* 14, 664–77; Hunt (1978), 82 ff.

of 332 meters per second for a temperature of zero centigrade by measuring the time delay between a cannon's firing and hitting.[51] Two years later Giovanni Lodovico Bianconi proved that sound velocity indeed depends on temperature.

The nineteenth century ushered in a fundamental improvement in the understanding of sound propagation and perception. Siméon-Denis Poisson studied the three-dimensional expansion of pressure waves (1820); the self-taught George Green solved the mathematical problem of reflection, refraction and transmission of a plane sound wave (1838); using rotating toothed wheels, the physician Félix Savart defined the minimum-audible frequency at eight vibrations per second and the upper limit at 24,000 vibrations per second (1830); by means of tuning forks, rods, strings and pipes, Karl Rudolph Koenig and Hermann von Helmholtz observed better values for lower and upper frequency limits (from 16 or 32 Hz up to 16,000 Hz or more);[52] August Toepler and Ludwig Boltzmann (1870), as well as Lord Rayleigh (1877) studied the auditory threshold, with Rayleigh also exploring the directional characteristic of vibrating objects. Georg Simon Ohm (1843), August Seebeck (1846) and Hermann von Helmholtz (1862) developed the basic theories of modern physiological and psychological acoustics, which were to be refined only in 1938 by Georg von Békésy (resonance theory, frequency analysis in the *cochlea* according to Fourier's theorem, theories of pitch and timbre perception).[53]

Ernst Florens Friedrich Chladni's traveled through Europe around 1800, during which he performed public demonstrations of vibrating plates sprinkled with sand to show the modal lines,[54] marked the onset of research to develop devices that would make sound vibrations visible.[55] Others used gas flames (John LeConte 1828, John Tyndall 1858), manometers (Karl Rudolph Koenig 1860), a membrane and a stylus (Leon Scott 1857, Scott und Koenig 1864) or a membrane and the registration of its movement on a light sensitive photograph plate (Francis Blake 1878). The next steps were the development of the telephone (Alexander Graham Bell 1876), the carbon microphone (David E. Hughes 1878), the registration of sound vibrations on wax cylinders with a horn, a membrane and a needle (Thomas A. Edison 1877) and on discs with a device called gramophone (Emile Berliner 1887).

Important studies in room acoustics were published during that time. In 1821 Chladni warned against elliptical and circular halls. His recommendation

51 The exact value is 331,8 meters per second without wind at 0° Celsius and a barometric pressure of 760 mm mercury.

52 The exact value for the lower limit is 16 Hz, see 1.5.1.

53 Lindsay (1968), 128 ff. The basic modern theories of the physiology and psychology of sound perception are presented in 1.5.

54 Ernst Chladni published in 1787 in Leipzig *Entdeckungen über die Theorie des Klanges*, and in 1802 *Die Akustik*.

55 Lindsay (1968), 182.

of the oblong rectangle as the best shape for good acoustics was successfully carried out in Otmer's new building of the *Singakademie* in Berlin in 1827.[56] John Scott Russel's *Treatise of Sightlines* published in 1838, has left its mark on the construction of auditoria and theaters with raked seating. In 1853 the Boston physicist J. B. Upham gave detailed descriptions of multiple sound reflections off room surfaces and of the effect of reverberation on speech clarity and intelligibility. He gave instructions for the reduction of reverberation time by means of curtains and furniture. The American physicist Joseph Henry published in 1856 a study on room acoustics where he described crucial physical factors. In 1877–78 Lord Rayleigh published the first edition of his *Theory of Sound,* which included a full survey on the then current theoretical knowledge in room acoustics. In 1895 Wallace C. Sabine developed an experimentally based formula for the calculation of reverberation time in a given space as a function of volume and total absorption of the surfaces.[57] Still, he was aware that this was only one of many determining factors in room acoustics. Consulted before 1900 on the project of the new *Boston Symphony Hall,* he recommended, contrary to the architect's proposals of a semicircular hall, to keep the rectangular shape of the old Music Hall and to copy not the dimensions but the proportions of the famous large hall of the new *Gewandhaus* in Leipzig opened in 1884.[58] Thus he laid the foundations for one of the best concert halls of the world.

The detection of the analogy between electrical and mechanical vibrations in the twentieth century ushered in the modern science of room acoustics (John Ambrose Fleming, Lee De Forest et al.). The improved application of theoretical knowledge in room acoustics on the basis of new experiments and measuring instruments had a powerful impact on the design of spaces that require appropriate acoustics, such as law courts, auditoria, theaters, opera houses, concert halls and cinemas. The development of certain devices – electro-acoustical microphones since 1917,[59] loud speakers,[60] the sound-on-film system in 1927 and, since 1932, electro-magnetic sound recording and reproduction, especially the high frequency magnetophone in 1940 – was groundbreaking for research

56 E. Chladni, *Beiträge zur praktischen Musik* (1821), 253; see D. Ullmann, *Chladni* (1996), 151 and *AMZ* 29 (1827), 44–5: "Aus einem Brief des Herrn Dr. Chladni, den neuen Saal der Berliner Sing-Akademie betreffend".

57 Lindsay (1968), 174 ff.

58 Forsyth (1985), 250; on the acoustics of concert halls, see 1.8.3 and 1.8.7.

59 E. C. Wente 1917: condensor microphone; E. Gerlach and W. Schottky 1924: ribbon tape microphone; E. C. Wente and A. L. Thuras 1931: moving-coil microphone; cf. F. Fischer and H. Lichte, *Tonfilm: Aufnahme und Wiedergabe* (1931), IV: Mikrofone und Lautsprecher; Lindsay (1968), 186.

60 E. W. Rice und E. W. Kellogg 1924: dynamic loud speakers; H. Riegger 1924: Blatthaller; cf. Fischer, Lichte, *ibid.,* and Wood (1944), 537 ff.

in acoustics and the quality of sound recording and reproduction. The visualization of the three-dimensional propagation of sound waves[61] refined the knowledge in geometrical acoustics and was applied to the design of acoustically effective shapes of reflectors, walls, ceilings, ground plans and finally of three-dimensional volumes.[62]

Changes in building construction, such as the use of iron introduced in the late eighteenth century and, in the second half of the nineteenth century, of iron supports and larger glass panels important for exhibition halls,[63] as well as the development of reinforced concrete,[64] offered new design options, in particular leaner supports and larger span without support or vaults. New materials and construction methods used to achieve traditional shapes with traditional proportions and decoration often yielded acoustically excellent halls. Nevertheless, these new shapes and proportions, mainly the concept of larger, shorter and, thanks to modern ventilation, lower halls, as well as the elimination of surface decoration, such as plaster ornaments, stills and columns, created ever increasing acoustical problems, which only slowly were solved by the new theoretical knowledge in acoustics. As mentioned, the acoustic consequences of this fundamental change in building design require further research that is beyond the scope of this study. Therefore, the following chapters refer only in specific cases to rooms built after 1918.

1.2.6 Acoustic knowledge available to architects since the seventeenth century

Knowledge in physical acoustics exerted a variety of influences on the design of rooms. In *Vollständige Anweisung alle Arten von Kirchen wohl anzugeben* (Augsburg 1718),[65] the architect and theologian Leonhard Christoph Sturm

61 *Spark-pulse method* developed by W. C. Sabine and others 1913; ripple-tank method used since Davies & Kaye 1926; the first description of three-dimensional ray tracing developed by F. R. Watson 1913; see Lindsay (1968), 186; Wood (1944), 536–358.

62 An introduction to geometrical acoustics is given in 1.6.1.

63 L. Benevolo, *History of Modern Architecture* (3rd ed. 1985), vol. 1, 21; Ch. Friemert, *Die gläserne Arche: Kristallpalast London 1851 and 1854* (1984).

64 The first known patent was registered in Paris in 1855 in the name of Lambot, see H. Straub, *Die Geschichte der Bauingenieurkunst* (4th ed. 1992), 269, footnote 37; see also Benevolo (1985), vol. 1, 117 and vol. 2, 398, and in general P. Collins, *Concrete. The Vision of a New Architecture* (1959).

65 See N. Goldmann, *Vollständige Anweisung zu der Civil-Baukunst* (1699), a comment on an earlier text by the same author (1611– before 1665).

(1669–1719) listed visual and acoustic reasons for designing Catholic churches ("Kirchen vor das Pabsthum") and "churches that do not follow Rome" ("Kirchen vor diejenigen Religionen, welche von der Römischen Kirche ausgegangen sind"):

> [...] Protestant churches aim primarily at allowing a large public to see and hear the preacher well. That is why it is impossible to find [all] this space down on the floor, because in very large churches those far from the chancel cannot hear anything; therefore that space has to be compensated above [with galleries] [...] That is why in Protestant churches the floor in the church must be free [of columns and pillars].[66]

Such general considerations helped to build many Protestant churches with good acoustics. It was at that time that some architects, such as Christopher Wren in the new church of *St. Paul's* in London (1666) or George Saunders while working on his *Treatise on Theatres* (1790), began experimenting with the radiating power of the human voice, the maximum distance reached and the maximum number of listeners and volume of space.[67]

Elements that proved helpful for a certain hall or theater were often copied for another without proper understanding of the reason and consequently did not achieve the expected results. For example, the mattresses lining all the walls and even the *Musick Room* above the rear stage in the *Court Theatre,* built in 1665 by Inigo Jones in the Great Hall in *White Hall* in London, failed to improve the clarity of acoustics, and the disturbing reverberation persisted because the playhouse with boxes had no ceiling, and the large hall in which it was located was not fitted with absorbing material.[68]

The assumption that a hidden water current flowing from under the stage to the back of the parterre at the *Teatro Argentina* in Rome (1732) would help to improve the acoustics of "this dead hall" *(sala sorda per le voci)* proved to be totally wrong.[69] The movement of the water meant to carry the sound and reach the listeners through openings under the seats failed to do so.[70] In other

66 "[...] in der Protestantischen Kirchen sihet man vornehmlich darauf / dass eine grosse Menge einen einigen Prediger wohl sehen und hören könne / daher man die Stellen unmöglich auf der Erden gewinnen kann / weil bey gar grossen Kirchen / die weit von der Cantzel zu stehen kommen / nichts hören können / sondern muss man sie übereinander [mit Emporen] zu gewinnen suchen /... / Dahingegen die Protestantischen Kirchen den Platz in der Kirche / so viel immer möglich ist / gantz frey [von Säulen und Pfeilern] haben wollen." Germann (1963), 147 f.

67 A. Favaro, *L'Acustica* (1882), 28 f.

68 J. Orrell, *The Theatres of Inigo Jones and John Webb* (1985), 168 ff. The measures of *Great Hall* in *White Hall* are given in table C in the appendix.

69 The reason for excessive sound absorption in this theater is the vaulted ceiling of thin wood.

70 "Sound", *Encyclopedia Metropolitana,* vol. 25, 752.

theaters the convection of the heated air channeled in certain directions was expected to improve the acoustics.[71] Since the late eighteenth century the installation of a semi-cylindrical resonance chamber under the orchestra was common mainly in Italy. This chamber could also be connected to the stage via acoustic tubes, as in Saunders' plan for the *Teatro Regio* in Turin.[72] While the resonance of such an installation can be audible, like the vibrations of any wooden orchestra podium, neither the resonance frequencies nor their loudness were predictable. In lowered orchestra pits of the nineteenth century the back wall was often built as a concave semi-cylindrical mirror expected to be sound "reinforcing". In practice, for musicians seated at the disturbingly loud sound focus this element was a disadvantage and on the audience it had no effect because it directed the sound horizontally, thus preventing it from reaching beyond the pit. In the late nineteenth century considerable money was wasted for completely useless wires or strings believed to absorb sound due to resonance.[73] Many theater ceilings were fitted with so-called resonance domes, with unpredictably good or bad results.[74] In 1808 the installation of a "sound dome" at the old *Burgtheater* in Vienna was an expensive failure in this acoustically intimate auditorium,[75] whereas the domed ceiling at the new *Covent Garden Theatre* in London, opened 1858 after the destruction of the old building in a fire in 1856, was a success.[76]

In the seventeenth century circular, semicircular, elliptic and other curved ground plans played an important, nearly mythical role in the building not only of churches and halls but also of opera houses with tiered boxes. Treatises of theater building from the sixteenth to the twentieth century have yet to be studied systematically, but many examples show that the ground plan's "curve" was chosen mainly to improve the sightlines from the lateral boxes. Sound distribution was usually planned for the sound source at the focus only.[77] A geometrical analysis of sound distribution from the singers and the orchestra players placed outside the focus indicates that sound reflections "glide" along the concave surfaces, creating a zone of sound concentration along the side

71 E. Tamburini, *Il luogo teatrale* (1984), 89; A. Niccolini, *Alcune idee sulla risonanza del teatro* Napoli 1816 (1816), 10.

72 Forsyth (1985), 95.

73 J. M. Allen, "Acoustics of Public Buildings, Defects and Remedy" (1874); A. E. Dolbear, *Il Telefono, descrizione dei fenomeni di elettricità, magnetismo e suono in esso implicati e norme per la costruzione del telefono parlante* (1878), 38, cf. Favaro (1882), 60 ff.

74 For the impact of the shape of the ceiling in theaters, see 1.8.8.

75 *Vaterländische Blätter* (1808), 39 f. For the old Burgtheater of Vienna see also 1.8.8: Theater acoustics and figure 61.

76 Forsyth (1985), 122.

77 G. Saunders, *A Treatise on Theatres* (1790), § 1: Optics and Phonics considered, as They Relate to Theatres.

walls and at the back of the parterre, as well as excellent acoustics in front of the boxes and in the first rows of the galleries, as shown in a ray diagram.[78] The sound concentration at the focus in the parterre's rear is usually not disturbing because the profusely decorated box fronts and the absorbing box openings create diffuse reflections of middle and high frequencies and absorb low frequencies. Echoes do not occur in rooms with limited dimensions.[79]

Without properly understanding this theoretical background, eighteenth- and early-nineteenth-century authors of many architectural treatises, mainly in France and England, recommended elliptical, semicircular or circular ground plans for theaters, as did Saunders in his *Treatise on Theatres* (1790), where he directly referred to Athanasius Kircher, or Claude Nicolas Ledoux in *L'architecture considérée sous le rapport de l'art, des moeurs, et de la législation* (1804). Ledoux proposed a circular ground plan for the building and a semicircular auditorium for the new theater of Besançon. In *Parallèle de plans des plus belles salles de spectacles d'Italie et de France* (1764) Gabriel Pierre Martin Dumont proposed a circular concert hall with spherical dome, and Sieglitz in *Encyclopädie der bürgerlichen Baukunst* (1792–1798) praised the oval shape for concert halls as excellent, "because the music will sound well" ("weil darin die Musik sich gut ausnehmen wird").[80] But in large, sparely decorated rooms without tiers these recommendations caused acoustic problems.[81]

Notwithstanding the progress in acoustic science, architects were not better informed in the nineteenth century. Many theater and concert hall architects used models that experienced musicians and experts deemed acoustically effective. Charles Garnier, the architect of the still extant and acoustically successful *Nouvel Opéra* in Paris opened in 1872, concluded, after a poll among directors of the main opera houses in Europe, that good acoustics are a matter of coincidence.[82] Fortunately, the condition at the architectural competition was that the *Palais Garnier* should be a somewhat larger version of the old, excellent *Opéra* at the rue Lepelletier (figure 53).[83]

78 D. Baumann: "Qualche riflessioni sull'acustica a soggetto della ricostruzione del teatro all'italiana 'La Fenice'" (1996), 81–84.

79 For the audibility of echoes, see also 1.1.8 and 1.8.7.

80 O. Schreiber, *Das Orchester* (1938), 209. On similar ground plans and ceilings of halls see 1.8.7.

81 H. Habel, *Das Odeon* (1967), 155 f. On the ellipse in theater building see 1.8.8. For a discussion of the echo in the Royal Albert Hall, see Beranek (1962), 317ff.; Bagenal and Wood (1931), 62 f.

82 Ch. Garnier, *Le nouvel opéra de Paris* (1876), 150.

83 M. Semper (1904), 485–9, especially 487.

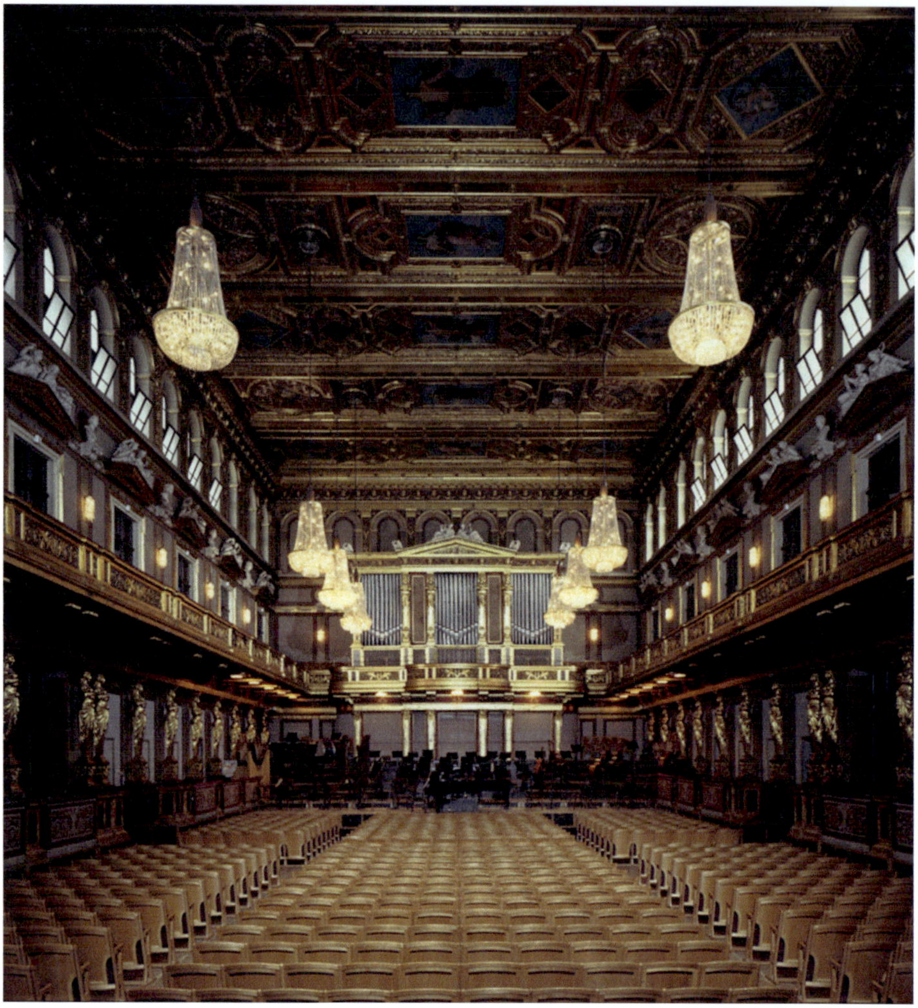

Figure 10: Vienna, *Musikverein* (1870), Great Hall, view to the podium.

The elliptical concert hall in the old *Schauspielhaus* in Berlin with its notoriously bad acoustics prompted Louis Catel in 1815 to recommend an oblong hall for excellent acoustics, referring to the hall at the opera house in Berlin, whose vaulted ceiling was said to be modeled after the *Capella Sixtina* in Rome. In 1827 Catel remarked that the new rectangular hall of Otmer's *Singakademie* was even better because it had a flat ceiling. Chladni had predicted it would have good acoustics even before construction began. Concert halls were not only compared with and built after temples and sacred rooms but also followed the models of excellent aristocratic halls. One of the most famous examples is the *Goldene Saal* of the *Musikverein* in Vienna (figure 10) built by Hansen in 1870, whose height-to-width proportions of 1:1 clearly follow the model of a *Sala grande,* as the great *Redoutensaal* at the Hofburg (figure 11) and the *Winterreitschule,* both also used for concerts.[84]

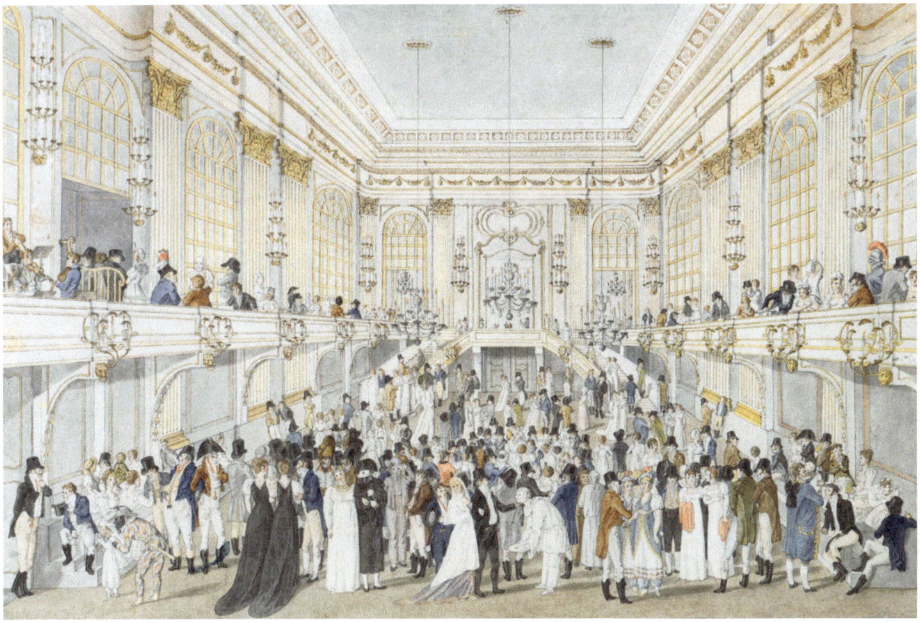

Figure 11: Vienna, *Hofburg,* Ball at the Great *Redoutensaal* (around 1810).

Since the eighteenth century, design after empirically improved models underpinned a series of excellent theaters and opera houses. The same approach produced a series of large concert halls with outstanding acoustics built during the second half of the nineteenth century and in the early twentieth century. With hindsight, this period deserves to be called the "golden age" in the history of opera houses and concert halls.[85]

1.2.7 Old and new rooms

Many excellent halls and theaters have disappeared for various reasons, such as inadequate size for the increasing number of listeners and musicians towards

84 See also 1.9.5.

85 F. Winckel, "Die besten Konzertsäle der Welt" (1955), 750–3; Forsyth: "The Shoebox and Other Symphony Halls", *Buildings for Music* (1985), 197 ff.; D. Baumann, "Ein Spätling im Goldenen Zeitalter des Konzertsaalbaus" (1995), 37–50.

the late eighteenth or during the nineteenth century and absence of stricter safety conditions (danger of fire and collapse, insufficient heating and ventilation, lack of staircases and bathrooms, etc.). Hence it is only partly true that mainly the acoustically best halls survived.[86] An excellent acoustic quality often penetrated awareness only when the hall was destroyed. This was the case when the old *Drury Lane Theatre* dating from 1775 was replaced by the new, twice-larger building designed by Henry Holland in 1794. With 3,600 seats it was the largest theater of Europe at the time. When it was destroyed by fire in 1809 Wyatt erected a circular, somewhat smaller hall, explaining:

> I was aware of the very popular notion that our theatres ought to be very small; but it appeared to me that if that very popular notion should be suffered to proceed too far it would in every way deteriorate our dramatic performances, depriving the proprietors of that revenue which is indispensable to defray the heavy expenses of such a concern.[87]

But the new house that opened in 1812 was no less disappointing and had to be changed due to inadequate sightlines and acoustics.[88] The old *Burgtheater* in Vienna, which had excellent acoustics but horribly bad sightlines, was equally missed by a whole generation of actors, musicians and theatergoers after it had been pulled down in 1888. Similar documents are extant for *St. James's Hall* in London, which was destroyed in 1905.[89]

Following the construction of larger theaters, opera houses, and concert halls, many of the old halls remained nevertheless in use. This variety of spaces enabled a separation between *Opera buffa* and *Opéra comique,* staged in smaller theaters, and *Grand opéra,* reserved for large houses, as well as between chamber music, more often played in small halls, and large symphony and choir concerts, usually performed in great halls. In the nineteenth century this separation became quite common, although not all music centers could afford both types of halls. Even today, small halls and theaters in smaller provincial towns have only partly been replaced by larger ones due to economical considerations.[90]

86 Forsyth (1985), 13, footnote 6.

87 Mackintosh (1993), 32 ff.; see also 2.5.3 and 2.5.4.

88 J. Gregor, *Geschichte des Österreichischen Theaters* (1948), 212; see also 1.8.8.

89 See also 2.7.5.

90 For a comment on the historical process of change, see table 11 (dimensions of selected opera houses) and table 12 (dimensions of selected concert halls). For music related quality of room acoustics of theaters and opera houses, see also 1.9.2.2.

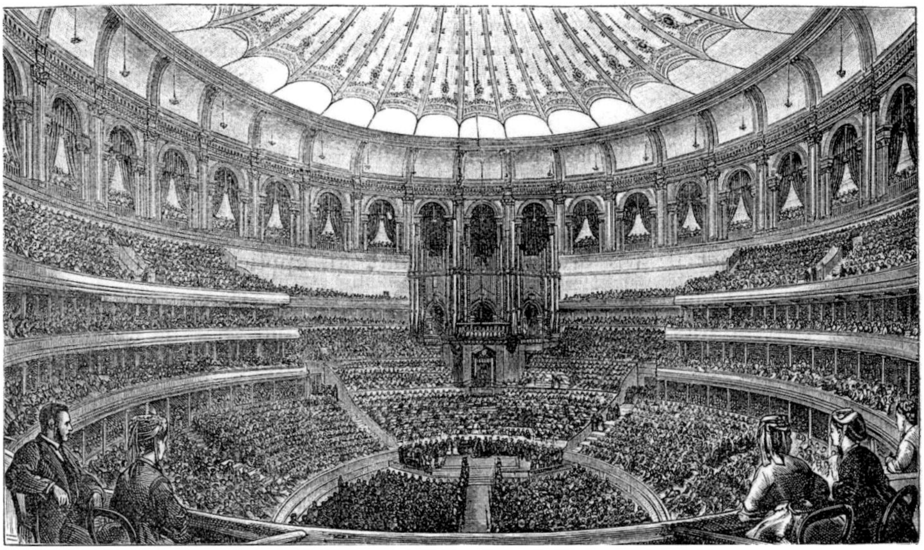

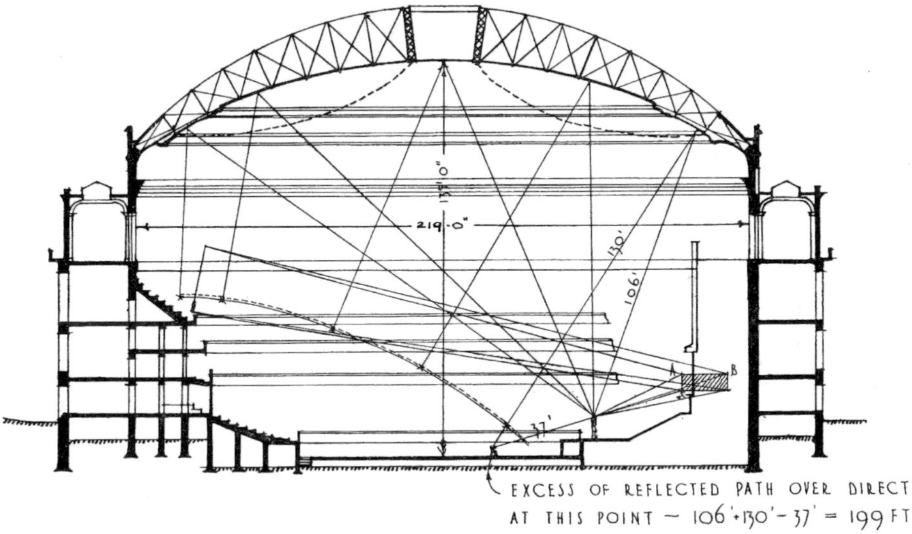

Figure 12: London, *Royal Albert Hall:* a) The grand opening by Queen Victoria, 29 March 1871; b) Longitudinal section with velum: on the podium the excess of reflected path via the ceiling is 60.65 m and the time delay of reflected sound is 0.176 sec (20° Celsius).

1.2.8 Festival halls

In the late eighteenth century and throughout the nineteenth the large choir festivals and public mass events in England, Europe and America entailed special architectural tasks. First held in music rooms, assembly halls, taverns and ball rooms, public musical performances were then moved into churches and cathedrals and finally took place in very large industrial exhibition halls, crystal palaces and ad hoc festival halls and buildings. Large cities could afford to erect permanent stone buildings as the *Royal Albert Hall* in London built in 1871 (figure 12) and the *Palais du Trocadéro* in Paris built in 1878 for the *Exposition universelle* (figure 52, page 132). Both spaces feature elliptical vaults and ground plans whose enormous dimensions caused severe acoustic problems.[91]

The echo generated in the *Royal Albert Hall* due to its dimensions of 67 × 56 × 42 meters and enormous time delay of reflections immediately sparked a public debate, whereas Parisians displayed lingering admiration for the technical achievement of the 60 × 50 × 32-meter hall with an oval *parterre* and circular upper floor, though the acoustics were practically a flop despite the small mirrors used during construction to verify the even distribution of sound reflections.[92] In many of the large festival halls raked platforms and large swaths of sound absorbing cloth were installed in the hope that they would improve the difficult acoustics.[93]

91 Forsyth (1985), 156; for more information on acoustic properties of festival halls, see 1.8.7.

92 Favaro (1882), 168; for ground plan of the hall at the *Trocadéro,* see *Handbuch der Architektur* IV/4/4/1-3 (1894), fig. 225.

93 For example, in the *Crystal Palace* in Sydenham, see 2.7.3: The *Handel Centennial* at *Crystal Palace* in *Sydenham* and 2.7.4: Documents on acoustics in the transept of the *Crystal Palace.* The dimensions of some of these mostly forgotten festival halls are listed in table 5 in 1.8.2.2.

1.3 Room acoustics and music: changing relations

The discussion on the performance practice in *San Marco* in Venice raises important questions: who selected the performance site and who decided where the musicians should be placed? When was this decision deliberately made in relation to the music to be performed? Were those responsible mindful of how these decisions affected the acoustical quality of the presentation?

1.3.1 The function of the performance site and its significance for room acoustics

Before the emergence of public opera in the seventeenth century and of public concerts in the eighteenth, Western music was widely determined by non-musical functions: as liturgy, courtly representation, dance, entertainment and spectacle – "it served these, notwithstanding its important role in each case."[1] Halls were built or used for, but also adapted to, certain purposes or functions. Whether and to what extent those responsible cared about the acoustical quality of these spaces calls for specific investigations. For each event we need documents on the number of musicians and their position, as well as on many other acoustically important details that would allow us to draw conclusions on the quality of room acoustics.

Before the seventeenth century the relation between music and room acoustics is sparely documented. There is evidence that those in charge aimed at worthy presentations that would be heard by the most important attendants. Documents show, for instance, that in 1436 Dufay's motet *Nuper rosarum flores* was sung at the consecration of *Santa Maria del Fiore* in Florence by ten Papal singers directly under Brunelleschi's huge dome in the octagonal presbyterium. Mainly the pope and the honorary guests sitting close by benefited acoustically from this placement, whereas the roughly 10,000 persons sitting or standing in the main nave and the side naves could hear the event only from far away, with many not even seeing anything.[2]

1 E. Lichtenhahn, *Musik und Raum*: Gesellschaftliche und ästhetische Perspektiven zur Situation um 1800 (1989), 10.

2 D. Baumann, "Musical Acoustics in the Middle Ages" (1990), 199–210.

Athanasius Kircher was probably among the first to address the quality of acoustics perceived by the listeners in a printed publication. In the chapter "On pathetic music" in his *Musurgia universalis* (1650) he describes, obviously on the basis of experience, the main characteristics of a good music room. This paragraph, inserted after the *Affektenlehre* and before the treatment of the different "harmonic styles", is titled "On the place for pathetic music":

> Not [too] filled with people and things [...] because voices break / bounce back and are stifled.
> Not too narrow [...] because the walls reflect too much / and the singers are standing too close together.
> In churches without people / or in halls without tapestries / music resounds better.
> An average space is necessary / not too narrow / nor too wide / the walls should be of heavy plaster / the ceiling either vaulted or plane with panelling / to render absolutely equal reflections.[3]

Nearly one hundred years later the German opera singer, composer and theoretician Johann Mattheson (1681–1764) discusses different musical styles and settings in his book *Der vollkommene Kapellmeister* (1739), focusing on the spaces for presentation, namely church, theater and courtly rooms, though without mentioning their specific acoustics. But his comments on the influence of room acoustics on the style of instrumental chamber music are astonishing:

> But although it is possible to perform church pieces and dramatic things in chambers and halls, their style changes no more than a sermon that does not become a poem if presented in a small room.[4]

According to Mattheson, then, it is still the function that defines the setting and style. This is a binding rule. However, referring to Corelli's *sonate da camera*

3 A. Kircher, *Musurgia universalis,* German ed., Schwäbisch Hall (1662), Liber IV, *Diacriticus,* Caput 3, §1, 154.: "Von der pathetischen Musik" – "Von dem Ort der Pathetischen Musik": "nicht [zu sehr] mit Leuten und Sachen erfüllt, [...] weil hierdurch die Stimmen gebrochen / reflectiert und erstöckt werden; nicht gar zu eng, [...] weil die Wänd gar starck reflectieren / auch die Singer gar zu dick in einander stehen; in Kirchen da keine Leut / oder in einem Saal da keine Tapezereyen / resonirt die Music viel besser; gehört also ein mittelmässiger Ort darzu / nicht zu eng / nicht zu weit / die Mauern sollen von starckem Gips gemachet / der [Ober-]Boden [d. h. die Decke] entweder gewölbt oder gleich [eben] getäfelt sein / damit also die reflexio gantz gleich sei." The author could find no evidence that the literature on the history of acoustics has ever discussed this description.

4 J. Mattheson, *Der vollkommene Kapellmeister* (1739), chapter 10, §103 (facs. ed. 1954, 91): "Denn ob man gleich in Sälen und Zimmern auch wol Kirchen-Sachen und dramatische Dinge aufführen kann; so werden doch durch die Veränderung des Ortes die Schreib-Arten eben so wenig verändert, als eine Predigt, die nimmer zum Gedichte wird, ob man sie gleich im Cabinet hält."

(available in print at that time), he explains that only good-quality compositions can cope with changes of space and room acoustics. He heard these sonatas

> [...] at the Netherlandish church not during service, namely at vespers or after their end, and played not only by the organ alone but also by a small violin consort, [and this] with much pleasure despite the fact that at church and in the theatre excellent voices are more important than instruments and their style, [...] which are used only for accompaniment and support.[5]

Thus, Corelli's sonatas could be played not only in the original setting of a trio sonata in halls and chambers but also in churches under completely different acoustical conditions because they are written "with diligence and care, [...] with pleasantly pure middle voices," therefore they could also be played on the organ alone (today we would say in a "version for organ solo") or by a small string orchestra.

These rather complicated explanations on the relation between musical style and function, which go back to Marco Scacchi and Giovanni Francesco Anerio and their definition of music according to function, are important for the second part of this study on Handel's *Messiah*. Let us, therefore, briefly summarize their meaning here: the function of music (sacred, secular or chamber) defines the music setting and the instrumentation, namely, as church style ("high"), theater style ("medium") and chamber style ("low"). The affect is clearly defined by style, namely, "solemn" in high or church style, "dramatic" in theater style and "natural" in low or chamber style. This means that besides the exterior aspect, which is always subject to the influence of room acoustics, a musical composition always includes an interior aspect that is more or less independent of exterior conditions. It is the setting structure that leaves a good musical composition independent of the exterior space, at least to a certain degree.

On the other hand, certain elements of the musical structure are affected by the acoustical properties of the functionally appropriate type of space. Mattheson does not say it, but such room-related elements – not only in Baroque music – may be perceptually related to spatial concepts, such as *topoi*. Even on the stage of an acoustically very dry opera house elements of church style may evoke symbolically a highly reverberating church, as does, for instance, the

5 *Ibid.*, §105–6: "in den Holländischen Kirchen, wiewol ausserhalb der zum Gottes-Dienst bestimmten Zeit, nehmlich in den Vespern, oder nach deren Endigung nicht nur vom Organisten allein, sondern auch von einem Violinen-Concert [d. h. Streicherensemble] ehemals mit viel Vergnügen gehöret habe [...wo es doch sonst] in Kirchen und auf dem Schau-Platz [d. h. Theater] immer mehr auf die Hervorragung der Menschen-Stimmen ankommt und der Instrument-Styl [...] nur zur Begleitung und Verstärkung da ist."

first scene of Puccini's *Tosca* performed in the church of *Sant'Andrea della Valle* in Rome. If the exterior acoustics are "wrong," it is the performing musician's task to adapt the playing or singing to actual room acoustics by changing the articulation, dynamics or tempo to achieve the composer's intended expression. A legato has to be played out fully in a theater, whereas in a church the same legato would sound blurred if exaggerated.[6] Such an adaptation can even exceed the usual interpretation and require a new version or arrangement (*Bearbeitung*) of the musical work.[7]

1.3.2 Public concerts and the change of performance practice

It was not fortuitous that Mattheson, as practicing musician, chose instrumental music to point to the effect of a change in the performance space.[8] Independent of words, instrumental music was not a separate category of style in Mattheson's function-based system of church, theater (or *Schauplatz* or hall) and chamber (also hall or room) but a subcategory of each. It was therefore expected to appertain to different categories of space. It could belong to church style (in *S. Petronio* in Bologna concerts for solo violin or solo trumpet were played during mass), theater style (as the *sinfonia* at the beginning of an opera or as music during the entr'acte) and chamber style. The new category of the concert hall, which developed in Catholic regions from oratorio halls and in Protestant regions from halls used by the church and *collegia musica* during the eighteenth century, became an important place for the performance of orchestral music, even though it was still played in church, princely chambers and in many theaters if these served also as concert halls.[9]

The new category of public music rooms and concert halls played a prominent role in the changing music life and the altered relation between space and music. In this context, publicly performed instrumental and vocal music was able to discard its sacred and courtly functions even before the French Revolution, relying increasingly on professional musicians. In Paris these changing relations became especially evident in the organization of music education: the *Conservatoire National de Musique et de Déclamation,* the first national school, was based on the *Ecole pour la Musique de la Garde Nationale* foun-

6 See 1.4.1: Music and space: conditions for perception – The situation of musicians.
7 See E. Lichtenhahn, Introduction to *Bearbeitung in der Musik* (1986), 14.
8 Lichtenhahn (1989), 14 f.
9 For other room categories used as concert halls, see H. Habel, *Das Odeon in München und die Frühzeit des öffentlichen Konzertsaalbaus* (1967).

ded in 1792 and in 1795 replaced the *Ecole Royale de Chant* founded in 1783. Already when it opened, the *Conservatoire* featured a concert hall,[10] which in 1811 received the still extant shape and décor. The hall with excellent acoustics and a seating capacity of 1,000 owed its fame to the *Concerts français* under the direction of Jean-François Habeneck (figure 13).

In music history the changes in performance practice are usually presented and explained as follows:

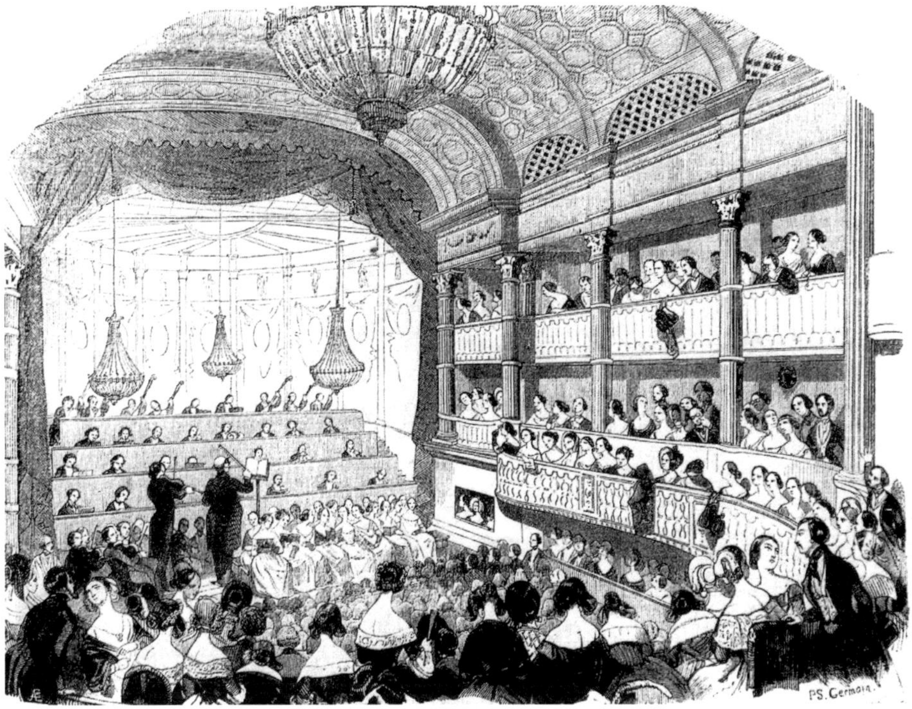

Figure 13: Paris, Hall of the *Conservatoire* (1811), concert in 1843.

1. The participation of the lower aristocracy, the middle class and, during the nineteenth century, also the working class (mainly in choir concerts) led to an increasing number of listeners.
2. Concert halls and opera houses therefore had to be enlarged.
3. This led to the enlargement of orchestras and choirs.
4. Louder instruments had to be developed in order to meet the new acoustic demands.

10 A. Donnet, *Architectonographie* (1837), 247 f.; D. Arnold, *Education in Music,* §V, "Conservatories", *NG2,* vol. 6, 19.

This argumentation is evidently too narrow. A more detailed analysis shows that the change was not a clearly aimed process but a transition consisting of various overlaying processes that evolved differently at different places. Thus the larger number of listeners during the first phase led only to an increase in the number of seats in the existing rooms, sometimes beyond the bearable limit.[11] In some opera houses the number of seats decreased following a period of increase. Each case, then, calls for a separate investigation. Sometimes old small rooms were still used, as existing larger halls were considered acoustically problematic due to their size.[12] Larger halls were often built quite some time after the need for them had become evident, partly because the financial means for musical events that had lost former sacred or courtly functions were hard to obtain. Another reason was the disappearance of patrons or their dwindling resources. In many cities the absence of appropriate rooms hindered the development of public concerts. The second part of this study presents documents on such a situation in eighteenth-century London,[13] and Vienna in the late eighteenth and early nineteenth century.[14] In Paris around 1835 traveling musicians had enormous difficulties in finding a publicly available hall that would provide enough space for an orchestra.[15]

While grand opera, public concerts and musical mass events took root an enormous surge in instrument making began to meet the need for the growing number of instruments. Instrument construction, too, changed, as it had to produce a more powerful sound and respond to new stylistic-aesthetic demands: a larger ambitus, the production of new sound color shades and the control of intonation of the entire chromatic scale. The increased dynamics of certain registers or of an instrument's full compass were an occasional side effect of such constructional changes. Causes and results cannot always be clearly distinguished. An impressive example is the improved upper register of the pianoforte in the early nineteenth century, which allowed the expansion of *cantabile* melodies up to the instrument's highest notes. Furthermore, the necessary reinforcement of the frame and chords increased the maximum sound power.

With the increased loudness, caused partly by the larger numbers of players and singers, room had to meet new acoustical standards of clarity and immediate acoustical "reaction".[16] As listeners began to perceive more conscious-

11 On the acoustics of the theaters in London, see 2.5.4.

12 For instance, the great Redoutensaal at the Hofburg in Vienna for Beethoven symphonies (*Allgemeine Musikalische Zeitung* 15.5.1805), see 1.9.5.

13 See 2.5 on Handel's performance practice of *Messiah* in London.

14 See 1.9.5 on the relation between type of space, music genre and orchestra size.

15 August Gathy, *Neue Zeitschrift für Musik* 30 (1835), 122; see O. Schreiber, *Das Orchester* (1938), 206, footnote 8.

16 See 1.4.1: Music and space: conditions for perception – the situation of the musician.

ly and subjectively,[17] an increased interest in architectural design for acoustically appropriate music rooms became evident since the late eighteenth century. Independence from non-musical functions made room for new architectural ideas, which, at the time of the French Revolution explored new stylistic references to Greek and Roman antiquity.[18] Yet the departure from traditional building models often yielded acoustically disappointing results, sparking, therefore, the need to understand and explain the relation between design and acoustical quality, as well as a growing number of personal and published documents on the relation between space and sound.

The music periodicals that had been appearing since the eighteenth century and proliferated during the nineteenth sometimes featured discussions on room acoustics.[19] The shift of criticism from the work itself to the musical performance clearly raised the importance of room acoustics for the sounding of the musical work. Traveling virtuosos, critics and music lovers began to compare music rooms from different cities. Towards the late nineteenth century all these important factors established certain standards of size and acoustical quality of spaces for the various musical categories.

1.3.3 New definition of chamber music around 1800 in relation to hall size

The small and large opera house and concert hall, the new types of space that emerged in the nineteenth century, generated new definitions of musical genres. The term "chamber music" is a telling example. From 1600 to the late eighteenth century it was defined by function: *musica da camera* meant music played in smaller or larger court halls and chambers or other rooms rather than in a church or theater.[20]

Already around 1795 Charles Burney offers a definition not by function but by number of parts: "Cantatas, single songs, solos and trios, quartets, concertos and symphonies of few parts."[21] In 1805 he rewrites the definition, adding hall size: "Compositions for a small concert room, a small band and a small audience; opposed to music for the church, the theater, or a public concert

17 Lichtenhahn (1989), 11.
18 E. Tamburini, *Il luogo teatrale* (1984).
19 Schreiber (1938), 202 f.
20 For an investigation of the size of these halls, see 1.9.2.
21 Ch. Burney, *A General History of Music* (1776–1789).

room."[22] Well half a century later Berlioz says that chamber music should be played "in a medium-size room without much furniture or carpets," in order to produce the desired effect. If the same music "a simple piano trio, for instance, the trio in b flat major by Beethoven" is played in an "ordinary theater the listener still hears it but does *not vibrate.* And one *must* vibrate *with* the instruments and voices and *because* of them in order to have genuine musical sensation." As a result he admires the music but "only with his mind, not his heart." In an even larger hall (Berlioz refers to the old *Opéra* at *rue Lepelletier*),[23] a listener "still *hears* but he has virtually stopped *vibrating;* he admires the work, but rationally, not any longer carried away be irresistible feeling. Enlarge the room even more, so that the listener is still further removed from the musical center. He is now as far away as he would be if the three performers were playing in the middle of the Opéra stage and he were sitting in the first tier. He still *hears,* he does not miss a note, but the *musical fluid* no longer reaches him."[24] This observation is based on Berlioz's experience not only as conductor and composer but also as a music critic who at that time was attending performances of both pure chamber music and mixed programs of all possible genres in various, more or less suitable, spaces in Paris and many other European cities.

1.3.4 Architectural space as part of the work's performance

As a composer and conductor Berlioz knew that the size of the music ensemble must be adapted exactly to the hall's size to achieve full sound.[25] In 1837 he had to trim his dream to fill the giant *St. Peter's* dome in Rome with sound to

22 Ch. Burney, "Chamber Music", *Ree's Cyclopaedia* (1802–1820).

23 The auditorium is about 2 m narrower and 1.50 m lower than the later *Opéra* by Charles Garnier; see 1.8.3, figure 58.

24 H. Berlioz: "Sur l'état actuel du chant", *A travers chant* (1862, repr. 1971), 115: "On entend, on ne vibre pas. Or, *il faut vibrer* soi-même avec les instruments et les voix, et par eux, pour percevoir de véritables sensations musicales. ... supposez ... que les salons ... puissent s'agrandir graduellement. ... voilà notre salon grand comme un théâtre ordinaire; notre auditeur ... il *entend* toujours, mais il ne vibre presque plus; il admire l'œuvre, mais par raisonnement et non plus par sentiment ni par suite d'un entraînement irrésistible. Le salon s'élargit encore, l'auditeur est éloigné de plus en plus du foyer musical. Il en est aussi loin qu'il le serait, si les trois concertants étaient groupés au milieu de la scène de l'Opéra, et s'il était, lui, assis au balcon des premières loges de face. Il *entend* toujours, pas un son ne lui échappe, mais il n'est plus atteint par le *fluide musical* qui ne peut parvenir jusqu'à lui."; English translation "The current state of the art of singing", *The art of music and other essays* (1994), 60–61.

25 For the theoretical background of this phenomenon in physical acoustics, see 1.6.4.

the dimensions of the *Dôme des Invalides* in Paris (figure 14). For the first performance of the *Messe des morts* in this more modest space (the central ground plan of the 60 × 60-meter church would have fitted snugly in the space below the dome of St. Peter's), he calculated very well "the power of means": the main orchestra with 210 singers and 190 instruments and four separate orchestras with 38 brass instruments and timpani were able to provide adequate power and quality of sound, and, at the most overwhelming moment in the *Tuba mirum,* "the horror produced by the five orchestras and eight pairs of timpani was indescribable."[26]

Seven years later Berlioz realized that acoustic quality and maximum space size have certain limits. After a spectacular concert at the *Salle des machines* during the 1844 *Grand Festival de l'Industrie* in Paris he was honest enough to admit that even 1022 musicians, two assistant conductors and five choir conductors could not fill the hall sufficiently with sound (figure 15).[27]

Only few moments were impressive, he later said, such as the *Andante* from Weber's *Freischütz-Ouverture* with 24 French horns, the prayer from Rossini's *Moses* with 25 harpists who played arpeggios not as usually but in four parts, Berlioz's *Hymne à la France* and the consecration of swords from Meyerbeer's *Les Huguenottes* with solo parts sung by a twenty-fold stronger group of singers, that is, 80 instead of four basses. Yet this strong impression was the privilege of only "the musicians and those listeners very close to the orchestra". According to Berlioz the *Marche au supplice* from his *Symphonie fantastique* and the *Scherzo* and *Finale* from Beethoven's *Fifth Symphony* sounded weak and dull despite the 36 double basses. These results indicate that the largest ensemble with 465 instruments and 360 voices recommended by Berlioz in his *Grand Traité d'Instrumentation* published that year was by no means utopian but calculated exactly for such special occasions, which the composer called "congrès musicaux dans les vastes locaux".[28]

Known since the legendary *Handel Commemoration* in 1784 at *Westminster Abbey* in London and the great musical events of the French Revolution, such large-scale events became increasingly fashionable. But Berlioz's idea "to convene all musicians in Paris to perform a work composed especially for such an event in a great hall built especially for this purpose and designed by an architect versed in acoustics and music" had to wait for a long time.[29]

26 "[...] au moment du *Jugement Dernier* l'épouvante produite par les cinq orchestres et les huit paires de timbales accompagnant le *Tuba Mirum* ne peut se peindre." H. Berlioz, letter to Humbert Ferrand, Paris 17 December 1837, *Correspondance générale* (ed. 1975), vol. 2, 391.

27 M. Forsyth, *Buildings for music* (1985), figure 4.20.

28 H. Berlioz, "Le diapason", *A travers chants* (1862, repr. 1971), 308.

29 "Il serait pourtant curieux d'essayer une fois, dans une composition écrite ad hoc, l'emploi simultané de toutes forces musicales qu'on peut réunir à Paris [...] dans un vaste local dis-

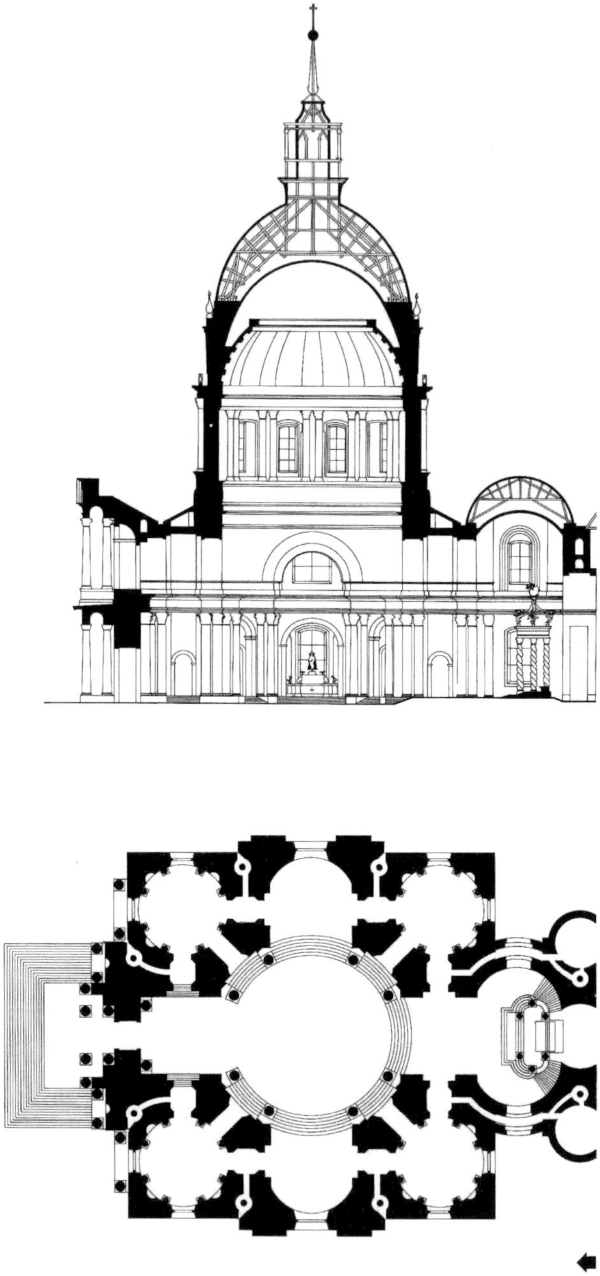

Figure 14: Paris, *Dôme des Invalides* (1675-1706), section and ground plan.

posé pour cet objet par un architecte acousticien et musicien [...]" – see H. Berlioz, *Grand traité d'instrumentation et d'orchestration moderne op.10* (1843, repr. 1971), 295.

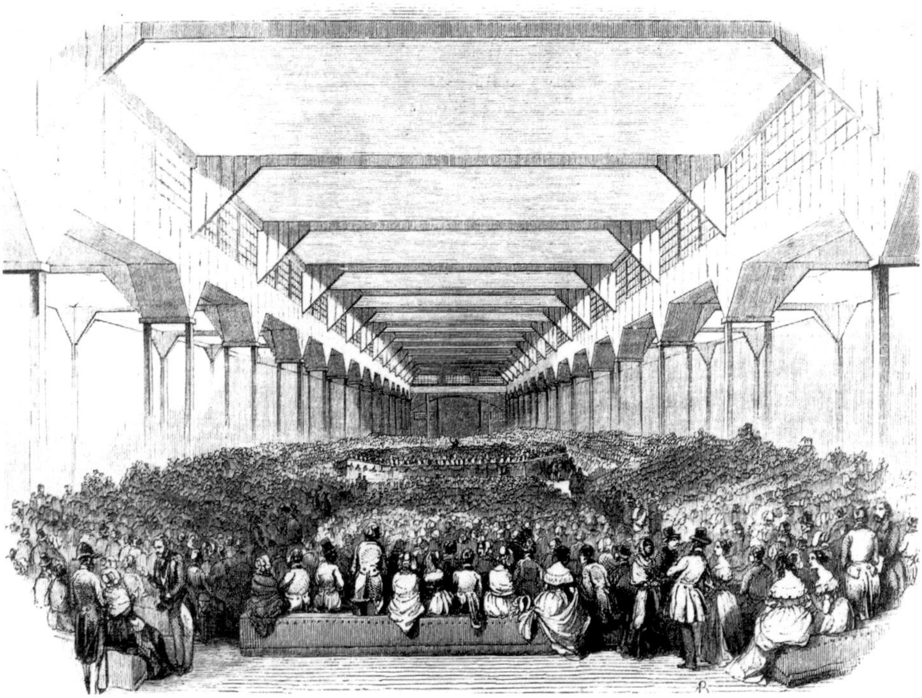

Figure 15: Paris, *Salle des machines*: Grand Festival de l'Industrie (1844).

Richard Wagner, who was no less aware of the impact of room acoustics on music performance, went even a step further. He not only exactly determined the ensemble's size but also demanded a special theater building and, following long struggles and against enormous financial difficulties, he finally realized his project in Bayreuth.[30] The long, difficult history of this theater reflects Wagner's pragmatic planning. The idea for a temporary wooden theater without boxes in Vitruvian style, developed first for Zurich and later for a floating platform on lake Lucerne, could not be implemented for financial reasons.

In the 1860s his friend Gottfried Semper, a Dresden architect and later professor at the *Eidgenössische Technische Hochschule* in Zurich, planned the Munich theater project. To test the architectural concept, Wagner insisted on a relatively small, temporary Vitruvian theater at the *Glaspalast* (figure 16a), where he could gain sufficient experience to build a "large stone theater in noble style" (figure 16b).

30 M. Kiesel, "Construction History", *The Richard Wagner Festival Theatre Bayreuth* (2007), 56–87.

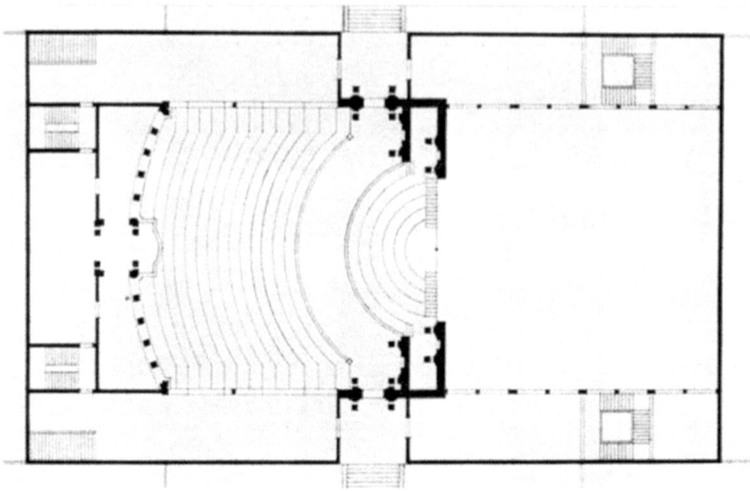

Figure 16: Munich: a) Project for the provisional theater in the *Glaspalast* (1865?).

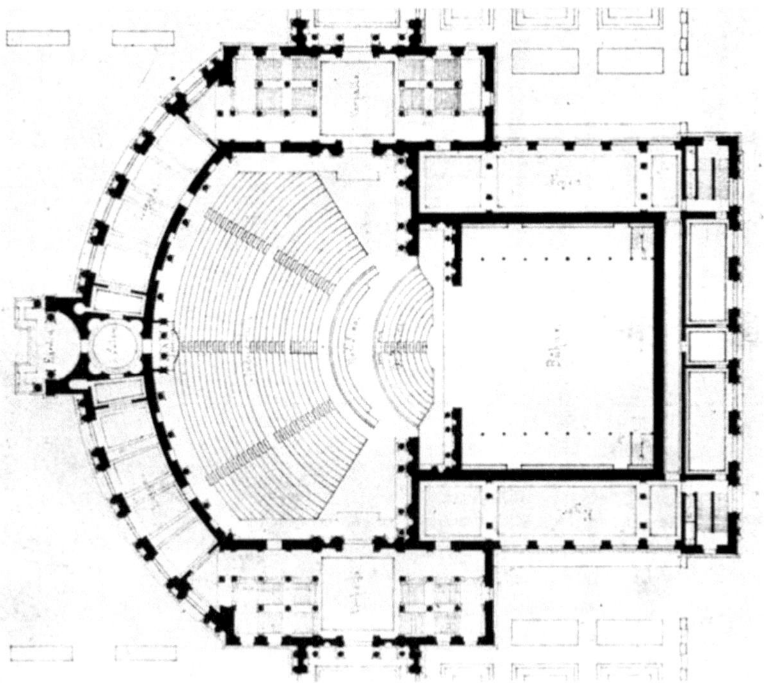

Figure 16: Munich: b) Project for the large theater (presumably 1867).

King Ludwig II rejected the provisional theater, and the expensive *Festspielhaus* failed not only due to financial reasons and the opposition of Munich's government and citizens but, as Semper surmised, because Wagner himself hesitated, as the auditorium, modeled after an ancient *odeion*, would have measured 38 × 44 meters, exceeding twice the width of the largest opera houses of the time, the *Opéra Lepelletier* and the *Teatro San Carlo* in Naples.

In 1876, when the *Festspielhaus* finally opened in Bayreuth, the auditorium length had been reduced to a realistic 28 meters (figure 17). Still, the 30-meter distance between the parallel side walls without boxes was nearly one third greater than the inner width of *La Scala* in Milan. For visual reasons the stage machinist Carl Brandt proposed side wings in the auditorium, which would not reach the ceiling, to fill the empty angles on either side of the proscenium and direct the view to the stage. Wagner probably hoped that these wings would enhance the acoustics too, like the back wall of the podium he had seen in the famous concert hall at the *Conservatoire* in Paris, which did not reach the ceiling and produced beautiful acoustics behind the stage (figure 13).[31]

Visual considerations lay also behind the Bayreuth orchestra's lowered position, which placed the musicians' instruments and movements in front of the stage below the sight lines. The first seating tests on 28 October 1874 revealed that the pit was too small for the large orchestra and had to be enlarged downwards under the stage and upwards at the expense of the first rows of raked parterre seats. Following acoustical tests the pit had to be partially covered because the sound of the instruments under the stage did not blend with those in the pit's open area. Wagner defined the result as "pure sound [...] transfigured by the acoustical sound wall, free from any extra-musical noise that inevitably accompanies the tone production of instrument playing." [32] Conducting in the "mystical abyss" [33] of the *Festspielhaus* in Bayreuth is not an easy task, as inside the pit neither the singers on stage nor the effect in the hall can be properly heard. *Parsifal*, composed for the special acoustic conditions of that theater, is fully adapted to the dark sound of the covered orchestra.[34]

31 M. Gregor-Dellin, *Wagner* (1980), 147; D. Baumann, "Wagners Festspielhaus – ein akustisch-architektonisches Wagnis mit Überraschungen" (1983), 123–50; K. Müller, "The democratization of acoustics" (2007), 174–94.

32 "... welchen verklärten, reinen, von jeder Beimischung des zur Hervorbringung des Tones der Instrumentisten unerlässlichem aussermusikalischen Geräusches befreiten Klang ein Orchester bietet, welches man durch eine akustische Schallwand hindurch hört." R. Wagner, *Ring des Nibelungen, preface* (orig. publ. 1863, ed. 1911), GS 6, 275.

33 R. Wagner, *Das Bühnenfestpielhaus zu Bayreuth* (orig. publ. 1873), GS 10, 338.

34 D. Baumann, "Wagners raumakustische Überlegungen" (1985), 117 ff.; D. Baumann, "Der Bayreuther Raumklang" (2009), 151–68.

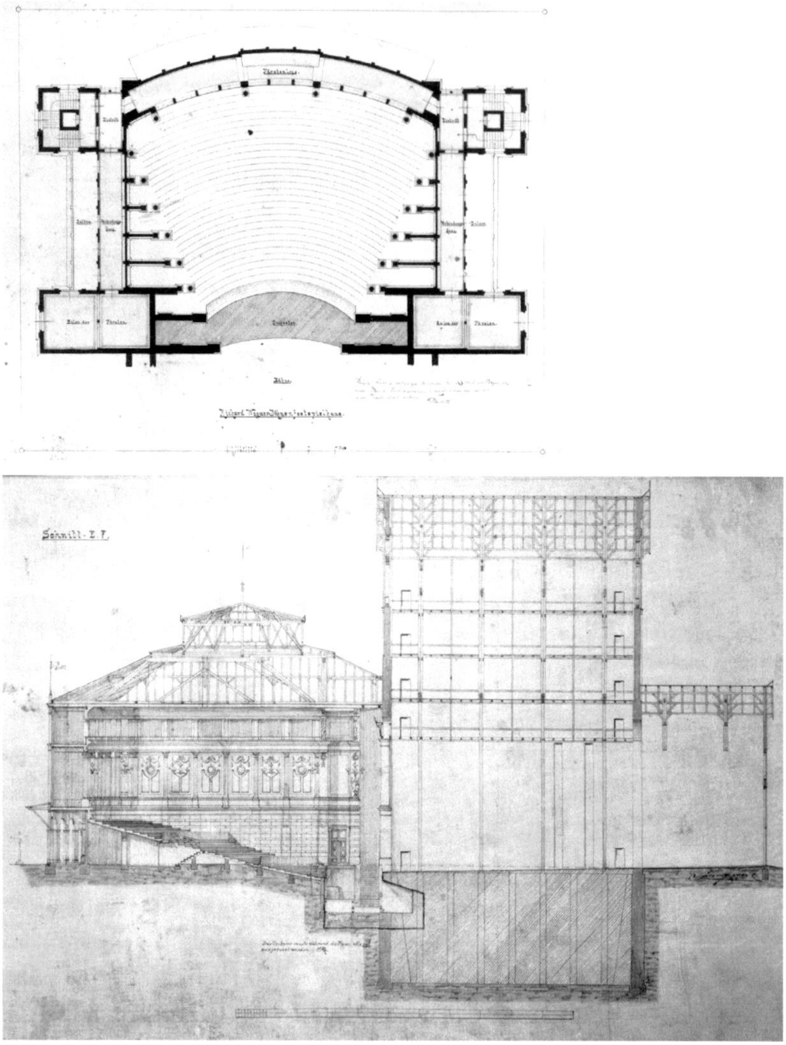

Figure 17: Bayreuth, *Festspielhaus*, ground plan and section with enlarged orchestra pit (1876).

Nevertheless Wagner created a room that protects singers from the enormous power of the very large orchestra, helps to present his music dramas in the best possible way and provides all seats with an excellent view of the stage. His modern concept of a theater auditorium without boxes, equally accessible to all, had the disadvantage that the income generated by only 1500 seats could hardly cover the enormous costs despite expensive tickets. Thus, music freed from non-musical functions at the "ideal" Bayreuth theater was now threatened by commercial limitations.

1.3.5 Electro-acoustical reproduction of space

The development of sound recording and playback laid the technical foundations for the reproduction of room acoustics. This possibility had been consciously perceived, discussed and applied long before the emergence of stereo, multi-channel and digital recording techniques. The comparison with "the original" posed a challenge from the very first days of the gramophone.[35]

The invention of microphones made recordings in large halls and churches possible. Since the 1920s loudspeakers and electro-acoustic recording reduced the perceived difference between real and electro-acoustically reproduced space. As stereo recording allowed various spatial representations of orchestra instruments and singers across the panorama between left and right channel and between virtual foreground and background, some critics came to regard the sound quality and even spatiality of the best recordings as better than those experienced in a concert hall. Sound recording was soon to exceed the documentary recording of a musical event and its room acoustical aspects, and today multi-channel electronic space is a feasible, controllable aspect of electronic sound art. A three-dimensional impression of space can be created electro-acoustically and it may differ from real room acoustics during the recording session in a hall or a studio. The task of sound recording engineers is, among others, to balance real and electro-acoustically created spatial elements.

If the real acoustics in the recording hall differ from the virtual space represented by the recording, the impression of playing quality can be surprisingly inadequate. Musicians control crucial details of musical expression, such as articulation, dynamics or tempo changes always related to room acoustics.[36] Only experienced musicians are able to adapt their playing during the recording session to later added artificial reverberation or other electronically produced changes of spatial sound. Yet the ability to control such acoustic discrepancies during the recording through experience never really replaces the musicians' direct sound control. While headphones and playback do allow for such a control of the final result, they also change the playing in several aspects. This is why they are still rarely used for recordings of "classical" music.

The more realistic the space reproduction of recordings, the more important are control and assessment of room acoustics from the viewpoint of music performance. But this is not the main topic here. Since this study includes sound recordings, the question is which recording rooms or studios were chosen and how various recording techniques affect spatial-acoustical representation.

35 See chapter 1.7.3: Spatial impression in sound recording: conclusions.

36 D. Baumann, "Musizieren im Raum – live" (1986), 49.

Chapter 1.7 will therefore elucidate the principles of sound recording techniques required for the analysis of the sound recordings of Handel's *Messiah* from different times in different halls and churches presented at the end of the second part in chapter 2.8.

1.4 Music and space: conditions for perception

1.4.1 The musicians' situation

Music is more than a mere succession of sounds. As a system of symbols that communicates emotions and musical thinking,[1] it has a certain meaning for a social group within a cultural context. It is therefore important to avoid any changes or losses in the temporal structure of sound as it travels from the musician to the listener. Some halls transmit elements of music as melodies, harmonies or rhythms without distortion. Other rooms effect such far-reaching changes in these basic elements that the music is unrecognizable when it reaches the listener. For example, a highly reverberant cathedral can unbearably blur melody, harmony and rhythmic patterns. Rapidly changing elements, such as dynamics (change of loudness), agogic, rubato (temporal fluctuation of melodic rhythm against the beat) and articulation, are more sensitive to room acoustics. The quicker the change and the wider its fluctuation, the more closely must room acoustics follow the signal. The intensity of reverberation has to diminish fast enough in order to leave musical units properly perceivable.[2] This does not mean that transparency requires a short reverberation time. Musicians often do not like acoustically dry rooms because they impede the blending of sound, which is crucial to good sound quality. In a dead hall a crescendo has to be forced, and a full and overwhelming fortissimo is impossible. In such rooms musicians not only feel hampered by an inadequate room reaction but experience the hall's "slow" answer as resistance against their playing. A good hall features both: proper reverberation characteristics for all frequencies and a sufficiently linear onset of sound in the hall down to the lowest frequencies.[3] The exact boundaries of these characteristics depend on music style. In good acoustics sound blends without any effort, and the slightest crescendo or diminuendo reaches the listener. As musicians feel supported, they explain that playing

1 H. de la Motte, *Handbuch der Musikpsychologie* (1985), 11 f.: "Musik als Sprache"; D. Baumann, "Musik von innen her empfinden und verstehen – Einsichten aus Susanne K. Langers Kunstphilosophie" (2008), 257–79.

2 Physical conditions are explained in detail in 1.5: Sound and sensation, 1.5.2.3: Masking, and 1.5.2.4: Temporal analysis of acoustical signals: blurring, blending, summation.

3 V. L. Jordan, "Einige Bemerkungen über Anhall und Anfangsnachhall in Musikräumen" (1968), 29–36; see also 1.8: Architecture related quality factors in room acoustics.

nearly happens by itself. They are able to control the unfolding of sound easily, despite the reverberation's duration.

If one of the sound parameters changes on its way from the musician to the listener, be it color (spectrum), loudness (volume), dynamics, articulation or weight of the different instruments, the sound's temporal structure and meaning can change as well. This warrants two conclusions: not every music style fits every space; playing practice is not merely technical control of the instrument but also adaptation to the hall's acoustics. Good musicians are able to cope with good or bad reactions of space by embedding its acoustical effects into the creation of sound.

For performers, the change of acoustics from an empty hall during rehearsal to a full house poses a major challenge. Furthermore, it is difficult to estimate the sound result in the hall from the podium if the acoustic conditions differ considerably between them.

Non-musicians are rarely aware of this subtle temporal coordination between playing and hall reaction. Excessively reverberant halls call for reduced dynamics or tempo or for very pronounced articulation. Dry halls prompt musicians to increase loudness and to accelerate tempi in slow movements or to slow down fast movements because of the impeded onset of sound, which musicians experience as their instruments' slow reaction. Musicians tend to compensate dead acoustics with increased expressiveness[4] and try to fill the hall with sound in order to sense the space volume and create the impression of envelopment.

This process of adaptation to room acoustics usually happens automatically. Excellent musicians are able to control it consciously, especially singers and players of instruments without resonance bodies, such as wind instruments or organs. A good hall or church thus becomes part of their instrument, acting like a resonance body that adds sound components with favorable or difficult characteristics. Instruments with resonance bodies like the piano or strings seem to be less sensitive to dry acoustics, but control of the hall's feedback is as important for them as for other instruments. In a hall with "dark sound" the musicians' playing tends to be brighter, in a dry hall fuller (a pianist with more pedal), in a reverberant hall shorter (a pianist with less pedal) because the hall fills the temporal gaps and binds notes and phrases together.

Playing music always means responding to the hall's influence on sound. It is the musician's task to play the piece in a way that transmits most clearly its structure and meaning. Room-dependent details of performance belong mostly to the non-written elements of music practice that are passed from master to pupil, be it selection and number of instruments, but also timbre, phrasing or

4 F. Winckel, "Optimum acoustic criteria" (1962), 81 ff.; J. Meyer, *Acoustics* (2009), 347 f.; J. Burghauser and A. Spelda, *Akustische Grundlagen des Orchestrierens* (1971), 105.

adaptation of dynamics of every single instrument in an ensemble. These elements are rarely discussed outside the teaching and rehearsal context.

Musicians are more or less sensitive to the influence of room acoustics. Organ players, conductors and soloists depend most on it. Great conductors know how to use the hall's effect and the buildup of sound and its blending in and through the hall. This capacity is part of their "aura", as illustrated by a recollection of the percussionist Werner Thärichen from the Berlin Philharmonic orchestra. During a rehearsal with an unnamed conductor he remarked a sudden change of sound in the orchestra. Looking up he saw Furtwängler standing in the doorway. At that moment musicians started to play differently because they began to listen to each other as he had told them to do:[5] they paid closer attention to the effect of music and thus, as far as possible for a single player within the orchestra, also to the sound's effect in the hall.

Improvised music permits a spontaneous reaction to room acoustics because sound imagination is not bound to written notation. The more detailed the notation, the more limited the musicians' possibilities to adapt their playing to the hall. In Renaissance music the freedom of *da cantare o sonare* and of *per ogni modo di strumenti* allows a choice between singing or playing a part with any instrument of the requested register. Given the more or less elaborate playing of the *basso continuo* and the choice of *continuo* instruments in Baroque music, the accompaniment can be adapted to both the soloists' needs and room acoustics. Embellishing the solo is also a way of adaptation to room acoustics. Thus, the performance of a Baroque solo piece or aria may be of greater importance than the score.[6]

Since the eighteenth century musicians have faced greater difficulties, as they have had to perform works from different times and of different styles in different types of space. They have often had to play in rooms that neither they nor the composer would have chosen. Modern instruments change the sound characteristics and require adaptation of playing if the music belongs to an earlier style. The definition of every single instrument and even of the instruments' number per part, as well as exact performance instructions, specify the execution so precisely that a change of room acoustics cannot be handled with mere "nuancing". In this case, adaptation may require a larger ensemble or even a change in orchestration by adding or eliminating parts and structural elements, to the point that musicians speak of a new version or of a "Bearbeitung". This term implies that a *work* has a stronger identity than a single *performance* because the composer imagines the sound, which he specifies as far as possible in a score. Based on this definition of the "work", musicology

5 Television interview, *Grosse Dirigenten,* emission by ARD, 9 March 1995.

6 R. Strohm, *Italienische Opernarien des frühen Settecento (1720–1730)* (1976); E. Selfridge-Field, "Dynamics of Performance" (1989), 8 f.

has introduced the term "original version" in the late nineteenth century and referred to "interpretation" as the musicians' performance.[7]

1.4.2 The composer's situation: interior and exterior spatiality of music

Richard Wagner rightly claims that the musical work of art *actually* comes into being during performance only. He addresses the role of the poet, the composer, the performer, the theater architect,[8] but also of the listener, as, for example, in the following passage from "Oper und Drama", which underscores the importance of this topic in his writings:

> Assuming [...] that in this performance the highest dramatic intention of [the work of art] would be perfectly realized, we would become keenly aware of the absence of whoever made the work of art possible, namely, the all-powerful collaborating public.[9]

In any case, space plays an important role in music performance. Compositions structurally adapted to the acoustics of a certain space of performance have existed throughout the history of music. Dufays' already mentioned motet *Nuper rosarum flores,* written for the consecration of *Santa Maria del Fiore* in Florence in 1436, is an impressive example from the early fifteenth century. Keeping in mind the reverberating acoustics of the then largest dome in Western culture, Dufay alternated slowly proceeding consonant chords for full choir that set on reverberation and fast duos with quickly passing dissonances practically without answer from the dome.[10]

Both written notation and playing instructions are fundamental for the composition's sounding aspect in space. With the beginning of public concerts and the shift of criticism from work to performance, a general increase of playing instructions in the eighteenth and nineteenth centuries would seem a logical consequence. But the composers' notation practice is individual and reflects various aspects of the musical work. Some composers, such as Beethoven, Schumann or Hindemith, left very few space-related comments and not many

7 E. Lichtenhahn, "Musikalische Interpretation – ein romantisches Konzept" (1999), 107–14.

8 See for instance R. Wagner, "Das Kunstwerk der Zukunft", *SS* 4 (1914), 115.

9 R. Wagner, "Oper und Drama" (1851, rev. 1868), *SS* 4 (1914), 225: "Setzen wir den Fall, dass [...] einer höchsten dramatischen Absicht in dieser Darstellung [des Kunstwerks] vollkommen entsprochen würde, so müssten wir dann erst recht lebhaft innewerden, dass uns der eigentliche Ermöglicher des Kunstwerks, das [...] allmächtig mitgestaltende Publikum, abginge". See also Baumann (1983), 123 f. and Lichtenhahn, *loc. cit.*

10 D. Baumann, "Musical Acoustics in the Middle Ages" (1990), 208.

instructions for these aspects of performance. Others, like Weber, Berlioz, Wagner or Mahler paid scrupulous attention to the sound quality of their works at the performance. It is not fortuitous that they were all excellent conductors. As a general rule, it seems that composers who give only few instructions related to room acoustics also pay less attention to the choice of the room for performance. For these, the space is like more or less adequate light for reading: walls that provide better hearing conditions. Such composers leave the interpretation of the score as sound to the performers. This does not mean, however, that their ideas of how their works should sound are less clear. They just trust the unwritten rules of musical practice. The others, such as Mahler, supervise the relation between notation, playing and room acoustics so closely that they revise scores and parts before and after each performance.

These different approaches are obviously not due to a composer's lack of practical involvement in the performance. Schumann and Beethoven were also excellent pianists and more or less successful conductors of their own compositions. The reason behind this difference is that their notation is coding another psychological level of the musical work. The spatial aspect of music has, in fact, two sides: music creates its own inner world with its own time, which is passing even if only in our imagination. A simple stream of sounds creates a sensation of space. But "musical" space is strangely ambiguous. Still, by means of thinking and sensation we can move within this virtual space, which has fullness and depth. Its ambiguity is reflected in the way various authors and composers described this musical-psychological phenomenon. The horizontal dimension can clearly be related to the work's immanent time, which is structured by meter, rhythm, phrase, accent, tempo, change of harmony, sequence of meaning units, density of the most frequent note values and other time-related parameters. The association between the vertical dimension and pitch in the melodic progression is clear, whereas the third dimension, which is less clear, gives the impression of depth in this virtual space: louder and more brilliant sounds with sharp articulation may appear closer, whereas soft, muted, dull, vague or blurred sounds seem more distant. Similarly, the main voice in a contrapuntal setting maybe sensed as closer, the secondary voice further away.[11] Those composers who give few playing instructions probably perceive interior and exterior spatiality as a whole and do less separate the impact of exterior space on the perception of the sound's spatiality.

11 H. de la Motte, *Handbuch der Musikpsychologie* (1985), 301 f.; H. Conradin, "Die Frage nach der Bewegung in der Musik" (1986), 224–33; K.-E. Behne: "Musik- und Raumwahrnehmung" (1989), 60 f.; Baumann (2008), 274.

1.4.3 The listener's situation: aim of perception and perceptual constancy

Listeners and even performers are rarely aware of the dependence of sound production on room acoustics. This astonishing fact is related to the nature of sensory perception. Since perception is generally conditional on change, and attention drops if nothing new happens, in a concert hall we are only briefly aware of its acoustics. It is during the first few seconds that we pay attention to the acoustic characteristics and notice whether the sound is dark or acute, reverberant or dry, blurred or clear. Attentive listeners take the opportunity to scan their surroundings before they concentrate on music. Eyes and ears help to adapt perception to the expected sound event and, nowadays, to the fact of staying put in one's seat during the concert. During this more or less conscious (or unconscious) moment of encounter with the hall, perception not only concentrates on sound characteristics of the hall but, due to spontaneous head movements, it also ascertains spatial orientation visually and aurally, as wells as the hall's size and the sound's direction ("from front or back", "from the side" or "from the ceiling"), the musicians' location in front on the podium and the distance to the next listener. Listeners thus "attune" to the hall: they become aware of its bright or dark sound, perhaps even of its effect on the timbre and clarity of voices and instruments. These characteristics of the surrounding space are registered in memory in order to back the analysis of the expected sound.[12]

Perception is usually aimed at the meaning of the sound signal rather than at its sounding aspect. Thus, during a talk we pay attention to the content and not to the sound quality of the speaker's voice, which may change completely when we move from an empty resounding corridor into a lounge with carpets, curtains and furniture. We ignore such a change as long as it does not impede or obstruct the understanding of speech. The process of auditory perception *eliminates* such change of room acoustics. In cognitive psychology this phenomenon is called *perceptual constancy*.

Euclid had already commented on the constancy of the perceptual size of objects at changing distances. Systematic observations show that even a considerable change of light intensity and, within certain limits, of its color has no effect on ordinary perception of color, although any change in illumination causes a measurable change in the spectrum of radiation that the external objects reflect back to our eyes. In fact, every change of illumination is accompa-

12 See also 1.5.3: Neuronal sound processing, 1.5.3.3: Experience and memory.

nied by a modification in retina stimulation.[13] A color film shows these different color shades produced by daylight, light bulb or a fluorescent lamp.[14] Persons interested in such details and experts, such as painters or photographers, are able to detect the difference.

Thus far, constancy of auditory perception has not been analyzed extensively.[15] But perceptual constancy is certainly one reason why untrained persons are hard put to memorize changes of timbre caused by changed room acoustics. The same is true for changes in color perception, although in both cases these persons do perceive the differences well in directly compared changes. In visual, tactile, olfactory, gustatory and auditory perception the analytical accuracy of such details is important for artists, professionals, experts and connoisseurs. Presumably a special talent, this accuracy is developed by intensive training in comparative analytical perception.

For the following research it is important to keep in mind that auditory perception functions in two different ways, analytically and synthetically. Inexperienced listeners can only briefly sustain analytical hearing. Synthetic hearing is a holistic process that usually dominates perception automatically after a short analytical phase if no extraordinary circumstances call for further analytical hearing. The type of hearing depends on the perceptual aim. Listeners usually forget a hall's acoustic properties if these are not called forth by special events, such as a moment of silence, a new instrument with a different sound or one that starts playing from a different position on the podium.[16] Predominant non-analytical hearing is one of the reasons for the striking scarcity of reports on musical performances that refer explicitly and in detail to a room's acoustic aspects. Inexperienced listeners often sway between cautious uncertainty and strongly confident judgment about the acoustics of a space. The individual trusts his own impression more than comments by other listeners, even if these are more experienced and base their judgment on more exact observation.[17] Individual and necessarily subjective, auditory experience has been

13 E. Cassirer, "The Concept of Group and the Theory of Perception", *GW* vol. 24 (2006), 209–50: 219 and footnote 15, referring to Adhémar Gelb, "Die 'Farbenkonstanz' der Sehdinge" (1929), 594–678: 596, and: "Zur medizinischen Psychologie und philosophischen Anthropologie" (1937), 195–271.

14 U. Eysel, "Sehen" (1993), 303.

15 C. Stumpf, *Tonpsychologie* (1883–1890); K. Mohrmann, "Lautheitskonstanz" (1939), 146–199: 145; A. Lang and R. Calmonte, *Klangkonstanz im Raum, Hören und Überhören von raumbedingter Schallvariation* (1986–1988); P. Knaus, *Studie zur auditiven Wahrnehmung auf der Grundlage der Theorie von J. J. Gibson* (1987).

16 See also 1.5.3: Neuronal sound processing and 1.5.4: Hearing and measuring: answers from psychoacoustics.

17 This fluctuation of judgment is known from witness reports at court.

written down in order to be remembered, mostly in diaries or letters exchanged between friends or persons with similar interests.

In cognitive psychology this categorization of sensory impression by constant norms and reduced perceptual complexity is also called "stereotype" or "prejudice".[18] Stereotypes of sensory perception are retained in memory for reference. Their definition is a complex process that depends on preference, repeated experience and judgment based on socio-cultural rules, education, and convention. Auditory experiences also produce stereotypes, such as "church acoustics", "violin sound", "violin in an orchestra" or "solo violin with orchestra". The quality of sensory perception depends strongly on one's experience with similar events. As we have a certain expectation for the average loudness of a singing voice, we are surprised if the perceived loudness is different. If our expectation is based on listening to sound recordings at high levels of loudness, the impression of an actual performance seems pale and faint. This misjudgment may elicit a correction of the stereotype but only if the new impression prevails.[19] As a result, stereotypes are only relatively stable, as they undergo a corrective process.[20]

Our electronic age is brimming with acoustic surprises and, thus, affords us a chance to observe these phenomena more deliberately than earlier generations. A telling stereotype stronger than momentary perception is that we are usually not bothered if an orchestra instrument visible on the television screen is barely audible. Even the total absence of its sound claims our attention only after a certain time, since at first we are convinced we hear it. These conditions for auditory perception call for careful consideration in any critical evaluation of all reports on music and experience of room acoustics.

18 de la Motte (1985), 159.

19 W. Lippmann, *Public Opinion* (1922).

20 See also chapter 1.5.4: Hearing and measuring: answers from psychoacoustics.

1.5 Sound and sensation: physical, physiological and psychological principles

Questions on music and room acoustics throughout history cannot be fully understood without applying the principles of the main disciplines, namely physical acoustics, physiology of hearing, neuropsychology, psychology of perception and related special fields. The following chapters present, therefore, the main theoretical foundations of these disciplines in so far as they are required to evaluate the collected data.

1.5.1 Sound and sound propagation

If an object is moved mechanically and the resulting variation of air pressure around the object is quick enough and lies between 0,02 μbar (auditory threshold) and 200 μbar (limit of pain), the human ear perceives sound. These values are equivalent to a force of 0.00002 to 20 Newtons per square meter, or Pascals. For convenience, loudness levels for 1000 Hz are indicated on a logarithmic scale of decibels ranging between the auditory threshold at 0 dB and the pain limit at 120 dB. Equal-loudness levels for all audible frequencies are indicated on a respective scale of 0 to 120 phons (see figure 22: equal-loudness curves).

Sound is always produced by some kind of mechanical movement. One part of the sound source is set in motion to produce periodic vibrations, for example, of the vocal folds or the lips of wind players that open and close periodically, of the air stream blown over the edge of a flute's mouthpiece and periodically moving into the tube, of chords activated by bowing or of struck membranes and plates. When these periodic movements oscillate at a rate of 16 to 20'000 per second, the resulting aural sensation is said to have a frequency of 16 to 20'000 Hertz (Hz). A regular movement is referred to as a sinusoidal vibration generating a sinus tone. Musical tones usually consist of a fundamental frequency and a number of more or less exactly harmonic partials. Noise as it occurs during the build-up of sound vibrations of musical instruments or during articulation in speech or singing is a blend of many closely laying frequencies within a certain bandwidth. If white noise, which contains all audible frequencies, is cut into a sequence of short bursts, we perceive clicks or crackles; if it starts suddenly and very loudly, we speak of bangs.

Such sound vibrations expand from the sound source through the air (or other elastic media) in narrower or wider bundles of longitudinal waves:

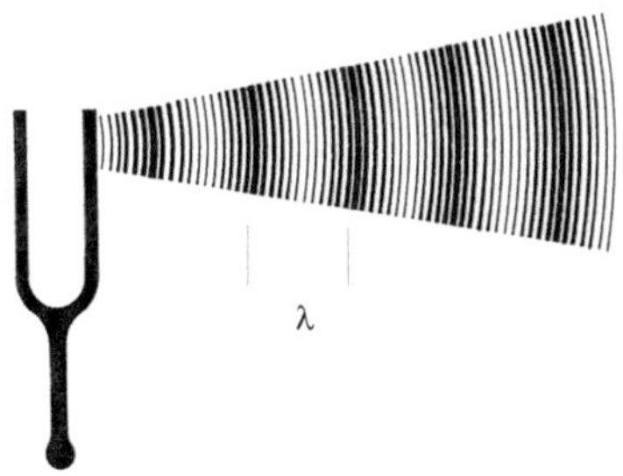

frequency	wavelength λ	
20 Hz	17.00 m	
100 Hz	3.40 m	
1'000 Hz	0.34 m	= 34 cm
10'000 Hz	0.034 m	= 34 mm
20'000 Hz	0.017 m	= 17 mm

Figure 18: a) Propagation of sound wave; b) Frequencies and wavelengths.

Air is compressed and expanded synchronically with the vibration of the sound source. The distance between zones of equal compression in a longitudinal sound wave is defined as wavelength λ, which is increasing at lower and decreasing at higher frequencies (figure 18a). These zones spread at sound velocity, which depends on the medium's temperature and elasticity: airborne sound waves propagate with a velocity c of about 340 meters per second at a temperature of 20° Celsius. The wavelength λ can be calculated as $\lambda = c / f$ because frequency f and wavelength λ are inversely proportional. The following table shows wavelengths for low, mid and high frequencies for c = 340 m/sec. They range from 17 meters to 17 millimeters (figure 18b). The human ear perceives noise and sound that often change direction, intensity, timbre and articulation within milliseconds.

1.5.2 The physiology of hearing and characteristics of auditory perception

Sound waves are propagated through the air. They reach the listener's ears, enter the outer ear canal and set the tympanic membrane at its end into vibration (figure 19).

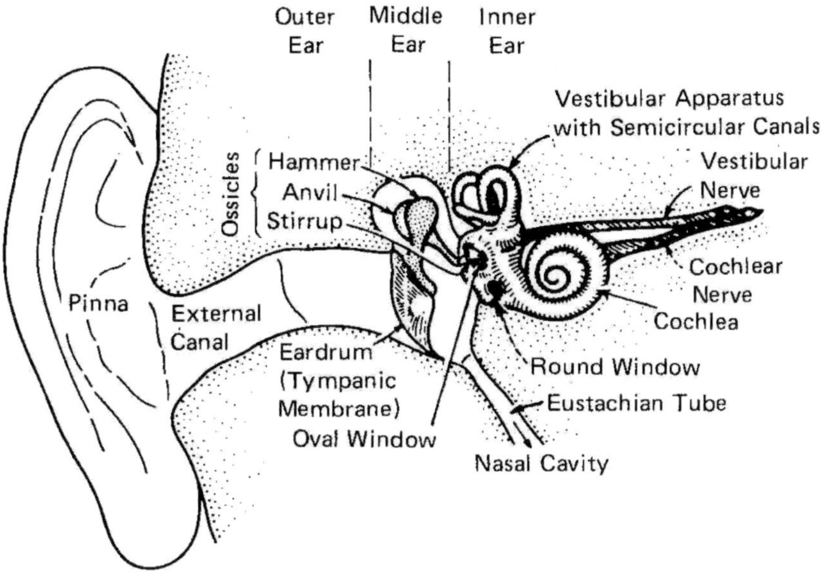

Figure 19: Section of the outer, middle and inner ear.

The movements of the *tympanum* are transmitted to the three auditory ossicles of the middle ear, the hammer *(malleus),* the anvil *(incus)* and the stirrup *(stapes).* The *stapes*, which is no larger than half a grain of rice, sets the membrane-covered oval window into movement and moves in and out like a record player stylus, creating pressure waves within the *cochlea,* which is filled with lymph liquid. The *cochlea* consists of three parallel ducts, the *scala vestibuli,* which runs upwards from the oval window to the top, the *scala tympani* which runs down from the top to the round window, and the *scala media,* which is separated from the two ducts by Reissner's membrane and the basilar membrane. At the *cochlea apex,* the *scala vestibuli* and the *scala tympani* communicate through a small opening, the *helicotrema*. The strongly damped pressure waves produced by the movement of the *stapes* travel along the *cochlea* ducts in the lymph liquid, and, as a result, the basilar membrane moves up and down. The displacements are synchronous with the amplitude of the traveling waves, reaching a maximum for high and low frequencies at the *cochlea's* lower and upper end respectively (figure 20).

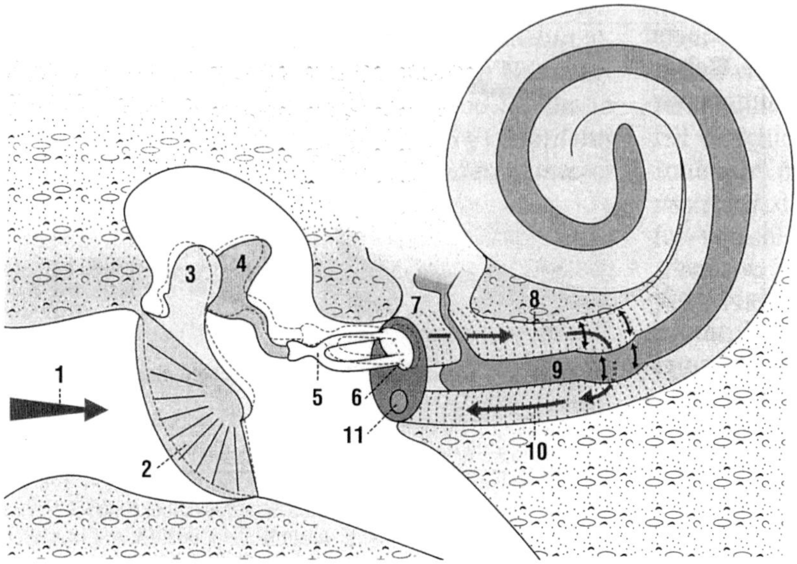

Figure 20: Sound conduction to the inner ear: 1 sound, 2 tympanum, 3 malleus, 4 anvil, 5 stapes, 6 oval membrane, 7 *vestibulum*, 8 *scala vestibuli*, 9 *scala media*, 10 *scala tympani*, 11 round membrane.

As shown in figure 20, these movements cause shearing movements between the basilar membrane (2) and the tectorial membrane (3), and, as a result, the outer hair cells of the acoustic nerves embedded in both membranes are bent. Bent hair cells send electrical impulses to the primary auditory centers in the cortex (see also figure 27 below). Transmitted via afferent neurons of the auditory pathway, these neural signals contain complex codes of frequency, intensity, direction, and other parameters of sound events. The exact details of the coding are not yet fully understood. The frequency perception based on the movements' location and amplitude can be sharpened by a brain-controlled feed-back sent down to the basilar membrane via efferent axons that set inner hair cells close to the maximum shearing shift into movement. Below about 5000 Hz frequency signals are also encoded directly by the movement of the basilar membrane in synchrony with the lymph's pressure variation. The other organs of the vestibular system, the utricle and the saccule, are also involved in the reception of sound vibrations due to resonance within the bony labyrinth.[1]

1 H. Fletcher, *Speach and Hearing* (1929), 117 f.; G. von Békésy, *Experiments in Hearing* (1960); E. Leipp, *La machine à écouter* (1977), 120 f.; W. Arnold et al., "Funktionelle Morphologie der äusseren Haarzellen des Menschen" (1990), 177–86.

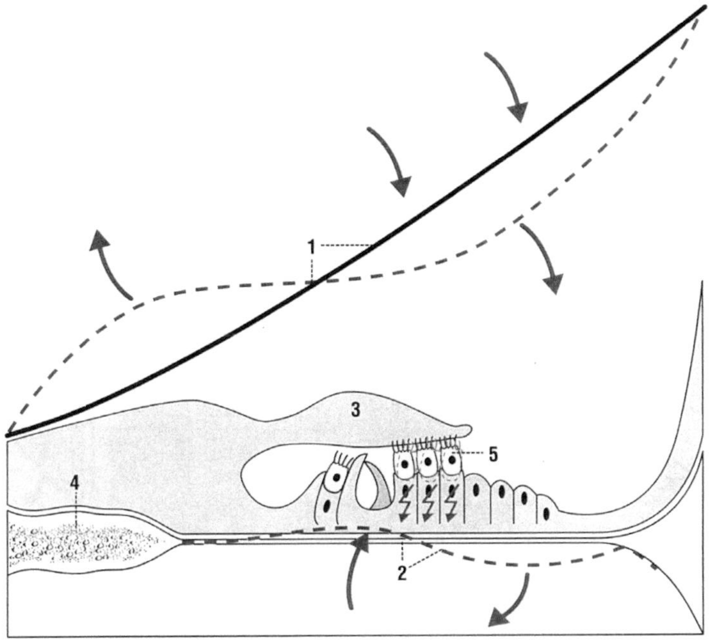

Figure 21: Sound perception through the outer hair cells: 1 Reissner's membrane, 2 basilar membrane, 3 tectorial membrane, 4 *lamina spiralis ossea,* 5 outer hair cells.

The sensitivity of the auditory system is not linear but varies according to frequency. For frontal binaural pure tones under normal free field hearing conditions it is represented in statistically determined curves of equal loudness for young persons with acute hearing (figure 22). Maximum sensitivity, which lies between 3000 Hz and 5000 Hz, strongly diminishes towards lower frequencies and less steeply towards higher frequencies. The discrimination of lows decreases for louder sounds.[2] The upper frequency limit for young persons with acute hearing lies between 16'000 and 20'000 Hz. With increasing age it falls to about 12'000 Hz. The perception of high frequencies is important for a reliable detection of sound direction.

2 The appropriateness of these curves for sound level and noise has been discussed since Fletcher-Munson's first measurements in 1933 which were based on ear phone signals. Values differ for side or frontal presentation of the test signals. A new experimental determination was made by Robinson and Dadson in 1956 which became the basis for the ISO standard 226. ISO revised these standard curves in 2003, in response to an international study coordinated by a research institute at Tohoku University in Japan. Audio equipment testers use special curves adapted to noise signals. Also curves for pure sound and diffuse sound differ, see E. Zwicker and H. Fastl, *Psychoacoustics* (1990), 181 f.

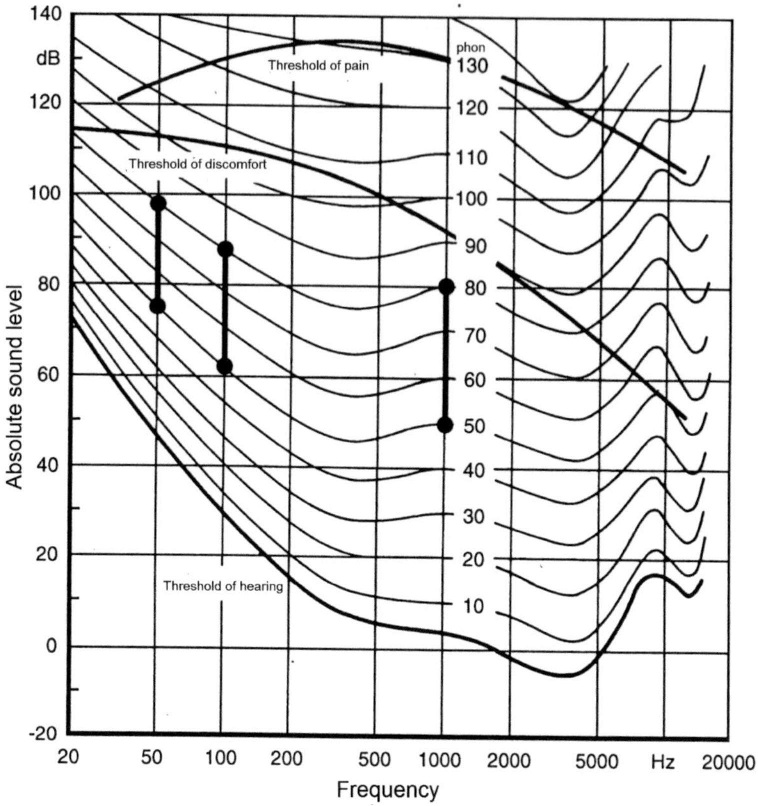

Figure 22: Equal loudness curves according to ISO-226:2003 revision.

As shown in figure 22, curves of equal loudness are not parallel. A rise of sound pressure level of 30 dB or phons from 50 to 80 dB at 1000 Hz is equivalent to a rise from 60 to 90 dB at 100 Hz and from 75 to 100 dB at 50 Hz. This means that at 100 Hz a sound pressure increase of 10 dB is necessary to cause the same loudness impression. The necessary sound pressure is even higher at 50 Hz, but an increase of 25 dB is sufficient to cause the same crescendo (see bold lines and bullets on the curves for 50 phons and 80 phons).[3]

Around 3000 Hz tympanum movements of less than 10^{-8} mm are sufficient to cause a sound sensation.[4] That is, sound perception at its most sensitive frequency range comes close to atmospheric noise because the amplitude of tympanum movements at threshold level is less than the diameter of a hydrogen molecule. The high sensitivity of the auditory system in this frequency region is caused by resonance in the outer ear canal and characteristics of the sound

3 J. Meyer, *Acoustics* (2009), 7, fig. 1.1, threshold of discomfort after F. Winckel (1969).

4 Békésy, *ibid.;* Leipp (1977), 66.

transfer by the middle ear ossicles. That is why the reach of a voice or an instrument depends not only on general loudness but also on sufficient loudness within this range of highest sensitivity of the auditory system.[5]

Perceived loudness also depends on the signal's duration: sound impulses shorter than about 250 milliseconds seem less loud than longer lasting sound at the same level. Furthermore, loudness impression depends on reverberation.[6]

1.5.2.1 The function of the middle ear muscles

The transmission of vibrations from the ear drum to the oval window by the ear ossicles is controlled by the middle ear muscles (figure 23). A contraction of the *tensor tympani* pulls the head of the hammer inwards and pushes its handle outwards. A contraction of the *stapedius* muscle pushes the *stapes* into the oval window in a slightly asymmetrical position. The combination of both muscle contractions causes a deformation of the ear drum and the membrane over the oval window. This process changes frequency characteristics and the time resolution of the perception system.

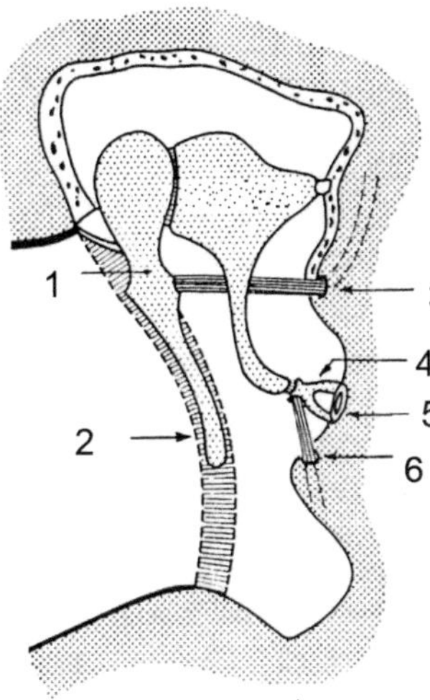

Figure 23: Cross section of the middle ear with middle ear muscles: (1) hammer, (2) ear drum, (3) tensor tympani, (4) stapes, (5) stapes on oval window, (6) stapedius.

The middle ear muscles are reflex-controlled and may also, though rarely, act on purpose. They protect the hair cells of the inner ear against damage caused

5 F. Winckel, "How to Measure the Effectiveness of Stage Singer's Voices" (1971), 228.
6 Zwicker and Fastl (1990), 315: fig. 16.32, 16.34; see also 1.6 Room acoustics.

by violent blows or lasting excessive loudness. The *stapedius* alone attenuates the transfer of low frequencies already at medium loudness levels: an inactive *stapedius* causes a hearing loss of up to 40 dB for frequencies between 500 and 4000 Hz after prolonged exposure to high levels of noise (figure 24, broken line against solid line). The *tensor tympani* is activated only at high loudness levels. The middle ear muscles begin to contract already several milliseconds before our own vocal chords are set into vibration and protects the inner ear from the high loudness levels in the head.[7] Contraction may also start in advance if we expect loud sound events or if the sound level is increasing slowly. When caught by surprise, the contraction reflex acts 100 to 200 milliseconds late in order to reach maximum attenuation and protection. That is why sudden sound clashes may cause irreparable hearing loss.

Tension adaptation to average loudness also amplifies the dynamic range of perception. A change of tension alters frequency characteristics of the transfer system like a selective filter. The attenuation of low frequencies reduces masking of speech. This is one of the reasons why we are able to understand speech despite excessive noise at a cocktail party *(cocktail-party-effect).* Selective filtering also helps to explain why conductors are able to identify single instruments within the full sound of an orchestra.

In order to detect the different directions of incoming sound selective perception is abetted by micro-analyses of time delay between the signals of both ears.[8] The higher the attenuation by the middle ear muscles, the better the impulse perception and temporal discrimination. On the other hand, analysis of frequency, sound color, and harmony, which is based on longer lasting sound signals, improves with low attenuation. That is, for optimum perception of frequency and time information, the perception system has to switch between these two characteristics.[9]

1.5.2.2 Directional hearing

The labyrinth and the *cochlea* provide information for motor control of the body. Both sensory systems may warn against outside danger. Precise directional hearing based on the comparison of signals from both ears is one of the most important warning processes. Due to the distance between the ears

7 Borg, Counter, Rösler, *Theories of middle ear function* (1984); Borg, Counter, "The Middle Ear Muscles" (1989), 65; more on hearing loss, see J. Frei, *Die Gehörbelastung des Orchestermusikers in der Konzert- und Opernformation der Tonhalle Zürich* (1979).

8 See also 1.5.2.4: Temporal aspects and 1.5.3: Neuronal sound processing.

9 Leipp (1977), 58.

and the head's screening effect signals reaching the brain differ laterally. Time, timbre and intensity difference conveys spatial information (figure 25).

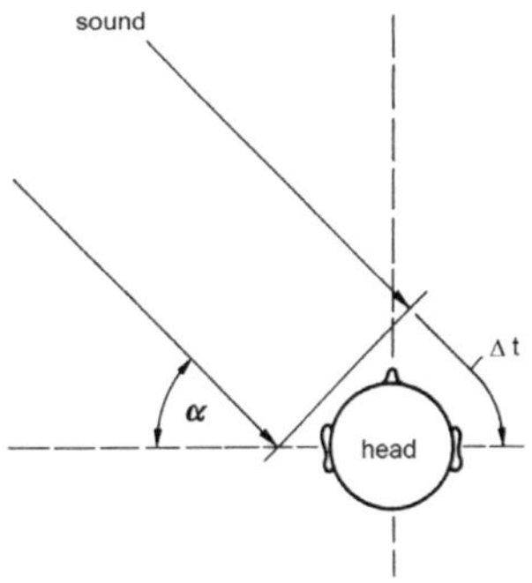

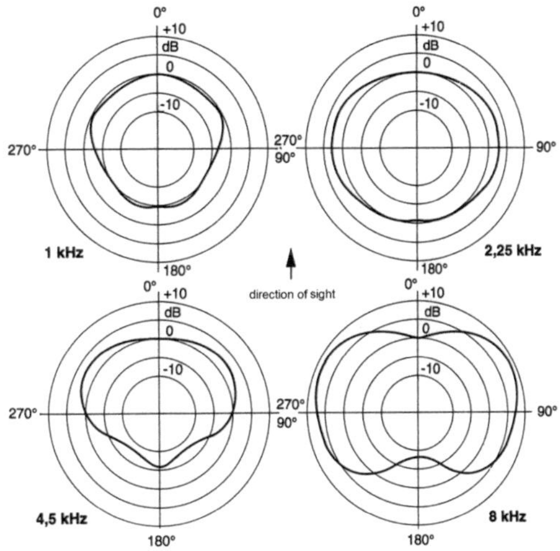

Figure 24: Time delay left ear – right ear. Figure 25: Binaural hearing, directional characteristics.

A comparison between frontal sound and a sound source circling around the head shows that perception of lateral sound is louder than of frontal sound at 2250 and 8000 Hz (figure 25, top right and below). At these frequencies the difference reaches a maximum of 10 dB and lateral sound appears to be twice as loud. For 1000 and 4500 Hz frontal-lateral sound is louder than sound from the back (figure 25, top left and below).[10]

The just noticeable *time difference* for frontal sound is 0.03 milliseconds only, which is equivalent to a minimal sound path difference of about 1 centimeter or an intensity difference of about 1 dB. The just noticeable *difference of sound direction* for frontal sound incoming at angles between 45° to the left and 45° to the right is 3°. For more lateral signals from 45° to 90° the just noticeable localization difference is 4.5°.[11] For angles between – 45° and + 45° intensity and timbre differences determine perception of a sound source in front or at the back. If several sound signals arrive within a time delay Δt of less than 3 milliseconds, direction is perceived as "from their center". For time delays of more than 3 milliseconds localization follows the direction of the first wave front even if later waves are somewhat louder.[12] Localization follows direct

10 Meyer (2009), 13 f.
11 J. Blauert, *Spatial hearing* (1997), 36 f., 93 f., 137 f., 312 f.
12 Hoeg and Steinke, *Stereofonie-Grundlagen* (1972).

sound as long as the intensity difference is less than 10 dB and the time difference is between 5 and 30 milliseconds.[13]

1.5.2.3 Masking

Quieter sound signals are partly or completely masked by louder sound. The closer the masking frequency to the signal frequency, the stronger is the masking effect. Low frequencies have a stronger masking effect on higher frequencies than vice versa. Masking is stronger in monaural hearing. As mentioned, binaural hearing may better avoid masking thanks to inter-aural time difference and different direction.

If the masking sound stops, the normal hearing threshold is reached again after 200 milliseconds. Lower sound levels are processed more slowly than louder levels. That is why low sound events may be pre-masked by louder events up to 20 milliseconds. Therefore, softly played music seems less clear because the characteristic noise of sound production is masked.[14]

1.5.2.4 Temporal aspects: blurring, blending, summation

The eye is able to separate 20 to 25 pictures per second. If the picture sequence is faster, the eye perceives continuous movement. The ear separates single sound pulses if frequency drops below about 16 to 20 Hz or if the time gap between sound pulses exceeds about 50 milliseconds. A quicker sequence of pressure pulses conveys a continuous sound. Fifty milliseconds is also the limit for just noticeable echoes if sound events are loud enough and have a clear onset. The 50-millisecond time difference, also called blurring limit, depends on incident angle of sound, loudness, number and time pattern of reflections.[15] The limit for conscious perception of echoes is usually above 100 milliseconds. As a result, clearly perceivable echoes caused by relatively hard sound events need sound path differences of more than 34 meters.[16]

Below a time difference of about 50 milliseconds direct sound and early reflections blend into a single sense impression, and early reflections with similar timbre are perceived, therefore, as coherent and reinforce loudness. During the arrival of coherent sound waves loudness is summed if time gaps

13 H. Haas, "Über den Einfluss eines Einfachechos auf die Hörsamkeit" (1951), 49–58.

14 Zwicker and Fastl (1990), 57 and 195.

15 See already Stumpp (1936); see also L. Cremer, *Die wissenschaftlichen Grundlagen der Raumakustik* (1976), vol. 2, 379.

16 W. Ahnert and W. Reichardt, *Grundlagen der Beschallungstechnik* (1981), 37.

between reflections are not too large. In a closed space with reflecting surfaces integration time usually exceeds 50 milliseconds and often 80 milliseconds.[17] Depending on sound characteristics and the aim of perception, it varies between 15 to about 200 milliseconds. Due to energy loss caused by distance and air and surface absorption, later reflections arriving with a delay of more than about 300 milliseconds after direct sound usually do not increase perceived sound intensity to any considerable level.[18]

As mentioned, reflections from different directions also help to reduce sound masking and pre-masking due to inter-aural signal comparison (see also figure 27 below). This explains why the cocktail-party effect derives not only from the attenuation of low frequencies by the middle ear muscles but also from binaural perception of direction. The same effect explains why reflections reaching both ears from different directions render perception more transparent. Coherent reflections arriving within 50 to 80, or even 200, milliseconds reduce masking thanks to time difference, the different angles of incidence and the timbre difference caused by binaural hearing. In neuronal processing coherence is defined by the similarity of the delayed signals and the limited time difference. Perception thus yields a total, more transparent and louder sound impression (see below figure 31 and 32).

1.5.2.5 Summary of physiological sound perception

The following characteristics of the physiology of hearing are relevant to the study of room acoustics and music:

1. Adaptation of the middle ear muscles causes a difference in perception between expected sound events (such as repetitions) and unexpected ones.
2. Signal transparency and clarity is diminished by masking and improved by coherent reflections arriving from different directions and within integration time.
3. The total energy of these reflections is integrated in order to produce a total sound impression. Integration time, which depends on the pattern of time delay and the relative intensity of reflections as well as on the listener's aim of perception, varies between 15 and 200 milliseconds. Early reflections arriving before 80 milliseconds are crucial. Very early reflections arriving with a delay of 5 to 30 milliseconds after direct sound and relatively loud reflections arriving after a gap of 50 milliseconds may be perceived separately (see 1.5.2.2 and 1.5.2.4).

17 Cremer, *op. cit.*, vol. 1, 382.
18 Ahnert and Reichardt, *ibid.*

4. Lateral reflections are important because lateral sound between 2000 and 5000 Hz appears to be 5 to 10 dB louder than frontal sound.
5. The highest ear sensitivity requires quiet conditions. To reach minimum tension middle ear muscles need a certain time without disturbing noise. Spaces with good hearing conditions need, therefore, sound insulation against outside noise. Silence and quiet thus contribute to the magic of excellent halls.[19] Contrary to anechoic chambers quiet halls characteristically have a non-disturbing very low-level background noise that helps to improve perception of very low sound.[20]

1.5.3 Neuronal sound processing

Important characteristics of sound perception result from neuronal processing in the brain. In humans the latter consists of two nearly symmetrical cerebral hemispheres of the *neocortex,* linked by the *corpus callosum* and connected to the next lower levels, the midbrain or *mesencephalon,* the *pons,* the *medulla oblongata* and the *cerebellum.* The *neocortex* consists of several lobes with different areas that are active during specific perception processes. Its nervous system contains about 10,000 million nerve cells (various types of neurons) that exchange signals via excitatory synapses, which activate nerve cells to discharge short electric impulses, and inhibitory synapses, which inhibit the discharge of impulses.

1.5.3.1 The auditory pathways

The auditory pathways connect the *cochlea* of each ear and the primary auditory area, which is situated in Heschl's *gyrus* in the superior temporal lobe *(gyrus temporalis superior)* of both hemispheres (figure 26).

19 D. Baumann, "Was ist Stille?" (1990), 981–3; D. Baumann, „Die Kunst des Konzertsaalbaus: Heute kann man gute Akustik im voraus planen. Interview mit Russell Johnson", *Neue Zürich Zeitung,* 25 June 1993, Nr. 144, 57.

20 F. Moss and K. Wiesenfeld, "The Benefits of Background Noise – Stochastic Resonance" (1995), 50–3.

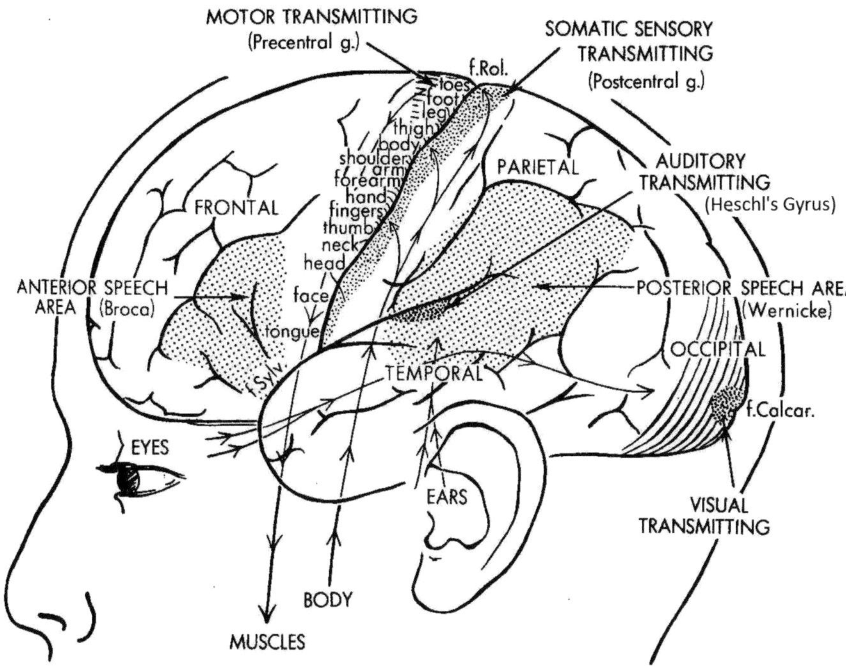

Figure 26: Human brain, sagittal section.

Seventy percent of the neurons of the auditory pathways cross from the *cochlea* on one side to the contralateral hemisphere of the cortex, 30 percent remain on the same side, projecting to the ipsilateral auditory area (figure 27). Some of the nerve fibers pass four or five *nuclei* (synaptic connections) on their way to the auditory cortex, some skip them partially. On the lowest level, from the *cochlea* to the *cochlear nucleus,* the nerve fibers of both ears remain completely separate. On the next level, at the superior olive, some nerve fibers from both ears meet, and from there upwards the number of nerve fibers connecting to both sides is increasing.

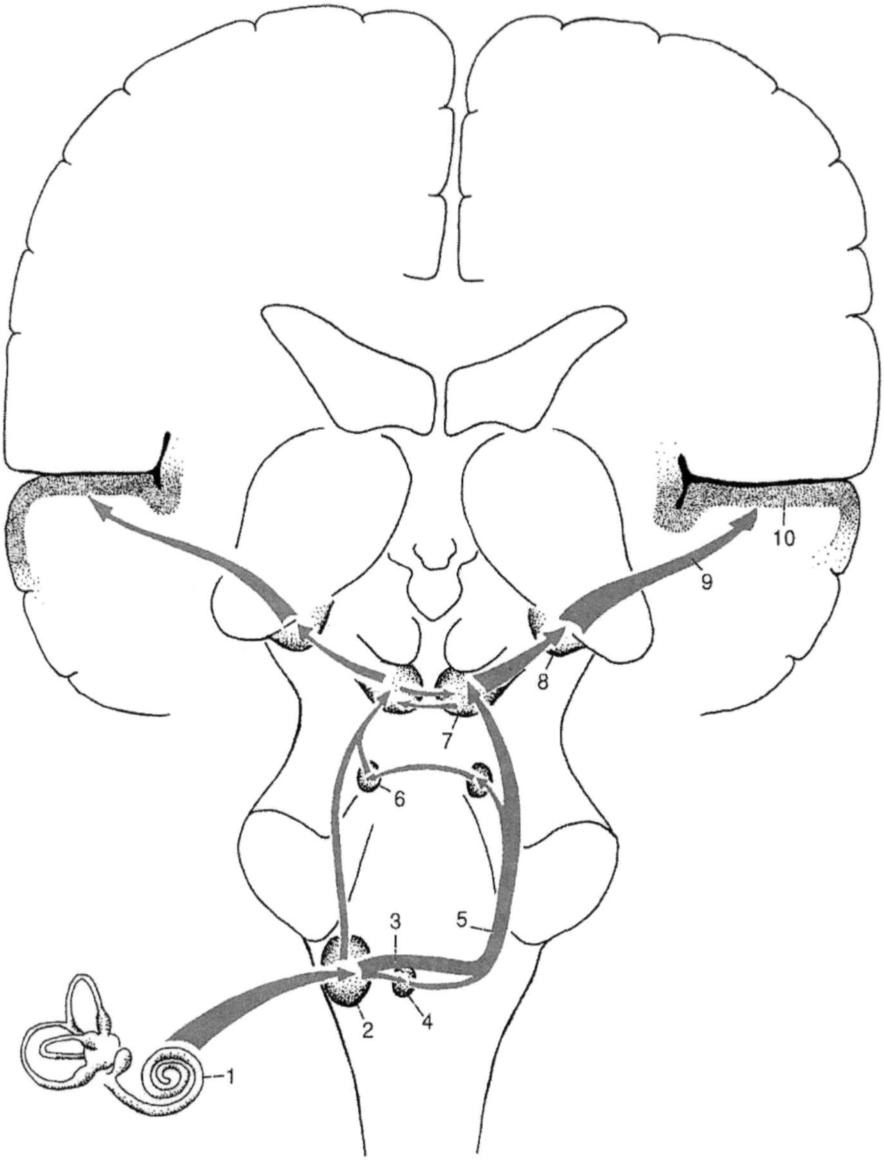

Figure 27: Auditory pathways from one *cochlea (1) via nucleus cochlearis* (2), *corpus trapezoideum (3), oliva superior (4), nucleus lemniscus lateralis (5), colliculus inferior (7), corpus geniculatum mediale (8),* to the primary auditory cortex in both hemispheres (10).

The electric pulses reaching the primary auditory cortex usually do not transmit codes of single frequencies but, rather, of more complex units, of Gestalten based on multiple processing that begins already on lower synaptic levels. The number of neurons in the human basilar membrane, which is only about 15,000 (compared to 150 million cells in the human retina), increases from synapse to synapse to finally reach about 100 million in Heschl's *gyrus* on either side of the cortex. The signal processing from the inner ear to the human brain is not yet fully known. But some functions of the *nuclei* are clear enough to give an idea of the principles of neural processing of frequency and direction detection in auditory signals. Animal testing, surgery and non-invasive imaging systems showing regional changes of brain metabolisms yielded a spate of results:[21] The cochlear nucleus focuses on frequency by suppressing weak signals; the ventral *cochlear nucleus* analyzes monaural information on direction; the superior olive compares frequency and time patterns from both ears with time differences of microseconds; the inferior *colliculus* combines and compares monaural and binaural direction information. Some nerve cells are sensitive to single tones but not to noise, some are capable of suppressing certain frequencies regardless of loudness; some react to phase shifts of partials, some to the onset of tones, some to signals from certain directions. Similar selections happen on the upper levels of the auditory pathway and in the auditory cortex, where signals from larger groups of nerve cells are gathered and compared. The results of processing are coded on each level into new pulse sequences and transmitted to new independent pathways. One may assume, then, that this is how comparisons are made also between earlier and later coherent signals from the same auditory "event" that result from reflections off more or less distant surfaces.

The classical model of auditory perception assumes that neuronal signals project from Heschl's *gyrus* to the other auditory areas of the cortex, that is, to Wernicke's area (posterior, sensory) with more analytical processing in the dominant hemisphere and more synthetic processing in the non-dominant hemisphere, and to Broca's area (anterior, motor) in the dominant hemisphere only (see figure 26).[22]

A parallel processing of sense signals occurs in different adjacent modules already in the primary auditory area of the cortex, and modules of one area project signals to modules of a remote area. Some processes are tonotopic (ordered

21 S. Handel, *Listening* (2/1991), 521 ff.; N. Wallin, *Biomusicology: Neurophysiological, Neuropsychological, and Evolutionary Perspectives on the Origins and Purposes of Music* (1991); Ch. Faessbender, "Funktionsteilung im Cortex" (1993), 622–30; H. Petsche, "Zerebrale Verarbeitung" (1993), 630–638; *Neuro- und Sinnesphysiologie,* ed. R.F. Schmidt (1993); R. F. Thompson, *The brain* (1993). 256; L. Jäncke, *Macht Musik schlau? Neue Erkenntnisse aus den Neurowissenschaften und der kognitiven Psychologie* (2008), 16–8: best resolution restricted to either temporal (MEG, EEG) or local information (MRT, fMRT).

22 On the differentiated interpretations of brain asymmetry, see Jäncke (2008), 304–5; on modifications of function and definition of language related zones due to modern neuroimaging technics, see M. Meyer, "Functions of the left and right posterior temporal lobes during segmental and suprasegmental speech perception" (2008), 108–115.

by frequency), some are not. Specialized in a certain way, each module is supplied by a different combination of lower-level neurons. The circuitry, or network, created by these interactions is partly genetic and partly stored from experience, yet the general tonotopic organization of the frequency analysis seems to be mainly inherited. Binaural direction discrimination seems to be learned mainly individually from experience. Hence the main conclusion that these auditory processes are based not only on inherited structures but also on those steadily developed and adapted throughout individual experience within daily spatial surroundings, which are strongly influenced by the social milieu.

Sound waves in the air are about one million times slower than light waves. The cascading manner of neuronal processing takes account of the relatively slow propagation of sound in space. Depending on the aim of perception, it can focus more on optimum resolution of time and analysis of quickly following short pulses (resulting in poorer frequency resolution) or, rather, on maximum temporal blending (resulting in better frequency resolution). Time-delayed but content-coherent sound events are gathered, and this process of time integration increases subjective loudness. Frequency selection and temporal resolution is supported by the adaptation of the middle ear muscles and the activated motion of the outer hair cells in the basilar membrane.[23]

1.5.3.2 Active perception

A general hypothesis in psychology has been that cortical processes can be separated into sensation, perception and higher-order processes. Yet neurology has shown that these processes, which are distributed throughout different areas of the brain, are not happening according to a strictly hierarchical order. Rather, different parallel ways of signal transfer operate, actively searching for information about events on which perception is focused.[24] Perception is affected by emotions, and emotions, in turn, influence our actions. A sequence of rhythmically structured sounds may deeply touch our sensation simply because the periodical processes of heart beating, breathing are influenced by such periodical sound processes.

Our behavior can be characterized by two dimensions: one cognitive, based on signals from the sense organs, the other affective, based on signals from the limbic system. Emotions, which activate the limbic system, are controlled in the prefrontal cortex, where they are combined with tactile, visual and auditory signals and produce patterns of reactions. Motor reactions can be roughly categorized into two tendencies: approach and escape. The mostly

23 H.-P. Zenner, "Hören" (1993), 324.
24 K. Popper, J. Eccles, *The self and its brain: an argument for interactionism* (1977), 250 ff.

unconscious vegetative processes controlled by the midbrain also depend on affective elements of perception. The change of corresponding vegetative functions can be measured as a change of blood pressure, pulse rate, electrical skin resistance, breathing rate and other parameters. For instance, a test subject ready to listen to the first movement of Johann Sebastian Bach's first *Brandenburg Concerto* showed an increased activity of the forehead and ankle muscles nearly one minute before the beginning of the music. With the beginning of the music a clear increase of pulse rate was measured, followed by fluctuations. The breathing rate changed completely and finally adapted to the tempo of the music.[25] Such characteristic changes induced by the effects of listening to music differ from person to person and from one presentation to another, as they depend on personal experience, the subjective interpretation of the situation and the emotional state of the test subject.

The still quite prevalent hypothesis that music is processed mainly in the non-dominant hemisphere relies on an incomplete definition of music and an inadequate understanding of the dynamics of brain processes. The analysis of all aspects of a sound event, be it language or music, needs many areas of both hemispheres of the cortex.[26] The areas' participation depends mainly on the signals' nature, on the individual's experience and on the momentary aim of perception. The same signals may be heard with different aims of perception. As explained before, the process of perception may be focused either on the "what" or on the "how". While listening to music one may focus one's attention on either the content or the interpretation, or on the influence of room acoustics. The extent of our ability to combine more than one focus or, rather, to switch between these different aims, is an important topic for further research.

Conscious perception is controlled by the "self-conscious mind",[27] which uses different regions and areas of the cortex.[28] The function of the self-conscious mind is interpretation and control. Through selective feedback (for instance, control of the middle ear muscles or head and body movements), it can influence the selection of the cortex modules used for analysis as well as the quality of the resulting information. Conscious perception is always directive. The self-conscious mind scans open modules for important and useful information. To reduce energy and time it tries to anticipate the future on the basis of former experience and the actual aim of perception. Events are thus compared to expectations. This continuous process of comparison is based on

25 G. and H. Harrer, "Music, Emotion and Autonomic Function" (1977), 212–13.

26 L. Jäncke, in his outline of the more differentiated view on asymmetry refers to the important fact of different structures in brains of musicians; see Jäncke *loc. cit.*

27 This concept has been developed and clarified by Popper and Eccles (1977), 552.

28 *Ibid.*, 250 f. and 440 f.; Thompson (1993), 388 f.: 409.

stored experience, with individually developed and genetically based stereotypes playing an important role.[29]

As far as we know today the modules of the non-dominant hemisphere communicate with the self-conscious mind of adults only via the corpus callosum and the dominant hemisphere. But processes in the non-dominant hemisphere can also be stored in unconscious parts of memory, from where they may affect behavior.[30]

1.5.3.3 Experience and memory

Memory can be defined in terms of content or time. The temporal aspect of memory is based on three different processes: the sensory register, short-term memory, and long-term memory (figure 28).[31]

The sensory register, also called "iconic" for visual aspects and "echoic" memory for acoustic events, keeps sensory information automatically and for a very short time (less than 1 second).

Only a minor part of this information is transferred to short-term memory, which keeps a limited amount of information for a few seconds. Short-term memory is based on repeated rehearsing, such as the repetition of words or movements in order to acquire practical skills. Success is based on alertness and active concentration.

Some information is also transferred directly from the sensory register to long-term memory. The latter is based on different brain circuits (among others, the spine synapses of pyramid cells of the cortex modules, the granule cells of the hippocampus in the limbic system and the Purkinje cells and the granule cells of the cerebellum, the latter with indefinite storage duration).[32]

The steady activation of the neuronal pathways used for perception and thinking increases signal intensity in these neural circuits. This process induces structural changes of the involved synapses, causing an increase of the number of synapses within half an hour to three hours. The pathways are reinforced such that they can be used with a minimum of energy. Information storage, which needs resources for protein synthesis to be ready within minutes,

29 H. de la Motte, *Handbuch der Musikpsychologie* (1985), 118; C. E. Tolman, "Cognitive Maps in Rats and Men" (1948), 189–208; see also 1.4.3: The listener's situation: the aim of perception and perceptual constancy.

30 Thompson (1993), 332 f.

31 Eccles (1977), 330, 334, 402 f.; E.R. Kandel and R.D. Hawkins, "The Biological Basis of Learning and Individuality" (1992), 53 ff.; H. J. Markowitsch, *Neuropsychologie des Gedächtnisses* (1992); Thompson (1993), 384.

32 Kandel (1992), 53 ff.

requires not only adequate alertness and concentration but also availability of short-term memory information.

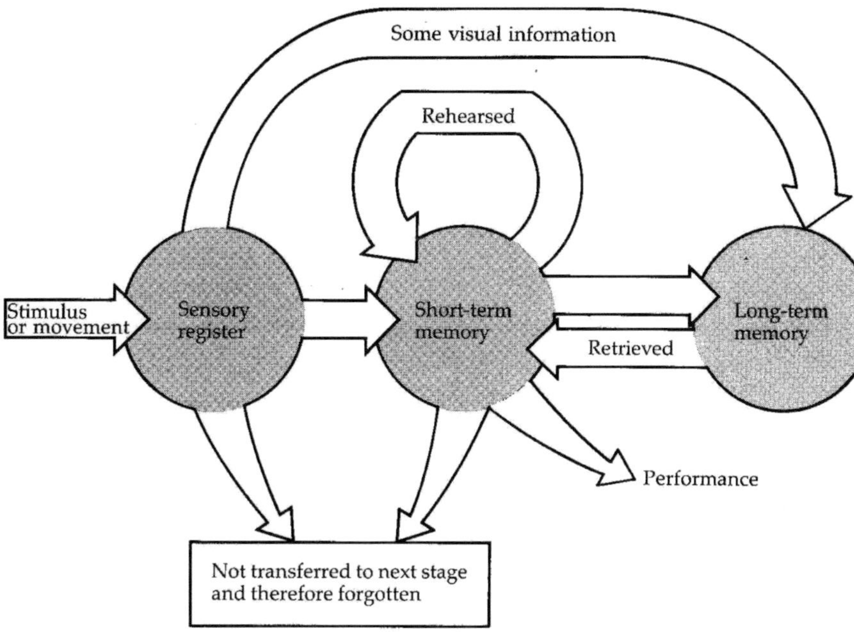

Long-term memory keeps information for hours, weeks, years or a lifetime. The transfer of information from short-term to long-term memory happens stepwise over an intermediary working storage that can keep information for seconds or hours based on ongoing and less intensive repetitions.[33]

Acoustical characteristics are memorized according to these procedures, also in terms that describe perceived quality such as "thin", "blurred", or "delayed bass", repeated constantly in order to store the impression of steadily incoming sound for a longer period. Information recall from short-term memory is relatively slow and needs full attention because the modules have to be activated in order to replay the experienced sequences of signals. On the other hand, the "compressed" information from long-term memory is quickly available as though an open safe were to display all its content.

A listener able to sing or play music usually has better memory and retrieval capacities than a passive listener because he or she learned certain processes of signal processing on the basis of motor activities and several other control mechanisms. These are deeply rooted in long-term memory and other

33 Eccles (1977), 402–3.

brain structures, can be recalled quickly and are available without change for a long period. Passive listening without motor action activates only limited capacity of memorizing. But any sound oriented activities, such as motor movements or the use of electronic music instruments or of measuring and recording equipment, support analytical experience that allows a more differentiated sound perception and cognition.

Analytical hearing of room-acoustic influence on the sound of music requires high sustained concentration and attention by the subject in order to compare sequences of information stored mainly in short-term memory. On the other hand, synthetic hearing of music is based on experiences stored in long-term memory. That is why musicians and recording engineers with special skills in the analysis of room acoustics are often able to remember the acoustics of a hall used for recordings even days later. Less experienced listeners retain a less stable impression that may be easily disturbed by new acoustical events. They have difficulties in maintaining focused attention to relevant acoustic features and store them in memory for comparing the acoustics of different rooms or even different positions in the same room.[34] Specialists, too, can achieve refined comparisons only if they immediately switch among sound recordings from different halls or positions thus concentrating on memorized perceptual contrasts. The following study uses descriptions and recollections of acoustical impressions. It should be noted, though, that memory-based impressions and information are not as reliable as measuring protocols. Still, if information from other sources provides a general acoustical background, these reports by experienced listeners are of high value because they supplement the acoustic measures and provide an expert's evaluation of room acoustics.

1.5.3.4 Music and the brain

Sound may warn us from imminent dangers but it transmits other information as well. Human language and music use auditory signals to transmit symbolic messages. Thanks to written symbols they can also be transmitted visually, and sound can exist in the imagination only without the production of any real sound.

Music in its various forms is generally very appealing to our perception, sensation and imagination. Music affects human beings in many ways, from the vegetative level, which transmits merely the feeling of life, to the highest level of abstraction, where humanly structured sound streams are analyzed in different neuronal networks in order to decode visual, auditory and spatial

34 I. Frei, *Raumakustische Qualität in komplexen Räumen. Theoretische und exemplarische Darstellung der Probleme,* masters thesis, University Zurich 2011.

information.[35] The spatial information results partly from imagination and partly from the real space of performance in which music is actually sounding. While listening to music in a holistic way concentrated on the work of art, we are in an existential situation free from exterior dangers that transports us to transcendental spaces.[36] That is why real exterior space and interior imagined musical space get mixed. The ability to differentiate between the perceptual conditions of the inner and the outer musical space is the result of a long term training that characterizes the status of experts.

1.5.4 Hearing and measuring: perspectives from the psychology of perception

Since Helmholtz the aim of research was to define the cause and effect of sound impressions by exactly measurable physical parameters. For example, values of sound pressure variations were assumed to cause an exactly defined loudness impression. Yet the psychology of perception clearly shows that such statical and causal relations between signal and sensation do not exist. These relations are more complex, as the characteristics of perceived sound are not defined by exact values but by bandwidths of values valid under certain conditions.[37] A value of physical sound intensity indicated in decibels (dB) is not sufficient to define perceived loudness, since sound impression derives from a temporal stream of sound events that are evaluated according to context and content and compared with memorized stereotypes. For instance, in complete silence a sudden sound of 40 dB seems loud, but during the performance of a piece played by a full orchestra it appears as a pianissimo. Furthermore, the eardrum's minimum displacement required for a sound sensation – the so-called threshold level – is not a fixed and constant value but depends on the listener's momentary condition and previous sound events. The reason for these complex definitions lies not only in the perception process itself but also in individual differences concerning physiological, neuronal and psychological conditions of sound processing.

35 O. Kühl, *Musical Semantics* (2007), 25; D. Baumann, "Musik von innen her empfinden und verstehen – Einsichten aus Susanne K. Langers Kunstphilosophie" (2008), 257 f.: 266.

36 H. de la Motte, "Der homo ludens – ein homo oeconomicus?" (1989), 15 f.; *Musikpsychologie, ein Handbuch,* ed. H. Bruhn et al. (1993), 263; D. Baumann, "Improvisierte Musik und Gehirn" (1994), 60.

37 Leipp (1977); W. Jauk, "Lautheit" (1995), 67–77.

The human ear does not measure sound events on a constant scale of values but compares sound impressions to immediately passed impressions or to the memorized impressions from previous experience. This is why acoustic quality (like other sensory qualities) cannot be absolutely quantified but must be qualified according to the relative values of contextualized sensory perception. Furthermore the auditory perception is affected by the individual biography and it also crucially depends on the momentary motivation, the level of attention and the *aim of perception* (see 1.5.2.4 and 1.5.3.2).

As a receptor of auditory signals, the ear does not only capture single sound signals or changing factors nowadays measurable with modern equipment. Rather, the ear transmits whole sound events comparable to pictures that are hard or even impossible to measure or verify objectively. Sound events are defined by a large number of intricately related parameters. Quality factors are scaled by ranges and bandwidths. Their values depend on certain perception conditions affected by use, the actual situation, experience and expectations. These auditory impressions depend on several psychological factors of the outer and inner world similarly to visual (also architectural) impressions and symbolic associations. Stereotypical expectations of loudness and of the sound of specific instruments are based on both short-term and long-term learning processes.

1.6 Room acoustics: fundamental concepts in physics

The intensity of sound from a sound source diminishes dramatically inversely to distance. In totally absorbing or "dead" surroundings, such as outdoors in fresh snow with sound not reflected by any surface, words scarcely reach beyond a distance of five meters. Sound intensity of an omnidirectional sound source decreases by four times if the distance is doubled and by a hundred times at a ten times larger distance. If I_1 is the intensity of the sound wave at point A_1, then I_1 a_1 is the amount of energy flowing through the surface a_1 during each second. Assuming that no energy is lost on the way, the same amount of energy will flow each second through the surface a_2 at point A_2, and I_1 a_1 is equal to I_2 a_2. Since the areas of the two surfaces a_1 and a_2 are proportional to the square of their respective distances r_1 and r_2 from the source, the intensity of a sound wave varies in inverse proportion to the square of the distance to the source:[1]

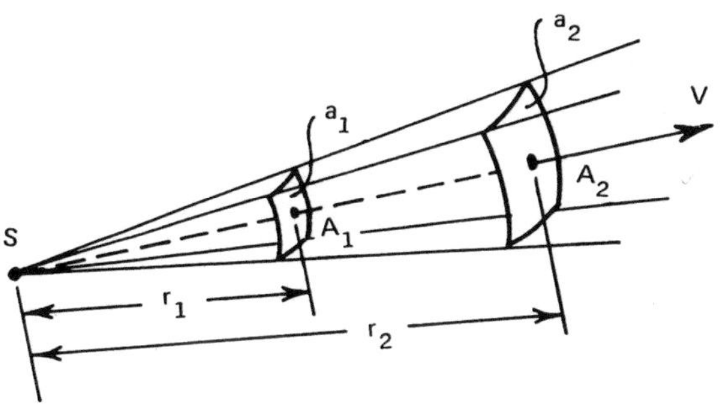

$$\frac{I_1}{I_2} = \left(\frac{r_2}{r_1}\right)^2$$

Figure 29: Decrease of sound intensity with increasing distance.

Sound level decreases by 6 dB at a double distance and by 10 dB at a tenfold distance. This inverse square law reveals the immense importance of supporting sound reflections.

1 J.G. Roederer, *Introduction to the physics and psychophysics of Music* (1975), 73–5.

1.6.1 Geometrical room acoustics

Geometrical room acoustics investigates sound distribution by tracing sound rays and their reflections off surfaces, which are similar to light rays reflected off mirrors.

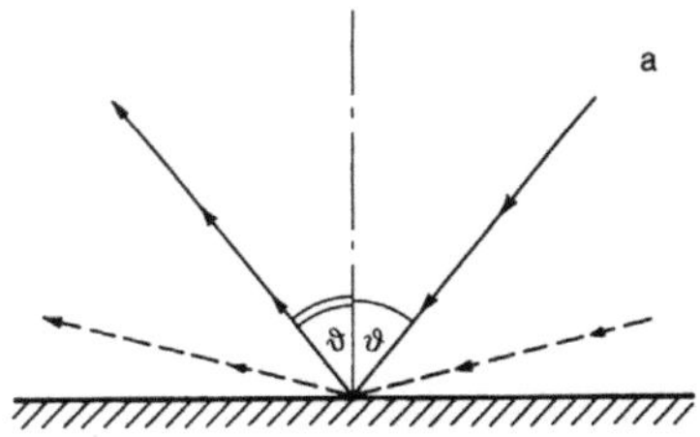

Figure 30: Law of mirrors: reflection off a plane surface.

Incoming and outgoing sound rays lie in a plane that is perpendicular to the reflecting surface. The angle ϑ between reflected and perpendicular sound is the same as the angle of the incoming sound. If the reflecting surface is bent sound rays lie in a plane that is perpendicular to the tangential surface.

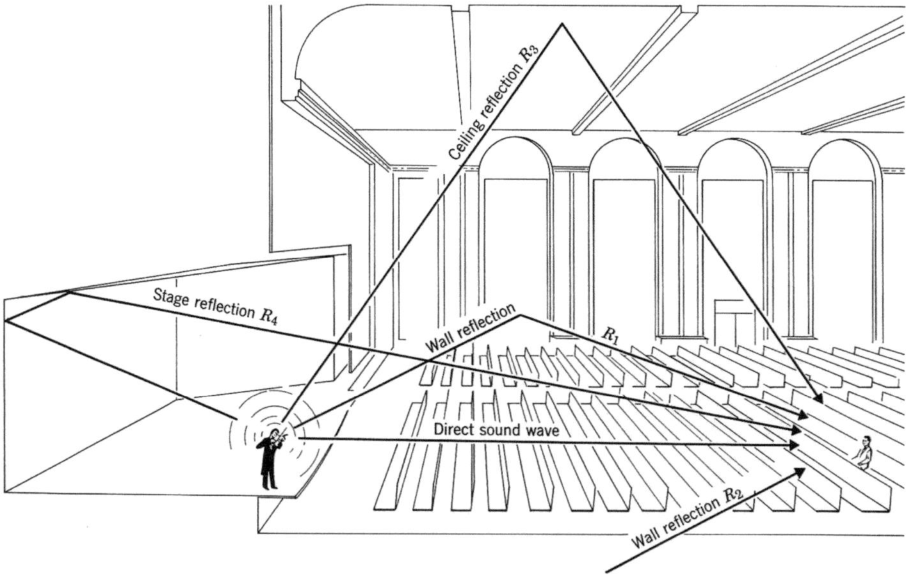

Figure 31: Sound reflections in a concert hall.

As shown in figure 32, a listener in a hall is situated in a field of time-delayed sound waves that spread at a sound velocity of about 340 m/sec from the sound source and are reflected off the enclosing surfaces and objects. A musical instrument or voice sends out sound waves or, in terms of geometrical acoustics, bundles of sound rays. Their directionality depends on frequency: the higher the frequency the narrower the bundle. Direct sound takes the shortest path from the sound source to the listener and arrives there first with a sound level higher than the following sound. So-called indirect sound hits one or more surfaces and is reflected one or more times.

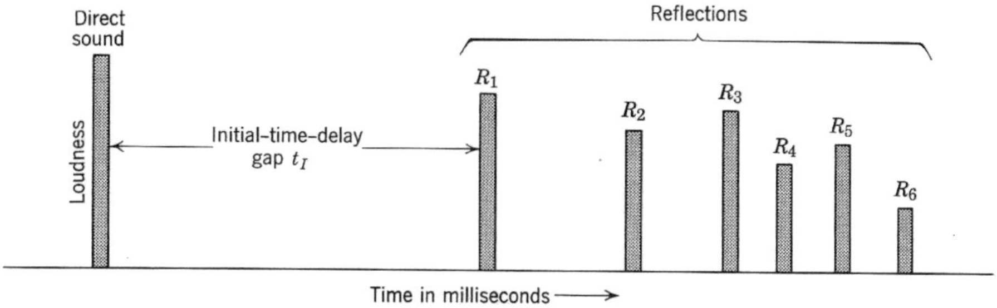

Figure 32: Reflectogram of direct sound and sound reflections at the position of the listener.

In a rectangular hall reflection R_1 from the closer side wall arrives first, followed by reflection R_2 from the other side wall and R_3 from the ceiling, if the height of the hall is more than half of its width. (With a lower ceiling R_3 would arrive earlier.) Later reflections follow after having passed more than one surface, like R_4 from the back wall and the podium ceiling. The pattern of this sequence of reflections depends on the room's shape. Any change in the latter causes a change in the reflection sequence, which changes also following any alteration in the position of the listener or the sound source.

As explained in chapter 1.5, our hearing process is able to cope with this complex situation of multiple reflections and their time delay. Ray tracing clearly shows that higher-order reflections off more than one surface also contribute to the enhancement of perceived loudness and transparency.[2] As long as sound reflections arrive in packages sufficiently coherent with direct sound, loudness and quality of sound impression are supported and enhanced by a brain process called time integration of sound signals. Integration time, which varies between 15 to about 200 milliseconds, is adapted to the number of reflections and their time pattern but also to the aim of perception. Reflections arriving within this time are integrated into one sound impression.

2 J. Webers, *Handbuch für Tonstudiotechnik* (1979), 177.

From the time pattern of incoming sound the brain produces a certain spatial impression, using direct sound to locate the sound source. The initial time delay gap between direct sound and the first reflection as well as further early reflections transmit information on the hall's size. (This capacity allows blind persons to gain an astonishingly exact impression of their spatial surrounding.) Later or much louder reflections may lose coherence with respect to the main package of reflections and cause an echo. They also mask a part of the earlier sound and decrease transparency.

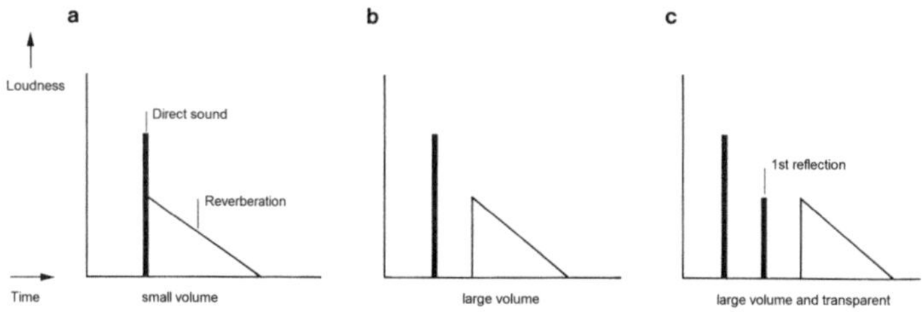

Figure 33: Early reflections and time integration.

As known from sound recording practice and the use of artificial reverberation, early reflections arriving within the first 50 to 80 milliseconds render sound more transparent and improve clarity. The closer the sequence of reflections within 50 to 200 ms, the better the intensity and transparency of sound perception. Thanks to the higher lateral sensitivity reflections from the side walls are crucial to improving the impression of spatiality.[3] Neuronal processing uses the multiple information to gain a clearer impression by accumulating signals up to a certain time delay and by suppressing disturbing signals.[4] As explained before, lateral reflections are especially important due to higher human ear's lateral sensitivity (6 dB to 10 dB above average) and, therefore, highly important in improving the impression of spatiality.[5]

Sound rays behave as if they came from a virtual sound source behind the wall. Geometrical ray tracing is based on the law of mirrors (figure 34). Rectangles generate the most regular distribution of sound rays (1): after two reflections the outgoing ray is parallel to the incoming ray. In narrow angles (2) sound is trapped but creeps along the surfaces when angles are wider than 90 degrees (3):

3 L. Cremer, *Grundlagen der Raumakustik* vol. 1 (1976); J. Meyer, *Acoustics* (2009), 15.

4 Leipp compared this process to visual "autocorrelation"; see E. Leipp, *La machine à écouter* (1977), 31.

5 Cremer (1978), 101; Meyer, *ibid.*

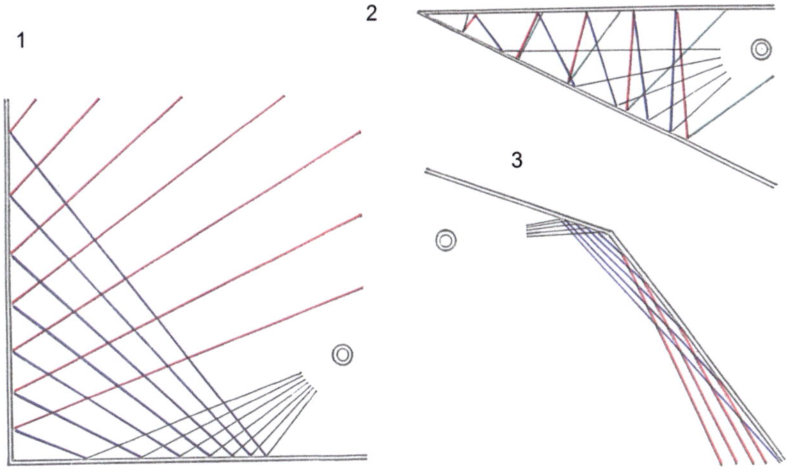

Figure 34: Reflections off surfaces joint at different angles.

Like light rays, sound rays are spread by a convex surface (1) and concentrated into a focus by a concave surface (2):

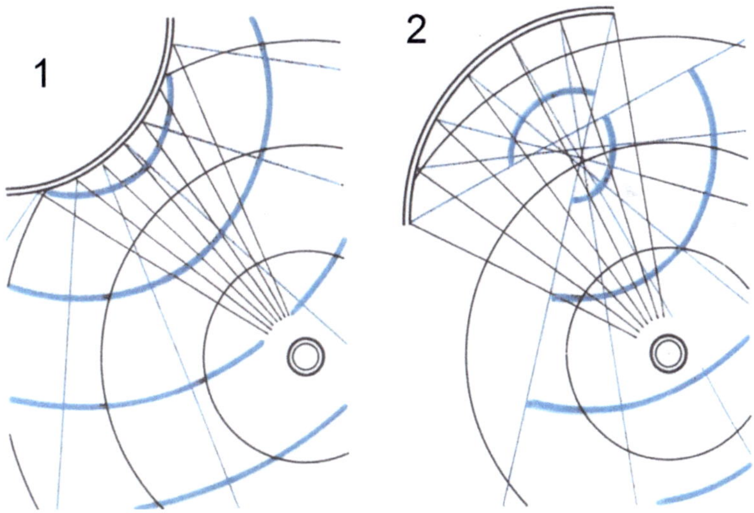

Figure 35: Reflections off convex and concave surfaces.

As in optics, dimensions are crucial (see figure 36). A change in the distance between source and surface may invert the effect: a convex ceiling (1) disperses sound to a large surface of the floor; a concave ceiling with a relatively small radius (3) has the same effect because sound rays spread after having passed the focus; a concave ceiling with a radius equal to the height above the floor reflects all sound back to the sound source placed exactly under the vault center – a speaker in this position speaks into his own ears with increased sound intensity (2):

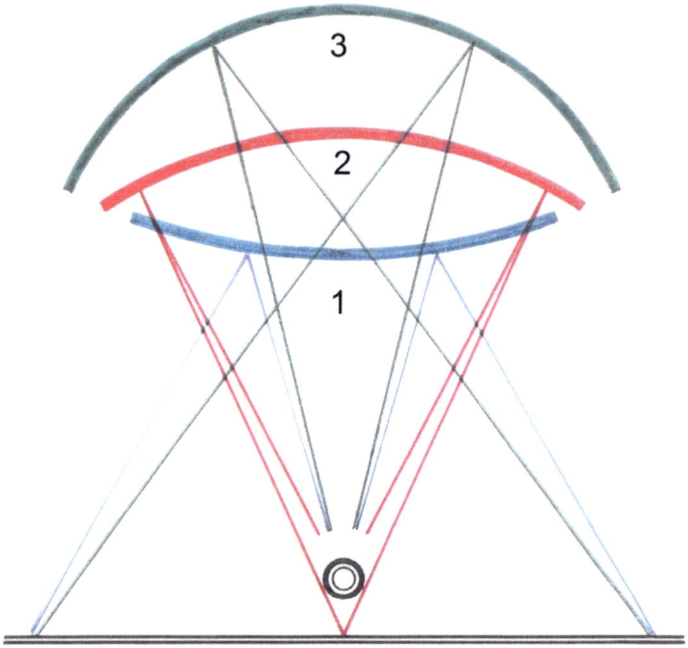

Figure 36: Convex and concave ceilings of different radiuses and with different distances to the floor.

As mentioned before, Athanasius Kircher described in 1650 a very refined geometrical construction of sound amplification by sound focusing (see figure 9). A similar whispering gallery with an elliptical vault, built in 1538–42, still exists below the octagonal chapel of the palace of Charles V in Granada (figure 37). When the speaker in one ellipse focus speaks to a listener at the other focus over a distance of nearly 14 meters,[6] the astonishingly strong sound amplification remains completely unnoticed by those outside the focus.

6 D. Baumann, "Whispering galleries and Arab features" (2001), 481–91.

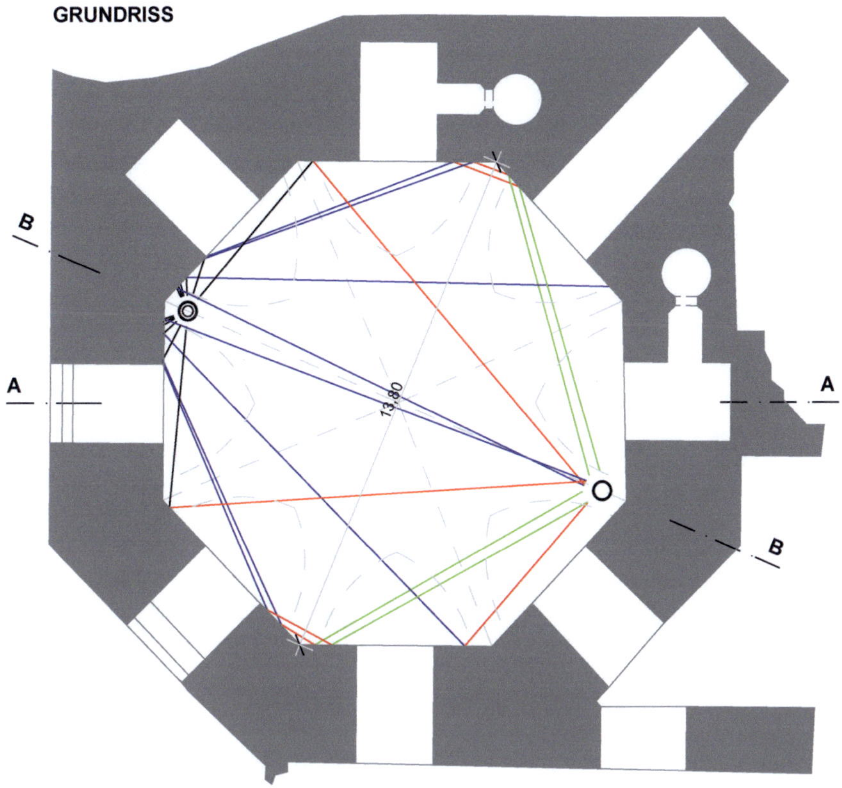

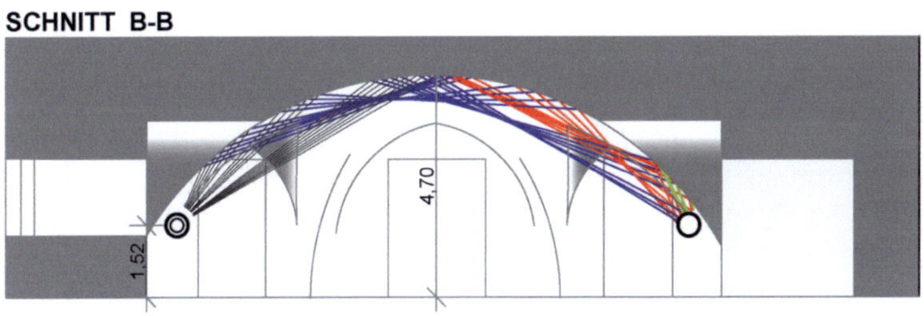

Figure 37: Whispering gallery below the octagonal chapel of the *Palace of Charles V* in Granada, (1538–42), ∅ 13.80 m, maximum height of vault 4.70 m.

Conclusion: The law of mirrors is the main principle of geometrical room acoustics. The room's shape determines the geometrical distribution of direct sound and reflections, provided that the reflecting surfaces are large in relation to the sound's wavelength.

1.6.2 Wave theoretical room acoustics

The geometrical law of mirrors is valid only if reflecting surfaces are clearly larger than the wavelength of incoming sound. Wavelengths of visible light are between 0.4 to 0.75 micro millimeters. In acoustics the range lies between 1.7 cm (for 20'000 Hz) to 17 m (for 20 Hz) – a difference of eminent importance for room acoustics.

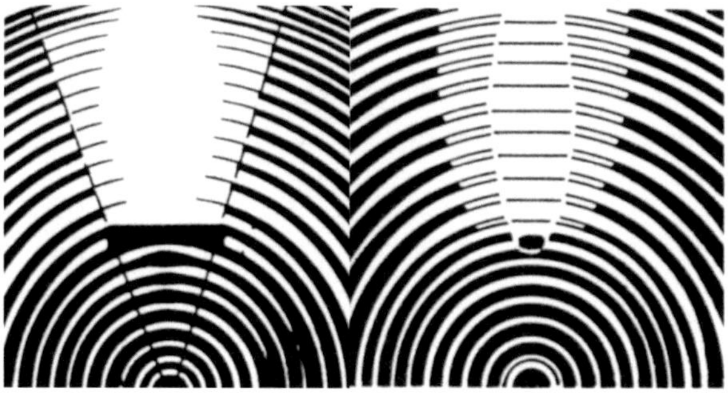

Figure 38: Obstacles: a) reflection ($\lambda << d$); b) diffraction ($\lambda >> d$).

Sound diffraction is a phenomenon related to wave theory. If wave length λ is clearly smaller than the obstacle's diameter d, a sound shadow will be formed behind the obstacle, whereas for higher frequencies normal reflection is occurring on its front side (figure 38, left). If wave lengths λ are clearly larger than the obstacle, sound waves "ignore" it and pass almost undisturbed (figure 38, right). As a result sound is darker behind the obstacle, since high frequencies are reflected, whereas low frequencies "creep" around. Openings cause diffraction if the wavelength is larger than the diameter. Lower frequencies with larger wavelengths spread beyond the sightline (figure 39, left), while higher frequencies with shorter wavelengths remain within the sightline (right).

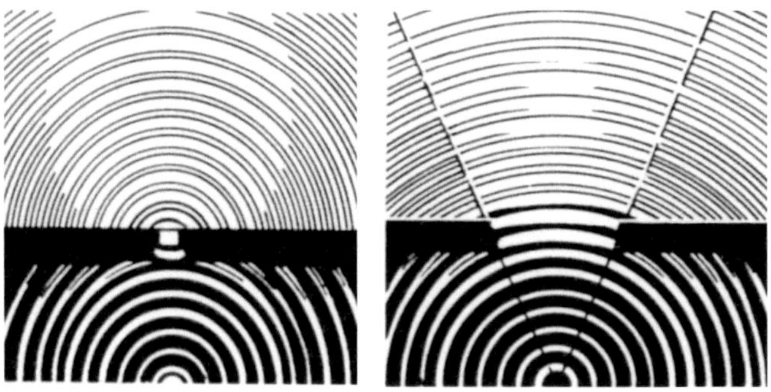

Figure 39: Diffraction through openings: $\lambda >> d$ (left) and $\lambda \approx d$ (right).

A special situation arises when wavelength and opening are of roughly the same magnitude. The opening then acts as a sound source, radiating sound of that wavelength in all directions (figure 39, left).

This explains why high sound frequencies passing easily through the door of a concert hall open into the adjacent corridor do not reach beyond the door sightline, whereas medium and low frequencies spread beyond. If the corridor is a long, narrow rectangular space, sequences of many reflections between the side walls and the ceiling occur. These phenomena produce a dark but surprisingly good sound. It was exactly this impression that fascinated Richard Wagner behind the stage of the *Conservatoire* in Paris. Thanks to diffraction, at the Bayreuth *Festspielhaus* middle and low frequencies also reach the auditorium from the covered orchestra pit.[7]

Reflections on structured surfaces can feature three different situations:

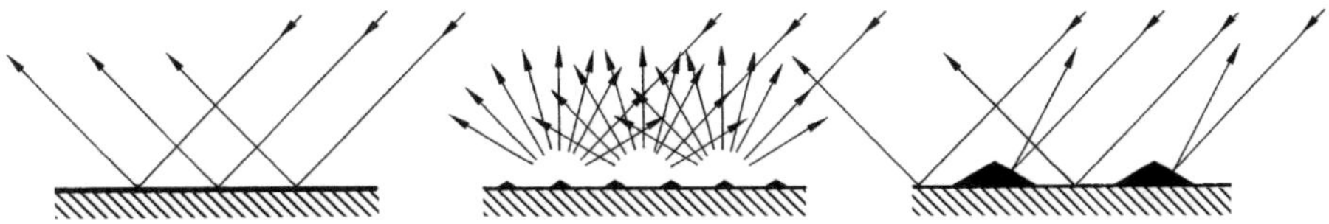

Figure 40: Reflections off wall structures:
from left to right: low frequencies, middle frequencies, high frequencies.

For proper reflection without diffraction, reflecting surfaces need to be larger than three wavelengths ($d > 3 \lambda$, see figure 40). This means that single reflectors may force high frequencies into certain directions. Reflection of low frequencies without diffraction requires very large plane wall surfaces: a violoncello's low C (66 Hz, wavelength 5.15 meters) needs a plane wall of a 15.45-meter diameter to be mirrored. For proper reflection of the lowest frequencies down to 20 Hz uninterrupted walls longer than 50 meters would be necessary. As a consequence, single reflectors cannot clearly direct very low frequencies because diffraction occurs. The distribution of low frequencies in a closed space is a mixture of diffraction and reflection off wall elements of different shapes joined at different angles. Low frequencies ignore smaller obstacles and structures (figure 40 left).

The main properties of reflectors depend on several parameters. The reflectors' effect improves as their dimensions increase relative to wavelength, the distance to the source and to the listener decreases, the angle of incoming

7 D. Baumann, "Bayreuth" (1984); K. Müller, "Die Demokratisierung der Akustik" (2007), 174–83; D. Baumann, "Der Bayreuther Klang" (2009). See also 1.3.4: Space as part of the work's performance.

sound becomes steeper, and their weight increases.[8] For normal speech and high singing voices reflectors need a minimum weight of 10 kilograms per square meter, whereas for low voices and bass instruments more than 40 kilograms per square meter are necessary in order to prevent sound penetration.

Surface structures such as profiles, cornices and coffered ceilings of roughly wavelength dimensions (figure 40, center) or that protrude and recede by a fourth or half a wavelength also cause diffraction. Since sound waves are dispersed widely from such elements with changing surface directions, the term "diffuse reflection" is used. The smaller surfaces of such structures act as mirrors for higher frequencies with shorter wave lengths (figure 40 right) but are "ignored" by low frequencies because their dimensions are small in relation to wavelengths. These surfaces then act as if they were smooth (figure 40 left).[9] That is why in a hall with richly structured surfaces only direct sound carries information on the direction of the sound source, whereas reflected and dispersed sound arriving later at the listener's position carries no such information.

Rough and patterned surfaces cause higher absorption than smooth surfaces of the same material. This effect, which is based on complex wave theoretical processes, is not exactly calculable because of penetration depth, phase shifts and diffraction. Too regular structures generate unpleasant sound decoloration due to selective filtering of certain frequencies (comb filter effects caused by interference). Sufficient variation in the dimensions of surface structures is therefore an important condition for good acoustics.[10]

Such surface structures with elements of various dimensions causing diffusion for a relatively broad frequency range are typical for halls in baroque, rococo and eclectic style richly decorated with marble, stucco or wood cornices and statues. One of the most famous examples is the *Goldene Saal* of the *Wiener Musikverein* with its acoustically favorable structures and statues of plaster (figure 41).

Smooth walls also create a certain amount of diffusivity if surfaces of different absorption alternate, as, for instance, stone wall, curtain, glass, stone wall, and so on. Impulse signals containing high intensities of very high frequencies reflected from plane and hard surfaces produce a shattering sound because our very fine analytic hearing capacity detects directions of reflections.

A further effect explained by wave theory is that the listeners' heads cause bending of high frequencies down into the space between the seat rows and

8 Meyer, *Acoustics* (2009), 182.

9 Cremer, *Raumakustik,* vol. 2 (1978), 310 f.; W. Fasold et al., *Bau- und Raumakustik* (1987), 270; Meyer, *ibid.*

10 Cremer (1976), 409; Fasold (1987), 271.

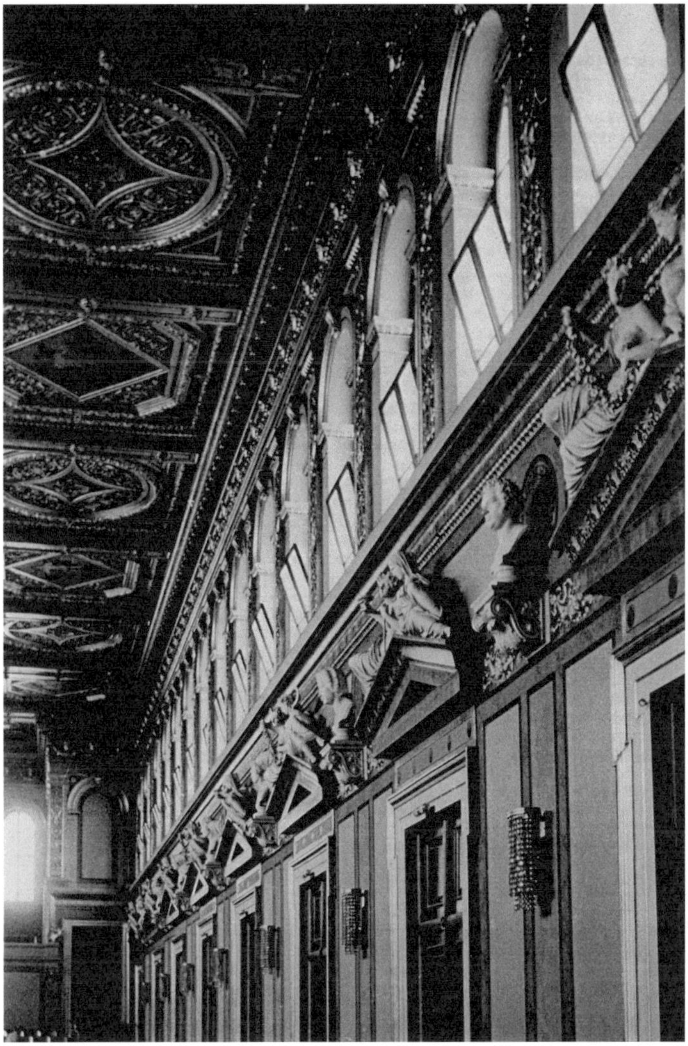

Figure 41: Wall structures in the *Goldener Saal* of the *Musikverein* in Vienna.

increase absorption of sound brilliance unfavorably. This is one of the reasons why a good view to the sound source does not always ensure sufficient quality of direct sound. In the air space closely above the heads a zone of very low sound pressure is generated for all frequencies. If the listeners' ears are exactly in this zone the sound impression is thin and weak,[11] but it may be improved significantly by stretching the neck. This effect is increasing if flat sweeping

11 G. von Békésy, "Interferenzerscheinung infolge Reflexion mit Phasenumkehr" (1933), 6; Cremer (1978), 117 f.

sound is coming from sound sources on a relatively low position. Already in the Greek theaters acoustics were improved by steep raking of seat rows. Vitruvius called for a step height of 1 ¼ to 1 ½ feet with a depth of 2 to 2 ½ feet, which corresponds to an inclination of about 0.6 (figure 3).[12] The same conditions hold for podium height in order to improve direct sound.[13] Rectangular halls with horizontal parterre, high ceiling and narrow parallel side walls cope better with flat sweeping sound because the weak sound pressure is compensated by lateral downward reflections of higher orders from the upper parts of the side walls.[14]

1.6.3 Reverberation time and absorption

When sound production is stopped, sound continues to reverberate until all reflections have reached the listener. The reverberation diminishes due to the increasing length of the reflections' sound path and absorption at the reflecting surfaces. The reverberation time T, an exactly measurable value, is defined as the time at which sound energy decreases to one thousandth of the initial value, which corresponds to a decay of 60 dB. If the initial value is not high enough, the so-called T_{30}-time for a decay from -5 dB to -35 dB is measured and multiplied by 2. The Early Decay Time (ETD) corresponds to -10 dB, the Initial Reverberation Time (IRT) to -15 or -20 dB. Measured values for EDT are often shorter than the calculation according to Sabine because sound is not yet distributed evenly. That is why ETD is closer to the subjective impression of reverberation during ongoing music than to the reverberation heard when music stops.[15] W. C. Sabine empirically determined the following formula for the calculation of T:

$$T = 0.163 \frac{V}{A}$$

T = reverberation time (seconds)
V = volume of the enclosure (m^3)
A = equivalent absorption area (m^2)

12 Cremer (1976), 90 ff.: Glanzwinkel.
13 E. Mommertz, "Einige Messungen zur streifenden Schallausbreitung über Publikum" (1993), 42; Meyer (2009), 192.
14 Cremer (1978), 118.
15 Meyer (2009), 190.

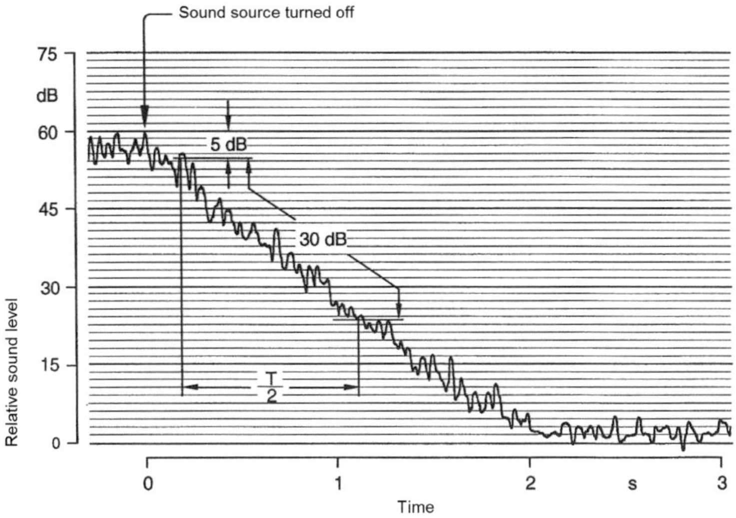

Figure 42: Reverberation time T_{30}.

The total equivalent absorption area A is the sum of all partial areas S with different absorption coefficients α:

$$A = \sum_{i=1}^{n} \alpha_i S_i$$

α = absorption coefficient
S = absorption area
n = number of absorption areas with different absorption coefficients

Calculations of T with Sabine's formula usually yield values quite close to on-site measurements if absorption coefficients of building materials are well known and low ($\alpha < 0.25$).[16] Since absorption and reverberation depend on frequency, values are either plotted as a frequency curve or indicated for a certain frequency, as, for instance, T_{125} for the reverberation time T for 125 Hz, or $T_{500-1000}$ for the average reverberation time for frequencies from 500 to 1000 Hz.

16 Eyring's formula corresponds better to measurements in the hall for higher absorption coefficients, see E. Meyer and E. G. Neumann, *Technische Akustik* (1974), 68. Differences between measured and calculated reverberation times are one of several reasons for yet insufficient exactness of computer auralization. The other is the difficulty to predict diffusivity in computer simulation, see 1.6.5.

The reverberation time T is one of several important factors in the description of a room's acoustics. It is not identical with the reverberation perceived when the music ends or is interrupted by a pause. Audible reverberation depends on the loudness of sound before the break, on noise and on the listener' actual sensitivity. Lower sound levels produce a shorter after ring. A musician can affect the amount of reverberation in several ways. Limitation of sound level, prolongation of breaks and adaptation of tempo and articulation are often the only means to play music in rooms with excessive reverberation.

Table 1: Characteristic standard values for absorption coefficients of some building materials:[17] total absorption: $\alpha = 1$, total reflection: $\alpha = 0$.

Absorption coefficient	*α for average frequencies (Hz)*					
Wall and ceiling surfaces	*125*	*250*	*500*	*1000*	*2000*	*4000*
Marble	0.01	0.01	0.02	0.02	0.03	0.03
Concrete, stucco, oil paint	0.02	0.02	0.03	0.04	0.05	0.05
Cement plaster, limewater color, wallpaper	0.03	0.03	0.02	0.04	0.05	0.08
Wooden board, chip board on firm ground	0.04	0.04	0.05	0.06	0.06	0.06
Glass window	0.25	0.15	0.10	0.05	0.03	0.03
Floors						
Concrete blocks	0.02	0.02	0.03	0.04	0.05	0.05
Wooden floor on bars	0.15	0.11	0.10	0.07	0.06	0.06
Wooden platform on cavity	0.12	0.08	0.07	0.06	0.06	0.06
Parquet flooring on firm subsoil	0.04	0.04	0.05	0.06	0.06	0.06
Parquet flooring on cavity	0.10	0.07	0.05	0.06	0.06	0.06
Carpet, less than 6 mm thick	0.02	0.04	0.06	0.20	0.30	0.35
Carpet, about 7 to 10 mm thick	0.04	0.07	0.12	0.30	0.50	0.50
Chairs 1.15 m distance between rows, occupied:						
Wooden chairs	0.01	0.15	0.30	0.40	0.55	0.60
Folding chairs, upholstered 20 mm thick	0.30	0.35	0.50	0..60	0.70	0.70

The equivalent absorption coefficients and surfaces of no longer extant historical rooms cannot be conclusively determined. Absorption coefficients depend on porosity, elasticity, vibrating mass or, in other words, on the material, its surface quality, and the substructure's attachment and construction. Exact values for α can be established only through on-site measuring or reverberant field tests of original materials.

Musicians know from experience that large surfaces of polished slates of marble or other stone cause excessive reverberation due to very low absorption ($\alpha = 0.01$ to 0.03). The large wooden surfaces in many famous halls gave rise

17 After Fasold, *Bau- und Raumakustik* (1987), 259.

to a nearly mythical belief in the acoustical benefits of this material. The most famous example in history was presumably the completely wooden concert hall at the old *Gewandhaus* in Leipzig.[18] Measurements have confirmed that wooden plates produce audible vibrations due to resonance at low frequencies, especially if directly set into vibration by such instruments as the violoncello, the double bass or the grand piano. Musicians have reported about this experience on the platform.[19] Another acoustical effect is caused by airborne sound that sets wooden plates into motion. Depending on their attachment, they cause quite high absorption if mounted over cavities with openings. The system then acts as a Helmholtz resonator absorbing low frequencies.[20] Plaster, also in the form of stucco or lime with sand, enhances acoustics due to its modest absorption that increases towards high frequencies and dampens hard reflections. Carpets, upholstered chairs or even simple wooden benches like those used in earlier concert halls and theaters have a high absorption coefficient for middle and high frequencies.

The zones occupied by listeners are a considerable absorbing surface in relation to a room's total inner surface, even more so in halls with low ceilings. Due to absorption caused by diffraction the surface occupied by listeners is more important than their number. Absorption caused by a certain number of listeners increases as the surface they occupy increases. In other words, the equivalent absorption surface A in relation to the number of listeners decreases with increasing seating density.

Further factors influencing absorption can be quantified only by measuring. They include the listeners' clothes, seat specifications (materials, cavities below the seat, height of the backrest and the distance of its lower end to the floor, the surface not covered by the seated person) and the sound's incident angle.[21] The surface occupied by listeners and the total number of seats are, therefore, among the most important factors in room acoustics.[22] An increased seating surface due to a larger number of seats does not only increase absorption but also the distance from sound sources to the reflecting walls and to the listeners, causing a further delay of reflected sound. This in itself reduces maximum loudness, and later reflections reduce also the density of incoming reflections per time.

18 H. Bagenal and A. Wood, *Planning for Good Acoustics* (1931), 91, 95, 101 f.

19 A. Askenfelt, "Stage Floors and Risers – Supporting Resonant Bodies or Sound Traps?" (1986), 43–61.

20 Cremer (1978), 231, 316.

21 Fasold (1987), figure 9.57, 9.59, 10; H. Kuttruff, *Room Acoustics* (3rd ed., 1991), table VI.3. For resonance-based absorption between 130–170 Hz for certain kinds of chairs, see Meyer (2009), 192 (Davies/Lam, 1994); acoustic problems due to change of chairs are documented for Zürich, *Tonhalle,* Grosser Saal 1968, and Basel, *Casino,* Grosser Saal 1989.

22 D. Baumann, "Die Kunst des Konzertsaalbaus" (Interview with Russell Johnson) (1993).

Since sound energy produced by music instruments and voices without electro-acoustical amplification is limited, the number of seats in halls without amplification must be restricted. For modern seating density the upper limit is at about 3000 seats. With a larger number of seats acoustics approach free-field conditions. Sound intensity decreases disappointingly with increasing distance, and sound becomes thin and devoid of low frequencies. Berlioz described this experience during his monster concerts.[23] For excellent acoustics with large orchestras and choirs the upper limit is 1800 to 2000 seats.[24]

On the other hand, sound loudness increases if halls are too small and the ceiling is too low, dead and without brilliance. A minimum required volume per seat has therefore been calculated to reach a certain reverberation time.[25] However, the disadvantage of this standard factor lies in its lack information on factors crucial for the quality of room acoustics, that is, the relation of room width to room height and the geometrical sound distribution as a function of time.[26]

1.6.4 Quantification of quality in room acoustics

The reverberation time T alone is not sufficient to define room acoustic quality. Some halls with relatively short reverberation and some of those with long reverberation are said to be excellent for music. On the other hand, imitations of the reverberation decay curves of acoustically famous halls are not sufficient to offer the same favorable experience. The quality of room acoustics obviously depends on further important factors. After many discussions with musicians and music critics, Leo Beranek selected 18 terms to describe the quality of room acoustics also in relation to music.[27] Table 2 lists these "subjective attributes" and, in the second column, relates them to the objective parameters (table 2).

J. S. Bradley proposed to use only five subjective attributes in relation to the following objective parameters (see table 3).[28] The parameters given in the second column of table 3 can be calculated or measured on site. We should remember, though, that physical definitions cannot be directly related to subjecti-

23 See 1.3.4: Architectural space as part of the work's performance.
24 J. Burghauser / A. Spelda, *Akustische Grundlagen* (1971), 152; Baumann, *ibid.*
25 Fasold (1987), 259.
26 For the importance of the relation of height to width, see 1.8.2: Room proportions.
27 L. L. Beranek, *Acoustics* (1979), 29, 64 and *Concert and Opera Halls* (1996), 22.
28 J. S. Bradley, "The Evolution of Newer Auditorium Acoustics Measures" (1990), 13–23.

ve impressions because the latter always depend on the listener's psychological background.[29]

Table 2: Quality factors (after Beranek).

Subjective attribute	Physical facts
Intimacy, presence	Early first reflection, initial-time-delay gap < 15 ms
Liveness, fullness of tone	T for f > 500 Hz
Warmth	T for middle & high frequencies (f = 20 – 250 Hz), boomy if low-frequency tones are exaggerated
Loudness of the direct sound	in comparison to the reverberant sound
Loudness of the reverberant sound	in comparison to direct sound
Definition, clarity	Early reflections until 80 ms
Brilliance	Relative prominence of the treble and the slowness of its decay
Diffusion	Diffuse reflections
Balance	within the orchestra and with soloists
Blend	Mixing of the sound
Ensemble	Co-ordination within orchestra and between stage and pit
Immediacy of response, attack	First reflections reaching the musician
Texture	Sequence of reflections
Freedom from echo	No annoying reflections after $\Delta t > 70$ ms
Quiet, freedom from noise	Isolation of all external noise sources
Dynamic range	Depends on the force of the orchestra and the acoustic characteristics of the hall
Distortion, tonal quality	Flutter echo, a loss of a band of frequencies
Uniformity	of sound in the hall

Table 3: Quality factors (after Bradley).

Subjective attribute	Objective parameter
Strength, loudness	Relative level G: impulse response at listener's position in relation to free-field conditions at 10 m
Reverberance, liveness	Early Decay Time EDT, T_{10} for the first 10 dB
Clarity, definition	C_{80}: impulse response at listener's position < 80 ms in relation 80 ms to ∞; centre time TS
Spatial impression, envelopment	Lateral Energy Factor LF: impulse response at listener's position for 25 – 80 ms (bidirectional microphone) in relation to response for 0 – 80 ms (omnidirectional microphone); Inter-aural cross correlation coefficent IACC (for artificial-head)
Timbre	Variation of EDT with frequency, EDT (f)

29 D. Baumannn, "Systematische Musikwissenschaft – eine Disziplin zwischen Kulturgeschichte und Naturwissenschaften" (2009), 40–51: 40.

Comparisons between these measured values and the subjective impressions therefore call for special care and experience. Still, this is the only way to relate objective factors to subjective attributes and these to the practical experience of musicians and listeners.

In a closed space with regular sound distribution there is a statistical sound field or diffuse field where sound energy decays with increasing distance from the sound source to a level at which direct sound and reflected sound become equally strong (diffuse-field distance). The energy absorbed by the enclosing surfaces is then equal to the energy produced by the sound source, and the sound level is equal everywhere in the hall.[30]

Based on the assumption of statistical sound distribution, the room damping index D_A may be used to compare different rooms used for music.[31] This numerical term objectively defines facts well known to musicians, namely, the influence of room acoustics (volume, reverberation time) on the relative sound level produced by a known sound source (all musical instruments and voices) in a certain hall. The sound pressure level of the diffuse sound field or the density of energy in a room is given by:

$$L_{p\ statist} = L_W - 10 \log \frac{V}{V_0} + 10 \log \frac{T}{T_0} + 14 dB$$

L_p	sound pressure level of the diffuse field = density of energy = energy per volume		
L_w	sound power level of the sound source		
V	volume of space	V_0	volume unit = 1 m^3
T	reverberation time	T_0	time unit = 1 sec

The sound pressure level L_p of the diffuse sound field depends on space volume, the power of the sound source (orchestra size) and reverberation time T (therefore also on frequency). In large rooms the energy density is lower than in small rooms. Berlioz left us an impressive description of this experience when listening to Beethoven's piano trio played in rooms of different sizes.[32] The decrease of sound energy in larger rooms is compensated only partly by a longer reverberation time T.[33]

30 Strictly only true in a cubic room which is not too large, see Meyer (2009), 350.

31 Meyer (2009), 191.

32 See 1.3.4, Architectural space as part of the work's performance.

33 Meyer, *ibid.*

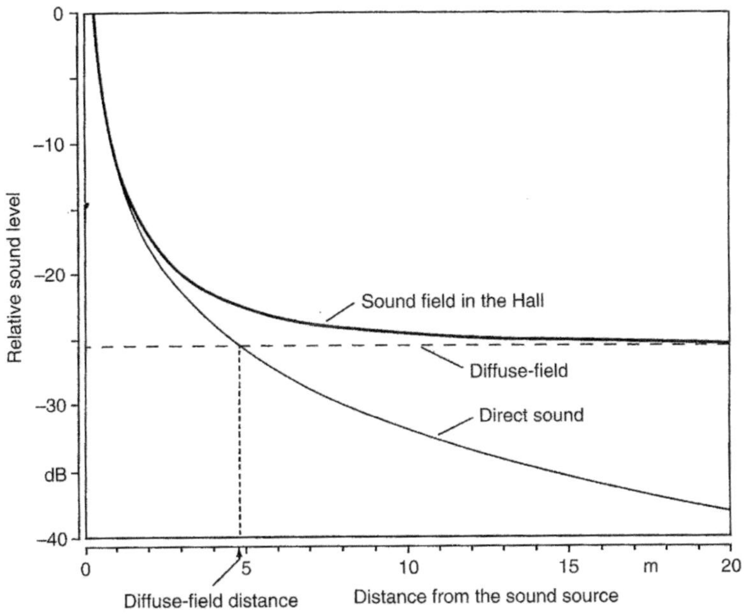

Figure 43: Sound level in dependence on sound source distance (omnidirectional).

The room damping index D_A is defined as the difference between the sound power level of the source L_w and the energy density L_p :

$$D_A = L_w - L_p = 10\log\frac{V}{V_0} - 10\log\frac{T}{T_0} - 14dB$$

The room damping index D_A permits to directly quantify the influence of room acoustics on the effect of a sound source. D_A is the numerical difference between the sound power level of the source (all musical instruments and voices), which is room independent, and the sound pressure level achieved in the room (based on diffuse-field conditions). The room damping index D_A is calculable if volume and reverberation time T are known. Data on the energy produced by the sound source are not needed. D_A permits to calculate the change of average sound level if the same sound source (an orchestra of the same size) is moved to different halls. For musicians this value indicates the change of the average level of sound necessary to reach an equivalent sound level of music in a room with different absorption (shorter or longer reverberation time) or of a larger or smaller size (different cubic space). Figure 44 shows a diagram of values for several empty and occupied spaces.[34]

Some of these halls and theaters are further discussed in this study, such as the hall of *Palais Lobkowitz* in Vienna (D_A calculated as 15 dB for the occupied hall), the great hall of *Hanover Square Rooms* in London and the hall of the old *Gewandhaus* in Leipzig (D_A occupied 20 dB), as well as the great halls of the *Musikverein* in Vienna, the *Philharmonie* in Berlin and the new *Gewandhaus* in Leipzig, which feature a higher D_A (D_A occupied 25–27 dB). To reach the same average sound level, an ensemble of the same size, such as the small orchestra that played Beethoven's *Eroica* in the private hall of Fürst Lobkowitz, would have to increase its sound level by 5 dB in the old *Gewandhaus* and by 10 to 13 dB in the great halls of the late nineteenth century. The task of musicology here is not only to interpret these values and keep in mind that for not extant halls they are based partly on uncertain information on room shape, materials and building construction, but also to study how musical practice adapted to such differences. Did musicians play louder in larger halls or was the size of the orchestra increased?[35] If not, were listeners aware that sound became thinner and sound level lower? If so, did they judge the size of the hall to be too large?[36]

If we compare the number of musicians playing in smaller halls and opera houses with their number in larger halls we have to be aware that we judge the size of the ensemble by stylistic and aesthetic attributes. We know from experience that the number of instruments affects sound quality (color, articulation, tempo etc.). Room acoustical considerations show how much these aesthetic assessments depend on the performance space and its acoustics.

Values for D_A are valid for a fully built-up statistical sound field only, or in musical terms, for long notes. In large or not cubic halls the statistical sound field is often incompletely built up, resulting in an energy density about 3 dB lower than expected. Sound level then decreases by about 0.85 dB for an increase of 10 meters distance.[37] The value depends also on the local reflections and the directivity of the sound source. That is why one must determine the strength G or relative level (in dB), which indicates the difference between the

34 Meyer (2009), 351.

35 The increase of level of power in relation to number of instruments in an orchestra has been measured by Burghauser and Spelda (1971). Doubling the number of instruments effects not only an increase of energy density by 3 dB but the so-called chorus effect sound becomes denser and fuller), see Meyer (2009), 347 f.

36 This was the case for the great hall at the Hofburg in Vienna which was thought to be too large for Beethoven's symphonies in 1805 *(AMZ,* 15 May 1805), see 1.9.5.

37 M. Barron and L.-J. Lee, "Energy Relations in Concert Auditoriums" (1988), 618; see Meyer (2009), 194.

power level of the sound source and the power level at the position of each seat. Since the exact calculation of G is very complex, it is measured for each situation on the basis of a so-called impulse response.[38] The sound pressure level is not identical to the perceived loudness of sound, which also depends on the momentary structure of the music. But sufficient loudness gives the general impression of fullness, spatiality and warmth. These differences between perception and measurements must be taken into account when physically defined values are used for the interpretation of musical situations.

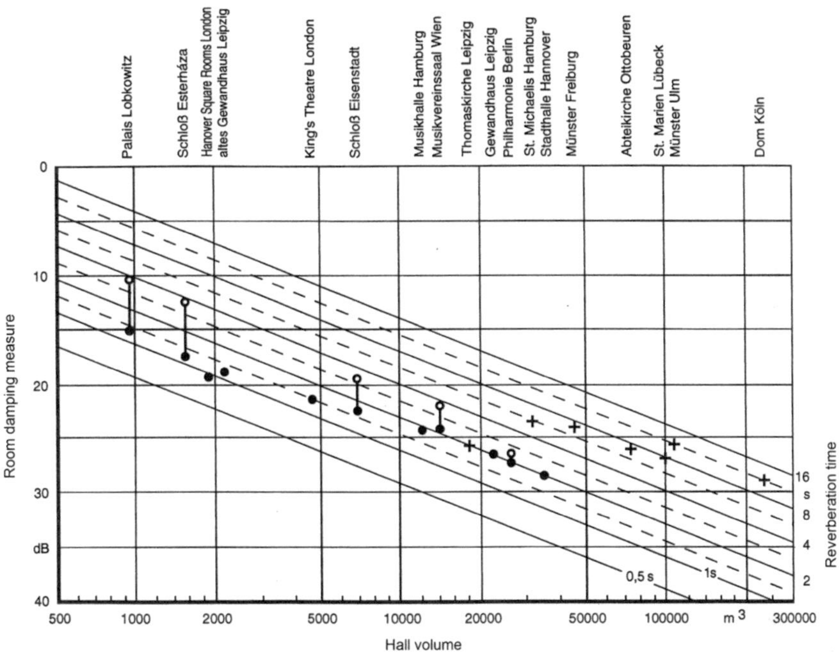

Figure 44: Dependence of room damping index D_A on the hall's volume of space and reverberation time: ● : fully occupied hall, ○ : unoccupied hall, + : empty church.

A further important value that can easily be heard by attentive listeners is the diffuse-field radius r_H. It depends on volume of space and reverberation time T but not on the power level of the sound source. r_H is the distance from the sound source at which loudness of direct sound and reflected sound are equal (see figure 43).

38 P. Lehmann, "Über die Ermittlung raumakustischer Kriterien und deren Zusammenhang mit subjektiven Beurteilungen der Hörsamkeit" (1976); see Meyer (2009), 192.

$$r_H = 0{,}057\Gamma_{st}\sqrt{\frac{V}{T}}$$

r_H	diffuse-field radius in m	V	volume of space in m^3
Γ_{st}	statistical directivity factor of sound source	T	reverberation time in s

For an omnidirectional sound source the diffuse-field distance, called diffuse-field radius, is the same in all directions. Within the diffuse-field distance sound is more brilliant because direct sound is stronger than indirect sound. Beyond the diffuse-field distance sound becomes darker and more diffuse because indirect sound prevails. The diffuse-field distance is an important factor in sound recording in order to determine best positions for microphones. The following examples of diffuse-field radiuses for the main frequencies of a trumpet indicate that in the highly prized long, relatively narrow concert halls of the eighteenth and nineteenth centuries brilliance and directness of sound reached many more seats than in modern halls of more complex shape, such as the new *Philharmonie* Berlin, a centered room built in 1963:

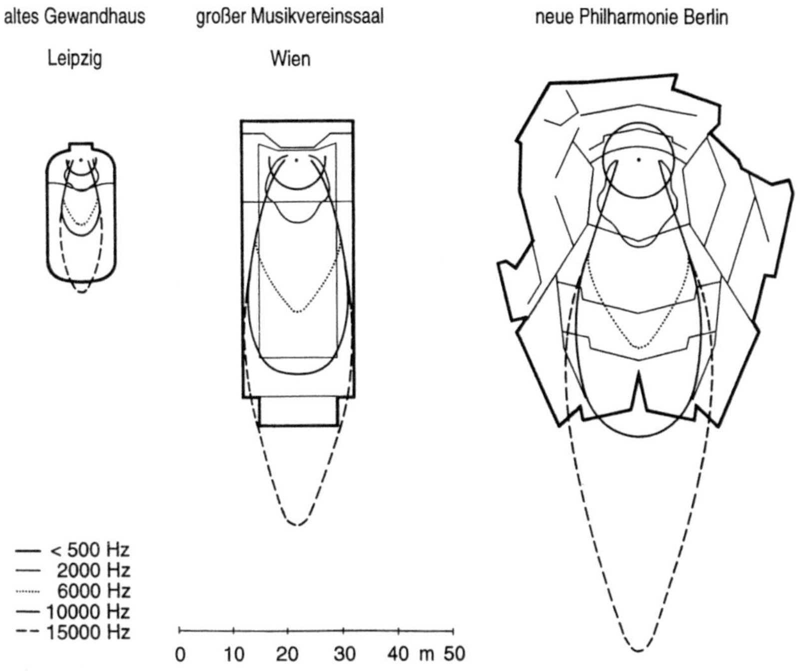

Figure 45: Diffuse-field distance of a trumpet playing facing the back wall.

Clearness of trumpet articulation and brilliance of sound already diminishes for lateral seats in a hall larger than 20 meters if the trumpet plays exactly towards the end of the hall. High singing voices with similarly pronounced directional characteristics for high frequencies elicit a similar impression.

1.6.5 General acoustic impression and acoustics at a certain place

Beranek's 18 attributes of room acoustics describe two main aspects important for musicians and listeners alike, as these judge both the general acoustic impression and the specific acoustics at their position in the hall. This double view manifests itself in the astonishing capacity of musicians on the podium to calculate the effect of their playing on listeners in the hall. The general acoustic impression (comparable to a room's general illumination) may be described best in terms based on the statistical sound field because experience and judgment are based on relatively slow aspects of sound events.

But in music relatively long quasi-stationary sounds are structured by very short, quickly passing articulation processes or onsets of sound vibrations. These micro-structures of sound events happen in a temporal range during which sound is not yet distributed evenly within the cubic space and before the statistical sound field is fully established. The appropriate transmission of these micro-temporal structures is crucial for the sound impression at a certain position in the room. As explained in chapter 1.5, micro-temporal auditory perception is so precise that these processes of the onset of sound vibrations play an important role for auditory perception. That is why geometrical room acoustics is more important than has long been assumed.[39] To determine the acoustics at a certain position it is not sufficient to merely follow the way of direct sound (somehow comparable to the direct light for a place of work).

It is crucial to select relevant positions for the sound source and the listeners in order to analyze reflections over one, two, three, and more surfaces (comparable to the buildup of indirect, general illumination in a room).[40] This allows the temporal buildup of reflections over a relatively extended period up to 80 ms or more to be followed on its geometrical way.

Today three-dimensional geometrical analyses can be plotted by computer ray tracing based on the construction of mirrored sources (see figure 46) that follow sound rays on their way from the source to the listener's position or vice versa (image method).

39 Cremer (1978); Y. Ando, *Concert Hall Acoustics* (1985).

40 Th. Baer-Loy, *Hören und Raumform* (1984); ARTEC, New York (1989); K. Heutschi, *Seminar Computersimulation in der Raumakustik* (1995).

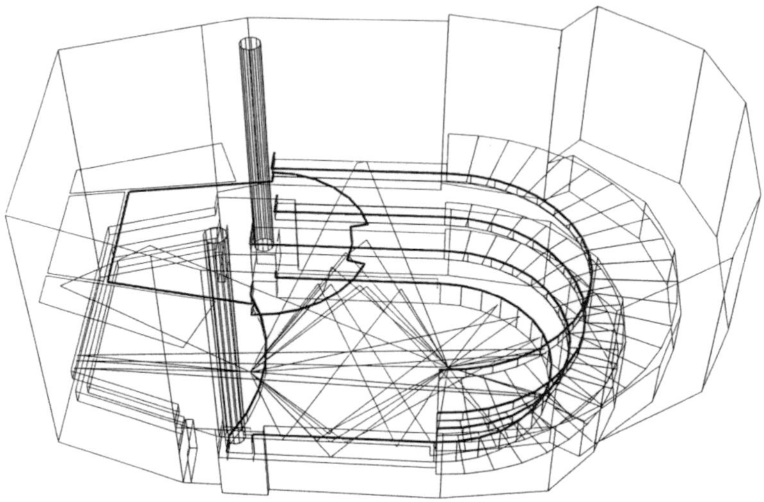

Figure 46: *McDermott Concert Hall,* Dallas, TX (1990), 3D ray tracing, canopy above the stage.

Unfortunately, the three-dimensional plots become quickly overloaded. Two-dimensional designs may give a clearer view on the distribution of direct and indirect sound in certain parts of a space if the sectional plane is carefully chosen. The following pictures show the distribution of lateral reflections from side walls in halls with two different ground plans. This analysis confirms the lack of lateral sound experienced in the middle part of fan shaped halls and the excellent distribution of lateral sound in rectangular halls:

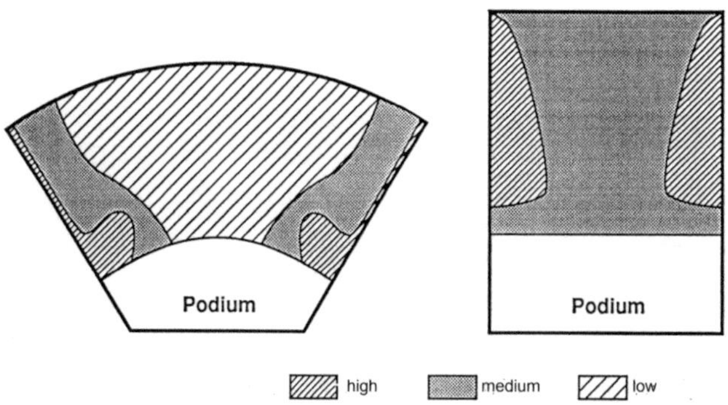

Figure 47: Early lateral sound in different ground plans with a surface of 600 m^2.

Another two-dimensional geometrical analysis of sound distribution confirms the result of several acoustic tests in the former *Philharmonic Hall* in New York (figure 48). The reflection distribution from the concave side walls in the ground plan shows that the center of the hall receives only side wall reflections from the first third of the hall length, while all lateral reflections from the middle part of the hall remain close to the side walls and are concentrated at the end of the hall. The longitudinal section shows that first-order reflections from the ceiling are well distributed. Reflections going sideward to the ceiling are inevitably directed to the side walls and from there also travel along the side walls and concentrate at the end of the hall. That is why the best seats in this problematic hall were close to the concave side walls and at the very end of the hall, while listeners in the center of the parterre complained about a lack of low frequencies.

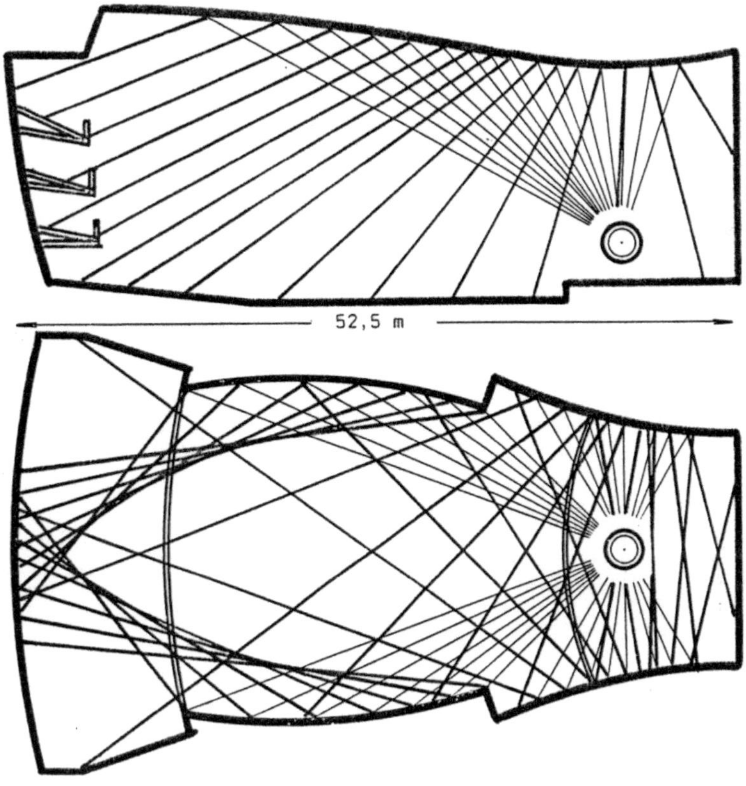

Figure 48: New York, *Philharmonic Hall* (1976).

As explained, the absence of low frequencies cannot be corrected by reflectors. During the acoustic tests in the *Philharmonic Hall* experts tried in vain to combine a great number of single reflectors in order to force low frequencies towards the hall's center. That is why in 1976 the *Philharmonic Hall* was torn down and replaced by the shoe-box shaped *Avery Fisher Hall* (architects Philip Johnson & John Burgee; acoustic consultant Cyrill Harris) to solve the problem of reflection distribution.[41]

On the other hand, the new *Philharmonie* in Berlin (1963 Hans Scharoun, acoustic consultant Lothar Cremer) lacked high frequencies on the podium due to the height of the convex ceiling. This problem was successfully corrected with convex reflectors hung by strings from the ceiling to improve clarity on the podium by adding early reflections:

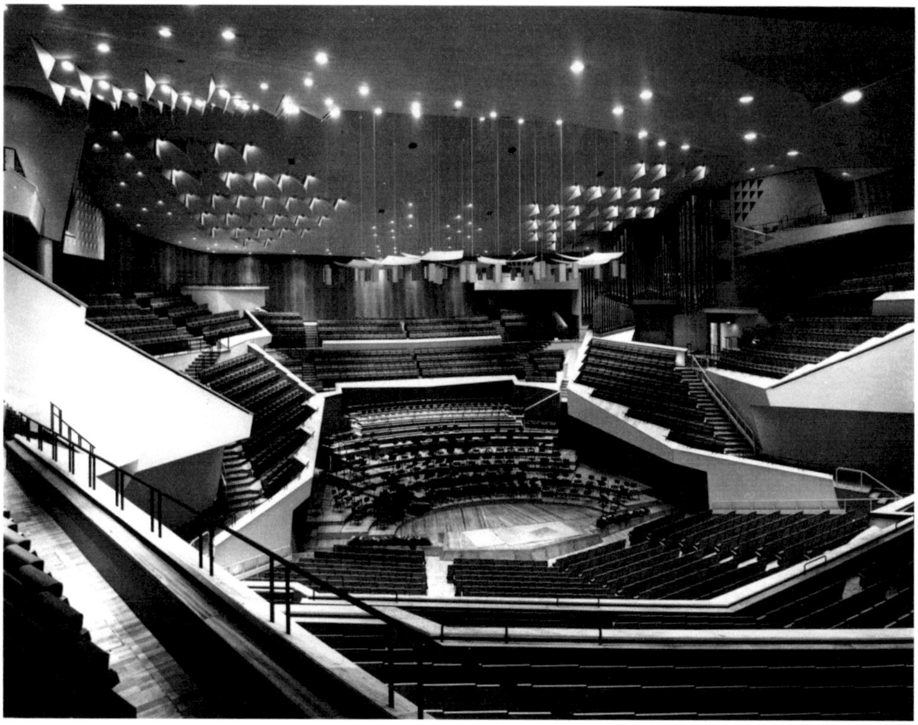

Figure 49: Berlin, *Neue Philharmonie*, great hall (1963): convex reflectors suspended from the ceiling above the podium.

Already in the eighteenth century theater architects and related specialists applied these geometrical facts by designing the proscenium's ceiling for more clarity, that is, by improving reflections towards the auditorium. For theaters this is especially important because lateral reflections are weakened by the absorbing openings of the boxes and the occupied gallery benches. In 1709 the

41 M. Forsyth, *Buildings for Music* (1985), 286 f.; J. Pierce, *The science of musical sound* (1992), 140 f.

acoustics of the *King's Theatre* in London were successfully improved with a lowered ceiling in the auditorium's front part. In 1732 the ceiling of the proscenium was inclined in order to direct reflections more towards the rear part of the auditorium.[42] The same measures were taken in many theater auditoria of the nineteenth century in order to improve sound transmission from singers and actors who were relegated behind the proscenium arch because the orchestra pit was enlarged and the proscenium stage eliminated.[43]

Three-dimensional computer ray tracing and the calculation of impulse responses require high processing capacity. This is why 3D models of halls and laser beams with little mirrors are still used to make reflections visible.[44] Modern software takes into account also absorption coefficients α of the respective surfaces for each reflection. Strictly regarded geometrical analyses show only reflections not submitted to diffraction. If surfaces of the enclosure are structured, ray tracing is valid only for middle and lower frequencies (according to the rule $d > 3\ \lambda$ for proper reflections). Higher frequencies reflected from the small surfaces of the structures or diffused omnidirectionally if structures are of about wavelength must be analyzed on the basis of larger scaled, exact models of the surface.[45]

Diffusivity and absorption of reflections do not depend on surface structure only but also on the penetration depth of the sound wave, that is, on the material, its porosity, fixation and elasticity, on cavities behind the surfaces, on materials below and on the acoustic behavior of closely coupled spaces. The quality of electro-acoustic auralization based on calculated impulse responses and anechoic sound recordings is, therefore, still insufficient for a fully reliable reproduction of room acoustics and sound quality, despite the enormous progress of computer programs in this field, as diffusivity and wave penetration and their effects are not yet predictable with sufficient precision.[46]

42 See 2.5.3 Theaters used by Handel.

43 D. Baumann, "Qualche riflessioni sull'acustica [del teatro 'La Fenice']" (1996), 81–3.

44 A famous example was the great hall at the *Trocadéro* in Paris, a recent example the *Opéra Bastille* in Paris.

45 Heutschi (1995), SP – 5.

46 M. Vorländer, *International Round Robin on Room Acoustical Computer Simulations* (1994); K. Nasshan, G. Schupp, U. Stephenson, "Echtzeitauralisation als Element optischer und akustischer virtueller Realität" (1995), 307–310; J. Blauert, *Spatial Hearing* (1997), 372 f.; M. Vorländer, "Room Acoustics in Virtual Reality" (2007); see also 1.6.3 Statistical room acoustics.

1.6.6 Principles for the analysis of a room's acoustic quality

Modern knowledge in room acoustics has come a long way from Sabine's formula (1895). Numerous factors determine room acoustic processes. On the one hand, the development of theoretical knowledge slowed down because twentieth-century construction techniques allowed nearly unlimited freedom in architectural design, and, as a result, the acoustic consequences of new room concepts could not be based on experience. On the other hand, defining criteria for the quality of room acoustics proved difficult. Factors must not only be clearly measurable and calculable; their relevance for auditory perception of sound quality is crucial. The difficulty in defining quality factors in room acoustics stems from our holistic perception. After a short adaptation time neuronal processes and their feed-back to the ears normally eliminate specific acoustic properties of a room in order to guarantee reliable perception of information. Furthermore, a sounding work of art is not a mere sound stream; it creates a virtual reality that prevents us from attentively analyzing the acoustic environment and the circumstances of spatial transmission. Progress in electro-acoustics and collaboration with experts in analytic hearing enabled the development of better methods and tests for the assessment of acoustic quality. An important means is the direct comparison of artificial head recordings from different halls.[47] Experiments, advanced theoretical knowledge in the physiology of hearing and the psychology of perception, and improved measuring methods and equipment for acoustic analysis provide today good conditions for the qualification of room acoustics on the basis of room shape, surface structure and material. Geometrical, wave theoretical, and statistical room acoustics, as well as the experience with new architectural designs during the last 100 years, yield the following basic rules for the analyses of room acoustic quality.

Seven basic rules

1. The intensity of the direct sound at the listener's position (depending on the distance to the sound source and its directivity), the number of reflections, their incoming direction, intensity, time sequence and frequency characteristics are the main quality factors of room acoustics at this position. The ear does not take a snapshot of the sound stream. Despite the quickly changing, constantly overlapping signals, auditory perception subjects this

47 H. Wilkens, "Kopfbezügliche Stereophonie – ein Hilfsmittel für Vergleich und Beurteilung verschiedener Raumeindrücke" (1972), 213; D. Gottlob, *Vergleich objektiver akustischer Parameter mit Ergebnissen subjektiver Untersuchungen an Konzertsälen* (1973); Pierce (1992), 150 f.; Blauert (1997), 372 f.

flow of direct sound followed by a number of delayed reflections to a complex integration process. The aim of perception determines integration time somewhere between 15 and 200 milliseconds. Few reflections or gaps in time sequence are negative quality factors. Early reflections arriving within less than 15 milliseconds and hard, rather late reflections with a delay of more than 50 milliseconds may be disturbing. Reflections delayed between 15 and 50 milliseconds are not perceived as single sound events.

2. The room shape determines the geometrical distribution of sound, the path of first, second and higher-order reflections as long as they are not absorbed. The wider and larger the hall, the more important are reflecting surfaces that provide early reflections to the listeners' zones. Rectangular halls higher than half their width feature a positive quality factor, as they provide lateral reflections coming down to listeners at a relatively steep angle.
3. Surface decoration sufficiently varied in structure provides diffuse reflections of higher frequencies. This is a positive quality factor, as information on sound direction is bound to direct sound only, whereas the darker colored reflections follow with increasing delay and without clear direction information. Diffraction off structures also causes absorption and, if structure patterns are too regular, dangerous comb-filter effects.
4. Surface material either increases the velocity of particles or slows it down: porosity and elasticity, but also coupling with open volumes, cause either absorption or amplifying resonance. Since these processes are frequency dependent, they influence the sound color of reflections. Sound perception usually adapts fast to the room specific sound color. Listeners who are not analytically attuned generally tolerate changes of sound color as far as these do not disturb perception of "information". Change of the frequency spectrum during the onset and decay of reverberation also affects the process of integration. If sound color is changing too much, reflections do not blend, because masking results instead of sound amplification.
5. Absorption and volume of space determine the process of reverberation, which is important for the general impression of room acoustics. Reverberation time is measured for a decay of 60 dB as a function of frequency and time. For musical instruments and voices without electronic amplification, exceeding absorption must be avoided.
6. The energy produced by instruments and voices, the space volume and the total absorption define the energy density in a hall.
7. Due to directional characteristics of hearing (greater sensitivity of lateral sound, direction analysis in neuronal sound processing) direction of incoming direct and reflected sound must be assessed as a function of time in order to determine perception quality at a certain place. This is why rules 2 and 3, which are based on geometrical information, are important for the forecast of the acoustics at a certain place. Rule 4 features also a geome-

trical component due to the placement of absorbing materials that affects the frequency characteristics of incoming sound. These objective physical values, which are bound to the position in a hall and are crucial for subjecttive perception, can be clearly defined for a subject at a certain place and are measurable for this position of the listener.

On the basis of these rules one can explain the main reasons for the excellent acoustics of many famous concert halls of the nineteenth century. For example, the Golden hall in the *Musikverein* in Vienna is a long, relatively narrow and high rectangular hall with a nearly 1:1 height-to-width relation. These properties generate a regular distribution of a high number of coherent reflections and lateral reflections that reach listeners via the ceiling and side walls. The surface structures of widely varying dimensions – statues, columns and niches of stucco plaster on the heavy wooden ceiling and stone walls – provide diffusivity for high frequencies (see figure 41) and constitute an excellent mixture of material that creates a frequency balanced reverberation of about 2.0 sec for mid-frequencies in the full hall.[48] The room width, which is about 19 m, provides excellent lateral reflections, and the narrow galleries along the side walls and at the back wall provide additional reflections going down to the listeners in the parterre.

These basic rules also provide qualification criteria for the acoustics of historical rooms with reference to their typical architectural properties or styles (see chapter 1.8), and the architecture-related quality factors lay the ground for the development of music-related criteria to assess room acoustics from the point of view of the history of musical performance (see chapter 1.9).

48 See Beranek (1996), 600; Fasold (1987), 314.

1.7 Spatial impression in sound recording

In this study sound recordings have been used for different purposes. Exactly controllable historical documents, they do not only transmit a piece of music but also a certain performance situation and, according to the technical standards of the time, a part of the acoustics of the recording room. Live recordings in particular transmit the former situation of listening at least up to a certain point. A recording reflects also the recording engineer's concept of space and the musician's adaptation to the acoustics of the recording room and even to the spatial impression of the recording itself, which may differ more or less from the musician's own spatial impression when playing for the recording. This is why we need knowledge of the main recording techniques, the possibilities of creating a spatial impression, the pick-up of acoustics in the recording room, the recordings' sound quality and historical development in order to use recordings as sources for the following analyses. Studied examples range from the earliest to recent times of recording history. Music related questions of sound recordings will be addressed in the second part of this study along with a discussion of the examples.[1]

1.7.1 Technical limits of recording

Since the development of electrical recording in the 1920s, the weakest part in the electroacoustical chain of transmission – from the microphone to the amplifier and from there to the electrical disc cutter and, via amplifier, to the loudspeaker – determines technical quality. Up to recent times the loudspeakers were a main source of concern. However, since historical sound reproduction is not part of this study, the following introduction presents the main recording systems without discussing the reproduction quality at the time of the recording. Instead, it concentrates, on technical data of microphones and further parts of the sound recording system that affected, directly or indirectly, the quality of music recordings. The year indicated in table 4 of technical characte-

1 See 2.8: Sound recordings of *Messiah*.

ristics of recording procedures and microphones is not the year of invention but of the beginning of standard use.[2]

Table 4: Technical characteristics of recording procedures and microphones.

Year	*Recording procedure or microphone*	*Frequency range*	*Dynamics, distortion, noise*
1888	Acoustic recordings (horn)	600–2000 Hz	30 dB, strong noise!
1920	Carbon microphone (Reiss)	60–10000 Hz	High sensitivity, low dynamics! Highly nonlinear frequency response!
1920	High-/Low-frequency condenser microphone	30– 10000 Hz	Omnidirectional up to 2000 Hz, few distortions
1920	Ribbon velocity microphone	50–10000 Hz	Bidirectional, harmonic distortion from 14,6% for 50 Hz to 0,27% for 800 Hz
1925	Electrical recording with microphone	100–5000 Hz	40 dB, frequency response highly nonlinear, resonance around 2500 Hz
1927	Sound-on-film system	50–7000 Hz	40 dB, high distortion above 1000 Hz!
1930	Dynamic microphone	50–10000 Hz	Omnidirectional up to 2000 Hz, long distance to amplifier possible
1933	Ribbon pressure microphone	50–10000 Hz	Unidirectional cardioid, relatively cheap, multimicrophone technique since 1941
1933	Electrical recording	50–9000 Hz	Improved frequency range
1940	High-frequency magnetophon	50–5000 Hz	65 dB, harmonic distortion 3%
1948	Standard 78-r.p.m. disc	50–14000 Hz	50 dB
1948	Long-playing 33- r.p.m. disc	40–15000 Hz	60 dB, less noise
1958	Studio magnetophon	40–15000 Hz	60 dB, harmonic distortion 2%
1958	Stereo LP	30–16000 Hz	60 dB
1953	Condenser microphone		Neumann M49, all directivity patterns
1978	FET condenser microphone	40–20000Hz	61–71 dB
1983	Studio magnetophon	30–20000 Hz	72 dB, harmonic distortion 0.3%–0.7%
1983	Digital recording, CD	16–20000 Hz	84 dB, harmonic distortion 0.03%

For instance, long playing discs were available since 1948, when normal 78-r.p.m. discs with a lower frequency range and 10 dB less dynamics were still produced and in use for quite some time, with many households playing these discs on horne gramophones, a much cheaper system that did not require an electrical amplifier.

Mechanoacoustical horn recordings were of limited quality due not only to high levels of noise and crackling but also to the very limited frequency range

2 H. F. Olson, "Microphones for Recording" (1977), 676–84; R. Flury, "Von Edison zur Compact disc" (1986), 25–40.

of 600 to 2000 Hz. Low sensitivity required the shortest possible distance between the sound source and the horn. Best results were achieved for voices, and singers developed special techniques to improve recordings. During a crescendo the distance to the horn was increased in order to avoid distortion, whereas a diminuendo required a closer approach to the horn. The recording of instruments and, even more so, of ensembles or complete orchestras was much less satisfactory because of the inevitable distance to the horn and the greater distortion caused by a more immediate change of dynamics and sound directivity. Usually the acoustics of recording rooms were strongly damped in order to render direct sound as clearly as possible.

The microphones developed for radio broadcasts in the 1920s introduced enormous improvements. In 1920, the Reiss carbone microphone could transmit a frequency range of 60 to 10'000 Hz, though with a highly non-linear frequency response. Its high sensitivity made long distances to amplifiers possible, but the dynamics were still very limited. If production costs were not limited, condenser microphones could be used with much less distortion and a frequency range of 30 to 10'000 Hz, though with diminishing sensitivity above 6000 Hz. Due to its structure the condenser microphone had omnidirectional characteristics below 2000 Hz and increasing directivity towards higher frequencies. The more expensive high-frequency microphone model allowed distances up to 30 meters.[3] Both models achieved a hitherto unknown sound quality with excellent low frequencies and low distortion if the dynamics were carefully controlled. The cheaper moving coil microphone available since 1930 had similar directional characteristics and a sensitivity comparable to a carbone microphone's but a much more linear frequency response from 50 to 10'000 Hz. The even cheaper ribbon microphones by Vente and Turas unfortunately produced non-linear distortion that increased towards low frequencies from 0.27% harmonic distortion at 800 Hz to 14,6% distortion at 50 Hz. The frequency range of these microphones was better than electrical recording on shellac discs available since 1925 with a frequency range of 100 to 5000 Hz and a resonance between 2000 and 3000 Hz. Still, an increase in dynamics by 10 dB and a greater frequency range brought crucial improvement for softer instruments and larger ensembles, making recording on disc an acceptable medium for classical music also in conductors' opinion.

Since 1927 sound-on-film had an even better frequency range of 50 to 7000 Hz, but strongly fluctuating noise and increasing distortion above 1000 Hz, with fluctuation of harmonic distortion between 5% and 20% at 6000 Hz. With carefully adjusted recording equipment dynamics reached 40 dB and the

3 H. Backhaus, Siemens-Zeitschrift 8 (1928), 298: transmission from *Victoria Hall* to the theater in Geneva, cf. F. Fischer, H. Lichte, *Tonfilm, Aufnahme und Wiedergabe* (1931), 167.

film negative a frequency range of 40 to 10'000 Hz. Since 1933 direct cut discs attained nearly the same frequency range. In 1940 high-frequency magnetic recordings achieved a more limited frequency range of 50 to 5000 Hz, but with enormously reduced noise, increased dynamics of 65 or 70 dB and harmonic distortions of only 3%. Recordings on tape brought crucial improvement, as they offered the possibility of any desired repetitions and assembly of the best results. The frequency range could be further improved, and in 1948 78 r.p.m.-shellac discs reached 50 to 14'000 Hz with dynamics of 50 dB, and long playing 33 r.p.m.-vinyl discs attained 40 to 15'000 Hz and 60 dB with less noise. The two-channel stereo long playing disc in 1958 further increased the frequency range to 30–16'000 Hz, which is nearly equal to the frequency range of the human ear.

Microphone construction improved as well. In the 1940s cheaper moving coil and ribbon microphones became available and, mainly, the unidirectional cardioid pattern microphones, which attenuated sound incident from the back. In 1959 the later legendary condenser microphone M 49 by Neumann with electronic vacuum tube amplifier and changeable directivity pattern was first produced. Despite its more compact dimensions, it preserved the excellent bass response of the old condenser microphones. In 1964 field effect transistors (FET) were even smaller and lighter, and their upper frequency limit exceeded 15'000 Hz. Progress in electronics allowed the construction of large mixing consoles and easy use of multimicrophone recordings, as well adaptation of dynamics and frequency characteristics (filtering) and the addition of artificial reverberation. Since the 1920s additional microphones were used. The new techniques permitted a larger number of microphones in order to catch the soloist or support weaker or badly placed instruments of the ensemble. The principle of one microphone per group or even per instrument and separate recording tracks per group opened new possibilities in sound mixing even after the recording sessions. On the other hand, this technique again required short microphone distance and more absorption of recording rooms in order to avoid throughput of undesired sound from other instruments. Thanks to the cheaper cardioid microphones this technique was first used for jazz and popular music in the United States in the 1940s, while classical music recordings remained traditional with one to three condenser microphones mainly in Europe.[4]

In the 1980s digital recording eliminated noise almost completely, highly increasing dynamics and decreasing distortion. This was much admired, but soon it became evident that human hearing is very sensitive to distortion caused by analogue-digital converters and too low resolution, which produced, already

4 P. K. Burkowitz, "Recording, Art of the Century" (1977), 873–9.

at low levels, artificial frequencies far below 20'000 Hz.[5] Listeners noted especially non-natural changes of noise and onset characteristics of instruments. With better A-D converters, higher oversampling and quantization of 24 bits these negative effects could be better handled.

The enormous progress in electronics, computer storage and sound processing today allows careful editing of historical sound recordings. Special filtering extracts details from horn recordings and electric recordings that were barely or completely inaudible at the time of the recording. Since early sound film and recordings from the 1930s and 1940s produced in large, acoustically excellent recording halls, such as the 18'000 m^3 concert hall of the Reichsrundfunkgesellschaft Berlin, built 1929 to 1932, or the EMI-Studios at Abbey Road No.1 in London, opened in 1931, or Studio A of the NBC in New York in the 1940s, sound recordings have been important for the research of room acoustic aspects.

1.7.2 Space in sound recording and reproduction

Since the introduction of electrical reproduction with loudspeakers the recording engineer has had to make recordings that are usually played back in rooms with completely different acoustics, mostly with a much smaller volume of space, a different room shape and different reverberation time. During listening spatial characteristics of the recording room and the playback room are superposed.[6]

1.7.2.1 Artificial head recording

This problem can be avoided with artificial head recordings, a process introduced in the 1970s featuring microphones at the position of a dummy head's ears in order to simulate a head's acoustic characteristics with respective time shift and intensity and frequency differences between left and right ear. Adequate listening is limited to headphones, which eliminate the acoustics of

5 P. Bienert, *Handbuch für die professionelle Studiotechnik: Digitale Audiotechnik* (1991), 36.

6 See J. Jecklin, *Musikaufnahmen* (1980) and "Mono – Stereo – Quadro" (1986), 81–178; T. Tryggvason, "Classical Music" (1976), 211–28; R. Auger, "Location Recording" (1976), 244–54; J. Culshaw, "The Role of the Producer" (1977), 255–61; M. Dickreiter, *Tonstudiotechnik* (3/1979) and *Mikrofonaufnahmetechnik* (1995).

the reproduction room. Since space characteristics of the recording system remain unchanged, different halls can be reliably compared. This is why artificial-head recordings became an important means for the scientific research in room acoustics. A crucial advantage lies in the possibility of placing the system at the best position in the hall and, thus, of achieving a natural sound balance between the ensemble and the room acoustic feed-back as perceived by the musicians during performance in order to reproduce their articulation and dynamic's control properly. This means that all room acoustic characteristics, both the good and the bad, are recorded.

In practice, the disadvantages of artificial-head recordings have so far impeded broader acceptance despite improved headphones and artificial-head constructions. Headphones are uncomfortable if used for an extended time, and low frequencies lack body vibrations perceived in the hall or with loudspeaker reproduction. Space reproduction also becomes inadequate if differences between the pinnas and the shape of the artificial and the listener's head are too large.[7] A further unnatural element is the fixed position of the artificial head and of the headphones, that is, the acoustic environment does not change according to the listener's movements.[8]

1.7.2.2 Monaural recording

Many musicians and music lovers still rate early monaural recordings highly because of the natural sound impression despite all the technical limitations. As mentioned, these recordings were made mostly with one microphone (usually a condenser microphone with excellent bass characteristics) at an acoustically excellent position in the hall and far enough from the sound source in order to provide adequate balance and spatial impression.[9] Such recordings do not allow exact localization within the recorded space but preserve a certain impression of room depth because closer instruments and voices are recorded with more direct sound than distant sources. Furthermore, they contain musically relevant spatial parameters that are rendered with astonishing precision. This qualification, often considered too subjective, has been confirmed by experienced sound engineers.[10] The recording procedure is adequate for the main aspects of the holistic process of space perception. A condenser microphone of the time with omnidirectional characteristics up to 2000 Hz and increasingly directional

7 For further information on the history of artificial head recording, see *25 Jahre RTM,* Hamburg, September 1981.

8 Recent technical developments permit head-related transfer functions.

9 Fischer and Lichte (1931), 167.

10 J. Borwick, "Studios and Studio Equipment" (1977), 694.

characteristics for higher frequencies correctly recorded the sequence of direct sound and room reflections according to the microphone's position. These recording methods turned even opera recordings of the 1930s with outstanding singers and musical quality, such as Fritz Busch's Mozart recordings from *Glydnebourne* or Otto Klemperer's *Salome* from the *Berliner Staatsoper,* into some of the most valuable documents of sound recording history.

Space impression of a monaural recording is comparable to listening "into" a hall through an opening and perceiving its depth but not width. Of course, listening to such recordings immediately after modern stereo or multi-channel recordings would be a mistake, as adaptation to new circumstances of perception takes some time. The human ear is able to cope with the noise and crackling of old records, to compensate by experience for the lacking frequencies above 5000 to 6000 Hz and to perceive the room acoustic atmosphere despite the masking of certain parts of the sound signal. If such early recordings are carefully cleaned of noise and played back by one neutral loudspeaker, an astonishing amount of the recording room's acoustics may surface (see 2.8, sound examples 16 and 17 from Handel's *Messiah).*

1.7.2.3 Stereophonic recording

Stereo recordings use the characteristics of our direction perception based on intensity and time differences. In natural hearing both effects are always combined. Technically it is possible to produce direction impression through either time difference or intensity difference only. Intensity stereophony (or x/y-stereophony) uses a coincidence stereo microphone or two unidirectional microphones, their two capsules closely placed and angled respectively to the left and right with an opening angle varying between 90° to 120° to give the desired stereo width or covering angle.[11] The instrument's position on the panorama can be selected according to the relative intensity of the right and left channel. Due to the lack of time difference a rather virtual impression of space is created, and characteristic elements of the recording room may be lost (figure 50a). These characteristics are better preserved with time delay stereophony (or A/B-stereophony) using two unidirectional microphones connected to the left and right channel placed at a 30–150-centimeter distance between them (figure 50b). In halls with excellent acoustics experienced sound engineers use a combination of both effects. Hearing tests and, if inevitable, also adjustment of the instrumentalist's placement determine the best microphone position.[12]

11 See Jecklin (1980), 37; Tryggvason, *ibid.*

12 Borwick (1977), 694.

Two-channel stereo recording based on time and/or intensity differences has both room depth and width.

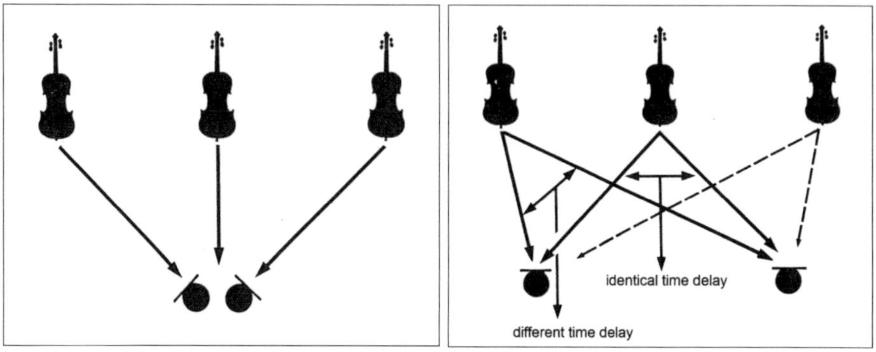

Figure 50: Stereophony a) intensity or x/y (left); b) time delay or A/B (right).

Instead of the relatively small opening produced by mono recordings, stereo recordings open a window into space that allows left-right orientation. The listener remains in that window but cannot enter the virtual room. In stereo recording natural perspective is best preserved if only one pair of microphones or a stereo microphone is used at a reasonable distance from the performance area.[13] Additional spot microphones always endanger natural perspective. But adequate level adjustment and time delay of the signal with respect to microphone distance allow excellent recordings that produce quite natural space impression. Controlled time delay, much easier thanks to modern digital techniques, is especially important for more distant stereo pairs.[14]

Since stereo reproduction uses two loudspeakers that must be placed at a certain distance, the acoustics of the playback room interfere more with the room image of the recording than in playback of a mono recording. The more the spatial characteristics of the recording room are picked up, and the more the acoustics of the playback room are activated by the playback, the stronger the overlay of both spatial components becomes. From this point of view recordings are situated between two extremes: either the recording room's acoustics are picked up as fully as possible in order to be played back in a dead room, or mainly direct sound is recorded in order to add the acoustics of the playback room. In practice neither solution is fully convincing. Pure direct-sound recording in classical music was not successful because musicians rarely give their best performances in dry recording studios. Furthermore, the effect of the playback room is out of their control because its feedback is lacking during the

13 Borwick, *ibid.*
14 See 2.8: Sound recordings of *Messiah,* example 2.

recording. On the other hand, recordings with full spatial information need either very dry playback rooms or headphone playback and the adaptation of frequency characteristics to the playback level.[15] This is why records usually present an adequate mixture of direct and diffuse sound with respect to the acoustics of an average playback room. Sound recording technique therefore tends to develop special esthetics based on a virtual, non-existent room. Sound engineers, listeners and, finally, also musicians adapt to these esthetics, which are influenced by several factors.[16]

The quality of recordings crucially depends on the microphone type, its position and its distance from the sound source. Directional microphones need more distance than omnidirectional microphones to pick up the same amount of direct and diffuse sound.[17] Omnidirectional microphones pick up most spatiality but also most disturbing noise. The mixing determines whether the recording transmits podium acoustics, hall acoustics or an enhanced mixture of both, which does not exist in reality. Stereo recordings that transmit details with clarity and left-right localization unheard even in excellent halls easily lose natural character. Limited dynamics caused by noise, saturation of the tape material and distortion by the transmission chain enticed producers to exaggerate presence. With digital techniques practically devoid of noise and distortion, many of these recordings appear cold, hard and not spatial enough. As a consequence, the number of recordings with more spatiality, even with added artificial reverberation, is increasing. This aspect indicates the sound engineer's responsibility for the adequate relation between outer and inner spatiality of music to be comparable to the musician's.

1.7.2.4 Quadraphonic and multichannel recordings

The problems of spatial perspective mostly disappear if recording is based on more than two channels. First used in sound film, multichannel recordings are increasingly used in television broadcasts and DVDs.[18] In 1969 systematic tests demonstrated that four channels are sufficient to transmit a spatial image of the sound source, as well as the main spatial characteristics of the recording room, to another room for reproduction. The loudspeakers' position is crucial for the quality of sound reproduction.

15 J. Jecklin, "Die notwendige Signalaufbereitung: Der Transdyn-Process" (1986), 156 f.

16 See also 1.9.7: Definition of room acoustic quality norms.

17 Dickreiter (1995), 28.

18 Actually the most common recording system is the 5.1 Dolby surround sound which is not discussed here, because no surround recordings are used for this study.

Quadraphonic recording technique is based on two stereo recordings. One pair of microphones takes a relatively close image of the sound source reproduced via two front loudspeakers, and a second stereo pair captures the acoustic environment from a greater distance, reproducing it with respective time delay over two lateral loudspeakers placed along the side walls at the first third of room length. During playback listeners have the impression that the actual room disappears to make room for another. For sound engineers the main difference from two-channel stereo recording is much greater freedom of microphone positions, for listeners it is the possibility of moving within this electroacoustically produced room from the position between the lateral loudspeakers towards the back wall. As in a real room, lateral sound (reproduced here by the two lateral loudspeakers) adds more transparence and clarity and attenuates exaggerated presence and harshness of the close stereo recording.

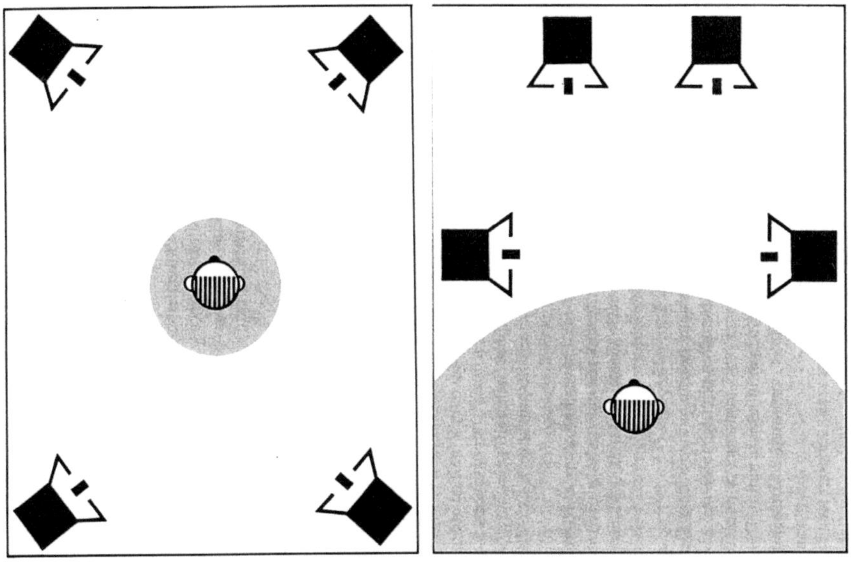

Figure 51: Quadraphonic playback in the 1970s (left); optimum position of loudspeakers (right).

The loudspeakers' position in the four angles of a rectangular room was the main reason for the failure of early quadraphonic recordings in the 1970s (figure 51, left).[19] Only recordings with a minimum of four channels and two front and two lateral loudspeakers approximate a real concert hall experience (figure 51, right). The convincing effect of this recording technique, also of carefully balanced multichannel recordings, confirms the importance of lateral reflections for room acoustics in general.

19 T. Nakayama et al., "Subjective Assessment of Multichannel Reproduction" (1969), 12.

1.7.3 Conclusions

As mentioned, a musician's playing always reacts more or less to room acoustics. Therefore, to preserve the complete information of the sound event, sound recordings should adequately transmit not only the instruments' direct sound but also the room's answer. Recordings have to balance direct and indirect sound in a way that allows musical units of meaning to be reproduced as intended by the musician.

The pick-up and reproduction of original or adequate sound was discussed whenever new technical possibilities were introduced. A questionnaire distributed after the transmission of the first artificial-head recordings in 1973 by ARD contained a considerable amount of hits for "transmission should be as close to perception in the concert hall as possible."[20] It seems obvious, though, that the possibility of direct comparison affected the answers. It is surprising, therefore, that already at the time of cylinder and horn recordings the immediate comparison between the machine and the original sound source yielded a majority of positive votes for the recording machine! Before the First World War many cities in the US offered so-called *tone tests* with one of the new *Edison Phonographs* playing in direct comparison with an instrument or a soloist in a great hall. Despite the enormous distortion, noise and limited frequency range and dynamics, the result was acclaimed to be "as loud and clear as the original."[21] In 1897 a *Lioret Phonograph* replaced the singer's position on the podium of the enormous hall of the *Trocadéro* in Paris, with equally positive results for the Phonograph (see figure 52). These are remarkable documents on our readiness to correct what we hear according to our aim of perception. But the answers underscore also the importance of systematic research on the role of questionnaires in the assessment of the sound quality of music and its room acoustic aspects.[22]

Once a new technique is in use and has become the norm, it is difficult to introduce new and even better techniques. This is why technical progress in sound recording was greeted, mainly at the beginning, with skepticism. Yet finally each new technique created new listening habits and expectations that influenced the listening to music of whole generations, as much as other typical room acoustic and performance related impressions and situations, especially when music started to be transmitted more often through sound recordings than live performances or music making. The possibility of listening more than once

20 G. Plenge, "Ausgewählte Aufsätze zum Thema Kunstkopfstereophonie", *25 Jahre Rundfunktechnische Mitteilungen,* September 1981, 3: foreword.

21 W. L. Welch, "Edison and His Contributions to the Record Industry" (1977), 660–65.

22 J. Blauert, *Spatial Hearing* (1997), 12 and 369.

to the same event, impossible in earlier music history, offers a new perception of music and a completely new approach to the analysis of room acoustics.

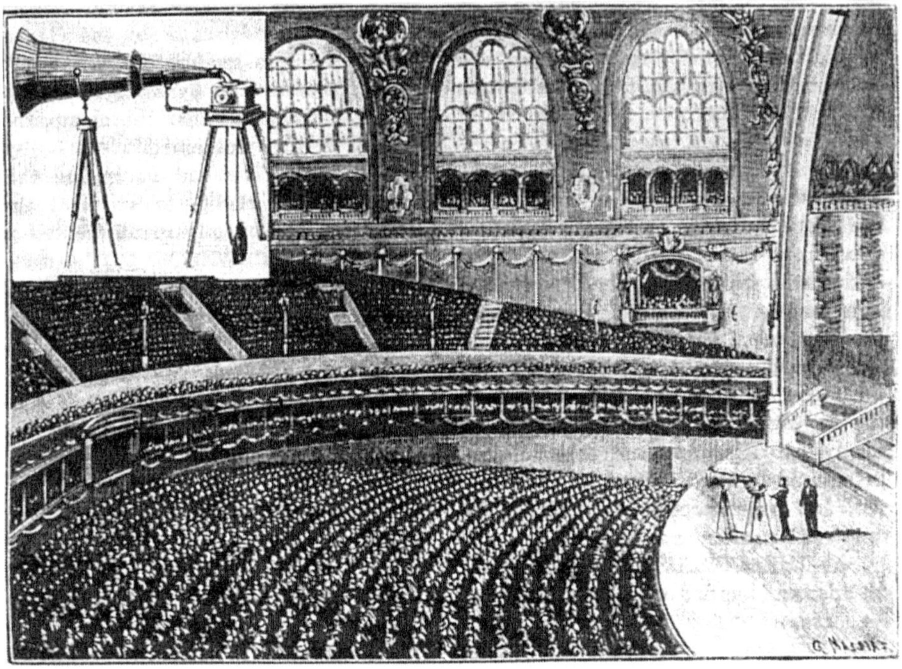

Figure 52: "Tone test": a Lyoret Phonograph and a singer at the *Trocadéro* in Paris (1897).

1.8 Architecture related quality factors in room acoustics

This chapter aims to describe the main elements of architectural design related to the definition of quality factors in room acoustics and to present well documented examples from the history of architecture.

As the physical facts compiled in 1.6 show, room acoustics are not only the result of interior decoration with appropriate materials but a whole, irreducible quality dependent on room shape, surface structure, material and acoustic properties of the construction. As outlined in 1.2, in earlier periods, too, room shape and other architectural elements were rarely chosen only in relation to acoustic properties, and the planning was not always as successful as in the antique theater, the *odeion,* or certain oratorio halls.[1] In architectural planning, acoustics are one factor among others and related to constructional, visual-esthetical and functional aspects (statics, span width, room proportions, decoration, stairs, fire safety, air condition, heating, etc.).

The seven basic rules for the qualification of room acoustics listed in 1.6.6 enable the analysis of historical rooms if reliable plans and detailed descriptions have been preserved. But what if elements of this documentation are lacking? For the following considerations the hierarchical evaluation of quality factors is important: (1) room shape determines the main geometrical distribution of sound; (2) surface structures affect the audibility of sound reflections; (3) materials determine absorption and, in conjunction with the cubic volume of space, the reverberation time T. This top-down sequence is non-reversible in the sense that a change of surface structure or, even less, of material alone cannot fully compensate for the effects of room shape on sound distribution. In other words, the qualification of a room's acoustics is defined stepwise from the top down. Information on room shape and proportions already predefines fundamental acoustic conditions. Hence the great importance of these parameters for studies. Such details as surface structure, interior decoration, material, the number of attendants and their placement allow a more detailed assessment of room acoustics, whose precision depends on the reliability of the available information.

But information on room shape is crucial. Sometimes it can be reconstructed from descriptions and a comparison with similar halls. If the room

1 Important examples of such room shapes are given in 1.2 Applied acoustics in construction and use of rooms.

shape is known, approximate or partial information on structure, interior design and the number of attendants may be sufficient to qualify room acoustics within certain boundaries. These considerations are the starting point for the following discussion of quality factors and also determine their order of presentation.

1.8.1 Room proportions; relation of room height to width

The rules for proportions with respect to the architectural style, along with the assessment of proportions of relevant parts of a building, sometimes permit the determination of a room's height-to-width relation, even if the measures of the room itself are not extant. We are referring to the rules given by Alberti, Palladio, Scamozzi, Vignola and other authors, based on Vitruvius and studies of antique buildings.[2] As explained in 1.5.2 and 1.6.6, the height-to-width relation determines the distribution of reflections, their time sequence and lateral components.[3] Thus we may conclude that room proportions are an important factor of acoustic quality.

Many of the studied *palazzi* from the sixteenth to the nineteenth century show such an application of esthetic rules for proportions. The great halls on the first floor or *piano nobile,* higher than the other floors, often show a height-to-width relation of 1:1, with vaulted ceilings of even 3:2. If longer than wide, a hall could also be less high, for instance, with a height-to-width relation of 3:4 or 2:3. With constant floor height, smaller adjacent rooms were either proportionally higher, or the great hall exceeded normal floor height and attained one and a half or two stories. If the hall was on the second floor, the height-to-width relation was mostly less than 1:1,[4] as, for instance, the music room on the second floor of the *Palazzo Barberini* in Rome, with a height-to-width relation of 3:4 (0.75), while the *salone grande* on the first floor had a relation of 1:1.[5] Most upper floors and basements were even lower. If halls with exceeding

2 See for instance Andrea Palladio, *Quattro libri,* I, 52 and VI, 3 (ed. 1983, 86); Vincenzo Scamozzi, *Dell'idea della archittetura universale* (1615), I, 310; see G. Germann, *Einführung in die Geschichte der Architekturtheorie* (1987), 164. Due the lack of investigations the influence of harmonic proportions on room acoustics cannot be resumed here. Reports on such effects exist but no systematic research; see H. Bagenal and A. Wood, *Acoustics* (1931), 216, § 6: 'Note' of the Church; P. von Naredi-Rainer, *Architektur und Harmonie* (1982).

3 See 1.5.2.2 on directivity of hearing and 1.6.6.2 on lateral sound; see also J. Meyer, *Acoustics* (2009), 15 f., 196 f.

4 Chr. L. Frommel, *Der römische Palastbau der Hochrenaissance* (1973), Bd. 1.

5 See 2.2 on Handel in Rome and table C in the appendix.

floor height were inserted (often due to later modifications), the rooms above and sometimes those below were lower.[6]

The indication of the floor for a *palazzo's* music room may therefore contain a precious hint on room acoustic conditions even if the room's exact dimensions are unknown. And even the general conclusion is warranted that halls on the first and second floor of *palazzi* in Renaissance or eclectic style have basic conditions for good or excellent acoustics because sound quality in rooms with a height-to-width relation of more than 1:2 (or > 0.5) begins to be excellent with sufficient length and appropriate interior decoration and outfitting. The attained room acoustic quality in such a room is determined by the surface structure, the material, acoustic properties of the construction and the number of attendants.

1.8.1.1 Halls with low ceilings

Up to the nineteenth century height-to-width proportions of less than 1:2 (< 0.5), a negative room acoustic precondition, were typical of taverns or modest private houses. Indeed, these halls were often reported to have bad acoustics, as for instance the ball room of the *Mehlgrube* in Vienna,[7] from 1740 to 1810 used as concert hall for mixed social classes and known as too low for a good effect of music and too small for the number of seats (17 × 8 × 4.75 meters).[8] The ball hall of *Ignaz Jahn's inn* at the Himmelpfortgasse in Vienna, between 1788 and 1806 used as concert hall for maximum 400 listeners, also had badly rated acoustics (Mozart performed here in November 1788 his version of Handel's *Acis und Galathea,* and on 4 March 1791, in the academy of the clarinetist Joseph Bär, he played his piano concerto KV 595). Documents speak of a "long, very narrow hall of three joined rooms. The pillars of the separating walls are still extant."[9] A report in the *Allgemeine Musikalische Zeitung* of 15 October 1800 says that "Jahn's hall is not high enough and also too narrow and thus inhibits the effect of music; furthermore, it provides space for

6 P. Waddy, *Seventeenth-Century Roman Palaces: Use and the Art of the Plan* (1990), 179 ff.; see also 2.2.2 on the halls at the *Palazzo Bonelli* in Rome.

7 See S. Weinzierl, *Beethovens Konzerträume* (2002), 93. In 1785 Mozart played here subscription concerts.

8 *Vaterländische Blätter für den österreichischen Kaiserstaat* (1808), 40; M. S. Morrow, *Concert life in Haydn's Vienna* (1989), 98, Plate 21.

9 "[...] ein langer, sehr schmaler Saal aus 3 Zimmern. Die Träger der Trennwände stehen noch." R. Klein, "Ein alt-wiener Konzertsaal: Das Etablissement Jahn an der Himmelpfortgasse," *AMZ* 11.4.1804, 407–71, (1963), 12–18.

no more than 400 listeners."[10] Other halls of this kind had astonishingly high ceilings, such as the greater hall of *Hickford's Rooms* in London (figure 72), 1729 to 1779 much favored for concerts, with an excellent height-to-width relation of 0.75 (15.20 × 9.10 × 6.70 meters).[11]

1.8.1.2 Festival halls

Festival halls of the nineteenth century, mostly ad hoc wooden or iron constructions, often suffered from inadequate acoustics because a lot of seats had to be provided for very low building costs. Halls were therefore often very wide and had low ceilings. As a result, the absence of sound transparence and the short reverberation were irremediable. For example, Abraham Mendelssohn, Felix Mendelssohn's father, reported in 1833 on the hall of the *Niederrheinische Musikfest* in Düsseldorf, with 1200 to 1300 seats: "At the Landstrasse to Berlin, in a large, well-shaded garden belonging to a restaurant, a hall 135 feet long and about 70 feet wide and unfortunately only 27 and a half feet high (obviously 10 to 15 feet too low) has been built, without any decoration and [...] whitewashed."[12] The hall's ground plan of 38.5 to 20 meters already met the standards of the later large rectangular concert halls. With a height of 7.85 meters, it featured a height-to-width relation of only 0.4. An increase in height of about 10 feet would have really improved the relation to an acoustically sufficient value of 0.54, and 15 feet would have achieved the excellent 0.6. It doesn't matter whether Abraham Mendelssohn himself or his son Felix, conductor of the festival in 1833, first talked about the deficient height. What matters is this letter, as it confirms that Felix Mendelssohn, who knew the main music centers in England and on the continent, was among the experienced people who were *aware* that sufficient room height is crucial for good acoustics, mainly in a full or overcrowded hall during concerts.

Problems increased with larger room dimensions. In 1872, on the occasion of the *Eidgenössische Schützenfest,* Zürich built "probably the largest hall ever for this purpose." The report mentions that the five speakers at the festival

10 "Der Jahnische Saal ist nicht hoch genug und auch zu schmal, so dass er die Wirkung der Musik beschränkt; überdies fasst er höchstens bis 400 Zuhörer." *AMZ, ibid.;* see Morrow, (1989), 11, table 1, 2, 101–2: Sittengemälde, 95–6.

11 W. Salmen, *Das Konzert, eine Kulturgeschichte* (1988), 23, Abb. 12.

12 "An der Landstrasse von Berlin, in einem grossen, schattenreichen, zu einer Gastwirthschaft gehörigen Garten, ist ein Saal von einhunderfünfunddreissig Fuss Länge, etwa siebenzig Fuss Breite und leider nur siebenundzwanzig und einen halben Fuss Höhe (offenbar zehn bis fünfzehn Fuss zu wenig) hineingebaut, ganz ohne alle und jede Verzierung und [...] geweisst." See S. Hensel, *Familie Mendelssohn* (1879), I, 353: letter by A. Mendelssohn, Whitsunday 1833.

could be understood well from the first five rows of the tables close to the podium, which was at the center of the hall. But, inevitably, a group of singers started performing on the stage although a brass band was still playing at the other end of the hall at a distance of 98 meters. Three years later, at the *Sängerfest* in Basel a 14-meter-wide one-nave festival hall with a vaulted wooden ceiling in "Moorish style" had better acoustics, but the great length of 112 meters caused several problems. In 1890 the hall for the fourth *Deutsches Sängerbundesfest* in Vienna attained the then enormous span of 56 meters. Comparable to the *Hallenstadion* in Zurich (1939), the hall, which seated 20,000 persons and featured a high vault, a long oval ground plan and gallery, was only half as high but nearly twice as long as the *Royal Albert Hall* in London (1871). The 36-meter-deep amphitheatrical podium covered more than one third of the hall's 95-meter length. The construction report does not contain any comment on the acoustics, but the velum visible in illustrations suggests acoustic problems.[13]

Table 5: Festival halls.

City	*Event*	Year	G	C	L(m)	W(m)	H (m)	H /W	Seats	m²
Düsseldorf	*Niederrhein. Musikfest*	1833	[ ]	__	38.50	20.00	7.85	0.40	~1'300	750
Zürich	*Schützenfest, Halle*	1859	[ ]	^	98.00	28.00	16.00	0.57	2'300	2744
Zürich	*Eidgen. Schützenfest*	1872	[ ]	^	115.00	45.00	21.00	0.47	5'300	5211
Basel	*Sängerfest*	1875	[ ]	∩	120.00	31.00	21.00	0.67	4'000	3780
Wien	*Deutsches Sängerfest*	1890	⊂⊃	∩	116.00	56.00	23.00	0.41	20'000	5867
Zürich	*Hallenstadion*	1939	⊂⊃	fV	112.00	62.00	24.00	0.38	11'000	10000

G = ground plan, L = length, W = width, C = ceiling: ^ = pitched, fV = flat vault, H/W = height-to-width.

1.8.2 Acoustic similarity of rooms

Interior decoration and materials determine the diffusivity of sound and the frequency-specific absorption of sound energy. This is why halls of similar shape, proportions, decoration and materials also have a similar micro sound structure. On the other hand, halls of exactly the same size may have a considerably different sound only because their construction, surface structures and interior décor are different. This became obvious after the Second World War when old halls and theaters were reconstructed without full decoration. Concerts in *palazzi* from the fifteenth and sixteenth century show that scarcely de-

13 J. Durm / H. Wagner, *Handbuch für Architektur* IV/4/4/1 (1894), 183 f.

corated halls (in most smooth halls only ceilings are coffered) appear more reverberating or even problematic than rooms decorated in baroque or eclectic style of similar proportions and dimensions. It is important to note that halls with excellent acoustics, as the mentioned *Oratorio del Gonfalone* in Rome, have head-high wooden paneling around the walls. Lack of surface structures and well distributed surfaces with changing absorption are always a negative factor for room acoustics.

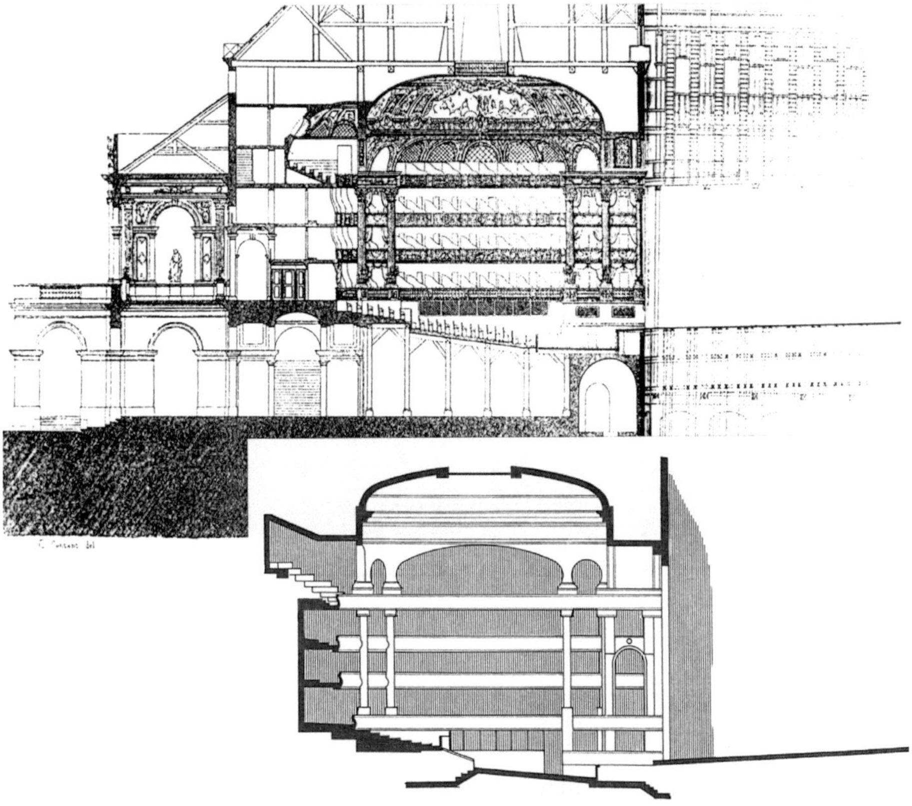

53: Paris, *Opéra Lepelletier,* 1821–73(above) and *Opéra Garnier,* 1875 (below), same scale.

With proper choice of materials, construction and decoration a hall's exact reconstruction preserves good acoustics. This is how the famous old *Gewandhaussaal* in Leipzig was preserved in 1882, when the small hall in the new *Gewandhaus* was built as a nearly identical copy (the acoustic consultant was Joseph Joachim). There were unanimous reports that the excellent quality of acoustics was preserved despite the right-side gallery with its deeply receding section. The success was attributed to the exact copy of the "resonating wooden construction."[14] However, the copy of the room's shape and dimensions was a highly important precondition (figures 54 and 55).

The enlargement of a hall while keeping the existent shape, proportions and respective decoration provides acoustic similarity because the pattern of reflections is preserved though with a relatively greater time delay. The construction of the new *Opéra* in Paris in 1875 was based on this principle, as may be seen from the comparison of plans with the former *Opéra Lepelletier,* which had been destroyed by fire. According to a report of its director Emile Perrin acoustics were "among the best known, the timbre was pleasant and the resonance perfect except in the rear part of some side boxes."[15] The architect Charles Garnier increased the new auditorium's width from 16.80 m to 20.60 m and maintained the height-to-width proportion, with the respectively higher flat dome reaching 22.50 m instead of 18.50 m. But he kept the greatest distance from the rearmost box to the platform unchanged at 22.50 m, probably for visual reasons (see figure 53).

Many of the great concert halls built in the late nineteenth century are based on models with good acoustics and feature similar proportions (see table 10 b), such as the great hall of the *Casino* in Basel (1876), of the *New Gewandhaus* Leipzig (1884) and of *Boston Symphony Hall,* the latter known to be an enlarged copy of the Leipzig hall (1900, acoustic consultant Wallace C. Sabine). All these halls have a height-to-width proportion of close to 3:4. Despite considerable differences in details, such as the exact shape of the rounded edges, the side galleries, the podium, and materials, we can feel the acoustic affinity of these halls. The great hall of the *Tonhalle* in Zürich (1895) also belongs to this series, but the deeply receding side galleries reduce and delay lateral reflections for the podium and certain parts of the parterre.[16]

Another group of related halls with similar cross-sections are the large Redoutensaal (see figure 11), the *Winterreitschule,* the *Musikvereinssaal* and the small *Brahms-Saal* in Vienna, all with height-to-width proportions close to 1:1.

14 R. Skoda, *Das Gewandhaus Leipzig, Geschichte und Gegenwart* (1984), 15, 49.

15 "L'acoustique est une des meilleurs connues, le timbre est agréable et la résonnance est parfaite, sauf dans le fond de quelques loges de côté." Ch. Garnier, *Le nouvel opéra de Paris* (1878–1881), vol. 2, appendix.

16 H. U. Glogau, *Der Konzertsaal* (1989), 53; R. Skoda, *ibid.*, 170; D. Baumann, "Glück und Unglück im Zürcher Konzertsaalbau" (1993), 115–28.

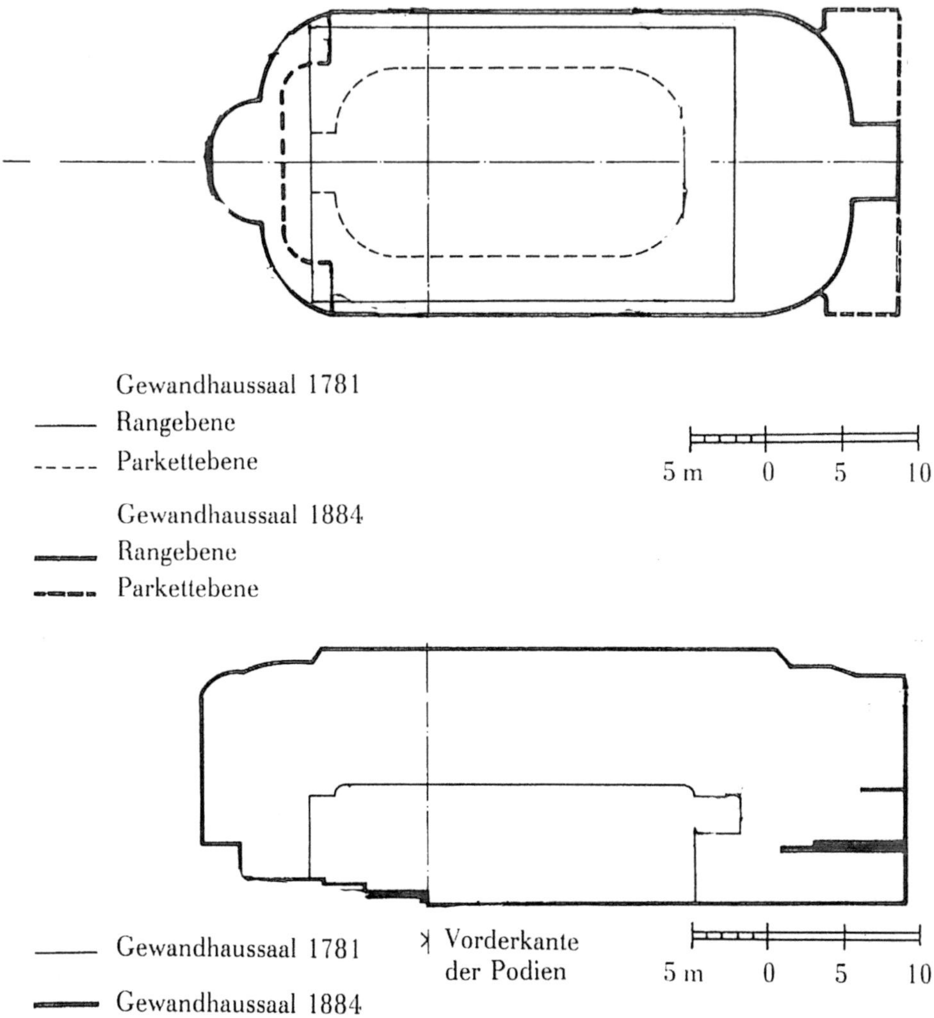

Figure 54: Leipzig, *Alter Gewandhaussaal* (1781–1894) and *Neues Gewandhaus*, Grosser Saal (1884–1944), comparison of ground plan and section (after Skoda).

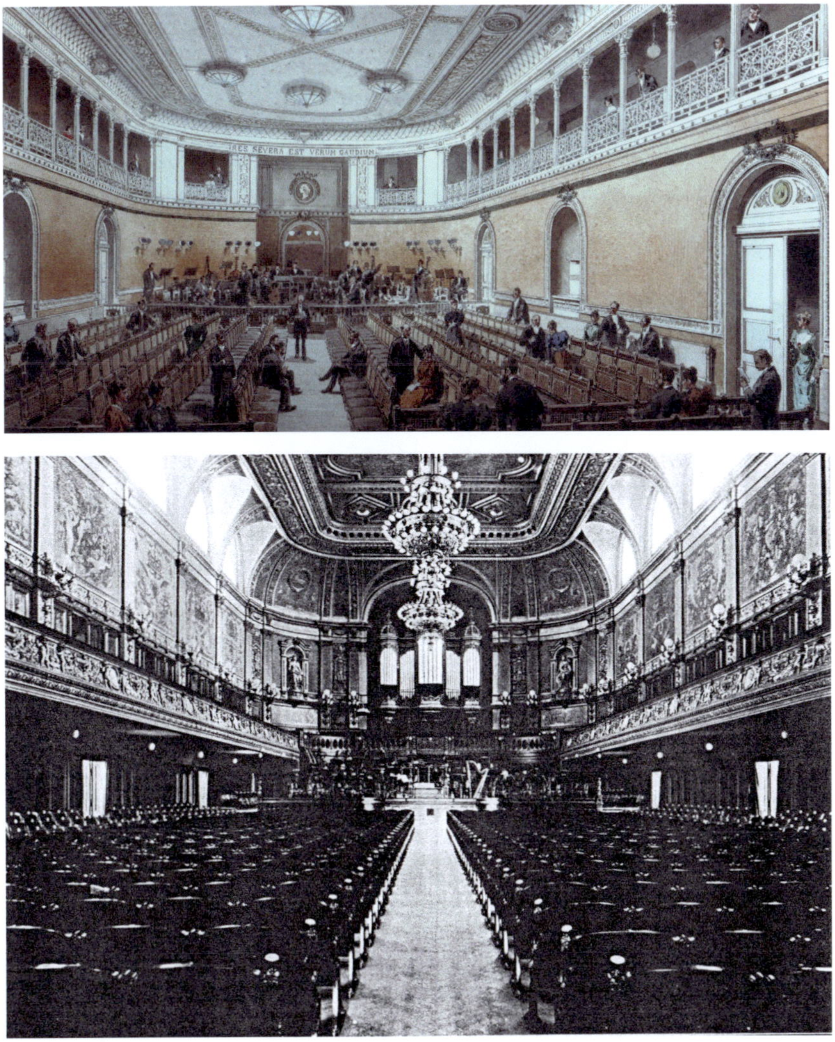

Figure 55: Leipzig: a) *Alter Gewandhaussaal* (above); b) *Neues Gewandhaus*, great hall (below).

1.8.3 Critical volume of space

The sound pressure level depends on total absorption and the volume of space. To preserve good acoustics, the enlargement of a model and the change of shape should, therefore, not exceed a critical limit. In 1966 this limit was reached (or, in the opinion of some singers, exceeded) at the new *Metropolitan*

Opera in New York, which, for visual reasons, featured a slightly fan shaped auditorium with a maximal width of 33 meters at the rear wall. Twenty-five meters high, this auditorium is already in the middle six meters larger than the 19-meter-wide old *Met,* which, in the opinion of singers, had more adequate measures. The total volume increased from 19,500 m^3 to 30,500 m^3. Despite the increase of reverberation time at mid-frequencies from 1.2 to 1.8 seconds, singers feel the decrease of sound pressure level in the new Met (see figure 56). The power required for voices is at the upper limit.[17] The auditorium of the new Met also exceeds the inner width of the acoustically excellent *Teatro Colón* in Buenos Aires by 3.50 meters (measured on the first floor), which has a space volume of 20,870 m^3 and a reverberation time for middle frequencies of 1.8 seconds.[18]

1.8.4 Critical room width

Besides the volume of space, room width should not exceed a critical value either, as the example of the new Met shows. An increase of distance between the main reflecting lateral walls to more than 20 meters noticeably reduces clarity. The consequences become all the more crucial as the distance of these surfaces from the listener increases and the sequence and number of incoming reflections decreases.[19]

In the history of room acoustics the free span of ceilings, which rarely exceeded 20 meters, was of utmost acoustic importance, although special constructions allowed wider spans in ancient and Renaissance architecture.[20] Among the widest rooms with difficult acoustics were vaulted brick, beton or stone buildings, such as the *Pantheon* in Rome built in 118 (diameter of the dome 43.30 meters), the *Hagia Sophia* in Constantinople (diameter of the central dome 32 meters and two lateral half domes above a ground of 64 × 32 meters, built 525–537), *S. Maria del Fiore* in Florence with Brunelleschi's dome (diameter 42 meters, finished 1436)[21] and *St. Peter's* in Rome with della Porta's dome on Michelangelo's tambour (diameter nearly 42 meters, finished 1500).

17 M. Forsyth, *Buildings for music* (1985), 172, 284 and 334; Meyer (2009), 235 f.
18 Dimensions see table 11b.
19 Meyer (2009), 196 f.; Baumann (1993), 115 f. and „Ein Spätling im goldenen Zeitalter des Konzertsaalbaus", *Hundert Jahre Zürcher Tonhalle* (1995), 37 f.
20 H. Schmidt: *Festigkeitsnachweis für eine von Dreiecksbindern getragene Satteldachkonstruktion* (1971/72), cf. R. Meinel, *Das Odeion* (1980), appendix.
21 H. Saalman, *Filippo Brunelleschi, The Cupola of Santa Maria del Fiore* (1980), 66 f.

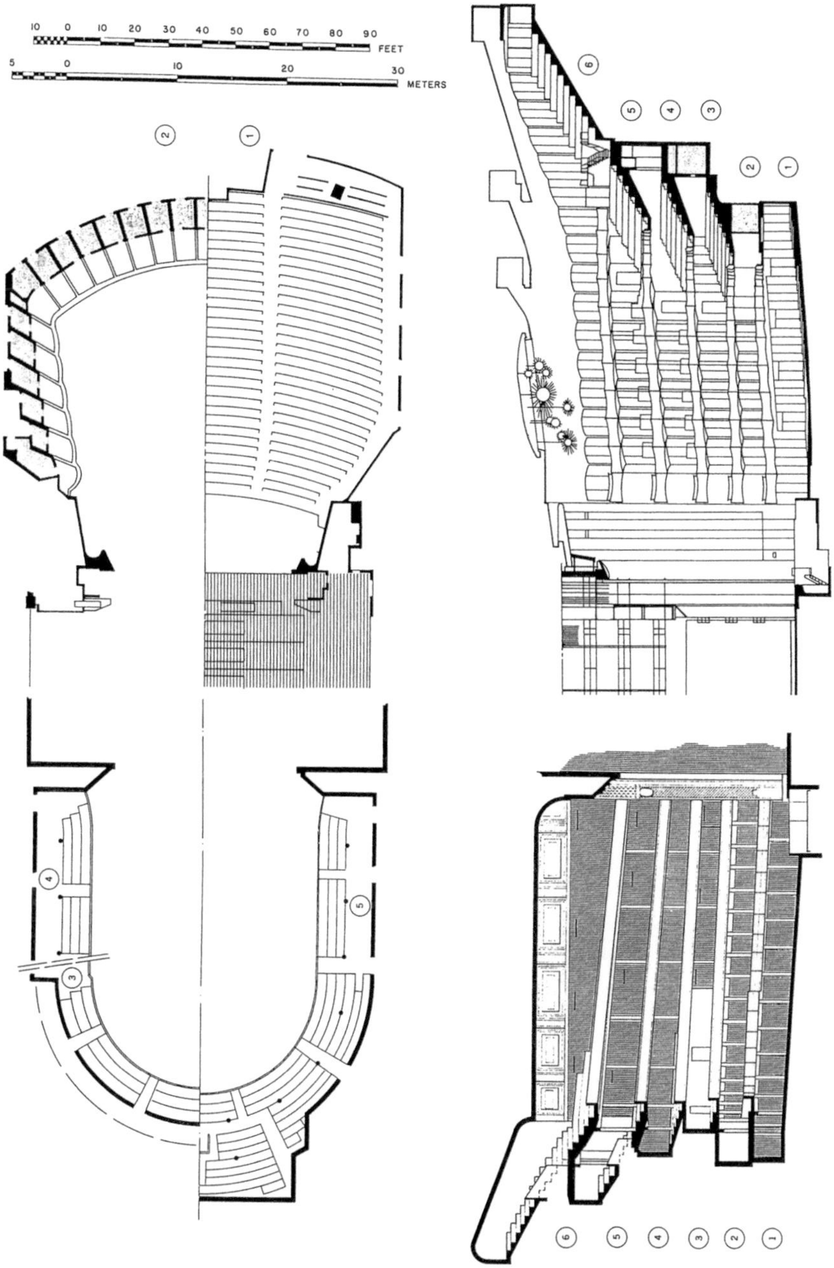

Figure 56: New York: *old Met* (1883–1966) (below) and *new Met* (1966) (above), same scale.

6a: Rooms with large span.

Stone ceilings	Typus	C	Sn	L(m)	L_{tot} (m)	W_i (m)	W(m)	H(m)	DH(m)	D ∅
Rome, *Pantheon*	O	D++	–	–	43.00	–	43.00	–	43.00	43.00
Constantinople, *Hagia Sophia*	Cross-domed	D	–	–	80.00	32.00	64.00	40.00	56.00	32.00
Florence, *S. Maria del Fiore*	Basilica, octag. D	gV+	1	78.40	153.00	19.00	40.80	40.80	86.40	42.00
Rome, *St. Peter's*	Basilica D	bV+	1	96.00	180.00	26.40	69.60	45.00	101.00	41.60
Wooden ceilings					L_i (m)	W_i (m)		H (m)		
Pompeji, *Th. tectum*	[ ))]	+++	–	–	28.60	26.60	–	?	–	–
Athens, *Agora-Odeion*	[ ))]	+++	–	–	24.60	24.75	–	22.50	–	–

C = ceiling shape: bV = barrel vault, gV = gothic vault, D = dome, + = structured, sN = number side naves, L = length, L_i = inside, W_i = width main nave, W = total width, H = height, HD = dome height, ∅ = diameter.

In ancient times beams of cedar trees also allowed for a span of more than 20 meters, as in the roofs of music halls *(odeion)* and the *theatrum tectum* at the time of the Roman Empire. A well-known example is the small theater in Pompeji, a typical *theatrum tectum* built around 80 BC with a span of 25 meters and exterior dimensions of 28.60 × 30 meters (figure 5). The nearly quadrate *Agora-Odeion* in Athens, built around 15 BC, had a span of nearly 24.75 meters (figure 4).[22] Basilicas with triangular trusses of cedar wood also exceeded spans of 20 meters, such as the *Basilica Ulpia,* built by Apollodorus around 100 AD in the *Forum Trajanum* in Rome with a main nave of 25 meters, or the Early Christian basilicas *S. Paolo fuori le mura* and old *St. Peter's* (around 324) with main naves 24 meters wide.

In modern Western architecture earliest wooden ceilings spanning more than 20 meters date from the Renaissance. In 1618 Aleotti covered the *Teatro Farnese* in Parma with a bold roof construction of 33.50-meter long wooden beams clamped with metal braces holding a flat horizontal ceiling 30 meters wide (figure 8).[23] In 1668 Christopher Wren used models by Serlio and ancient examples to build the *Sheldonian Theater* in Oxford with a ceiling span slightly over 21 meters (figure 73). Even in the nineteenth century, after the develop-

22 It collapsed in the middle of the second century AC and was reconstructed later with a 7.65 m shorter hall. Based on archeological results and static calculations certain semicircular *odeia* shaped like large antique theaters probably also were roofed with wooden beams as the *odeion* of Herodes Atticus described by Philostratus, according Pausanias with a wooden beamed roof, a semicircular cavea with 5000 seats and 81 m diameter, see R. Meinel, *Das Odeion* (1980), 44–59, fig. 5–13. A roof without columns would span 49 m demanding triangular cedar trusses and metal braces. For the calculation of span in wooden constructions, see W. Mönck, *Holzbau* (1995), 276.

23 On the rediscovery of the ancient theater, see also 1.2.3.

ment of iron and steel constructions, 22 meters were exceeded mainly industrial and festival halls, as well as special hall and theater constructions, and only after the First World War did wider rooms become generally more common. Due to new air conditioning techniques ceilings could be kept lower, yet both factors increasingly caused acoustic problems.

Table 6b: Rooms of large width.

City	G	L_i (m)	W_i (m)	$Wi_{1st\ rank}$ (m)	H (m)
Parma, *Teatro Farnese*	U	42.25	30.00	–	20.67
Oxford, *Sheldonian Theatre*	U	24.10	21.90	–	14.60
Amsterdam, *Concertgebouw*	( ]	43.15	28.15	20.50	16.40
Bayreuth, *Festspielhaus*	[ ))]	28.20	30.00	–	15.90
Buenos Aires, *Teatro Colón*	U	34.30	30.00	21.50	26.80
New York, *Metropolitan Opera*	U	32.00	20.65	18.75	23.45
New York, *Met Lincoln Center*	/\	31.00	33.00	23.00–27.00	25.00

G = ground plan, L_i = length inside, W_i = width inside, H = height.

Thus, one may conclude that before the twentieth century large room spans were an exception. Concert and theater halls with large reflecting lateral surfaces very rarely attained widths of 30 meters, as did the mentioned Teatro Farnese, which prompted Monteverdi to be remarkably prudent.

For a listener the hall's acoustic quality is crucially determined by early reflections. This is why clarity is diminished if these early reflections have to travel too far. Sound waves lose intensity with increasing distance, and high frequencies are damped due to dissipation in the air. If the time sequence is good and the number of reflections high enough or, in other words, if the height-to-width relation is big enough, transparent sound is possible also in wide halls, though sound color tends to become rather dark. Therefore, rectangular halls nearly 30 meters wide also belong to the excellent halls if their ceiling is high enough, such as the great hall of the *Concertgebouw* in Amsterdam. The wider a hall, the more important are surfaces that provide additional lateral reflections, as do the side galleries in the *Concertgebouw* hall with a width of 20 meters between the balustrades, or the protruding walls not fully reaching the ceiling in the Bayreuth *Festspielhaus* (figure 17). The new *Met* was probably the first important opera house with boxes or galleries exceeding an average inner width of 22 meters on the first floor (figure 56).

1.8.5 Architectural quality factors and type of space

1.8.5.1 Music rooms and concert halls

Stylistic preference for certain ground plans and ceiling shapes has marked the acoustic characteristics of concert halls throughout history. Historical architectural data show that most halls from the sixteenth to the early twentieth century had rectangular ground plans, partly with more or less rounded edges, added sectors in front, at the rear or on the sides and with a varying number of galleries of varying depth, more or less jutting out or receding from the main walls of the hall. Ceiling shapes were mostly flat and horizontal, often with quite deep wood or plaster coffers. The mentioned halls in Vienna, Basel, Leipzig, Zürich, Boston and Amsterdam belong to this type of *shoe box halls* with flat ceilings.

Yet another type is the oblong rectangular hall with a high or flat barrel vault. Rectangular halls with coffered barrel vaults were much in favor in the late eighteenth and during the nineteenth century, mainly in England, but also on the continent. If room height is sufficient, the sound focus of the vault is far above the listeners' heads without causing problems. If dimensions are too large and surfaces too smooth, such halls can have a booming after ring. With the exception of the mentioned hall in the opera of Berlin,[24] such halls as the *Hanover Square Rooms* in London (figure 93) at the time of Haydn's visit, known for its favorable acoustics, the flat-vaulted hall of the Paris *Conservatoire* (figure 13), or large English halls of the nineteenth century with barrel vaults, such as *St. James's Hall* in London (figure 95), are said to have excellent acoustics. *St. George's Hall* in Liverpool (figures 98) and the town hall in Leeds belong to the problematic examples.[25]

Table 7a: Halls with vaulted ceilings.

Hall	G	C	L(m)	W(m)	H(m)	Seats
London, *Hanover Square Rooms*	[ ]	flV	24.10	9.80	[8.50]	800
Paris, *Conservatoire*	( )	flV	24.45	11.10	10.40	1021
London, *St. James's Hall*	( ]	BV	42.40	18.30	18.30	2500

G = ground plan, C = ceiling: flV = flat vault, BV = barrel vault, L = length, W = width, H = height.

A third common type has been developed from theaters with two or more galleries, such as the just mentioned still extant and surprisingly small hall of the old

24 See 1.2.6: Acoustic knowledge of architects.
25 Forsyth (1985), 144 and 147.

Conservatoire in Paris from 1811, with slightly over 1000 seats and an oblong ground plan with rounded edges, *Queen's Hall* in London, destroyed during the Second World War, with roughly 2000 seats and a shorter, wider ground plan approaching a semicircle (figure 96) or the much larger *Carnegie Hall* in New York, based on a similar ground plan with 2760 seats. Like theater halls, these halls have the advantage of seating many listeners on a relatively small ground plan within relative short distances from the podium.[26] Clarity is preserved also in larger halls of this type, but too low galleries and too large balconies may excessively reduce direct sound for the seats below.

Table 7b: Halls with several galleries.

Hall	G	C	L (m)	W (m)	W_i (m)	H (m)	Seats
Paris, *Conservatoire*	U	flV	24.45	11.10	10.25	10.40	1021
Geneva, *Victoria Hall*	U	flV	36.00	19.50	11.50	17.00	1754
London, *Queen's Hall*	[ ))]	__	32.10	26.45	17.45	17.45	2026
New York, *Carnegie Hall*	[ ))]	__	41.85	31.00	24.00	24.00	2760

G = ground plan, C = ceiling: flV = flat vault, L = Length with podium, W = total width, W_i = width between galleries, H = height.

Besides these much favored rectangular halls, Italy in the seventeenth century and France, England and other countries in the eighteenth and nineteenth century preferred elliptic and semicircular halls, as they were considered elegant and acoustically excellent.[27]

Table 7c: Elliptic and circular halls.

Hall	G	C	L (m)	W (m)	H (m)	Seats
Paris, *Concerts rue Cléry* 1791	O	D	8.00	8.00	6.25	–
Edinburgh, *St. Cecilia's Hall* 1762	0	flV	19.20	10.70	5.20	500
London, *Ranelagh Rotunda* 1742	0	fl	> 46.00	45.70	?	3000
London, *Royal Albert Hall* 1871	0	flV	66.80	56.40	41.80	8000

G = ground plan, C = ceiling shape: D = dome, flV = flat vault, L = length, W = width, H = height.

Early-nineteenth century reports on performances increasingly referred to the acoustic problems caused during loud music due to sound focusing in larger halls. The acoustics of the elliptic concert hall *Felix Meritis* in Amsterdam, opened in 1788, prompted a critical report in the *Allgemeine Musikalische Zei-*

26 D. Baumann, "Die Kunst des Konzertsaalbaus" *NZZ* 25. June 1993, Nr.144, 57: Interview with Russell Johnson; see also A. Favaro, *L'Acustica* (1882), 96.

27 See 1.2.6: Acoustic knowledge of architects.

tung in 1814: "The hall is long and oval; the music reverberates too much, especially if the hall is not full enough. Perhaps this is the reason why symphonies never cohere well and are often unstable." [28] The same periodical noted even worse conditions in the oval concert hall of the old Berlin *Schauspielhaus* (1803–1816): "Due to the same circumstances the capricious Echo so often plays her indecent jokes here, mainly with louder instruments that cause more reverberation, such as tympanis, trumpets, trombones etc." [29] A change of the orchestra's position to the lateral wall and the canvas stretched over the proscenium brought no remedy. Similar problems with delayed echoes appeared in circular or elliptic halls with domes, such as the Rotunda at *Ranelagh Gardens* in London (1742–1805) with a diameter of 45.70 meters and a circumference of nearly 170 meters (figure 86). At Mozart's time, the floor was completely covered with straw mats, and the public adored the hall because promenading was permitted during concerts, as it was at the *Royal Albert Hall* in London (1871), where this tradition has been preserved on the uppermost floor in the acoustically excellent standing area (figures 12 and 97).

1.8.5.2 Theaters

Since the sixteenth century theater building has produced a great number of different models to be discussed here only in reference to those addressed in the second part of this study. The auditorium of theaters with boxes had quadratic, rectangular, oval, bell, horseshoe, semicircular, u-shaped and fan-shaped ground plans (figure 57). For the acoustics it is important whether this shape is created by the walls of the hall or the balustrades and boxes' front walls.

In a theater and opera house the acoustics are determined by the position and dimension of reflecting surfaces, such as proscenium arch, ceiling, balustrades, front walls of boxes and main walls of the auditorium. An exact analysis of the geometrical distribution in such spaces is technically extremely demanding. The mostly concave surfaces of the auditorium cause sound concentrations that reach listeners in the rear parterre as well. Echoes are rather scarce because of the limited room length and the absorbing box openings and galleries. Reflections causing sound concentrations are soft and arrive within a short delay, therefore producing a very intense sound and, closer to the wall, also very clear acoustics.

28 *AMZ* 16 (1814), Sp. 417: "Der Saal ist länglich rund; die Musik hallt daher zu stark wieder, besonders wenn er nicht gut gefüllt ist. Dies mag wohl auch die Ursache seyn, warum die Symphonien nie gut zusammengehen, und man ein öfteres Schwanken bemerkt."

29 *AMZ* 12 (1810), Sp. 617: "Diesem Umstand schrieb man es zu, dass die launige Echo so oft ihr unartiges Spiel hier trieb, besonders mit den stärker hallenden Instrumenten, den Pauken, Trompeten, Posaunen etc."; see also O. Schreiber, *Das Orchester* (1938), 209.

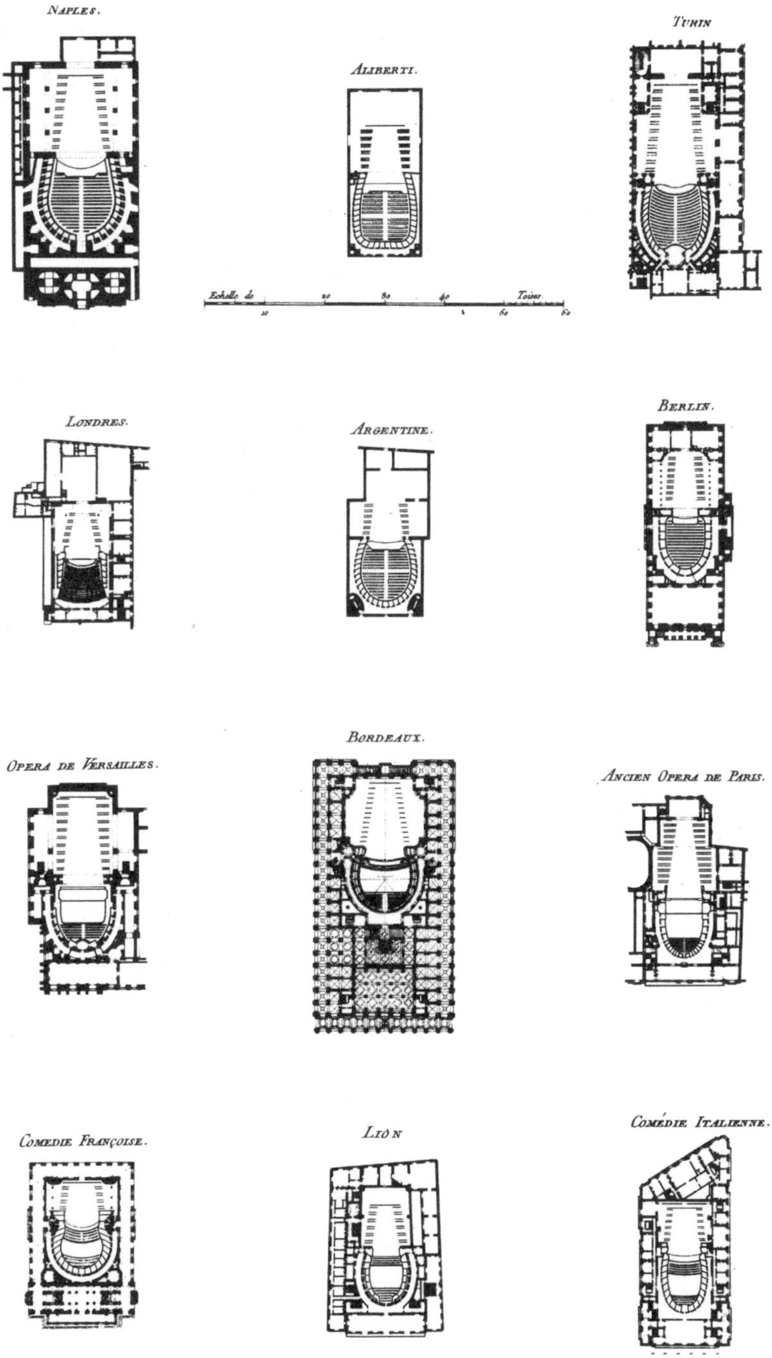

Figure 57: Ground plans of modern theaters, same scale (Durand, *Receuil,*1801).

The dignitaries' box on the first floor lies in this zone of increased sound intensity, combining acoustic and visual advantages. The standing area on the parterre directly below usually has a dark sound due to the absence of direct sound but is still outstandingly transparent, as may be experienced in *La Scala* in Milan. The cheap seats under the roof, the so-called *olympus,* are often excellent due to sound concentrations from the rounded balustrades and the ceiling.

A further important factor in theater acoustics are the dimensions of the box openings, as the surface of the boxes' front wall determines the percentage of reflecting lateral walls. Accordingly, the part of direct sound reaching the box decreases the smaller the box openings and the steeper the viewing angle to the stage. Sound entering the box hardly gets back to the auditorium. In an Italian style opera house the time delay of main reflections is therefore determined mainly by the distance to the front rather than the rear walls of boxes. French style opera houses feature open boxes and gallery balustrades with reflecting front walls clearly less than 50% of the lateral walls (figure 58).

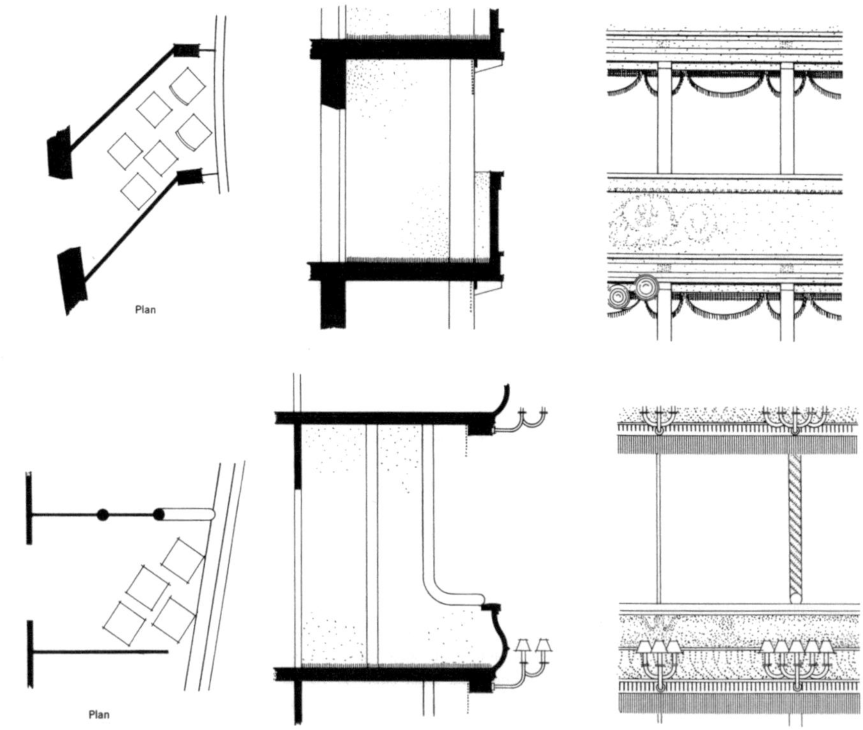

Figure 58: Boxes: Italian style: Venice, *La Fenice* (above); French style: *Covent Garden* (below).

Still, distances between these relatively small surfaces are acoustically important because they produce early lateral reflections. The main walls of the hall become main reflecting surfaces if galleries are very high and wide open, and steep ranking seat rows are highly absorbing. In theater acoustics, too, well structured reflecting front walls of the boxes are important for diffusivity, which increases on the ranked galleries and because reflecting and absorbing zones alternate between the box openings.

Compared to large concert halls, opera and theater acoustics are often very dry. Even large opera houses may have mid-frequency reverberation times below 1.2 seconds (see table 12b).[30] To determine acoustic quality, reverberation time is even less significant in theaters than in concert halls.

A theater with boxes has a much smaller ground surface than a hall with the same number of seats, and its acoustics are therefore relatively clearer and more intense.[31] The sound in good opera houses and theaters is warm and close to the boxes' openings and balustrades it is full, transparent and clear. Another typical property of theaters is that good voices are not covered by the full sound of the orchestra. This is based not only on the strong directivity of voices and on the elevated position of singers on the stage, but also on the exactly calculated position of reflecting surfaces above the apron within the proscenium arch. Many experienced singers know what the best acoustic position for the big aria is.

The stage behind the proscenium, which is more or less strongly coupled to the auditorium, has its own acoustics depending on stage size and height. Acoustics vary considerably depending on the stage setting. Scenery has an important influence on acoustics, especially when the stage plafond is opened for raising backdrops. Scenery and the position of the scenery's reflecting surfaces are therefore important for historical studies of acoustics. In case of excessive acoustic loss on stage and insufficient contact between stage and auditorium only the apron stage provides good acoustic conditions for singers and actors. For concerts the stage is often closed with a concert shell to prevent backstage sound loss.

The increased dimensions of theaters since the eighteenth century prompted, for visual reasons, discussions on the maximum distance between spectators and stage. The limitation of room length caused an increase in the width of auditoria. The acoustic problem worsened when the orchestra increased and the apron stage was sacrificed in nineteenth century opera houses, as in *La Fenice* in Venice where the ceiling above the orchestra had to be inclined in order to improve reflections from singers behind the proscenium arch (figure 59). Today the orchestra plays in more or less lowered pits between the proscenium opening, and singers are restricted to acoustically less favorable positions more backstage.

30 See Meyer (2009), 6.2, 234 f.

31 Bauman (1993), Interview with R. Johnson.

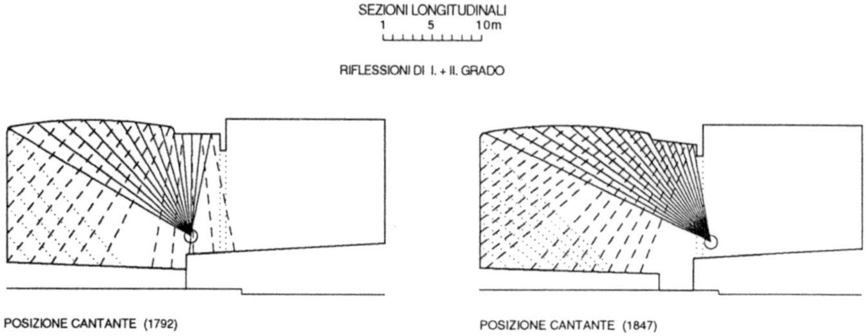

Figue 59: Venice, *La Fenice* (1792) (left) and after elimination of apron stage and change of the ceiling above (1847) (right).

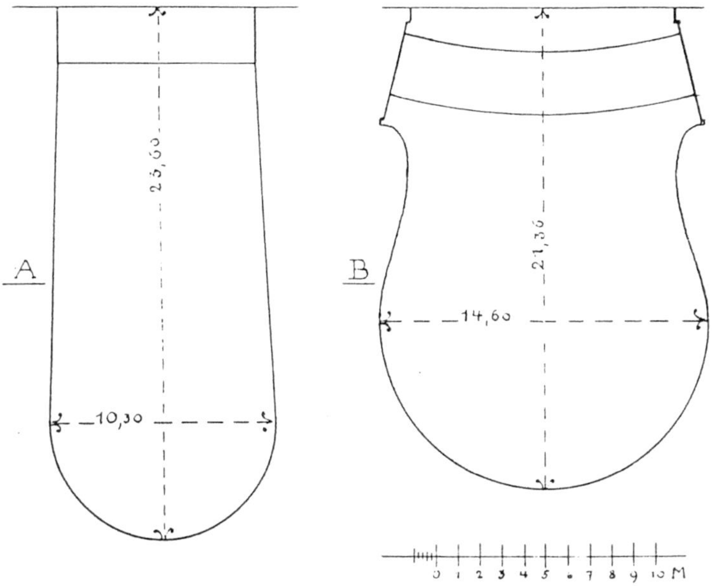

Figure 60: Vienna, a) old (1748) and b) new *Burgtheater* (1888), comparison of ground plans.

The price of improved sight lines in shorter and wider halls was a loss of acoustic presence and intimacy compared to the longer and narrow baroque theaters. This is one reason why the new *Burgtheater* in Vienna of 1888 had a bad reputation (figure 60).[32] With 174 additional seats it was nearly 50% wider between box openings and higher than the old *Burgtheater,* since 1748 famous

32 See also 1.2.2.7 on old and new halls.

for its acoustic presence and transparency, and, furthermore, its lyre-shaped ground plan caused bad sight lines for many seats.

Table 8: Old and new *Burgtheater* Vienna.

Vienna	Ranks	L (m)	W_i (m)	W (m)	H (m)	Vol. (m^3)	Seats	T_m (s)
Old *Burgtheater* (1748)	4	23.60	10.30	15.00	12.00	4'250	1300	1.3
Burgtheater (1888)	4	21.40	14.60	17.50	17.50	5'500	1474	–

L = Length with podium, W_i = Width between galleries, W = total width, H = height, T_m = reverberation time for mid-frequencies (500–1000 Hz)

Joseph Gregor wrote the following obituary on this theater which had seen the first public performance of Haydn's *Creation* with 180 musicians in 1799:

> This long and narrow hall with four galleries without decoration, these old comfortable chairs, [...] the four galleries were as tuning forks of an incredibly clear and attentive devotion, and these produced an acoustics never heard again and even the acoustics of the theater in the Josefstadt seem to be but a weak after ring! [...] In the much too large auditorium [of the new theater] it becomes soon obvious that from the more distant seat rows and boxes little and from most of the galleries' practically nothing is heard and only rarely a good and undistorted view to the stage is possible. The distance of the actors even from the first row is much too large and this dead hall also kills the playing because it makes contact so incredibly difficult. Not the [new] Burgtheater has to be admired but so many artists able to find this contact in spite of the impediments of this magnificent-capricious house.[33]

The loss of intimacy in theater buildings since the late eighteenth century, which was caused not only by larger volumes of space and lower energy density but mainly by the increase of hall width and the delay of lateral reflections, was analyzed first in spoken theater. But this loss changed also the sound impression of opera, choir and orchestra concerts, often performed in theaters and

33 Cf. J. Gregor, *Geschichte des Österreichischen Theaters* (1948), 212: "Dieser lange und schmale Saal, den vier Ränge schmucklos umliefen, dieses alte, unbequeme, rote Gestühl, [...] diese vier Ränge waren wie Stimmgabeln einer unglaublich scharfen und hingegebenen Aufmerksamkeit, und auf ihnen tönte eine Akustik, die niemals wiedergekehrt ist und gegen die selbst jene des Theaters in der Josefstadt nur ein schwacher Nachhall sein soll! [...] In dem viel zu grossen Zuschauerraum [des neuen Theaters] aber merkt man bald, dass man von weiter liegenden Sitzreihen und Logen wenig, von grossen Teilen der Galerie aus aber überhaupt so gut wie nichts hört, und dass man nur selten einen guten, unverzerrten Blick auf die Bühne geniesst. Die Distanz der Schauspieler selbst von der ersten Reihe ist eine viel zu grosse, und dieser tote Raum tötet auch das Spiel, weil es den Kontakt unendlich erschwert. Nicht das [neue] Burgtheater ist zu bewundern, zu bewundern sind so viele Künstler, die diesen Kontakt trotz der Widerstände des grossartig-eigenwilligen Hauses dennoch gefunden haben."

opera houses.[34] This change becomes obvious in comparisons of the dimensions of selected theaters and opera houses given in table 12b. More documents on Handel's performances of *Messiah* in London theaters of the eighteenth century are presented in the second part of this study.

1.8.5.3 Churches

Regardless of social context most listeners in Western culture owe their most impressive sound experiences to church acoustics and to the composers, organ builders and musicians who know how to handle them. Church acoustics, too, can be described only incompletely by reverberation decay curves. Since no other room category used for music and speech shows such a variety of shapes and dimensions, only basic characteristics can be given here with respect to the analyses in the second part of this study.

Church acoustics are extremely complex, especially in large cathedrals. They are determined by the shape and height of vaults, width and height of the main nave, the choir, the crossing and the transept, the number, height and width of side naves, the height and diameter of pillars, which are the closest lateral reflecting surfaces on the floor level, the dimensions of openings to all further acoustically coupled spaces, such as side chapels, galleries, triforia, the total length of the main nave and the choir, and the volume of space of all these parts. A comparison of reverberation decay curves in churches of various architectural styles shows differences that depend not only on total volume, surface structure, different interior elements (rough stone walls or plaster absorb high frequencies; leaded glass windows, wooden ceilings, vaults, and installations absorb low frequencies) but also on room shape and on the way these different parts of space are acoustically coupled.

This is why, for example, in the huge cathedral of *St. Paul's* in London, which features barrel vaults and a relatively high percentage of reflecting lateral surfaces, a very loud organ chord is audible for about 12 seconds. In the full church this is due mainly to the long reverberation time for low frequencies. Since persons absorb mainly middle and high frequencies, the measured reverberation time T for middle frequencies decreases from 12 to 6.5 seconds as the number of attendants increases. In the gothic cathedral of *Westminster,* on the other hand, a respective organ chord in the crossing is audible only for about five seconds. Here, the measured reverberation time T for middle frequencies is about five seconds in both the empty and the full church (figure 61).

34 See also 2.5.4 documents on acoustics of London theaters and 2.6.5 overview on number of instruments and hall size.

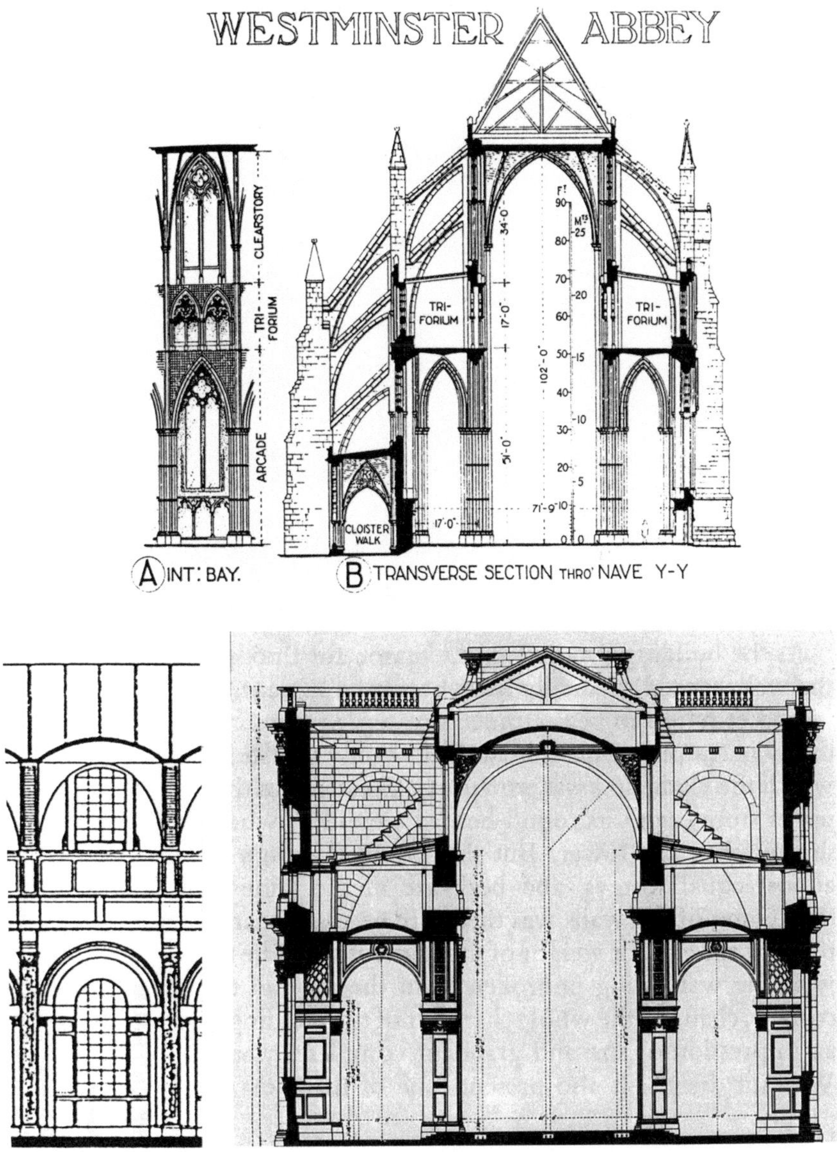

Figure 61: London, *Westminster Abbey* (above) and *St. Paul's Cathedral* (below), same scale.

In gothic cathedrals the reverberation time depends much less on the number of persons because a large part of the sound remains up in the vaults and in the connected parts of space, where it is absorbed.[35] *St. Peter's* in Rome has surpri-

35 H. J. Purkis, "The Reverberation Time of Some English Cathedrals" (1963), 8; see also Bagenal / Wood (1931), VIII § 11.

singly dry acoustics despite its enormous dimensions, the barrel vaults and the more than 100-meter-high dome (if fully occupied the measured reverberation time T is only about five seconds!),[36] because the many side chapels absorb a lot of sound, and reflection paths in this enormously wide and high space are long and multiplied. Visitors listening to the guide in the nearly empty church close to other groups with guides, have an acoustic impression similar to outdoors.[37]

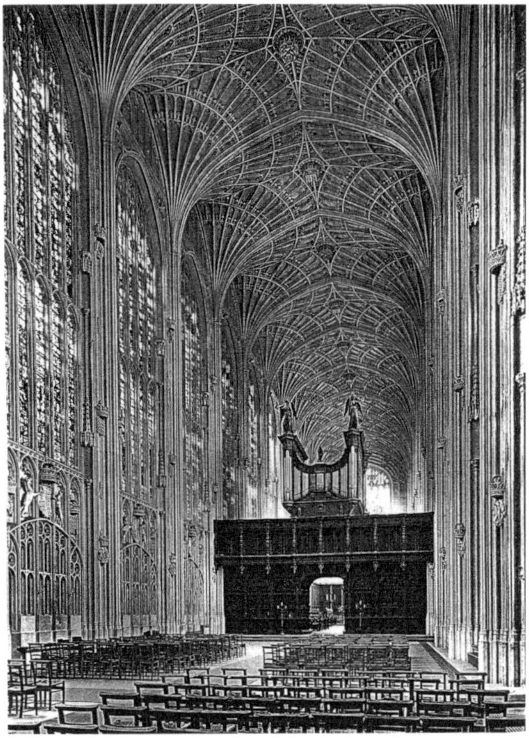

Figure 62: Cambridge, *King's College Chapel* (1446-1515), view to rood screen.

In empty Romanesque and Gothic churches with stone vaults maximum reverberation time T in general is at low frequencies. For high frequencies T strongly depends on surface roughness. For middle frequencies T in the empty church depends mainly on the space volume but very rarely exceeds 11 seconds due to absorption in coupled spaces. Smaller Gothic churches and chapels with space volumes of 50,000 m^3 or less and fewer coupled parts can have similar or even

36 R. S. and H. K. Shankland, "Acoustics of St. Peter's and Patriarchal Basilicas in Rome" (1971), 2.

37 F. Winckel, "Das Stimmverhalten des Sängers in Abhängigkeit von den Musikgattungen und den Aufführungsstätten während der Florentiner Renaissance" (1976), 11 f.

longer reverberation times than large cathedrals, as, for instance, *King's College Chapel* in Cambridge (figure 62) and *Westminster Abbey* in London, both of which have a reverberation time of five seconds for middle frequencies when fully occupied. The volume of both *King's College Chapel* without side chapels and of the main nave of *Westminster Abbey* up to the crossing is, in fact, about 26,000 m³.

Table 9: Churches: volume and reverberation time (occupied).

Church	D	sN (m)	LmN (m)	total L (m)	WmN (m)	W (m)	HmN (m)	HD (m)	∅	T_m	Vol.
Rome, *St. Peter's*	bV D	1	96.00	180.00	26.40	69.60	45.00	101.0	41.60	5.0	550'000
London, *St. Paul's*	bV D	1	56.00	140.00	14.40	30.00	28.00	65.60	32.80	6.5	155'000
Westminster Abbey	goth V	1	71.85	113.00	11.50	22.00	31.00	–	–	5.0	72'500
Cambridge, *KC Chapel*	goth V	1/ sC		84.00	12.80	20.80	26.40	–	–	5.0	46'500

D = dome, C = ceiling shape: bV = barrel vault, goth V = gothic vault, sN = number of side naves, sC = side chapels, LmN = length main nave, WmN = width main nave, W = total width, H = height, HmN = height main nave, HD = dome height, T_m = reverberation time for mid-frequencies (500–1000 Hz).

Baroque churches have longer reverberation times for high frequencies because of the mostly smoothly plastered walls. In empty churches maximum reverberation occurs therefore at lower frequencies,[38] if these are not absorbed by a lot of wooden installations, such as galleries, altars, benches or flat vaults. Richly structured stucco provides good diffusivity. With acoustically favorable shapes, such as a not too wide one-hall church,[39] Baroque churches provide excellent acoustics for music.

38 Meyer (2009), 245.

39 For an overview on special formal aspects of protestant one-hall churches such as the lack of the choir or the importance of galleries, and its relation to hall and theater architecture, see G. Germann, *Der protestantische Kirchenbau in der Schweiz von der Reformation bis zur Romantik* (1963); see also 1.2.6 on the knowledge of architects.

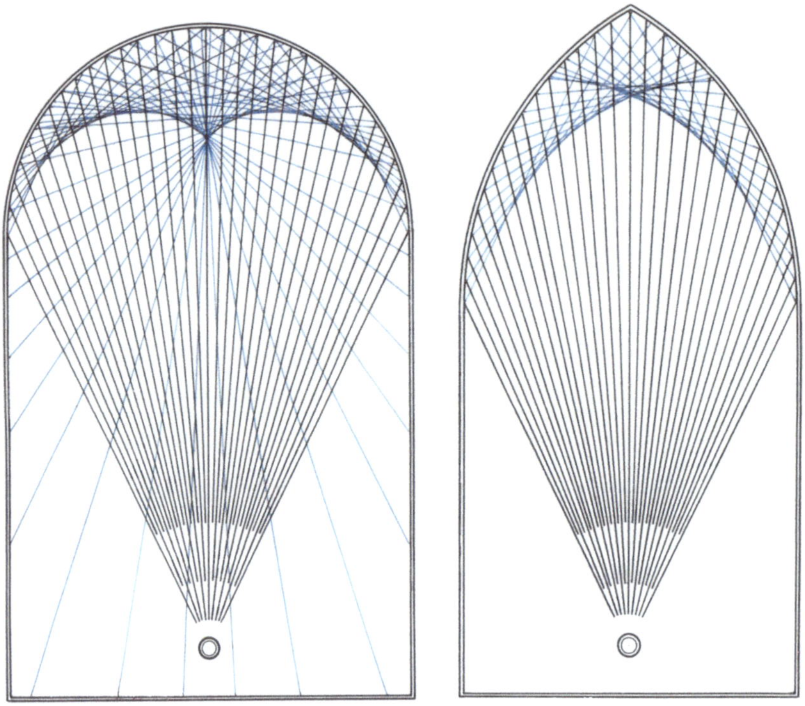

Figure 63: Reflections off barrel vault (left) and Gothic vault (right).

Geometric studies reveal further characteristics perceivable by attentive listeners: Romanesque vaults reflect larger amounts of sound back to the floor level than gothic vaults (figure 63 and 64). They send first-order sound reflections getting directly into the vault and second-order reflections reaching the vault via a lateral surface back to the floor, whereas Gothic vaults give the impression that sound sources on the floor level directly under the vault center send sound upwards, where it is lost as though it were reaching "heaven." In fact, none of the first- or second-order reflections come back to the floor level, and only a small part of loud sound directly sent to the vault reaches the listeners, such as the low and later reverberation of high voices, trumpets or organs, with vault-to-wall reflections coming before wall-to-vault reflections. Lateral reflecting surfaces, such as the walls of the side naves, are often far removed from the listeners in the main nave and hindered by pillars. Hence the important role of the floor surface as reflecting surface. In the main nave acoustics depend therefore on the number of listeners on the floor.

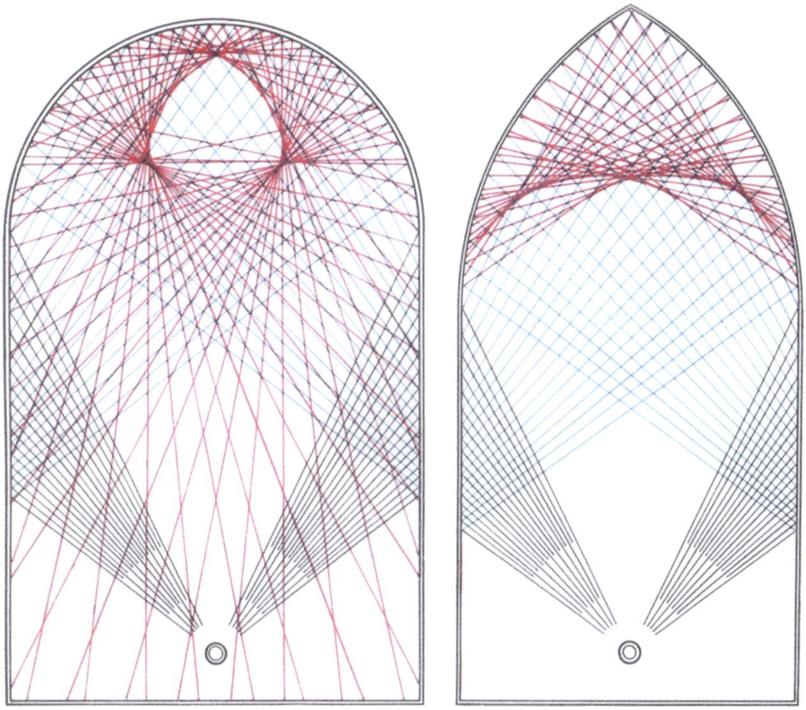

Figure 64: Reflections off sidewalls into barrel vault (left) and Gothic vault (right).

Sound getting into the side naves is reflected there several times mainly in case of a completely or nearly unoccupied floor and therefore reaches back to the main nave with considerable delay, which is disturbing at high sound levels (this is the effect of Vitruvius' *reflexiones circumsonantes).*[40]

These observations indicate that the position of the sound source is one of the main factors of sound quality in large churches. Below vaults, even small horizontal or vertical shifts of the sound source's position can change reflection paths completely. Church musicians often find best places due to their long experience. Already in the Middle Ages those in charge spared no effort to install high galleries, balconies, tribunes, rood lofts, and swallow nests for organs, singers and later even for whole orchestras. This did not only improve direct sound going down to the listeners but also provided better support by lateral and ceiling reflections due to shorter path ways.[41]

40 D. Baumann, "Acoustics in Gothic Cathedrals, Theory and Practical Experience in the Middle Ages" 2001, 37–48.

41 D. Baumann, "Musical Acoustics in the Middle Ages" (1990), 199–210.

The less wide choir behind the crossing or in the nave with relatively high walls closed on three sides (as used in Mediterranean and Anglo-Saxon church architecture) has an important acoustic function: these relatively closed spaces protect musicians from the reverberation in the main church volume and provide excellent lateral reflections, as in the choir of *Westminster Abbey* or in the no longer extant choir of *St. Paul's* in London, which was used as concert hall with boxes and without ceiling when Handel played his organ concertos (figure 65).[42]

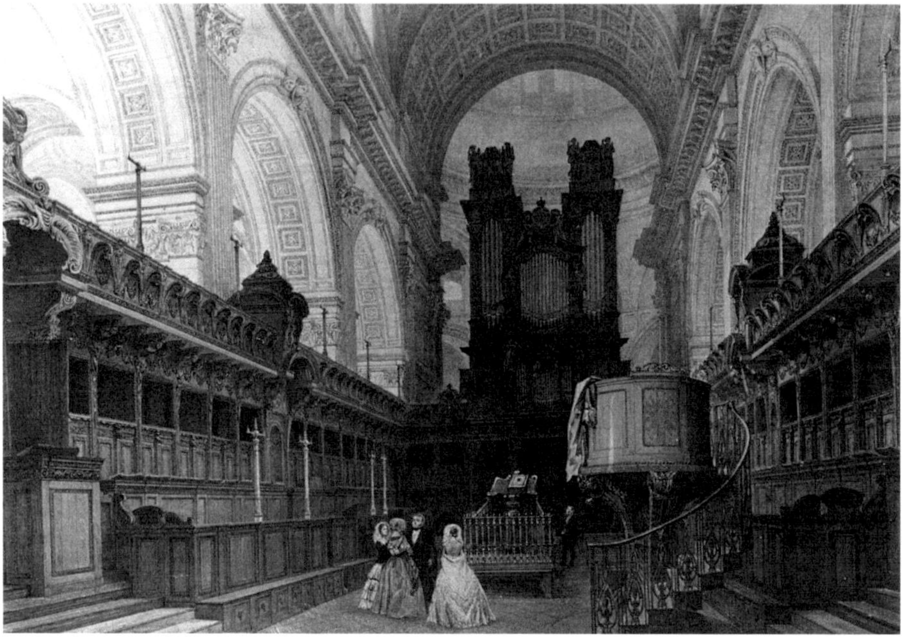

Figure 65: London, *St. Paul's Cathedral,* choir (1830) with Bernhard Smith's organ from 1695–97.

Rich decorations with cloth and tapestries as used in many churches on feast days had and still have a favorable influence against excessive reverberation but can also hinder the effect of music in churches with dead acoustics, such as *St. James's Palace* in London, described by the *Chapel Royal's* organist and composer in 1928 as follows: "The hangings and coverings here are a very real hindrance to vocal music."[43] This is an example of how important exact knowledge of details is for reliable information on acoustics during a certain performance.

42 D. Baumann, „Whispering Galleries“ (2001), 481–491.

43 E. S. Roper, "Music at the English Chapels Royal c. 1135 – Present Day" (1927/1928), 19–33, mainly 28–9; see also 2.6.2 on Handel's performances with large ensembles.

Composers, organ builders and musicians have at their disposal important means to improve acoustics. Since only loud sounds fully activate reverberation, a reduction of sound level and dynamics reduces excessive blurring at least for close listeners. The sound onset is usually about 1/20 of reverberation time.[44] This is why short articulation with few low frequencies and respective shorter or longer breaks between the notes and a *messa di voce* increasing from piano to higher sound levels also help to avoid full reverberation. Therefore, baroque instruments with their typical slow onset of low frequencies and the baroque playing techniques, such as the *non legato* and the *messa di voce,* are especially well adapted to church acoustics. It is also known that highly directive sound sources, such as trumpets and high sopranos, reach far in difficult cathedral acoustics and sound better than the omnidirectional low frequencies of instruments and voices.

1.8.6 The number of attendants

Special care is required to calculate the number of listeners on the basis of the dimensions of seating or standing space. Detailed historical sources indicate that in the eighteenth century the distance between seat rows and the surface per seated person were much smaller than today. In 1793 London theaters seat rows were only 53 centimeters apart, with four persons per square meter seated on benches without back rest (0.3 m^2 per person),[45] whereas today the norm in Europe is only two seated persons per square meter (0.5 m^2 per person) or 6.6 standing persons per square meter (0.15 m^2 per person).[46] In the United States the norm for seat density is even lower. Besides the increasing number of listeners, the need for security and commodity also finally demanded larger halls that brought a principal change of room acoustics and sound impression.

Shortly after the disappearance of many intimate small halls and theaters at the end of the nineteenth century theater goers realized that good acoustics were determined also by dense seating.[47] In small concert halls the situation worsened due to the increased space needed by the musicians. At first, standing

44 Meyer (2009), 250; see also V. L. Jordan, "Einige Bemerkungen über Anhall und Anfangsnachhall in Musikräumen (1968), 29–36.

45 See 2.5.3.1: The *King's Theatre* at the Haymarket; G. Saunders, *A Treatise on Theaters* (1790), 30 ff.

46 After W. Fasold et al., *Bau- und Raumakustik* (1987), 276; for density of seats, see L. Beranek, *Acoustics* (1971), 485.

47 "[…] dem dichten Aufeinanderhocken", see M. Semper, *Handbuch der Architektur* IV/4/4/1 (1904), 204.

string players, as in the old *Gewandhaus* in Leipzig, were able to remedy the change, but, finally the permanent podium had to be enlarged.

Table 10: Density of seated persons.

Hall	Year	L (m)	W (m)	H (m)	m^2	Seats parterre	Seats/m^2	m^2/seat	Vol. m^3
Oxford, *Holywell Music Room*	1748	19.85	9.80	9.15	150	300 – 500	2–3.38	0.3	1660
London, *Hanover Square Rooms,* great hall	1775	24.10	9.80	[8.50]	236	800–1500*	3.38	0.3	1875
Leipzig, *old Gewandhaus*	1781	23.00	11.50	7.40	259	430 –860	1.6–2.32	0.3	1800
Vienna, *Musikverein, Brahmssaal*	1870	32.80	10.60	10.60	348	625	1.8	0.5	3500
Zürich, *Tonhalle,* small hall	1895	26.80	12.00	9.00	322	429	1.33	0.75	2900

L = length with podium, W = width inside, H = height. *Attendance during a Haydn concerto in 1792.

Oratorio performances needed the largest platforms, as documented in an impressive overview published in 1859 in the *Musical Times,* which clearly showed the relation between large choir festivals and the dimension of the ensembles and the stage (figure 66). The special platforms used for the *Handel Commemorations* since 1784 in *Westminster Abbey* and for the Handel Festivals since 1854 in the *Crystal Palace* at Sydenham are described in more detail in the second part of this study. Here we concentrate on the relation between podium surface and the hall's ground plan and its influence on room acoustics. The great hall of *Hanover Square Rooms* in London, yearly used for 11 Oratorio concerts between 1783 and 1793, had been used for small concerts only (figure 93) since 1859.[48] At the time, the podium was 88.55 m^2 (953 ft^2) or more than a third of the hall's ground surface of 236.20 m^2. In London, from 1831 to 1860 large choir concerts were held in *Exeter Hall,* one of the earliest large public halls built for the concerts of the *Sacred Harmonic Society,* whose *podium* of 338.60 m^2 (3645 ft^2) covered more than 28% of the parterre of this large hall with 3000 seats and a ground surface of 1215 m^2 (figure 94).

The enlarged podium of the old *Gewandhaussaal* in Leipzig (1781) covered 63 m^2 or 25% of the parterre of 264.5 m^2, the podium size in the great Hall of the new *Gewandhaus* (1884) was nearly double with 116 m^2 but only 16% of the parterre of 722 m^2.[49] A podium of this size had to be installed in the old *Gewandhaus* for choir concerts before the opening of the new hall. On such

48 R. Elkin, *The Old Concert Rooms* (1955), 93 f.
49 R. Skoda, *Das Gewandhaus Leipzig, Geschichte und Gegenwart* (1984), 177.

occasions it encompassed more than 44% of the parterre! In 1866 the definitive enlargement of the podium to 9 × 13 m or 117 m^2 in the hall of the Paris *Conservatoire* required the removal of 150 parterre seats (figure 13).

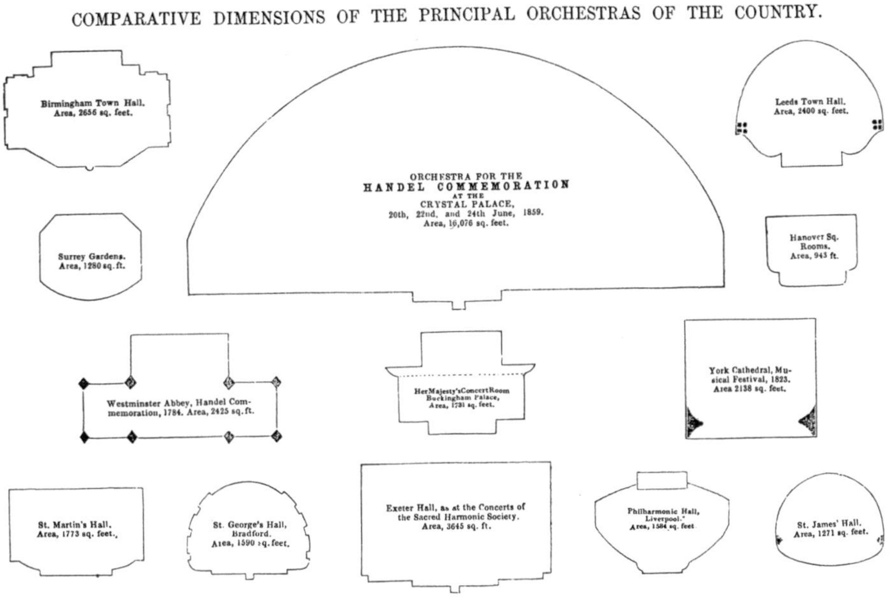

Figure 66: Podium's dimensions in English concert rooms (Musical Times, April 1859).

Although the reverberation time T can be calculated only if exact plans and descriptions on the interior decoration and material are extant, in certain cases a rough estimation of the number of attendants may be sufficient to calculate T for the occupied hall because the occupied surface is more relevant than the exact number of persons on this surface. Indications on the number of attendants and their seating density sometimes permit also to estimate a hall's size if plans are not extant. In private halls and theaters relatively few persons per square meter were often seated in fauteuils and single chairs. In the London theaters of the eighteenth century galleries often remained empty if reserved only for subscribers.[50] The sociological level is reflected here in the number of listeners. Excellent acoustics obviously belonged to aristocratic and bourgeois privileges. But the opera houses and concert halls offered excellent acoustics also to lower classes on the *Olympus,* though only as standing places or seats with bad sight lines below the ceiling. Not only the increase of listeners but

50 See 2.5.3 on the theaters used by Handel.

also the endeavor to provide convenience and security finally led to an increase of room dimensions in many cities, thus introducing a fundamental change in room acoustics and music perception.

Today video and audio techniques offer nearly everyone the privilege to see and hear an event from “the best seats”, although the playback of a film can never replace a live performance in real space. But the acoustic difference is fully perceivable only if the recorded event can be compared directly with the real sound in the hall. Supported by visual information, our perceptual system is ready to use all information and experience to build a three-dimensional sound impression of the whole event.[51]

51 On neuronal sound processing and active perception, see also 1.5.3.2. A recent example for new audiovisual experiences of space was the performance of Verdi’s *Aida* in Basel in the rooms of a hotel and on a ship on the Rhine, emission by Swiss Television on 1 October 2010, edited on DVD *Aida am Rhein, Das Live Opernerlebnis mitten aus Basel,* SF 2010.

1.9 Music related quality factors in room acoustics

1.9.1 Assessment of quality factors

The assessment of a music room's acoustic quality never is absolute but always depending on the style of the performed music, the musical instruments, the number of musicians, their position in the room, the room's size and the number of listeners. A description of room acoustic quality has to take into account that since the eighteenth century more and more genres in different styles of music from different times used to be played in the same hall and even during the same concert. Musical dynamics (the difference between piano and forte), the variety of articulation also in the forte, the used ambitus of instruments, the frequency range in a piece and the variety of sound colors presented during a concert further increased during the nineteenth and twentieth century. That is why demands on room acoustic properties increased and composers and musicians had to take into account room acoustics more consciously.

In the history of concert hall and theater architecture the number of buildings is too small, documentation too incomplete and unique solutions too frequent in order to permit a purely statistic assessment of music related room acoustic quality. That is why two steps are taken:[1] first a representative selection of halls and theaters with known dimensions is grouped chronologically (group 1a). Then spaces used for music in certain music centers are selected in order to document a certain period of time as fully as possible, even if dimensions for some rooms are lacking or have to be based on estimation (group 1b). Well known rooms and music centers and such investigated in the second part of this study are particularly regarded. This broad collection of information then has to be related to documents on musical compositions (group 3) and music performance (group 2). Together with information on playing practice (group 4) and reports on the music's reception (group 5) these documents lay ground for the definition of music related room acoustic quality factors. But acoustic quality only becomes evident if this information is regrouped in order to define certain types of spaces (characteristic for certain periods of time), typical programs and genres presented in concerts and finally the period of use of certain types of spaces, where desired, grouped by music centers.

1 See 1.1.4: Methods of research: a study in two parts.

1.9.2 Types of space

The systematic collection of information for the investigation of spaces used for music provided more or less complete data for over 1400 spaces. For many spaces in this study the main dimensions gained from different sources are published for the first time. Special care was taken to assess room height so important for acoustics. Where ever possible historical plans were used. The main sources used for the indication of measures in the tables commented in the following paragraphs (see table 11b for music rooms and concert halls, 12b for opera houses and 13b for churches) are listed in the full alphabetical list of rooms (see table C in the appendix). These sources and sources for citations are fully listed in the bibliography.

Spaces partly were adapted to the musical purpose, or built mainly or completely for music. Musical events happened in a certain space only once (as a performance during a festival) or sometimes or regularly.[2] With the increased number of spaces exclusively used for music during the seventeenth, eighteenth and nineteenth century, partly based on ad hoc adapted spaces, certain types of music rooms and theaters developed. The result is a list with different types of spaces characteristic for certain periods of time and bound to certain social functions. From this certain information on the social provenance of listeners, their number and distribution in the auditoria could be gained.

1.9.2.1 Music rooms and concert halls

Since the Renaissance rooms and halls in aristocratic and high bourgeois buildings were used or rebuilt for music performance, such as music rooms or *stanzas* on the second floor but also the great hall or *sala grande* or *stanzione*[3] on the first floor or *piano nobile* or garden pavillons. In the eighteenth century also ball halls or *redoutes* in opera houses or theaters could be used for music with the consequence that only exclusive social groups had access. Theaters and opera auditoria themselves were and still are used as concert halls, either with the orchestra in front of the closed main curtain or on stage with the orchestra pit raised, often with closed scenery (Musikzimmer) in order to improve acoustics.

2 An important survey on the characteristic types of halls is published by H. Habel, *Das Odeon in München und die Frühzeit des öffentlichen Konzertsaalbaus,* Berlin (1967). His list given above has been enlarged by further research.

3 *Stanzione* means hall or entrance hall.

Music societies which had no access to such rooms used halls of the city as the town hall, the guildhall, the aula of the university, museum halls, the pit of trade halls, or halls belonging to churches as oratorios, assembly halls or music rooms, larger rooms and several connected rooms in private houses, halls in inns, ball houses, sanatoria and, already since the seventeenth century, halls for sports such as tennis, riding, skating halls.[4] In the eighteenth century music societies started to become independent from church and erected their own concert halls increasingly open to public.

A similar role played the chapels and halls of conservatories since the seventeenth and eighteenth century in Italy and since the nineteenth century in France spreading over whole Europe. In some cities piano builders had their own, mostly smaller halls important for music life, as the old *Salle Pleyel* (1839–1927) in Paris or the *Bösendorfersaal* (1872–1913) in Vienna. Public concert halls of the nineteenth century often follow aristocratic models (as the great hall of the *Wiener Musikverein)* or sacred prototypes (as Schinkel's project for the Berlin *Singakademie* following the model of the *Cappella Sixtina* or Wyatt's *Pantheon* in London imitating the Roman *Pantheon* and the *Hagia Sophia).*

Relations between music life, music rooms and social background are complex. The important influence on the sounding appearance of music for instance becomes evident in Beethoven's situation in Vienna: In 1805 concert goers could hear his third symphony *Eroica* for the first time in the fully occupied *Theater an der Wien* (between the balustrades 18 × 20 m, 15 m high) in clear, dry but not unfavorable acoustics played by an ensemble which would be called today a chamber orchestra (strings: 6-5-3-2-2, woodwinds: 2-2-2-2, 3 French horns and 2 trumpets) sitting in the auditorium in front of the closed main curtain. Prince Lobkowitz and his guests had the privilege to hear the same composition played by an orchestra of practically the same size in the hall of the *Palais Lobkowitz* since more than one year.[5] This hall on the *piano nobile,* in modern terms a small chamber music hall (15.20 × 7.15 m, 8.25 m high)[6] for about 150 persons, then was mostly occupied by the roughly 30 musicians and a few listeners only. Such social background conditions and limitations have to be considered in looking at the following definition of types of space classified by dimensions, since the number of listeners often depended more on this background of an event than of a room's dimensions.

4 Such an example is the *Philharmonie* in Berlin, a skating gall which in 1882 was changed to a concert hall, or the 1914 destroyed *Bösendorfersaal* in Vienna, a former riding hall of Graf Starnhemberg with excellent acoustics.

5 See facsimile edition by O. Biba, *Gesellschaft der Musikfreunde,* Wien (1996).

6 Measured by the author in 2004 during renovation. Marble slabs on the walls probably are added later replacing wooden paneling at Beethoven's time.

As explained in chapter 1.6, for a general evaluative background for a historical survey, categories of size based on the main dimensions relevant for acoustic conditions are sufficient as parameters for the definition of groups of music rooms and concert halls in spite of their great variety of design if special ground plans and ceiling shapes as well as details of the interior décor are taken into account for a more detailed acoustic analysis. These categories of size are divided into 3 levels. Collected data show that the dimensions for each type of space and category lay within characteristic band widths marked in bright or dark grey in the following tables.

Table 11a: Music rooms and halls – categories.

Category	Hall	Very small	Small	Large	Very large	Even larger
	Width	W < 9 m	W 7–13 m	W 12–20 m	W 15–25 m	W > 25
	Length	L ≥ W	L 15–30 m	L 30–45 m	L > 45 m	
	17th/18th century					
A	Stanza					
B	Oratorio, Salone					
C	Salone grande Gallery					
and	***18th century***					
A	Room, tavern hall					
B	Music room					
C	Festival hall					
→	***~1830–1880***					
A	Music room, salon					
B	Small hall					
C	Great hall					
D	Festival hall					

For some categories it is appropriate to speak of a standard due to the great number of halls with similar dimensions (see fields marked in dark grey) such as for the nineteenth century the small hall (category B) and the great hall (category C), which were developed from the respective hall categories of the seventeenth and eighteenth century, the smaller music room (category B) and the larger festival hall or *salone grande* (category C). Table 11b shows that the typical width of a "small hall" in the nineteenth century varied between 7 and 13 m with a length between 15 and 30 m, while halls wider than 12 m and longer than 30 m normally were used as "great halls".

Table 11b: Music rooms and halls – examples.

Hall	Year	Gr.	C	L	W	H	Vol.	S.	T_m
Category A				m	m	M	m³		sec
Rome, *Palazzo Barberini,* Salotto	1632	[ ]	V	17.15	12.45	9.35	2550	-	
London, *Hickford's Rooms*	1729–1779	[ ]	----	15.20	9.10	6.70	934	300	
Paris, *Concerts rue Cléry*	1791–1801	O	D	8.00	8.00	6.25	627	-	
Category B									
Vienna, *Palais Lobkowitz*	~1700	[ ]	flV	15.20	7.15	8.25	900	150	[1.45]
Oxford, *Holywell Music Room*	1748	[ )	----	19.85	9.80	9.15	1660	300	[1.5]
Edinburgh, *St. Cecilia's Hall*	1762	0	?	19.20	10.70	5.20	1394	500	[0.8]
Esterháza, *Musiksaal*	1766	[ ]	----	15.50	10.30	9.20	1530	200	[1.2]
London, *Hanover Square*	1775–1874	[ ]	D	24.10	9.80	[8.50]	1875	800	[0.95]
Leipzig, *old Gewandhaus* with gallery	1781 1842	( )	-----	22.85	11.50	7.40	2396	430 570	[1.9] [1.2]
London, *King's Theatre, Hall*	1792	[ ]	+++	29.60	14.60	11.90	4550	1000	[1.55]
Paris, *Conservatoire*	1811	( )	flV	24.45	11.10	10.40	2800	1021	
Berlin, *Singakademie*	1829–1945	( ]	+++	32.40	12.80	9.70	~3300	1200	
Vienna, *Musikverein,* Brahmssaal	1870	[: :]	⌒	32.80	10.60	10.60	3500	625	
Category C									
London, *Whitehall* , Banqueting Hall.	1621	[ ]	+++	32.00	16.00	16.00	8192	400	
Vienna, *Grosser Redoutensaal*	1748	[ ]	+++	39.90	16.95	15.60	10440	1500	1.4
Vienna, *Aula,* old University	1753	[ ]	flV	28.00	14.00	10.00	3920	750	1.7
London, *Exeter Hall*	1831–1907	[ ]	+++	45.00	27.00	14.60	17739	3000	
London, *St. James' Hall*	1858–1904	[ ]	V	42.40	18.30	18.30	13000	2500	
Vienna, *Musikverein,* great hall	1870	[ ]	+++	45.70	19.10	17.70	15650	1600	2.0
Amsterdam, *Concertgebouw*	1888	[ )	+++	43.15	28.15	16.40	18700	2206	2.0
London, *Queen's Hall*	1893–1941	[ )	---	32.10	26.45	17.45	12000	3000	1.3
Category D									
Rome, *Palazzo Doria,* Galeria	1684	[ ]	V	51.00	15.00	20.40	12675	-	
London, *Royal Albert Hall*	1871	0	V	66.80	56.40	41.80	86600	6080	2.5

Gr. = ground plan, C = ceiling, V = vault, flV = flat vault, D = dome, L = length with podium, W = total width, H = height, S. = seats, T_m = reverberation time for middle frequencies (500–1000 Hz) in the occupied hall, values are measured or calculated on the basis of measures in the empty hall, values in [ ] are calculated on the basis of room dimensions and descriptions. For sources, see appendix C.

But in the seveneteenth and eighteneeth century halls belonging to category C were mainly used for other purposes and only on special occasions for music performances.[7] On the other hand in many music centers small halls of category B continued to be used as the main concert hall, as for instance the great room of the *Hanover Square Rooms* (1775–1884) in London (figure 93) or the old *Gewandhaus* (1781–1886) in Leipzig (figures 54 and 55). The width of very large halls could reach 15 to 25 m with a length of more than 45 m.

Table 11b gives dimensions of examples for these categories of halls from the seventeenth to the twentieth century, within each category in chronological order. Well known halls and halls used in this study were selected in order to provide information for the comparison of halls also in relation to their social background.

For the study of performance practice in relation to room acoustics further documents are needed, and rooms must be grouped by city or region. One can thus reconstruct a general room acoustic background for a music center during a given period. The second part of this book presents such documentation on Rome around 1700 (chapter 2.2) and on England in the eighteenth and nineteenth centuries (chapters 2.6 and 2.7, respectively).

1.9.2.2 Theaters and opera houses

In theater and opera house building ground plans and ceiling shapes are even more varied than in music rooms and concert halls. As explained in chapter 1.8.8,[8] not only the ground plans but also the number and design of boxes and galleries, the proscenium shape, the ceiling's inclination and shape, the position and level of the orchestra pit, as well as other factors have an important influence on theater acoustics. Since theaters were often destroyed by fire, theater design changed faster than concert hall architecture and could therefore adapt more rapidly to new needs. In this respect, too, theater building in the eighteenth and nineteenth centuries introduced standardization of design, which allows the following definition of types.[9]

The indication of the number of listeners in a theater with boxes and standing areas is especially difficult.[10] Due to the enormous change in seating density, the number of attendants has to be assessed carefully in each case. As in

7 See also 2.2: Handel's encounter with the Italian *Oratorio volgare*.

8 See 1.8.8: Acoustics in theaters.

9 The semicircular theater with raked seats developed from the ancient *odeion* and theater as built since the Renaissance, are not listed here because they play no part in the second part of this study.

10 See also 1.8.10: The numbers of attendants, and L. Beranek, *Acoustics* (1971); I. Mackintosh, *Architecture, Actor, Audience* (1993).

concert halls, in theaters and opera houses, too, the dimensions of size categories show a shift from smaller to larger rooms (see table 12a). But size had begun to increase considerably earlier. The usual size of public theaters changed already in the late eighteenth century from category B to category C. In the nineteenth century the small houses of important music centers could be reserved for more intimate genres, such as spoken theater, Italian opera, *opéra comique* and *Singspiel,* while operas with a large orchestra were performed in the large house. Provincial towns often made do with one small theater only.

The orchestra size was first increased with extra musicians, and only during the nineteenth century and in important music centers could the regular number of musicians be increased. It is noteworthy that for oratorio performances in theaters, larger orchestras than for operas were used quite early, also due to the choir's participation. Documents confirm 150 and more musicians for late eighteenth-century oratorio performances.[11]

Table 12a: Theater and opera houses – categories.

Category	Type of hall	Very small	Small	Large	Very large	Even larger
	Width	W < 9 m	W 9–12 m	W 12–15 m	W 15–20 m	W > 20–29
	Length	L ≤ W	L 10–15 m	L 12–20 m	L 15–25 m	L = W
	17th / 18th century					
A	Court theater					
B	Public theater					
C	Festival theater					Hall theater
	18th / 19th century.					
A	Court theater					
B	Small house					
C	Large house					

Table 12b lists dimensions of examples of the different categories of theaters and opera houses, to be compared in the second part of this study:

11 See table A in the appendix and also 2.8.2: Sound recordings of *Messiah:* Sound balance and number of instruments per part.

Table 12b: Theaters and Opera houses – examples.

Theater	Year	Gr.	C	L	B	H	Vol.	S.	T_m
Category A				m	m	m	m3		sec
Rome, *Palazzo della Cancelleria,* Teatro Ottoboni	1690–1740	U	Conc	9.20	8.95	10.60	856	?	
Munich, *Cuvilliés-Theater*	1753	Gl.	-----	14.00	10.00	10.85	1512	436	0.90
Category B									
Venice, *SS. Giovanni e Paolo*	1639	U	-----	17.00	13.75	11.65	2400		
Vienna, old *Burgtheater*	1741–1889	U	-----	23.60	10.30	12.00	2900	1300	[1.20]
London, *Covent Garden*	1732	/\	_---	13.20	10.00	9.90	3200	1400	
Rome, *Teatro Argentina*	1732	U	flV	18.15	15.15	15.00	5120	1872	
Category C									
Florence, *Teatro Mediceo*	1585	[ ]	+++++	*44.00	20.40	14.00	11424	2000	
Parma, *Teatro Farnese*	1619	U	-----	42.25	30.00	20.70	32000	2500	
Naples, *San Carlo*	1737	U	-----	25.00	22.50	22.50	12375	2550	
Milan, *La Scala*	1778	U	flV	24.85	21.50	20.00	11250	3000	1.25
Venice, *La Fenice*	1792	U	flV	22.00	18.00	14.00	5420	2000	
Vienna, *Theater an der Wien*	1801	U	----	18.00	20.00	15.00	5200	1060	1.15
London, *Covent Garden*	1858	U	D	23.05	18.90	19.25	12240	2180	1.10
Paris, *Opéra Garnier*	1875	U	flV D	25.60	20.00	20.00	9960	2156	1.10
Dresden, *Semper Oper II*	1878	U	-----	20.25	17.25	18.50	7400	2000	1.50
Zurich, *Stadttheater*	1891	U	flV	19.00	13.50	14.00	6800	1139	
Buenos Aires, *Teatro Colón*	1908	U	flV	34.30	21.50	26.80	20870	2487	1.80
New York, *Metropolitan Opera*	1883–1966	U	—	32.00	20.65	23.45	19500	3639	1.20
New York, *Metropolitan Opera*	1966	/ \	—	31.00	22.00 33.00	25.00	30500	3816	1.80

Gr. = ground plan, C = ceiling, V = vault, flV = flat vault, D = dome, L = length (curtain – back balustrade) or *length (with platform), W = width (between front walls of boxes), H = medium height, S. = seats, T_m = reverberation time for middle frequencies (500–1000 Hz) in the occupied hall, values are measured or calculated on the basis of measures in the empty hall, values in [ ] are calculated on the basis of room dimensions and descriptions. For sources, see appendix C.

1.9.2.3 Churches

According to chapter 1.8.9, the assessment of church acoustics must be based on the following main elements: Ground plan; distance of reflecting surfaces: width and height of the main nave, total width of the main nave and side naves, height of the side naves, width and height of the choir; shape, structure and material of the ceiling; the acoustically relevant cubic volume of the main space; the volumes and position of connected and acoustically coupled spaces; the position of the sound sources in relation to the acoustical interaction between the main space and the connected spaces.

The relatively high main nave is characteristic for churches in general, reaching 20-30 m or more in large churches, as in *S. Maria del Fiore* in Florence (41 m) or in *St. Peter's* in Rome (45 m). The total cubic volume of space shows an enormous variety. In churches with side naves the volume of the main nave is acoustically more independent if pillars are large and side naves low. Acoustics are then determined mainly by the height-to-width relation, the ceiling shape and the volume of the main nave. From this point of view it is striking how few of these main naves exceed 12–14 m.

Table 13 below lists dimensions of the main churches investigated in this study, grouped by the shape of the ground plan; therefore, only the following types of churches are represented: chapels and small churches; Gothic cathedrals; hall churches; centered churches.

A more detailed analysis of churches must open with a comparison of the listed elements in order to find acoustic similarities and relations. Gothic cathedrals with main naves of roughly the same width and the same number of side naves can have similar acoustics, even if their shape differs in other respects. Thus, the cathedrals of York, Exeter and Salisbury offer similar sound impressions despite the different heights of their main naves. The similarity is based on the acoustic effect of Gothic vaults and side naves, which withhold large amounts of sound. An even closer acoustic similarity is evident in Gothic cathedrals with main naves of nearly the same width and height, such as *Westminster Abbey* in London and the *National Cathedral* in Washington D. C. But due to this close similarity, the importance of the sound source position for the sound distribution and impression becomes especially evident.[12]

12 See 2.7.1.2: The reconstruction of the 1784 Handel *Commemoration* in *Washington Cathedral* (1984).

Table 13: Churches – examples.

Church	Typus	C	Sn	L mN	L tot.	W mN	W tot.	H mN	H D	∅
Rome, *S. Lorenzo in Damaso*	Basilica	+++++	1	22.00	48.00	16.00	26.00	24.00	-	-
Venice, *San Marco*	byzant.	bv 5 D	1	46.40	48.00	13.00	26.00	29.60		
Florence, *S. Maria del Fiore*	Gothic	gV D	1	78.40	153.00	19.00	40.80	40.80	86.40	42.00
Rome, *St. Peter's*	Ren.	gV D	1	96.00	180.00	26.40	69.60	45.00	101.00	41.60
London, *Westminster Abbey*	Gothic	gV	1	67.00	113.00	11.50	22.00	31.00	-	-
London, *Chapel of Henry VII*	Chapel	gV	1	18.40	39.00	11.50	22.00	22.00	-	-
London, *St. Paul's*	Ren.	gV D	1	56.00	140.00	14.40	30.00	28.00	65.60	32.80
Cambridge, *King's College Chapel*	Chapel	gV	(1)	-	84.00	12.80	20.80	26.40	-	-
London, *Chapel Royal*	Chapel	+++	-	-	[20.00]	-	[10.00]	[10.00]	-	-
York, *Minster*	Gothic	gV	1	61.60	116.60	12.80	32.80	34.40	-	-
Exeter, *Cathedral*	Gothic	gV	1	44.80	118.40	12.00	22.40	18.40	-	-
Salisbury, *Cathedral*	Gothic	gV	1	58.40	136.00	10.40	23.20	24.00	-	-
Winchester, *Cathedral*	Gothic	gV	1	76.80	142.40	10.40	24.00		-	-
Washington, *Cathedral*	Neo-gothic	gV sCh	1	55.55	120.30	12.50	32.40	31.70	-	-
Paris, *Les Invalides*	Central	D	–	16.00	64.80	12.00	53.60	28.00	64.00	28.00

C = ceiling, bV = barrel vault, gV = gothic vault, sCh = side chapels, D = dome(s), L = length, sN = side nave(s), W = width, mN = main nave; T and volume see appendix C.

1.9.2 Music genres and program types

To assess the acoustic quality of rooms in relation to music performance it is helpful to group compositions by genre and style. The descriptions of music performances from the seventeenth to the early twentieth century yield the following characteristic program types:

a) oratorio (with solo concerto during the intermission)
b) opera (with solo concerto during the intermission)
c) instrumental music for ensemble
d) mixed program with symphony (also single movements), solo concerto, aria, lied, music for choir, chamber music (up to 4 hours)

e) chamber music for small ensemble, lied
f) church music for small ensemble
g) church festival with music for choir, instrumental music, solo concerto

The evaluation of data with known orchestra size shows a strong relation between the performance space, ensemble size and program type.[13] Since research on musical life has thus far covered mostly single music centers, further research is necessary to obtain a broader view on the different program types. Nevertheless, collected data on performances permit the following conclusions:

1. Even at the time of the first performance there is not only one typical performance of a specific genre, such as opera in the opera house or the symphony in the great concert hall or the sonata in the chamber music hall. The relation between space and genre is more complex and was subjected to changes.
2. Genre, selection of the performance space and even the ensemble size depended in most cases also on non-musical factors. For performers this could cause serious problems in the sounding presentation of a work.
3. Composers and musicians obviously had a clear evaluation scale for good and bad room acoustic conditions.

1.9.4 Use of space types by period

Architectural spaces can be grouped by building type as theaters, halls and churches, and then classified according to shape, volume and other relevant criteria that enable comparisons between spaces of the same class or type used at the same time or even at different times. The information for this study had, then, to be collected systematically, with such comparisons in mind.[14]

The assessment of the period spaces were used also requires broader research. Studies show that in small provincial towns spaces often remained small and longer in use than in larger cities.[15] Importantly, the period of use of certain theaters, halls and churches provides the background of the then *general acoustic experience* at that place. More detailed information on certain performances, also from personal recollections, makes this acoustic background of

13 A. M. Hanson, *Musical Life in Biedermeier Vienna* (1985); D. J. Koury, *Orchestral Performance Practices in the Nineteenth Century* (1986); W. Weber, *The Rise of Musical Classics* (1992); on numerous further sources, see bibliography in the appendix.

14 See 1.1.2.

15 Mackintosh (1993), 33.

musical life clearer and more concrete. Such information is crucial because, as explained, the judgment of the sound impression depends on the acoustic experience of individuals and is assessed in relation to it.

Since subjective comments can be better evaluated when based on this general background, the value of reports by travelling virtuosos, critics and music lovers familiar with many music spaces at various places is different, as they are based on broader experience rather than comments from listeners who know only the local conditions. Data for a certain period and certain music centers (Rome, London, Vienna, etc.) are listed in the second part of this study. They are used also for the selection of sound recordings at the end of part two.

1.9.5 Relations between space type, program type and size of the ensemble (numbers per part)

Information on type of space and program can be related to the information on the orchestration and the number of musicians in the ensemble (see 1.1.2, group 2) in order to find typical ensemble sizes for certain types of space at certain periods. Collected data show that the information obtained from different sources must be used very carefully: the numbers of musicians for a certain performance may be estimated on the basis of personal recollections or exactly known from contracts with musicians, lists of payment (probably not containing unpaid musicians or those otherwise paid), instrumental parts and their number (possibly obtained from other performances), etc.[16] Despite the relatively short period of research in this field, the assessment of information on space and the number of musicians per part yields surprisingly clear results that can be summarized as follows:

The design of characteristic space types often depended on social and non-musical functions as well as on the performance of musical genres related to these functions. Still, in all periods different genres (or parts of them) were played in different types of space. Already before Mattheson's comments[17] but even more so in the late eighteenth century, music genres had been increasingly gaining independence from their strong ties to non-musical functions and spac-

16 Koury (1986), 13 f., 117 f., 327 f.; O. Schreiber, *Orchester* (1938), mainly 202 f.; N. Zaslaw, J. Westrup, "Orchestra", *NG* 13, 679–90, and *NG Instruments* 2, 823–36; N. Zaslaw, J. Snitzer, "Orchestra", *NG Opera* 3, 719–35; E.W. Galkin, *A History of Orchestral Conducting* (1980), 127 f.; for special studies, see bibliography.

17 This is reflected in the repeated prohibition of playing secular song on the organ or the use of opera arias in church or the performance of church music in theaters.

es. With the emergence of public music life special types of spaces appeared, whose size shows a relation to the main music genres, such as opera and opera house, symphony and concert hall. But genres related to a certain size of space in the late nineteenth century, such as symphonies written for great halls or chamber music for smaller halls, are still played in spaces of different size.

A selection of orchestras of typical size reflects also the change of music style (see table 14a), such as the minimal ensemble of 5 strings with several instruments for *basso continuo* for the Italian baroque opera around 1658 in Venice, Lully's opera orchestra with 25 strings its group of woodwinds around 1670 in Paris, the opera orchestra in Vienna between 1781–1808 with 22 strings, double woodwind and two trumpets, or Wagner's orchestra for the *Ring* with 64 strings, fourfold woodwind and brass as indicated in the score, compared to the setting used in 1876 in the new *Festspielhaus* in Bayreuth with 87 strings, sevenfold woodwind and 26 brass instruments.

Table 14a: Typical size of orchestras – examples.

Year, place, *source*	vn	va	vc	db	fl	ob	cl	bn	hn	Trbn	tpt	bc	instr.
1658, Venice, *T. Giovanni e Paolo*	3	1	1	0	0	0	0	0	0	0	0	5	**10**
1670, Paris, *Opéra,* Lully	10	8	6	0	2	2	0	1	0	0	2	1	**31**
1781-1808, Vienna, *Burg-theater,* orchestra	12	4	3	3	2	2	2	2	0	0	2	0	**35**
1876, Bayreuth, *Ring score*	32	12	12	8	4	4	4	3	8	5	4	0	**104**
1876, Bayreuth, *Ring orchestra*	44	16	16	11	7	7	7	7	12	7	7	0	**147**

Such authors as Johann Joachim Quantz (*Versuch einer Anweisung die Flöte traversière zu spielen,* 1789), Francesco Galeazzi (*Elementi teorico-pratici di musica,* 1791), Heinrich Christoph Koch (*Musikalisches Lexikon,* 1802), Ferdinand Simon Gassner (*Dirigent und Ripienist für angehende Musikdirigenten, Musiker und Musikfreude,* 1844) and Hector Berlioz (*Grand traité d'instrumentation et d'orchestration modernes op. 10,* 1843) indicate common or desirable numbers of instruments per part (table 14b). Still, in each case one must assess where such information is appropriate and define general limits regarding the time and place of performance. Sizes of ensembles for certain music centers are obtained mostly from documents on specific performances and are bound to certain music genres, such as the usual orchestra size for operas at the *Burgtheater* or the much larger number per part for oratorios with choir at the same theater, for instance, Haydn's *Ritorno di Tobia* (28 March 1784), *Sieben Worte* (April 1798) or *Schöpfung* (19 March 1799). These large ensembles were not tied to Haydn's choir compositions only.

Table 14b: Orchestras, common or desirable numbers per part – examples.

Year, place or source, *event*	vn	va	vc	db	fl	ob	cl	bn	hn	trbn	tpt	bc	instr / v.
1752, Quantz	4	1	1	1									**7**
1752, Quantz	6	1	1	1				1					**10**
1752, Quantz	8	2	2	2	2	2		2					**20**
1752, Quantz	10	2	3	2	2	2		2					**23**
1752, Quantz	12	3	4	2	4	4		3	1			3	**36**
1781–1808, *Vienna Burgtheater* (VBT)	12	4	3	3	2	2	2	2			2		**35**
1784.03.28, VBT *Ritorno di Tobia*	40	6	5	6	2	7		6	6	2	2	1	**88 / 65**
1791, Galeazzi	4	1	1										**6**
1791, Galeazzi	8	2	1	1									**12**
1791, Galeazzi	16	4	3	3									**26**
1798.04.01, *VBT Sieben Worte*	41	9	8	8	2	6	2	6	4	4	4		**95 /57**
1799.03.19, *VBT Schöpfung,* 1stperf.	36	12	11	11?	6	6	6	6	6	4	4	1	**113 /63**
1802, Koch, *church, theater*	8	2	2	2									**14**
church, theater	10	3	3	2									**18**
Symphony	12	4	4	3									**23**
1844, Gassner	8	2	2	2									**14**
1844, Vienna, *Concerts. Spirituels*	16	4	4	4	2	2	2	2					**59/ 80**
1844, Berlioz,Traité, *(earlier for Mozart)*	17	6	7	6	2	2	2	2	2		2		**48**
today (f. Beethoven)	32	12	12	8	2	2	2	2	4	3	2	3	**92**
orchestra, best	41	18	15	10	4	3	3	4	4	4	2	4	**115**
orchestra, festival	120	40	45	36	8	12	15	12	16	14	8		**465/ 360**

The size of these ensembles is only partly related to room size. Data clearly show that the size of the symphony orchestra during the nineteenth century started to increase independently of the hall's size. In small halls the podium had to be enlarged, as, for instance, in the old *Gewandhaussaal* in Leipzig until the great hall in the new *Gewandhaus* was opened in 1884, finally covering over a third of the parterre. The growing need for space is reflected in the orchestra's numbers per part: a total of 29 in 1807, 39 in 1831, 70 in 1865 and 72 in 1881.

The construction of the great public halls in the nineteenth century was prompted mainly by the increasing number of concerts with large choirs and the need for halls for more than 1500 listeners. *Exeter Great Hall* in London (1831) and Birmingham *Town Hall* (1834) belong to the earliest examples of this category. In many cities the construction of great halls of category C was launched with considerable delay despite already evident necessity, as in Boston in 1863, in Vienna in 1870, in Leipzig in 1884 (see table 11b). In the absence of large halls churches had to be used with more or less adequate acoustics.[18]

The construction of large opera houses started earlier. In Italy the first great theaters with boxes of category C with more than 2000 seats already appeared in the eighteenth century: the *Teatro San Carlo* in Naples in 1737, *La Scala* in Milan in 1778 and *La Fenice* in Venice in 1792 (see table 12b). Similarly, opera orchestras were first enlarged in response not to hall size but to the change of orchestration. In opera history, too, the problem of too large orchestras for too small theaters and orchestra pits is well-known. Long before Wagner additional instruments had to be placed in the proscenium boxes or in the parterre.[19] In the nineteenth century orchestra pits started to be enlarged at the expense of the proscenium stage. Many large houses continued to hire small music ensembles only, with additional players if necessary. Oratorio performances in theaters made use of the stage as a large podium with raked risers, also taking advantage of the fact that theater boxes and galleries provided space for many listeners.

The greater number of listeners in the existing concert halls and opera houses and the subsequent increase of the hall's size changed sound perception due to changed room acoustics. This change is reflected in reports on musical events and in books on playing techniques of instruments and on instrument building. To cite just one important example, in 1803 a critic of the *Allgemeine Musikalische Zeitung* recommended violin virtuosos to have two violins, one by Stainer and the other by Stradivari, in order to better adapt to different halls. Further documents from that time show the increased distribution of violins based on Stradivari's model, which coped better with larger halls and theaters by producing more sound energy when the instruments were fitted with longer

18 As for oratorios in Zürich, see D. Baumann, *Ein Spätling im Goldenen Zeitalter* (1995).

19 D. Baumann, *Richard Wagners Festspielhaus* (1983); on the size of opera orchestras, see *NG Opera* 3, 726 f.; in *La Scala* the elimination of the proscenium stage to provide more space for the orchestra only happened in 1908.

bass bars, thicker sound posts, stronger chords and heavier bows.[20] The tension of strings increased until 1850. Sound energy also increased due to bow construction and, since 1914, to the spreading use of steel e-strings on violins.[21] At the same time, opera required better trained voices, first to cope with overcrowded small houses, then with the new larger houses. The use of the full voice in all registers is documented since the 1830s and a generally increased power of sound was much discussed after the First World War and became common after the Second World War. This was concomitant with changes in instrument building and playing techniques. The increase in the average playing level in concert and opera performance during the last 100 years has been preserved in sound recordings and confirmed by reports of musicians and concert goers.[22] The choir festivals in large churches and special halls, at some places performed only once, at others regularly repeated, played an important role in this complex process.

Collected data permit the conclusion that a balanced relation between ensemble size and performance space is bound to the development of public opera and concerts starting in the seventeenth but standardized only in the late nineteenth century. Before and even afterwards, relatively small ensembles continued to play in large spaces with strong reverberation, just as relatively large ensembles did in small, incredibly overcrowded halls by modern standards. These facts, thus far only partly addressed in research, shed new light on many aspects of the history of performance practice. The two examples given here introduce the topic, to be treated more extensively in the second part of this study.

The numerous, well documented performances under Corelli in the late seventeenth and the early eighteenth century in Rome in the acoustically favorable church *S. Lorenzo in Damaso* (figure 67) show ensembles of widely differing sizes, from small to middle to large: the total number of strings varied between 6 and 50, and violins between 4 and 30, with 1 to 6 basso continuo instruments and on special occasions 2 trumpets and sometimes 1 trombone (see table 14c). Corelli's numbers per part were obviously standardized neither in relation to room size nor to the event. The ensemble's size was determined by

20 *AMZ* 4 (1802/03), 773: "Jeder Virtuos sollte eigentlich Geigen von beiden Gattungen [Stainer und Stradivari] haben, damit er jederzeit diejenige, die für das Lokale passt, zu seinem Gebrauch wählen könnte."

21 D. Gill, *Book of the Violin* (1984), 42.

22 We are aware that these general considerations call for further investigations based on early sound recordings and personal recollections going beyond the limits of this study. We only refer to important evidence given by a former viola player of the orchestra of the *Tonhalle Zurich* recollecting that rehearsals were always played at reduced loudness in order to avoid hearing stress for the musicians; cf. J. Frei *Die Gehörbelastung des Orchestermusikers in der Konzert- und Opernformation der Tonhalle Zürich* (1979).

different factors, such as the feast's liturgical level, the structure of the composition, available musicians, but also the social standing of listeners.[23]

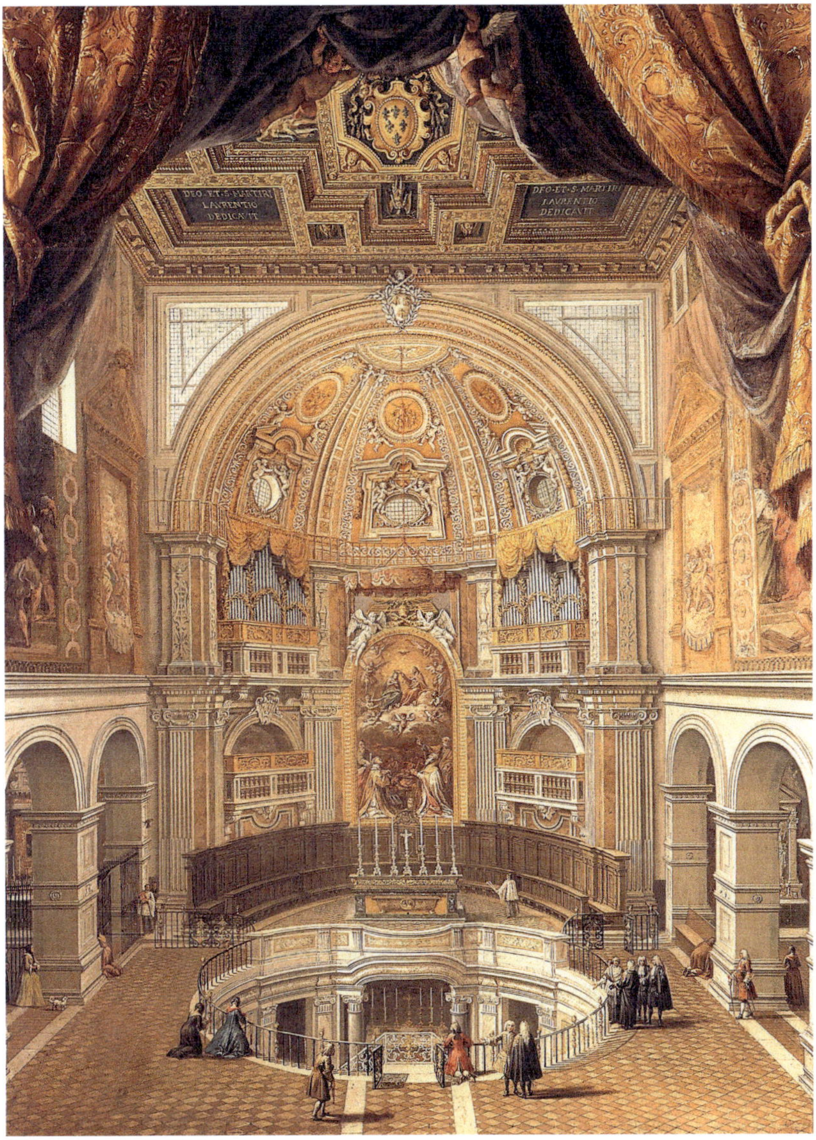

Figure 67: Rome, *Palazzo della Cancelleria, San Lorenzo in Damaso* (15th century): Choir with musician's benches, demolished in the 19th century.

23 See 2.2.3 for Corelli's performances in Rome; see H. J. Marx, *Corelli Gesamtausgabe* V (1976), preface, and "Die Musik am Hofe Pietro Kardinal Ottobonis unter Arcangelo Corelli" (1968), 104–77.

Table 14c: Performances under Corelli in Rome, *San Lorenzo in Damaso.*

Year, place, *event*	vn1	vn2	va	vc	db	tbn	tpt	bc	instr	v.	total
1690.02. *concerto*	9	9	7	7	7	0	2	5	46		
1690.02,	6	6	4	4	4	0	0	4	28		
1691	10	9	5	8	5	1	2	2	42	16	58
1692, *40 hore*	2	2	0	1	1	0	0	1	7		
1692.08.10,	9	8	4	7	4	0	2	3	37	19	56
1693, *oratorio*	6	6	4	6	3	0	0	0	25		
1693	8	8	4	5	5	1	2	2	33		
1694, *40 hore*	3	3	2	2	2	0	0	1	13		
1694, *40 hore*	8	7	4	3	4	0	0	1	27		
1693, *festa S. Lorenzo*	2	2	2	0	0	0	0	1	7		
1695, *40 hore*	9	16	4	6	4	0	0	1	42		
1696.08.10. *festa S. Lorenzo*	2	2	0	2	2	0	0	1	9		
1697, *40 hore*	7	7	4	5	5	0	0	1	29		
1698, *oratorio, Chiesa nuova*	14	13	6	8	9	0	0		50	38	88
1699.12.25. *la notte di Natale*	15	15	6	7	6	0	2	6	57		
1700, *40 hore*	15	14	7	6	8	0	0	1	51		
1704.08.10. *festa S. L.*	8	7	5	7	5	0	0	0	32		
1705.08.10. *festa S. L.*	11	10	3	8	4	0	0	0	36		

Another very impressive example from early nineteenth-century Vienna concerns Beethoven and the relation between the performance rooms and the numbers per part used for his symphonies (table 14d).[24] Up to the *Seventh Symphony* Beethoven's choice of rooms and numbers of musicians was very restricted partly because the few adequate public halls and the best musicians and dates were reserved for the performance of Haydn's *Creation* from 1799 to at least 1810, since 1801 alternating with the *Seasons*.[25] The size of the orchestra used for the *Eroica* (see table 14c) in the private performances in the hall of the *Palais Lobkowitz* since 1804 (cubic space 870 m^3) and in the first public performance at the *Theater an der Wien* (cubic space 5200 m^3) with 17 to 21 strings was only slightly larger than the minimal size Beethoven indicated in a letter of October 1811 to archduke Rudolf of Habsburg for the performance of his symphonies in his palace: "Concerning the number of instruments for the sympho-

24 See also D. J. Koury, *Orchestra* (1986), 118 and 327.

25 See appendix, table B; H. C. R. Landon, *Haydn,* vol. 5, (1977), 288.

nies I ask for at least 4 violins, 4 secund, 4 prim, 2 contrabasses, 2 violoncellos," in other words a total of at least 14 strings (4-4-[2]-2-2).[26] This minimal number of strings is still indicated by Gassner in 1844, Hofmusikdirektor in Karlsruhe, while in the same year Berlioz in Paris already refers to 36 strings as "earlier normal" and to 64 as "necessary for Beethoven".

During Beethoven's lifetime only *Wellington's victory* was performed with a large orchestra in concerts organized by Johann Nepomuk Mälzl in 1813/14 in Vienna. These concerts were given in halls of category C, namely in the *Aula* of the old University (cubic space 3920 m^3)[27] and in the great *Redoutensaal* (10440 m^3),[28] then thought to be too large for music (for measures, see table 11b). In this case Beethoven asked for "2 Harmonies" or double woodwinds and French horns (4-4-4-4) and a respectively larger number of strings, namely "18 first violins, 18 second, 14 violas, 12 violoncellos, 7 contra-basses, 2 contra-bassons," as Beethoven noted,[29] according to the principle: "The larger the hall the more players."[30] The same number of players was also used for symphonies number 7 and 8 performed in the same concert. Yet one cannot conclude that since the Seventh Beethoven would have insisted on great halls and a great number of strings with "double harmony." It is interesting to read Prince Lichnowsky recommendations for the first performance of the Ouverture *Die Weihe des Hauses,* the *Kyrie, Credo* and *Agnus* from the Mass in D and the *Ninth Symphony* given in 1824 to the deaf composer, who had relied on his friends for such questions at least since the beginning of his Conversation Books in1818:

> Lichnowsky: "It is right that the orchestra be doubled, but superfluous to engage more than are necessary; after Schuppanzigh and Umlauf know what is at their service at the *Wiedener Theater* we can tell what is needed" – Schindler: "Lichnowsky says that a smaller orchestra is more effective at the Theater-an-der-Wien than a large one in the Redoutensaal. [...] You will not have to pay the forces at the Theater-an-der-Wien at all – so that may be deducted."[31]

26 "Zu der Besetzung der Sinfonien wünsche ich wenigstens 4 Violinen, 4 Secund, 4 Prim, 2 Contrabässe, 2 Violonschell," see A.C. Kalischer, *Beethovens sämtliche Briefe* (1906), 1, II, 46; F. Prelinger, *Ludwig van Beethovens sämtliche Briefe und Aufzeichnungen* (1907), I, 253, Nr. 191. Beethoven does not mention the number of violas, but it seems clear that this would also be two.

27 *Zeitung für die elegante Welt,* 25. 12. 1807: "Die Stimmen verlieren sich in dem Saal."

28 *Allgemeine musikalische Zeitung,* 15. 5. 1805: "Ungünstig gross für Musik."

29 E. Forbes, *Thayer's Life of Beethoven* (1967), 576.

30 "Je grösser der Saal, desto mehr Spieler." See L. van Beethoven, *Wellington's Sieg,* ed. by H.-W. Küthen (1974), *SW* II, Bd. 1, 124: Beethovens "Bemerkung für die Aufführung"; see also A. Riethmüller, "Wellingtons Sieg op. 91," *Beethoven, Interpretationen seiner Werke* 2 (1996), 39.

31 English translation Forbes, *Thayer's Life of Beethoven* (1967), 902. German original, see A. W. Thayer, *Ludwig van Beethovens Leben,* H. Riemann (ed.), vol. 5 (1908), 75: *Lichnowsky:*

Table 14d: Performance of Beethoven's symphonies – examples.

Year, place, *event*	vn	va	vc	db	fl	ob	cl	bn	hn	trbn	tpt	timp	instr. /v.
1804.06, *P. Lobkowitz, Eroica* reh.	7	2	2	2	2	2	2	2	3	0	2	1	27
1805.01.23, *P. Lobkowitz, Eroica*	10	3	2	2	2	2	2	2	3	0	2	1	31
1805.04.07., *Th. a.d. Wien, Eroica*	11	3	2	2	2	2	2	2	3	0	2	1	32
1807.12, aula old University, *Eroica, Coriolan*	25	7	6	4	2	2	2	2	3	0	2	1	56
1813.04.30., archduke Rudolf, *7th* rehearsal	8	4	2	2	0	0	0	0	2	0	0	0	18
1813.12.08., aula old University *7th, Wellington's Vict.*													100
1814.02.27., gr. *Redoutensaal, 7th, 8th, Wellingt.*	36	14	12	7	4	4	4	4	4	0	4	1	94
1815, GdM, *2nd, 4th, 7th, Egmont*	40	12	10	8	4	4	4	4	4	0	4	1	105
1824.05.07., *Kärtnerthortheater, 9th, Missa Solem.*	24	10	6	6	4	4	4	4	4	2	4	1	73 /84
1824.05.23., gr. *Redoutensaal, 9th, Missa Solem.*	28	10	6	6	4	4	4	4	4	2	4	1	77 /84
1828, Paris, *Conservatoire, Eroica* (Habeneck)	32	8	12	8	4	3	4	4	2	3	0	4	85

Undecided between the different options, Beethoven later even proposed the 500-seat *Landständische Saal,* but Lichnowsky warned: "Concerned about your well-being I have to ask you why you could accept to give your Academy in such a small hall where every effect will be lost."[32] Further discussions be-

"Das Orchester doppelt besetzen ist recht, aber überflüssig mehr zu nehmen als nöthig ist, wenn Schuppanzigh und Umlauf wissen was Ihnen am Wiedener Theater zu Gebote steht, so ergibt sich daraus was noch gebraucht wird – dann um – *Schindler:* Lichnowsky sagt, dass an der Wien ein weniger besetztes Orchester mehr effektuirt, als ein grosses im Redoutensaal. [...] Das Personal an der Wien überhaupt bezahlen Sie ja nicht, also das fällt schon weg."

32 Thayer, *Beethoven* (1908), vol. 5, 77: "Bloss für Ihr Wohl besorgt muss ich Sie doch fragen, wie Sie denen Gründen Ihre Akademie in einem so kleinen Lokale zu geben nachgeben konnten, wo aller Effekt verloren gehen muss."

tween Beethoven and his consultants show the complex reasons for the change of performing conditions in Vienna in these years: The *Theater an der Wien* provided the small but professional orchestra of the house; the public *Landständische Saal,* opened in 1822, was too small, had insufficient acoustics, and the aristocratic listeners did not come.[33]

This example calls for cautious evaluation of information on acoustics. In this respect a comparison with other music centers is helpful. The usual performance practice of Beethoven's symphonies with large orchestra and double woodwinds since the late nineteenth century, often referring to these Vienna performances in the great *Redoutensaal,* dismisses the fact that the most discussed performances of these works were given at the *Conservatoire* in Paris by a large symphony orchestra with "double harmony" (in 1828 with 1+15-16-8-12-8 strings, 4-3-4-4 woodwinds, 2-3 french horns, 4 trumpets and timpani). The excellent hall belongs to category B and in modern terms is a chamber music hall.[34] In Leipzig and London halls of the same category not only symphonies were performed but also works nowadays considered utterly inappropriate for a "small hall." For example, Brahms' *Deutsches Requiem* and his violin concerto were performed in the old *Gewandhaus* in Leipzig, in 1879 and 1869 respectively; the performance of Schumann's *Das Paradies und die Peri* with Jenny Lind in 1856 in London's *Hanover Square Rooms* (figure 93) was highly problematic due not only to the large number of musicians (in addition to the usual orchestra there were 6 soloists and 80 choir singers) but also to the large public attracted by the presence of Queen Victoria and Prince Albert.[35] London critics complained about the dead acoustics as well as the lack of space that even impeded the bow movements of the string players.

1.9.6 Music and space: ideal and reality

Halls for music performance often were and still are not perfect. Composers and musicians have to adapt to conditions that are often unfavorable and determined by non-musical factors. Since we tend to compare our actual impression with our habitual experience, the evaluation of performances in earlier periods requires careful study of all the important parameters explained in chapter 1.5.4 on perception psychology. Discussions of historical performances tend to neglect the fact that earlier instruments with their relatively weak

33 Hanson (1985), 96.

34 A. Elwart, *Histoire de la Société des Concerts du Conservatoire* (1864), 98.

35 R. Elkin, *The Old Concert Rooms* (1955), 101–2; Koury, *Orchestra* (1986), 328.

ground tones, bright onset, and concise, rich sound, which were used for baroque, classical and even early romantic music and their respective playing techniques, cope better with intense reverberation than modern instruments with their quicker onset of lower partial tones. Especially quieter instruments, such as the lute, cembalo and viola da gamba, need the support of good quality reverberation. In the overcrowded small halls of the late eighteenth and nineteenth centuries these instruments reached the limits of their capacity. This was certainly a reason for changes in instrument building and performance practice. The use of these instruments dropped, and others, better adapted to the increased bandwidth of dynamics, such as the pianoforte and violoncello, gained territory. But another fact should not be forgotten, namely, that performances in great halls for many listeners and performers did not emerge in the nineteenth century but had taken place at any earlier time on special occasions. Without discussing details here it is clear that on such occasions acoustic conditions were often far from ideal. But up into the electronic age the very occurrence of such events is obviously more important than the question whether really all listeners can adequately hear the music. In this sense experience and social norms determine acoustic expectations.

These explanations warrant the question whether it is possible to define the ideal room for a work's performance. An appropriate answer can be given only in terms of holistic perception. After all, sound is experienced in our imagination. The evaluation of auditory impressions according to the constancy phenomenon plays an important part in the analysis of sense impressions. In this regard the definition of room acoustic conditions for a musical performance is simple: real space should not disturb but support this process of perception. The detailed acoustic conditions are determined by the structure of the work, the instrumentation and the number per part of the ensemble. But the assessment of these conditions depends also on the experience of the composer, the musicians and the listeners. Therefore, the history of performance practice must also assess the usual sounding appearance at the time of the composition in an adequate room with respective dimensions and décor, the then used instruments and usual number per parts. However, although the former sounding appearance in the original hall can never be replicated exactly, a good musician gains precious experience from such historical performance conditions in order to perform a work appropriately. Experience with historical instruments in a room that corresponds to past conditions also allows musicians at least to approximate the sounding imagination of the composer with modern instruments in modern halls.[36] Such experiences are important partly because only scores writ-

36 Actual performance practice shows all levels of approach to historical conditions such as exact copies of historical music instruments in historical rooms for instance in Norrington's performance and recording of Berlioz's *Symphonie fantastique* in the hall of the old *Con-*

ten since the nineteenth century contain precise information concerning the performance room's acoustics and the instrumental aspects of the work's sounding appearance. An adaptation to changed room acoustics may therefore be required, such as the increase of the total number of music instruments and of the numbers per part for a larger room or, possibly, only a change of room related playing details. Finally, this decision depends on artistic and esthetic factors. But musicians always take room acoustics into account, even if unconsciously. The best interpretation is not the closest reconstruction of past conditions but a masterly control of the instrument and playing based on artistic sensitivity and experience, adapted to the best possible hall and the structure of the composition with the appropriate number of instruments per part.

1.9.7 Room acoustic quality norms and their practical importance

From the viewpoint of architects and room acoustic consultants a question still remains: is it possible, despite the caveats and complex relations under discussion, to give recommendations for room acoustic conditions? In building construction physical norms are defined by clearly determined measuring procedures in order to attain certain values. In room acoustics, too, the recommended values have been based on measurable parameters since Sabine. Depending on the test method, these recommendations have shown considerable differences.[37] As explained, since Beranek's investigations research has expanded in order to better relate room acoustic quality to the assessments of the psychology of perception, as well as to music style and playing conditions.[38] Acoustic publications recommend values for different uses, such as speech, chamber music, symphonic music, opera, church music (oratorio, organ music), which are based on reverberation time T and the hall's cubic space (figure 68). Some authors give recommendations for certain styles of music, such as longer reverberation times and larger volumes of space (*Concertgebouw* Amsterdam) for

servatoire in Paris, or Harnoncourt who started with exact reproductions and later began to use historical playing technics on modern instruments in usual concert halls.

37 See Bagenal/Wood (1931), 116; W. Kuhl, *Acustica 4* (1954), 618 and *Acustica* 8 (1958), 280 (based on 4 sound recordings with classic, romantic and "modern" music); G. C. Izenour, *Theatre Design* (1977); W. Gabler, *Historische Räume* (1989); Fasold (1987), 259.

38 Beranek (1971), M. Forsyth, *Buildings for Music* (1985), 250 f.; S. Bradley, "The Evolution of Newer Auditoriums Acoustics Measures" (1990), 13–23; J. Meyer, *Acoustics* (2009), 203 f.; L. L. Beranek, *Concert and Opera Halls* (1996).

Romantic orchestral music, and halls with shorter reverberation time *(Altes Gewandhaus Leipzig)* for baroque music and classical orchestral music.[39]

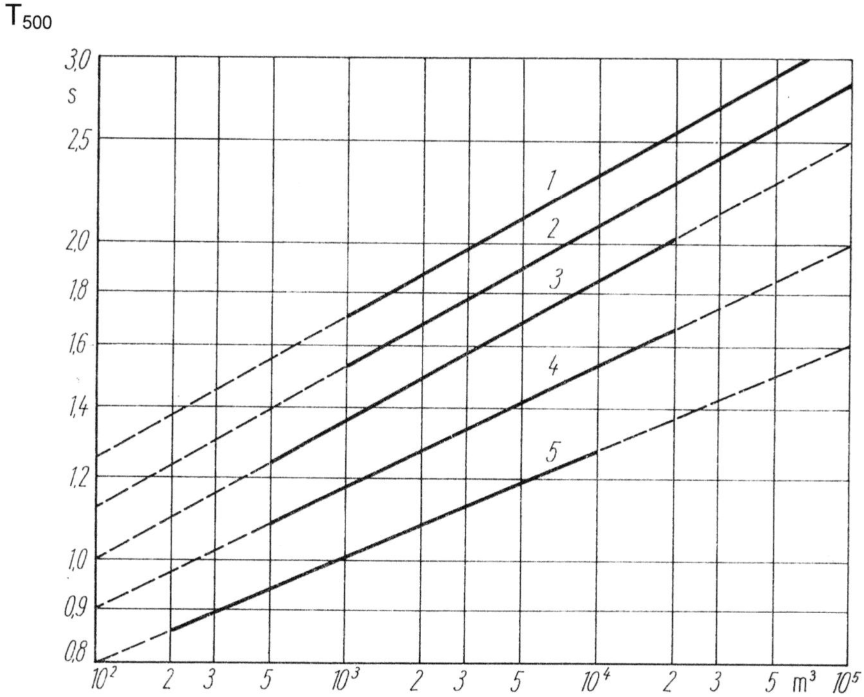

Figure 68: Recommended reverberation time T_{500} for different room categories according to volume of space (after Fasold): 1) hall for oratorios and organ music (1.7 to more than 3.0 seconds), 2) symphony halls (1.5 to more than 2.8 seconds), 3) halls for solo and chamber music (1.3 to 2.0 seconds), 4) opera, theater, multipurpose halls (1.1 to 1.7 seconds), 5) spoken theater, assembly halls, sports (0.85 to 1.3 seconds).

From the musician's and music historian's point of view the definition of standards remains problematic even if based on correct psychological and physical principles. Research has shown how varied the conditions of performance practice have been throughout history. Since the construction of public opera houses in the seventeenth century opera has developed certain standards, such as typical shapes and sizes of auditoria and orchestras, but these differed by region and changed at different periods. For concert life, too, certain standards of orchestra size and typical halls have existed since the late nineteenth-early twentieth century. At that time earlier music was adapted to the then current standards, if necessary even with changed orchestration. Today, when these works are successfully reset within past conditions, such as baroque music

39 Forsyth (1985), 17; Fasold (1987), 305; Meyer (2009), 250 f.

played on instruments of that period in historical halls, these performances are often based on insufficient information on the previously used rooms and their acoustic conditions.

Besides the question whether the given definitions sufficiently determine room acoustic properties in relation to the music played, the fact remains that such standards strongly affect our way of music perception if they are used in the construction of new halls, the renovation of old halls, the selection of rooms for sound recordings and the engineers' treatment of spatial aspects in sound recording. Standards applied to room design do not only influence the perception of single musical events but change the acoustic background of whole groups of listeners. These standards are based on tests, but the tests' results inevitably depend on the individuals' experience, which relies on their habits and comparison with their latest acoustic impressions before the test.

Our perception tends to create norms in order to be more reliable. Norms are usually based on the most important and frequent impressions. We tend to qualify unusual situations as "wrong" or – consciously or unconsciously – at least as "worse".

Acoustic recommendations for the design of music rooms must be based on physical, perceptual and architectural parameters. Current scientific knowledge clearly indicates that in the design of appropriate halls or the choice of historical halls for performance or sound recordings and films the values for reverberation time T related to the cubic volume of space must be accompanied by recommendations for room shape, mainly width and height, and interior décor. To analyze how room acoustics were perceived this study investigates not only these physical and architectural elements but also performance rooms and conditions for specific musical works.

Each architectural space is unique. Using shape, surface structure and materials, the architect and specialists create a whole. The design and construction process of a music room is so complex that the successful result is comparable to a work of art. In music performance and recording, too, the selection of a room and the approach to room acoustics is based, finally, not only on rational rules; it is also a creative process informed by the holistic perception of those involved.

From the musicians' viewpoint the best performance in bad acoustics is not impossible but rare. On the other hand, it is beyond dispute that good acoustic conditions support good musical results. If perceptual information and music history are included in the assessment of room acoustic quality, recommended values are helpful. Room acoustic characteristics obviously admit boundary conditions and band widths within which musicians can work well. Musicians use their own language in order to define these conditions. This study aims to find more ways to translate these terms into real, visible, maybe even measurable room acoustic characteristics and, thus, to assess the spatial aspect of a musical performance and the composer's view on appropriate rooms for his works. In the second part of this book Handel's oratorio *The Messiah* will serve as reference point for such investigations from the perspective of music history.

Part 2

Handel's *Messiah*

2.1 The change of performance practice and room acoustics

2.1.1 Handel's *Messiah:* an uninterrupted performance tradition

The English oratorio *Messiah* holds a special position both among Handel's compositions and in the history of performance practice. During Handel's lifetime *Messiah* was among his best-loved works. The list of his choir works' performances (see table B in the appendix) shows only six oratorios and the *Masques* with over 15 performances conducted by Handel himself: *Saul* (17), *Deborah* (20), *Acis and Galathea* (22), *Samson* (26), *Esther* (30) and *Messiah* (41, of which five by the blind composer).[1] This is impressive even compared to Handel's most successful operas – those with 19 or, if resumed later, with over 30 presentations,[2] considering that the *Messiah* performances were distributed over the years 1742 to 1759. During Handel's lifetime some 40 *Messiah* performances were documented by other organizers in England and Ireland. The number of such performances of Handel's choir compositions and especially of *Messiah,* often documented only by chance,[3] was certainly higher.

Messiah belongs to the few late Baroque compositions that continued to be performed practically without interruption from the first performance until the twentieth century. None of Handel's other oratorios and no other contemporary choir composition garnered so much success over such a long period. Handel's *Messiah* outdoes even Johann Sebastian Bach's choir works, which were performed mainly within the circles of Bach's students and their students but started to be more widely performed only since Mendelssohn's performance of *St. Matthew's passion* in 1829.[4] Following the first performance of selected parts of *St. Matthew's Passion* in 1837 at the *Birmingham Festival* and in London and a complete performance in the London *Hanover Square Rooms* in

1 For a survey, see W. Dean, *Handel's Dramatic Oratorios and Masques* (1959), 640. Table B in the appendix lists also performances not yet known at the time of Dean's publication, see also *HHdb.* 2 and *HHdb.* 4.

2 A. Hicks, "G. F. Handel, work list", *NG* 8, 114–6: for instance *Admeto:* 19 performances in January 1727, 6 in September 1727, 3 in May 1728 and 6 in December 1731.

3 See O. E. Deutsch, *A Documentary Biography of Handel* (1955) and enlarged ed.: *Händel: Dokumente zu Leben und Schaffen, Händel-Handbuch,* vol. 4, (1985), cited as *HHdb.* 4.

4 See H.-J. Hinrichsen, "'Urvater der Harmonie ?' Die Bach-Rezeption", *Bach-Handbuch,* ed. Konrad Küster, Kassel (1999), 31–67, with bibliography.

1854, a Bach tradition emerged in England only in 1884, with regular performances of the Passions in *St. Paul's, St. Anne's* and *Christ Church* in London.[5]

During Handel's lifetime *Acis and Galathea* was also often performed in public concerts by other organizers. Another group of works brought the composer early immortal fame – his sacred compositions for state ceremonies: the *Utrecht Te Deum* and *Jubilate* for soloists, choir and orchestra, which, after its splendid first performance in *St. Paul's* in July 1713, was performed there yearly on St. Cecilia's day during the charity service of the *Festival of the Sons of the Clergy* and regularly in other English cathedrals at similar charity services; the famous *Coronation Anthems,* 1727, which was composed for the coronation of King George II as church anthems and parodied parts of Handel's oratorios *Esther* and *Deborah,* belonged to the favorite pieces for charity and subscription performances; the *Dettingen Te Deum,* which since 1743 replaced the *Utrecht Te Deum* once or twice a year in *St. Paul's* and at other charity festivals in London and other cities of the kingdom. The following study of performances of Handel's choir compositions offers evidence of the exceptional performance tradition of *Messiah* during Handel's lifetime and after his death, a tradition directly related to church charity festivals that were so important for the development of public music life in eighteenth-century England.

2.1.2 Important performances

A survey on important performances of Handel's *Messiah* from its first performance to the twentieth century shows that different types and categories of rooms have been used.[6] Each category has its characteristic room dimensions, design, social background and organization of the event. All these parameters influence the musical performance directly or indirectly.

The first performance of *Messiah,* with the best singers of the choirs of Dublin's three main churches, was on 13 April 1742 in one of the first public concert halls in the English speaking part of Europe, *Neal's Music Hall,* opened in 1741 (figure 77).[7] Following the great success Handel organized another performance on 3 June 1742. It has since been a lasting tradition to perform *Messiah* each December in the same hall.

5 N. Temperley, "Bach Revival", *NG* 1, 883 f.; P. A. Scholes, *The Mirror of Music* (1947), vol. 1, 87.

6 For sources see table B in the appendix.

7 M. Forsyth, *Buildings for Music* (1985), 51, figure 2.23.

After its first performance in 1749 under William Hayes in Oxford's *Holywell Music Room* (figure 74), *Messiah* was performed there yearly and since 1754 alternately at the University's *Sheldonian Theatre* (figure 73). *Holywell Music Room* of the *Oxford Musical Society* opened in 1748 with a performance of Handel's oratorio *Esther*.

Back from Dublin, Handel performed *Messiah* in London and his other oratorios with singers of the choirs of the *Chapel Royal, St. Paul's* and *Westminster Abbey* either at *Covent Garden Theatre* (in 1743 and yearly during lent since 1749, figure 80) or at the *King's Theatre* (two performances in April 1745, figure 79). The performance was on stage but not acted, and the choir, orchestra and soloists were placed on amphitheatrically raked risers reaching from the stage down into the orchestra pit. This tradition was preserved up to the nineteenth century (figure 85).[8]

With the building of the *Foundling Hospital* and its chapel, Handel had access to an appropriate room for the performance of his choir compositions and instrumental works. On 1 May 1750 an overwhelming charity performance of *Messiah* was held to inaugurate this 1200-seat chapel, although neither the interior decoration nor the organ were ready (figure 78).[9] Since then *Messiah* was regularly performed at Easter as charity for the hospital, mostly with the same choir singers and musicians as at the opera. Until 1753 the performances where conducted by Handel, and subsequently, because of his blindness, as well as after his death in 1759, by his pupil John Christopher Smith Jr. until 1768. The tradition was continued by John Stanley and later by Felice de' Giardini, until it stopped for unknown reasons in 1777. John Christopher Smith Jr. also conducted the yearly performances at *Covent Garden Theatre* from 1754 until 1774.

Already during Handel's lifetime *Messiah* performances were organized by church choirs in other English cities, at first in the church's assembly hall. In Salisbury, where the *Anniversary Music Festival* in October and St. Cecilia's feast in November were combined with Handel performances, the first performance of *Messiah* was initiated by Handel's friend James Harris as an *Evening Concert* on 4 October 1750 for the consecration of the organ in the cathedral's new *Assembly Hall*. On 27 and 28 September 1752, during the *Festival,* the organist John Stephens dared to perform parts from *Messiah* as *Anthem* at the two morning services after Handel's *Te Deum* and *Jubilate,* with the Corona-

8 W. Salmen, *Konzert* (1988), fig. 95: oratorio performance in *Covent Garden Theater* 1808.

9 The concert should be held for the consecration of the organ which was not yet ready because the chapel was not finished. *HHdb.*4, 438; see also 2.5.1: Handel's concerts at the *Foundling Hospital.*

tion Anthem *Zadok the Priest* sung at the end of the service.[10] Further performances took place during the *Three Choirs Festival,* since 1716 held alternating in Gloucester, Worcester und Hereford.[11] Every year the choirs of the three Cathedrals sang Purcell's *Te Deum* and Handel's *Utrecht Te Deum* (later the *Dettingen Te Deum)* "in the cathedral manner", after the model of the London festival, in *St. Paul's* as charity for the orphans of the poor clergies' dioceses.[12] The regular performance of Handel's oratorios, odes and serenades in den *Evening Concerts* of the *Festival* is documented since 1738 (*Esther,* 1738 in Worcester, *Alexander's Feast,* 1739 in Gloucester, *Athalia,* 1743 in Worcester, *Acis and Galathea,* 1745 in Gloucester).[13] *Messiah,* first performed at the *Three Choirs Festival* in September 1750 in the *College Hall* of Hereford, was performed since 1759 each year during service in one of the three cathedrals. The performance was held in the church choir with more choir singers than usual but with an orchestra of normal size. Thus, the music could benefit from the choir stalls and the organ screen that shields this part of the church from the often problematic acoustics of the rest.[14] In April 1758 and on 10 Mai 1759, still before Handel's death but in his honor, the *Hallelujah* from *Messiah* (then "For the Lord God omnipotent") was performed for the first time in the same manner in *St. Paul's* choir in London during the *Feast of the Sons of the Clergy* (figure 65)[15] – after the ouverture of *Esther*, the *Dettingen Te Deum* and *Jubilate* and the Coronation Anthem *Zadok the Priest.* Such events, not attended by Handel and slowly slipping out of his control, show an increasing distance from his own performance practice of *Messiah.*

For these performances not only the number of choir singers but also of instruments was increased. At Winchester *Cathedral,* where *Messiah* was played every year from 1770 to 1783, the participation of 50 choir singers and 60 instrumentalists is documented.[16] The *Handel-Commemoration* in 1784 on the occasion of the composer's 100st birthday, held in *Westminster Abbey* with permission of the King,[17] exceeded all former performances: the architect James Wyatt built a platform for the more than 500 performers (figures 88 and

10 *HHdb.* 4, 461. The performance tradition of Handel's *Te Deum* und *Jubilate* at this festival started in November 1714.

11 Scholes (1947), 151.

12 *HHdb.* 4, 129.

13 Not 1739, see *HHdb.* 4, 311.

14 W. Shaw, *Messiah,* 1965, 61.

15 *HHdb.* 4, 515: in April 1758 also at *St. Margaret's* and *St. Anne's* charity performances were held with the same program; Deutsch, *Handel*, 1955, 825; the tradition of performances in the choir of *St. Paul's* lasted until 1854, see Scholes (1947), 177.

16 Dean (1959), 103.

17 According to the English calendar old style new year started on 24 March. Handel's birthday 23 February therefore in England belonged to 1684 and not 1685.

89).[18] This event was repeated yearly until 1791 with a steadily increasing number of performers,[19] and other *Handel-Festivals* also started to use larger ensembles with a growing number of amateur instrumentalists and choir singers. The climax of this *gigantomania* was reached at the *Handel-Memorial* in *Crystal Palace,* which was transferred to Sydenham in 1854.[20] In 1857 a rehearsal for the *Handel Centennial* of 1859 was held under Michael Costa with 2000 singers and 397 instrumentalists. These numbers increased in 1859 to 2765 singers and 453 instrumentalists (figure 92).[21] Since then this festival was held every three years until 1926, with the number of performers reaching the maximum of 4,500 in 1883.

In the meantime the ensemble size also increased at choir festivals, and the performances consequently had to be held in the cathedrals' main nave, as shown on a picture of the *Yorkshire Festival* on 29 September 1823 in *York Minster:* 465 musicians (180 instruments and 285 singers, among them many working people) performed before 5,000 listeners.[22] At *Hereford Cathedral, Messiah* was performed in 1834 for the first time in the main nave.[23]

On the continent Michael Arne conducted the first performance of *Messiah* in Hamburg in 1772 in English, and Carl Philipp Emanuel Bach in 1775 in German. The latter marked the beginning of a specific performance tradition in German and sometimes in Italian, first in halls and soon also in churches, also with increasingly larger ensembles, mainly at choir festivals. For the *Niederrheinisches Musikfest,* since 1812 held alternately in Elberfeld, Düsseldorf, Cologne and Aachen, and since 1819 regularly with choir compositions by Handel and mainly *Messiah* in German, the number of performers was usually 200 to 600.[24] Abbé Vogler reported on his own performance of *Messiah* in Italian on 1 November 1777 in Mannheim, which was critically commented by W. A. Mozart.[25] Hiller also reported on his own performance in Italian at the

18 Ch. Burney: *Account of the Musical Performances in the Westminster Abbey and the Pantheon May 26th, 27th, 29th and June the 3rd and 5th, 1784: in Commemoration of Handel,* London 1785, 8, further cited as Burney, *Account* (1785).

19 Ch. Hogwood, *Handel* (1984), 249.

20 1851 constructed by Joseph Paxton at Hyde Park, 1854 dismantled and reconstructed in Sydenham with an additional transept and with its own train station.

21 Scholes (1947), 178.

22 See Hogwood (1984), 151, Plate IX; "York"; *NG,* vol. 20 (1980), 575 by mistake gives 1829 as festival year.

23 W. Shaw, *A textual and historical companion to Handel's Messiah* (1965), 61.

24 R. Schaal, "Feste und Festspiele", *MGG* 4 (1955), 104–28; H.-Chr. Mahling, "Zur Aufführung und Stellung der Oratorien Händels im 19. Jahrhundert" (1987), vol. 1, 264 f.; J. Müller-Blattau, "Händel", *MGG* 5 (1956), 1274 f. with survey on Handel performances in Germany.

25 M. Seiffert, "Die Mannheimer 'Messias'-Aufführung 1777" (1916), 61 f.; W. A. Mozart, letters from 31 October and 4 November 1777, *Briefe und Aufzeichnungen* (1962–63): vol.

Berliner Dom on 19 Mai 1786 with 308 performers and Carl Friedrich Zelter as violinist. In 1786 and 1787 he conducted the German version in Leipzig and in 1788 at the *Maria-Magdalena Church* in Breslau with double orchestra and a total of 259 performers.[26]

Vienna, too, developed a Handel and *Messiah* tradition, beginning in 1778 with the performance of selected choir numbers from *Messiah, Alexander's Feast, Israel in Egypt, Acis and Galatea, Samson* and *Ode for St. Cecilia's Day,* in Italian translation, by the *Tonkünstlersocietät* at the *Kärtnertortheater* with about 100 performers. This trend was initiated by the Bach and Handel admirer Baron van Swieten, former ambassador in Berlin, who was called back to Vienna in late 1777.[27] On his request Mozart prepared versions in German with revised orchestration of *Acis und Galatea* in 1788, *Messias* in 1789, *Das Alexandersfest* and the *Caecilien-Ode* in 1790. The first complete performance of Mozart's version of *Messias* in Vienna in German translation by Ebeling, with revised orchestration according to contemporary taste, was held in March 1789 at the hall of *Palais Esterházy* at Schenkenstrasse.[28] Further performances followed in other palaces of the *Associierten Cavaliere.* The edition of Mozart's version, printed in 1803 by Breitkopf & Härtel in Leipzig (though with some changes in text and music), became during the nineteenth century the model for new performance versions in German speaking countries. Important events were the Vienna performances of *Alexandersfest, Samson* and *Messias* from 1812 to 1815 at the *Winterreitschule* (*Messias* in Mai 1815)[29] with 600 to 900 performers, explicitly following the model of the London *Handel Commemorations.* Not long before, in 1806, the correspondent of the Leipzig *AMZ* had referred to the great *Redoutensaal* as being too large for Salieri's performance of *Messiah.*[30]

In Romanic countries the *Messiah* reception began slowly and late: in 1784 La Mara,[31] the admired soloist of the London *Handel Commemoration* held the same year, sang *I know that my redeemer liveth* in Paris with modest success. In 1827 selected numbers of Mozart's version were performed, of which

2, 94f.; L. Finscher, "Zur deutschen Händel-Rezeption im 18. Jahrhundert" (1987), 271–83; idem: "Mozart und Mannheim" (1992), 72.

26 W. Siegmund-Schultze, "Über die ersten Messias-Aufführungen in Deutschland" (1960), 51–110, especially 64 f.

27 B. Edelmann, "Händel-Aufführungen in den Akademien der Wiener Tonkünstlersozietät" (1984), 178.

28 Not at the Hofbibliothek, see *Mozart und seine Welt in zeitgenössischen Bildern, NAW* X 32, ed. by M. Zenger and O. E. Deutsch (1961), 93.

29 A. M. Hanson, *The concert life in Biedermeier Vienna* (1985), 93; H. W. Schwab, *Das Konzert* (1971), fig. 61, 62.

30 *AMZ* 9 (1807), 336.

31 Gertrud Elisabeth Mara-Schmeling (1749–1833), not to be mistaken with the singer Marie Lipsius (1837–1927) also called La Mara.

mainly the *Hallelujah* won acclaim. A complete performance of *Messiah* is documented only in 1873. In 1900 the performance of Handel's compositions was so common that *Messiah* was chosen for the opening of the Paris World Fair.[32] In the U.S.A. the Handel tradition started in 1812 in Boston with the performance of several oratorio choirs conducted by George K. Jackson, a former choir member of the London *Handel-Commemoration* in 1784, who co-founded the *Handel and Haydn Society of Boston* in 1816.[33] In the late nineteenth century *Messiah* and other choir works (such as Haydn's *Creation* and Beethoven's *Ninth Symphony)* became festival music *per se*. In many cities several new great concert halls were opened with Handel performances, such as the *Neue Gewandhaus* in Leipzig (1884), the *Concertgebouw* in Amsterdam and the *Philharmonie* in Berlin (both in 1888).[34]

2.1.3 Aim of research

Handel's *Messiah* initially remained tied to the English speaking countries. The performances in German speaking countries, which first opened in the eighteenth century, and the those in Romanic countries, which began only towards the late nineteenth century, were related to considerably different social conditions and performance traditions. A more detailed study of these traditions would exceed the limits of this research.[35] Therefore, since the assessment of each performance must be based on the history of concert life and choir societies in the respective cities and on the general development of these institutions in the eighteenth and nineteenth centuries, the following research will concentrate only on the English speaking countries.

The uninterrupted performance tradition of Handel's *Messiah* from its first performance in 1742 to the twentieth century reflects the change of performance practice within this period. This change becomes evident already in the number of performers and in the selection of the performance room. In this respect it is important to know how far Handel's own idea of his work was based on a room category and a certain number of performers and how he determined

32 Hogwood (1984), 254.

33 H. Earle Johnson, "Handel and Haydn Society of Boston", *MGG* 5 (1956), col. 1291–4.

34 Forsyth (1985).

35 On the relation between choir compositions and music festivals in German speaking countries, see S. Weibel, *Die Deutschen Musikfeste des 19. Jahrhunderts im Spiegel der zeitgenössischen musikalischen Fachpresse* (2006); E. Lichtenhahn, "Das bürgerliche Musikfest im 19. Jahrhundert" (1987), 161–79; G. L. Mosse, *Die Nationalisierung der Massen* (1976); D. Düding, *Öffentliche Feste* (1981); Mahling (1987), Bd. 1, 264 f.

these elements. The answer to this question requires an analysis of some additional compositions by Handel, using comparisons of ensemble size and numbers per part and further investigations on the performance background (local practice, Handel's performance characteristics related to the music genre, the organizational background of the performances, adaptation to representational conditions of the performance, available soloists and instrumental players, and room size). The performance rooms of selected English performances from the eighteenth to the twentieth century and their influence on the change of performance practice are investigated through a comparison with Handel's own performance practice.

2.1.4 The transformation of performance practice

Preserved sources reveal that Handel himself did not consider any of the *Messiah* versions as final (see list of sources in the appendix, 3.4.1). The score was adapted to each performance by changing, eliminating or adding certain parts, mainly for soloists.[36] Handel's *Messiah* performances took place in music rooms, theaters and in the chapel of the *Foundling Hospital* in London. He himself never performed *Messiah* in a large cathedral or during church service. Only three choirs were adapted to liturgical works, one of them the unchanged *Hallelujah* as part of the *Foundling Hospital Anthem* HWV 268.[37] The following investigations will show when and how Handel adapted the ensemble size to the size of performance spaces.

Messiah performances in churches, which developed out of the tradition of the *Te Deum* at the festivals in *St. Paul's* and at the *Three Choir Festivals* with respective sizes of choirs, showed differences from Handel's own performance practice already during his lifetime. The sounding characteristics of *Messiah* changed principally with the *Westminster Abbey* performance during the *Handel Commemoration* in 1784. The choir and instrumental ensemble were not only five times larger than Handel's largest ensemble, but more instruments were added to fill the church with sound. This tendency to gigantic ensembles continued up to the performance in the huge transept of *Crystal Palace* in 1857, where thousands of listeners barely could hear the music properly. The work became an object of national pride and magically transmitted some of its symbolic grandeur to both musicians and the audience. The attraction was such

36 For a survey on the different versions and soloists, see J. P. Larsen, *Handel's 'Messiah'* (1957, 2 / 1972) and D. Burrows, *Handel: Messiah* (1991).

37 On the size of ensembles and spaces in the nineteenth century, see 2.7.5.

that the triennial performances ran up to 1926 in that building which in 1931 was destroyed by fire. This background also explains why this "incredible" hall and an oratorio by Handel were selected for one of the first live recordings in 1888.[38]

In this fight against mass excitement based on the development of choir concerts in England and on the continent a return to Handel's modest ensemble could happen only step by step: in 1868 a facsimile edition of the autograph of *Messiah* was published by the *Sacred Harmonic Society,* headquartered in *Exeter Hall* in London,[39] where most of London's oratorio performances took place in 1831–1860 (figure 94). Another facsimile of the autograph by the Deutsche Händel-Gesellschaft in Hamburg, with a foreword by Friedrich Chrysander, was published in 1892 to prepare the *Messiah* performance at the 1895 *Händel-Festspiele* in Mainz.

In 1874 the study of the autograph sources and early copies of *Messiah* and lists of payments from the archives of *Foundling Hospital* by William G. Cusins, music master of Queen Victoria, offered the first evidence on the surprisingly small ensemble for the 1759 performance at the *Hospital* – 5 soloists, 34 instrumentalists and 18 choir singers.[40] A 1899 performance at the London *Royal Albert Hall* for 6,000 listeners was billed as the "original version" despite the 100 instruments and 800 choir singers. In this new version based on the autographs and early printed editions, Bridge at least dispensed with the additional accompaniments" *favored* since 1784.[41] A 1902 performance in London's *Queen's Hall* (figure 96) for 2,000 listeners with an orchestra of 100 instruments featured a reduced choir of 65 singers. In 1906 August Manns finally performed *Messiah* in *King's College Chapel* in Cambridge (figure 62), using the exact numbers per part of the *Foundling Hospital* materials, but no other such performances followed.[42]

The movement of historical performance practice of old music began in the nineteenth century with the collection, reconstruction and use of old instruments. In Paris of 1832 François-Joseph Fétis was among the first dedicated to early music, followed since 1877 by Victor-Charles Mahillon in Brussels and since 1883 by the Belgian-born Arnold Dolmetsch in England. Since the 1920s and after the Second World War specialized soloists and ensembles have

38 *Israel in Egypt,* recorded on 30 June 1888 in the transept of *Crystal Palace,* conducted by Sir August Manns.

39 *HHdb.* 2, 193.

40 W. G. Cusins, *Handel's Messiah, An Examination of the Original and of some Contemporary MSS.* (1874), 14.

41 J. P. Larsen, "Georg Friedrich Händel – Oper und Oratorium, Gattungstypologie und Aufführungspraxis" (1988), 23; Scholes (1947), 181.

42 *Musical Times* 47 (1907), 608.

increasingly called attention to the importance of early music,[43] which got a further boost since the 1950s and 1970s.[44] Yet astonishingly, the performance practice of *Messiah* remained for a long time outside this return to small ensembles with old instruments.

After the Second World War England offered new conditions of music education that allowed further steps toward Handel's own work idea.[45] Thus, the reorganization of the study of musicology at English Universities significantly influenced music practice, and the availability of the autograph version and other sources of *Messiah* was improved after Queen Elizabeth II donated the *Royal Music Library* to the nation in 1957 on the occasion of the bicentennial of Handel's death.

In 1952 Otto Erich Deutsch published a documentary biography of Handel based on important contemporary documents on the composer's life that he collected during his exile in England since 1939. On the occasion of the Handel celebrations in 1959 Winton Dean published an important book on Handel's oratorios, followed by Jens Peter Larsen's study on *Messiah* and the new edition of score and parts by Watkins Shaw, who also published articles mainly on Handel's conducting score. John Tobin's edition of *Messiah* for the Hallische Händel-Ausgabe in 1972 does not include all the results of this research. From 1978 to 1985 the *Händel-Handbuch (HHdb.)* published the most complete documentation of Handel's life and works with a thematic index of his works ordered by genre and a revised and augmented edition of Deutsch's documentary biography in the fourth volume.

After the Second World War radio and sound recordings became important media for historical performance practice. In 1947 the first complete recording of *Messiah* conducted by Sir Thomas Beecham on 21 shellac discs claimed authenticity.[46] Only in 1980 did Christopher Hogwood present a *Messiah* recording based on actual knowledge of Handel's performance practice and with an ensemble corresponding to the well-documented *Foundling Hospital*-performance of 1754.[47] In 1984 Hogwood also published a well-documented Handel biography (1992, in German translation). In 1984, the year of the Handel commemoration, further publications appeared, among them H. C. Robbins Landon's biography with not yet analyzed documents of the time and a spate of

43 The performance of Handel's operas during the *Göttinger Händelfestspiele* with baroque instruments and playing practice since 1920 had an influence on the performance practice of his oratorios, see Larsen (1988), 23 ff.

44 More information on the development of performance practice, see H. Haskell, *Early Music Revival* (1988) and D. Gutknecht, "Alte Musik heute, Geschichte und Stand der Aufführungspraxis der Musik des 18. Jahrhunderts" (1989), part 2, 9–28.

45 D. Burrows (1991), 53.

46 In 1905 selected parts of *Messiah* were published on 20 discs.

47 Chr. Hogwood, Comment to *L'Oiseau Lyre* D189D3 (Florilegium Series).

illustrations. There were also further recordings with historical or reconstructed instruments in "original" numbers per part, such as the recording of the reconstructed 1784 *Handel Commemoration* in *Westminster Abbey,* performed in 1984 in the cathedral of Washington, D.C.

Widely indebted to musicological research and sound recordings, the renewed enthusiasm for Handel, is ongoing. Besides the still existing tradition of festival performances with large amateur choirs and large or very large orchestras, historically informed performances, especially of *Messiah,* are still the province of very specialized ensembles. They are presented mainly at special festivals and, since the introduction of the compact disc in the 1980s, as increasingly successful sound recordings. But concerts and even recordings of such historical performances rarely take place in spaces fully equivalent to those Handel used. The question why organizers, musicians and sound recording engineers rarely address the issue of appropriate rooms for the performance of Handel and of *Messiah* in particular, despite their extensive knowledge of baroque performance praxis, will be discussed in the last chapter of this study.

2.1.5 Handel's own performance practice of *Messiah* and other works

Like many other topics of historical research, Handel's own performance practice and its relation to room acoustics is embedded in partly preserved, partly even lost, documentation. Written evidence of Handel's comments on his own works and performance practice is extremely scarce. The exact dimensions of some of the rooms he used for the performance of his works remain unknown. They could be partly reconstructed, partly only categorized on the basis of room shape and the number of listeners. This is why Handel's own approach to room acoustics and their impact on the sounding appearance of his works called for a broad comparative assessment of documents on works comparable to *Messiah* in genre or size of ensemble, performance rooms, structural elements of these compositions related to room acoustics, and of Handel's few comments on acoustic questions. The following analyses include, therefore, Handel's general experiences in Rome and with the Italian *oratorio volgare,* the later development of the English oratorio and well documented performances of such compositions, as well as the performance tradition of church compositions, such as the *Utrecht* and the *Dettingen Te Deum* and *Jubilate* and the famous *Coronation Anthems.* A further element is the role of the musicians of the London opera houses and the singers of the *Chapel Royal* and other church choirs, as all these constitute the background for a proper understanding of Handel's own performance practice of *Messiah.*

2.2 Handel, the oratorio volgare, and Arcangelo Corelli

2.2.1 *La Resurrezione,* oratorio for Rome (1708)

Handel spent the years 1706 to 1710 in Italy, where he met Alessandro and Domenico Scarlatti, Caldara, Corelli, Pasquini, Lotti and Gasparini in Rome, Perti in Florenz and Vivaldi and Albinoni in Venice. He composed instrumental music, catholic Latin church music for soloists, choir and orchestra, more than 100 important chamber cantatas, in which he developed the art of writing for a singer's voice, an opera for Florence and one for Venice, which are counted among the masterworks of their time, the serenata *Aci, Galatea e Polifemo* and the oratorios *Il Trionfo del Tempo* and *La Resurrezione* for Rome.

Written in spring 1708 for the *Accademia Arcadia,*[1] Handel's sacred oratorio *La Resurrezione* is especially important for his experience in performance practice. Unlike the secular *Il Trionfo del Tempo,* whose circumstances of creation and performance are scarcely known, this oratorio by the 23-year-old Handel is well documented. We chose it, therefore, as a starting point for this study.

Since 1690 the *Accademia Arcadia,* founded in Rome in memory of the 1689 deceased patron of the arts Christina of Sweden, met weekly for discussions and concerts and counted among its members Cardinale Benedetto Pamphilij, Cardinale Pietro Ottoboni and Marchese Ferdinando Ruspoli, but also poets and musicians, such as Arcangelo Corelli, Bernardo Pasquini and Alessandro Scarlatti. Not least, in these circles Handel developed his own musical language inspired by vocal music (mainly of Alessandro Scarlatti) and instrumental music (mainly of Arcangelo Corelli) performed partly by very large string ensembles. He thus combined Italian art with German and French elements in an unique way[2] and later, in his own parodies, often made use of his splendidly orchestrated, highly inventive Italian works.

In Rome's halls, theaters and churches Handel met the best singers and instrumentalists working for the great churchly and private patrons and became

1 U. Kirkendale, "The Ruspoli Documents on Handel" (1967), 222–73 and 518, rev. publ. in *Music and meaning* (2007), 287–346 (if information unchanged, cf. 1967); S. H. Hansel, "Orchestral Practice at the Court of Cardinale Pietro Ottoboni" (1966), 398–403, especially footnote 11.

2 R. Strohm, "Händel und Italien: ein intellektuelles Abenteuer" (1993), 7; as for the orchestra size, see H. J. Marx, *Corelli, Gesamtausgabe* V (1976), preface.

acquainted with their performance practice. Due to the Papal prohibition of public opera from 1689 to 1709 the *oratorio volgare* was of special importance in Rome. During Lent performances took place not only in oratorio halls but also in the rooms of the *palazzi* and *collegia.* The works were not staged but played without acting on a decorated stage. Private performances in the *palazzi* were held without reading from the bible and with refreshments during the entr'acte. Cardinals *(purpurati)* were not only hosts and guests but often also poets of the oratorio libretti. Uffenbach left a vivid description of such an event:

> In the evening we went to Prince Ruspoli's palace to listen to the regular Sunday concert and attend the gathering. The oratorio or the music was performed at that place in the long gallery that I described earlier, and this time we found a much larger crowd than the last time. The music was also better, no doubt incomparable, but especially the inimitable voice of Mariotgi. [...] The concert wasn't so long as last time and very pleasant, and refreshments, preserves, etc. were served. Cardinal Ottoboni, Imperiali and others were also present, as was a large number of people of high standing and strangers, yet everyone was so focused and delighted that one could have heard a fly in flight, so quiet was everyone, though the passionate Italians constantly rolled their eyes and gesticulated in admiration and also applauded when something was over, though with the multiply folded coat between their hands so as not to be heard, as that would have been disrespectful, being allowed only in the theater.[3] (7 April 1715)

Handel's Italian oratorio *La Resurrezione* was performed on Easter Sunday and Monday 1708 for the *Accademia Arcadia* at *Palazzo Bonelli* on *Piazza dei SS. Apostoli* in an ad hoc hall theater:[4]

> Easter Sunday the 8th. This evening marchese Ruspoli presented at the Palazzo Bonelli near SS. Apostoli a most beautiful oratorio with music in the salone, which was decorated as a theater for the public. Many noblemen and some cardinals attended.

The account books of Marchese Ruspoli, the organizer of the performance who resided at the time in *Palazzo Bonelli,* list the singers' names: Filippo (soprano castrato), Matteo (soprano castrato), Pasqualino (alto castrato), Vittorio (tenor), Cristofano (bass).[5] A Vatican reprimand regarding Handel's work addressed to

3 Cf. E. Preussner, *Die musikalischen Reisen des Herrn von Uffenbach* (1949), 78, referring to the performance of an oratorio by Caldara at the *Palazzo* of Marchese Ruspoli at the Corso.

4 Francesco Valesio, *Diario,* cf. *HHdb.* 4, 34: "Dom. 8 Pasqua di Resurrezione / Questa sera il marchese Ruspoli fece nel Palazzo Bonelli a'SS. Apostoli un belliss.mo Oratorio in musica havendo fatto nel salone un ben'ornato teatro per l'Uditorio, si intervenne molta nobiltà et alcuni porporati." See also Kirkendale, "Ruspoli" (1967), 236.

5 Kirkendale (1967), 231 f.; U. Kirkendale, *Antonio Caldara, Sein Leben und seine venezianisch-römischen Oratorien* (1966), 74 and 222 f.; *HHdb.* 2, 38 f.

Marchese Ruspoli indicates that the first performance on Easter Sunday included a female singer, replaced by the castrato Filippo in the second, as the public performance of women was prohibited at the time in Rome.[6] The female singer was most probably the soprano Margherita Durastanti, then in the service of Ruspoli, who also performed in Handel's *Agrippina* at *Teatro S. Giovanni Grisostomo* in Venice in 1709. Handel composed several chamber cantatas for her and later engaged her for the opera in London, where she sang in 1720, 1722–24 and 1733–34.[7]

The orchestra consisted of the regular musicians in Ruspoli's service and 50 additional instrumentalists as follows:[8] 22 violins with Corelli as leader and soloist, 4 *violette* (violas), 6 *violoni* (violoncelli), 6 *contrabassi,* 1 *viola da gamba* (not in the payrolls, played by the virtuoso Ernst Christian Hesse from Darmstadt, then in Rome),[9] 2 trumpets, 1 trombone,[10] 4 oboes, 2 recorders, 1 *flauto traverso* (oboes and flutes alternating and probably played by the same musicians), 2 bassons and Handel at the harpsichord.[11] There were three public rehearsals and two performances. After the first public rehearsal in the *Stanzione delle Accademie* on the second floor of the *Palazzo,* the performance was moved to the *Salone grande al piano nobile,* the great hall on the first floor, in anticipation of a great number of listeners. After the second rehearsal the podium and the whole installations had to be moved within six days from the second to the first floor and adapted to the *Salone grande.*

6 Kirkendale (1967), 236, footnote 51: "Lunedì 9 [Aprile, 1708] ha fatta S. B. fare una ammonizione per haver fatto cantare nell'Oratorio della sera precedente una Cantarina." ("On Monday, the 9th, S. B. sent a reprimand because the previous night a female singer had been allowed to sing in the oratorio.")

7 W. Dean, "Margherita Durastanti", *NG of Opera* 1, 1283.

8 All indications on the orchestra and number per parts are listed again in table A in the appendix.

9 A. Wenzinger: "Die Viola da gamba in Händels Oratorium 'La Resurrezione'" (1987), 80–3.

10 The participation of a trombone is not as improbable as Hicks remarks (s. booklet for the recording of *La Resurrezione* without trombone for DECCA 1982). Ewerhart mentions Caldara's oratorios used trombones, see R. Ewerhart, "New Sources for Handel's 'La Resurrezione'" (1960), 127–35, footnote 3; trombones are also listed several times for performances by Corelli in Ottoboni's account books, see H. J. Marx, "Die Musik am Hofe Pietro Kardinal Ottobonis unter Arcangelo Corelli" (1968), 128, 139 and Marx (1968), 208. In large ensembles they played the bass *colla parte* with the bassoons, see Zaccaria Tevo, *Il Musico Testore,* Venezia (1706), 360 and H. J. Marx, *Corelli, Gesamtausgabe* V (1976), 23, footnote 72.

11 Kirkendale (1967), document 11, 256 ff.; Marx (1986), 86; Marx, "Instrumentation" (1988), 496 –505. The bassoons notated in the aria *Risorge il mondo* are not listed in the account books.

2.2.2 The halls at *Palazzo Bonelli*

Almost none of the *palazzi* of Rome remained unchanged over the years. Mainly those used by the ever changing churchly and diplomatic representatives as apartments and offices were often radically altered.[12] *Palazzo Bonelli* (today *Palazzo Valentini* or *Palazzo della Provincia),* from 1707 to 1713 Marchese Ruspoli's residence during the reconstruction of *Palazzo Ruspoli* at the Corso, has also been altered since. The not extant halls in which the rehearsals and performances of Handel's oratorio took place, the *Stanzione delle Accademie* on the second floor and the *Salone grande* on the *Piano nobile,* can be described more in detail thanks to the restorations undertaken during the last hundred years,[13] and the payments to craftsmen recorded in Ruspoli's 1708 account books.[14]

2.2.2.1 The *Salone grande*

From the carpenter's bill we know the width of the podium installed in the *Salone grande* on the first floor. It was 55 *palmi romani* or barely 12.30 m.[15] When the stage was adapted to the *Salone grande* a wooden beam had to be mounted above the stage opening. To install the beam the mason had to break through a bricked-up window above the vault cornice that formerly opened to the attic and then fix the other end in the transom window in the main facade.[16]

12 In 1675 the number of the city's inhabitants was 150'660, among which "650 cardinals, bishops and abbots with about 8'000 *cortigiani,* about 10'000 officials with entourage and about 30'000 moving strangers." For this steadily changing upper class the *palazzi* were constantly altered; see A. Ademollo, *Teatri di Roma* (1888, reprint 1969), 231.

13 G. C. Argan (ed.), *Palazzo Valentini* (1984); G. Farina (ed.), *Palazzo della Provincia* (1985); G. Eminente, "Dopo cento anni radicali restauri al'palazzo Valentini" (1973), 7.

14 Kirkendale (1967), 260 f.

15 R. Mattia, "Feste Musicali a Palazzo Bonelli" (1985), 213–30, especially 223: "Lo scenario si presentava con la forma di teatro a scalinata, con quattro file di sedie per l'orchestra, avente un'estensione in larghezza di cinquantacinque palmi." See also Kirkendale, *ibid.* (1967), 261–2, abbreviated citation ("Fatture da Mas.o Crespineo Pavone falegniame ... di 11 aprile 1708, no.168: „per formare li no. 4 seditori ..."); the dimensions are confirmed by the bill of the painter, see *ibid.*, 258. 1 p[almo romano] = 0.2234 Meter, see P. Waddy, *Seventeenth Century Roman Palaces: Use and the Art of the Plan* (1990).

16 Kirkendale (1967), documents 12 f., 258 f., especially 264, [Page 3]: "Lauori Fatti nel Salone al Piano Nobile doue fù trasportato il Teatro, e Cantato il sudetto Oratorio per maggior Capacità di Persone – Per hauer rotto il muro di una finestra smurato sopra il Cornicione dell'Imposta della uolta di detto Salone che corrisponde al piano dell'apparti-

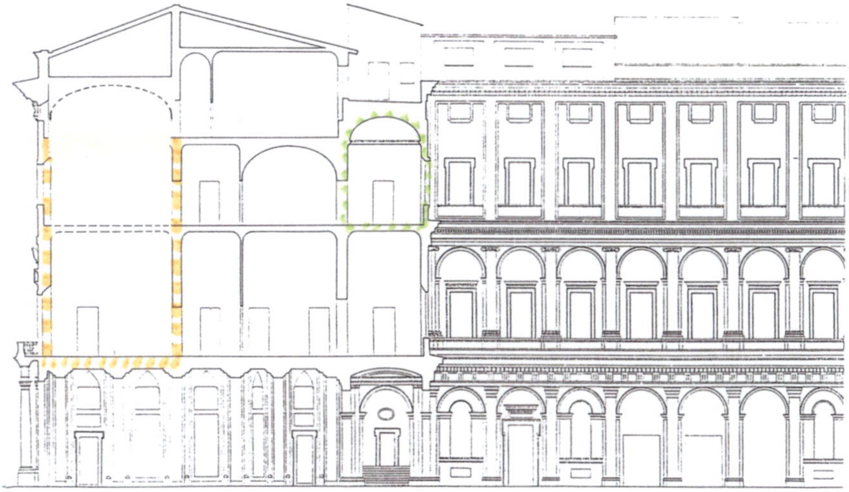

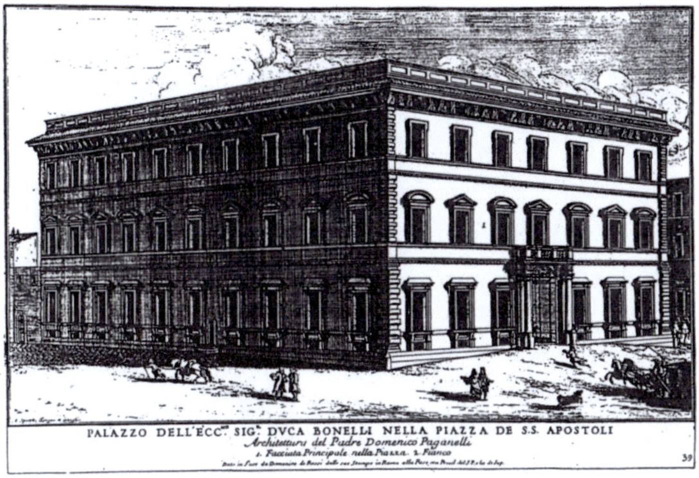

Figure 69: Rome, *Palazzo Bonelli,* modern section with probable position of the *Salone grande* and the *Stanzione* (above) and view to the main façade (17th century, below).

mento di Cima [...] per mettere in opera l'armatura de legni che furno fatte e cauate fuori delle dui uani delle finestre ... per reggere il filagnone sopra la bocca del Teatro." – ("The work done in the hall on the first floor to which the theater was transported and the mentioned oratorio was sung because of the greater seating capacity, the latter being due to the removed wall of a walled-up window above the base cornice of the vault of that hall, which reaches up to the level of the uppermost apartment [...] in order to fix the wooden beams inserted through the two openings of the windows [...] to erect the beam above the theater's stage opening.")

From the mason's bill it is evident that the *Salone grande's* vault then reached into the *2° Piano* – at the time the attic. The beam's wall-to-wall length was 64 *palmi romani* or 14.30 m.[17]

An eighteenth-century view of the *Palazzo* and a modern section permit the assessment of the following boundary measures (figure 69): based on the building shape at the time, the main façade facing the *Piazza dei SS. Apostoli* was 31 m long.[18] The *Salone grande* on the *Piano nobile* therefore had a maximum length of 31 m and a width between 12.30 and 14.30 m (these values are given by the podium's width and the length of the beam resting in the windows of both lateral walls, above the stage). The first load-bearing inner wall still stands today at a distance of 12.50 m from the outer wall. The shape and height of the vault is unknown but the hall's crown height was at least 19 m (up to the upper border of the second floor windows) but less than 21 m (the total height of both floors minus thickness of the vault). These measures yield a presumable height-to-width relation of 3:2 (19.00 / 12.65 m). The volume of the *Salone grande* did not exceed 6750 m. The floor area was slightly larger than in the small hall at the *Tonhalle* in Zürich (26.5 × 12 m) or the *Brahms-Saal* at the *Wiener Musikverein* (23.40 × 10.60 m) but slightly smaller than the Berlin *Singakademie* (32.40 × 12.80 m). Today a hall of this size would be defined as a chamber music hall. But with a height of 19 m it was about twice as high as the hall in Zürich (9 m) and Berlin (9.70 m) and also considerably higher than the chamber music hall at the *Wiener Musikverein* (11 m). The acoustics were comparable to those in a small chapel.

2.2.2.2 The *Stanzione delle Accademie*

The carpenter's bill contains also important information on the *Stanzione delle Accademie,* here named the "Gallaria doue (...) si uoleua Fare l'Oratorio."[19] This hall may correspond to the still extant porch on the second floor, measuring 21.20 m × 7.70 m and 10.60 m in height (volume slightly below 1700 m^3).[20] In any case, room height and volume were considerably less than those of the *Salone grande*. From the bills it becomes evident that the hall was

17 Kirkendale (1967), 262, bill of the carpenter, [Page 4]: "Per ... li due Filagnioni aggiunti insieme posti pertrauerso sopra la bocca di detta sala long'asieme p[almi] 64." – ("For [...] the two beams connected and placed over the [stage] opening of the mentioned hall measuring a total length of 64 palmi.")

18 Documents concerning a festive meal offered by the French ambassador in this hall in 1739 confirm that it filled the whole façade's length of 31 meters: 3 tables of minimum 27 m length were necessary to serve 140 sitting guests; see Mattia, *ibid.,* 223.

19 Kirkendale (1967), 262–3.

20 *Stanzione* means hall or entrance hall.

so narrow that parts of the podium had to be placed in an adjacent room, connected to the *stanzione* after a wall was torn down. The carpenter's bill indicates that the space for the extraordinarily large orchestra had been miscalculated and that the podium had to be enlarged after the first rehearsal.[21] Following the second rehearsal it was dismantled and moved to the hall on the first floor.

2.2.2.3 Decoration, seating, number of listeners

For the Handel performance the *Salone grande* was decorated after designs by the famous church decorator Contini from the Bernini school. Velvet, red and yellow taffetta and damask covered the ceiling and walls as well as the podium, which featured four steps for the orchestra *(quattro file di sedie per l'orchestra)* with 28 music stands, and a separate riser for the *Concertino di Violini* with four music stands (solo strings and basso continuo).[22]

The exceptionally high number of listeners in the *Salone grande* can be estimated on the basis of libretti: 1500 were printed for *La Resurrezione,* but only 200–300 for the other events at Ruspoli palace, at the time all held on the second floor in the *Stanzione delle Accademie*. Assuming that libretti were also given to the listeners at the first two public rehearsals in the small hall on the second floor, still more than 300 libretti remained for each performance in the great hall. The number of chairs and of eventually standing listeners is not known. The above named halls of comparable size from the nineteenth century actually provide 300–400 seats for 1.3 persons per m^2 (*Kleine Tonhalle,* Zürich: 429 without gallery; *Brahms Saal* at the *Musikverein,* Vienna: 321 without gallery).[23] Single chairs at the time would have allowed for a maximum of 2.5 persons per m^2 or 600 listeners on an available floor area of about 240 m^2.

In fact, seating was offered on hired chairs,[24] probably in rows parallel to the side walls, as described by Uffenbach regarding a performance of an oratorio by Caldara in 1715 in *Palazzo Ruspoli* at the Corso. He provides a detailed picture of such an event:

> In the evening the big weekly concert was held at the palace of Prince Ruspoli, it was the best one can find here because of the prince's lavish yearly expenses in this respect. And because he gladly welcomes any stranger without any reference, we all went there and were led through a large number of superbly furnished rooms to a huge long gallery, where, as in the rest of the house, there were plenty of incomparable paintings and silver works, the

21 Kirkendale (1967), 263.
22 Kirkendale (1967), 262.
23 See also 2.4: The first performance of *Messiah* in Dublin.
24 Kirkendale (1967), 238.

lighting was sumptuous, and both sides of the entire gallery were lined with seats, but the space above was left free for the musicians – a large number of virtuosos, with three female singers and a little castrato in the service of the ambassador Gallas sitting up front. They gave such an excellent concert or so-called oratorio, that I was immensely delighted and convinced that I've never heard such perfection in my life. The composition is always completely new and it's by Caldara, the Pope's well-known music director, who also conducted.[25] (31 March 1715)

2.2.3 Other concerts under Corelli in Rome

2.2.3.1 Scarlatti's Oratorio *La Passione* at the *Palazzo della Cancelleria*

The circumstances of Handel's work must be related to other concerts within the same circles in Rome, mainly Alessandro Scarlatti's oratorio *La Passione del Nostro Signor Giesù Cristo,* performed a few days earlier on 5 April 1708 (Wednesday of Passion Week) in a similar manner at the *Palazzo della Cancelleria.*[26] Related to Christ's death and resurrection,[27] both Handel's and Scarlatti's oratorios belong to a cycle held by the *Accademia Arcadia* during Easter 1708. The orchestra for Scarlatti's work, also conducted by Corelli, consisted, in typical Italian manner, of strings, 2 trombones, 4 trumpets, timpani and harpsichord, but no woodwinds as for Handel's works.

On the basis of an undated scenery draft by Juvarra marked "Oratorio in the Holy Week in the theater of Signore Cardinale Ottoboni" Smither assumed that the performance of Scarlatti's *Passione* took place in the *Palazzo* theater (figure 71a).[28] Yet the theater in the very high hall with vaulted ceiling and four ranks of boxes at the *Palazzo della Cancelleria* was undergoing renovation by Juvarra at the time and did not reopen until January 1710 (figure 70).[29] Recent research has provided evidence that in this 900 m^3 hall, which measured 8.95 × 9.20 m and about 10.60 m in height, a theater with boxes had existed already

25 E. Preussner, *Die musikalischen Reisen des Herrn von Uffenbach* (1949), 77 and 78.

26 Unfortunately the original version is lost. On 25 March 1708 A. Scarlatti's *La SS. Annunziata* was performed as first oratorio of the cycle at *Palazzo Ruspoli*, but no further details are known so far.

27 H. E. Smither, *A History of the Oratorio*, Vol. 1 (1977), 274.

28 Torino, Biblioteca Nazionale Ris. 59.4, f. 23 [1]: "Oratorio nella Settimana Santa in Teatro del Signore Cardinale Ottoboni," see Smither, *ibid.*

29 M. Viale Ferrero, *Juvarra* (1970), tav. 176, 182–185 (groundplan, longitudinal and cross section of the theater); *NG of Opera* 4, 26: fig. 2 and *NG* 13, 670: stage view of Costanzi's *Componimento sacro* (1727). Viale Ferrero mentions a draft for a larger theater of 10.30 ×13 m which was not realized (*Juvarra* (1970), 74 and tav. 180).

since 1690/91. In December 1690 Alessandro Scarlatti's opera *Gli equivoci in amore* was performed there and in winter 1690/91 also *La Statira* (libretto by Pietro Ottoboni, first performance in January 1690 at the *Teatro Tordinona).* The theater was closed at least from June 1707 to January 1710.[30]

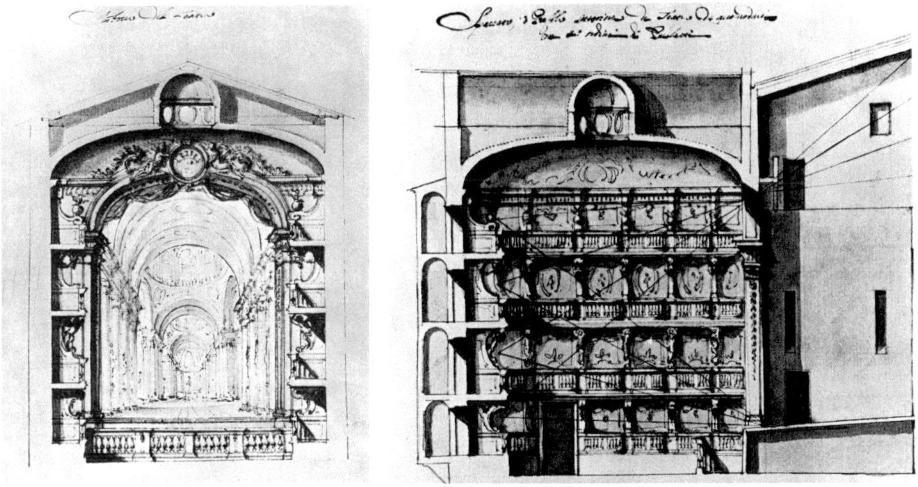

Figure 70: Rome, *Palazzo della Cancelleria, Teatro Ottoboni* (1690–1740), sections.

If a further undated scenery draft by Juvarra annotated "His Excellence Ottoboni, the oratorio. Holy Week in his great hall" (figure 71b)[31] refers to this performance, it would prove that the great hall at the *Palazzo della Cancelleria* was used for Scarlatti's oratorio.

The still extant and unchanged *Sala Riaria* with the beautiful coffered ceiling has a rectangular ground plan of 26.90 m × 15.80 m comparable to the *Salone grande* at the *Palazzo Bonelli,* but with a much lower flat ceiling (12.40 m); the hall volume is 5,300 m^3. Juvarra's drawing shows that the musicians were not seated on a stage but on a balcony.[32] This was a common orchestra placement, as may be seen in the interior design of the *Oratorio della Chiesa nuova*[33] and also in the 1708 published *Descrizione di Roma moderna,* where the *Anticamera,* the porch or a corridor of the *Palazzo della Cancelleria,* is described as the usual space for the performance of oratorios, comparable to

30 Ferrero, *ibid.* (1970), 77.

31 Torino, Biblioteca Nazionale Ris. 59.4, f. 81 [1]: "Eo Ottoboni .. l'oratorij La Settimana Santa nella sua gran Sala."

32 Kirkendale (1966), 150; Ferrero (1970), tav. 177; see also Smither (1977) I, 267 f., fig. VI-4: Marx (1968); S. H. Hansell, "Orchestral Practice at the Court of Cardinal Pietro Ottoboni" (1966), 398–403; Ch. L. Frommel, *Der römische Palastbau der Hochrenaissance* (1973), 67 and t. 163a.

33 J. Connors, *Borromini and the Roman oratory* (1980), 34.

the *Stanzione delle Academie* on the second floor of the *Palazzo Bonelli,* which also was described as gallery:

> The antechamber of the Palazzo della Gancelleria [...] where [the presentations] are usually given several times during the year, especially the oratorios during Lent [...], is well designed with gilded balconies for musicians.[34]

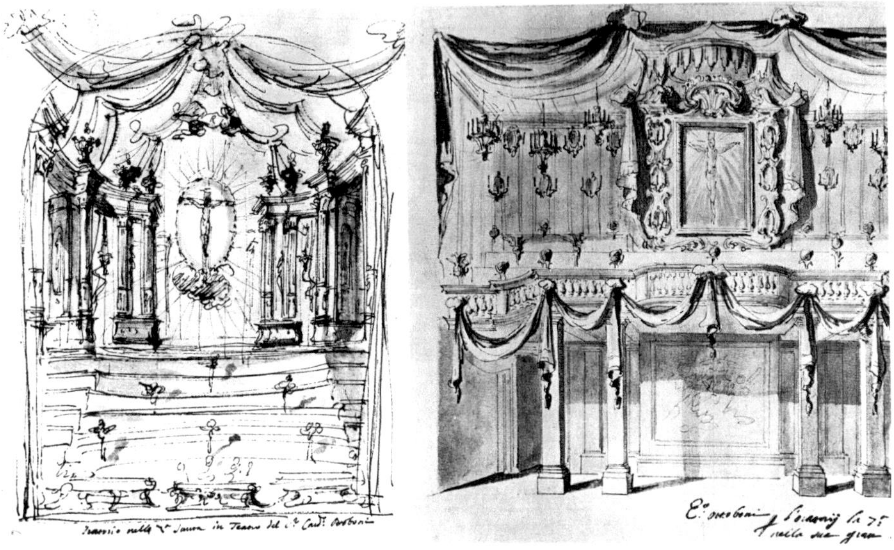

Figure 71: Rome, *Palazzo della Cancelleria,* oratorio stage design by Juvarra (1708?): a) in the theater, b) in the *Salone grande*.

Juvarra's first draft shows a podium "a scalinata"– with raked steps similar to the one described for Handel's *Resurrezione.*[35] Handel later used such an orchestra placement for his oratorio performances in the London theaters and in 1727 for the coronation of King George II in *Westminster Abbey,* where it caused a great sensation. He had certainly seen such raked platforms for large orchestras in Italy, both outdoors and in churches.

34 "Anticamera del palazzo della Cancelleria [...] dove suole fare alcune volte durante l'anno, e particolarmente la Quadragesima gl'Oratorii [...] ben accomodata con balconi dorati per i Musici." See *Descrizione di Roma moderna,* Roma (1708), cf. Ferrero (1970), 73, footnote 5.

35 See the above cited bill of the carpenter for the Handel performance at *Palazzo Bonelli,* which mentions 3steps and 4 rows of seats for the musicians.

2.2.3.2 Orchestras' size under Corelli

The enormous prestige of large string orchestras in Rome becomes evident in the documents on oratorio performances under Corelli, such as *La S. Beatrice d'Este* 1689 (by Lulier with an *Introduzione* by Corelli) with 74 strings, 2 trumpets and basso continuo in the famous gallery of Carlo Fontana in the *Palazzo Doria-Pamphilij* at the *Piazza Navona* – a large and very long hall of nearly 15,000 m^3 (51 × 15 × 20.4 m), 5 m narrower and 2 m higher than the *Goldene Saal* of the Vienna *Musikverein* (51 × 20 × 18.5 m).[36] Even more strings played on 23 August 1705 at Scarlatti's Oratorio *Il regno di Maria Virgine* in the *Palazzo della Cancelleria,* but this performance with 100 strings was held in the outdoor courtyard.[37] At *Palazzo Riario* the *Accademia per Musica* by Pasquini took place on 2 February 1687, organized by Christine of Sweden on the occasion of James' II coronation as king of England, for which 150 strings and 100 singers are reported.[38]

A survey of Corelli's performances for Cardinale Pietro Ottoboni from 1689 to 1705 in the rooms of the *Palazzo della Cancelleria,* in the palazzo's *Cortile* and *S. Lorenzo in Damaso* church and in other places shows, however, that such great string groups were used only on special festive occasions (see table 14c):[39] of 57 documented events only 4 further performances had 50 and more strings; 25 concerts had 25 and more strings, the remaining 28 had a group of only 3 to 24. Ottoboni's account books show varied numbers of performers also for the same rooms because the ensemble size depended on the event's importance. During Lent the number of strings in *San Lorenzo in Damaso,* a basilica with a coffered flat ceiling and favorable acoustics (48 × 26 × 24 m, figure 67), was between 7 (1692 "p[er] le 40 hore [per sette servitij] 4 violini, violone, contrabasso, leuto") and 50 (1700 "Per le 40 hore 29 violini, 7 violette, 6 violoni, 8 contrabassi, leuto").[40] The orchestral forces obviously conformed to two rules: the first was "secondo il bisogno del luogo, " defined by Emilio de Cavalieri in the printed edition of the *Rappresentazione di anima e di corpo* in 1600:

36 Marx (1976), 19.
37 Smither (1977), 274.
38 M. Talbot, "Corelli", *NG* 4, 769.
39 Marx (1968), Giustificazione, 122–61; Marx (1976), preface; Hansell (1966), 400–403.
40 Marx, (1968), 129 [27]: number of instruments. The first information on these documents in the account books of the Cardinal was given by M. Lütolf, *Corelli GA* I (1987), preface.

> The instruments must be well played and their number more or less adjusted to the place, be it a theater or a hall, which, in order to be well proportioned to this [kind of] recitation in music, must not exceed a capacity of thousand persons.[41]

The second rule could be called "secondo il bisogno dell'occasione" – according to the need of the occasion – because performances for more important occasions had larger ensembles than everyday music, even if played in the same room.

2.2.4 Handel's *Resurrezione*: size of the orchestra and hall

This second rule explains why Handel's oratorio was carefully planned for the small *Stanzione delle Accademie* on the second floor: the performance was organized as an event of the *Accademia,* which normally took place here if not in the Esquilin garden or in a hall of another *Accademia* member's Palazzo.[42] Only the unexpected high number of guests and problems with the musicians' placement finally opened access to the great hall on the first floor.

Corelli's account for *La Resurrezione* is so detailed that it lists not only the number of musicians paid for the performance in the *Salone grande* but also the smaller number rehearsing in the *Stanzione delle Accademie.* The two rehearsals in the small hall had 8 strings less than the third rehearsal, and the two performances in the *Salone grande* (the additional musicians were not paid "per due funtioni, e trè Proue" but "per due funtioni, e una Proua").[43] The ensemble for the smaller *Stanzione delle Accademie* had 30 strings with 18 violins, 4 violas, 4 violoncellos and 4 double basses that played at the first and second public rehearsal. Remarkably, this group was already larger than usual, obviously following the rule "secondo il bisogno dell'occasione": oratorios on the second floor at Marchese Ruspoli's usually had 12 strings only with 8 violins, 2 violas, 1 violoncello and 1 double bass.[44] For the great hall the string group was increased "secondo il bisogno del luogo" to 38 with 22 violins, 4

41 "Gli stromenti siano ben sonati, e più, e meno in numero secondo il luogo, ò sia Teatro, overo Sala, quale per essere proportionata à questa recitatione in Musica, non doveria esser capace al più, che di mille persone." Cf. Quellentexte zur Konzeption der europäischen Oper im 17. Jahrhundert, *ed. by H. Becker et al.* (1981), 13.

42 Also in other Roman palaces music was performed in smaller rooms on the second floor and not in the great hall on the first floor, as for instance at the *Palazzo Barberini* (see table A in the appendix).

43 Kirkendale (1967), 256–7: additional 4 violins, 2 violoncellos and 2 double basses.

44 Kirkendale (1966), 52, footnote 61: 1710 *Oratorio per Santa Francesca Romana.*

violas, 6 violoncellos and 6 double basses. The additional instruments were, as mentioned above, the solo viola da gamba, 2 trumpets, 1 trombone, 4 oboes, 2 recorders, 1 flute, 2 bassoons,[45] with Corelli playing the first violin as concert master and Handel at the harpsichord, which makes a total of 53 instruments played by 50 musicians. The ensemble of Handel's *Resurrezione* remained slightly below the splendid events at the time of Christina of Sweden, but after 1689 it certainly was among the largest, and the effect was certainly not less sensational thanks to the hall's favorable acoustics, the excellent musicians and Handel's original composition.

Handel used the ensemble in a new way. Besides the exceptional sound effects, such as the subtle use of the solo viola da gamba, an instrument rarely heard in Rome at the time, the special importance of the middle voices is striking, thus allowing a rich and full sound of strings. Instead of Corelli's usual relation of 2:1 between violins and lower strings (viola, violoncello and double bass)[46] Handel uses a relation of 3:2. For some movements the violins are divided into three or even four parts, and in several arias the orchestra is divided into *concertino* and *tutti,* sometimes explicitly indicated in the score as *concerto grosso.*[47] In the choirs at the end of the first and second part sung by the soloists the soprano is mostly doubled by *oboe col soprano* or *oboi e violini col soprano,* a technique not common in Italy at the time and later used by Handel also in his English anthems and many choir numbers of his English oratorios. In this respect the autograph of *Messiah* will offer a surprising exception.[48]

With 39 strings the oratorio by the 23-year-old Handel no doubt produced a splendid sound in the *Salone grande* of the *Palazzo Bonelli*. The 22 violins, 4 violas, 6 violoncelli and 6 double basses achieved a size nearly equal to a modern symphony orchestra that would play today in halls larger than 10,000 m^3.[49] In the long, narrow and high *Salone grande* with nearly 6750 m^3 the tutti of Handel's orchestra must have been extraordinarily intense and warm. The description of the interior design shows excellent acoustic conditions. The richly decorated ceiling, the good lateral reflections in this relatively narrow and long hall provided a clarity and a transparence undiminished by the great number of listeners. The installations and decoration material used on stage, on the ceiling and walls provided good diffusivity and prevented booming

45 For the use of bassons, see D. Möller, *Besetzung und Instrumentation in den Opern Georg Friedrich Händels* (1989), 66, footnote 80; H. J. Marx, "Zur Kompositionsgeschichte von Händels Pastoralkantate Apollo e Dafne HWV 122" (1984), 77–92: the score demands oboe I, II and *Bassons* (tutti and soli).

46 Marx (1976), preface.

47 In Maddalena's aria "Per me già di morire" and "Se impossibil".

48 For the *Messiah* performances in Dublin, see 2.4.3.

49 See 1.9, table 14c and table A in the appendix.

resonance from the vault, conditions especially important in regard to the large number of instruments. The orchestra's rich sound was well supported by the favorable acoustic conditions in the not too large, long and narrow hall.

2.3 The performance practice of the early English oratorio

After a first visit during the 1710–11 season Handel lived from 1713 until his death in 1759 mostly in London, with only temporary stays outside the capital, a few journeys on the continent and a stay in Dublin during the winter of 1741–42. His main activity until 1741 was the composition of *opere serie* in Italian for famous Italian singers. In 1732 he developed the English oratorio, writing a new version of his masque *Esther,* a new genre to which he committed practically his full creative power from 1741 onward, and to which he owes his later reputation more than to any other genre.

2.3.1 Two forerunners from 1718: *Acis* and *Galathea* and *Esther*

Handel's compositions for the Anglican church (*Anthem, Te Deum* und *Jubilate)* and of English masques were a crucial prerequisite for the development of the new genre. In 1717–18, when the opera in London was closed until the foundation of a new *Royal Academy of Music,* which appointed him permanent *Master of the Orchestra* in 1719, he entered the service of James Boydges, Earl of Carnarvon, who owned a Palais in London on Albemarle Street as well as a country residence in Cannons near Edgware. The Earl, since April 1719 first duke of Chandos, employed several singers and a small orchestra, the *Cannons Concert,* with Johann Christoph Pepusch as music director. Handel composed for Cannons Italian cantatas, instrumental works, the eleven so-called *Chandos Anthems* and a *Te Deum,* integrating the style of English choir music (mainly of Purcell) in the same original manner as the Italian style in his secular vocal music 10 years earlier. He also composed two English masques in a new form with recitatives, arias and choirs. These works actually belong to the genre of the English opera, which, despite successful performances by several composers, failed to establish itself in the years 1715–18 in London against the then more favored Italian opera. In 1718 the charming two-act masque *Acis and Galathea,* with a libretto by John Gay,[1] immediately sent Handel to the top of this

1 Gay is mainly known as the author of *Beggar's Opera,* the successful English parody on Italian opera and its star singers, played in London since 1728 during many years.

English movement. The orchestra for *Acis,* modest compared to Handel's works for Rome, presumably contained 4 violins, 1 viola, 2 violoncellos, 1 double bass, cembalo, 2 oboes alternating with recorders and, in Polyphem's aria Nr. 12, with *flauto piccolo ottavo* and in a second performance also 1 basson.[2]

Esther, a masque in six scenes, was supposedly composed also in 1718. The libretto of this piece of religious and uplifting entertainment was written in the circles of Cannons after Thomas Brereton's *Faith, Triumphant, A Sacred Tragedy* (London 1715), an English translation of Jean Racine's 1689 tragedy *Esther.* The music for which Handel reused and altered 16 numbers from former works (among them several numbers from the *Brockes Passion,* which could not be used in its original version in England) is excellent in parts, the strange libretto notwithstanding. Six solo parts, important choirs and a larger orchestra than usually available for Cannons, with strings, 2 oboes, 2 bassons, 2 French horns, trumpet, basso continuo and harp, suggest that the performance was staged either in the London palace of the Earl or in the Cannons palace after August 1720, when the music room designed by Edward Sheperd had been completed.[3]

2.3.2 The singers of the Chapel Royal and the 1732 version of *Esther*

The *Chapel Royal*, whose composer Handel became in 1723, offered him the opportunity to write several works that became highlights of his English period. For this choir he did not write only the *Anthems* for the coronation of George II (1727) and the *Dettingen Te Deum* (1743) but, already in 1713 on behalf of Queen Anne, also the *Utrecht Te Deum* and *Jubilate.* The latter were rehearsed in *Queen's Chapel* of *St. James's* and performed in *St. Paul's Cathedral* by the *Chapel Royal* on the occasion of the celebration of the Utrecht contract and since then were performed regularly on *St. Cecilia's Day* and on other days in

2 *Acis and Galatea,* ed. by W. Windszus, preface to *HHA* I/ 9, 1.

3 Regarding HWV 50a, see *HHdb.* 2, 81 and 97 f. Handel's original genre term is not known, because the autograph title page has been lost. The earliest version termed *oratorio* is an undated, later copy from the property of Elizabeth Legh's with 1718 given as year of composition. Chrysander kept the title of a copy now at the Bayerische Staatsbibliothek Munich with the number B/1667: *Haman and Mordecai, a Masque.* See also preface to *Esther, HHA* I/ 8, VII, ed. by Howard Server, also in respect of the use of a viola.

the choir of *St. Paul's,* alternating with Purcell's *Te Deum.*[4] In this function Handel got acquainted with the important English cult of the religious sublime, some of whose adherents described religion as "the most entertaining thing in nature" and praised the combination of religion and entertainment as "a practical means of increasing the devotion of people."[5] Reacting against the strict style of church music under William III, a tradition of devotional singing at home and a ceremonial style of anthem- and Te Deum-compositions developed. Such choir compositions became part of the early concerts in the churches' assembly halls, the new music rooms and rooms of inns. The music referred to the Old Testament temple music as a model of sacred music that included musical instruments.[6]

Bernard Gates, singer of the *Chapel Royal* and conductor of the boys' choir,[7] soloist in the *Utrecht Te Deum* and *Jubilate* in 1713 and in the 1727 performance of *Coronation Athems for* George II in *Westminster Abbey,* performed in 1732 Handel's *Esther* three times with singers from the *Chapel Royal* and other church choirs. These staged performances took place on 23 February on Handel's 47th birthday and on 1 and 3 March in the *Crown and Anchor Tavern* with the *Philarmonic Society* [sic!], an orchestra of *gentlemen* (aristocratic amateurs), and a few additional professionals,[8] as explained in the introduction of the libretto:

> Mr. Bernard Gates, Master of the Children of the Chapel-Royal, together with a Number of Voices from the Choirs of St. James's and Westminster, join'd in the Chorus's, after the Manner of the Ancients, being placed between the Stage and the Orchestra; and the Instrumental Parts (two or three particular Instruments, necessary on this Occasion, excepted) were performed by the Members of the Philarmonick Society, consisting only of Gentlemen.[9]

Soli and choir sopranos were sung and played by the boys of the *Chapel Royal.* The first two events were organized by the *Philarmonic Society,* the last by the *Academy of Ancient Music* conducted by Pepusch, which worked regularly with Gates and the *Chapel Royal Children* until 1734.[10] It was the first time

4 Deutsch (1955), 63; *HHdb.* 4, 367. 1743 it was replaced by the *Dettingen Te Deum.* The tradition continued not only until 1843, but until 1854 nearly without interruption (see P. A. Scholes, *Mirror of Music,* 177, footnote 283).

5 A. H. Shapiro, "'Drama of an Infinitely Superior Nature': Handel's Early Oratorios and the Religious Sublime" (1993), 215 f., footnotes 114 and 115.

6 Shapiro (1993), 224 and 232.

7 W. Dean: *Handel's Dramatic Oratorios and Masques* (1959), 656.

8 The duke of Chandos and the Earl of Egmont were members, see Dean, *Oratorios* (1959), 83.

9 *HHdb.* 4, 98.

10 Dean (1959), 83.

that this academy, which succeeded the *Academy of Vocal Music* in 1726, sang works by Handel. Like the Italian Renaissance *accademie,* the society was committed to the performance of "chiefly ... grave and ancient vocal music" or, as in this case, new music in old style. This explains the position of the choir between stage and orchestra after the manner of the ancient theater. From then on the *Academy* regularly performed Handel's works. His enthusiasm for the activities of the "indefatigable society, the gropers into antique music, and hummers of madrigals"[11] and for the amateur orchestra of the *Philarmonic Society* seems to have been limited. Handel himself always sought the best soloists, choir singers and instrumentalists for the performance of his compositions and never held any larger performance in any of London's acoustically inadequate taverns or clubs, although several of the well-known singers and musicians performed there on their own account.

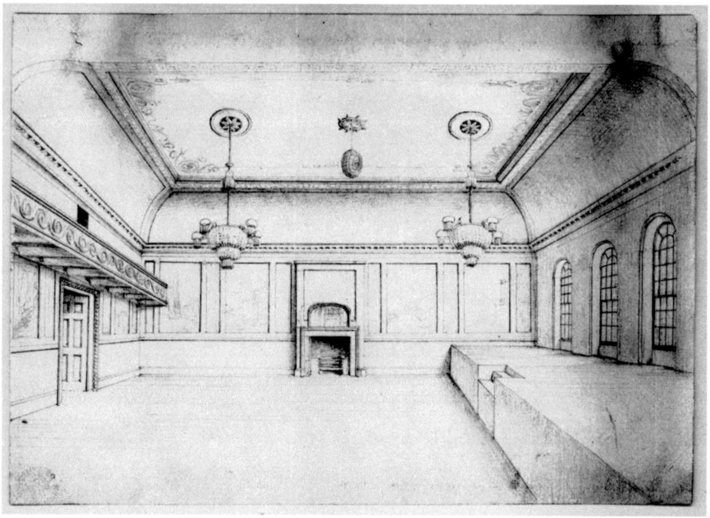

Figure 72: London, *Hickford's Concert Room* (1729–79).

Unfortunately we know only the names of most of these rooms, where various private music societies used to meet on different weekdays. About *Crown and Anchor Tavern* (not to be mistaken for the new 1790 building with the same name on *Arundel Street)* we know only that it was inaugurated on St. Cecilia's day, 22 November 1727, with a concert by the violinist Dubourg and the best Italian singers of the opera for a public consisting of the members of the

11 Deutsch (1955), 352: *Harmony in an Uproar,* 1734; see also W. Weber, "Intellectual Bases of the Handelian Tradition 1759–1800", (1981–82), 107; Chr. Hogwood, *Handel* (1984), 234.

Academy of Ancient Music and 200 women.[12] This space seems to have been one of the larger rooms of that kind in London. Measures and seating capacity are known for the great hall of *Hickford's Rooms* on Brewer Street (figure 72): with 300 seats and a ground plan of 15.20 × 9.10 m it was clearly smaller than the Oxford *Holywell Music Room* (19.85 × 9.80 m), and with 6.70 m also about 2.50 m lower. *Hickford's Rooms* were used for concerts from 1729 to 1779, and in 1764 Wolfgang and Nannerl Mozart performed there.[13]

Contemporary comments complained of the insufficient room height of these rooms that caused problems of air conditioning and acoustics.[14] Fougeroux, who presumably referred to the concert on 12 April 1728 in the *York Buildings* and to a concert of the *Castle Society* in *Castle Tavern,* wrote in his travel report:

> While talking about music I have to tell you about the public concerts of London, which aren't much compared to ours. We heard one that was held in a low hall, all painted by very dirty, usually serving as a ball hall; it has one platform at the end where you climb a few steps and there the music is set up. They played some sonatas and sung English and German vaudeville: you have to pay five shilling for these bad concerts, which is 5 fl. and 10 s. We heard also another concert on the first floor of a caffé, where the violins of the opera practice every Thursday. There were only Germans, who played very well but very hard.[15]

In the *Great Room* of the *York Buildings* on *Villars-Street* the fourth unauthorized performance of *Esther* took place, publicly announced as "Never performed in Publick before," which prompted the following reaction by the composer in the *Daily Journal* on 19 April 1732:

> By His Majesty's Command, At the King's Theatre in the Hay-Market, on Tuesday the 2d Day of May, will be performed, The Sacred Story of Esther; an Oratorio in English. Formerly compos'd by Mr. Handel, and now revised by him, with several Additions, and to be performed by a great Number of the best Voices and Instruments. NB. There will be no

12 *HHdb.* 4, 156: letter from Mary Pendarves to her sister Ann Granville from 25 November 1727.

13 W. Salmen, *Das Konzert* (1988), 26.

14 Similar complaints are known from Vienna, see 1.8.2: Room proportions; relation of room height to width.

15 "Pendant que nous sommes sur la musique, il faut vous parler des concerts publics de Londres, qui sont peu de chose en comparaison des nôtres. Nous en entendimes un qui se tint dans une salle basse, toute peinte mais fort noircie, que sert ordinairement de salle à danser; il y a une tribune au bout où l'on monte quelques marches, c'est où se met la musique. On y joua quelques sonates et l'on y chanta des vaudevilles anglois et allemands: on paye pour ces mauvais concerts cinq schelings qui valent 5 ff 10s. Nous entendimes encore un autre concert au premier étage dans un caffé, où les violons de l'opéra s'exercent tous les jeudys. Il n'y avoit que des allemands qui exécutent fort bien, mais qui jouent durement." cf. HHdb. 4, 166: P. J. Fougeroux, *Voiage d'Angleterre d'Holland et de Flandre fait en l'année 1728,* fifth letter.

> Action on the Stage, but the House will be fitted up in a decent Manner, for the Audience. The Musick to be disposed after the Manner of the Coronation Service.[16]

The King's permission put the *Chapel Royal* and the *King's Theatre* with its professional orchestra at Handel's disposal. Hence the "great number of best voices and instruments" mentioned explicitly in the announcement. Forgoing acting and the entailed definition "Oratorium in English language" was a condition to ensure collaboration of the *Chapel Royal,* as the puritan London Bishop and the Chapel Royal's dean had first denied permission for the participation of the Royal Singers in a public theater, despite the Royal request. The Bishop did not object against the performance of a Biblical play but feared that the *Chapel Royal* Boys' performance on stage could bring them in contact with bad people, against whom even the proposed measure of wearing their habit and holding the psalm book would not protect them.[17]

The reference to the coronation of George II in *Westminster Abbey* in 1727, for which Handel had demanded the construction of an amphitheater-like tribune – a great sensation at the time – effectively explained to the public the placement of the choir on raked steps on stage and elicited the association to a solemn ceremony.[18] These promises were made good musically too, as Handel's new version included the Coronation Anthem *My Heart is inditing* and the first part of *Zadok the Priest.* Handel used the occasion to perform these highly appreciated works, which could not be presented to the general public in their original version in the past, in the form of a parody and under completely different acoustic conditions.[19] These famous pieces had the effect of ceremonial *topoi* even in a theater. Handel expanded the 1718 masque *Esther,* which featured six scenes, seven choirs, 14 recitatives, 11 arias, a duet and an arioso, into a three-act oratorio with 16 choirs, more than 20 recitatives and arias, three ariosos and four duets. Four Italian star soloists and two English countertenors also joined the choirs sung by the boys and a few singers of the *Royal Chapel.* This was new for the theater public, as in Italian operas the few choirs were sung by the soloists alone. In order to please Italian bel canto experts Handel wrote *Solo-Allelujas* for the castrato Senesino inserted into the final six-part choir, reinforcing once again the allusion to a sacred ceremony. Even after six successful performances the press printed long vehement condemnations by puritan critics against this indecent transfer of the religious sublime to the theater.

16 *HHdb.* 4, 99.
17 Burney (1785), 100.
18 Actually a picture of the coronation of King George II is not publicly available.
19 On the acoustics, see 2.6: The impact of room acoustics on Handel's compositions.

Esther in this form contained all the characteristic elements of Handel's future English oratorios: recitatives and arias in the manner of the Italian opera seria, more or less virtuoso and dramatic solo parts demanding high-quality singers, a large number of important choirs in the manner of the Anglican *Anthem* and ceremonial church music, whenever possible with an *Alleluja* for choir (sometimes even more than one, sometimes with solo part), an *ouverture* often in French style, sometimes instrumental pieces and at least since 1734 virtuoso solo concertos during the entr'actes between the three parts (mostly organ concertos, sometimes for other solo instruments or *concerti grossi).* The detailed sequence of events shows the development of the new genre of the English oratorio, which was to become a new longstanding field of activity for the singers of the *Chapel Royal.*

2.3.3 Further oratorio performances in London

The emergence of the English oratorio and of odes and *serenatas* with English text happened at the time of the difficult competition between the *Opera of Nobility* under the music director Porpora (founded in 1733 with the support of the Prince of Wales) and the *Second Academy of Music* directed by Handel (founded in 1729 with the King's support), leading in 1741 to Handel's retirement from the Italian opera, which for some time thereafter played only a modest role in London.

At first Handel's oratorio performances were reactions against performances by the *Academy of Ancient Music* under Pepusch or by the *Apollo Society* under Maurice Greene.[20] These societies themselves and others, such as the *Castle Society* or the *St. Cecilia Concert,* later became regular performers of Handel's works.[21] Evidence for these private events, often preserved by chance in diaries or letters, is rarely confirmed by press announcements. In December 1732 Handel reacted against an unauthorized performance of *Acis and Galatea* with a new version, named *Serenata in tre parti* despite its English text. In autumn 1732 Maurice Greene composed the oratorio *Deborah,* to which Handel responded in March 1733 with his *Deborah, an oratorio in three acts,* performed six times until 10 April. Composed mainly from parts of former works, the oratorio requires no less than 10 soloists, a double choir and a large orchestra. Here too Handel included choirs from the *Chandos-* and *Coronation-Athems* and the *Brockes-Passion.*

20 W. Dean, *Handel's Dramatic Oratorios and Masques* (1959), 81 f.; *NOHM* VI, 32.
21 See table B in the appendix.

Until 1750 Handel's London oratorio performances were held exclusively at the theater, where the whole professional infrastructure was at his disposal. An additional advantage was the amphitheatrical podium erected from the orchestra to the stage, providing more space than the usual opera orchestra room, which accommodated only 25 to 30 instrumentalists. The most splendid oratorio score, *Deborah* calls for three-part violins, viola I/II, violoncello and double bass, travers flute I/II, oboe I/II, bassoon I/II, 3 French horns, 3 trumpets and timpani. In addition to this orchestra of 75 musicians Handel used five of his best Italian soloist and a choir with 25 singers.[22]

The exclusive connection to the theater had both good and bad sides. For the large orchestra and Italian stars, Handel was forced to sell single tickets to cover the costs, which were barely lower than for opera, because oratorios could not be sold by subscription. Despite the not staged concert form of these performances, the puritan segment of the public, which usually attended sacred choir concerts at the *Stationer's Hall* (the booksellers' stock exchange)[23] and at *Crown and Anchor Tavern,* continued to criticize the participation of church choirs and the use of biblical topics in the theater. The number of oratorio lovers remained so restricted that in 1740 Handel could still not imagine surviving without Italian opera – neither artistically nor financially.

2.3.4 Handel's 1733 oratorio performances in Oxford, the building of *Holywell Music Room* in 1748 and the beginning of a Handel tradition

Fortunately London did not go so far as Cambridge or Oxford, where the universities occasionally prohibited theater and opera performances. In Cambridge academics who attended theater performances were punished. In 1737 a parliamentary decree forbade actors to be within a distance of less than five miles from universities.[24] A similar rule was enforced in Oxford – one more reason for the success of Handel's oratorios, which were regularly performed there since 1733.[25] *Esther, Deborah* and *Athalia,* the latter composed for Oxford in July 1733, were reportedly performed at the ceremony for the award of

22 Diary of the Earl of Egmont, 27 March 1733, see Deutsch (1955), 309, see *HHdb.* 2, 117, footnote 3.

23 *HHdb.* 4, 60.

24 Dean (1959), 143, footnote 1.

25 D. Burrows, "Sources for Oxford Handel performances in the first half of the eighteenth century" (1980), 177 f.; S. Wollenberg, "Music in 18th-Century Oxford" (1981–82), 69 f.

Handel's honorary doctorate from Oxford University, which, for unknown reasons, the composer did not accept. He was nonetheless invited to conduct the ceremony in order to restore its former dignity after a period of neglect. Handel met the expectations with six concerts at the University's *Sheldonian Theatre,* among which the serenata *Acis and Galatea* and, on Sunday morning, 8 July 1733, the *Te Deum* and *Jubilate* with *Anthems* in *St. Mary's Church.* With hindsight, this concert series appears to be the first Handel Festival. Seventy voices and instruments are said to have performed his new oratorio *Athalia* for 3700 listeners. The U-shaped theater designed by Christopher Wren in 1669 (24.10 × 21.90 m), still extant in its original form, had 1,000 fixed and 1,000 additional seats; this warrants the conclusion that the reported number of 3700 listeners refers to the total of the two performances (figure 73).[26]

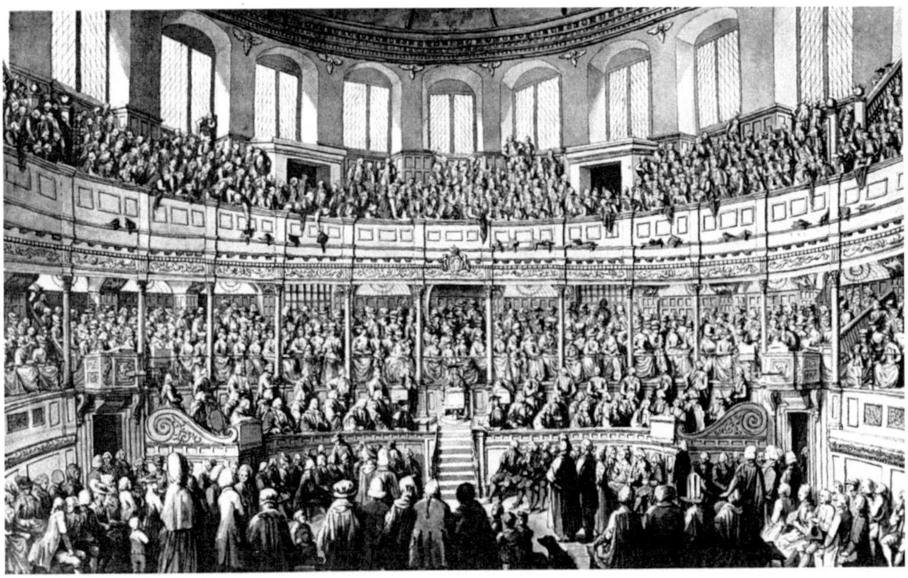

Figure 73: Oxford, *Sheldonian Theatre,* view from the organ gallery: University Commemoration (1781).

The next Handel performances of the following years in this city are insufficiently documented, as closed events held by the different music societies in inns, at *Christ Church Hall,* in other College halls and in private homes were rarely announced in the press. Preserved performance material from Oxford, which includes also as yet unpublished works, indicates ongoing direct relations with Handel and a lively Handel tradition.[27]

26 G. Beard, *The Work of Christopher Wren* (London) 1987, 16; see appendix, table A.
27 Burrows (1980), 177 f.

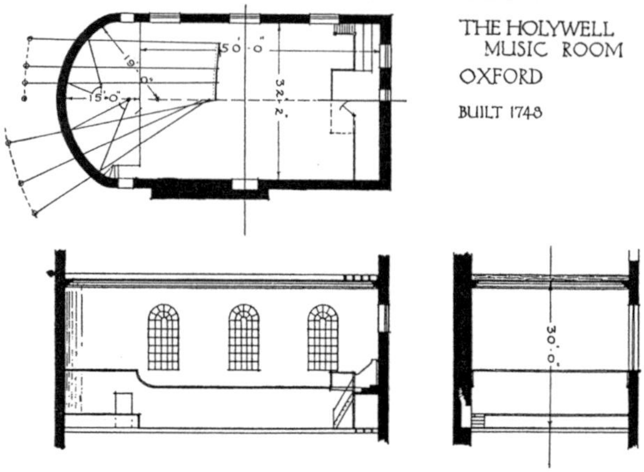

Figure 74: Oxford, *Holywell Music Room* (1748).

In 1742 the *Oxford Musical Society,* then directed by the organist and music professor William Hayes, a staunch Handel admirer, began to plan the *Holywell Music Room* (figure 74). The money was collected through the performance of oratorios and choir and instrumental music by Giovanni Bononcini, Geminiani, Hasse and Handel at *Christ-Church Hall* and *King's Head.* The still extant, then 300-seat, *Music Room* (20 × 9.80 m, 9.15 m high) was inaugurated six years later, on 9 July 1748, with a performance of Handel's *Esther.*[28] More oratorio performances had to benefit the completion of the interior construction. Directed by Hayes, the first performance of *Messiah* was held in 1749 at the *Sheldonian Theatre* with 50 instruments and a "large choir".[29] Since 1754 performances alternated between the theater and the much smaller *Holywell Music Room.*[30] Though no public concerts by Handel are documented, he may still have played the organ or cembalo in this hall.[31]

Unfortunately the acoustics changed so much due to the installation of steeply raked seating rows at the back of the hall and to the removal of stucco decorations that the curtain now covering the side walls does not help restore the former much better conditions. Nevertheless, *Holywell Music Room* is one of the very few preserved small music rooms of category B from Handel's time. The number of seats offers evidence that the hall is smaller than 450-seat

28 Deutsch (1955), 653.
29 *HHdb.* 4, 424; Dean (1959), 106; W. Shaw, *Messiah* (1965), 56.
30 See table 14, chapter 2.6.4.3.
31 H. Bagenal and A. Wood, *Acoustics* (1931), 99 f.

Neal's Music Room, the hall used by Handel in winter 1741–42 for his oratorio Saison in Dublin.[32]

2.3.5 Handel's oratorio performances in London until 1739

In the 1734–35 season, after the move from the *King's Theatre* to the *Covent Garden Theatre* with its large apron stage and good acoustics, Handel started to play organ concertos and other solo concertos between the acts in his oratorios. These additions became an even greater attraction than his harpsichord concerts and improvisations played in his operas since *Rinaldo* in 1711.[33] For this purpose Handel provided the theater with a new large organ.[34] In the meantime the situation of Italian opera once again deteriorated. On 11 June 1737 the *Opera of Nobility,* at the time playing at *King's Theatre,* was forced to close for financial reasons. Queen Caroline's death delayed the opera season. Handel's *Funeral Anthem* was rehearsed at *Banqueting Hall* in *Whitehall* in the presence of the Royal family (figure 75) and on 17 December 1737 was performed at *King Henry VII's Chapel* in *Westminster Abbey* (figure 76) with no less than 100 instruments and 80 singers.[35]

After his badly sold opera performances, held again at *King's Theatre,* Handel finally reaped great personal and financial success during the holy week 1738 with the oratorio pasticcio from *Esther, Athalia, Deborah,* which included an organ concerto and the anthem *Zadok the Priest* as final choir. The Earl of Egmont counted over 1,300 listeners without the *Gallery* and *Upper Gallery*. Benches for 500 listeners were placed on the stage behind the orchestra and the choir in order to accommodate the waiting crowds.[36]

The 1738–39 opera season closed early due to poor subscription. The successful performance of Handel's new oratorio *Saul* in January 1739 at the rented *King's Theatre* featured trombones, the big *Tower* timpani, a carillon and a new organ combined with harpsichord (a so-called *claviorganum).* It was followed by *Alexander's Feast* and the Italian three-part oratorio *Il Trionfo del Tempo e della Verità.* The series concluded with another performance of *Alexander's Feast,* Handel's first documented charity "for the Fund for the

32 See 2.4.2: The Handel-Saison 1741–42 in Dublin.

33 *HHdb.* 3

34 *HHdb.* 2, 117, footnote 5. Handel bequeathed the organ to the *Covent Garden Theatre* in his will. Unfortunately it has been destroyed by fire in 1808.

35 Hogwood (1984), 145.

36 Deutsch (1955), 455: Diary of the Earl of Egmont, 28 March 1738.

Support of Decayed Musicians and their Families." In the entr'actes of all performances organ concertos or other solo concertos were played. In April *Israel in Egypt,* the first oratorio before *Messiah* based exclusively on texts from the Bible, which contained considerably more choirs than arias (19 choirs, four arias and four duets), again triggered critical discussions among the puritans because biblical texts were performed in a theater. This oratorio, which used many pieces from other works, among which the newly texted *Funeral Anthem* for Queen Caroline, garnered only lukewarm success.

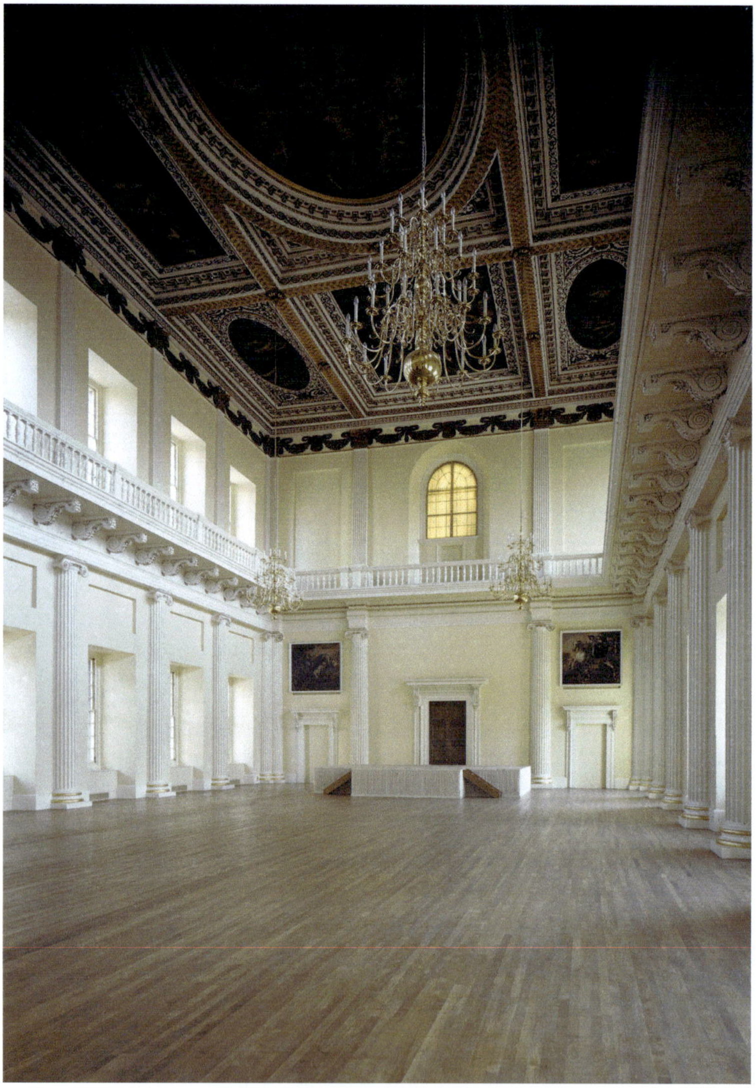

Figure 75: London, *Whitehall, Banqueting Hall* (1621).

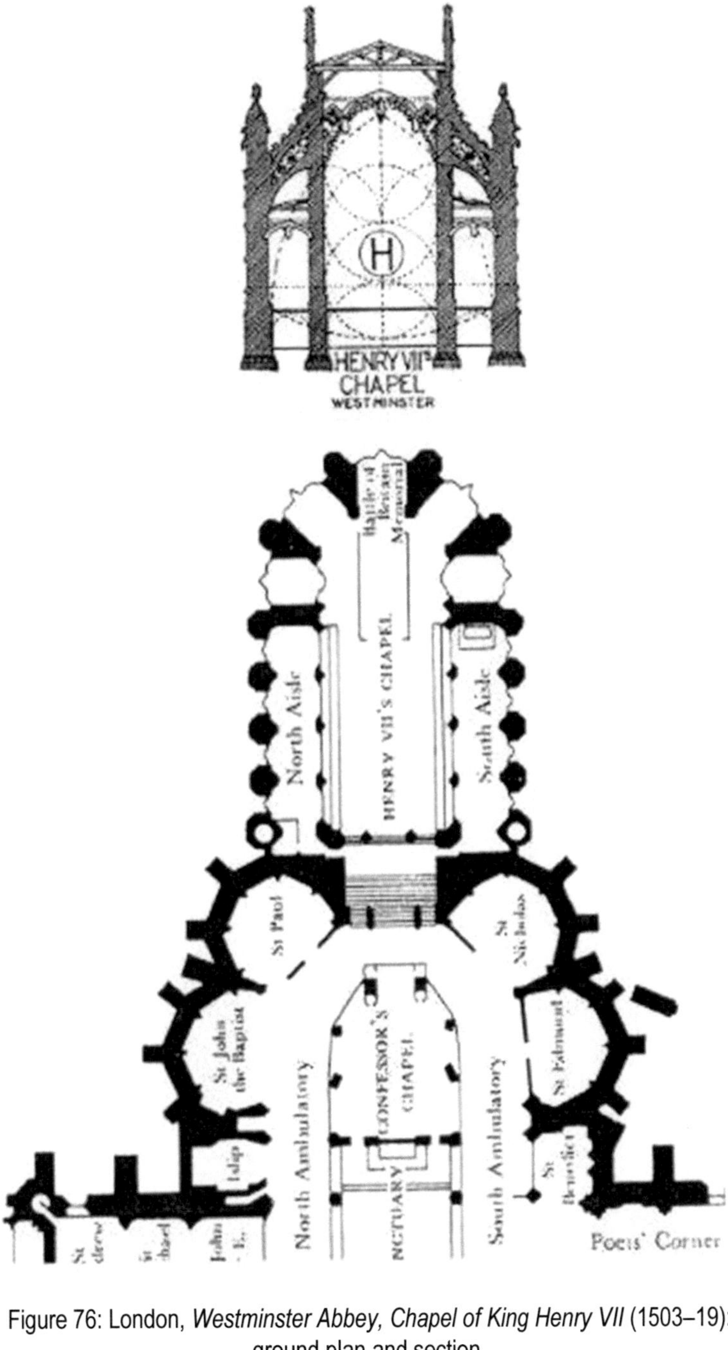

Figure 76: London, *Westminster Abbey, Chapel of King Henry VII* (1503–19): ground plan and section.

2.4 The Oratorio *Messiah*

2.4.1 Introductory note

During the summer of 1740 Handel visited the continent, looking for opera singers though unsuccessfully. The following winter the performance of his last Italian opera *Deidamia* at *Lincoln's Inn Fields Theatre* was a failure. In April Handel did not protest against the announcement of his "last concert". It was rumored in London that he would go back to Germany.[1] He received from Charles Jennens the libretto for *Messiah* in July 1741, at the latest, and composed the work in about 20 days, from 22 August to 14 September.[2]

Messiah is the first oratorio that describes the whole life of Christ on the exclusive basis of texts from the Bible. Charles Jennens, who had already written the libretti of *Saul* (1739), *Israel in Egypt* (1739) and *Il Moderato* for *L'Allegro ed il Penseroso* on a text by Milton, used the Anglican "Authorized Version" of 1611 and partly the *Great Bible* as it appears in the 1539 *Book of Common Prayer.* Christ does not appear in person. Described mostly in direct narration, the events of the New Testament are illustrated and commented by passages of the Old Testament. The founder of the Christian faith is identified with Messiah, the king promised by the old prophet. In the autograph the tenor as main narrator is reminiscent of the evangelist in the German passion oratorio. Handel's later changes reduce his special position in favor of a more balanced distribution of the narration to all soloists and the choir in a manner more akin to the composer's odes, mainly *Alexander's Feast* and *L'Allegro.* Soloists and choir are less participants in a drama than common believers transmitting the Messiah's message. Against this background the few moments of direct discourse, such as the angel's "Fear not: for behold, I bring you good tidings," are of special importance. The text taken from the Bible contains all the important feasts of the Anglican liturgical year, the prophecy of Messiah and his birth, crucifixion, resurrection, ascension and salvation. Handel's musical composition is a synthesis of elements from the Italian *oratorio volgare*, the Italian opera, the English *masque,* the northern German passion oratorio and Anglican church music. The participation of church choirs and English soloists brings the work especially close to the listeners.

1 O. E. Deutsch, *Handel* (1955), 518: diary of Earl of Egmont, 8 April 1741.
2 *HHdb.* 2, 194 and 4, 336.

2.4.2 The 1741–42 Handel-Season in Dublin

Extant documents do not confirm whether Handel received an official invitation from Ireland's governor William Cavendish for a Dublin season in winter 1741–42 before or after the autograph was written. Nor is it certain that the Dublin *Charitable Music Society* commissioned Handel to write *Messiah*. But the circumstances hint to his careful planning of the season in Dublin and of the performance of the new oratorio. For *Messiah* Handel needed an excellent professional choir. Unlike the oratorio *Samson,* which was also composed before the Ireland journey but finished only afterwards and used 11 soloists (four soprani, alto, four tenori, basso) representing different roles, *Messiah* in the autograph version has only five solo parts not tied to any personal roles (soprano, alto I, II, tenore, basso) and assigned by Handel to several soloists. The orchestra is reduced to a minimum (strings, basso continuo, trumpet and timpani),[3] whereas the orchestra of *Samson* requires, in addition to trumpet I/II, traverse flute I/II, oboe I/II, bassoon I/II, French horn I/II, obligate organ. Handel seems to have conceived *Messiah* for an eventual performance outside of London that would be easily adaptable to more modest performance conditions. A hint to his initial intention of composing the work for a charitable performance appears in the announcement of a delayed first performance in *Faulkner's Dublin Journal* on 10 April 1742, where the "many well-wishers to this Noble and Grand Charity for which this Oratorio was composed" are mentioned.[4] The same publication welcomed Handel on his arrival in Dublin as the composer of the *Dettingen Te Deum* and *Jubilate,* which, already before his visit, were performed yearly in Dublin in *St. Andrew's* to benefit *Mercer's Hospital* after the model of the performances in *St. Paul's* in London.[5]

Shortly after his arrival Handel agreed to participate in a charitable concert for *Mercer's Hospital* in the *Round Church St. Andrew's* with the performance of a *Te Deum* and *Jubilate* and one of the Coronation anthems. A new anthem by William Boyce, composer of the *Chapel Royal* in London, was also played. Another charitable concert with a similar program was held on 8 February 1742 at the same church.[6] Besides Handel's sincere commitment to charity benefits, his obvious intention was to work with the choirs of Dublin's two main churches, *Christ Church* and *St. Patrick's Cathedral*, in order to try them out

3 For the eventual participation of woodwinds, see 2.4.3 and 2.6.5.2: Orchestra size and volume of space.

4 Deutsch (1955), 454.

5 H. C. R. Landon, *Handel and His World* (1984), 175 f.

6 *HHdb*. 4, 342 and 345: minutes of Mercer's Hospital from 22 January 1742, libretto of the performance of 8 February 1742.

for his later oratorio performances and to eventually obtain permission for their participation in his own oratorios.[7] The soprano Christina Maria Avoglio and the mezzosoprano Susanna Maria Cibber (the sister of the composer Thomas Arne and the daughter-in-law of the actor Colley Cibber), the organist Mr. Maclaine and his wife, a contralto, came with Handel from England. The male soloists were members of the Dublin cathedral choirs, now carefully prepared by Handel for the task. The violinist Matthew Dubourg, *Master of the State Music,* was leader of the orchestra. Dubourg, a pupil of Geminiani who was a pupil of Corelli, knew Handel from London and had performed already on 8 April 1736, partly with the same male soloist and more than 70 participants, Handel's *Te Deum, Jubilate* and *Anthem.*[8] Later he launched the Dublin tradition of the yearly performance of *Messiah* in December.[9] The orchestra for the Handel season comprised the best musicians of the two large partly amateur music societies, the *Academy of Music* and the *Charitable Music Society.*

Handel's satisfaction with the Dublin conditions is evident in a letter he wrote on 29 December 1741 to Charles Jennens after the success of the first six concerts:

> The Nobility did me the Honour to make amongst themselves a Subscription for 6 Nights, which did fill a Room of 600 Persons, so that I needed not sell one single Ticket at the Door. and without Vanity the Performance was received with a general Aprobation. Sigra Avolio, which I brought with me from London pleases extraordinary, I have form'd an other Tenor Voice which gives great Satisfaction, the Basses and Counter Tenors are very good, and the rest of the Chorus Singers (by my Direction) do exceeding well, as for the Instruments they are really excellent, Mr Dubourgh beeng at the Head of them, and the Musick sounds delightfully in this charming Room, which puts me in such Spirits (and my Health being so good) that I exert my self on my Organ with more than usual Succes.[10]

The first series of six subscription concerts was performed at the *New Music Hall* (also *Neal's Music Hall* in acknowledgement of the generous donation by the music sellers John & William Neal to cover building costs),[11] which was opened by the *Charitable and Musical Society* shortly before Handel's arrival. Designed by the architect Richard Castell, the hall had more than 450 seats and

7 Documents by the Dean Jonathan Swift temporarily cancelling the permission for the participation of the choirs of *St. Patrick's*. With the help of friends it was conceded again. See Deutsch (1955) 536 f.

8 *HHdb*. 4, 262: *Pue's Occurence,* critic of 10 April 1736.

9 The first performance was held on 7 February 1744 after postponements, see: W. G. H. Flood, "Fishamble Street Musical Hall, Dublin, from 1741 to 1777" (1909/10), 51–7, especially 55; *HHdb.* 4, 367.

10 *HHdb*. 4, 341.

11 *HHdb*. 4, 341; William Neal was treasurer of the *Charitable Music Society.*

excellent acoustics.[12] Besides organ concertos and instrumental works, Handel performed there *L'Allegro, il Penseroso & il Moderato, Acis and Galatea* with the Ode for St. Cecilia *From Harmony* and the oratorio *Esther,* and in a second series *Alexander's Feast, L'Allegro, Esther* and the serenade *Hymen* (a concert version of the opera *Imeneo),* when necessary in adapted or reduced versions.

2.4.3 The performances of *Messiah*

Handel planned the *Messiah's* first performance for the end of his stay in Ireland, namely, April 1742. Such a late date was not only close to Easter but, after the performance of other works by Handel, it also allowed a trained ensemble to perform the new oratorio as well as possible.

For Dublin it was especially important that Handel's *Messiah* was performed in the vernacular, bound to the Bible but not to liturgy, and definitely not in a church. The work thus attracted the new middle-class public, whose economic rise and prosperity allowed it to seek not only at material goods but also education and culture. This new social, cultured group in Ireland and England fostered a puritan attitude that the open-minded Handel disliked though he shared the need for Christian love expressed in the foundation of numerous *charitable societies.* These pious societies organized charity concerts for the poor and poorest of a quickly growing population increasingly concentrated in the cities. Not least, for these events new church assembly halls, music rooms and inn's halls were built. Handel and the concert organizers proved to be very sensitive to this tendency: as a charity concert, the performance of *Messiah* ensured not only the participation of the choirs[13] but also a full hall and the appropriate atmosphere for the work's reception. On 27 March 7142 the first performance was announced in the *Dublin Journal* as follows:

> For Relief of the Prisoners in the several Gaols, and for the Support of Mercer's Hospital in Stephen's Street, and of the Charitable Infirmary on Inns Quay, on Monday the 12th of April, will be performed at the Musick Hall in Fishamble Street, *Mr. Handel's new Grand*

12 The first series of subscriptions sold 150 tickets giving access to three persons. With 600 listeners for Handel's concerts the hall was much more occupied than usual, see Flood (1909/10), 52.

13 The participation of the cathedral's choirs was permitted only on the condition of charity performances, see *HHdb.* 4, 346: minutes of the *Mercer's Hospital* of 4 March 1742.

Oratorio, call'd the MESSIAH, in which the Gentlemen of the Choirs of both Cathedrals will assist, with some Concertoes on the Organ, by Mr. Handell.[14]

The announcement of the postponement to 13 April due to organizational reasons published in *Faulkner's Dublin Journal* included the request ladies should forgo crinolines to allow space for more listeners.[15] The review of the first performance published in the same journal on 17 April 1742 indicates that the direct relation between the history of Messiah and the practice of Christian love was reflected not only in the number of attendants but also in the sum of collected money:

> The Sublime, the Grand, and the Tender, adapted to the most majestick and moving Words, conspired to transport and charm the ravished Heart and Ear. It is but Justice to Mr. Handel, that the World should know, he generously gave the Money arising from the Grand Performance, to be equally shared by the Society for relieving Prisoners, the Charitable Infirmary, and Mercer's Hospital, for which they will ever gratefully remember his Name. [...] There were about 700 People in the Room, and the Sum collected for that Noble and Pious Charity amounted to about 400 l. out of which 127 l. goes to each of the three great and pious Charities.[16]

On 3 June 1742 Handel was offered the occasion for a second performance for his own profit.[17] As a farewell gift he donated to the *Charitable Music Society* a new organ consecrated at the beginning of the next season on 8 October 1742.[18] This donation confirms Handel's never realized intention to return to Dublin, as he provided with organs also other halls where he intended to play concerts (in 1735 the *Covent Garden Theatre* and in 1750 the *Foundling Hospital Chapel).*

2.4.4 First performance: ensemble size and hall size

The dimensions of the no longer extant *New Music Hall* in Dublin are not known. After a collapse in 1755 the hall was reconstructed and in 1777 converted into a theater with two ranks of boxes. Closed around 1800, it was conver-

14 Deutsch (1955), 542.
15 Deutsch (1955), 545.
16 Deutsch (1955), 546.
17 Deutsch (1955), 550.
18 Flood (1909/10), 53.

ted in 1866 into an industrial hall and later pulled down.[19] Pictures of the façade (figure 77a)[20] show a two-story building clearly higher than the neighboring houses. A picture from the *Hibernian Magazine* of March 1794 shows the hall as a theater (figure 77b).[21]

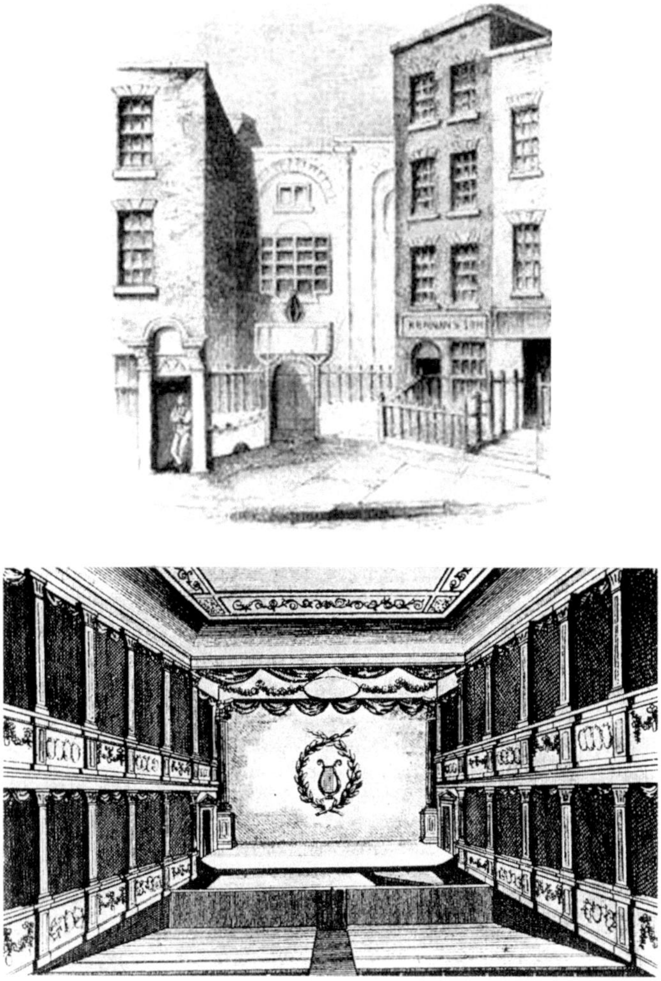

Figure 77: Dublin, *Neal's Music Hall* (1741): a) façade, b) later view of theater

19 Dublin had also other halls, such as *Mr. Johnson's Hall* in Crow Street, 1731 seat of the *Musical Academy for the Practice of Italian Musick,* 1732 of the *Anacreontic Society,* 1758 changed into a theater, see Flood (1909/10), 55.

20 Landon (1984), 179 (undated) and W. Shaw, *Messiah* (1965), print from 1840, after an older colorated pencile design.

21 M. Forsyth, *Buildings for Music* (1985), 51, figure 2.23; besides these pictures and the mentioned view of the facade now further designs or plans are known.

The rectangular hall was high enough for the insertion of two lateral galleries and a slightly inclined parterre. The room was long and narrow even before the addition of the lateral galleries. These facts confirm sufficiently the favorable acoustics praised by Handel. With 450 seats (or 600 to 700 seats and standees when overcrowded) that hall is comparable to small halls of category B, such as the old *Gewandhaussaal Leipzig* (23 × 11.50 × 7.40 m), then without galleries, also barely two stories high, which featured excellent acoustics and admitted 860 listeners at the then usual 0.3 m^2 per person.[22] The Dublin hall was presumably somewhat smaller than the hall in Leipzig (figure 55a) but larger than *Holywell Music Room* in Oxford (19.85 × 9.80 × 9.15 m) with 500 seats at 0.3 m^2 per person, where regular performances of Handel's oratorios had been held since 1748 (figure 74).[23]

The exact numbers of musicians for the Dublin *Messiah* performances is not documented. Presumably, Handel directed from the harpsichord, and Mr. Maclain played the organ. The choirs of both cathedrals numbered between 20 and 32 boys' and men's voices.[24] Except the Italian first soprano Christina Maria Avoglio, the soloists were English: Susanna Maria Cibber (contralto), William Lamb and Joseph Ward (countertenor), James Baileys and John Church (tenor), John Hill and Joseph Mason (bass).[25] The eight soloists participating in the choirs bring the total of voices to 28–34.

The autograph score calls for strings, basso continuo, 2 trumpets and timpani but has no separate oboe and bassoon parts and no indications for the participation of woodwinds in the tutti.[26] It is impossible to draw final conclusions about the delicate question of woodwinds in the Dublin *Messiah* orchestra. The pay lists for Handel's later *Messiah* performances at the *Foundling*

22 See 1.8.3: The acoustic similarity of rooms, and 1.8.10: The number of attendants.

23 See 1.9.1, table 10b.

24 *HHdb*. 2, 195: 20 [these 20 voices probably do not count the male soloists]; Chr. Hogwood, *Handel* (1984), 176: "26 boys of both cathedrals" – [probably by mistake 26 instead of 16?]; Tobin in the preface of *HHA* I, 16 says: "16 men's and 16 boys' voices". None of the authors give direct sources for the numbers. The only confirmed number according to our knowledge is Jonathan Swift's permission for 8 singers, "six Vicars Choral and two Choristers", which could have been no more than half of the alto, tenor and bass. See Flood (1909/10), 52.

25 *HHdb*. 2, 195; Burrows mentions Mrs. McLean [sic] instead of Christina Maria Avoglio for the soprano solos annotated in the only preserved libretto (British Library K.8.d.4), a source which in all other cases is confirming Handel's score annotations and the information in the press. If the libretto owner was right, either Mrs. McLean was identical with Christina Maria Avoglio and married to the organist Mr. MacLain, or the soprano McLean really replaced Avoglio in the second performance on 3 June 1742, see D. Burrows, *Handel: Messiah* (1991), 22.

26 Only exception is the in 1745 added new choir "Their sound is gone out into all lands" with woodwinds in the score, see also 2.5.2: Handel's *Messiah* performances at the theater.

Hospital include oboes and bassoons, and the performance tradition continued with woodwinds. A performance without these instruments thus seems very unusual.[27] A hint in the score to an original conception without woodwinds is the *Pifa* (No. 12), an imitation of the bagpipe that seems to call for woodwinds. But in the preserved *Foundling Hospital* material this piece remains without woodwinds and is played by strings only in the impressive setting for *violino I, II, III e viola all'ottava bassa* and *continuo* – in fact a trio with all parts in octaves.[28] Furthermore, the technically demanding choirs need neither a large number of singers nor the support of woodwinds for a convincing sound. The magnificent effect is already set in the music's structure.[29]

In this small and, according to Handel's own comment, acoustically excellent hall, his work must have reached an unusually moving intensity notwithstanding the small ensemble. The balance between the sublime, the great and the tender obviously corresponded to Handels wish.[30] This is explicitly confirmed by reverend Dr. Delany, the later husband of Mary Granville-Pendarves who, like her brother Bernard Granville and her sister Ann Granville-Dewes, was among Handel's closest English friends. After the aria "He was despised", overwhelmingly sung by the contralto Susanna Maria Cibber, the reverend is said to have exclaimed: "Woman, for this, be all thy sins forgiven!"[31] The meaning of this comment becomes fully clear only if one

27 For the *Foundling Hospital* version, see 2.5.1, for the orchestra at *King's Theatre* and *Covent Garden Theatre,* see 2.5.2.

28 All scores of the other works performed in Dublin ask for woodwinds: double oboes and bassoons for *Acis and Galatea,* three bassoons for *Alexander's Feast,* not doubled bassoon for *From Harmony*. Even if we have not evidence of how these works were played at the time (it is known that Handel adapted the works to the more modest circumstances), it is unlikely that all were performed without woodwinds. In the *Foundling Hospital* parts tutti oboes always play *colla parte* with violino 1 or 2 or with the choir's soprano, bassoons *colla parte* with the bass. The lack of indication in the autograph and in the conducting score is neither an argument for a *basso continuo* without bassoon nor a proof for the lack of woodwinds. Performances on baroque instruments without oboes in a small hall are absolutely convincing. The orchestra then follows Italian baroque manner. Handel's Italian oratorio *La Resurrezione* makes evident that the size of the hall cannot be used as argument for or against woodwinds because already the rehearsals in the small *stanzione delle Academie* of *Palazzo Bonelli* were done with 4 oboes and 2 bassons. Without paying lists or original parts the question of the participation of woodwinds remains an artistic decision of the conductor.

29 See 2.6.4: Space, genre and music structure.

30 On the relation between room acoustics, compositional technique and ensemble size, see 2.6 and especially 2.6.5: size of ensemble and room in comparison to other performances by Handel, and the sound examples in 2.8.

31 V. Schoelcher, *The Life of Handel* (1857), 249. Reverend Delany and maybe also the listeners knew Mrs. Cibber's personal fate, that in London she had left her husband, the actor Theophilus Cibber, because of his alcohol problems and had secretly taken a separate

bears in mind that a woman, being a "sinful" wife, would have never been permitted in Anglican liturgy to sing this description of Jesus being jeered. The model of the German passion oratorio with women soloists is unmistakable here. Indeed, this aria looses much of its effect if sung by a countertenor, mainly in the middle part in C minor "He gave his back to the smiters," which needs the dramatic power of a mezzo-soprano, a contralto or a castrato with a good *voix mixte* for these relatively low notes.[32]

The Irish Reverend's compassion touches us because it vividly transmits the atmosphere and the acoustics of the Dublin hall: everyone could easily understand his exclamation, and it is hardly conceivable that in a larger, less intimate hall the Reverend would have dared such a comment, as its necessary loudness would have interrupted the performance. Interjections were usual at the time in the theater and at the opera in halls of similar size.[33] The intimate acoustics are confirmed by an anecdote about Handel's reaction to a very long and risky solo cadence by the concertmaster Dubourg. When the violinst ultimately reached the final trill Handel, though facing the orchestra, said clearly enough for everyone to hear: "Welcome home, Mr. Dubourg!."[34]

apartment and that her husband in 1738 went to court against her lover. In fact these Dublin concerts were Mrs. Cibber's first public come-back before the later successful performances after her return to London in 1742.

32 Handel obviously wrote this aria for a woman alto. The use of a countertenor according to our opinion is a misunderstanding of "historic" performance practice not meeting Handel's own practice. In lack of a woman alto he used to give the part to an alto castrato. The opposite practice to use a woman's voice instead of a castrato in opera is documented in a letter by Handel from 27/16 October 1730 to the British ambassador in Florence, Mr. Colman, seeking women singers able to sing and play men's parts instead of castratos. See *G. F. Händel, Biographie von John Mainwaring, Briefe und Schriften,* ed. by H. and E. H. Mueller von Asow (1949), letter XII (in French).

33 *The London Theatre World,* ed. by R. D. Hume (1980), 52 ff.; on the acoustics and dimensions of the London theaters, see 2.5.3 and 2.5.4.

34 Burney, *Account* (1785), cf. Hogwood, *Handel* (1985), 176.

2.5 Handel's performances of *Messiah* in London

After Handel's return to London, a new wave of Puritanism generated an atmosphere inimical to a performance of *Messiah*. Following the model of the Dublin season, Handel organized a first subscription series at the *Covent Garden Theatre* that offered six very successful concerts of the oratorio *Samson,* composed shortly before his journey to Ireland, and a second series consisting of *L'Allegro* and the *Ode for St. Cecilia.* At the end of the season, from 23 March 1743, three performances of *Messiah* were held under the title *A New Sacred Oratorio* with an organ and a violin concerto (Dubourg was briefly in London at the time and had already taken part in *Samson).* Despite the innocuous title Puritans were fiercely critical, thus thwarting the work's complete success.[1] Only from 1749 was *Messiah* performed regularly at the end of the season at the *Covent Garden Theatre*.[2]

2.5.1 Handel's concerts at the *Foundling Hospital*

Among the donors of the *Foundling Hospital,* founded in 1739 by the retired captain Thomas Coram as a "Hospital for the Maintenance and Education of Exposed and Deserted Young Children", were not only aristocrats and members of the middle class but also artists, such as the painter Hogarth and the music publisher Walsh. The construction in *Lamb's Conduit Fields* began in 1742. How Handel made contact with this charitable institution, which enjoyed also the King's support, is not known. On 27 May 1749 he directed a charitable concert to raise money for the chapel's completion. At this matinée, held in the bare brickwork of the chapel[3] with "more than 100 musicians for more than 1,000 listeners," the following works were performed: the *Fireworks Music* HWV 351 in a version with strings added to the wind instruments of the original version ordered by King George II for the performance during the fireworks on 27 April 1749 at *Green Park,*[4] parts of the oratorio *Solomon* and

1 *HHdb.* 4, 359.
2 For further analyses of these performances, see 2.5.3: The theaters used by Handel and 2.5.4: Documents on the acoustics of the London theaters.
3 *HHdb.* 4, 428.
4 Handel then had in vain insisted to have strings, see *HHdb.* 3, 123.

the newly composed *Foundling Hospital-Anthem* HWV 268 "Blessed are they that considereth the poor" (taken partly from the *Funeral Anthem* and *Susanna,* the music ended with the *Halleluja* from *Messiah).*[5] At the suggestion of King George II, who supported the *Foundling Hospital* with considerable sums, after this impressive and profitable performance Handel was elected board member of the foundation, although he was at first reluctant to accept this honorary position. Not subject to church control, the board could decide independently on the chapel's use.[6] This was how Handel secured the first appropriate space in London for the performance of his oratorios without the interference of Puritan circles.

2.5.1.1 *Messiah* at the *Foundling Hospital*

On 1 Mai 1750, planned as the chapel's inauguration date, then announced as the organ consecration in the still unfinished chapel and, finally, without the still unfinished organ, *Messiah* was performed for nearly 1,400 listeners. The five soloists were a boy's soprano,[7] Caterina Galli (mezzo soprano), Gaetano Guadagni (alt castrato), Thomas Lowe (tenor) and Thomas Reinhold (bass). The concert was repeated on 15 May for those who had been denied entrance on 1 May due to lack of seats. This was, in fact the *Messiah's* big breakthrough in London. As in Dublin, a new group of listeners was attracted, who welcomed the presentation of the work outside the compromising theater. From now on the oratorio was played yearly on Easter for the hospital's benefit and on 18 April 1751 for the consecration of the organ donated by Handel, which was in use until a new organ replaced it in 1769. The enthusiasm was so outstanding that from now on Handel also performed *Messiah* at the *Covent Garden Theatre* once or twice at the end of nearly every spring season. The Puritan ban had been broken. Thanks to the charitable performances at the *Foundling Hospital* the oratorio achieved an importance comparable to the *Utrecht Te Deum* (of 1713) and the *Dettingen Te Deum* (of 1744), which were performed yearly

5 *HHdb.* 4, 428; Deutsch, *Handel*, 669; W. Dean: *Handel's Dramatic Oratorios and Masques,* Oxford 1959, 526; Hogwood, *Handel,* London 1984, 217, by mistake gives 19 May. The concert was postponed on the wish of the Prince of Wales from 24 to 27 May. The preserved librettos are dated with 25 May.

6 N. Temperley: "London", § I, 5: Charities and proprietary chapels, *NG* 11, 149.

7 Giulia Frasi, the soprano of the second performance, does not seem to have sung at the first performance at the *Foundling Hospital;* s. D. Burrows, *Handel: Messiah,* 1991, 38.

at the *Festival for the Sons of the Clergy* in *St. Paul's Cathedral*.[8] But unlike these liturgical works, *Messiah* was not performed during the service.

2.5.1.2 *Messiah* after Handel's blindness

In February 1751 Handel had to interrupt the composition of *Jephta* because of problems with his left eye. Despite a treatment in Bath and Cheltenham the sight of his right eye also deteriorated. In August 1752 a stroke caused blindness. A cataract operation brought temporary relief, but since 1754 Handel could no longer direct his oratorios himself.[9] Thanks to his art of improvisation he regularly played organ concertos between the acts. The minutes of the *Foundling Hospital* continued to list him as director of the *Messiah* performances, but John Christopher Smith junior took over the direction and John Christopher Smith senior the organization. The *Foundling Hospital* archives of 1754 to 1759, when Handel did not pay the musicians himself, have preserved four bills with lists of the participating singers and musicians that indicate the exact setting of these performances.[10] In his last will Handel left a score and a set of parts of *Messiah* to the *Foundling Hospital*. A set of parts written by Handel's long-time copyist in 1760, one year after Handel's death, confirms the 1754 lists and shows that Handel had not made or permitted any score changes since he went blind.

2.5.1.3 Size of orchestra and room at the *Foundling Hospital*

The exact measures the *Foundling Hospital Chapel*, destroyed in 1926, are not known. John Sanders' etching of 1773 (figure 78)[11] shows a high rectangular hall with a flat, round-edged ceiling and a narrow gallery with regressing walls and supported by columns. The ticket sales of the first *Messiah* performance suggest that the maximum number of seats and standees must have been slightly less than 1,400. By modern standards the hall was certainly overcrowded (the ladies were again asked to come without or with only small crinolines, and the men to dispense with swords). On 15 May 1754, 1,219 tickets were sold. On the basis of the picture, the gallery seating can be estimated at 150–200 persons. This seating capacity suggests that the chapel must have had a some-

8 *HHdb.* 4, 189; Deutsch, *Handel* (1955), 270 f. At the beginning of this service from about 1720 to 1743 the Ouverture of *Esther* was played.

9 *HHdb.* 1, 35.

10 See 2.6.5.2, table 14.

11 H. C. R. Landon, *Handel and His World* (1984), 199.

what smaller ground plan than the later *Concert Hall* at the *King's Theatre* from 1792, a long rectangular hall without gallery with a flat ceiling (29.60 × 14.60 × 11.90 m), whose 4,500-m³ volume at the time admitted no more than 1,200–1,400 persons.[12] The ground plan must have been larger than the old *Gewandhaus* (23 × 11.50 × 7.50 m), which, without a gallery, accommodated 860 listeners and, once the gallery was added in 1842, 1,000 listeners. Since the *Foundling Hospital Chapel* featured an organ gallery with raked steps for the choir and the orchestra, the room height can be estimated at 12 m and the volume at about 4,000 m³. With these dimensions the hall belonged to category B.[13]

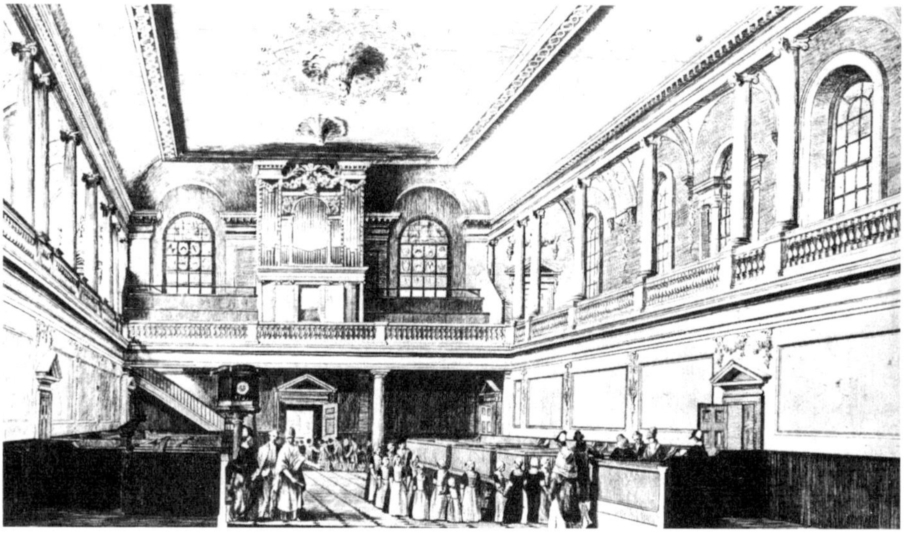

Figure 78: London, *Foundling Hospital Chapel* (1750), engraving 1774.

Given the visible interior installations, the acoustics must have been good or even excellent. Mary Delany-Granville, like her sisters among Handel's closest friends, reported on the *Foundling Hospital* performance of May 1754:

> [...] the music was too fine, I never heard it so well performed. The chapel is fine, and the sight of so many poor children brought up (I hope to good purpose), was a pleasant sight.[14]

For this performance the preserved *Foundling Hospital* payrolls list five soloists (two sopranos, one mezzo soprano, one tenor and one bass), 39 instrumentalists and 19 choir singers (13 men and six boy sopranos) and for three diffe-

12 See table 9.

13 See table 10a and b.

14 HHdb. 4,481, entry of 16 Mai 1754; Mary Delany: *Autobiography and Correspondence of Mary Granville, Mrs. Delany,* London 1861–62, Vol. III, 27.

rent years an ensemble of five or six soloists (in 1750 a boy as soprano solo and the alto castrato Guadagni, in 1759 the alto castrato Ricciarelli) with 34 instrumentalists and 17 or 18 choir singers (1756, 1758 and 1759). The strings played in 8-6-6-3-2 (1754) or 6-6-3-3-2 (for the other years) and, as seen from the bills, 4 oboes played in the Ouverture with the violins and in the choirs with the soprano (sung by six boys and the solo soprano)[15] and 4 bassoons supporting the bass. Double reed instruments increase loudness and provide clearer articulation due to the short onset of sound. In *colla parte* their playing is therefore perceived as mixed sound color that renders the boys' sopranos more focused and intense. Similarly, Handel used the 4 bassons to make the bass line clearer, as stringed bass instruments in a large hall are easily blurred due to the slow onset of sound.[16] This is why bassoons are often used for recitatives and arias to support the *basso continuo*. In addition to the 2 trumpets and timpani (kettledrums), the *Foundling Hospital's* bills list 2 French horns, which appear neither in the score nor in the *Foundling Hospital* parts. Modern recordings with baroque instruments use them in the choir finale of the second and third part, in no. 39 "Hallelujah" and in no. 47 "Worthy is the Lamb – Amen", playing an octave below the trumpets[17] and thus lending the *tutti* more volume without covering the brilliance of the trumpets.[18]

15 Chr. Hogwood, *Handel* (1984), 228.

16 J. Burghauser and A. Spelda, *Akustische Grundlagen des Orchestrierens*, 1971, 147–8. J. Meyer, *Acoustic and the Performance of Music* (2009), 70, 79, 85 ff. The onset time depending on the way of playing and the frequency region has been measured for modern instruments as follows: violin 30–60 ms, oboe 20–40 ms; violoncello 60–300 ms, double bass 100–400 ms (pizziccato 15–35 ms), bassoon 20–60 ms. For baroque instruments no such measured values have been published yet. With baroque bows and gut strings onset and articulation are quicker and easier because the onset of the relatively low frequencies is slower. *Colla parte* playing with wood winds is blending more than with modern instruments.

17 For instance Christopher Hogwood's *Messiah,* recorded 1980 in London at *St. Jude-on-the-Hill*. For an analysis of this recording see 2.8.

18 In the choir no. 15 "Glory to God" the remark in the autograph "[trombe] da lontano e un poco piano" speaks against a participation of the horns with the trumpets which sound here for the first time. The horn's task has to be investigated further based on a complete analysis of Handel's use of horns in other works and a detailed analysis of *Messiah* which goes beyond this study. Interesting is Handel's genre related use of horns: in his Anglican church music he never uses horns (the liturgic *Allelujas* included). Also the *Brockes Passion,* which uses oboes and bassoons not only in choirs and chorals but also in some arias and also obligate, Handel does not use horns. Since *Radamisto* (1720) horns belong to Handel's opera orchestra and already in the first version of *Esther* (1718) he uses horns, shortly after the first documented written horn parts of the second suite of the *Water Music* (1717). (The

2.5.2 Handel's performances of *Messiah* at the theater

Most of Handel's performances of *Messiah* were held at the theater, twice at the *King's Theatre* (April 1745) and at the *Covent Garden Theatre* three times in 1743 and since 1749 twenty times. Burrows lists the soloists of these performances and Handel's score changes.[19] The orchestra settings of these performances are not exactly known, but for March 1749 Handel put *ripieno* and *solo* marks into the orchestra accompaniment of the choirs (as indicated in the Hallische Händel-Ausgabe). At that time Handel had additional strings for the double orchestra of the oratorio *Solomon,* which he obviously used also for the *tutti* of *Messiah.* No other *Messiah* performance by Handel featured such a large string *tutti,* neither in other London theater performances nor at the *Foundling Hospital.* This string group remained a one-off experiment. Other changes can most probably be ruled out both for 1749 and the other theater performances, as Handel would have entered them into the preserved conducting score.[20] The string group for the theater performances of *Messiah* was presumably the same as for Handel's operas. Until today only two exact lists of the opera orchestra's size are known, one for the *King's Theatre* in 1720 with 25 string players and a total of 34 instrumentalists and a somewhat larger group in 1728 with 30 string players and a total of 41 instrumentalists.[21] In 1754 the *Foundling Hospital* orchestra had 25 string players.

Handel's theater orchestra usually included oboes and bassoons (the 1720 list mentions 4 oboes and 3 bassoons). Evidence for the use of wood winds can be found in the choir no. 35 "Their sound is gone out", composed in 1745 for the *King's Theatre,* which, in the appendix of the autograph score, has systems for oboe I and II, the latter on a separate line, sometimes doubling oboe I,

dates for the early suites HWV 348 are not certain.) Also for his English oratorios since the second version of *Esther* (1732) he repeatedly used horns. Still horns and trumpets are rarely used together. One such example is the richly set first choir of *Deborah* (1733) with double choir, oboe I/II, bassoon, trumpet I/II/III, horn I/II/III, timpani, strings with double bass, organ I/II and harpsichord I/II, (the piece is a contrafactura of the first choir of the *Chandos-Anthem* HWV 254 (1718), only for soli, choir, oboes, violine I/II and basso continuo). Even less Handel used horns and trumpets in oratorios for the *Alleluja,* for example in the final choir of *The Triumph of Time and Truth* (1757). As far as we do not have authentic documents on the function of the horns in *Messiah,* the way of their use remains a artistic decision of the conductor. The above mentioned used to re-inforce the trumpets at least is not against Handel's parctice in other oratorios.

19 Burrows (1991), 23 ff.

20 D. Burrows (1991), 36; A. Hickford, text for L'Oyseau Lyre D189D3, London 1980.

21 D. Burrows, "Handel's London Theatre Orchestra" (1985), 349 f.: the list for 1720 in 1983 was found in the collection of the Duke of Portland (s. Burrows, footnote 11), the list for 1727 belonged to Pierre-Jacques Fougeroux (see Burrows, footnote 14).

sometimes the alto or tenor, sometimes running independently. Oboes and bassoons presumably participated also in the other choirs, perhaps in this differentiated way or simply *colla parte*, as in the *Foundling Hospital* material. The use of French horns in analogy to the *Foundling Hospital* performances is not documented but possible, as horns, like trumpets and timpani, belong since the first *Royal Academy* to Handel's opera orchestra,[22] which also played oratorios since 1732.

To sum up, we can thus conclude that except in 1749, Handel used, in the theater as in the chapel of the *Foundling Hospital,* an ensemble of more or less the same size and setting and, as far as possible, even with the same instrumentalists and singers.[23]

2.5.3 Theaters used by Handel

2.5.3.1 The *King's Theatre* on Haymarket

The *King's Theatre* on Haymarket (until 1715 *Queen's Theatre),*[24] built in 1705 by the architect Sir John Vanbrugh, was refurbished in 1709 in the manner known by Handel. Its original form with semicircular seat rows in the parterre[25] was similar to a not theater draft by John Webb[26] and to Serlio's theater plan of 1539.[27] Even more striking is the relation to another theater draft by Christopher Wren,[28] inspired by the *Teatro Olimpico* in Vicenza (Scamozzi / Palladio 1580–85) (figure 8)[29] and Roman theaters known at that time.[30]

22 D. Möller, *Die Besetzung und Instrumentation in den Opern Georg Friedrich Händels* (1988), 208.

23 See *HHdb.* 4, 481, 498, 516 and 532, with the identification of several London musicians mentioned in the books of *Foundling Hospital.* Burrows also thinks that Handel's London orchestra with the exception of some special cases was relatively stable. See Burrows (1991), 350.

24 Not to be mistaken for the 1766 built *Little Theatre* on Haymarket, later also named *Haymarket Theatre.*

25 G. F. Barlow, "Vanbrugh's Queen's Theatre in the Haymarket 1703–9" (1989), 515–21, especially 519, fig. 7–10.

26 J. Orrell, *The Theatres of Inigo Jones and John Webb* (1985), 162.

27 Orrell (1985), 27: Sebastiano Serlio, *Della architettura, secondo libro di perspettiva,* Paris 1545, f. 66b.

28 D. Thomas, *Theatres in Europe: Restoration and Georgian England* (1989), 67: No. 56; Hume, *London Theatre World* (1980), 59, fig. 12.

29 R. Leacroft, *Development of the English Playhouse* (1988), 45, fig. 45.

Vanbrugh does not mention Wren but the Palladio's more famous example despite the latter's half oval form.[31] In the 1705 *Queen's Theatre,* the columns of the proscenium, which had a flat vaulted ceiling, created three quarters of a circle (as in Wren's draft). The lateral walls without boxes and the very high ceiling made the voice sound "like the gabbling of so many people in the lofty aisles in a cathedral," as the actor Colley Cibber 1740 recalled:

> [...] they had not yet discovered that almost every proper quality and convenience of a good theatre had been sacrificed or neglected to show the spectator a vast triumphal piece of architecture! And that the best play for the reasons I am going to offer, could not but be under great disadvantages, and be less capable of delighting the auditor here than it could have been in the plain theatre they came from. For what could their vast columns, their gilded cornices, their immoderate high roofs avail, when scarce one word in ten could be distinctly heard in it? Nor had it then the form it now stands in, which necessity, two or three years after, reduced it to. At the first opening it, the flat ceiling, that is now over the orchestra, was then a semi-oval arch that sprung fifteen feet higher from above the cornice; the ceiling over the pit, too, was still more raised, being one level line from the highest back part of the upper gallery to the front of the stage. The front boxes were a continued semi-circle to the bare wall of the house on each side: this extraordinary and superfluous space occasioned such an undulation from the voice of every actor that generally what they said sounded like the gabbling of so many people in the lofty aisles in a cathedral. The tone of a trumpet or the swell of an eunuch's holding note, 'tis true, might be sweetened by it, but the articulate sound of a speaking voice were drowned by the hollow reverberations of one word upon another.[32]

In 1709 the acoustics were improved by the construction of proscenium boxes and the lowered ceiling, horizontal above the proscenium and orchestra, then stepwise higher and inclined upwards to the back wall of the auditorium. Cibber comments on this refurbishment of 1709 as follows:

> They contracted its wideness by three ranges of boxes on each side, and brought down its enormous high ceiling within so proportionable a compass that it effectually cur'd those hollow undulations of the voice formerly complain'd of. The remedy had its effect; their audiences exceeded their expectation.[33]

Neither plans nor pictures remain of the theater as it existed from 1719 to 1732 after its rebuilding, and where Handel played on 2 May 1732 the three-part oratorio *Esther* with organ concerto as the last event before another rebuilding.

30 George Saunders, *A Treatise on Theaters,* London 1790, Plate III.

31 T. Oosting, *Teatro Olimpico* (1970).

32 Colley Cibber, *Apology,* 183, cf. Thomas (1989), 75: No. 65, s. a. Barlow (1989), footnote 7. These comments by the actor Cibber are remarkable not only because of the extremely exact description of the room but also because of the clearly defined difference between acoustics for speech and music.

33 Colley Cibber, *Apology*, 242, cf. M Forsyth, *Buildings for Music* (1985), 341, footnote 20; s. a. Barlow (1989), footnote 26.

The theater had 322 parterre seats, 182 in the boxes, 182 on the middle gallery, 250 on the upper gallery and about 64 in the proscenium boxes – a total of about 1,000.[34] In 1732 the theater was renovated as shown in Dumont's plans (figure 79).[35]

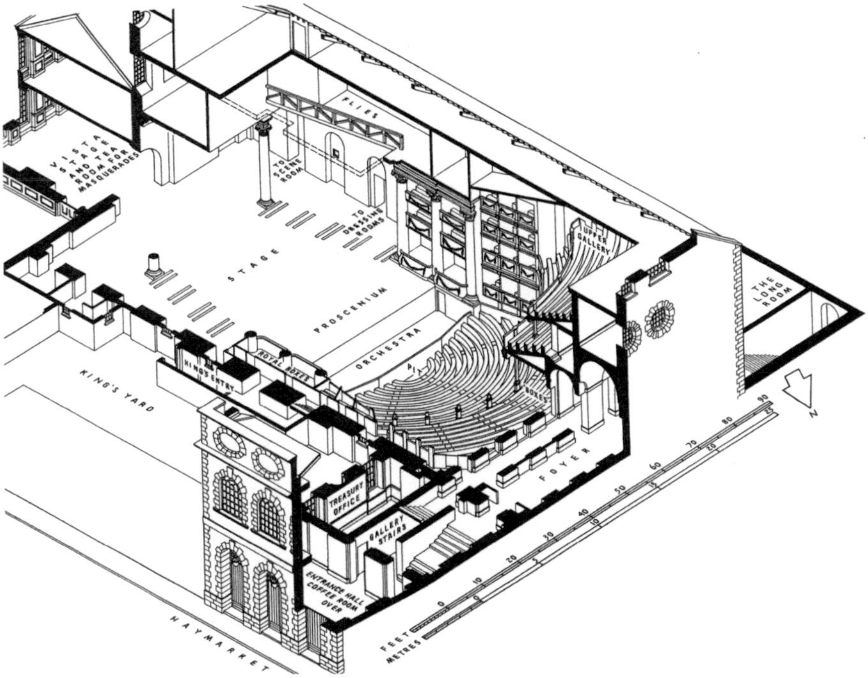

Figure 79: London, *King's Theatre* (1733), perspective view.

Due to incredibly close benches in the pit and on the galleries the number of seats was increased to 1,400.[36] From Cibber's recollections it is evident that already in 1709 the proscenium columns were set in parallel lines, behind which the proscenium boxes were inserted.[37]

34 Barlow (1989), footnote 22; Hume (1980), 64, estimates the number of seats to maximum 950.

35 Various reproductions, for instance *NG* 11, "London", 168.

36 Hume (1980), 242; in table 6, 65 the maximum number of seats for 1735 is indicated with 2000; these must include seats on the stage and in the orchestra as installed for plays. George Saunders in his treatise on theater construction 1790 gives for the distance between the benches in *Covent Garden Theatre* only *1 foot 9 inches* (ca. 53 cm), in spite of the obviously necessary seating distance of 1'10" (ca. 56 cm)! These benches were in use until 1792.

37 Hume (1980), *ibid.*

The fan-shaped widening did not start at the proscenium but after the orchestra pit in the parterre. The side wall and proscenium boxes were closed after the Italian model, whereas the parterre boxes under the gallery were open. If these so-called *front boxes* and the first and second gallery were fully occupied, the acoustics were very dry. Furthermore, the ceiling raised stepwise from the proscenium produced a bad reflection distribution in the *front boxes.* These seats with an excellent view therefore lacked acoustic clarity and transparence, which was obviously annoying, especially in spoken theater, as the repertoire indicated that the space was used primarily for operas and oratorios.[38] For these the *King's Theatre* was preferred to the slightly larger *Covent Garden Theatre* also because of the former's larger apron stage (for a comparison of the dimensions of both theaters, see table 11c).

2.5.3.2 The *Covent Garden Theatre*

In 1732 Edward Shepherd, a theater lover and important patron of Handel, designed the new *Covent Garden Theatre*. On 23 March 1743 the first London performance of *Messiah* took place, and from 1749 until after Handel's death the oratorio was played here yearly at Easter. The plans, also printed by Dumont, show a similarity between Shepherd's design and the *King's Theatre* of 1732. The inner width between the main walls was 56 feet (17.10 m) in both theaters, the height up to the roof beams about 45 feet (13.70 m), the length including the *Upper Gallery* 112 feet (34.15 m).[39] In both theaters the number of seats was about 1,400.[40] The similarity extends considerably also to the interior design, as shown in the following comparison of the theaters' main dimensions:

Table 15: London theaters.

Theaters	Vol. m^3	L m	L_i m	W_i m	Wmax. m	H apron m	H m
King's Theatre 1709/32	3250	20.50	15.80	11.60	13.20	8.25	10.50
Covent Garden 1732	3200	22.70	13.20	10.00	12.50	7.00	9.90
Drury Lane 1775	?	18.30	15.60	9.60	12.50	6.55	?
Drury Lane 1794	?	30.50	22.60	?	24.50	11.60	[14.00]

L = length from curtain to the upper gallery's back wall; L_i = length from curtain to the front of the lower gallery; W_i = width of the apron stage; W max. = maximum width; H apron = height of the apron stage; H = medium height of the parterre.

38 Hume (1980), *ibid.*

39 The reason for these equal ground sizes is a law installed after the big fire in 1666, which did not allow a change of the outer fire walls.

40 Hume (1980), table 3, 61 and 46 for the *King's Theatre*.

But Shepherd made several important improvements in the *Covent Garden Theatre* (figure 80). Unfortunately we do not know whether he discussed his ideas with Handel, whom he had known at least since 1719, as he was the architect of the country residence of the Duke of Chandos in Cannons. He dispensed with the "ancient" semicircular seat rows in the *Pit.* From the 10-m wide apron stage, which was 1.10 m narrower than at the *King's Theatre,* the balustrades of the boxes and galleries widen immediately and steadily up to the rear wall, thus improving sight lines. This negative acoustic factor for the seats in the rear parterre and under the first gallery, which in the *King's Theatre* also lack ceiling reflections due to the raised ceiling above the first row of the *Pit*, is more than compensated for in the *Covent Garden Theatre* by an ingenious change of the ceiling's shape, whose slight inclination, following the Italian model, starts already above the apron stage and the orchestra (figure 81a).

The following, much smaller, stepped rise affects only reflections for the lower gallery. Despite its inclination, the ceiling provides contact between orchestra and singers even in positions at the rear part of the apron stage (figure 81b). With an upwards and backwards opening to the outer wall of the building Shepherd gained seats with adequate acoustics and view, although the greatest distance from the last row of seats on the upper gallery to the curtain is 22.70 m – about 2 m more than at the *King's Theatre*. Still, the lower class gallery public had good reason to complain about the dangerously steep stairs and the stagnant heat. In winter ventilation was nearly impossible because of the icy draught, the burning candles and a freezing temperature in the pit.[41] The last seats under the lower *Gallery* at *Covent Garden* were too heavily covered and hardly received any useful reflections. For the first time in a London theater the King's boxes were installed after the French manner on the left and right side of the apron stage, in the event's immediate proximity, where the acoustics were very clear.

The worst seats in *Covent Garden* were in the rear part of the boxes and in the last seat rows of the *2 Shilling Gallery,* whereas in the *King's Theatre* bad acoustics affected all listeners in boxes at the rear end of the parterre and those in the center of the *Pit* (figure 81c).

41 Hume (1980), 46.

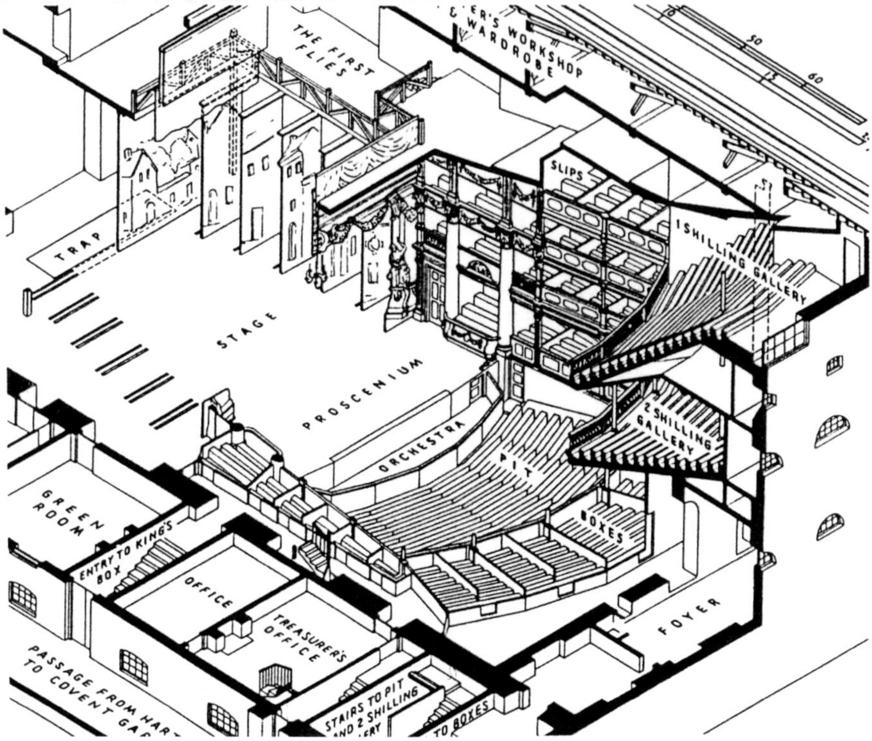

Figure 80: London, *Covent Garden Theatre* (1732), perspective view.

Shepherd was a plasterer by profession. The wooden construction with iron reinforcements of *Covent Garden Theatre* was completely covered with plaster also for fire protection. The construction's stability, the plaster's absorption characteristics, which were not too strong and increased slightly towards high frequencies (see table 1, line 2), the structured decoration of the surface, and the already described good elements of room shape – all these explain why the acoustics, contrary to the *King's Theatre,* were praised as favorable. But Handel's comment, transmitted by Burney, on the badly sold performance of the oratorio *Theodora* refers also to the *Covent Garden Theatre:* "Nevre moind;

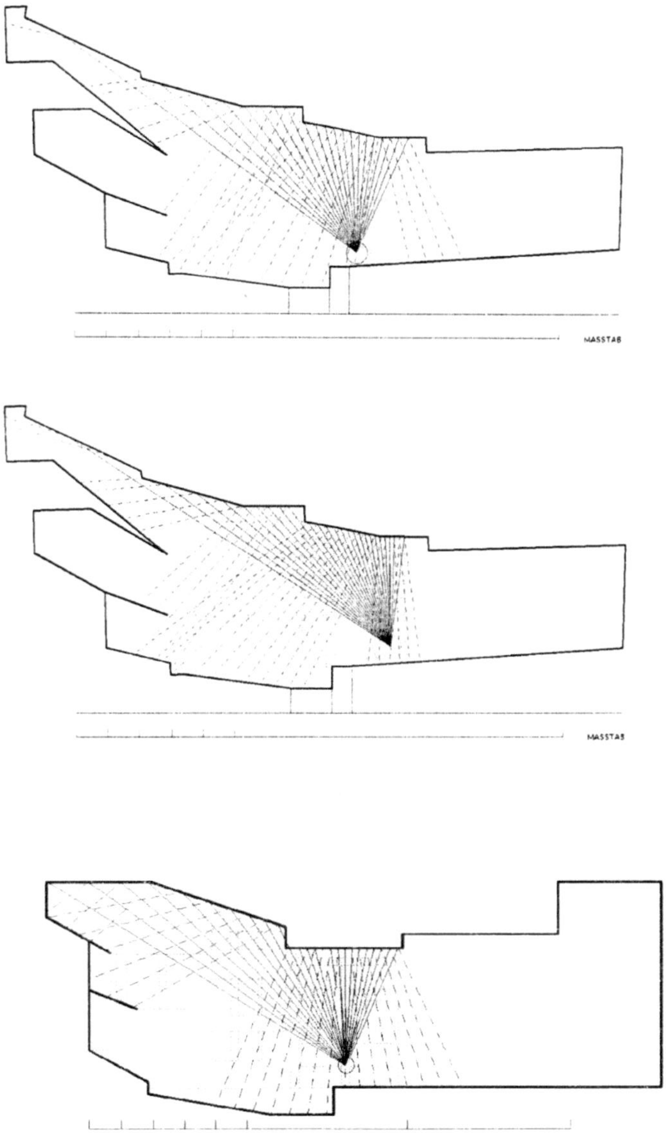

Figure 81: *Covent Garden Theatre* (1732): a) sound source on apron stage, b) sound source behind apron stage, c) *King's Theater* (1733): sound source on apron stage.

de moosic vil sound de petter." [42] The dry acoustics in a full house, even dryer than in the *King's Theatre,* and the better situation at *Covent Garden* become

42 Ch. Burney, *Account* (1785), 29; see also Hogwood (1984), 219.

evident in a comparison of the interior design of both theaters: in the *Covent Garden Theatre* the space volume behind box openings and below galleries is about 1,600 m^3 or roughly half of the cubic space of the auditorium and the apron stage, whereas in the *King's Theatre* it is 1,250 m^3 or only about one third of the respective cubic space. The lateral boxes in the *Covent Garden Theatre* have only low balustrades and cover the side walls along the entire length of 8.30 m. The reflecting lateral surface in the full house is therefore smaller than in the *King's Theatre,* where the lateral boxes have higher fronts and cover the side walls only along a length of about 5.20 m. But the disadvantage of larger absorbing surfaces in *Covent Garden* counts less for hearing than the better design of the regressing box fronts and the ceiling that offer better clarity.

During oratorio performances in both theaters, the apron stage reaching far into the parterre and the steps installed down into the orchestra provided space for more than 100 participants all the way up to the curtain line. This excellent configuration placed the whole ensemble with choir and soloists close to the listeners under the same ceiling, that is, in the same acoustic space, all the more so as the stage was closed backwards and sidewards with walls and upwards with a ceiling. This setup was described in a *General Advertiser* announcement of 21 March 1748 for Cibber's concertante *Masque of Music in two Interludes 'Venus and Adonis',* played between the acts of Shakespeare's *The Merry Wives of Windsor* in *Covent Garden Theatre:* "The Stage (for the better accomodation of the ladies) will be form'd into an Amphitheatre, illuminated, and enclos'd as at an Oratorio."[43] For ladies who in such performances also sat in the parterre at such performances (unlike at opera performances) these seats on the stage were more convenient and comfortable because they were not exposed to draughts. As mentioned, this eighteenth-century practice is still in use today for concert performances at the theater (1.8.5.2).

2.5.4 Documents on the acoustics of the London theaters

As we are looking for still extant comparable theaters, we should bear in mind that at the time all London theaters (including the smaller *Drury Lane Theatre* and the *Lincoln's Inn Fields Theatre)* were built within outer walls of nearly the same size and belonged to the described mixed type of category B unknown on the continent. They featured an inclined parterre, an *apron stage* reaching far into the auditorium (the so-called "Shakespeare stage" in the manner of the Elizabethan theater), proscenium boxes (in the manner of the French Baroque

43 *HHdb.* 4, 413.

theater), rising seat rows in the rear of the parterre (the so-called *amphitheater* similar to the Italian Renaissance theater, itself modelled after the ancient manner) and closed or half-closed boxes and/or open galleries. The acoustics were therefore not comparable to any extant French nor Italian baroque theaters.

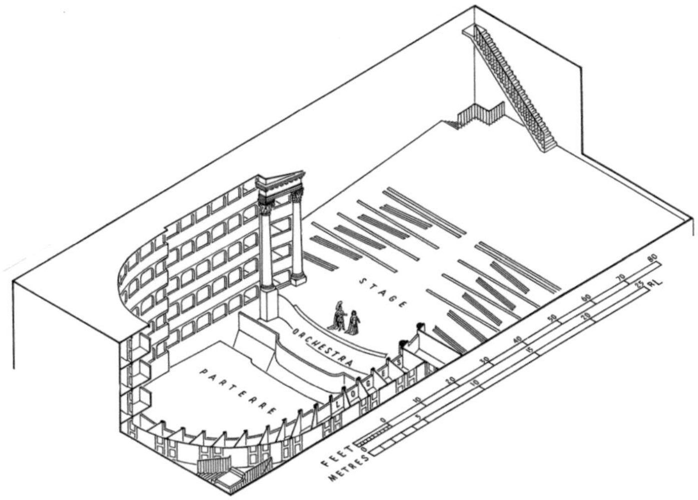

Figure 82: Venice, *Teatro SS. Giovanni e Paolo* (1639).

These London theaters were larger than the intimate Baroque castle theaters of category A with one gallery (if any at all) or single boxes, such as *Drottningholm* or *Potsdam*. In size they are comparable to smaller Italian theaters with closed boxes of the seventeenth and early eighteenth century, such as the *Teatro SS. Giovanni e Paolo* (figure 82) or to the still regularly used *Cuvilliés-Theater* in Munich, said to be ideal for Mozart's operas, though its reverberation times were shorter due to the fan shaped widening of the gallery balustrades, the open galleries and the low ceilings rising towards the rear end. The London theaters used by Handel were also considerably smaller than the acoustically completely different London theaters of the late eighteenth and early nineteenth century of category C, which featured a U-shaped ground plan and a domed auditorium after the French manner (dimensions, see table 12b, 1.9.2.2).

An idea of the spatial experience in a London theater of this type of category B is conveyed by two documents on the *Drury Lane Theatre* with an auditorium that was enlarged in 1775 nearly to the same dimensions as the *Covent Garden Theatre* of 1732 (figure 80). Its acoustics were very close to the *Covent Garden* used for *Messiah* between 1749 and 1759, mainly because the ceiling over the apron stage had been raised by 12 feet and inclined towards the rear wall,[44] extending over the third rank above the *Upper Gallery:*

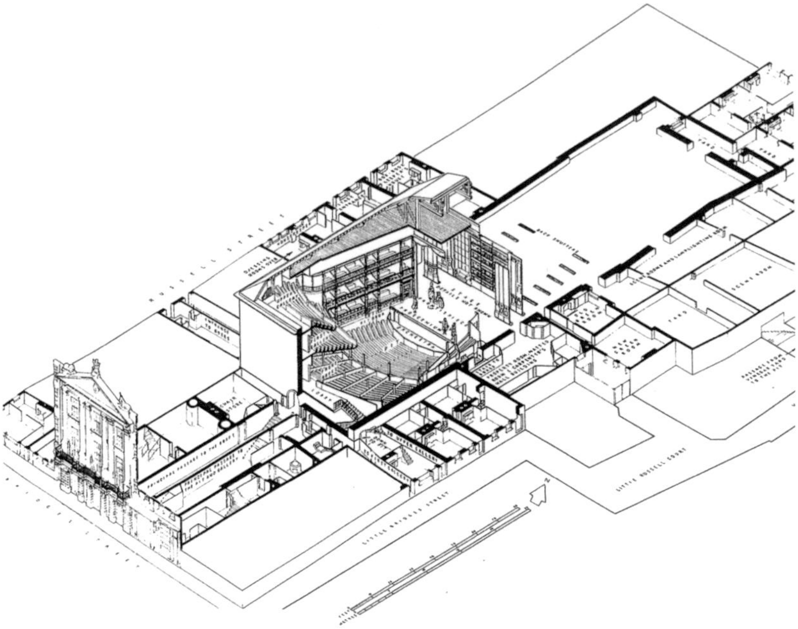

Figure 83: London, *Drury Lane Theatre* (1775).

> I [...] observed, that the Sounding Board was much raised on the Part next the Stage [...] and what astonished me greatly was to find that the Sound of the Music and Actors Voices both improved by this additional Height.[45] (*Public Advertiser*, 30. September 1775)

After a fire in 1791 this theater was replaced by a new building nearly twice as large designed by Henry Holland with a horseshoe ground plan and dome (figure 84): the auditorium's width was enlarged from 16.30 m to 24.50 m, the length from the curtain line to the balustrade of the first rank from 15.60 m to 22.60 m, the greatest distance between the rearmost seats on the upper gallery to the stage from 18.30 m to 30.50 m. A category B theater was replaced by one of category C (see 19.2.2).

The deep regret at the disappearance of the old house, already too small in greater London, is documented in the diary of John Byng, later Viscount of Torrington, in his entry of 14 May 1794:

44 *The London Stage 1660 – 1800,* G. W. Stone (ed.) 1962, xli.
45 Hume (1980), 50.

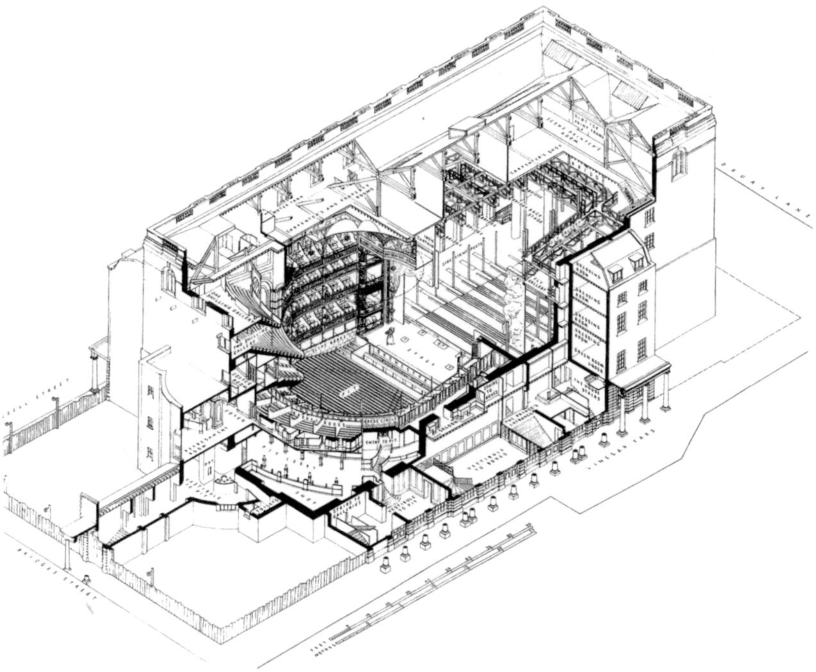

Figure 84: London, *Drury Lane Theatre* (1794), perspective view.

I adjourned to Drury Lane Playhouse where I enjoy'd the highly wrought exhibition of Mrs S[iddon]'s performance in Catherine in Henry 8th, altho' lost and sent to waste in this wild wide theatre, where close observation cannot be maintain'd, – nor quick applause received! – Restore me, ye overuling powers to the drama, to the warm close, observant, seats of Old Drury where I may comfortably criticise and enjoy the delights of scenic fancy: These now are past! The nice discriminations, of the actors face, and of the actors feeling, are now all lost in the vast void of the new theatre of Drury Lane. [...] – 'Why here is a fine theatre,' say they? 'Aye, it may be fine, it may be magnificent; but I neither hear, nor see in it!!' 'That's your misfortune' – 'So it is I allow, but not yet my failing.' – 'Does it proceed from the narrowness of my faculties; or the width of your new stage? Answer me that? Is my decrease equal to your increase?' No; No; fill your stage with monsters – gigantic cars, and long train'd processions – whilst the air vibrates with the sound of trumpets, and kettle drums: These will beat all your actors, and actresses out of the field. Who will listen to, or who can hear the soliloquies of Shakespeare, the inward terrors of the mind – perturbed imaginations and the strugglings of a guilty conscience – ? – To see a 'fellow hunting a dagger about the stage –; or an old 'princess' wasting in a great chair? Who will go here after to see their tiresome attitudes? To hear them none will attempt, so let us have the battlements, – the combat, the sulphur, the torches, – the town in flames, – and the chorus.[46]

46 *The Torrington Diaries*, C. Bruyn Andres (ed.) (1938), IV, 18–19, cf. Hume (1980), 52 and D. Thomas, *Restoration and Georgian England* (1989), 269.

Byng not only regrets the disappearance of the building but the decline of a tradition of acting, best represented by the *Natural Style* of the actor David Garrick. The actor John Philip Kemble is said to have responded as follows to protest from the public in this new theater with four ranks, three *Galleries* and a distance of 30.50 m from the last seats on the uppermost gallery to the stage: "I will raise my voice, and the Galleries shall hear me!"[47] The new way of speaking modeled after Ancient Greek theater was called *Classical Style*. It became fashionable to include procession scenes in each play to provide the opportunity for a great choir number,[48] which could be produced in the large house with overwhelming effect.

The change of performance practice in plays and operas followed an earlier change outside the theater, at the choir festivals and in the performances of *Messiah*: the steady enlargement of the choirs, due also to an increasing number of amateur singers and an ever growing public, carried the change from the *Music Room* to the *Assembly Hall,* from here to the choir and finally to the church nave as a logical consequence. Only few sensitive listeners were aware that the enthralling large choirs resulted in a loss of subtleties. Actors and musicians were forced to develop a new way of performance in order to achieve the highest expression and best transmission of information under the new demanding and often difficult conditions. This is also expressed in the following review by Richard Cumberland, which refers not only to the new *Drury Lane Theatre* but also to the new *Covent Garden Theatre* in 1792:

> Since the Stages of Drury Lane and Covent Garden have been so enlarged in their dimensions as to be henceforward theatres for spectators rather than playhouses for hearers, it is hardly to be wondered at if their managers and directors encourage those representations, to which their structure is best adapted. The splendor of the scenes, and the ingenuity of the machinist and the rich display of dresses aided by the captivating clamours of music now in a great degree supercede the labours of the poet.[49]

These short but important remarks on the old *Drury Lane Theatre* (1791 destroyed by fire), and on the new theaters at *Drury Lane* and *Covent Garden* give an idea of how intimate Handel's oratorios appeared in the much smaller old theaters, which were similar in size to the old *Drury Lane Theatre*.

47 Cf. Leacroft (1983), 94.

48 *The London Stage 1660 – 1800* (1962), cxxxi.

49 Richard Cumberland, *Memoirs*, London 1807, II, 384–5, cf. Hume (1980), 54. For the dimensions of the new *Covent Garden Theatre,* see table C in the appendix.

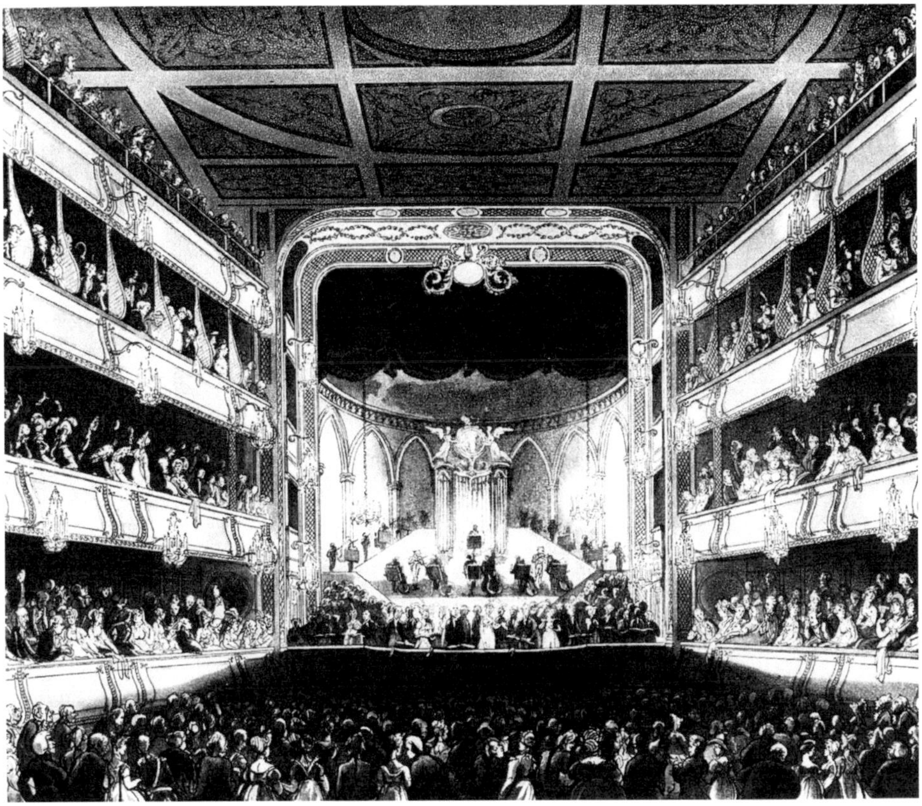

Figure 85: London, *Covent Garden Theatre,* oratorio performance 1808.

2.6 The impact of room acoustics on Handel's compositions

2.6.1 Room acoustics and the process of composition

Handel's compositions were written mostly on commission or at least for a specific or eventual performance. The performance's significance and social function, its musical genre and often the space, too, were specified in advance. Handel's influence on the circumstances of the performance was restricted for several reasons. If he worked on commission he could at most choose among the halls at his disposal. The late change of hall for *La Resurrezione* at *Palazzo Bonelli* in Rome, caused by the large, unexpected number of guests and the large orchestra, shows that Handel knew how to take advantage of possibilities. On the other hand, he had to take into account the available musicians (soloists, choir singers, orchestra musicians) in order to adequately present the compositional structure in the hall at the orchestra's given position.

A list of performance factors relevant to Handel's compositional technique shows several parameters related in various, sometimes complex, ways:

Composition depending on:	Ensemble size and number of instruments:
Patron, organizer, finances	*secondo il bisogno dell'occasione*
Social aspect of the event	
Available rooms for performance	*secondo il bisogno del luogo*
Genre, music structure	*secondo il bisogno del opus*

For Handel, rejection of a room could mean loss of a commission or a performance. No relevant documents could be found for such cases but there are indirect indications, such as the lacking oratorio performances by Handel in London's inns and other narrow rooms, even though his theater soloists, choir singers and instrumentalists regularly performed in these rooms.[1] In such rooms (for instance, *Queen's Arms Tavern*)[2] Handel only played harpsichord or, if the hall was large enough, organ, and he used such rooms for rehearsals. At the *Gothic Pavillon* of *Vauxhall Gardens* and in the huge elliptic *Rotunda* of *Rane-*

1 See 2.3.3.

2 Hawkins mentions a harpsichord concert by the young Handel in this hall, see Sir John Hawkins, *A General History of the Science and Practice of Music,* 5 vols., London 1776, cf. Chr. Hogwood, *Handel* (1984), 71.

lagh Gardens, a space for 2,000–3,000 listeners much favored by the public despite problematic acoustics, Handel performed only as organist (figure 86).[3]

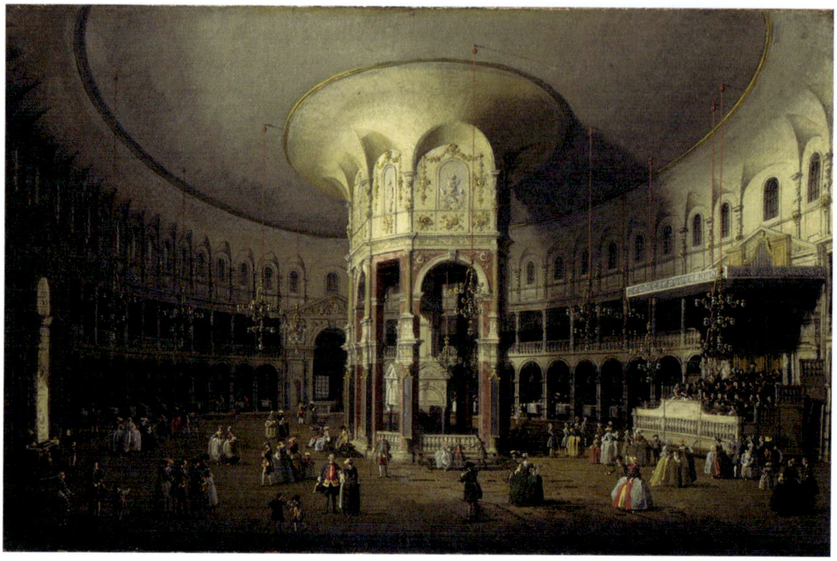

Figure 86: London, *Ranelagh Rotunda,* interior (1742–1805).

If offered a good room Handel was ready to make a supreme effort to improve a performance. Such was the case of his dispute with the archbishop of London in 1733 in order to receive permission for the participation of the children of the *Chapel Royal* in the performance of *Esther* at the *King's Theatre.* Even more impressive examples of Handel's commitment are the concerts at the *Foundling Hospital* chapel, which enjoyed his financial help for the event, the donation of the organ and his personal search for musicians for the yearly performance of *Messiah* until 1753.

If Handel did not know the hall in advance (like the new Dublin hall), he depended on information from others (such as the reliable report by the local concertmaster Dubourg in Dublin). If the circumstances were unexpectedly worse or changed (different listeners, musicians or hall), Handel followed the Baroque practice of adapting the music with changes of instrumentation or even cuts of single movements. If necessary, he enriched or reduced the composition's structure or wrote new parts or completely new pieces. The comparative analyses in the next two chapters aim to explore these relations between room acoustic conditions and the structure of Handel's music, especially with respect to his oratorio *Messiah.*

3 P. M. Young, *The Concert Tradition* (1965), 140; M. Forsyth, *Buildings for Music* (1985), fig. 2 and 2.2; see 1.8.7: Groundplans and ceiling shape in halls.

2.6.2 Handel's performances with large ensembles

Performance conditions did not always permit a perfect relation between music structure, number of musicians and room acoustics (see table 15). In this respect Handel's readiness for compromises or experiments shows certain limits. Documents indicate that for several London performances he used an exceptionally large number of performers *secondo il bisogno dell' occasione,* for a special occasion, such as the following ceremonial events:

– the *Coronation Anthems,* 1727, for the coronation of George II, with 36 choir singers, 4 soloists, 12 choir boys or a total of 52 singers and about 90 instrumentalists on a amphitheatrical podium in front of the main altar in *Westminster Abbey;*[4]
– the *Funeral Anthem* for Queen Caroline, 1737, in *Henry VII's Chapel* at *Westminster Abbey,* with nearly 80 singers and 100 instrumentalists on a special balcony;[5]
– the *Fireworks Music* on the occasion of the Peace of Aachen, performed in *Green Park* in 1749 according to Handel's score, with 55 wind players and 3 timpanists (the press spoke of 100 musicians).[6]

In these cases the large ensemble was also necessary because of the room acoustic conditions, that is, *secondo il bisogno del luogo.* Preserved documents reveal Handel's awareness of the difficulties such occasions presented: before the performance of the *Coronation Anthems* he therefore warned of failure due to the dangerous position of the choir on two separate tribunes, with the organ in the middle behind the main altar.[7] Furthermore, he insisted on holding the main rehearsal of the *Funeral Anthem* in the presence of the entire Royal family in the acoustically excellent *Banqueting Hall* in *White Hall,* probably because he knew about the less favorable conditions in *Henry VII's Chapel*[8] at *Westminster Abbey.* He also insisted, to no avail, on strings, 16 trumpets and 16

4 *HHdb.*4, 154; *HHdb.* 2, 702; *MT* 188 (1977), 469 f.
5 *HHdb.* 4, 486; *HHdb.* 2, 735.
6 *HHdb.* 4, 423, 426; *HHdb.* 3, 123.
7 *HHdb.* 4, 154–5; H. C. R. Landon, *Handel and His World* (1984), 117.
8 The Duke of Chandos to his nephew, Reverend Theophilus Leigh: "Lond.n 18 Decem.r 1737./ The Solemnity of the Queen's Funeral was very decent, and performed in more order than any thing I have seen of the like kind. [...] the Anthem took up three quarter of an hour of the time, of which the composition was exceding fine, and adapted very properly to the melancholly occasion of it; but I can't say so much of the performance." (Huntington Library, San Marino, California)

horns (instead of only 9) for the outdoor performance of his *Fireworks Music* at *Green Park* because he knew that the music would thus sound much better.[9]

After failures he did not stop looking for better solutions for repeated performances under better acoustic conditions with improved settings. In May 1749 he repeated the *Fireworks Music* with string players in the *Foundling Hospital Chapel* in the King's presence. The annotations to the autograph score list 36 strings (12-12-4-4-4), 37 wood winds (24 oboes, 12 bassoons and 1 contra bassoon), 9 French horns, 9 trumpets and 3 timpani. The total of 94 instruments at this performance is confirmed, at least broadly speaking, by the announcement of 26 May 1749 in the *General Advertiser:* "that above One Hundred Voices and Performers have engaged to assist upon that Laudable and Charitable Occasion."[10] The *Foundling Hospital* version with strings certainly had a much greater effect than the outdoor version of "military music" alone, as requested by the King at the first performance.

During the same concert at the *Foundling Hospital* in May 1749, parts from *Solomon*, the *Anthem of the Peace* of Aachen (originally composed for the *Chapel Royal),* the *Foundling Hospital Anthem* with the *Hallelujah* from *Messiah* and parts from the *Funeral Anthem* were played. The exact number of singers and instrumentalists for the anthems is not documented, but Handel presumably used the occasion to separate the available 36 strings into *ripieni* and *soli,* as he had done in the *Covent Garden* performance of *Susanna, Solomon, Samson, Hercules* and *Messiah* in spring 1749. The orchestra was much greater than at all known *Messiah* performances in the same room since 1750, which had featured at most 25 strings. The number of strings in this case was large not for room acoustical reasons but because the King's presence allowed an especially magnificent presentation in order to raise money for the still unfinished chapel of the charitable foundation. Once again Handel used the occasion to exceptional musical effect.

The performance of the *Fireworks Music* at the *Foundling Hospital* points to a further reason for the large number of players: *secondo il bisogno del opus*

9 *HHdb.*4, 422: John, 2nd Duke of Montague to Charles Frederick, 28. März 1749: "I think Hendel now proposes to have but 12 trumpets and 12 French horns; at first there was to have been sixteen of each, and I remember I tolt the King so, who, at that time, objected to their being any musick; but, when I told him the quantity and nomber of martial musick there was to be, he was better satisfied, and said he hoped there would be no fidles. Now Hendel proposes to lessen the nomber of trumpets, &c. and to have violeens. I dont at all doubt but when the King hears it he will be very much displeased." *Cf. HHdb.* 4, 286.

10 *HHdb.* 4, 429; O. E. Deutsch, *Handel* (1955), 672; we do not see any reason to follow Nicholas Kenyon's opinion on these numbers, see booklet to the recording with historic instruments by the English Baroque Soloists directed by John Eliot Gardiner (Philips CD 411 122–2). Handel also used smaller ensembles, but at the *Foundling Hospital Chapel* he aimed at the best possible performance.

as Winton Dean said, that is, when the music's structure required a certain number of performers.[11] An earlier example was *Athalia,* performed in 1733 at the *Sheldonian Theatre* in Oxford with 70 performers, a number appropriate not only to the celebration for the honorary doctors and the large theater hall but also to the topic of the oratorio Handel had chosen, as the occasion allowed writing splendid music for an exceptionally large group of musicians. The other large ensembles, for *Deborah* in 1733 with 75 instrumentalists, 25 choir singers and 7 soloists and the not exactly known ensemble for *Solomon* in 1749 at the *Covent Garden Theatre,* can be explained only by the textual requirements of the sacred oratorio.[12] In these cases the large ensemble *secondo il bisogno del opus* exceeded not only financial limits (for *Deborah* most subscribers had refused to pay the doubled entrance fee in 1733) but also the room's acoustic conditions: at least part of the listeners complained about excessive loudness at the *Covent Garden Theatre,* as evident in the following letter by a Lady Irwin to Lord Carlisle dated 31 March 1733:

> The subscribers forced into the House. [...] I was at this entertainment on Tuesday; 'tis excessive noisy, a vast number of instruments and voices, who all perform at a time, and is in music what I fancy a French ordinary in conversation.[13]

Like the Italian opera, which exceeded financial limits due to the expensive Italian star singers, Handel exceeded his own financial limits with these oratorios, as insufficient patronage at that time forced him to count on paying listeners. Indeed, the possible income could not cover the costs for such a large number of performers.

2.6.3 Virtual and actual space in Handel's compositions

2.6.3.1 Virtual space

Handel was admired as a composer because of the splendid sounding effect of his works, as confirmed by Charles Burney at the end of his report on the 1784 *Handel Commemoration:*

11 W. Dean, *Handel's Dramatic Oratorios and Masques* (1959), passim.

12 For *Solomon* Handel used 8 soloists (4 soprani, 1 alto, 2 tenori and 1 basso, double choir and and orchestra with strings (violas *divisi*!), flauto traverse I/II, oboe I/II, bassoon I/II, horn I/II, trombone I/II, timpani, organ and basso continuo.

13 *HHdb.* 4, 210.

> That Handel was superior in the strength and boldness of his style, the richness of his harmony, and complication of parts, to every composer who has been most admired for such excellencies, cannot be disputed. And, while FUGUE, CONTRIVANCE, and a FULL SCORE, were more generally reverenced than at present, he remained wholly unrivalled.[14]

As a musician Burney was able to explain that splendor is not necessarily the result of a large number of instruments and voices but of the music's main structure and Handel's incomparable art of counterpoint.[15] But in view of the overwhelming event he nonetheless got carried away with the following bold generalization:

> Indeed, Handel was always aspiring at 'numbers' in his scores and in his orchestra; and nothing can express his grand conceptions but an omnipotent band. The generality of his productions in the hands of a few performers, is like the club of Alcides, or the bow of Ulysses, in the hands of a dwarf.[16]

Similar hymns of praise could be heard already during Handel's lifetime. Thus Benjamin Victor wrote from Dublin on 27 December 1752 to William Rothery, Reverend in Chelsea:

> If HANDEL'S MESSIAH should be performed in London, as it undoubtedly will in the lent season, I beg it as a favour to me, that you will go early, and take your wife with you, your time and money cannot be so well employed ; take care to get a book of the oratorio some days before, that you may well digest the subject, there you will hear GLAD TIDINGS and truly divine rejoicings at the birth of CHRIST, and feel real sorrows for his sufferings – but oh! when those sufferings are over, what a transporting full chorus! where all the instruments, and three sets of voices are employed to express the following passage, which I must quote – 'Lift up your heads ...' [...] I would ride forty miles in the wind and rain to be present at a performance of the Messiah in London, under the conduct of Handel – I remember it there – He had a hundred instruments, and fifty voices! O how magnificent the full choruses.[17]

Burrow's list of Handel's own performances indicates that Victor is referring to a performance of *Messiah* with more than 25 strings and some 20 to 30 choir singers in March 1749 at the *Covent Garden Theatre,* presumably the only performance with such a large ensemble. We cannot confirm that Handel actually used at the time 100 instruments and 50 voices for the performance of *Messiah* and *Solomon* (see table B in the appendix). But the charitable performance on 27 May 1749 at the *Foundling Hospital,* for which the press reported 100 performers at least for the *Fireworks Music,* suggests that this does not seem im-

14 Ch. Burney, *Account* (1785), 39.

15 Burney was organist and played in Handel's orchestra.

16 Burney (1785), 106.

17 Deutsch (1955), 729, cf. Myers, *Moral Criticism,* 35 f.

possible. Besides the ensembles for the great Royal occasions, the *Coronation anthems* for George II (1727) and the *Funeral Anthem* for Queen Caroline (1737), this would be Handel's largest documented orchestra and choir ensemble for *Messiah*, even larger than the theater performance of *Deborah* with 75 instruments, 25 choir singers and 6 soloists mentioned above. Perhaps the round numbers of 100 and 50 are not to be taken literally here but in the sense of "very large", as they often are. Overwhelmed by this London performance, Victor praised in the same letter also the Dublin performances by singers of the cathedral in the *New Music Hall:*

> Mr. Handel, when he was here, composed this excellent oratorio, and gave it to a charitable musical society; by whom it is annually performed, for the relief of poor debtors, and very well, as we have good cathedral singers, to whom this music is chiefly adapted.[18]

These lines confirm Handel's capacity to reach magnificent effects with more modest forces thanks to structural means, such as obligate middle voices, which, despite the relatively small hall, evoked a magnificent space in the listeners' imagination. When based on this principle of "virtual space", Handel's scores don't lose much effect as long as the main voices of the music structure are presented appropriately.

An example of a masterly *downsizing* due to performance circumstances is the reworking of the *Utrecht Jubilate* HWV 279 (1713) into the anthem "O be joyful in the Lord" HWV 246 (1717) for the Earl of Carnarvon, later Duke of Chandos. A comparison of the scores shows how Handel adapted the splendid *Jubilate* for 3 soli, double choir, strings with triple violins, double oboes, bassoon, double trumpets, basso continuo and obligate organ to the much smaller ensemble with only 4 soli, three-voice choir, one oboe and bassoon, first and second violin and *basso continuo* for the small St. Lawrence church in Whitechurch near Canons: the overwhelming fortissimo at the beginning of the "Glory be to the father" (*Utrecht Jubilate* HWV 279, no. 6), set in 8-part chords spread over three and a half octaves for the enormous vaulted nave of *St. Paul's,* is reduced for the small church to 3-part chords supported only by oboe and bassi (HWV 246, no. 7). The figurative ground moving in eighth notes, played by first and second violin and viola in the *Utrecht Jubilate*, is distributed in the *Chandos-Anthem* between first violin, oboe and second violin, as violas were lacking. The quality of the score and its "virtual space" permit such a reduction of parts to the bare main structure without any further change (music example 1).

18 Deutsch (1955), 729. In Dublin it was not known that Handel composed *Messiah* in London before his trip to Ireland.

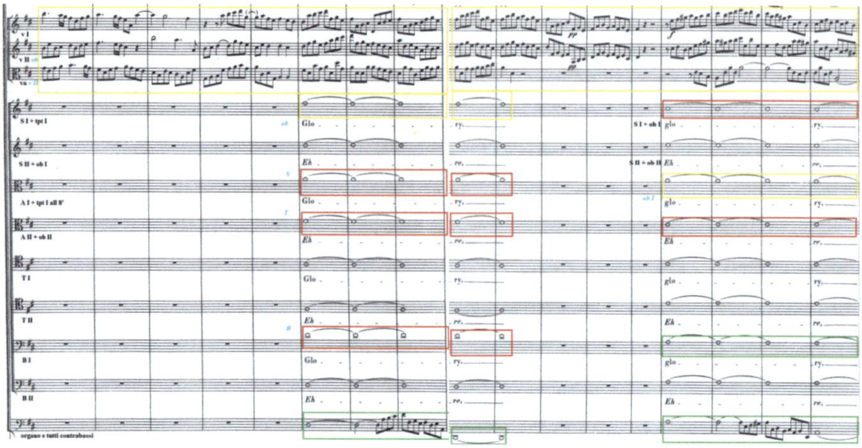

Example 1: *Utrecht Jubilate* HWV 279 (colla parte instruments entered in black), and *Chandos Anthem* "O be joyful" HWV 246 (strings and oboes marked yellow, choir parts marked red).

The temporal structure of the *Utrecht Jubilate,* too, reflects the acoustics of *St. Paul's* with several seconds of reverberation mainly for low frequencies: after a *fortissimo* D major chord of five measures Handel inserts a break of three measures. Wyatt presumably erected the tribune for the musicians between the no longer extant choir screens (see figure 65). Thanks to this position, partly protected from the reverberation of 14-meter-wide main nave of the cathedral, the quicker melodic lines and the figurative ground could benefit from the lateral walls of the choir stalls. In the *Chandos-Anthem* this figurative ground in eighth notes is reduced to *pianissimo* with a long final *fermata* in order to compensate for the short reverberation of the small village church.

2.6.3.2 Exterior acoustic conditions

On the other hand, a small score could not be performed in a larger space without changes. In such situations Handel did not only double the parts in octaves or add woodwinds *colla parte* and enrich chords but also inserted additional contrapuntal lines if necessary. This increase in volume and richness was reportedly characteristic also for his harpsichord and organ playing.[19] Such additions are limited by the main structure of the score. If moved to too large halls, compositions for small halls always lose a part of the intimate expression due to the necessary adaptations of structure and playing, as the "virtual spatiality" of the score is enlarged by the acoustics of the performance space.

19 The increase of power of sound on the organ is not only achieved with more and louder stops, but also with fuller chords and improvised obligate middle voices.

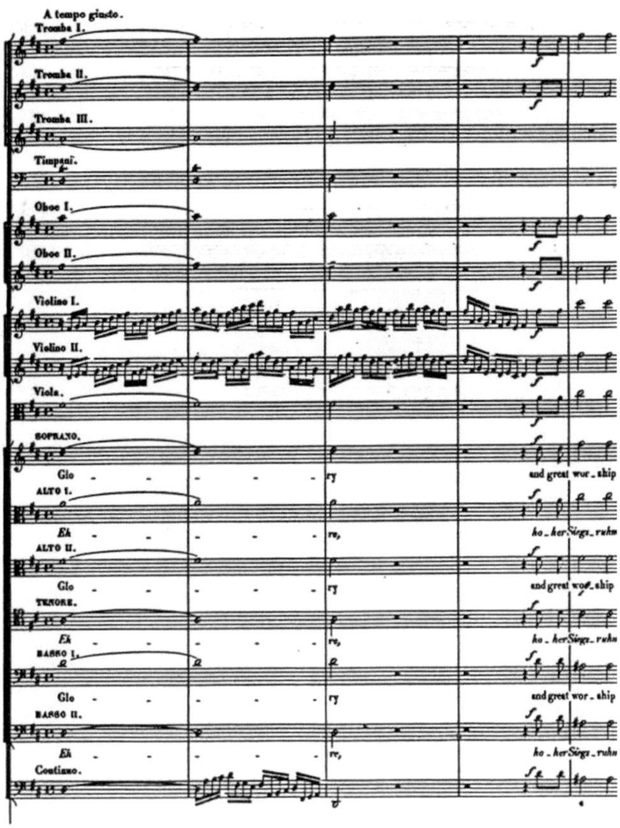

Example 2: "Glory and great worship" HWV 260.

Handel's special way of composition for very large spaces becomes evident in a comparison of *Glory* exclamations, such as the magnificent "Glory – and great worship" from the *Anthem* "The King shall rejoice", composed in 1727 for George's II coronation in *Westminster Abbey,* and the "Glory to God in the highest, and peace on earth" from *Messiah* (see examples 2 and 3 and the sound recordings series VII).

The single *Glory* exclamation for the act of coronation, a D major chord of 2½ measures above a ground moving in sixteenths in third parallels and followed by a break of 1½ measures, is written for the long reverberation in the crossing of *Westminster Abbey.* On the other hand, the *Glory* from *Messiah,* composed for much smaller, acoustically transparent rooms and for a much smaller ensemble, creates an extraordinary "virtual spatiality": the angel's choir, announced in the *accompagnato* of the solo soprano, sings the "Glory to God" vividly with quickly changing harmonies. The trumpets "da lontano e un poco piano", supporting mainly soprano and alto, add not only a secret background to the choir's *forte* but also an impressive spatial perspective. The *Glory* exclamations are accompanied by sixteenth notes, then breaks and a sequence of equal eigth notes (the dots here instruct to avoid the baroque *jeu inégale* on the different beats within the measure) followed by a rhetorical octave fall of the melody on the words "And peace on earth".[20]

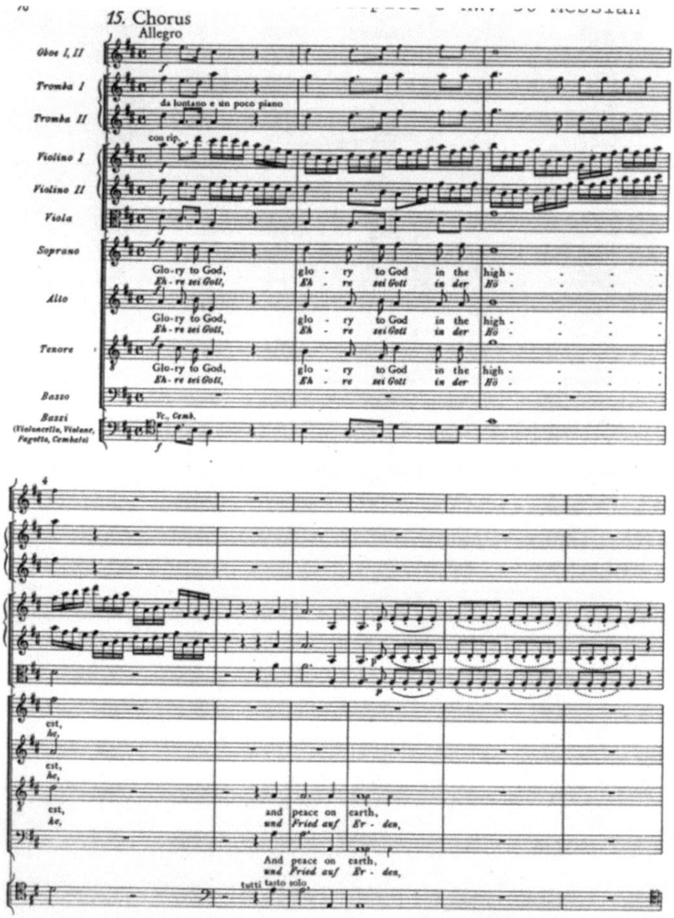

Example 3: "Glory to God" from *Messiah.*

The dynamically shaded trumpets, the structural fading of the instrumental closing section and the decreasing dynamics evoke increasing distance – a process that produces the desired effect only in transparent acoustics. Too large

20 E. R. Jacobi, Notes inégales, *Riemann Lexikon,* 12th ed., *Sachteil,* Mainz 1967, 642–3. The rule still treated by Quantz, Leopold Mozart and J. J. Rousseau is valid as long as the practice of *jeu inégale* in French style is used.

spaces with excessive reverberation would impair also the preceding scene of the shepherds in the fields and of the terrifying closeness of the angel bathing in light, whose recitative should be joyful but neither too loud nor to quick: "Fear not, for behold, I bring you good tidings of great joy."

2.6.4 Space, genre and structure of music

Handel's parody practice, which spans over related motifs, rewriting and literal quotes of complete parts from his own or borrowed works, where necessary with new text and a reduced or amplified score, offers further examples of adaptation to changed acoustic conditions.

Although composed within a short time, *Messiah* contains only few borrowings from other compositions. But the parodied parts offer clear evidence on such specific structural elements in music genres and certain room acoustics.

2.6.4.1 From chamber duet to oratorio choir

The choirs no. 7, 11, 18 and 23 from *Messiah* are based on chamber duets written immediately before the oratorio (HWV 189 und 192, London, 3 and 1 Juli 1741) or earlier (HWV 193, London, ca. 1720/24). A comparison between choir no. 18 "His yoke is easy" from *Messiah* and "Quel fior che all'alba ride" from the chamber duet HWV 192 for soprano, alto and basso continuo shows how Handel elaborated the famous choir "His yoke is easy" for soprano, alto, tenor, bass, strings, oboe and basso continuo using the relatively simple procedure of adding two solo voices and increasing the density with the insertion of tutti parts (see music example 4 and sound recordings VI, 21 / 22).

The duet consists of three developments of the main theme A (5 measures). To create the two additional entrances required for the four-part expansion, Handel uses octave transpositions and changes the *tessitura,* distributes the theme to two separate voices and adds extensions or abbreviations. The insertion of *tutti* cadences at the end of the developments in measure 11, 15, 19, 23, 29, 36 and 40, which are always marked *forte con ripieni* in the 1749 version (see framed fields in music example 4), produces a *concerto grosso* structure of increasing density.

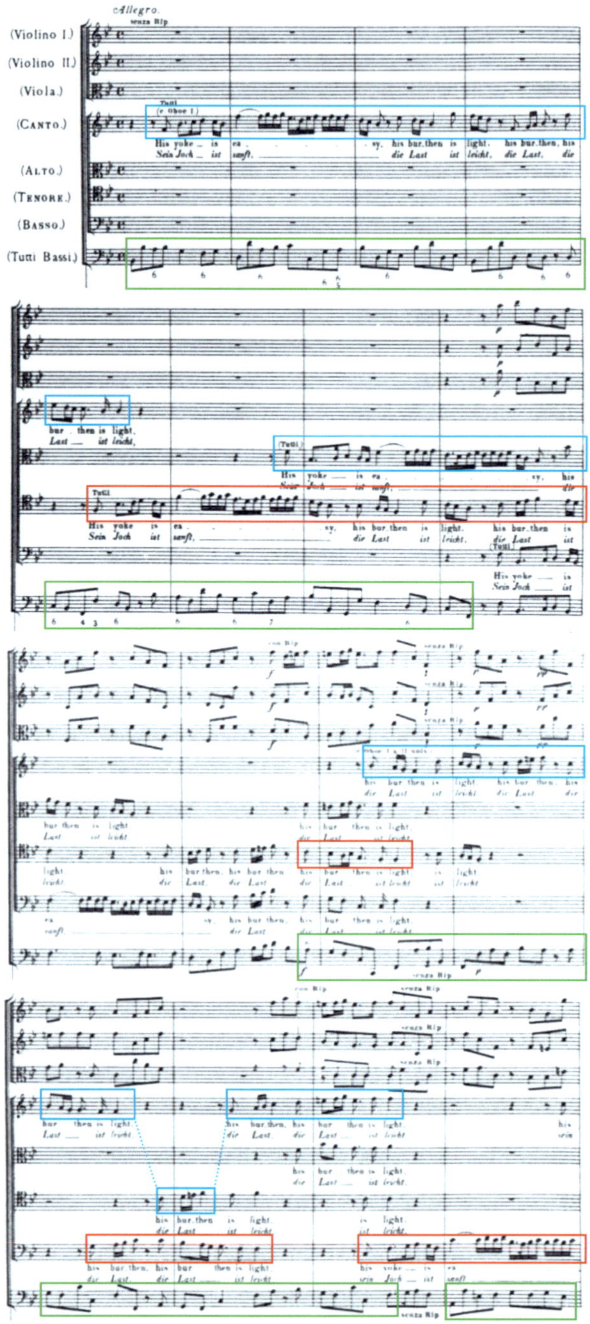

Example 4: comparison between choir no. 18 "His yoke is easy" from *Messiah* and chamber duet "Quel fior che all'alba ride" HWV 192 (parts of duet score marked grey).

The change of genre notwithstanding, Handel keeps the chamber duet's bass line with quick harmonic changes practically unchanged with the exception of the inserted measures 8–10, the added final measures 41–51 and small but important changes in the *tutti* cadences, such as a chord in the main position instead of a sixth chord. A comparison of the *basso continuo* numbering shows a few harmonic simplifications in the choir version. Figures have a richer embellishment in the solo voices in the duet, while the choir parts are limited to the usual appoggiaturas and trills. With such an enriched structure Handel developed a choir that changes the bitter-sweet chamber duet "Quel fior che all'alba ride, il sole poi l'ucide, e tomba nella sera" to a mysteriously joyful yet sad choir "His yoke is easy, his burthen is light". The condition for the parody was the possibility of transferring the rhetorical figure "ride" to the word "easy" in the new text. The duet's *Andante larghetto* is changed into the choir's *Allegro.* In terms of baroque style, these are less tempo instructions than indications of the expressive character, namely, a fluent movement for the duet and serene joy for the choir to be sung more or less in the same tempo.

In England, too, room acoustic conditions for the performance of an Italian chamber duet were determined by halls of limited size with rather few listeners. Usually ordered by an aristocratic patron, these pieces were performed in elegant, rather high, sparsely furnished rooms, such as the London residences of the Earl of Burlington or the Earl of Carnarvon. The change of genre from a chamber duet to an oratorio choir brought along a change from the *drawing room* (salon) to a small music room or hall, an English baroque theater or a chapel like the one at *Foundling Hospital,* which were used by Handel for the performances of *Messiah.* These rooms had a larger cubic volume than the small private halls then used for chamber music, but their transparent and intimate acoustics allowed a clear presentation of the bass line with quick harmonic changes literally taken from the chamber duet. Developed from chamber duets, these scores are technically so demanding that only excellent choir singers are able to sing them properly. The acoustics of the *Foundling Hospital Chapel,* the room furthest away from contemporary chamber music conditions, were also transparent enough to follow the quickly changing harmony without difficulties thanks to good lateral reflections.

With the decision to use these chamber duets and keep their bass lines for *Messiah,* Handel set room acoustic limiting conditions. Exceeding them would mean a change of musical expression. In larger, more reverberant, halls with blurring acoustics, which were never used by Handel, a larger choir and adapted playing techniques destroy the original chamber-music transparence and flexibility. Execution becomes slower and heavier if musicians do not hear each other well (see the analyses of sound recordings in chapter 2.8). The text declamation in these choirs, which is bound to the rhythm and tempo of natural speech, is eminently understandable when executed in appropriate room acoustic conditions.

2.6.4.2 From oratorio choir to anthem choir

Handel used three choirs from *Messiah* as parts of liturgical anthems with different performance conditions. Two of the newly composed choirs of *Messiah* were incorporated unchanged or with minor text changes into the *Anthem on the Occasion of the Peace* HWV 266 (for the peace of Aachen 1748/49), rehearsed on 22 April 1749 in the *Royal Chapel* of *St. James's Palace* and performed there on 27 April 1749. Under normal conditions the Royal choir had 8 to 12 *Boys* and about 32 *Gentlemen*[21] and was larger than the usual *Messiah* ensemble at the theater or in the *Foundling Hospital Chapel* since 1750.

Handel instructed the copyist to indicate on an additional leaf, which contained only the changed text, the insertion of the choir "How beautiful are thy feet" (no. 34b) from *Messiah* as no. 1 and the choir "Blessing and Honour" (no. 47) from *Messiah* as no. 4 "Blessing and Glory" into the score of the anthem.

Whoever imagines that the *Royal Chapel* at *St. James's Palace* had reverberant church acoustics should take note of the complaints of a twentieth-century Royal organist about the all too dry acoustics:

> In spite of its limited space [the Chapel] has been the scene of many great and brilliant ceremonies [...] We are denied the space, and those architectural features of our Cathedrals so conductive to resonance and blend and shading of vocal tone. The hangings and coverings here are a very real hindrance to vocal music. On the other hand, it is possible to obtain perhaps a greater degree of neatness and finish and clear diction in this comparatively small chapel.[22]

Handel used the acoustic clarity of this small chapel (figure 87), in his lifetime perhaps less decorated with hangings, also for rehearsals in order to prepare the performers well for the difficult acoustics of *Westminster Abbey* or *St. Paul's*. We wonder whether in 1720 the Earl of Carnarvon, later Duke of Chandos, took into consideration not only architectural but also acoustic reasons when he built a copy of the *Royal Chapel* at his country residence in Cannons.[23]

21 J. C. Bridge, "Music at the English Chapels Royal c. 1135 – Present Day" (1927–28), 19–33; HHdb. 4, 425; D. Burrows, Handel's Peace Anthem" (1973), 1230.

22 Bridge, 28–9.

23 Bridge, *ibid.* The construction only started after Handel's activity in 1717/18. See Gerald Hendrie, foreword to vol. III/4 of the *Hallische Händel-Ausgabe: Anthems für Cannons I* (1985), VIII.

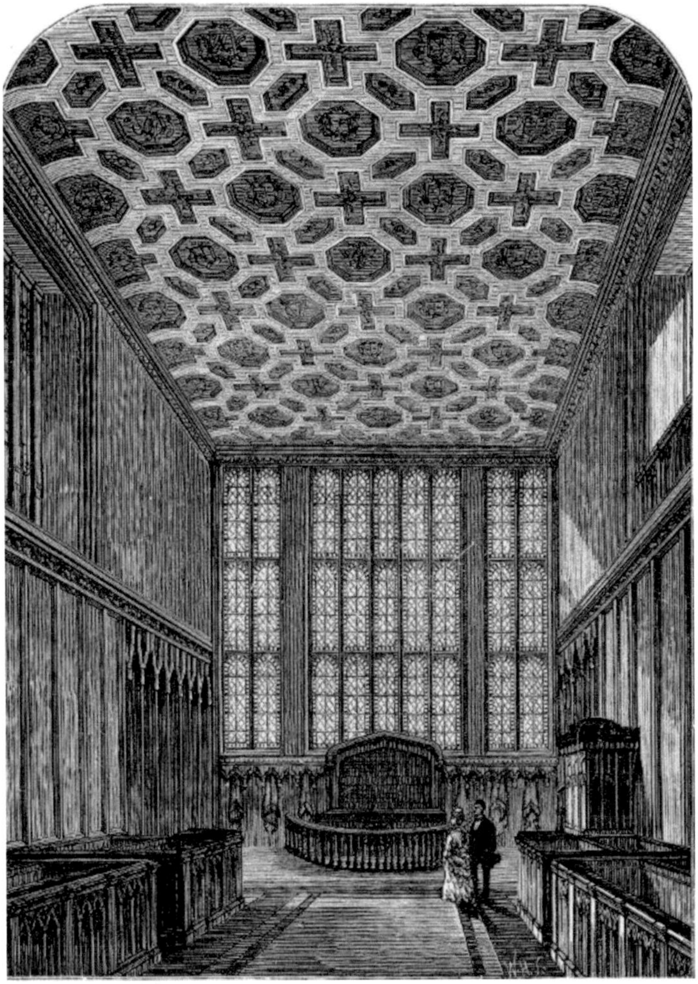

Figure 87: London, *St. James's Palace,* interior of the *Chapel Royal.*

The *Anthem on the Occasion of the Peace* was also performed with a very large ensemble on 27 May 1749 during the charity concert at the *Foundling Hospital Chapel*.[24] On the same occasion the *Foundling Hospital Anthem* HWV 268 was performed with the *Hallelujah* from *Messiah* (No. 39) as final number,[25] whose only change was the increased number of performers.[26] The other pieces of this concert were parts from *Solomon*.

24 Deutsch (1955), 669; Dean (1959), 526.

25 This Anthem also contains the choirs no. 5 and 7 from the *Funeral Anthem* for Queen Caroline HWV 264 (1737), Scene II, Recitative before no. 6 from the first version of *Esther* HWV 50[a] and no. 12 from the oratorio *Susanna* HWV 66 (1749).

26 Similar to the *Messiah*-performance of March 1749 in *Covent Garden.*

2.6.4.3 Limits in music structure for larger space

It is impossible to define exact boundary values for the change in musical expression due to an increase of space and ensemble, as this change depends not only on the score's structure but also on the manner of singing, the articulation in instrumental playing, the dynamics, the musicians' placement, the manner of conducting and the characteristics of room acoustics. For his *Messiah* performances Handel obviously remained within certain limits also with respect to the cubic space of the performance space: the size of the small *Chapel Royal* at *St. James's Palace* (figure 87) lies between the smaller *New Music Hall* in Dublin and the larger *Foundling Hospital Chapel.* In size and shape the *Foundling Hospital Chapel* resembled acoustically less a church than a good "shoe box concert hall". Considering the number of musicians per part, confirmed by the *Foundling Hospital* pay rolls, these performances came closer to Handel's sound ideal than those in the dryer *Covent Garden* or *King's Theatre.* This is confirmed by the anecdote that in the empty theater the acoustics were better for the music,[27] and, on the other hand, by Mary Delany-Granville's enthusiastic report on the *Foundling Hospital* performance in 1754: "the music was too fine, I never heard it so well performed. The chapel is fine [...]."[28] But the performances in the small London theaters with relatively dry acoustics had a special intimacy and spontaneous expression thanks to the ensemble's position on the *Apron Stage* under the same ceiling and close to more than 1,000 listeners. In such theaters or chapels or halls of category B with a dense seating of about 1,200 listeners, Handel used for the complete oratorio a choir of 25 to 30 singers and an orchestra with 25 strings, 4 oboes, 4 bassons, 2 trumpets, 2 French horns, timpani and basso continuo. An increase of these numbers to more than 100 is confirmed only for 1749 (and perhaps only for the mentioned choir numbers, including the *Halleluja* from *Messiah).*[29] Regarding the later changes in performance practice the most important fact is that *Messiah* was performed exclusively in relatively small spaces of the category B.

Still, the performance in the Dublin *Music Hall* for about 600 listeners must have been a special experience. In such an excellent "chamber music hall" the oratorio probably reached its most intense effect thanks to compositional reasons and the excellent musicians (even if performed without oboes and bassoons).

27 See 2.5.3.2: The *Covent Garden Theatre.*

28 *HHdb.* 4, 481, 16 May 1754; *Autobiography and Correspondence of Mary Granville, Mrs. Delany,* London 1861–62, III, 272; see 2.5.2: Handel's *Messiah*-performances at the theater.

29 See also 2.6.5.2: Size of ensemble and space.

2.6.4.4 Is Handel's *Messiah* a "chamber oratorio"?

Astonishingly, the question whether Handel's *Messiah* is a "chamber oratorio" had already been discussed in 1874 by William G. Cusins, the music master of Queen Victoria: "except for the trumpets [Messiah] [...] is scored as if it were a mere drawing-room cantata."[30] But what does chamber oratorio mean in this context? As mentioned, Burney was probably among the first to define in *Ree's Cyclopedia* (1805) chamber music not only by the ensemble's but also by the hall's size: "Compositions for a small concert room, a small band and a small audience; opposed to music for the church, the theater, or a public concert room."[31]

The use of the term "chamber oratorio" for *Messiah* would have been appropriate already around 1805 with regard to the relatively small orchestra and the modest size of the early concert halls in Dublin, Oxford and the theaters in London. But Handel himself, who composed music in relation to function and genre, considered *Messiah* not chamber music but a paraliturgical choir composition with soloists for a large public rather than as an exclusive work for the private use of an aristocratic patron.

Cusins' definition underlines Handel's small *Messiah* ensemble, which was modest compared to the ensembles used for Royal ceremonial works in very large spaces. It was very small also in comparison to later *Messiah* performances in even larger spaces. But for the halls of Handel's *Messiah* performances in Dublin and the London theaters the ensemble was large, whereas for the *Foundling Hospital Chapel* it was appropriate. The term "chamber oratorio" refers to the full effect of the choirs admired by the composer's contemporaries, which did not depend on church-like acoustics nor on large numbers of singers. Even more important is the fact that Handel's *Messiah* performances mark the beginning of a concert tradition that offered in these small and excellent halls a special acoustic presence rarely heard today: the relatively large ensemble in a small volume of space in narrow and high rooms of this size produced an intensity, brilliance and clarity never attained later by larger ensembles, be it in the famous large concert halls of the nineteenth century or in large cathedrals or crystal palaces. The listener had the impression of being inside the event and of participating in the music. The further development of the performance tradition shows that many centers, such as London, Vienna, Paris, Berlin, Leipzig or Zürich, kept these relatively small halls for more than a hundred years for the performance of instrumental music and choir works if no large churches were available, since in these cities large concert halls appeared only in the nineteenth century, often not before the last quarter of the century (see tables 11a and b).

30 W. G. Cusins, *Handel's Messiah, An Examination of the Original and of some Contemporary MSS.* (1874), 14.

31 Cf. M. Tilmouth, "Chamber Music", *NG* 4, 114.

2.6.5 Size of orchestra in relation to the size of architectural space

2.6.5.1 Size of architectural space

A comparison between the halls Handel used for *Messiah* and those he used or visited in Rome reveals his unmistakable orientation after the Italian models. Music life in England at Handel's time was based on the activity of several academies founded by the aristocracy with the explicit aim of bringing Italian culture to England and, like the Italian academies, they supported science and education. The *Royal Academy of Music* (founded in 1718), the *Second Academy of Music,* founded in 1729 for the performance of Italian opera in London, and the *Academy of Ancient Music,* founded in 1726 for the performance of vocal music in the old style, belong to this tradition. On the other hand, national and Christian charitable societies, supported also by the middle class and committed to sacred music, especially the English oratorio developed by Handel, founded such societies as the Dublin *Charitable Music Society,* the *Oxford Musical Society,* the London *Fund for the Support of Decayed Musicians* and the *Committee of the Foundling Hospital.*

The theaters and concert rooms built by these societies were also, at least partly, inspired by Italian models: the theaters from the second third of the eighteenth century clearly display, besides French elements, Italian influence, and the new *Music Rooms* and *Assembly Halls* of the churches undoubtedly follow the tradition of the Italian oratorio halls and small music rooms in the *Palazzi,* where the concerts of the *accademie* took place. England and Ireland had to make do with more modest conditions, mainly outside the theaters, since building costs had to be covered mainly by the middle-class co-founders of the societies. If the latter could not afford their proper halls they used halls in inns, public halls of cities or the *Assembly Halls* of churches.

In terms of seating capacity the Dublin *New Music Hall* with its 450 to 600 listeners was comparable to the *Salone grande* at *Palazzo Bonelli,* were *La Resurrezione* was performed in 1708 for 400 to 600 listeners (figure 69). The *Foundling Hospital Chapel,* which accommodates about 1,200 listeners, belongs to the same category as the *Oratorio della Chiesa nuova,* where regular oratorio performances were held at the time of Handel's stay in Rome. The extant documents on the Dublin hall, the *Foundling Hospital Chapel* and the great hall at *Palazzo Bonelli* are insufficient for exact calculations. But a comparison of these halls shows principal differences with acoustic impact. The more modest English conditions become evident in less room height and the probably higher seating density. But the shape, height, volume and interior decoration of all these English halls were nonetheless favorable enough to provide good acoustics.

The performance traditions of oratorios during Lent in London theaters goes back also to Roman tradition. In Rome oratorios were performed not only in private theaters, which in some cases even had several ranks with boxes, as at the *Palazzo della Cancelleria* (figure 70), but since 1710 again also in public theaters, after the prohibition had been lifted.[32] Handel's London theaters were comparable in size to early baroque Italian theaters of category B, such as the *Teatro SS. Giovanni e Paolo* in Venice (1639–1715) or the *Teatro Valle* in Rome (1727). The more modest conditions prevalent also in England's theater auditoria at Handel's time are reflected in less room height. In Italy singers at that time were already confronted with larger houses: the *Teatro Argentina* in Rome from 1732 with 5,120 m^3 shows the first step towards a larger volume of space, and *San Carlo* in Naples (1737) with 22.50 m width and a volume of more than 12,000 m^3 is a forerunner of the large nineteenth-century opera houses of category C. In London this change happened only around 1791–94 with the new *Drury Lane;* in 1809 the *Covent Garden Theatre* was about the same size as the *Zürcher Stadttheater* (1891).

Table 16: Theaters.

Theaters	L m	W min m	W m	H m	Vol. m^3	Seats
Venice, *Teatro Giovanni e Paolo* 1639–1715	17.00		13.75	11.65	2400	1000
Rome, *Teatro della Valle* 1727	14.50	12.00	12.00	12.00	1740	1100
London, *King's Theatre* 1709/32	15.80	11.60	13.20	10.50	3250	1400
London, *Covent Garden* 1732	13.20	10.00	12.50	9.90	3200	1400
Rome, *Teatro Argentina* 1732	18.15		15.15	15.00	5120	1872
Naples, *Teatro San Carlo* 1737–1817	25.00		22.50	22.50	12375	2550
Drury Lane 1775	15.60	9.60	12.50	?	?	2000
Drury Lane 1794	30.50		24.50	[14.00]	9000	3611
Covent Garden 1809–1847	19.20		15.50	?	?	2800
Zürich, *Stadttheater* 1891	19.00		13.50	14.00	6800	1139

L = length from curtain to front of lower gallery, W = medium width of parterre, H = height at the middle of the parterre. For sources, see appendix table C.

Despite different ground plans all theaters of category B had an acoustic intimacy thanks to the narrow proscenium and parterre in the first third of the room length, which was lost in the later, wider houses. None of the theaters had a forestage and an adjacent parterre larger than 12 m. These auditoria close to

32 H. M. Smither, *A history of the oratorio in the baroque era,* vol. I.

the stage were up to 3.50 m less wide than the *Stadttheater Zürich* (1891) and not larger than such chamber music halls as the old *Gewandhaussaal* in Leipzig or the small hall at the *Tonhalle* Zürich.

2.6.5.2 Orchestra size and volume of space

Documents show that Handel increased the strings by one third from 30 to 40 for the move from the small to the large hall at *Palazzo Bonelli* in Rome. The total number of instruments for the large hall at *Palazzo Bonelli* was the same as for a festive oratorio performance by Corelli at the *Oratorio della Chiesa nuova* in 1698, but instead of Corelli's pure string orchestra with 50 players Handel used a mixed ensemble with 40 strings and 9 wind instruments.

Presumably, Handel adapted also the *Messiah* ensemble somewhat to room size. If we consider the small ensemble with 20 strings, as documented for 1756, 1758 and 1759 at the *Foundling Hospital,* the number of strings for *Messiah* in Dublin probably did not exceed 20 in order to fit the smaller volume of space. In addition to the two trumpets, a bassoon and 2 oboes could be used according to the practice of the time. For the *Foundling Hospital Chapel* this ensemble was increased by 2 French horns, 4 oboes, 4 bassoons (!) and, if available, 5 additional strings. The total number of instruments at the Dublin performance was probably only 24, whereas at the *Foundling Hospital* it was 34 to 37.

At the theater an increase of strings up to 30 is possible, which would compensate for the greater absorption by the boxes and galleries.[33] More strings were certainly used in 1749 at *Covent Garden Theatre.* If the number was the same as at the *Foundling Hospital* that year, the orchestra had 36 strings. With the performance in the chapel of *Foundling Hospital,* where in 1749 only the *Hallelujah* from *Messiah* was played as part of the *Foundling Hospital Anthems,* Handel's orchestra was even larger in relation to the space volume than the festival performances directed by Corelli at the *Palazzo Doria-Pamphilij* in 1689 (about 80 instruments for 15,000 m^3) or the Christmas performance in 1798 at the *Oratorio della Chiesa nuova* (50 instruments and 38 singers for 11,000 m^3).

Truly Roman conditions also in terms of room height were at Handel's disposal only for the main rehearsal of the *Funeral Anthem* for Queen Caroline on 14 December 1737 in the *Banqueting Hall* at *White Hall,* a rectangular hall in Palladian style by Inigo Jones, with a total of 180 instruments and singers for a

33 Such a string ensemble is documented for an opera performance in 1728, see *HHdb.* 4,167. On numbers per part, see also Dean (1990), 102 f.

volume of about 8,500 m^3. This extraordinary large ensemble for a small hall of category C with 32 × 16 × 16 m (height-to-width relation of 1:1), undoubtedly enhanced the high level of the performance.

Table 17: Orchestra and volume of space.

Place, hall, event	strings	wood-winds	brass	voices	b. c.	soli	total	vol. m^3
Dublin, *New Music Hall, Messiah* 1742	20?	1–3?	2	20–30?	2	8	44–56?	[>1'660]
London, *Foundling Hospital, Messiah* 1752–1759	20–25	8	4	17–19?	2	5–6	53–58?	[~4'000]
London, *Foundling Hospital, Anthem (Hallelujah)* 1749	36	>8?	>4?	50?	2	5	>100?	[~4'000]
London, *King's Theater, Covent Garden Theatre, Messiah*	20–30?	8?	4?	20–30?	2	5?	56–62?	3'200
Rome, *Palazzo Bonelli,* stanzione, *Resurrrezione* 1708	30	(3)/4+2	3	4	2?	4	46	[<1'600]
Rome, *Palazzo Bonelli,* great hall, *Resurrezione* 1708	40	(3)/4+2	3	4	2?	4	56	[<6'750]
Rome, *Oratorio della Chiesa nuova, Oratorio* (Corelli) 1698	50	–	–	38	?	?	89	11'000
Rome, *Palazzo Doria- Pamphilij*, Galeria, Lulier *S. Beatrice* 1689	74	–	3		1	?	>78	15'000
London, Whitehall, *Banqueting Hall, Funeral Anthem* 1737				80			180	8'500

For sources, see appendix, tables A and C.

2.6.5.3 Conclusions

In view of the small rooms and Handel's elite ensemble it is surprising that *Messiah* did not only survive the changing performance conditions after his death but resisted them to such a degree despite a certain change in musical expression. This resistance is based on the quality of the composition and of the libretto with texts from the Bible. The content consists of simple, general Christian ideas, and Handel's music maintains its effect thanks to basic *topoi* that function also under acoustically "distorted" circumstances. Thus, an idea may survive all exterior changes if the content remains understandable.

2.7 The enlargement of the musical ensemble after Handel's death

In the yearly *Messiah* performances held since 1759 during liturgy in one of the three cathedrals of Gloucester, Worcester and Hereford, the oratorio's choir pieces gained increasing importance compared to the arias. With the move from the *Assembly Hall* to the choir in the main nave of the Cathedrals, and even more so with the expansion of the performing forces since 1770 in Winchester *Cathedral,* magnificence was increasingly engaging the attention at the expense of the tender and sublime. Yet hardly anyone noticed the loss of delicate sound effects caused by this shift. Indeed, after the 1784 *Handel Commemoration* in *Westminster Abbey* ostensibly critical comments on loudness were published only to stress the importance of the overwhelming event:

> The immense volume and a torrent of sound was almost too much for the head or the senses to bear – we were elevated into a species of delirium.[1] (European Magazine, March 1784)

2.7.1 The 1784 *Handel Commemoration* in *Westminster Abbey*

In his *Account of the Musical Performances in the Westminster Abbey,* published one year after the centenary celebration of Handel's birthday in 1784 on behalf of King George III,[2] Charles Burney gives a reliable, detailed report, notwithstanding the King's wish to see uncritical praise of the performance of Handel's music. Burney's letters reveal that this Royal instruction annoyed him. According to Burney three London music lovers, among them Joah Bates, director of the *Concerts of Antient Music,*[3] drafted early in 1783 a plan to convene a great number of these musicians living in London "on such a scale of magnificence, as could not be equalled in any part of the world." The tradition was to be launched in 1784, on the occasion of the 25th anniversary of Handel's

1 Cf. Chr. Hogwood, *Handel* (1984), 239.

2 Charles Burney: *Account of the Musical Performances in the Westminster Abbey and the Pantheon May 26th, 27th, 29th and June the 3rd and 5th, 1784: in Commemoration of Handel,* London 1785.

3 Hogwood (1984), 45. Not to be confounded with the earlier mentioned *Academy of Ancient Music,* founded in 1726.

death in April (according to the old calendar 1784 was also the centenary of Handel's birthday in February). The board of the *Concerts of Antient Music,* which had included Handel's music in its concerts since 1768, took charge of the organization of the concerts, and the treasurer of the *Musical Fund* promised support. It was King George's III idea to assume the patronage of the second part with two concerts in *Westminster Abbey,* the place of the coronations and graves of the English kings and heroes of the nation, among whom Handel is buried. A third concert was to take place in the *Pantheon,* the popular concert and exhibition hall built by James Wyatt after the model of the *Hagia Sophia.*[4] A part of the festival's income was charity for the *Westminster Infirmary,* which missed its yearly charitable event due to this event, the rest went to other charitable societies. James Wyatt received the instruction to install the necessary tribunes, boxes and benches, "without using one nail." Etchings of the orchestra tribune, the Royal boxes and an exact plan of the orchestra's disposal are enclosed in Burney's *Account* (figure 88a, b, 89b) with the following comment:

> The general idea was to produce the effect of a royal musical chapel, with the orchestra terminating on one end, and the accommodations for the Royal Family, the other. [...] The orchestra was so judiciously contrived, that almost every performer, both vocal and instrumental, was in full view of the conductor and leader; which accounts, in some measure, for the uncommon ease with which the performers confess they executed their parts.
> The whole preparations for these grand performances were comprised within the western part of the building, or broad aisle [...] At the east end of the aisle, just before the back of the choir-organ, some of the pipes of which were visible below, a throne was erected in a beautiful Gothic style, corresponding with that of the Abbey, and a center box, richly decorated and furnished with crimson satin, fringed with gold, for the reception of their Majesties and the Royal Family; on the right hand of which was a box for the Bishops, and, on the left, one for the Dean and Chapter of Westminster. [...]
> The orchestra was built on the opposite extremity, ascending regularly from the height of seven feet from the floor, to upwards of forty, from the base of the pillars; and extending from the centre to the top of the side aisle.
> The intermediate space below was filled up with level benches, and appropriated to the early subscribers. The side aisles were formed into long galleries, ranging with the Orchestra, and ascending, so as to contain twelve rows on each side: the fronts of which projected before the pillars, and were ornamented with festons of crimson morine.
> At the top of the Orchestra was placed the occasional organ, in a Gothic frame, mounting to, and mingling with, the saints and martyrs represented in the painted glass on the west window. On each side of the organ, close to the window, were placed the kettle-drums [...].
> The choral bands were principally placed in view of Mr. Bates, on steps, seemingly ascending into the clouds, in each of the side aisles, as their termination was invisible to the audience. The principal singers were ranged in the front of the Orchestra, as at Oratorios, accompanied by the choirs of St. Paul, the Abbey, Windsor, and the Chapel Royal.[5]

4 See M. Forsyth, *Buildings for Music* (1985), 34, Figure 2.9 and 2.10.

5 Burney (1785), 9–11.

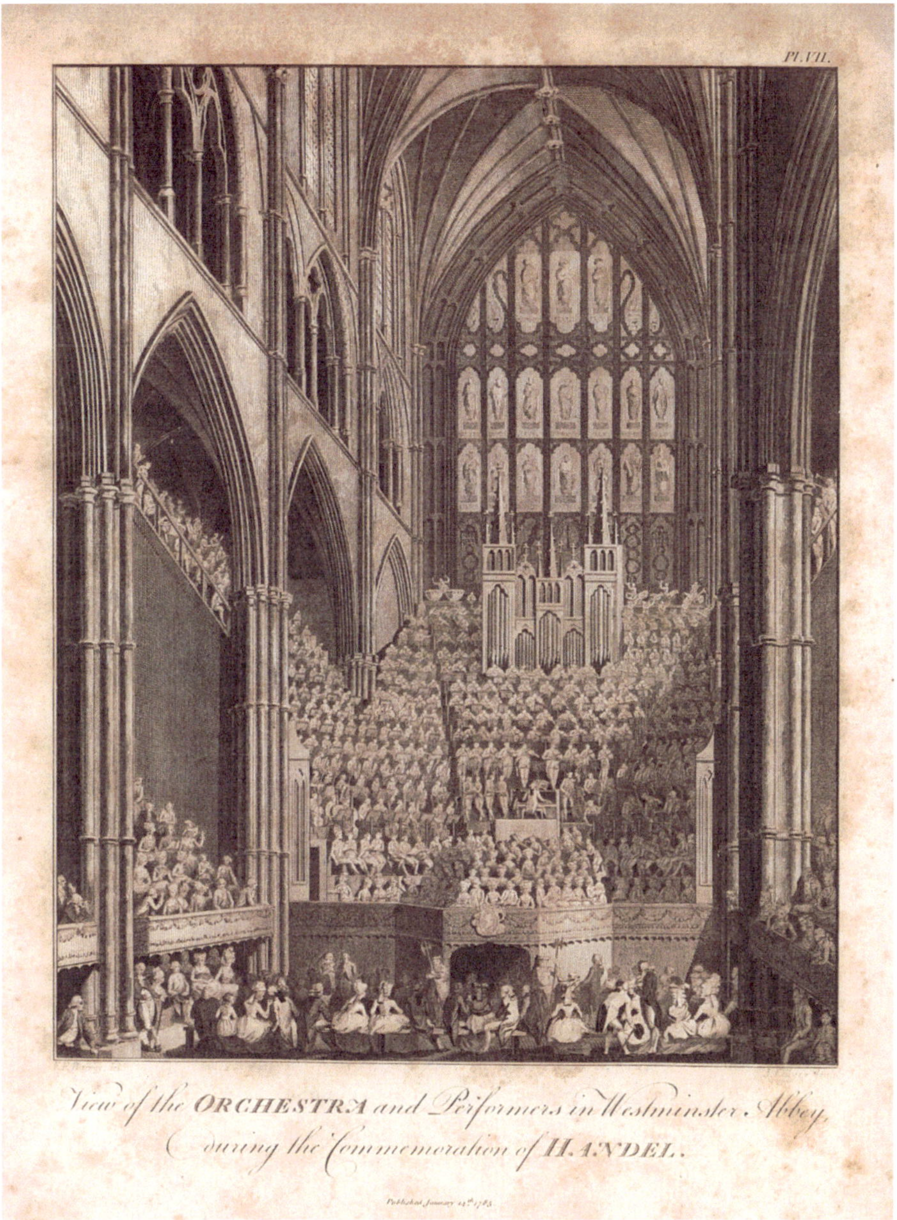

Figure 88a: London, *Westminster Abbey,* orchestra tribune (1784).

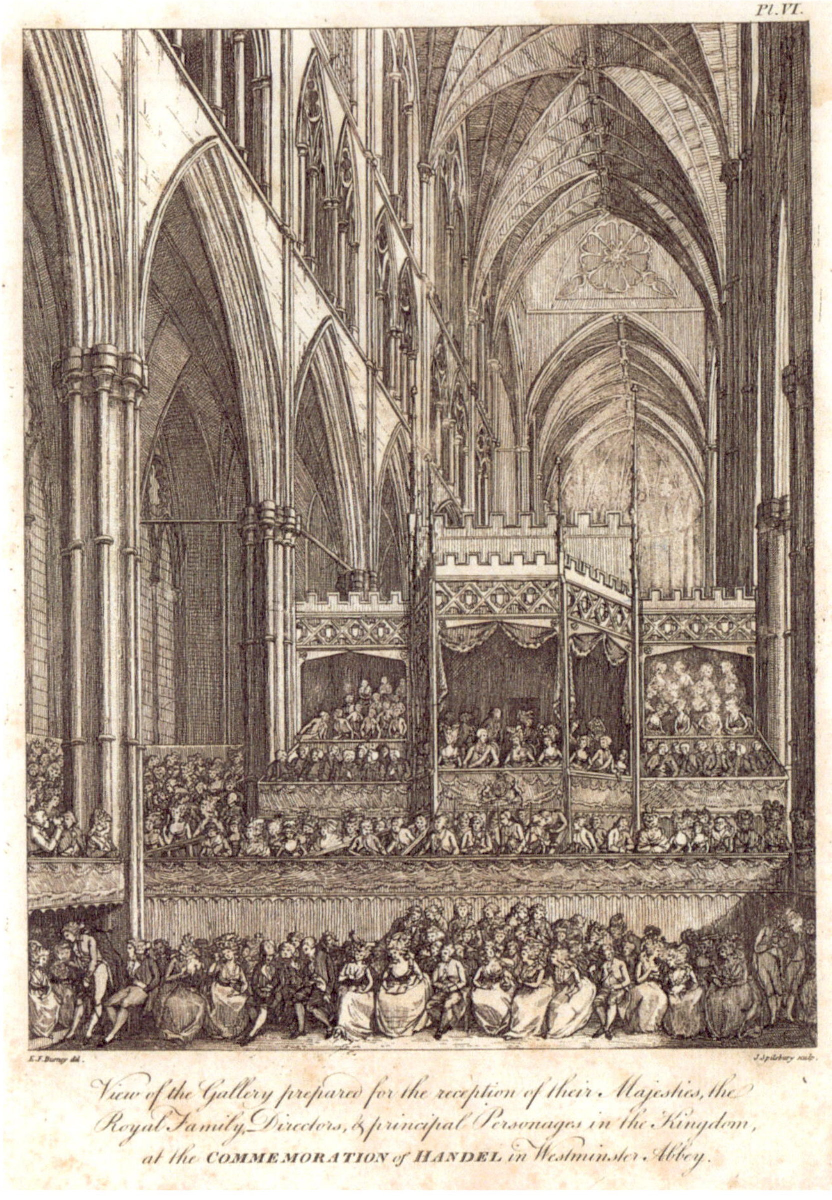

Figure 88b: *Westminster Abbey,* view from the orchestra tribune to the Royal boxes (1784).

This orchestra disposal had a long tradition in London: already in 1713 Christopher Wren installed the tribunes for the performance of the *Utrecht Te Deum and Jubilate* in *St. Paul's*. In 1727 two amphitheatrical tribunes for more than 100 musicians, installed for the coronation of George II in *Westminster Abbey,* were so close to the crossing that the organ had to be placed behind the altar because Handel's request to move the altar had been, understandably, rejected.[6] The confusion during the *Coronation Anthem* "My heart is inditing" was certainly due not only to the short rehearsal time but also to the lack of view to the organ, which was highly important in this acoustically difficult place at the crossing. William Wake, archbishop of Canterbury noted on the margin of his missal: "The Anthem s in Confusion: all irregular in the Music." This embarrassing incident was not forgotten, and in 1784 the tribune was erected at the west end of the main nave. The space here offered all performers a good view to the conductor, who on several keyboards could play the harpsichord and the large organ mounted 19 feet behind and 27 feet higher.[7]

Great efforts were invested in 1784 to gather the orchestra:

> In order to render the band as powerful and complete as possible, it was determined to employ every species of instrument that was capable of producing grand effects in a great orchestra, and spacious building.[8]

Instruments with sufficient power at low frequencies were especially sought, such as trombones ("Sacbut, or Double Trumpet, [...] which the Italians call Trombone, and the Germans Posaune"), for years not in use in the kingdom but finally found in the Royal military chapel. A clumsy *Double Bassoon,* presumably built earlier for Handel, was used, the large *Tower-drums,* already borrowed earlier by Handel with his Majesty's permission (this time the orderly's drums used by Duke of Marlborough in 1709 during the battle of Malplaquet were provided), were supported by especially large drums with copper kettles copied after those built for the *Drury-Lane orchestra.*

Only one "drilling rehearsal" was held one week before the first concert in a hall on Tottenham-Street, with more than 500 musicians, among them the regular members of the *Concerts of Antient Music* and other orchestras, as well as local and foreign professionals, good musicians from the province and *dilettanti.* Of the 120 non-famous musicians examined only two were denied participation. The list of instruments published in Burneys *Account* by category and in alphabetical order of musicians contained 250 instrumentalists (157 strings,

6 H. C. R. Landon, *Handel and His World* (1984), 117.

7 The organ built by Samuel Green from Islington ordered for the cathedral of Canterbury was installed temporarily in *Westminster Abbey* for these concerts.

8 Burney (1785), 9 f.

59 woodwinds and 30 brass) and 283 voices (among them six sopranos, three male altos, three tenors and five basses), or a total of 533 performers:[9]

Instruments:
48 first violins (Principals: Mr. Hay, Mr. Cramer)
47 second violins (2 Principals)
26 violas or Tenors (4 Principals)
13 first oboes (4 Principals)
13 second oboes
6 flutes
21 violoncelli (4 Principals)
26 bassons (4 Principals)
1 double basson (Mr. Ashley)
15 double basses (4 Principals)
12 trumpets (4 Principals)
6 trombones (performers played on other instruments, when sacbuts not wanted)
12 French horns
3 timpani (Kettle-Drums)
1 large drum (Double Kettle-Drum: Mr. Ashbridge)

Singers:
60 sopranos (7 Trebles Principals, among them 2 Italian castratos, one singing only at the second concert at the Pantheon; the choir with 6 ladies, 34 boys, 13 men's trebles or soprano counter tenors)
58 altos (3 Counter Tenors Principals and 55 men's altos)
83 tenors (3 Principals)
83 basses (5 Principals)

The conductor Joah Bates at the *Claviorganum* was supported by three *Subdirectors*. The soloists stood in the first row, in front of the choir's soprano, divided into two symmetrical sections on the steps that reached high up into the side naves (see figure 88a). The number of listeners was reportedly 3,000 to 4,000.

These numbers are impressive, and according to Burney there were no comparable performances either in the kingdom or on the continent. We will quote some of his interesting comments before proceeding to the analysis of the acoustic conditions. Astonishingly, only two of the soloists are described in detail. The German soprano La Mara (1749–1833)[10] was graced with an beautiful and virtuoso coloratura voice. Born Gertrud Elisabeth Schmeling, she was

9 Burney (1785), 16-21. The exact use of the instruments not indicated in Handel's score is not known. See table A in the appendix and analysis in chapter 2.8 and also the comment to the CD: *Handel Messiah,* Recorded Live at the Maryland Handel Festival in Washington Cathedral (1984), Commemorating the Westminster Abbey Performance of 1784.

10 J. Rosselli "La Mara", *NG of Opera* 3, 198.

primadonna of the Royal Opera in Berlin until 1779, then a guest at different opera houses and since 1784 in London:

> Indeed, she had not only the power of conveying to the remotest corner of this immense building, the softest and most artificial inflexions of her sweet and brilliant voice, but articulated every syllable of the words with such neatness, precision, and purity, that it was rendered as audible, and intelligible, as it could possibly have been, in a small theatre, by meer [!] declamation.[11]

Burney also praises her aria from *Messiah* "I know that my redeemer liveth" and the recitative "There were shepherds abiding in the field" following the *Pastorale:*

> These Recitatives, as delivered by the sweet voice and articulate pronunciation of Madame Mara, had an effect far beyond what might be expected from such few and simple notes, without air, or measure: they were literally made 'melting sounds', to every hearer of sensibility present.[12]

A second soprano, Miss Abrams, is mentioned because the acoustics of *Westminster Abbey* had a surprising effect on her rather small voice:

> The happy construction of Westminster-Abbey for cherishing and preserving musical tones, by a gentle augmentation without echo or repetition, was demonstrated by no part of the performance more clearly than in that of Miss Abrams; whose voice, though sweet and of a good quality, is not regarded as Theatrical, but such as the Italians denominate Voce di Camera. Yet, in singing the pleasing Air, "But thou didst not leave his soule in hell", [...] her voice was rendered more audible in every part of that immense building, than it has ever been in any Concert-Room in London.[13]

However, as we look for precise depictions of loud moments in Burneys *Account* the stereotypical effusiveness expressed in recurring terms becomes tiresome over the more than hundred pages filled with exact descriptions of every piece. At the end of the first part on the first concert Burney reports about the public's general reaction:

> I have long been watching the operations of good Music on the sensibility of mankind; but never remember, in any part of Europe, where I attended Musical exhibitions, in the Church, Theatre, or Chamber, to have observed so much curiosity excited, attention bestowed, or satisfaction glow in the countenances of those present, as on this occasion. [...]

11 Burney (1785), 36.
12 Burney (1785), 77.
13 Burney (1785), 80.

> The Choral power of harmonical combinations affected some to tears, and fainting; while others were melted and enrapt, by the exquisite sweetness of single sounds.[14]

It is not surprising that for *Messiah* Burney selected two choirs as climaxes: "For unto us a Child is born" and the moment "Wonderful! Counsellor! the mighty God! the everlasting Father! the Prince of peace!," where the reverberation of the Cathedral became part of the experience thanks to the pauses:

> [...] the idea and effect are so truly sublime, that, assisted by the grandeur and energy of this band, I never felt the pression of words, so strongly before.[15]

And then the expected *Hallelujah,* repeated in both performances on the King's request:

> [...] but I hasten to speak of the Alleluja, which is the triumph of Handel, of the *Commemoration,* and of the musical art. [...] Now as the Orchestra in Westminster Abbey, seemed to ascend into the clouds and unite with the saints and martyrs represented on the painted glass in the west window, which had all the appearance of a continuation of the Allelujah, to imagine that this Orchestra, so admirably constructed, filled, and employed, was a point or segment of one of these exhibited a more respectable appearance to the eye, or afforded a more extatic and affecting sound to the ear, than this.[16]

2.7.1.1 The acoustic conditions in the main nave of *Westminster*

A good observer despite his effusive tendencies, Burney noted that *Westminster Abbey* was "preserving musical tones, by a gentle augmentation without echo or repetition." As mentioned, the reverberation of a loud organ chord lasts about five seconds in this gothic cathedral whose ground plan is based on a symmetrical cross with 67-meter-long arms.[17] Still, for lower sound the acoustics are surprisingly dry, especially in the main nave, which is firmly separated, both visually and acoustically, from the other parts by the high choir rood. This effect is enhanced as the number of listeners increases.[18] Even more important was the advantageous placement of the ensemble, which, for the *Commemoration* and the performances repeated yearly until 1791, was indeed at the best place in the main nave: against the west wall above the main entrance facing towards the rood screen (figure 89a).

14 Burney (1785), 40.
15 Burney (1785), 76.
16 Burney (1785), 83.
17 See 1.8.9: The acoustic conditions in church and 1.9.1: Music related quality factors in churches.
18 The dry acoustics are clearly audible in the film of the 1956 coronation of Elisabeth II.

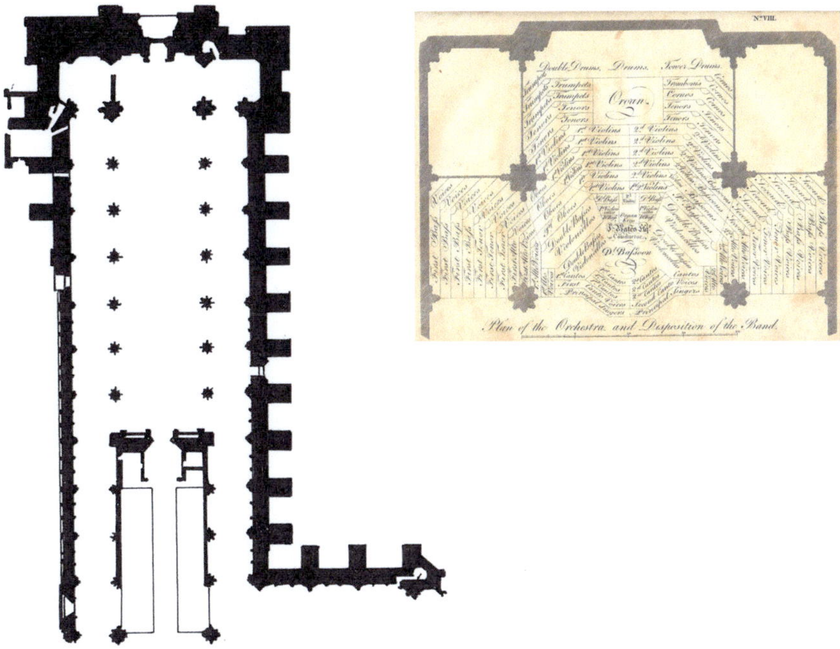

Figure 89: London, *Westminster Abbey:*
a) ground plan; b) plan of the orchestra (1784).

This 31-meter-high section reached a width of 11.50 m between the pillars and of 22 m between the outer walls of side naves (for the section of *Westminster Abbey,* see figure 61). The length from the west wall to the back screen of the choir roods was about 48 m. The reflecting surface of lateral walls in *Westminster Abbey* is only about 50%. The rest is 13% glass surface of the windows and openings. A larger part of sound energy is lost in the Gothic vaults because reflections either don't return to the floor at all or arrive late and weakened, thus neither reinforcing sound nor producing much reverberation. Further sound energy is lost through openings into the side naves and the *triforia*. During the Handel concerts of 1784 to 1791 the side naves were filled with steeply raked tribunes. This setup eliminated the disturbing effect of sound accumulations that increase late reverberation in an empty church. With the occupied tribunes much of this energy is absorbed, and the acoustics in the main nave become clearer. The sound is slightly reinforced by reflections off the pillars and off the wooden installations. In addition to this slight lateral support, listeners in the main nave perceived mainly direct sound (as confirmed by Burneys description). This is why strongly directed sound sources, such as the trumpet and high singing voices, produced the best effect. A fuller and louder sound in such a room can be produced only by a very large ensemble and only if placed such that direct sound radiates best from above down to the listeners, as it did from the musician's tribune. The audible reverberation diminished so quickly that it gave the solo soprano a melting sweetness. The forte of the full choir activated the full reverberation to such a level that tempo, articulation and dynamics had to be adapted, as in the mentioned break after the exclamation "wonderful! Counselor!" Compared to the bad acoustic experiences at the 1727 coronation, sung by the choir in the crossing, the situation had improved considerably. The fear of deafness that might be caused by the high sound level of so many musicians proved to be completely unfounded. Even Burney had wrong expectations regarding the loudness of the large ensemble.

2.7.1.2 The reconstruction of the *Commemoration* in *Washington Cathedral* 1984

At the 1984 reconstruction of the *Handel Commemoration* in the *National Cathedral* in Washington D. C. (completed in 1976), a cathedral in Gothic style with dimensions similar to those of *Westminster Abbey,* the orchestra and choir were not placed at the west end but in the acoustically problematic zone of the crossing on the main floor, on a podium with rising steps for the choir (figure 90; for the number of instruments, see table 15). Reports from members of the orchestra with 100 Baroque instruments confirm that the excessive and incoherent reverberation at this position caused acoustic difficulties. The sound recording engineers had problems finding a good mixture of direct and reverberant sound. The unmistakable difficulties of synchronization between the large orchestra and the choir, caused by the musicians' and the conductor's hearing difficulties, were one reason for the dragging and partly mechanical tempi. The other was the emphatic interpretation of the conductor, who was not a specialist in Baroque performance. Still, the recording conveys an idea of the acoustic situation in a cathedral comparable in size and shape to *Westminster Abbey* (the Gothic vault of the main nave in Washington is about 1 m wider, some 70 cm higher and roughly 11.50 m shorter) and of the reverberation effect caused by a choir with 400 singers – even larger than at the 1784 performance in London. The recording also justifies Joseph Haydn's admiration for Handel: during the *Messiah* performance at *Westminster Abbey* in 1791 he reportedly said after the *Halleluja:* "He [Handel] is the master of all of us."[19] (See chapter 2.8.3 and sound recordings 6, 12, 13, 18.)

19 Hogwood (1984), 243.

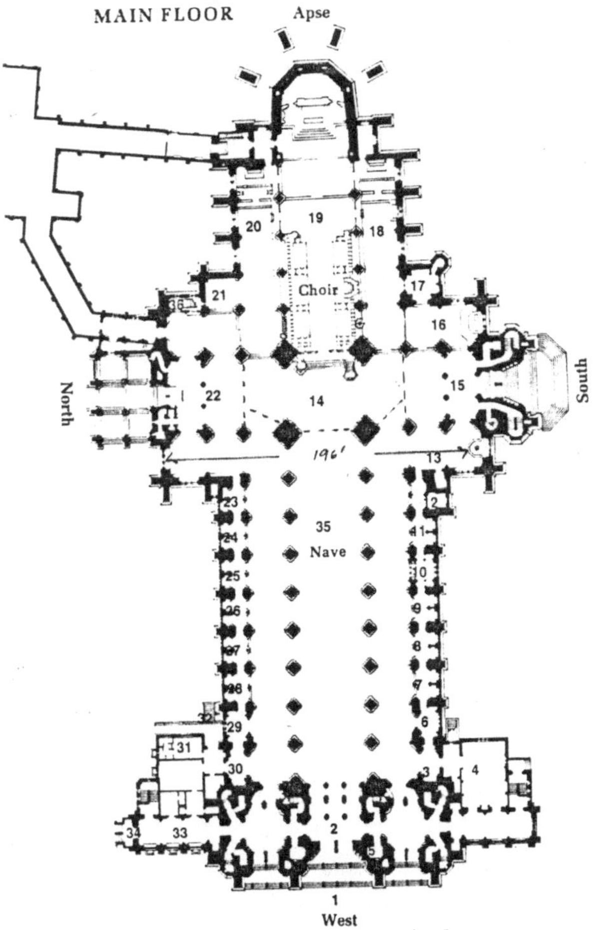

Figure 90: *Washington Cathedral* (1976), ground plan.

In his 1784 report Burney criticized the *Ouverture* of *Messiah* as weak compared to the *Processional Symphony* of the *Coronation Anthem* "Zadok the priest" or the famous Ouverture of *Esther*, often played in *St. Paul's* before the *Te Deum* already during Handel's lifetime. At the beginning of the first part of the *Commemoration* these works conveyed well Handel's grandeur.[20] In fact, the cathedral's acoustics were problematic for the Ouverture of *Messiah,* as its music structure was composed to create a solemn atmosphere in intimate acoustics. (See chapter 2.8.3.1 and sound recordings series I.)

Some of the most impressive numbers of the Washington recording are the forte-parts in the choir "For unto us a Child is born" and the *Halleluja,* thus

20 Burney (1785), 26 and 74.

confirming Burney's comments on the 1784 performance in *Westminster Abbey.* Indeed, the extraordinary forces with the huge drums activating the long reverberation of *Washington Cathedral* are clearly audible on the sound recording (see chapter 2.8.3.4 and sound recordings series IV).

2.7.2 The large choir festivals in England until the *Handel Centennial* in 1859

A further change in the performance practice of Handel's oratorios and *Messiah* in English speaking countries is directly related to the social changes brought about by the industrial development in the nineteenth century. Since the yearly performances in *Westminster Abbey,* held until 1791 (the last time with 1,068 performers), various choir festivals appeared outside London with performances of *Messiah* and other oratorios held by large ensembles, such as the *Edinburgh Festival* in 1815 with more than 120 performers or the *Yorkshire Festival* in 1823 with 180 orchestra musicians and 285 singers. In Birmingham, too, where six Handel Festivals were held between 1759 and 1784, Triennial Festivals were introduced, and in 1834 a larger concert hall was built with more than 2,300 seats, presumably the first concert hall outside London belonging to category C (figure 91).

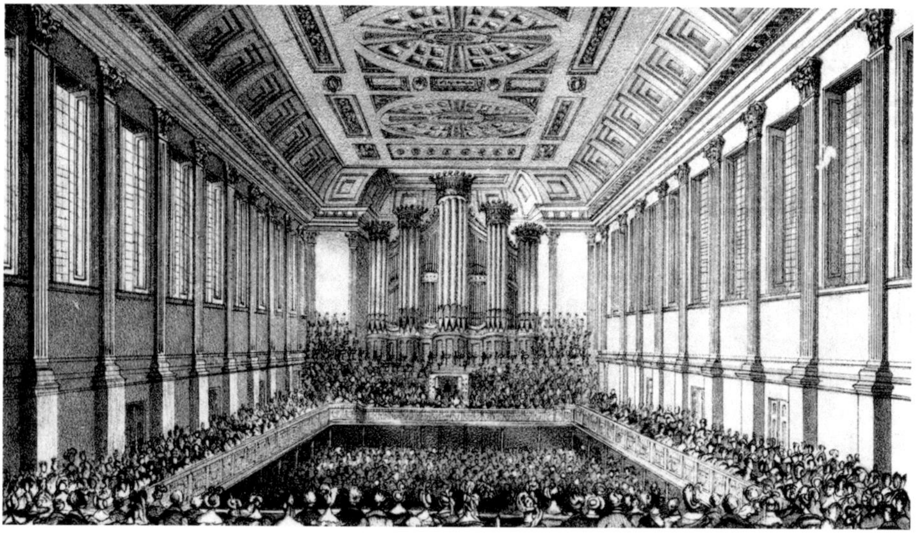

Figure 91: Birmingham, *Town Hall,* organ by William Hill (Music Festival 1834).

The church choirs were not only assisted by choirs from other churches but increasingly enlarged by amateur singers. After 1800 important festivals emerged in the new industrial cities, such as Leeds (since 1858), with choir singers coming mainly from the working class.

In 1834 *Messiah* was again performed in *Westminster Abbey,* now in the Mozart version with a professional choir assembled from different cathedrals in England, for 3,000 listeners of the upper class in the Queen's presence. The number of performers was 644. The organizers of this *Royal Musical Festival* had turned down the *Sacred Harmonic Society's* offer of collaboration. As a consequence the amateur choir founded in 1832 began to offer its own oratorio performances in *Exeter Hall* as *Amateur Music Festivals.* Following the big success the concerts were moved in 1836 from the small hall to the great hall, which featured 3,000 seats and a platform for 500 performers.[21] The building had been erected in 1831 for religious, charitable and scientific purposes, and until 1880 the great hall (45 × 27 × 14.50 m) was the most important concert hall in London. At first only sacred music was performed, and applause was prohibited. At Christmas 1837 the fifth yearly *Messiah* performance took place with 400 amateur singers and some professionals, for the first time in the Queen's presence. Throughout the year several repeat concerts were held, and in the 1840s and 1850s the *Sacred Harmonic Society* garnered increasing success. In 1848 the Neapolitan Michael (Michele) Costa was appointed director of the choir and the Society. The choir and orchestra were increased to 700 members, and in 1850 the podium had to be adapted. At that same time the ceiling was raised to improve the hall for music, and the organ was moved further back on the podium (figure 94).[22] In the meantime the oratorio performances still held in *Covent Garden Theatre* had reached a deplorable level. The upper-class public returned to the Italian opera, while the middle class was the main force behind the ongoing oratorio tradition. Recalling the year 1855, Richard Wagner commented on the remarkable level of the oratorio concerts in *Exeter Hall:*

> In the province of music, with which I was more concerned, I have still to mention several of the Sacred-Music Society's concerts, which I attended in the large room at Exeter Hall. The oratorios given there nearly every week have, it must be admitted, the advantage of the great confidence which arises from frequent repetition. Neither could I refuse to recognise the great precision of the chorus of seven hundred voices, which reached quite a respectable standard on a few occasions, particularly in Handel's Messiah. It was here that I came to understand the true spirit of English musical culture, which is bound up with the spirit of English Protestantism. This accounts for the fact that an oratorio attracts the public far more than an opera. A further advantage is secured by the feeling among the audience that an

21 H. W. Schwab, *Konzert* (1971), 112–13, fig. 82 and comment.

22 Cf. Forsyth (1985), 200.

evening spent in listening to an oratorio may be regarded as a sort of service, and is almost as good as going to church. Every one in the audience holds a Handel piano score in the same way as one holds a prayer-book in church. These scores are sold at the box-office in shilling editions, and are followed most diligently – out of anxiety, it seemed to me, not to miss certain points solemnly enjoyed by the whole audience. For instance, at the beginning of the 'Hallelujah Chorus' it is considered proper for every one to rise from his seat. This movement, which probably originated in an expression of enthusiasm, is now carried out at each performance of the Messiah with painful precision.[23]

Since 1841–42 amateur singing expanded thanks to the foundation of several singing schools where anyone could study for modest fees. The former catholic priest Joseph Mainzer had already founded singing schools for workers in Germany, Belgium and France. In 1841 he looked for music teachers with access to large public halls in England and founded centers in Bristol, Manchester and other industrial cities. In 1842 reports already mention 20,000 pupils taught by teachers from Mainzer's school. Shortly before Mainzer's arrival John Pyker Hullah began to teach choir masters in *Exeter Hall,* and by July 1842 these had 50,000 pupils. Money was raised to build *St. Martin's Hall,* which opened in 1847 despite the unfinished interior design, which was completed only in 1850. The long rectangular hall in simple *Elizabethan style,* a center for *Sight-Singing* until 1860 when it was destroyed by fire,[24] had 3,000 seats and a steeply raked podium for 500 singers and 70 instrumentalists. The third competitor with his own network in England was John Curwen, who used the Tonica-Sol-Fa-method of Miss Sarah Ann Glover. He, too, addressed mainly the middle and working classes. Cheap textbooks and music editions were printed, and in 1842 Mainzer founded his own periodical, the *Musical Times,* which was bought in

23 R. Wagner, *Mein Leben,* M. Gregor-Dellin, ed. (1977), 538 f.: "In das mir näherliegende Bereich der Music fallen noch mehrere Concerte der 'Sacred Music Society', welchen ich in dem grossen Saal von Exeter Hall beiwohnte. Die Oratorien-Performanceen, welche dort fast allwöchentlich stattfinden, haben wirklich den Vorzug einer grossen Sicherheit, wie sie durch sehr häufige Wiederholungen gewonnen wird. Ausserdem konnte ich dem 700 Köpfe zählenden Chore meine Anerkennung seiner sehr präzisen Leistungen nicht versagen, welche besonders im Handelschen 'Messias' einige Male zu respektabler Bedeutung sich erhoben. Ich lernte hier überhaupt den eigentlichen Geist des englischen Musickultus kennen. Dieser hängt wirklich mit dem Geiste des englischen Protestantismus zusammen, daher denn auch eine solche Oratorien-Performance viel mehr als die Oper das Publikum anzieht; wobei sich noch der Vorteil herausstellt, dass ein solcher Oratorienabend zugleich als eine Art von Kirchenbesuch [...] angerechnet wird. Wie man in der Kirche mit dem Gebetbuch dasitzt, trifft man dort in den Händen aller Zuhörer den Handelschen Klavierauszug, welcher in populären Schillingausgaben an der Kasse verkauft und in welchem eifrigst nachgelesen wird, das letztere, wie mich dünkte, auch um gewisse allgemein gefeierte Nuancen nicht zu versäumen, wie z. B. den Eintritt des 'Halleluja', wo es für schicklich gefunden wird, dass alles sich von den Sitzen erhebt." English translation by A.Gray and M. Whittal, *My Life by Richard Wagner* (1983), 525.

24 Schwab (1971), 112–13, fig. 83 and comment.

1844 by the music editor Novello. *Messiah* belonged to the standard repertoire of these amateur choirs at Christmas and on Ash Wednesday and Easter Sunday.

2.7.3 The *Handel Centennial* in *Crystal Palace* at Sydenham

In 1851 Joseph Paxton built the *Crystal Palace* at *Hyde Park* in London for the first World Fair. The huge cast-iron and glass construction measuring a total length of 563 m, a maximum width of 124 m and a maximum height of nearly 41 m was dismantled when the fair closed and rebuilt in 1854 in *Sydenham* in the south of London as an enlarged permanent exhibition hall.[25] The north section was planned as a school of the arts, architecture, botany, geography and ethnology, the south section as a "living display of world trade."[26] The transept in-between was enlarged to an enormous 117-meter-long, 66-meter-wide and 51-meter-high hall with a 45 m × 18 m amphitheatrical concert podium covering a surface comparable to a large concert hall. Thirty-five steps rose from 1.80 m to 15 m above floor level. In the middle of the raked podium a huge Gray & Davidson organ with 4,500 pipes and a steam engine for the organ belts was installed. It was used not only for the accompaniment of the large choir concerts but also for organ competitions[27] and organ concerts.

In 1856 Michael Costa was asked to organize, with the *Sacred Harmonic Society*, a *Great Handel Commemoration Festival* for the centenary of Handel's death in 1859. Already in 1857 a three-day preparatory festival was held with 2,000 singers and about 400 instrumentalists form all over England. The first public rehearsal was held on Saturday. *Messiah* followed on Monday, *Judas Maccabaeus* on Wednesday and *Israel in Egypt* on Friday. Costa himself enriched the score of *Messiah* with "additional accompaniments" developed from Mozart's version.[28] The festival's aim was not only to honor Handel but also to show Britain's love of music:

> [...] it indicates the existence on the part of the British public of a love of Music to a much greater extent, than it has hitherto been supposed to possess. [...] Never yet in the world's

25 C. Friemert, *Die gläserne Arche: Kristallpalast London 1851 und 1854* (1984), 7 ff.

26 Gottfried Semper, responsible for the décor of the South part made a project for an open theater in antique style which later became the starting point for the provisory theater at the Glaspalast in Munich see M. Fröhlich, *Gottfried Semper (*1991), 97.

27 P. A. Scholes, *The Mirror of Music* (1947), 591. Anton Bruckner 1871 played here six organ recitals, see A. Göllerich and M. Auer: *Anton Bruckner* (1936), IV/1, 159 ff.

28 Scholes (1947), 178.

history has choral music been sung to greater perfection, or its vast powers in affecting the human mind been more strongly felt.[29]

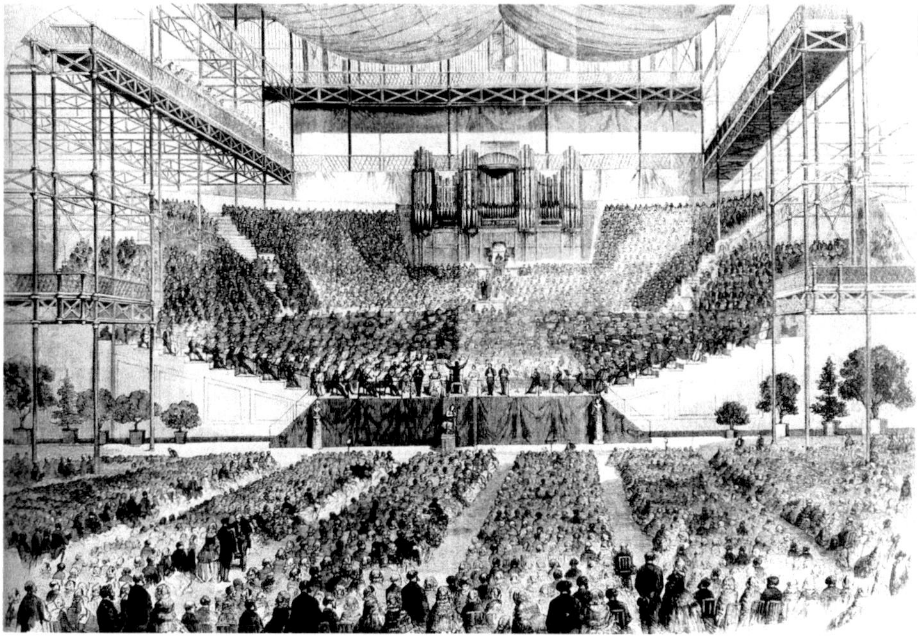

Figure 92a: Sydenham, *Crystal Palace,* Handel Memorial 1857, view to the podium with the grand organ.

The event demonstrated the nation's capacities in the fields of education and art in an exhibition palace displaying everything that represented progress and success at the time:

> Where, except in England, could there have been collected such an array of choral and instrumental talent? Where could there have been found solosingers, who, taking all in all, more capable of satisfactorily executing the responsible task confided to them, and where, lastly, can there be pointed out a public who will listen more devoutly to the music of the great Saxon composer, or who entertain a more profound, a more enthusiastic, and a more intelligent appreciation of its great and manifold beauties?[30]

In 1859 the number of musicians was increased to 2,765 choir singers and 453 orchestra players. Once again, the newspapers published enthusiastic reports, and critical remarks by a Scottish visitor offered an opportunity to boost the praise:

29 *Musical World* 35 (1857), 388.
30 *Musical World ibid.*

> Handel's choral music cannot be obscured by numbers – that is supposing the singers to sing and sing well, as they did at the Crystal Palace. Many listened to the 'Hallelujah' in the garden outside, and declared (which we can readily believe), that the effect surpassed anything they had heard before – whether at Exeter Hall, or at the country Festivals, or at Cathedrals – in solemnity, in grandeur, and in precision.[31]

The visitor from the country, on the other hand, had interpreted the crowds' awesome devotion differently:

> The Crystal Palace was not built for music [...]. This enormous mass of sound in some degree helped us over the difficulty, for a reasonably sized, or even a grand, band and chorus would have been powerless, and only the mammoth noises of the orchestra could have had a chance. But, as it was, except to a very small portion of the audience, the solos and duets were all but utterly lost, and half the grandeur of the great effects was wasted in empty air. [...] But the real Handel effects – the things that set the throat choking, and the eyes swimming – there were none of these. I have experienced them in Exeter Hall, and ten times more in one of the old cathedrals, Gloucester, or Hereford, at the festivals of the choirs. At Sydenham the music could not do its work, and the whole atmosphere was gay, fashionable, and morning-concerty.[32]

Such critical voices remained marginal, at least in England, and could not prevent success. The *Handelomania* was so overwhelming and lasting that until 1926 a *Triennial Handel Festival* was held with the same works. The number of performers was increased to 4,500 in 1883, and at the last concerts in the 1920s was nearly 4,000. Choir singers came from different provinces, mainly from the Middlands and the North with their "strong lunged and well-trained singers."[33] The number of listeners increased from about 12,000 to 87,763 in 1883, then slowly dropped to about 67,000. In 1871 the *Musical Times* replied to growing criticism with an astute reference to the event's non-musical importance and a remarkable observation on the problematic imbalance between the magnificent choirs and the unchanged arias and recitatives:

> We are not of those who believe that the greater the number of executents in a musical work, the greater must be the effect, nor are we by any means disposed to endorse the assertion, so often advanced, that Handel would have been in an ecstasy of delight could he have lived to hear his compositions with every part, vocal and instrumental, multiplied ad infinitum in the choral portions, whilst the solos are sung precisely as he wrote them; but such a magnificent demonstration as that presented at this triennial festival has perhaps scarcely a right to be judged purely on artistic grounds, for there is something to be seen as well as to be heard when upwards of four thousand performers assemble, under the experienced direc-

31 *Musical World* 37 (1859), 441.
32 *Musical World* 37 (1859), 440–1.
33 *MT* 35 (1891), Juni.

tion of one mind, to render homage to a genius, the admiration for whose works in the land of his adoption has deepened almost into a worship.[34]

2.7.4 Documents on the acoustics in *Crystal Palace*

The acoustic difficulties in this hall, so enormous it was comparable to a railway station, did not surprise experts.[35] Berlioz declined to write the music for the opening ceremony.[36] With 117 m the new *Transept* of 1854 was comparable to a cathedral. But its 66-meter span was more than twice the main nave of *St. Peter's* in Rome and four times the main nave of *St. Paul's* in London. The barrel vault was 6 m higher than *St. Peter's* in Rome (table 8a). Besides the glass walls, the glass roof and the balconies high up affixed to the iron construction, the hall had practically no surfaces that produced any helpful reflections. To loud sound the glass responded with masses of reverberation delayed by more than 2 seconds. The managers tried to improve the situation as much as possible. The picture of the 1857 *Preliminary Festival* shows heavy cloth over the stage (a so-called *velarium).* The stepwise raised podium was closed to the rear with a wooden wall. The correspondent of the *Niederrheinischen Musikzeitung* recalls in 1859 that these means were insufficient:

> At the concert in 1857 the public noticed that the mass of performers in the choir and orchestra still did not produce the expected overwhelming impression and, especially, that the sound of the choirs did not fill the transept space completely. It was therefore decided to increase the number of performers for the actual festival to 4,000. [...] However, we stick to our opinion that just as the theory has not yet solved the enigmas of architectural acoustics, so the sound effect of certain masses in certain spaces is hardly calculable. Experience seems to show that up to a certain limit the sound volume increases in relation to the number of voices and instruments: above this limit the same proportion is not maintained. Of course, it makes a big difference whether a choir has 20 or 200 voices; but between 20 and 2,000 voices there is by no means an arithmetic relation in terms of sound effect, as we often noticed at large male-chorus festivals, especially at the international Belgo-German festival in Cologne in 1846, where 2,000 singers in the Gürzenich hall did not produce a greater effect than 300–400 normally would.[37]

34 *MT* 15 (1871), 138.

35 The widest span had 1868 *St. Pancras Station* with 74 m (243'), see N. Pevsner, *History of Building Types* (1976).

36 Forsyth (1985), 149 f.

37 *Niederrheinische Musikzeitung* 7 (1859), Nr. 27, 198: "Bei dem Concerte im Jahre 1857 hatte man beobachtet, dass die damalige Masse der Ausführenden in Chor und Orchester doch nicht den erwarteten übermächtigen Eindruck machte, dass namentlich der Klang der

A comparison of the illustration published in the *Illustrated London News* shows that the "canopy of heavy reflecting material," in 1857 mounted about 45 m above ground (figure 92a), was lowered to about 36.50 m in 1859:

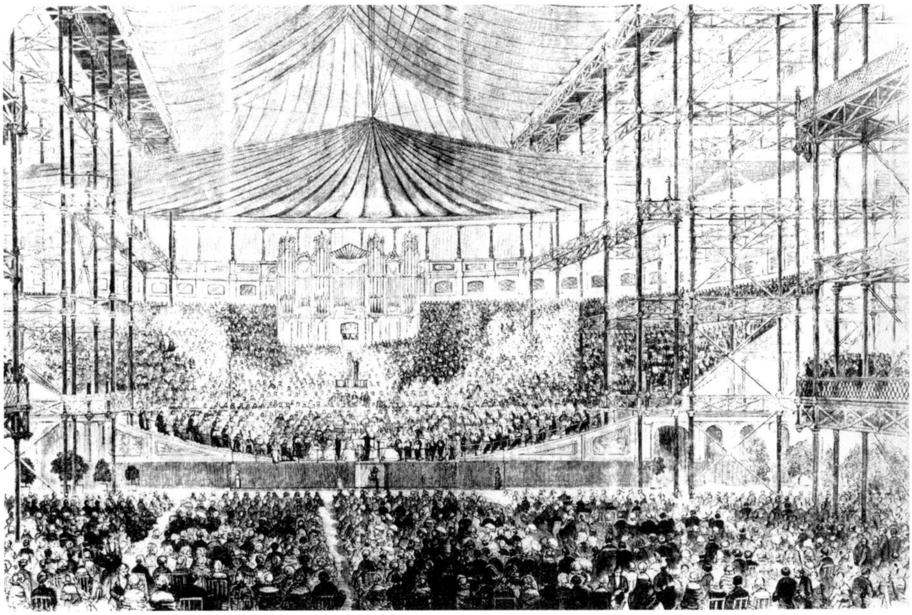

Figure 92b: Sydenham, *Crystal Palace,* Handel Centennial 1859.

The correspondent of the *Niederrheinischen Musikzeitung* confirms a certain improvement obviously supported by some changes at the rear wall of the orchestra:

> Notwithstanding all the improvements in the Palace one cannot deny that the immense hall is still lacking a lot to be suitable for musical performances. It is impossible to organize a

Chöre den Raum des Querschiffs nicht vollständig ausfüllte. Man beschloss deshalb, die Zahl der Mitwirkenden für das gegenwärtige Fest auf 4000 zu steigern. [...] Wir bleiben indess der Ansicht, dass, so wie der akustische Bau seine durch die Theorie noch nirgends unfehlbar gelösten Räthsel hat, auch die Klangwirkung der Massen in bestimmten Räumen sich nicht berechnen lässt. Die Erfahrung scheint zu lehren, dass es eine gewisse Gränze gibt, innerhalb welcher sich die Stärke des Klanges im Verhältniss zu der Zahl der Stimmen und Instrumente steigert: über diese Gränze hinaus findet nicht dieselbe Proportion statt. Es ist allerdings ein grosser Unterschied, ob ein Chor von 20 oder von 200 Stimmen singt; aber zwischen einem Chor von 200 und von 2000 Stimmen entspricht das Klangverhältnis keineswegs dem arithmetischen, was wir besonders bei grossen Männer-Gesangsfesten häufig beobachtet haben, namentlich bei dem internationalen belgisch-deutschen Feste in Köln im Jahre 1846, wo im Gürzenichsaale die 2000 Sänger keine grössere Wirkung machten, als sonst drei- bis vierhundert."

large enough mass of musicians that would be able to fill these wide open galleries, these endless wings of the main nave, this sky-high vault, whose enormous abyss swallows up the thunder of twenty thousand voices and instruments without any reverberation – On the first day we sat conveniently in row no. 55, directly above the podium, below the south-gallery. The view was magnificent and wonderfully impressive, and one would have expected this to be one of the best seats for seeing and hearing. But we soon started measuring the distance between our seats and Costa's conductor stand and estimated it at three times the distance between the last gallery and the stage at Her Majesty's theater.
The orchestra, set up like an enormous shell, seemed perfectly appropriate to send the sound directly into the transept, and we thought that this [shape] would spare us sensing the great distance. But this thought was refuted already at the God save the Queen, although this same piece was said to have made an enormous impression on many of the reserved seats. The high notes of Mrs. Novello were clearly audible also at our seats but did not suggest any powerfulness, and everyone knows that a soprano is better heard from far away than any other voice. [...] But when the quartet of Miss Dolby, Sims Reeves, Belletti and Weiss sang the second verse we were forced to listen with utmost attention in order to catch a tone here and there, and sometimes the singing was not audible at all.[38]

The editorial department of the *Musical World* closed its report of that year, partly reprinted in the *Niederrheinische Musikzeitung* in German translation, with a great eulogy followed by a reply to critical voices:

Regarding the critical view we cannot avoid to admit that never before such an affect was produced by any group of singers. Some fluctuations of the masses, the lack of stability, the

38 *Niederrheinische Musikzeitung* 7 (1859), Nr. 28, 219: "Bei aller Anerkennung der baulichen Verbesserungen in dem Palaste lässt sich doch nicht läugnen, dass noch viel fehlt, den ungeheuren Raum für musikalische Aufführungen geeignet zu machen. Es ist nicht möglich, eine Masse von Musikern aufzubringen, die im Stande wäre, diese weithin offenen Galerieen, diese endlosen Flügel des Langschiffes, dieses himmelhohe Domgewölbe zu füllen, dessen ungeheurer Schlund den Donner von zwanzig Tausend Sängern und Instrumentlisten verschlingt, ohne einen Wiederhall davon zurückzugeben. – Wir sassen am ersten Tage günstig, in der Reihe Nr. 55, gerade über der Tonbühne, unter der Süd-Galerie. Der Überblick war prächtig und wunderbar imposant, und man konnte erwarten, dass dies einer der besten Plätze zum Sehen und Hören sein müsse. Bald nachher fingen wir indess an, den Abstand unserer Sitze von dem Directionspulte Costa's zu ermessen und mussten ihn auf das Dreifache der Entfernung der letzten Galerie von der Bühne im Theater ihrer Majestät schätzen. Das Orchester, wie eine riesigen Muschelschale aufgebaut, schien vortrefflich geeignet, den Klang gerade aus in den Raum des Querschiffs auszuwerfen, und wir dachten, dass dies wohl die grosse Entfernung nicht fühlbar machen werde. Aber gleich durch das God save the Queen wurde dieser Gedanke widerlegt, obwohl dasselbe, auf einem grossen Theile der reservirten Plätze angehört, einen gewaltigen Eindruck gemacht haben soll. Die hohen Töne der Frau Novello waren auch da, wo wir sassen, deutlich zu hören, allein sie brachten auch keine Idee von Stärke hervor, und Jedermann weiss, dass eine Sopranstimme besser als jede andere in der Ferne gehört wird [...]. Als aber das Quartett Miss Dolby, Sims Reeves, Belletti und Weiss den zweiten Vers sang, sahen wir uns genöthigt, mit der grössten Aufmerksamkeit zu lauschen, um gelegentlich einen Ton aufzufangen, und zuweilen war der Gesang ganz und gar nicht zu vernehmen."

more frequent lack of subtlety of performance, impossible under such circumstances, and nearly complete perdition of all more delicate effects of instrumentation in this horrible sea of sound (in a way that Messiah would have been as well or even better without Mozart's additions, just to give an example), up to a certain amount were really disturbing, but could not count again this magnificence and sublime, surprising the ear and filling the soul with admiration for the power of Music.[39]

The managers did not rest and in 1862 installed new improvements:

> Despite the four-to-five-hour duration the public's attention did not flag, and thanks to the new installations, mainly a huge canopy, for two thirds of the listeners no tone was in fact lost. One can boldly claim that Handel's masterpiece has never been performed with such dignity, majesty and impressive power. [...] The overall impression was tremendous, not marred by any failure. The orchestra's volume of sound, revealed immediately in the Ouverture, was the strongest element of effect, especially when joined by the organ. [...] The solos were excellent.[40]

It seems that this helpful *canopy* of massive wood installed directly above the stage served its purpose until a fire destroyed the *Crystal Palace* in 1937. The management now organized up to 10 weekly concerts, such as the popular brass concerts directed by August Manns, the *Concerts of the National Brass Band Festival* (since 1860) and the famous Saturday concerts with program books by Sir George Grove, which offered countless first performances and first presentations in England (among others, works by Schubert, Schumann, Brahms and Dvořák). Bernhard Shaw, a feared critic, commented repeatedly

39 *Niederrheinische Musikzeitung* 7 (1859), Nr. 28, 220: "Was den kritischen Standpunkt betrifft, so können wir nicht umhin, zuzugeben, dass noch niemals durch irgend eine Vereinigung von Stimmen und Instrumenten ein solcher Effect hervorgebracht worden ist. Das jeweilige Schwanken der Massen, der Mangel an Stätigkeit, der noch häufigere Mangel an Feinheit der Ausführung, die unter solchen Verhältnissen unmöglich ist, und das schier gänzliche Untergehen aller zarteren Instrumentation in das furchtbare Klangmeer (so dass, um nur Ein Beispiel anzuführen, der Messias ohne Mozart's Zuthaten eben so gut, wo nicht besser, gewirkt haben würde), waren in gewissem Grade allerdings störend, aber konnten nicht von Gewicht sein in der Waagschale gegen die Grossartigkeit und Erhabenheit, die das Ohr in Erstaunen setzte und die Seele mit Bewunderung über die Macht der Musik füllte."

40 *Niederrheinische Musikzeitung* 10 (1862), Nr. 30, 233 f.: "Trotz der vier- bis fünfstündigen Dauer blieb doch die Aufmerksamkeit des Publikums gespannt, und durch die neuen baulichen Einrichtungen, deren Hauptbestandtheil ein ungeheurer Schalldeckel [aus Holz] ist, hatte man es in der That erreicht, dass wenigstens für zwei Drittheile der Zuhörer kein Ton verloren ging. Man kann dreist behaupten, dass Handel's Meisterwerk [der Messias] noch nie mit so viel Würde und Hoheit und imponirender Kraft aufgeführt worden ist. [...] im Ganzen war [der] Eindruck ein gewaltiger, den kein Misslingen störte. Die Klangfülle des Orchesters, die sich gleich in der Ouverture offenbarte, bildete das mächtigste Elemente des Effectes, zumal wenn sich die riesige Orgel mit ihm vereinigte. [...] Die Soli waren ausgezeichnet."

around 1900 on the *Handel-Festivals.* In 1894, towards the end of a hardly enjoyable concert season, he wrote:

> Like all our national institutions, the Handel Festival has a great deal of nonsense about it, and is applauded for its nonsense much more than for its sense. As, for example, the size of the thing, though, after all, it is only about a quarter as large as the London police force, about which nobody makes any fuss. Mr. Manns is much praised for the Napoleonic feat of conducting nearly four thousand performers, as if that were fifty times as difficult as conducting the seventy-five or eighty who constitute his band at an ordinary Saturday concert.[41]

2.7.5 Size of the orchestra and the concert halls in nineteenth-century London

2.7.5.1 Size of the concert halls

A list of London halls used for *Messiah* and other oratorio performances during the nineteenth century allows us to exactly trace, by period, the increase in dimensions described in chapter 1.8 (table 10). The number of seats and standees, the ground surface, the height-to-width relation show the shift from category B to category C during the century: the great hall of the *Hanover Square Rooms* (1775–1874), until 1869 the seat of the *Philharmonic Society*, whose symphony concerts for 500 to 800 *upper-class* listeners sometimes included also choir music, belonged to category B (figure 93).

In 1862, after careful renovation, the acoustics were commented as follows:

> [...] the admirable manner in which the Concert-room has been constructed for sound: the smallest piano is distinctly audible, and the most powerful forte produces no confusion, whilst the absence of galleries gives an air of comfort to the audience part which cannot be realised in any other Hall built for musical performances.[42]

Exeter Small Hall, where the *Sacred Harmonic Society* held its concerts from 1831 to 1835, also belonged to the smaller halls of category B. Thus far no exact information on forces used for oratorios in these two halls could be found. The previously mentioned performance of Schumann's *Paradise and Peri* at *Hanover Square* strongly suggests that the ensembles used for choir compositions reached the upper limit of the podium's size.

41 G. B. Shaw, *Music in London 1890–94* (1932), vol. 3, 252–3.
42 Scholes (1947), 1, 205.

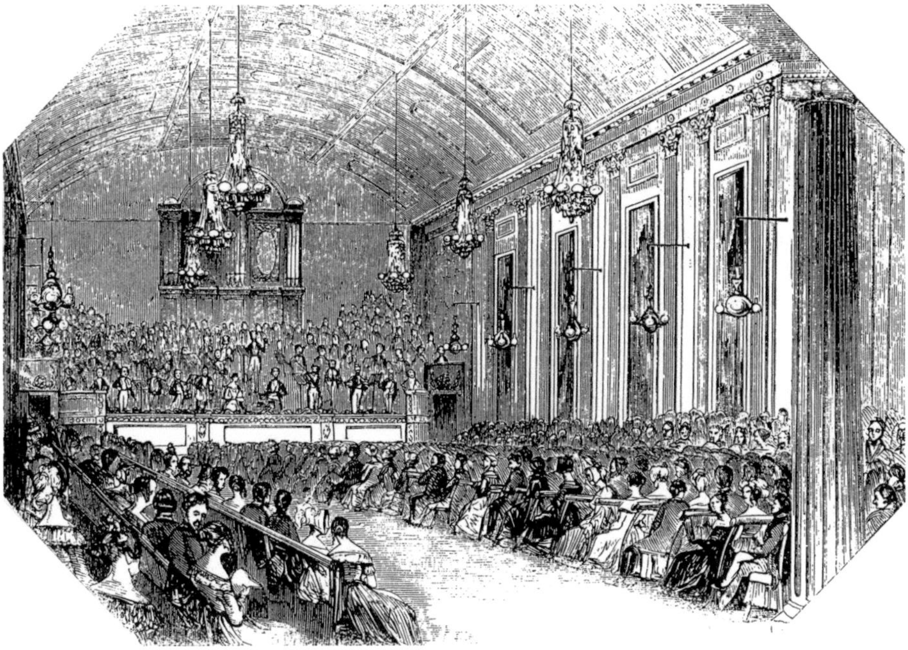

Figure 93: London, *Hanover Square Rooms* (1775–1874).

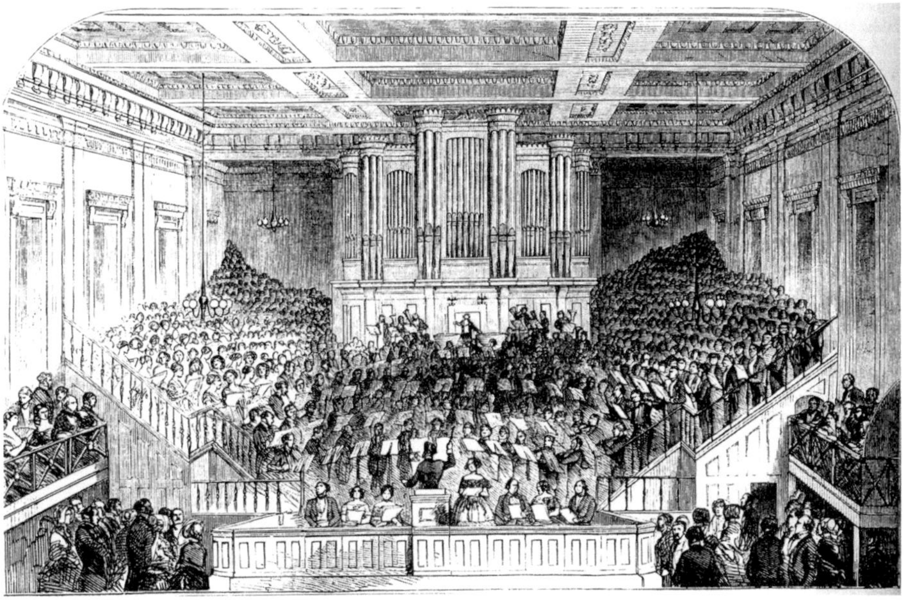

Figure 94: London, *Exeter Hall* (after 1850).

This tendency to larger forces is confirmed by the *Sacred Harmonic Society's* move from *Exeter Small Hall* to *Exeter Great Hall* and by the required podium enlargement in the latter after 15 years (figure 94).

According to the *Musical Times* the 1850 reconstruction and the now raised ceiling transformed this too low hall of category C, where most London oratorio concerts were held between 1836 and 1860 from "the musically worst space [in]to England's perhaps best concert hall." It then certainly was London's largest and most convenient hall.[43]

After the renovation the hall's size (45 × 27 × 14.60 m) was comparable to the large hall at the *Concertgebouw* in Amsterdam (this hall is 2 m shorter but also 2 m higher). *St. Martin's Hall,* which opened 1847, the same year as the seat of John Pyke Hullah's *First Upper Singing School,* was until 1867 an important category C hall for choir concerts.[44]

Choir concerts were held since 1858 also in the acoustically excellent *St. James's Hall*, seat of the *Philharmonic Society* from 1869 to 1894 (figure 95). On Good Friday *Messiah* was a popular oratorio. In size the hall was comparable to the great hall at the Vienna *Musikverein:* with a length of 42.40 m and a width of 18.30 m it was only 3 m shorter and somewhat narrower than the 19-meter-wide Vienna hall; its proportions were also nearly 1:1, but instead of a flat ceiling it featured a coffered barrel vault. Due to very narrow and uncomfortable chairs the hall had 2,500 seats, over 900 more than the *Goldene Saal* in Vienna! Thanks to its restricted width and exceptionally present acoustics the hall was regularly used for chamber music despite its size. The musicians used to sit in the center of the hall.

For this hall, too, with its highly popular *Monday Pops* since 1859 and the memorable concerts of the *Joachim-Quartet,* homages were written when it was pulled down in 1904 for a new building. In 1944 Helen Henschel still remembered:

> Those who ever heard music in St. James' Hall will always mourn its loss, though it wasn't very comfortable and certainly not beautiful. I do not suppose that any concert hall has existed with more perfect acoustics, and I wonder if it can be entirely imagination that endows it with a unique atmosphere of intimacy and charm, a warmth of welcome to those who came there to make music and to listen to it.[45]

The *Queen's Hall* (figure 96), from 1894 to 1941 the seat of the *Philharmonic Society,* initially had difficulties prevailing against *St. James's.* But thanks to its good acoustics and comfortable interior design – two unsupported balconies and the U-shaped galleries at the rear wall – it became a very successful hall

43 Scholes (1947), 205.
44 Schwab (1971), 112–13, figure 83.
45 Cf. Scholes (1947), 211.

with nearly 3,000 seats, which were later reduced to 2,000, until it was destroyed during the Second World War.

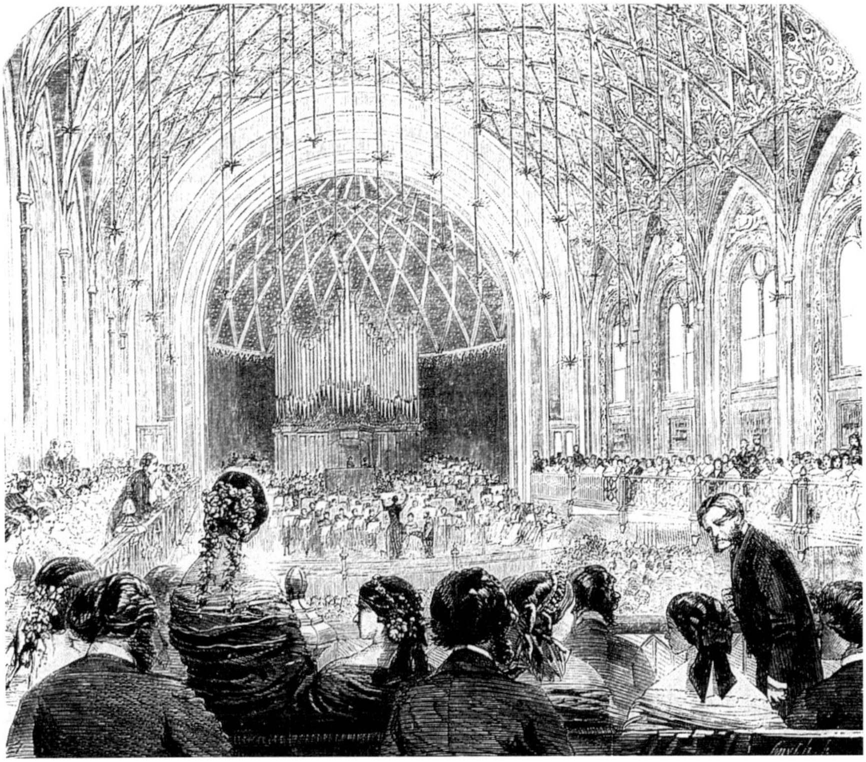

Figure 95: London, *St. James's Hall* (1858–1905).

The *Queen's Hall*, from 1894 to 1941 the seat of the *Philharmonic Society,* initially had difficulties prevailing against *St. James's*. But thanks to its good acoustics and comfortable interior design – two unsupported balconies and the U-shaped galleries at the rear wall – it became a very successful hall with nearly 3,000 seats, which were later reduced to 2,000, until it was destroyed during the Second World War.

Despite its rather dry acoustics and the disadvantaged seats under the balcony the hall enjoyed an excellent reputation since the opening concert:

> Every detail was heard with the utmost clearness, and in the loudest passages no trace of obscurity was apparent. A large audience had assembled, and its approval both of the hall and the performance was emphatic and unanimous.[46]

46 *MT* 35 (1894), 27. For a detailed analysis of the acoustics, see A. Bagenal and H. Wood, *Acoustics* (1931), 109 f.

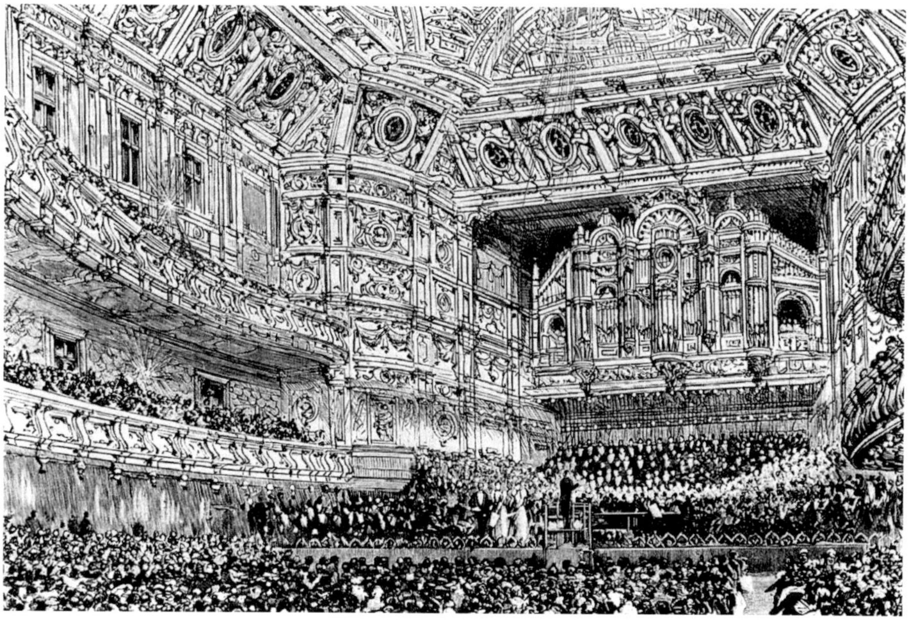

Figure 96: London, *Queen's Hall* (1893–1941).

2.7.5.2 Seeing and listening

With *Queen's Hall* a new theater-like building type with a shorter and larger hall, fortunately with good acoustics, replaced the old long, narrow halls. The architectural design met the desire for a better view to the podium, which was crucial to a full experience of the performance of virtuoso soloists and conductors and the breathtaking functioning of the large orchestra. At the overcrowded hall of *Hanover Square Rooms* only listeners in the hall's front section had a free view to the podium. For a long time this flaw obviously seemed sufficiently compensated by the good acoustics and the relatively free comings and goings of the upper-class public.

The visual aspect of the musical event was important already in 1784 in *Westminster Abbey*. Not only the decorated Royal box at the rear wall of the rood screen but also the amphitheatrical podium for the performers and the special organ were part of a spectacular presentation. A seat in row no. 55 prompted the critic of the *Niederrheinische Musikzeitung* to comment that the breathtaking view was an essential element of the event in *Crystal Palace* as well. No wonder, then, that the architects of the *Royal Albert Hall* chose an elliptic hall with a dome (figure 12) when a proposal came up in 1854 to replace the provisory *Crystal Palace* with a permanent exhibition hall. The popularity of this huge exhibition and concert hall, built after the model of the Roman *Colosseum,* is visually justified. However, the inevitable echo caused by the enormous dimensions came as a bad surprise at the first choir rehearsals. The musicians somehow tried to cope with the problem, which was especially disturbing on the podium, and the specialists tried to eliminate the echo with heavy cloth applied to the whole dome. But only in 1931 did Bagenal & Wood reach effective results by placing over the podium a large reflector that diminished the sound energy getting into the dome, thus directing early reflections to the listeners.[47] After the *Queen's Hall* was destroyed by a bomb in 1941 and concerts were transferred to the *Royal Albert Hall,* the acoustic situation was regarded as inadequate. In 1971 the podium's and the reflector's position could be optimized, and the diffusers mounted in the dome finally reduced the echo to an acceptable level.[48]

The envelopment by sound, so impressively described by Berlioz as typical of the good small halls in the eighteenth century and of the great nineteenth-century halls where correspondingly larger ensembles played, was replaced in the large festival halls by the spectacular event because here only the exceptional and super-human had an effect. This change is reflected also in the organ's importance. The instrument playing the basso continuo in the *tutti* and the solo instrument used by Handel for his virtuoso entr'acte improvisations later turned into the magnificent, overwhelming, space-filling miracle of sound. In the great oratorio performances the organ was the widely visible and audible attribute of a solemn ceremony. This significance dates back to the big State ceremonies of the seventeenth and eighteenth centuries, when special organs were installed in *Westminster Abbey* for each coronation ceremony, from James II (1685) to Victoria (1838). For the coronation of George II in 1727 an organ by Shrider was installed, in Handel's opinion "a very curious instrument."[49] The organ by William Hill for the *Town Hall* in Birmingham was the first large organ in England in 1834 (figure 91).[50] Both the organ at *Crystal Palace* by Gray & Davidson (1857) and the organ at *Albert Hall* by Henry Willis, completed 1871, were considered the largest organs of the world at the time (figure 97).[51]

47 Bagenal and Wood (1931), 62 f.

48 L. L. Beranek, *Music, Acoustics, and Architecture* (1979), 325.

49 *HHdb.* 4, 182.

50 St. Bicknell, *The History of the English Organ* (1996), 228.

51 Scholes (1947), 590. Bruckner in 1871 also played 17 recitals on the "Handel-Organ in the *Royal Albert Hall,*" see Göllerich / Auer (1936), IV/1, 143 ff, and figure after p. 144.

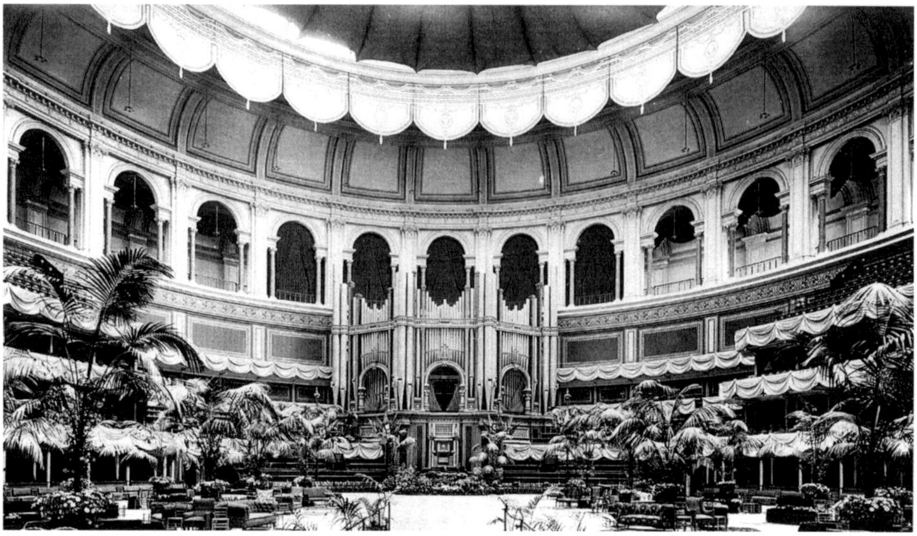

Figure 97: London, *Royal Albert Hall* (1871), view to the organ.

2.7.5.3 Enlargement of the ensemble and the "Additional Accompaniments"

With the increase of forces the conductor's task was to write the parts for the additional instruments. Their use at the performances in *Westminster Abbey* from 1784 to 1791 is not exactly known. Presumably they were mainly doubling Handel's instrumental parts.[52] Unfortunately, Michael Costa's version for his performances in *Crystal Palace* since 1857 was lost in a fire at the London publishing house Novello. It was most probably based on the then very popular so-called Mozart version published by Peters in Leipzig 1803, with considerable additional changes to Mozart's score, which was first performed in England on 21 March 1805 at the *Covent Garden Theatre* in London.[53]

Written in 1789 for Baron van Swieten and the *Gesellschaft der associierten Cavaliere* with a German translation after Klopstock and Ebeling, Mozart's *Messiah* version was based on the English scores published by Walsh, Randall and Randall & Abel and adapted Handel's score to the conditions in Vienna (for instance, the absence of Clarintrompete) and to contemporary taste. Elements no longer used and, therefore, not understandable had to be "translated". Mo-

52 CD booklet to *Handel Messiah,* Recorded Live at the Maryland Handel Festival in Washington Cathedral, Commemorating the Westminster Abbey Performance of 1784 (Pro Arte 232).

53 *MT* 40 (1899), 18 and footnote 545.

zart added "harmony" with complete woodwinds instead of Handel's Baroque score with woodwinds (oboes and bassoons) and trumpet moving mostly *colla parte* (with the exception of the trumpet solo). Mozart used two-part flutes, clarinets, bassoons, horns, trumpets and three-part trombones, the latter *colla parte* with alto, tenor and bass of the choir. At some points Mozart included woodwinds in the motivic development. Motives are not only rhetoric but also tone painting, and instruments are sometimes used for their characteristic sound color. The dynamic indications are changed and added according to the "empfindsame Stil."[54]

The first performance of Mozart's version was presented on 6 March 1789 in the *Salon* of the no longer extant *Palais of Baptist Graf Johann Esterházy* at Schenkenstrasse in Vienna (formerly *Palais Palffy*),[55] a narrow, high rectangular hall of category B. The concert followed the performances of Philipp Emanuel Bach's *Auferstehung und Himmelfahrt Jesu* in 1788 in the same hall and in the *Burgtheater* (figure 60) and Mozart's version of Handel's *Acis and Galatea* in 1788 in the same hall, for which an ensemble with about 36 instruments and 30 voices is documented,[56] whereas *Messias,* according to the extant parts, was performed by 45 instruments, 12 choir singers and four soloists (see table A in the appendix). With his changes Mozart evidently did not mean to increase sound power in order to adapt it to the hall's size. Larger halls of category C, such as the great *Redoutensaal* or the *Winterreitschule,* which would have required such an adaptation, were used only since 1801 for oratorio performances, sometimes with 200 performers for the *Redoute* and with about 900 performers for the *Reitschule,* as in 1814 for Handel's *Samson*. In 1807 the *AMZ* correspondent found the great *Redoutensaal* too large for Mozart's *Messias* version.[57]

In London the adaptation of sound power to larger spaces was clearly the aim of the revised versions, at least since the 1784 *Handel Commemoration* in *Westminster Abbey.* Amateur singers needed the support of woodwinds. The organ had to be loud enough to support and lead the large ensemble. Several revised versions of Handel's *Messiah* were in use. The re-orchestration ranged from simple doubling of Handel's parts to rich elaborations of the score in the

54 A. Holschneider, preface to *W. A. Mozart, Neue Ausgabe sämtlicher Werke* X/28/1: *Bearbeitungen von Werken Georg Friedrich Händels,* Bd. 2: *Der Messias*; Chr. Wolff, "Mozart's 'Messiah' – 'The Spirit of Handel' from van Swieten's Hands" (1989), 1–15; D. Baumann, "Georg Friedrich Händel" (1986), 119–33 (the numbers of musicians given for Dublin and Wien are corrected according to the latest research in table A in the appendix).

55 *Mozart und seine Welt in zeitgenössischen Bildern, Neue Ausgabe sämtlicher Werke* X / 32, M. Zenger and O. E. Deutsch, ed. (1961), 93.

56 The performance of Haydn's *Schöpfung* 1808 in the Aula of the old University was an exception.

57 *AMZ* 9 (1807), 336.

then current style. At the end of the nineteenth century, when comparisons with Handel's original score were again undertaken, this enrichment with sound effects was called "gingering up."[58]

The large number of singers and the participation of amateurs brought into the choir women's voices that increasingly replaced the boys' sopranos and men's altos. The earliest documents on women's participation in London presumably go back to a 1773 performance of Thomas A. Arne's oratorio *Judith*. The *Musical Times* reported that at the *Handel Festivals* in *Crystal Palace* the men's altos decreased from 418 in 1859 (of a total of 800 altos, among which 45 boys) to 23 in 1903 (of about 400 altos). In 1784 in *Westminster Abbey* the soprano part was sung by six women, 34 boys, one castrato and 13 men; among the 58 altos there was no woman with the exception of the solo contralto.[59] For how long soloists sang with the choir has not yet been investigated.

The influence of opera singing on soloists became evident in *Messiah* as well. The large opera houses, the increasingly larger orchestras and the acoustic disadvantage caused by the removal of the proscenium stage since the early nineteenth century had furthered a singing technique that required register compensation, coloraturas with full voice and a large range with all dynamic nuances. How and when this stylistic change evolved into the direct expression of emotions in sound nuances has not yet been investigated sufficiently. For Handel's *Messiah* the singing style changed relatively late. Reports in newspapers indicate that soloists continued to sing in the old style with instrumental voice and without too much pathos, though the Baroque embellishments (*appoggiatura, messa di voce,* additional embellishments) became gradually simpler. This is still preserved in the sound recordings of the early twentieth century in the singing style of singers who followed a tradition going back to the nineteenth century.[60] Baroque arias and recitatives sung in this manner were increasingly lost amid the larger, more dramatic choirs, which displayed big dynamic contrasts and enriched orchestration and were even overstressed by varying tempos in Wagnerian style. The shift to operatic singing style was therefore a logical upshot of these developments.

The documented orchestrations of the Handel performances follow the general relative trimming back of wood winds in relation to brass instruments (see table A in the appendix). Overall, the sound power of the wind, mainly brass, instruments increased, as in addition to trumpet and horn, trombone and low instruments, such as ophiklëide, serpent and bass horn, and later also tuba, were introduced. On the stage of *Crystal Palace* the large group of 300 to 363

58 Scholes (1947), 68.

59 Scholes (1947),180 and *MT* 41 (1900), 462 (see table A in the appendix).

60 CD *Georg Friederic Handel: A Collector's Messiah, Historic Handel Oratorio Recordings, 1899–1930* with several acoustic recordings from 1899 to 1907.

strings and the relative increase of low strings was indispensable. Critics in the *Musical Times* reported that the hall's volume was itself part of the overwhelming effect. Even in *Exeter Hall* Costa used 14 double basses.[61] The number of choir singers in relation to the orchestra was even more overpowering, with a maximum relation of 6:1 at *Crystal Palace!* The fortissimo was still dominated by the organ and the especially constructed timpani by Distin.[62]

As shown in chapter 2.6.4 (Space, genre and structure of music), the opposite relation was characteristic of Handel's large ensembles. Handel always expanded the orchestra more than the choir. Until 1759 the orchestra and choir for *Messiah* at the *Foundling Hospital* had a 4:3 relation. The Oxford ensembles of *Deborah* (1733) and *Solomon* (1749) showed a relation of 2:1 because the orchestra was enlarged. Handel sought not loudness but richness of sound. Moreover, the number of available good and professional choir singers was limited.

The answer to the question what role the space played in the performances of the nineteenth and twentieth centuries is not as simple and focused as expected. Critics like Shaw complained about the slowed down tempos and the stubborn, mechanical singing:

> [...] The multitudinous choristers, singing what is to them the easiest and most familiar music in the world, taken so slowly that there is plenty of time for them to consider what they are about, have only to observe the simplest mechanical conditions of their art: that is, accuracy as to the notes, clearness of pronunciation, and an occasional reduction of their chronic forte to piano.[63]

Wagner had already remarked that the performances in *Exeter Hall* were repeated so often that the choirs showed a surprising virtuosity. For the great performances at *Crystal Palace* the carefully selected choir singers came also from other cities with their own performance traditions and large, often excellent, halls with large stages, such as the *Town Halls* in Birmingham or Leeds or *St. George's Hall* in Liverpool (see figure 98).[64]

On the other hand, the tradition of performances in small churches and large cathedrals with its typical style of choir singing continued: boy's sopranos, men's altos and a special non-legato with instrumental solo voices developed for reverberant church acoustics. Certain conductors practiced this tradition also in *Crystal Palace*. Sound recordings reveal another group of opera and concert conductors who produced rather emotional, brilliant and dramatic

61 Scholes (1947), 178 f.

62 Schwab (1971), figure 97.

63 Shaw (1890–94), Vol. 3, 253.

64 see Forsyth (1985), 144, figure 4.11: *Liverpool, St. George's Hall* (1854) and 4.13: *Leeds, Town Hall*.

interpretations, mainly in *Albert Hall.* But quicker tempos were possible not only in *Albert Hall* but also in *Crystal Palace,* if the choir sang loud enough and before piano passages took breaks long enough to allow the reverberation time to decrease.

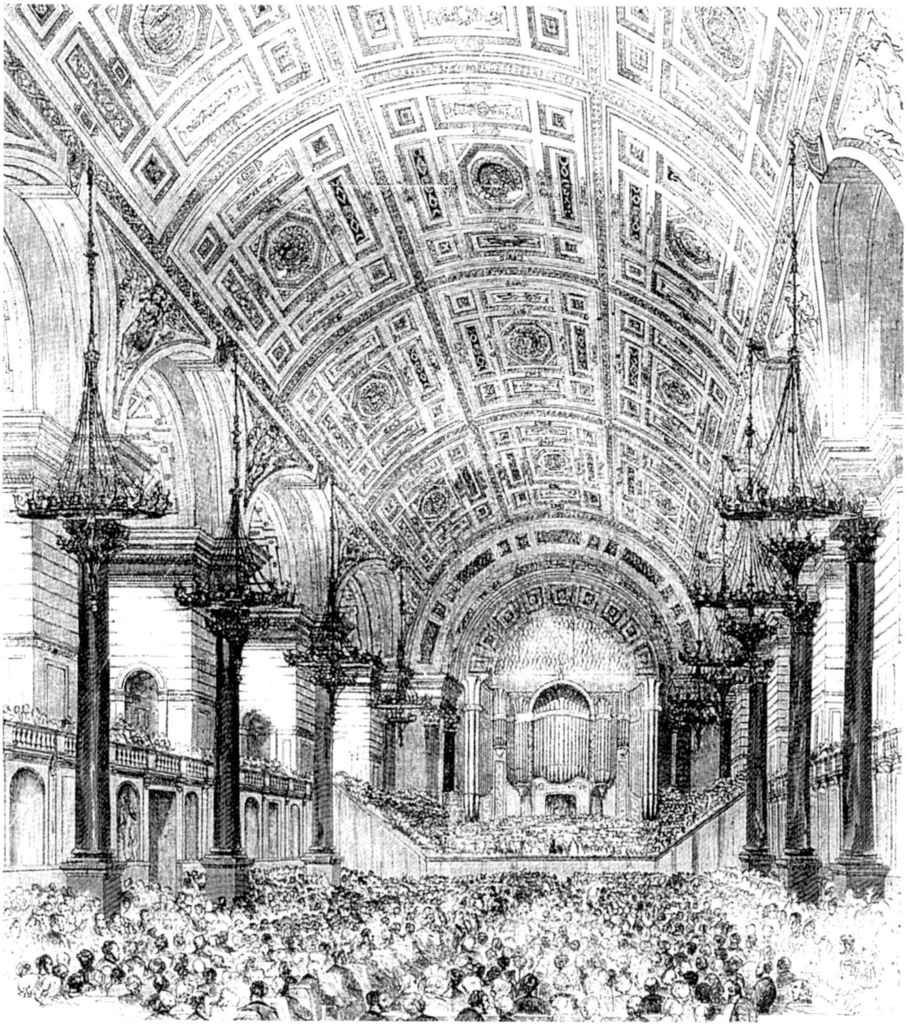

Figure 98: Liverpool, *St. George's Hall* (1854).

2.7.6 Back to the "original" size of the ensemble

With the discovery of the *Foundling Hospital* account books in 1784, which listed the exact numbers of musicians for the 1759 *Messiah* performance, certain groups initiated a return to Handel's score and a reduction of forces. In 1890 Bernhard Shaw proposed sarcastically to make a law that would prohibit the performance of a Handel oratorio with more than 80 musicians or 48 singers and 32 instrumentalists.[65] In 1894 Dr. A. H. Mann performed Handel's *Messiah* for the first time with the oboe and bassoon parts that had been discovered several months earlier by Davan Wetton behind the organ gallery of the *Foundling Hospital Chapel*. Using this and the facsimile of Handel's autograph and his *Conducting-Score* (the *Tenbury Score),* he revised the score with Ebenezer Prout but performed the oratorio with 63 instrumentalists and 200 choir singers – a performance that did not respect Handel's practice in either size or the choir-orchestra relation. Still, the *Musical Times* critic reported that the effect of the performance in the *King's College Chapel* in Cambridge was much more pleasing than expected:

> Though the hearer constantly caught himself missing familiar phrases – here a Mozartean viola part, there a phrase for clarinet, the general effect was more satisfying than might have been expected from a score which appears so thin to nineteenth century ears.[66]

He proposed to repeat the experiment with the exact numbers of Handel's practice in the chapel of the *Foundling Hospital* in London. No documents have been found that would attest to such a performance, but in February 1899 Frederick Bridge presented in the *Royal Albert Hall* a *Messiah* performance based on the "original sources, without Mozart's wind parts." For acoustic reasons Bridge was forced to compromise: he used 842 choir singers and 100 instruments which, according to the report in the *Musical Times,* still felt completely lost in the enormous hall, especially during the rehearsals:

> [...] the gigantic Royal Albert Hall presents a very different appearance, especially internally, from that of a concert night. The building is so vast that even the presence of the 800 or more singers, and of their friends who occupy seats in the arena, fails to change the atmosphere of desolation and weirdness caused by the shadowed auditorium and the echoes that come and go [...]. The accompaniment used at the rehearsals is that of the organ, most efficiently played by Mr. H. L. Balfour, organist to the [Royal Choral] Society. Having an instrument of such immense resource as that provided by Mr. Willis for the building, an or-

65 Cf. Hogwood (1984), 267
66 *MT* 41 (1900), 464.

> ganist might be strongly tempted to assert himself too prominently, but Mr. Balfour commendably exercises artistic restraint in his admirable accompaniments.[67]

The 1902 performance in *Queen's Hall* with 65 choir singers and 100 instruments was closer to Handel's ensemble, thanks also to the relation between choir and orchestra, reversed for the first time, and to acoustic conditions that made details of the score audible.

In 1906 Dr. A. H. Mann performed *Messiah* exactly in the manner of the 1759 performance in the *King's College Chapel* in Cambridge (figure 62). In 1900 he had presented the oratorio *Alexander Balus* with the same ensemble at the *Conference of the Incorporated Society of Musicians* in the "relatively small hall" of the Grand Hotel in Scarborough for an interested public of professionals.[68] The performance of *Messiah* in Cambridge was also organized for academic listeners interested in history. The large scale *Messiah* performances continued in *Crystal Palace, Royal Albert Hall, Queen's Hall* and several other halls and churches, but conductors now had the choice between Handel's score with oboes and bassoons or with the usual "additional accompaniments." The typical performance by large choirs for many listeners has resulted in the still observed tradition of large-scale performances. The roots of this tradition, which is related to industrial development and the positivistic admiration of progress, go back to the late eighteenth century. The wider public was extremely slow in reaching the awareness that large forces in great halls obliterated precious nuances, especially in *Messiah.*[69] Musicians interested in historical performance practice studied, rather, instrument building, playing technique and the rediscovered Baroque embellishments and basso continuo, as evident in different periods of historical performance practice, in particular also in the large number of sound recordings since the 1980s. Such recordings allow us to present at the end of this study several important room acoustic aspects of the historical changes in performance practice.

67 *MT* 40 (1899), 23.

68 *MT* 41 (1900), 115.

69 For the background on the change of performance practice, see 2.1.5: Handel's own performance practice of *Messiah* and other works.

2.8 Sound recordings

Since many elements determine the quality of a sound recording, special recordings would be necessary for a systematic comparison of room acoustic parameters. For example, the same ensemble would play in different rooms with microphones and microphone positions that record the halls' acoustics in a comparable way. As explained in chapter 1.9,[1] such an analysis of recordings would have to be assessed in a research project conducted by specialists from various related disciplines who would use artificial head and ambiophonic recordings with detailed recording protocols. Such a study would assess the effect of space shape, different sound source positions, the listeners and the microphones. However, the present study aims not at a quantitative analysis of such parameters but at a qualitative description of the relation between music performance and room acoustics. The following series of sound recordings were selected for a direct comparison of different phases in the performance history of the oratorio *Messiah* and their characteristic room acoustic aspects.

The last *Messiah* performance in the transept of the *Crystal Palace* in Sydenham during the last *Handel Festival* in 1926, with about 4,000 singers and instrumentalists conducted by Sir Henry Wood, has been preserved in one of the first electro-acoustic live recordings. In the same year single choir numbers were recorded in the *Royal Albert Hall* in London with nearly 1,000 performers. The London *Handel Commemoration* of 1784 in *Westminster Abbey* with more than 500 performers is indirectly documented by a recording in similar acoustics in *Washington Cathedral* in 1984, though with a larger choir and a smaller instrumental ensemble. A 1992–93 recording conducted by Stephen Cleobury conveys a picture of Dr. Arthur Henry Mann's 1906 historical reconstruction of Handel's *Foundling Hospital* forces in the *King's College Chapel* in Cambridge. From the many relatively recent recordings we selected three whose historical instruments and playing technique were most similar to Handel's first performance in Dublin, a London theater performance and a performance in the *Foundling Hospital* chapel: the 1992 recording by the *Scholars Baroque Ensemble* directed by David van Asch in the small *Rosslyn Hill Chapel* in London, the 1991 recording by the *U. C. Berkeley Chamber Chorus* and the *Philharmonia Baroque Orchestra* under Nicholas McGegan in the *Hertz-Hall* of UCLA in Berkeley, Los Angeles, and the 1980 recording by the *Academy of Ancient Music* under Christopher Hogwood in *St. Jude's-on-the-Hill* in

1 For methodology, see 1.1.4.

London.[2] Table 18 lists the number of instruments and voices for all sound examples and the respective historical ensembles of the eighteenth and nineteenth century. Table 19 lists the sections from Handel's *Messiah* taken from the different sound recordings and their duration, according to their sequence on the CD enclosed with this book.

Table 18: Messiah ensembles.

	Vl1	Vl2	Vla	Vc	Kb ***Str.***	Ob	Fg	Trp	Hr	Blech ***Bl.***	Timp. ***Orch.***	B c	Chor +Soli	**total**	**Vol.** **m³**
Crystal Palace, 1859	92	90	60	60	61 ***363***	10 +10	10 +10	6	9	12+15 ***82***	10 ***453***	1	2765	**3221**	847855
Albert Hall, 1899	19	19	12	12	12 ***74***	12	6	2	2	4 ***26***	2 ***103***	1	842	**945**	86600
Westminster *Abbey,* 1784	48	47	26	21	15 ***157***	6 +26	26 +1	12	6	12 ***89***	3+1 ***252***	2	284	**535**	72500
Washington *Cathedral,* 1984	18	18	10	10	6 ***58***	12	10	5	6	3 ***36***	2 ***98***	2	405 +4	**511**	~90000
Cambridge, *King's Coll. Ch.,* 1992/3	6	6	4	4	2 ***22***	2	1	2	–	– ***5***	1 ***30***	2	?+4		46500
Foundling *Hospital,* 1759	6	6	3	3	2 ***20***	4	4	2	2	– ***12***	1 ***35***	2	17+6	**57**	~4000
Foundling *Hospital,* 1754	8	6	6	3	2 ***25***	4	4	2	2	– ***12***	1 ***40***	2	19+5	**61**	
London, *St. Jude's,* 1980	8	7	6	3	2 ***26***	4	4	2	2	– ***12***	1 ***40***	2	31 +4	**74**	
Berkeley, *Hertz-Hall,* 1991	6	6	4	3	2 ***21***	2	2	2	2	– ***8***	1 ***32***	2	42+6	**80**	
London, *Rosslyn* *Hill Ch.,* 1992	4	3	2	1	1 ***11***	–	–	2	–	– ***2***	1 ***16***	2	14	**30**	

The sound recordings should be listened to first with headphones of the best quality in order to compare the spatial sound of the different examples. The ear will thus be prepared for systematic comparisons of room acoustic aspects by means of a stereo equipment with excellent loud speakers in a good playback room. The sequence of the examples in the five series begins with the largest forces in *Crystal Palace* and goes back in performance history to the smallest ensemble and room acoustic conditions similar to the first performance in Dub-

2 The few acoustic recordings of *Messiah* from the very beginning of sound recording have been excluded because of their restricted room acoustic elements. Furthermore they would need a detailed analysis of the singing technique at the time. Also the excellent 1959 stereo recording under Sir Thomas Beecham, available on CD, a *Messiah*-version in late Romantic style by Eugene Goosens, has been excluded. As studio recording it would need an assessment of the sound engineer's concept of room acoustics and of Beecham's interpretation which would go beyond the limits of this study.

lin. The sequence also moves stepwise forward from the old electro-acoustic sound recording technique to modern digital techniques in order to avoid direct confrontation of too large technical contrasts in sequential listening. Direct comparisons of single examples make details audible that are usually not perceived consciously during listening to the complete oratorio. To exclude the influence of one example on the next, longer breaks should be taken between them. A comparison of the examples from the same room requires listening to the numbers given in the horizontal lines of the following table, as, for instance, 1, 6, 12, 13 and 18 for the Washington recording.

Table 19: Sound recordings and timing.

Recording / Series	***no.***	**I**	***no.***	**II**	***no.***	**III**	***no.***	**IV**	***no.***	**V**	***no.***	**VI**	***no.***	**VII**
Messiah HHA 1/17		**1**		**10**		**38**		**39**		**30**		**18**		**15**
Sydenham, *Crystal Palace,* 1926 (Wood)		–		–		–		–	***16***	*4:02*		–		–
London, *Albert Hall,* 1926 (Balfour)		–		–		–		–	***17***	*3:08*		–		–
Washington *Cathedral,* 1984 (Dorati)	***1***	*3:15*	***6***	*2:09*	***12***	*2:18*	***13***	*3:52*	***18***	*3:02*		–		–
Cambridge, *KCC,* 1992/3 (Cleobury)	***2***	*3:09*	**–**	–	**–**	–	**–**	–	**–**	–		–		–
London, *St. Jude's,* 1980 (Hogwood)	***3***	*2:54*	***7***	*1:57*	***11***	*2:02*	***14***	*3:54*	***19***	*3:13*		–		–
Berkeley, *Hertz-Hall,* 1991 (McGegan)	***4***	*2:49*	***8***	*1:38*	**–**	–	**–**	–	**–**	–		–		–
London, *Rosslyn Hill Ch.,* 1992 (van Asch)	***5***	*3:09*	***9***	*2:41*	***10***	*2:09*	***15***	*3:38*	***20***	*3:05*	***22***	*2:14*	***24***	*1:53*
Duetto HWV 92, 15 London, *Rosslyn HC*											***21***	*2:11*		
Glory from HWV 260 Cambridge, *KCC*													***23***	*0:30*

Series	*HHA*	*I/17*
I	1	"Symphony" (Adagio – Allegro): example 1–5
II	10	Basso: Recitative "For behold, darkness shall cover the earth" (Larghetto): example 6–9
III	38	Tenor: Air "Thou shalt break them" (Andante): example 10–12
IV	39	Chorus "Hallelujah" (Allegro): example 13–15
V	30	Chorus "Lift up your heads" (a tempo ordinario): example 16 –20
VI	–	Duetto No. 15 "Quel fior che all'alba ride" (Andante larghetto) from HWV 92 example 21
	18	Chorus "His yoke is easy" (Allegro): example 22
VII	–	Chorus No. 3 "Glory" from HWV 260 "The King shall rejoice": example 23
	15	Chorus "Glory to God" (Allegro): example 24

2.8.1 Some general remarks on recording techniques and musical analysis

As explained, the density of sound power depends on the cubic volume of space and the size of the ensemble. If an ensemble, its size unchanged, is moved to a larger hall, the sound becomes thinner, whereas in a smaller hall its density will be higher. A similar change to a richer and denser sound occurs when a larger ensemble plays in the same hall. This effect depends also on the type, distance and number of microphones. A very present recording can appear richer and denser than a more distant recording, which conveys a more general impression of space. However, musicians and experts are able to perceive these different elements that separately affect the impression of room acoustics mainly for the following reasons:

1. Sensitive musicians adapt dynamics, articulation, phrasing, embellishment, breathing, and even intonation and vibrato character to the hall's reverberation. The slower the decrease of reverberation and the longer it lasts, the shorter are tones cut and the more "air" has to be left between the tones in order to produce a *non legato.* With very long reverberation a *non legato* is impossible because tones are bound together or even blurred by one or several following notes.
2. The effect depends on the directivity and frequency spectrum of each instrument. Room acoustics may be excellent for oboes, good for the soprano but problematic for the bass due to different frequency characteristics of the reverberation. The amount of the reverberation can depend also on the position of the sound source in the room.
3. To avoid blurring, a slower tempo may be necessary, or the musicians may be forced to avoid breaks, play loud enough and drag the music forward. On the other hand, in dry acoustics musicians have to produce volume and sound power in order to achieve enough blending.
4. The average playing volume and playing dynamics, too, must be adapted to the desired blending. This element is crucial for room acoustics, as in good halls the musicians sense the other instruments of the ensemble, whose total sound and sound blending they can control. The poorer the acoustics, the more difficult the blending and its control. A larger orchestra makes this process more difficult. This element can be traced in the comparison of the different recordings of the same section in each series, especially in the live recording of 1926 with 4,000 performers on the immense podium at *Crystal-Palace.* Under these conditions even outstanding conductors could not avoid coordination problems among the different groups, mainly between choir and orchestra. In a good hall the separation between

choir and orchestra, such as in "Lift up your heads" conducted by Sir Henry Wood, undoubtedly would have prompted the musicians to stop and repeat the recording (example no. 16, series V/1). Yet in a live performance in the chaotic acoustics of *Crystal Palace* it had to be accepted. A large part of the public obviously did not even notice the problem. In critical situations only the organ could help thanks to its sound power, which is clearly audible in the last measures of this section.

5. All these factors affect also the tempo, as reflected in the duration of the sections that are marked in a separate column of table 15. There is a clear difference among experienced tempo, measured tempo and intended expression. The correct assessment of these elements is easily lost if a section is taken out of its context within the work. On the other hand, the direct confrontation with other recordings makes the creation of "virtual space" audible as an element of the holistic perception process. The sequence of examples in table 16 proceeds from the largest to the smallest ensemble. On the other hand, the sequence 1 to 24 follows an order in which changing room acoustics are as clearly audible as possible.

2.8.2 Some remarks on ensemble size and sound balance

The critic of the *Niederrheinische Musikzeitung* in 1859 was right to note that an increase from 20 to 200 musicians makes a big difference, but a further increase to 2,000 a relatively small one in a proportionally larger hall. Already the greater distance of the reflecting surfaces due to the increasing seating space for musicians and listeners plays a crucial role. Burghauser and Spelda measured an increase of sound power by 2–3 dB if the number of instruments is doubled.[3] This is reflected in the characteristic setting of strings in symphony orchestras with either 6-6-4-4-2 or, rather, 15-14-10-12-8, as a smaller increase would not make a big difference.[4] A smaller increase may be sensed under certain circumstances because it still affects sound color and blending. Loudness does not increase linearly with the number of instruments but is bound especially to the maximum sound level of the instruments. How this factor depends on room acoustics, especially on the hall's width, has not yet been systematically assessed. In this respect, too, relations are more complex than simple numeric proportions. Similar room acoustic effects can be heard in sound mixture, that is, instrumentation, and in organ stops. Depending on the frequency spectrum

3 J. Burghauser and A. Spelda, *Akustische Grundlagen des Orchestrierens* (1971), 146 f.
4 *Op. cit.,* 147 and table A in the appendix.

the sound may be either audible in clear contours or blended with the total sound. The audible frequency spectrum depends also on the directivity of the sound source and on the influence of room acoustics on the frequency spectrum.[5]

The power of voices must be higher than the total volume of the orchestra in order to achieve text intelligibility. But in this respect, too, the crucial elements are the frequency spectrum of the voice and, mainly, the intensity of the singing formant, but also the position of the sound source and room acoustics.[6] The sound effect of wind instruments playing *colla parte* is difficult to estimate. Even with a well-balanced number of instruments and playing technique the sound engineer may have difficulties in creating the correct mixture of spatial sound. Too much presence or incorrect time delay can turn the supporting effect into masking. The same effect can result in playback even as a result of a slight signal distortion by the amplifier or loudspeakers. The delicate recording of the boys' voices, counter tenors and reed instruments requires, in particular, good acoustics because well-sequenced reflections from different directions help transparence and playing control.

2.8.3 Comparative analyses of sound recordings

2.8.3.1 Series I: "Symphony" (Grave, Allegro moderato)

Due to its pregnant articulation and breaks, the instrumental French ouverture of *Messiah* transmits information on room acoustics that may be perceived during neither arias nor choir sections (example 5).

According to contemporary rules the punctuation in a French ouverture must be sharpened, as described in theoretical treatises, such as Quantz' *Versuch einer Anweisung die Flöte traversiere zu spielen:*

> In this metre, as well as in three-four time, the quavers that follow the dotted crotchets in the loure, sarabande, courante, and chaconne must not be played with their literal value, but must be executed in a very short and sharp manner. The dotted note is played with emphasis, and the bow is detached during the dot. All dotted notes are treated in the same manner if time allows; and if three or more demisemiquavers follow a dot or a rest, they are not always played with their literal value, especially in slow pieces, but are executed at the extreme end of the time allotted to them, and with the greatest possible speed, as is frequently

5 *Op. cit.,* 138.

6 J. Sundberg: The Acoustics of the Singing Voice, *Scientific American* 236 (1977), 82; J. Meyer, *Acoustics* (2009), 123 and 403.

the case in overtures, entrées, and furies. Each of these quick notes must receive its separate bow-stroke, and slurring is rarely used.[7]

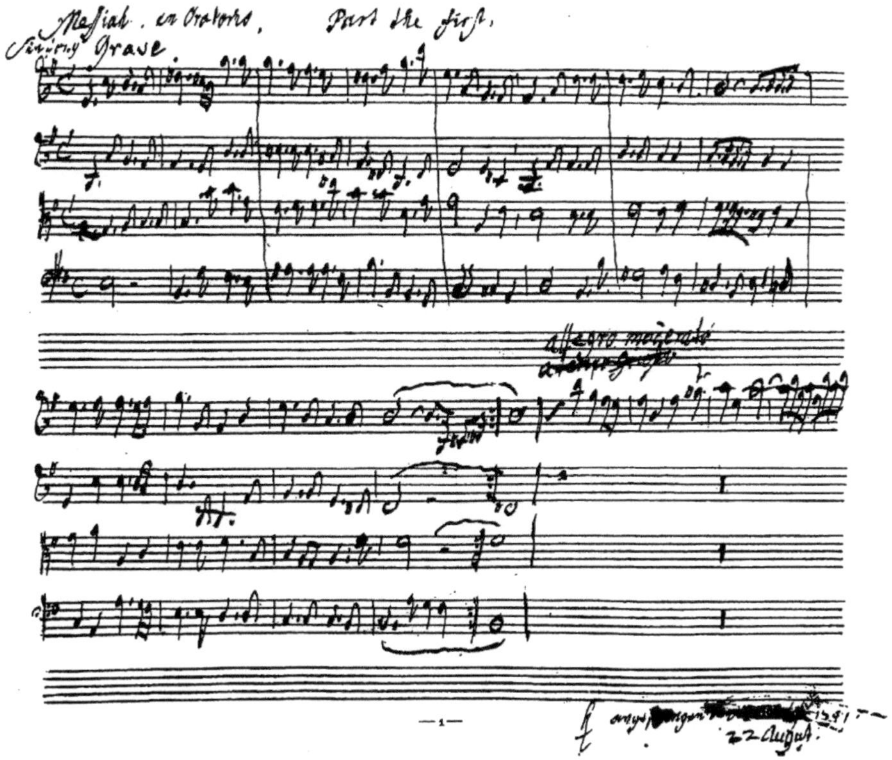

Example 5: *Messiah,* "Symphony", facsimile of the autograph score, first page.

7 Johann Joachim Quantz, *Versuch einer Anweisung die Flöte traversiere zu spielen,* Breslau 1789, XVII, VII, § 58: "[...] die Achttheile, so auf punctierte Viertheile folgen, [müssen] nicht nach ihrer eigentlichen Geltung, sondern sehr kurz und scharf gespielet werden. Die Note mit dem Puncte wird mit Nachdruck markieret, und unter dem Puncte der Bogen abgesetzet. Eben so verfährt man mit allen punctierten Noten, wenn es anders die Zeit leidet: und soferne nach einem Puncte oder einer Pause drey oder mehr als dreygeschwänzte Noten folgen; so werden solche, besonders in langsamen Stücken, nicht allemal nach ihrer Geltung, sondern am äussersten Ende der ihnen bestimmten Zeit, und in der grössten Geschwindigkeit gespielet; wie solches in Ouvertüren [...] öfters vorkömmt. Es muss aber jede von diesen geschwinden Noten ihren besonderen Bogenstrich bekommen: und findet das Schleifen [Legato] wenig statt." English translation Johann Joachim Quantz, *On Playing the Flute,* transl. by E. R. Reilly (2001), 290–1. See also XVII, IV, § 10 regarding the violoncello; and R. Donington, *The Interpretation of Early Music* (1963), 382 f.

The French ouverture is undoubtedly a topos of a festive ceremony whose special articulation would evoke the idea of a large and reverberant hall even if it is played in dry acoustics.

1. The maximum duration of 3 minutes 15 seconds is reached by the extremely slow *Grave* in the 1984 recording at *Washington Cathedral* with Antal Dorati. Despite the 100 instruments the ensemble appears to be far away due to the bad acoustics in the crossing of the neo-Gothic Cathedral. The musicians play without supporting lateral reflections. The large space volume, the relatively long reverberation and a small diffuse-field radius caused by the distant reflecting surfaces become evident in the recording: even at a small distance from the sound source the microphones receive a lot of blurring reverberation. To achieve a more or less proper sound the acoustics of the Cathedral had to be considerably masked out. Still, the reverberation is audible after loud passages with pauses. Despite these difficult circumstances (confirmed by reports of participating musicians) the tempo of the *Allegro* is surprisingly fast thanks to the short articulation, the quickly decreasing reverberation and the low sound power density in the large space volume. The multi-microphone recording could not fully compensate for the evidently bad blending of sound reported by musicians. The baroque instruments with their bright sound are less powerful than modern instruments. The conductor, not a specialist in historical performance practice, does not strictly apply baroque articulation, yet despite the difficult acoustic situation he remarkably controls the enormous ensemble seated on a very large podium. The soft punctuation of the very slow *Grave* (the over-punctuation is very modest) is well adapted to the decreasing reverberation. Thanks to excellent dynamic balance the difficult synchronization between strings and woodwinds is nearly always perfect. Only in the *Allegro* are slight shifts perceivable due to the insufficient blending of sound, though they were probably not audible in the Cathedral itself. As explained in chapter 2.7.1.2,[8] only the placement of the ensemble on a steeply raked tribune at the end of the main nave would have created the much better acoustic conditions of 1784 in *Westminster Abbey* in London. But the different sound in Washington is the outcome not only of the different placement in the cathedral but also of the much smaller orchestra with only 102 musicians as opposed to 150 in 1784 (see table 15).
2. The second and longest recording of the *Ouverture* – 3 minutes and 9 seconds – was made in the *King's College Chapel* in Cambridge, a Gothic room with 5 seconds of quite slowly decreasing, beautiful reverberation. The *Grave* is 20 seconds shorter and the *Allegro*-Fugato 7 seconds longer

8 On the reconstruction of the *Commemoration* in *Washington Cathedral* in 1984, see 2.7.1.2.

than in the *Washington Cathedral* recording of 1984. With 22 strings, 2 Oboes, 1 bassoon, 2 trumpets, timpani and basso continuo, Stephen Cleobury uses practically the same ensemble as Dr. A. H. Mann in 1906 in the same space, with wind players reduced from 12 to 5 compared to the *Foundling Hospital* ensemble of 1759 (2 oboes, 3 bassoons and 2 French horns are lacking). Since the musicians are placed at an excellent position behind the rood screen and within the choir stalls, the acoustics are surprisingly present despite the long, intense reverberation. The ensemble does not seem lost in the large volume of space, which is probably ten times larger than the *Foundling Hospital Chapel*. The spatial sound is excellently recorded thanks to digital time delay carefully adapted to microphone distances. The correctly sharpened punctuation of the *Grave* is well adapted to the decrease of the room's reverberation (the *crescendo détaché* is rounded to a *messa di voce* (< >) by the *diminuendo* added by the reverberation). Due to the reverberation the *Allegro*-fugato appears rather too fast despite the slightly longer duration than in the Washington recording. The tempo intended by the conductor according to Handel's score becomes slightly rushed and the expression becomes less brilliant. (Nota bene: this impression may be produced only by the recording or under certain playback conditions and may not have existed in the chapel itself.)

3. In direct comparison the acoustics in the London church *St. Jude's-on-the-Hill* appear much brighter than in the *Chapel* of *King's College* in Cambridge. The relatively long reverberation is audible only in forte passages and with excellent playback equipment. The sound power of the ensemble appears appropriate. After repeated listening to the whole work the recording seems slightly overpresent. The relatively limited amount of indirect sound on the recording is probably due, among others, to the difficulty of controlling the playing dynamics in the empty church, which increased too rapidly as loudness increased. The reproduction of the articulation appears therefore rather shorter than intended by the musicians, the sound of the baroque instruments brisker, the *messa di voce* nearly too strong due to the increase of the *crescendo*. These effects may be best heard with headphones. With only 57 seconds the duration of the *Grave* is the shortest of the five recordings. The properly executed over-punctuation appears too sharp with bad playback equipment but is exactly adapted to room acoustics if listened to with excellent playback. The *colla parte* playing of the 4 oboes and bassoons with the strings in the *Allegro* appears slightly blurred if the playback quality is bad, but if listened to several times with good equipment these effects, which are caused partly by the room acoustics, lose their negative character. The impression of the quality improves, and the recording belongs without any doubt to the best reconstructions of Handel's performance practice.

4. With 2 minutes and 49 seconds McGegan's recording in the very dry *Hertz-Hall* in Berkeley has, surprisingly, almost the same duration as Hogwood's in the rather reverberant church of St. Jude's-on-the-Hill. Both conductors use baroque playing practice but offer completely different interpretations of the ouverture: after an energetic *Grave* Hogwood presents a fully sounding quick *Allegro;* McGegan begins with a dramatic *Grave* in extremely short articulation that lasts 5 seconds longer than Hogwood's, then adds a very quick, almost nervous *Allegro,* which does not fully produce the intended brilliant result because the recording hall does not feature very pleasant acoustics. This evaluation may be too severe, as the recording is compared with much more reverberant earlier recordings of the series. Indeed, McGegan's recording appears more convincing if heard in the context of the complete work, as the listener then adapts to the dry acoustics. These and the orchestra in McGegan's recording transmit a sound impression quite similar to the London theaters at Handel's time despite the not doubled wind instruments. Due to less room width Handel's theater presumably had a warmer, more pleasant and intimate sound that lent his orchestra with 4 oboes and 4 bassoons a more splendid sound.
5. The recording of the *Scholars Ensemble* in the small, relatively dry but excellent acoustics of *Rosslyn Hill Chapel* in London, with the smallest string group of only 11 instruments and without woodwinds, has the same duration (3 minutes and 9 seconds) as Cleobury's in the much more reverberant *King's College Chapel* in Cambridge. This identical duration is due to the solemnly played *Grave* in soft *détaché* and with lightly played embellishments. The rippling *Allegro*-fugato, easily enhanced by the light onset of the acoustics, has a special transparence with especially clear and beautiful middle voices because oboes and bassoons are absent. Space, ensemble and recording are very present, and the listener has the impression of being involved in the sounding event. If listened to as a whole, this recording provides a sound impression similar to the intimacy of the Dublin first performance. Such a small ensemble in such an excellent small room enjoys much creative freedom and flexibility. Brilliance and solemn spatiality can be produced with the most simple means. Musicians define the property of this hall as "inspiring".

2.8.3.2 Series II: "For behold, darkness shall cover the earth", bass recitative

To analyze the relation between room acoustics and the solo bass and to enable analytic listening in a quick sequence of short examples we will focus on the *accompagnato* HHA I/17, no. 10 "For behold, darkness shall cover the earth" preceding the bass aria "The people that walked in darkness" (music example 6). The main points addressed in series I are confirmed: the excellent bass soloist eminently copes with the difficulties in the crossing of *Washington Cathedral,* and thanks to its bright timbre, flexibility and beautiful legato the bass voice is well heard despite the absence of room acoustic support. The effect of his message remains at a noble distance due to room acoustics. Hogwood's quicker interpretation in *St. Jude's-on-the-Hill* appears much more present. Only after *forte* moments is the reverberation slightly audible. Adequate playing and singing technique with appropriate embellishments, few vibrato and convincing *messa di voce* transmit Handel's rhetoric expression. If compared directly, McGegan's recording in the *Hertz-Hall* of UCLA again appears extremely dry. This impression is reinforced by the very light playing of the strings and the slightly overemphasized dynamics. The thin, therefore seemingly further distant, sound suggests a strong bass absorption in the hall. Every noise is audible, creating a slightly uneasy atmosphere compounded by strange pulse and intensity changes that are obviously caused by the musicians' difficulty in maintaining auditive control. Nevertheless, the excellent solo bass uses the room acoustics to create highly subtle nuances. In *Rosslyn Hill Chapel,* the smallest room of the series, the sound appears fuller and more substantial. Still, it allows the same nuancing as in *Hertz-Hall,* though with better blending as a result of low frequency resonance. The beautifully singing solo bass shapes the *accompagnato* very slowly, with 12 seconds more than McGegan's and 17 seconds more than Hogwood's. His illuminating rhetoric creates a very concentrated atmosphere and excellent intelligibility. The very limited string ensemble (4-3-2-1-1) is in a delicate situation, as the slightest deviations of intonation or phrasing are audible. The basso continuo carefully played by the organ provides additional sound volume.

2.8.3.3 Series III: "Thou shalt break them", aria for tenor (Andante)

The tenor aria on psalm II, verse 9: "Thou shalt break them with a rod of iron. Thou shalt dash them in pieces like a potter's vessel" (HHA I/17, no. 38, music example 7) offers the room acoustic answer to the sharp voice articulation and to the strings with their large melodic jumps in the high violins I and II in unison over the non legato bass. This aria makes the onset of room acoustics audible (see music example 8). The series, which comprises only the churches no. 1 *(Washington Cathedral),* 3 *(St. Jude's)* and 5 *(Rosslyn Hill Chapel),* starts with 5. This sequential change from series II to III allows a direct comparison of two pieces in the same room (bass accompagnato and tenor aria). On the other hand, a room's increasing impedance due to its larger cubic volume and the increasing distance of the side walls reaches highest audibility.

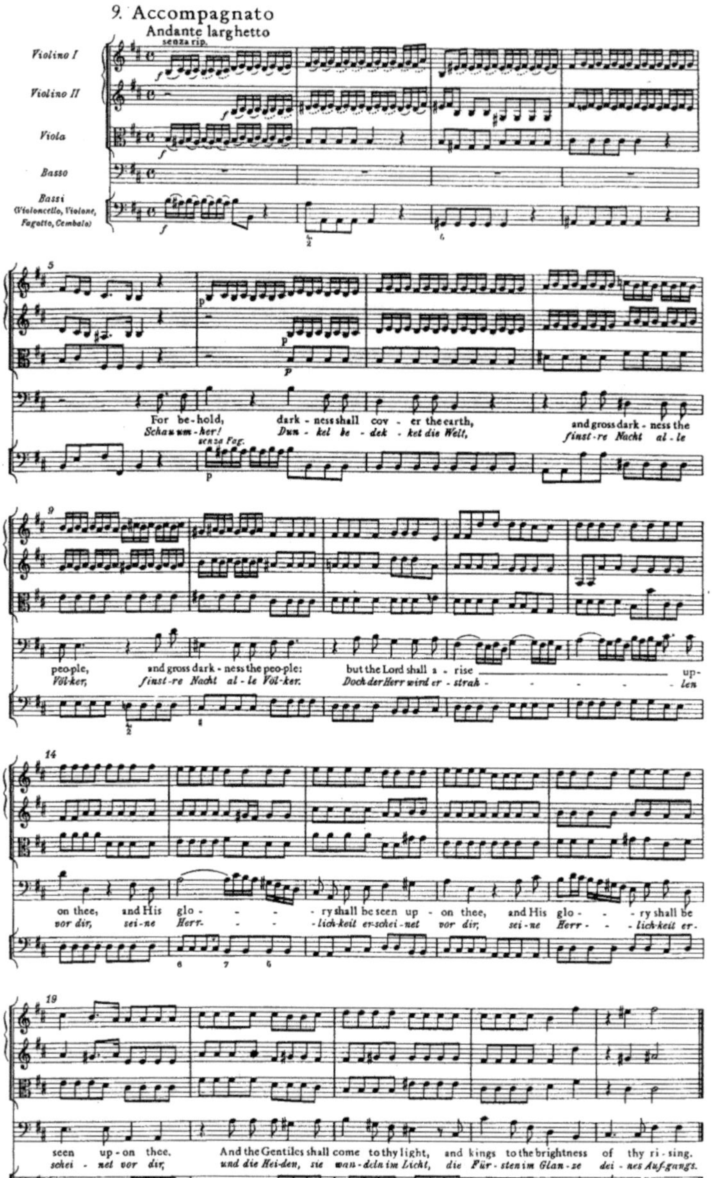

Example 6: "Darkness shall cover the earth" from *Messiah,* bass recitative no. 10.

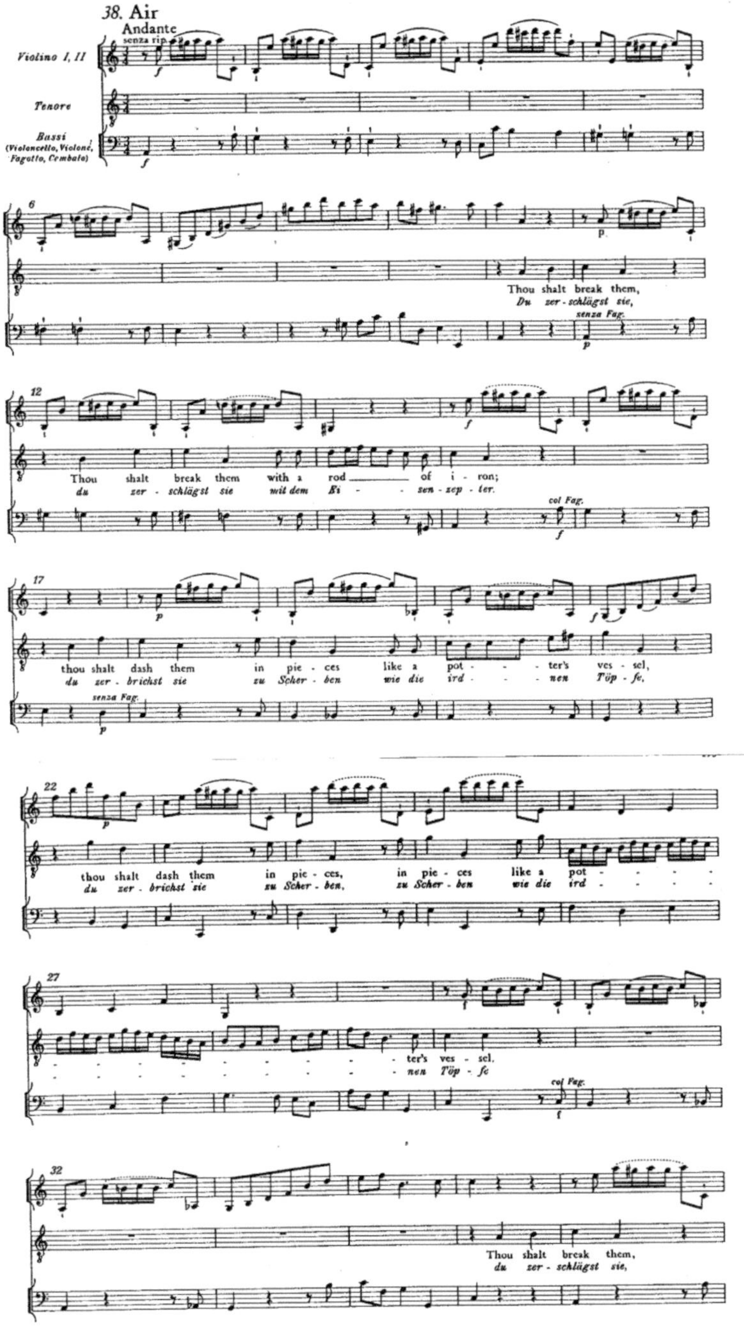

Example 7: “Thou shalt break them”, from *Messiah,* tenor aria no. 38.

It is not surprising that the recording in *Rosslyn Hill Chapel* with 1 minute and 21 seconds is by 6 seconds shorter than in *Washington Cathedral* (1 minute 27 seconds), but it is rather surprising that the recording in *St. Jude's* has exactly the same length (1 minute and 20 seconds) as the recording in the smaller, more present chapel.

In *Rosslyn Hill Chapel* the tenor tells the story like a comment devoid of strong emotions. The expression is created indirectly by the rhetorical figures (large jumps, quick coloraturas, staccato rhythm with violent punctuations). The equally convincing presentation by the tenor in *St. Jude's* is even more impressive thanks to the jolting room resonance of the low basses and the same quick tempo carried against the room's slightly dragging onset. In direct comparison the following instrumental introduction in *Washington Cathedral* appears almost harmless. The slower tempo of the aria is due to the long reverberation, and the heavier tenor presents less fluent coloraturas. The tenor had to produce also enough sound power and cope with room acoustic resistance caused by the very slow onset of sound.

2.8.3.4 Series IV: *Hallelujah,* chorus (Allegro)

These room acoustic characteristics are clearly audible in the directly following *Hallelujah* from the same recording (HHA I/17, no. 39). The full ensemble with about 100 instruments and 409 voices transmits an impression of the enormous space volume in *Washington Cathedral* and of the podium's size. Compared to the 1784 *Westminster Abbey* performance the sound difference must have been considerable due not only to the problematic placement but also to the much larger choir without boy's trebles and men's altos (405 voices in Washington compared to 284 in *Westminster Abbey*) and less than half the number of instruments (98 compared to 252 in *Westminster Abbey*).

In the much smaller barrel-vaulted church of *St. Jude's* the intense reverberation is surprising, as is the clearly audible sound increase of the men's altos and the silvery boys' trebles in the choir due to supporting woodwind and brass instruments. The even smaller ensemble in *Rosslyn Hill Chapel* convinces thanks to its natural presence and outstanding flexibility in the easily following acoustics. The excellent, relaxed, as though exclaimed *forte* lends splendor to the ensemble. Without boys' voices and men's falsettos the sound is darker, warmer and fuller. Handel's elaborated counterpoint with instruments participating in thematic entrances provides the allegro-fugato its appropriate magnificence in the modest space, the small ensemble notwithstanding. The finely woven structure of the dense score with flexible voice leading and quick harmonic changes produces its full effect only in transparent acoustics.

2.8.3.5 Series V: "Lift up your heads", chorus (a tempo ordinario)

On the other hand, in too large spaces with excessive reverberation the choirs of *Messiah* become heavy and coarse. Although they "survive" this change of character, the splendor disappears because details are lost either in the reverberation or due to heavy, overemphasized execution. This is highly striking in the choir "Lift up your head", recorded in 1926 during the last Handel Festival at *Crystal Palace* in Sydenham (HHA I/17, no. 30). The lack of coordination within the huge ensemble, the insufficient sound power and the sheer distance between choir and orchestra and even between the different parts of the choir cause recurring loss of coordination, until the entire ensemble is dragged forward by the organ and the increasingly louder wind instruments, finally almost drowning in a grotesquely excessive, blurred sea of trumpets, organ and 10 timpani!

At the *Royal Albert Hall,* too, the organist in residence H. L. Balfour cannot fully prevent coordination problems, but the choir with its shorter *non legato* copes better with the then not yet fully eliminated echo on the podium of the elliptic hall. This was all the more surprising, as Balfour required his over 800 singers to perform in the same tempo as Hogwood's 35 baroque specialists in *St. Jude's.* In *Albert Hall* the final chord is also dominated by the organ and overemphasized in romantic espressivo. The sweeping tempo helps mask the strong reverberation. Unfortunately, the piece ends abruptly because the long reverberation is cut off completely.

The choir with 405 singers in *Washington Cathedral* is moving slightly slower and activating the reverberation of the cathedral to such a level that the choir's flexibility is hindered despite the short and quickly kicked off *non legato.* The modest baroque playing could not remedy the dragging, pounding impression.

In *St. Jude's* the reverberation also appears to slightly interfere with the brilliant splendor of the *forte,* while the small ensemble in the transparent acoustics of *Rosslyn Hill Chapel* can freely expand its technical qualities and flexibility. In the latter intelligibility is excellent, the *concerto grosso* structure increases the intensity, and the intonation of the choir in pure thirds profits from the absence of the woodwinds.

2.8.3.6 Series VI: chamber duet "Quel fior che all'alba ride" (Andante larghetto) and choir "His yoke is easy" (Allegro)

The same flexibility and virtuosity is required for the *Messiah* choirs based on chamber duets. The comparison between the first part of the duet "Quel fior che all'alba ride" (HWV 192 no. 15) with the final choir of the first part of

Messiah (HHA I/17, no. 18), shows not only the sound difference between the slightly wider swinging solo duet and the joyful and slightly quicker oratorio choir (see music example 4, 2.6.4.1, p. 274) but also the advantage of the small ensemble in the small *Rosslyn Hill Chapel*. Performances in very reverberant halls with too large forces often lend this difficult section a heaviness that tempts even excellent conductors into mechanical drilling. Mozart remarkably solved this delicate problem of performance by presenting the first phrase with the solo singers only.

2.8.3.7 Series VII: "Glory to God", chorus (Allegro) and "Glory" from the *Coronation Anthem* HWV 260 "The King Shall Rejoice"

The last example presents a comparison between the *Glory* exclamation from the coronation anthem "The king shall rejoice", composed for *Westminster Abbey* and 100 performers and recorded at *King's College Chapel* in Cambridge (music example 2, chapter 2.6.3.2, p. 271) with adequate reverberation for a splendid sound, and the "Glory to God" from *Messiah* with the solo trumpets playing "da lontano e un poco piano" (music example 3, p. 272). This part from *Messiah* actively creates the spatial perspective if sung by a small ensemble in transparent acoustics. Under such conditions the message of the music is perceived as directly as an eye blink in a small baroque theater, and the musicians do not have to interpret Handel's exactly composed baroque rhetorics any further. Natural good declamation is sufficient.

Part 3

Appendix

3.1 Bibliography

3.1.1 Georg Friedrich Handel's *Messiah:* selected sources

For further information, see HHdb. 2, 178 f., mainly 193 f.; HHA I/17, especially the critical report, 10 f.; John Tobin, *Handel at work* (London, 1964), 80 f.; Watkins Shaw, *A textual and historical companion to Handel's Messiah* (London, 1965); Peter Larsen, *Handel's 'Messiah': origins, composition, sources* (New York, NY, 1972); Donald Burrows, *Handel: Messiah* (Cambridge, 1991), 103 f.; Donald Burrows and Martha J. Ronish, *A catalogue of Handel's autographs* (Oxford, 1994); Donald Burrows and Watkins Shaw, "Handel's 'Messiah', supplementary notes on sources", *Music and Letters* 76 (1995), 356–68.

Autograph (HHA source A)
London, British Library, R.M.20.f.2 (22 August – 14 September, 1741). Contains autograph amendments made during composition, others that antedate the first performance, still others added in preparation for the London first performance in 1743 (annotations and the insertion of singers' names). Later not used by Handel but only for further copies. Additional movements composed in 1742–45 presumably included with the first binding in the 1790s. Cover of the Royal Library.

Conducting Score (Tenbury Score) (HHA source B)
Oxford, Bodleian Library MSS Tenbury 346, 347; 3 vols. bound in 2. Copy by J. C. Smith senior, probably September – October 1741 before Handel left for Dublin; used by Handel in the preparation of his performances (except 1743, see above). Many autograph annotations and additions for different performances (cancellations, new compositions etc.). Several layers of singers' names, transposition directions by Handel and his immediate successors. The manuscript was probably used also to prepare performance material. It contains some detailed autograph musical revisions, including the orchestral *ripieno*-directions added in 1749.

Foundling Hospital Material (HHA source C)
London, Thomas Coram Foundation. Written in 1759 as bequeathed in Handel's will, third codicil dated 30 July 1757 by J. C. Smith senior and two other hands; score in 3 volumes, 13 vocal part books (9 choir and 4 solo voices), 15 orchestral part books. The principal interest lies in the part books, presumably copied from another set of parts that originated in 1754, including oboe and bassoon parts presumably copied from the appropriate lines of the score. With the accounts from 1754, 1758 and 1759 these sources provide precious information on the Foundling Hospital performances of *Messiah.*

Printed word-books *(libretti)*
Extant from 1742, 1743, 1749, 1750, 1755, 1758, 1759 with some undated editions probably from the 1750s. If approached with critical caution these often hastily produced books permit important conclusions. Some entries of singers' names, such as for the first London performance in

1742. Libretti are extant also for the 1784 performance (see CD booklet for the reconstruction by the Smithsonian Institution in 1984 at *Washington Cathedral* (see 3.4: Discography).

Early musical copies, manuscript and printed

The following copies from the eighteenth century preserve original material contained neither in the autograph score A nor in the *conducting score* B:

The Granville Copy (HHA source E)

London, British Library, Egerton MS 2937. Copy by J. C. Smith senior for Bernard Granville, ca. 1743–44. One of the earliest copies.

The Coke Collection Copy (HHA source S)

Collection Gerald Coke, Jenkyn Place, Bentley, Hants.

The Matthews' Copy (HHA source G)

Dublin, Archbishop Marsh's Library, St. Patrick's Cathedral, Z 1.2.26. Prepared by John Matthews, an amateur singer, from musical sources found in Winchester, Salisbury and Durham, written between 1761–65; contains oboe parts and vocal ornamentation giving insight into contemporary practice, though not by Handel.

The Goldschmidt Score (HHA source O)

Copy from about 1743–45, owned by William Hayes, Oxford; possibly prepared in relation to his performances with soloists who performed also with Handel; actually privately owned.

Jennens' Copy (HHA source F)

Manchester, Public Libraries, Newman Flower Collection, Henry Watson Music Libary, MS 130 Hd4, vols. 198–200: score in 3 volumes; *Manchester, Public Libraries, MS 130 Hd4, vols. 142–149, 247–248, 353:* parts copied directly from Handel's autograph textually independent of his score. *London, British Library, in RM 19.a.2:* independent copies of later composed movements.

Jennens' version for keyboard (HHA source R)

London, British Library, R.M. 19.d.1. Incomplete, richly embellished.

The Schoelcher collection score (HHA source D)

Staats- und Universitäts-Bibliothek Hamburg, Sign. 221. Conducting score, copied after 1759, despite of performing annotations it is not a reliable source for Handel's own performance practice.

Facsimiles

Autograph: London 1868, publ. by the Sacred Harmonic Society; Hamburg 1892, publ. by the Deutsche Händelgesellschaft, introd. by Friedrich Chrysander, repr. New York, NY: Da Capo 1969.
Conducting Score (Tenbury Score): for the Royal Music Association London: Scolar Press 1974.

Early printed editions

John Walsh (London, 1749): Arias, first edition based on a pre-1745 copy; contains versions probably never used for performance by Handel; some particularities probably related to Handel's 1743 performances.

Randall & Abell (London, 1767): first printed score; contains several variants. The edition may be directly related to the Lennard copy, *Cambridge, Fitzwilliam Museum, Mu. MS 844* (HHA source M).

H. Wright (London, 1785): printed score; nearly identical with the first print of 1767.

S. Arnold (London, 1787–88); printed score; contains some variants; probably not related to the copy owned by Arnold; today in the *Glasgow University Library, Euing Music Collection R.d.20*).

Goulding & D'Almaine (London, 1831): print of score, ed. J. Addison; with numerous variants and Mozart's additions of instrumentation.

E. F. Rimbault (London, 1850): printed score for the English Handel Society (The Works of Handel, Vol. 10/11); Mozart's additions of instrumentation, more than 30 mistakes.

Breitkopf & Härtel (Leipzig, 1803): edition of W. A. Mozart's 1789 version with changes and additions.

3.1.2 Editions of Handel's works

F. Händels Werke: Ausgabe der Deutschen Händelgesellschaft, (HG), ed. F. W. Chrysander (Leipzig and Bergedorf bei Hamburg, 1858–94): *Esther* (vol. xl & xli), *Messiah* (vol. xlv), *Oratorio per la Resurrezione* (vol. xxxix), *Funeral Anthem* (vol. xi), *Dettingen Te Deum* (vol. xxv), *Utrecht Te Deum und Jubilate* (vol. xxxi), *Italienische Duette und Trios* (vol. xxxii), *Anthems* (vol. xxxiv–xxxvi).

Hallische Händel-Ausgabe, im Auftrage der Georg Friedrich Händel-Gesellschaft, (HHA), ed. M. Schneider, R. Steglich et al. (Kassel, 1955f.): Series i: Oratorios: *La Resurrezione, oratorio in due parti* (vol. i/3); *Acis and Galatea,* first version (vol. i/9), *Esther,* first version (vol. i/8); *Deborah* (vol. i/11); *Saul* (vol. i/13); *Israel in Egypt* (vol. i/14); *Messiah* (vol. i/17); Series iii: Church Music: *Anthems für Cannons I, II, III* (vol. iii/4–6); *Anthems für die Chapel Royal* (vol. iii/9).

Messiah, An Oratorio for Solo Voice, Chorus and Orchestra, Eleven Versions 1742–54, ed. Donald Burrows (London: Hinrichsen and Peters, [1987]).

Foundling Hospital Anthem "Blessed are they that considereth the poor", ed. Donald Burrows (London: Hinrichsen and Peters, s.a.).

3.1.3 Editions of works by other composers

Arcangelo Corelli: *Historisch-kritische Gesamtausgabe der musikalischen Werke,* ed. Hans Oesch (Cologne, 1973 ff.): vol. i: *Sonate da chiesa, Opus I und III mit Francesco Geminianis Concerto grosso-Bearbeitungen von sechs Sonaten aus Opus I und III,* ed. Max Lütolf

(Cologne, 1987); vol. iv: *Concerti grossi, opus VI,* ed. Rudolf Bossard (Cologne, 1978); vol. v: *Werke ohne Opuszahl,* ed. Hans Joachim Marx (Cologne, 1976).

Ludwig van Beethoven: *Werke,* ed. Beethoven-Archiv Bonn, Joseph Schmidt-Görg (Munich-Duisburg, 1961 ff.): Series ii, vol. 1: *Wellingtons Sieg oder die Schlacht bei Vittoria op. 91,* ed. Hans-Werner Küthen (Munich, 1974; critical comment Munich, 1991).

Symphonie Nr. 3, Es-Dur, op. 55 "Eroica", facsimile edition of the autograph score, the orchestra parts of the first performance and early performances with corrections by Beethoven; commented ed. Otto Biba, 4 vols. (Vienna: Gesellschaft der Musikfreunde, 1996).

Wolfgang Amadeus Mozart: *Neue Ausgabe sämtlicher Werke,* ed. Internationale Stiftung Mozarteum Salzburg (Kassel, 1955 ff).: Series x/28/1: *Bearbeitungen von Werken Georg Friedrich Händels*, ed. Andreas Holschneider: vol. 1: *Acis und Galatea* (Kassel, 1973; critical comment Kassel, 1995); vol. 2: *Das Alexander-Fest* (Kassel, 1962; critical comment Kassel, 1963); vol. 3: *Messias* (Kassel, 1961; critical comment Kassel, 1962); vol. 4: *Ode auf St. Cecilia* (Kassel, 1969; critical comment Kassel, 1969).

Joseph Haydn: *Sämtliche Werke,* ed. Eugen Mandyczewski (Leipzig: Breitkopf & Härtel, 1907 ff.): series 16, vol. v: *Die Schöpfung;* vol. vi–vii: *Die Jahreszeiten.*

3.1.4 General bibliography

Acoustics for choir and orchestra: papers at a seminar organized by the Music Acoustics Committee of the Royal Swedish Accademy, ed. Sten Ternström, Publication of the Royal Academy of Music 52 (Stockholm, 1986).

Acoustics → Lewcock, Ronald

Adam, Max: *Raum- und Bauakustik* (Blauen, 1985).

Ademollo, Alessandro: *I teatri di Roma nel secolo 17°* (Rome, 1888; repr. Rome, 1969).

Ahnert, Wolfgang and Walter Reichardt: *Grundlagen der Beschallungstechnik* (Stuttgart, 1981).

Ahnert, Wolfgang and Frank Steffen: *Beschallungstechnik* (Stuttgart, 1993).

Alberti, Leon Battista: *De re aedificatoria libri X* (Florence, 1485), *L'architettura (de re aedificatoria),* Latin with Italian translation ed. Giovanni Orlandi and P. Portoghesi (Milan, 1966).

– *Zehn Bücher über die Baukunst,* German translation by Max Theuer (Vienna and Leipzig, 1912; repr. 1975).

Alf, Julius: *Geschichte und Bedeutung der Niederrheinischen Musikfeste in der 1. Hälfte des 19. Jahrhunderts* (Düsseldorf, 1987).

Algarotti, Francesco: *An Essay on the Opera (Saggio Sopra L'Opera in Musica)* by Francesco Alagarotti: anon. English translation (s.l., 1768); ed. Robin Burgess (Lewinston NY, 2005).

Allen, J. M.: "Acoustics of public buildings: defects and remedy", *Scientific American* 30 (1874), May 23.

Aloi, Roberto: *Architetture per lo spettacolo* (Milan, 1958; 2nd ed. Milan, 1972).

Alte Musik als ästhetische Gegenwart: Bericht über den internationalen musikwissenschaftlichen Kongress Stuttgart 1985, ed. Dieter Berke and Dorothee Hanemann, 2 vols. (Kassel, 1987).

Ando, Yoichi: *Concert hall acoustics* (Berlin,: Springer 1985).

Argan, Giulio Carlo (ed.): *Palazzo Valentini* (Rome: Editalia, 1984).

Argentieri, Domenico: "Die Optik Leonardos", *Leonardo* da Vinci Ausstellung, ed. S. Piantanida and C. Baroni (Berlin, 1939), 405–36.

Aristotle: *Problemata,* XI, 23, 901/b → Ch. B. Boyer (1945), 93.

Arnold, Denis: "Education in music", § V: "Conservatories", *NG,* vol. 6, 19 f.
Arnold, W. et al.: "Funktionelle Morphologie der äusseren Haarzellen des Menschen", *Laryngo-Rhino-Otologie* 69 (1990), 177–86.
Arteaga, Stefano: *Le rivoluzioni del teatro musicale italiano dalla sua origine fino al presente,* 3 vols. (Bologna: Trenti, 1783–88).
Aschoff, Victor: "Phantasie und Wirklichkeit in der Frühgeschichte der Akustik", *Acustica* 42 (1979), 121–32.
Askenfelt, Anders: "Stage floors and risers – supporting resonant bodies or sound traps?" *Acoustics for choir and orchestra,* ed. Sten Ternström (Stockholm, 1986).
Auer, Max → Göllerich, August
Auger, Robert: "Location Recording", *Sound recording practice* (London, 1976), 244–54.
Authenticity and early music, ed. Nicholas Kenyon (Oxford, 1988).
Avery Architectural Library: *Columbia College Catalogue,* ed. New York Academies (New York, NY, 1995).
Bach, Carl Philipp Emanuel: *Über die wahre Art das Clavier zu spielen* (Berlin, 1753, 1759, 1762; Leipzig, 1787).
Backhaus, H.: "Transmission from *Victoria Hall* to the *Opéra* in Geneva", *Siemens-Zeitschrift* 8 (1928), 298.
Baer-Loy, Thomas: *Hören und Raumform,* 2nd ed. (Dübendorf, 1984).
Bagenal, Hope: "The Leipzig tradition in concert hall design", *Journal of the RIBA* 36 (1929), 756–63.
– "Bach's music and church acoustics", *Journal of the RIBA 37* (1930), 154–63.
– and Alexander Wood: *Planning for good acoustics* (London, 1931).
– "Concert hall", *Chamber's Encyclopedia,* vol. 3, (s.l., 1950).
Barlow, Graham F.: "Vanbrugh's Queen's Theatre in the Haymarket, 1703–09", *EM* 17 (1989), 515–21.
Barnes, Carl F. jr.: *The portfolio of Villard de Honnecourt, a new critical edition and color facsimile* (Farnham, 2009).
Barron, Michael: "The subjective effects of first reflections in concert halls: the need for lateral reflections", *Journal of Sound Vibration* 15 (1971), 475–94.
– and A. Harold Marshall: "Spatial impression due to early lateral reflections in concert halls", *Journal of Sound Vibration* 77 (1981), 211–32.
– and L.-J. Lee: "Energy relations in concert auditoriums", *JASA* 84 (1988), 618.
– "Subjective study of British symphony concert halls", *Acustica* 66 (1988), 1–14.
– *Auditorium acoustics and architectural design* (London, 1993).
Baselt, Bernd: "Händels englische Oratorien und ihre Anfänge", *Festschrift Günther Fleischhauer zum 60. Geb.,* (Halle, 1990), 5–11.
→ *Händel-Handbuch*
Bauer-Reinhold, M.: *Theater des Barock* (Munich, 1966).
Bauer, Anton: *150 Jahre Theater an der Wien* (Vienna, 1952).
Baumann, Dorothea: "Richard Wagners Festspielhaus: ein architektonisches Wagnis mit Überraschungen", *Festschrift Hans Conradin* (Berne, 1983), 123–50 (Publ. SMG II, 23).
– "Qualitätskriterien für Streichinstrumente – Aspekte historischen Wandels", *Glareana* 32/2 (1983), 7–20 and 33/1 (1984), 3–11.
– "Wagners raumakustische Überlegungen in der Verwendung von Blasinstrumenten", *Kongressbericht Seggau 1983* (Tutzing, 1985), 117–28.
– "Musizieren im Raum – live", *Mono – Stereo – Quadro,* ed. Radio DRS Basel (Basel, 1986), 47–60.
– "Georg Friedrich Händel", *Schweizerische Musikpädagogische Blätter* 74 (1986), 119–33.

- "Veränderungen des Hörens im elektronischen Zeitalter", *Musikerziehung im Zeitalter der Elektronik,* ed. AGMÖ (Vienna, 1989).
- "Musical acoustics in the Middle Ages", *EM* 18 (1990), 199–210.
- "Was ist Stille? ", *Bauwelt* 81 (Berlin, 1990), 981–83; *Schweizerische Musikpädagogische Blätter* 79 (1991), 71–73.
- "Hören im Raum: Qualitätskriterien der Schallwahrnehmung", *Neue Zürcher Zeitung* (1991) 23 January, no. 18, 67 (Forschung und Technik).
- "Performance practice and architectural acoustics: bibliographic sources in related disciplines, *Fontes Artis Musicae* 38 (1991), 104–09.
- "Acustica", *Enciclopedia dell'arte medievale* (Rome, 1991), vol. 1, 23–26.
- et al.: "Exkurs über die akustischen Gefässe im Jöchelsthurn", *Der Jöchlsthurn in Sterzing* (Innsbruck and Vienna, 1992), 345–46.
- "Glück und Unglück im Zürcher Konzertsaalbau", *Festschrift Ernst Lichtenhahn* (Winterthur, 1993), 115–28.
- "Die Kunst des Konzertsaalbaus: Heute kann man gute Akustik im voraus planen. Interview mit Russel Johnson", *Neue Zürcher Zeitung* (1993), 25 June, no. 144, 57.
- "Improvisierte Musik und Gehirn", *Improvisation II, Tagungsbericht Luzern 1993,* ed. Walter Fähndrich (Winterthur, 1994), 37–61.
- "Ein Spätling im goldenen Zeitalter des Konzertsaalbaus", '*Ein Saal in dem es herrlich klingt', Hundert Jahre Tonhalle Zürich,* (Zurich, 1995), 37–50.
- "Qualche riflessioni sull'acustica a soggetto della ricostruzione del teatro all'italiana *La Fenice*", *La Fenice verso la ricostruzione,* (Venice, 1996), 81–84.
- "Whispering galleries and special places for music: Arab and Iberian acoustical sources and practices", *Fuentes en la Peninsula Ibérica (ca. 1250–ca. 1550), Actas del Coloquio Internacional, Lleida 1996,* ed. Universitat Lleida (Lleida, 2001), 481–91.
- "Acoustics in gothic cathedrals: theory and practical experience in the middle ages", *Colloque sur l'orgue gothique, Royaumont 1995* (Paris: Créaphis, 2001), 37–48.
- *Musik und Raum. Eine Untersuchung zur Bedeutung des Raumes für die musikalische Aufführungspraxis. Mit einer Studie zum Wandel der Aufführungspraxis von Händels Messiah.* Habilitationsschrift, Universität Zürich (2000).
- *Vom Musikraum zum Konzertsaal: Auf den Spuren von Zürichs Musikleben* (Zurich, 2002). (186. Neujahrsblatt der Allgemeinen Musikgesellschaft Zürich auf das Jahr 2002).
- "Konzertsäle und Opernhäuser des 20. Jahrhunderts im Spannungsfeld zwischen Umbruch und Tradition", *Brüche Torsi, Unvollendetes,* ed. Kurt Schärer and Erwin Sonderegger (Zurich, 2002), 147–68.
- "Geometrical Analysis of Acoustical Conditions in San Marco and San Giorgio Maggiore in Venice", *Architecture and Music in Renaissance Venice*, ed. Deborah Howard and Laura Moretti (Milan, 2006), 117–46.
- "Die Wechselwirkung zwischen architektonischem Raum und musikalischer Aufführungspraxis im 18. und 19. Jahrhundert am Beispiel von Händels Messiah". Symposium *Espaces et lieux de concert en Europe, 1700–1920: architecture, musique, société* (Berlin, 2008), 331–60. (*Musical Life in Europe 1600–1900, Circulation, Institutions, Representation,* European Science Foundation, Strassburg).
- "Musik von innen her empfinden und verstehen – Einsichten aus Susanne K. Langers Kunstphilosophie", *Intimität,* ed. Ingrid Tomkowiak and Werner Egli (Zurich, 2008), 257–79.
- "Systematische Musikwissenschaft – eine Disziplin zwischen Kulturgeschichte und Naturwissenschaften", *Musicology today: problems and perspectives, Scientific Researches* 80,

Ukrainian National Tchaikovsky Academy of Music, ed. Olena Zinkevych (Kiev, 2009), 40–51.
– "Der Bayreuther Raumklang", *Wagnerspectrum* 2009, H. 1, 151–68.
Beard, Geoffrey: *The work of Christopher Wren* (London, 1987).
Becker, Heinz → Quellentexte
Beethoven, Ludwig van: *Beethovens sämtliche Briefe,* ed. A. C. Kalischer (Leipzig, 1906–08).
– *Ludwig van Beethovens sämtliche Briefe und Aufzeichungen,* ed. F. Prelinger (Vienna and Leipzig, 1907–11).
– *Ludwig van Beethovens Konversationshefte,* ed. Karl-Heinz Köhler (Leipzig, 1972 f).
– *The letters of Beethoven,* ed. Emily Anderson, 3 vols. (London, 1986).
– Music editions → see 3.1.3
Begault, Durand R.: *3D sound for virtual reality and multimedia,* (Boston, 1994).
Behne, Klaus-Ernst: "Musik- und Raumwahrnehmung", *Musik und Raum* (Mainz, 1989), 60–81.
Beijer, Agne: *Slottsteatrarna på Drottningholm och Gripsholm* (Malmö, 1933 and 1937).
Békésy, Georg von: *Experiments in hearing,* ed. E. G. Wever (New York, NY: Mc Graw-Hill, 1960; repr. from *Scientific American* August 1957).
Bell, James F. → Dostrovsky, Sigalia
Benevolo, Leonardo: *History of Modern Architecture,* 3rd ed. (Cambridge Mass.: MIT Press, 1985), 2 vols.
Beranek, Leo L.: *Music, acoustics, and architecture* (New York, NY, 1962; 2nd ed. 1979).
– *Concert and Opera Halls: how they sound* (Woodbury: ASA, 1996).
– and T. Hidaka: "New subjective and objective data on 20 opera houses of the world", *Proceedings Forum Acusticum. Joint Meeting with ASA and EAA* (Berlin, 1999).
Berlioz, Hector: *Grand traité d'instrumentation et d'orchestration modernes op. 10* (Paris, 1843, 2e éd. 1855).
– *Traité d'instrumentation.* German translation, commented and with additions by Richard Strauss (Leipzig, 1904).
– *A travers chant* (Paris, 1862; repr. Paris, 1971); English translation: *The Art of Music and Other Essays (A Travers Chant)*, trans. and ed. Elizabeth Csicsery-Rónay (Bloomington and Indianapolis: Indiana UP, 1994).
– *Correspondance générale,* ed. Pierre Citron (Paris, 1972–2003).
– *Mémoires de Hector Berlioz* (Paris, 1870; repr. Paris, 1969); German translation ed. Wolf Rosenberg, (Munich, 1979).
Bernhardt, Reinhold: "Händel's Judas Maccabaeus", *Zeitschrift für Musikwissenschaft* 17 (1935), 516.
Bertelli, Sergio and Franco Cardini, Elvira Zorzi: *The Courts of the Italian Renaissance,* translated from the Italian, (New York, NY, 1986).
Bianchi, Lino: "Dall'oratorio di Alessandro Scarlatti all'oratorio di Händel", *Kongressbericht Händel-Scarlatti, Rome 1985* (Florence, 1987), 79–91.
Bianconi, Lorenzo → Storia dell'Opera Italiana
Biba, Otto: "Beispiele für die Besetzungsverhältnisse bei Aufführungen von Haydns Oratorien in Wien zwischen 1784 und 1808", *Haydn-Studien* 4/2 (1978), 94.
– "Beethoven und die 'Liebhaber Concerte' in Wien im Winter 1807/1808", *Beethoven-Kolloquium 1977* (Kassel, 1978), 82–93.
– "Concert life in Beethoven's Vienna", *Beethoven Congress, Detroit 1977,* ed. Robert Winter (Detroit, MI, 1980), 77–93.
– "Beispiele für die Besetzungsverhältnisse bei Aufführungen von Haydns Oratorien in Wien zwischen 1784 und 1808", *Haydn-Studien* 4/2 (1978), 94–104.

Bibliography of the History of Art, ed. J. Paul Getty Trust, Institut du CNRS, S. and F. Clark Art Institute (Williamstown, MA, 1991 f).
Bicknell, Stephen: *The History of the English Organ* (Cambridge: Cambridge UP, 1996; paperback 1998).
Bieber, Margarete: *The History of the Greek and Roman Theatre,* 2nd rev. and enl. ed. (Princeton, N.J.: Princeton UP, 1961).
Bienert, Peter: *Handbuch für die professionelle Studiotechnik: Digitale Audiotechnik* (Oberhausen, 1991).
Blaser, Werner → *Drawings of great buildings*
Blauert, Jens: *Räumliches Hören* (Stuttgart, 1974); *Nachschrift: Neue Ergebnisse und Trends seit 1972* (Stuttgart, 1985); *2. Nachschrift* (Stuttgart, 1997).
– *Spatial Hearing: the Psychophysics of Human Sound Localization,* (Cambridge MA: MIT Press, 1999).
Blume, Friedrich → *Musik in Geschichte und Gegenwart*
Böhme, Gerhard and Kunigunde Welzl-Müller: *Audiometrie: Hörprüfungen im Erwachsenen- und Kindesalter,* 3rd rev. ed., (Berne, 1993).
Borg, Erik, S. Allen Counter and G. Rösler, *Theories of middle ear muscle function* (Orlando, FL: Academic Press 1984).
Borg, Erik and S. Allen Counter: "The middle-ear muscles", *Scientific American* 267 (1989), no. 8, 62–68.
Borris, Siegfried: *Die grossen Orchester* (Hamburg and Düsseldorf, 1969).
Borwick, John (ed.) → *Sound recording practice*
– "Studios and Studio equipment", *JAES* 25, no. 10/11 (1977), 689–95.
Boullée, Etienne-Louis: *Essai sur l'art de construire les théâtres, leurs machines et leurs mouvements* (Paris, 1799).
Boyer, Charles B.: "Aristotelian references to the law of reflection", *Isis* 36 (1945), 92–95.
Bradley, John S.: "The evolution of newer auditorium acoustics measures", *Canadian Acoustics* 18 (1990), No. 4, 13–23.
– "Comparison of concert hall measurements of spatial impression", *JASA* 96 (1994), 3525–35.
Bräm, Thüring → *Musik und Raum*
Braun, Joachim: "The sound of Beethoven's orchestra", *Orbis Musicae* 6 (1978), 59–90.
Braunbehrens, Volkmar, Karlheinz Jürgens: *Mozart: Lebensbilder* (Bergisch-Glattbach, 1990).
Brauneis, Walther: "Theater an der Wien", *Österreichische Zeitschrift für Kunst- und Denkmalpflege* 36 (1982), 33–38.
Breh, Karl: "Quadrophonie gestern und heute", *Mono – Stereo – Quadro* (Basel, 1986), 61–80.
Bridge, Joseph C.: "Music at the English chapels Royal c. 1135 – present day", *PRMA* 54 (1927–28), 19–33.
Brinkmann, Reinhold: "Kleine 'Eroica'-Lese", *ÖMZ* 39 (1984), 634–38.
Brown, Clive: "The orchestra in Beethoven's Vienna", *EM* 16 (1988), 4–20.
Brown, Howard M. → *Peformance practice*
– and John W. McKinnon: "Performing practice", *NG,* vol. 14, 370–94.
Bruhn, Herbert et al.: *Handbuch der Musikpsychologie* (Reinbeck b. Hamburg, 1993).
– *Das neue Handbuch der Musikpsychologie* (Reinbeck b. Hamburg, 2008).
Bruyn, Andres → Torrington, Lord
Bryant, David: "The 'cori spezzati' of St. Mark's: myth and reality", *EMH* 1 (1981), 165–86.
Bucky, Gerhard: *Die Rezeption der Schweizerischen Musikfeste (1808–67) in der Öffentlichkeit.* Dissertation, Universität Zürich (Zurich, 1933).

Burghauser, Jarmil and Antonín Spelda: *Die akustischen Grundlagen des Orchestrierens,* German transl. by Adolf Langer (Regensburg, 1971).
Burkowitz, Peter K.: "Recording, art of the century", *Centennial Issue JAES* 25 (1977), 873–79.
Burney, Charles: *A general history of music* (London, 1776–89).
– *Account of the musical performance in the Westminster Abbey and the Pantheon 1784: in commemoration of Handel* (London, 1785).
– "Chamber music", *Ree's Cyclopaedia* (1802–20).
Burris-Meyer, Harold: *Theatres and auditoriums* (New York, NY, 1949).
Burrows, Donald: "Handel's Peace Anthem", *MT* 114 1973), 1230–32.
– "Handel's performances of 'Messiah': the evidence of the conducting score", *ML* 56 (1975), 319–34.
– "Handel and the Foundling Hospital", *ML* 58 (1977), 269–84.
– "Handel and the 1727 Coronation", *MT* 118 (1977), 469–72
– "Sources for Oxford Handel performances in the first half of the eighteenth century", *ML* 61 (1980), 177–85.
– "Handel's London theatre orchestra", *EM* 13 (1985), 349–57.
– "The autographs and early copies of 'Messiah': some further thoughts", *ML* 66 (1985), 201–19.
– "Handel in England: sacred music", *Kongressbericht Stuttgart 1985,* Vol. 1 (Kassel, 1987), 213–17.
– "Some thoughts concerning the performance of Handel's oratorios", *Händel-Jb.* 35 (1989), 63–68.
– "Dramatic music and scenic elements in Handel's English church music", *Händel-Jb.* 37 (1991), 87–102.
– *Handel: Messiah* (Cambridge, 1991) (Cambridge Music Handbooks).
– and Martha J. Ronish: *A catalogue of Handel's autographs* (Oxford. 1994).
– and Watkins Shaw: "Handel's 'Messiah', supplementary notes on sources", *ML* 76 (1995), 356–68.
Antonio Caldara: essays on his life and times, ed. Brian W. Pritchard Gower (Aldershot, 1987).
Calmonte, Roland → Lang, Alfred
Carse, Adam: *The orchestra in the 18th century* (Cambridge, 1940).
Cassirer, Ernst: *Gesammelte Werke (GW),* Hamburger Ausgabe, ed. Birgit Recki (Hamburg, 1995–2009).
– *Die Philosophie der symbolischen Formen* (orig. publ. 1921, 1925, 1929; GW 11–13).
– *An Essay on Man. An Introduction to a Philosophy of Human Culture* (orig. publ. 1944; GW 23).
– *Determinismus und Indeterminismus in der modernen Physik* (orig. publ. 1937; GW 19).
– *Studien zur Kulturwissenschaft: Dritte Studie: Naturbegriffe und Kulturbegriffe* (orig. publ. 1942; GW 24).
– *The concept of group and the theory of perception* (orig. publ. 1944; GW 24).
– *Structuralism in modern linguistics* (orig. publ. 1945; GW 24).
Cavalieri, Buonaventura: *Lo specchio Ustorio overo Trattato delle Settioni Conice, ed alcuni loro mirabili effetti intorno al lume, caldo, freddo, suono e moto ancora* (Bologna, 1632).
Emilio de' Cavalieri, *Rappresentatione di anima e di corpo,* ed. M. C. Bradshaw, AIM, *Miscellanea,* vol. 5 (2007), preface.
Chamber's Journal of Popular Literature, Science, and Arts (Edinburgh and London, 1854).
Chatillon, André-Marie: "Projekt eines runden Konzertsaales 1803", *The architecture of the école des beaux-arts* (Cambridge MA, 1977), 138 f.

Chladni, Ernst: *Entdeckungen über die Theorie des Klanges. Beyträge zur praktischen Akustik* (Leipzig, 1787. repr. Leipzig, 1980).
– *Die Akustik* (Leipzig, 1802).
– *Neue Beiträge zur Akustik* (Leipzig, 1817).
– *Beiträge zur praktischen Musik* (Leipzig, 1821).
Cibber, Colley: *An apology for the life of Mr. Colley Cibber,* ed. R. Lowe (London, 1740; repr. London, 1869).
Clark, Ronald W.: *Royal Albert Hall* (London, 1958).
Cobb, Gerald: *London city churches* (London 1977).
Coeyman, Barbara: "Theatres for opera and ballet during the reigns of Louis XIV and Louis XV", *Early Music* 18 (1990), 22–37.
Colins, Howard Montagu: *The Sheldonian Theatre and the Divinity College* (Oxford, 1981).
Collins, Peter: *Concrete. The Vision of a New Architecture* (London, 1959; 2nd ed. Montreal, Quebec, 2004).
Concert music (1630–1750), ed. Gerald Abraham (Oxford, 1986), (*NOHM* VI).
Connors, Joseph J.: *Borromini and the Roman oratory* (Cambridge, 1980).
Conradin, Hans: "Die Frage nach der Bewegung in der Musik", *Orbis Musicae* 9 (1986), 224–33.
Contant, Clément et Joseph de Philippi: *Parallèle des principeaux théâtres modernes de l'Europe et machines* (Paris, 1840).
Corelli, Arcangelo: Music editions → see 3.1.3
Counter, S. Allen → Bork, Erik
Cremer, Lothar: *Structure-borne sound* (Berlin and New York, NY, 1973).
– "The different distribution of the audience", *Applied Acoustics* 8 (1975), 173–91.
– *Die wissenschaftlichen Grundlagen der Raumakustik,* 2 vols., 2nd ed. (Stuttgart, 1976–78).
Cremer, Lothar and Helmut A. Müller: *Principles and applications of room acoustics,* 2 vols. (London and New York, NY, 1982).
Creuzburg, Eberhard: *Die Gewandhauskonzerte zu Leipzig 1781–1931* (Lepzig, 1931).
Crombie, Alistair Cameron: *Science in the Middle Ages* (London, 1959, repr. 1979).
Culshaw, John: "The Role of the Producer", *JAES* 25, no. 10/11 (1977), 255–61.
Cumberland, Richard: *Memoirs,* 2 vols. (London, 1807).
Curzon, Henri de: "History and glory of the concert-hall of the Paris conservatory (1811–1911)", *MQ* 3 (1917), 304–18.
Cusins, William George: *Handel's Messiah: an examination of the original and of some contemporary mss.* (London, 1874).
Daidalos: Der hörbare Raum / The audible space: Berlin Architectural Journal, ed. Wolfgang Meisenheimer et al. (Berlin, 1985).
Dandelot, Arthur: *La Société des concerts du Conservatoire 1828 à 1923* (Paris, 1923).
De l'Orme, Philibert: *Premier tome de l'architecture* (Paris, 1567–68; Rouen 1648: repr. Ridgewood 1964).
De la Motte, Helga: *Handbuch der Musikpsychologie* (Laaber, 1985; 2nd ed. 1993).
– "Der homo ludens – ein homo oeconomicus?", *Musik – Gehirn – Spiel, 4. Karajan Symposium 1988* (Basel, 1989).
Dean, Winton: "A French traveler's view of Handel's operas", *ML* 55 (1974), 172–78.
– *Handel's dramatic oratorios and masques* (Oxford, 1959).
– "Margherita Durastanti", *New Grove Dictionary of Opera,* vol. 1 (London, 1992), 1283.
Delany, Mary → Granville, Mary
Delvedez, Édouard-Marie-Ernest: *La société des concerts, 1860 à 1885* (Paris, 1887).
Denkmäler des Theaters: *Denkmäler des Theaters: Inszenierungen, Dekorationen, Kostüme des Theaters und der grossen Feste aller Zeiten* (Vienna: National Library, 1925).

Deutsch, Georg → Springer, Sally
Deutsch, Otto Erich: *A documentary biography of Handel* (London, 1955; repr. New York, NY 1974); new edition see: *Händel-Handbuch,* vol. 4.
– "Das erste Wiedner Theater", *Sternstunden im Theater an der Wien* (Vienna, 1962), 4–19.
– "Festkonzerte im alten Universitätssaal", *ÖMZ* 18 (1963), 428–32.
Dickreiter, Michael: *Tonstudiotechnik,* 6th ed. (Munich 1997 and 2008).
– *Mikrofonaufnahmetechnik* (Stuttgart, 1995).
Dilthey, Wilhelm: *Gesammelte Schriften,* ed. Bernhard Groethuysen (Stuttgart and Göttingen, 1973 f.) (GS); English translation: W. Dilthey, *Selected Works,* ed. Rudolf A. Makkreel & Frithjof Rodi (Princeton: Princeton UP, 1989 f.).
– "Einleitung in die Geisteswissenschaften: Versuch einer Grundlegung für das Studium der Gesellschaft und der Geschichte (orig. publ. 1883)", *GS* 1 (1973), p. xv–xx;
– "Ideen über eine beschreibende und zergliedernde Psychologie" (1894); "Beiträge zum Studium der Individualität" (1895/96); "Die Entstehung der Hermeneutik" (1900); "Die geistige Welt: Einleitung in die Philosophie des Lebens", *GS* 5 (1968);
– "Der Aufbau der geschichtlichen Welt in den Geisteswissenschaften" (1895, 1907), *GS* 7 (1973).
Dittersdorf, Karl Ditters von: *Lebensbeschreibung, seinem Sohne in die Feder diktiert* (Leipzig, 1800; repr. Munich, 1967).
Doelle, Leslie L.: *Acoustics in architectural design: an annotated bibliography on architectural acoustics,* ed. Nat. Research Council of Canada, Div. of Building Research (Ottawa, 1965).
Dolbear, Amos Emerson: *Il Telefono, descrizione dei fenomeni di elettricità, magnetismo e suono in esso implicati e norme per la costruzione del telefono parlante,* (Florence, 1878).
Donington, Robert: *The Interpretation of Early Music* (London, 1963; revised ed. 1992).
Donnet, Alexis: *Architectonographie des Théâtres de Paris ou parallèle historique et critique de ces édifices* (Paris, 1821; 2nd ed. Paris 1837).
Dörffel, Alfred: *Geschichte der Gewandhausconcerte zu Leipzig vom 25. November 1781 bis 25. November 1881* (Leipzig, 1884).
Dostrovsky, Sigalia, James F. Bell, Clifford Truesdell: "Physics of music", *NG,* vol. 14, 664–77.
Drawings of great buildings, ed. Werner Blaser (Basel, 1983).
Drexler, Arthur: *The architecture of the Ecole des Beaux-Arts* (London, 1975; New York, NY, 1977).
Düding, Dieter: *Öffentliche Feste.* Habilitationsschrift, Universität Köln (1981).
Dumont, Gabriel Pierre Martin: *Parallèle de plans des plus belles salles de spectacles d'Italie et de France* (Paris, 1764–65; repr. New York, NY, 1968).
Durm, Josef and Heinrich Wagner: *Handbuch der Architektur,* Teil IV/4/4/1–3: "Öffentliche Vergnügungsstätten und Festhallen", ed. Josef Durm et al., 2nd ed. (Darmstadt, 1894).
Durand, Jean-Nicolas-Louis: *Recueil et parallèle des édifices de tout genre anciens et modernes, remarquables par leur beauté, par leur grandeur. ou par leur singularité, et dessinés sur une même échelle* (Paris: J. G. Legrand 1801).
Eccles, John → Popper, Karl
Edelmann, Bernd: "Händel-Aufführungen in den Akademien der Wiener Tonkünstlersozietät", *Göttinger Händel-Beiträge* 1 (1984), 172–99.
Eggenschwiler, Kurt: "Erfahrungen mit zwei kommerziellen Raumakustik-Simulations-Programmen [dBRay und EPIDAURE] ", *DAGA '93,* separatum EMPA, (Dübendorf, 1993).
Eichhorn, Albert: *Die Akustik grosser Räume nach altgriechischer Theorie,* 3rd ed. (Berlin, 1903).
Eisen, Walter and Margret → Händel-Handbuch
Elkin, Robert: *The old concert rooms of London,* 2nd ed. (London, 1955).

Elwart, Antoine: *Histoire de la Société des Concerts du Conservatoire* (Paris, 1864).
Eminente, Gianfranco: "Dopo cento anni radicali restauri al palazzo Valentini", *Rassegna di Latio,* nuova serie, 20 (1973), 7.
Enciclopedia dello Spettacolo, 9 vols. (Rome, 1954–62, suppl. 1963, 1966).
Encyclopedia Metropolitana, ed. Samuel Taylor Coleridge et al., 30 vols. (London, 1817–1845).
Encyclopédie ou Dictionnaire raisonné des sciences, des arts et des métiers, éd. Denis Diderot et al., 28 vols. (Paris, 1751–72); [vol. 10, appendix: *Théâtres,* 38 planches].
Ewerhart, Rudolf: "New Sources for Handel's 'La resurrezione'", *ML* 41 (1960), 127–35.
– "Die Santini-Bibliothek in Münster", *Händel-Jb.* 6 (1960), 111–18.
Eysel, Ulf: "Sehen (betr. Farbkonstanz)", *Neuro- und Sinnesphysiologie* (Berlin, 1993), 303.
Faessbender, Christoph: "Funktionsteilung im Kortex", *Handbuch der Musikpsychologie* (Hamburg, 1993), 622–30.
Falda, Giovan Battista and Alessandro Specchi: *Palazzi di Roma nel '600,* repr. (Rome: Audino n.d.), (Collana Vedute d'Italia 3).
Farina, Gennaro → *Palazzo della Provincia*
Fasold, Wolfgang, Wolfgang Kraak, Werner Schirmer: *Taschenbuch der Akustik* (Berlin, 1984).
– Ernst Sonntag, Helgo Winkler: *Bau- und Raumakustik* (Berlin and Köln-Braunsfeld, 1987), (Bauphysikalische Entwurfslehre).
– Eva Veres: *Schallschutz und Raumakustik in der Praxis* (Berlin, 2003)
Favaro, Antonio: *L'acustica* (Padua, 1882; repr. Bologna, 1969).
Fenlon, Iain: "The Performance of *cori spezzati* in San Marco", *Architettura e musica nella Venezia del Rinascimento,* ed. Deborah Howard and Laura Moretti (Milan, 2006), 79–98.
Ferrerio, Pietro: *Palazzi di Roma* (Rome, s.a.).
Filippi, Joseph de: *Parallèles des principaux théâtres modernes de l'Europe et des machines théâtrales françaises, allemandes et anglaises.* Dessins par Clément Constant (Paris, 1859).
Finscher, Ludwig: „Zur deutschen Händel-Rezeption im 18. Jahrhundert", *Kanon und Zensur: Beiträge zur Archäologie der literarischen Kommunikation* II, ed. A. and J. Assmann (München, 1987), 271–283.
– "Mozart und Mannheim", *Die Mannheimer Hofkapelle im Zeitalter Karl Theodors,* ed. L. Finscher (Mannheim: Palatium, 1992), 71–96.
→ *Musik in Geschichte und Gegenwart, MGG²*
Firenze e la Toscana dei Medici nell'Europa del Cinquecento: la scena del principe (Florence, 1980).
Fischer, Fritz and Hugo Lichte: *Tonfilm: Aufnahme und Wiedergabe* (Leipzig, 1931).
Fletcher, Harvey: *Speach and hearing* (New York, NY and London, 1929).
Fletcher, Banister: *History of Architecture*, ed. by Dan Cruickshank, Oxford: Architectural Press, © Elsevier cg: Butterworth-Heinemann, 1987
Flood, William Henry Grattan: "Fishamble St. Music Hall, Dublin, from 1741 to 1777", *SIMG* 11 (1909/10), 51–57.
– "Crow St. Music Hall, Dublin, from 1730 to 1754", *SIMG* 14 (1912/13), 442–46.
Flury, Roman: "Von Edison zur Compact Disc", *Mono – Stereo – Quadro* (Basel, 1986), 25–40.
Forbes, Elliot (ed.): *Thayer's life of Beethoven,* rev. ed. (Princeton, 1967).
Forsyth, Michael: *Buildings for music* (Cambridge, 1985).
– *Bauwerke für Musik: Konzertsäle und Opernhäuser, Musik und Zuhörer vom 17. Jahrhundert bis zur Gegenwart.* German translation by R. and M. Dickreiter (Munich, 1991).
Forner, Johannes: *Die Gewandhauskonzerte zu Leipzig 1781- 1981* (Leipzig, 1981).
Fougeroux, Pierre Jacques: *Voiage d'Angleterre d'Holland et de Flandre fait en l'année 1728,* Coke Collection, Foundling Museum, London.

Francesco di Giorgio, Francesco Giorgi: *Memorandum di S. Francesco della Vigna* (1535); see Giannantonio Moschini, *Guida per la città di Venezia,* vol. I (Venice, 1815), 55–61; English transl. Rudolf Wittkower, *Architectural principles in the age of Humanism* (London, 1949), appendix I, 137; Italian ed. *Principi architettonici nell'età dell'umanesimo* (1964), 149–50; German transl. Wittkower, *Grundlagen der Architektur* (Munich, 1983), 125.
Frei, Isabelle: *Raumakustische Qualität in komplexen Räumen. Theoretische und exemplarische Darstellung der Probleme,* Lizentiatsarbeit, Universität Zürich (2011).
Frei, Jürg: *Die Gehörbelastung des Orchestermusikers in der Konzert- und Opernformation der Tonhalle Zürich.* Dissertation, Universität Zürich (1979).
Friemert, Chup: *Die gläserne Arche: Kristallpalast London 1851 und 1854* (Dresden, 1984).
Fröhlich, Martin: *Gottfried Semper als Entwerfer und Entwurfslehrer.* Dissertation, ETH Zurich (1974).
– *Gottfried Semper: zeichnerischer Nachlass an der ETH Zürich. Kritischer Katalog* (Basel, 1974), (gta 14).
– *Gottfried Semper* (Zürich: Verlag für Architektur, 1991).
Frommel, Christoph Luitpold: *Der römische Palastbau der Hochrenaissance* (Tübingen, 1973).
Furrer, Willi and Anselm Lauber: *Raum- und Bauakustik, Lärmabwehr,* 3rd ed. (Basel, 1972).
Furttenbach, Joseph d. Ä.: *Architectura recreationis,* (Augsburg, 1640).
– *Mannhafter Kunst-Spiegel* (Augsburg, 1663).
Gabler, Werner: *Der Zuschauerraum des Theaters* (Leipzig, 1935).
– "Historische Räume", *Räume zum Hören* (Cologne, 1989).
Gade, Anders Chr.: "Acoustics of the orchestra platform from the musician's point of view", *Acoustics for choir and orchestra* (Stockholm, 1986), 23–42.
– "Investigations of musicians' room acoustic conditions in concert halls", *Acustica* 69 (1989), 193–203, 249–62.
Galeazzi, Francesco: *Elementi teorico-pratici di musica,* 2 vols. (Rome, 1791 and 1796).
Galkin, Elliott W.: *A history of orchestral conducting in theory and practice,* (New York, NY, 1988).
Gallavotti Cavallero, Daniela: *Palazzi di Roma dal XIV al XX secolo* (Rome: NER, 1989).
Galli-Bibiena, Ferdinando: *L'architettura civile preparata sulla geometria e ridotta alle prospettive* (Parma, 1711).
Gallo, F. Alberto: *La prima rappresentazione al teatro Olimpico* (Milan, 1973).
Garnier, Charles: *Le nouvel opéra de Paris,* 2 vols. (Paris, 1876–81).
– *Le théâtre* (Paris, 1871).
Gasbarri, Carlo: *Lo spirito dell'oratorio di S. Filippo Neri* (Brescia, 1949).
– *L'oratorio filippino (1552–1952)* (Rome, 1957).
– *L'oratorio romano dal '500 al 1900* (Rome, 1962).
Gasparoni, Francesco: "Sulle curve dei teatri", *L'architetto girovago* 1 (Rome, 1841), 102–07.
Gassner, Ferdinand Simon: *Dirigent und Ripienist für angehende Musikdirigenten, Musiker und Musikfreunde* (Karlsruhe, 1844).
Gathy, August: *Neue Zeitschrift für Musik* 30 (1835), 122.
Gelb, Adhémar: "Die 'Farbenkonstanz' der Sehdinge", *Handbuch der normalen und pathologischen Physiologie,* ed. Albrecht Th. J. Bethe et al.,vol. 12/1 (Berlin, 1929), 594–678.
– "Zur medizinischen Psychologie und philosophischen Anthropologie", *Acta Psychologica 3* (1937), 195–271.
Germann, Georg: *Der protestantische Kirchenbau in der Schweiz von der Reformation bis zur Romantik* (1963).
– *Einführung in die Geschichte der Architekturtheorie,* 2nd enl. ed. (Darmstadt, 1987).
Geschwind, Norman: *Die Grosshirnrinde,* 4th ed. (Heidelberg, 1984).

Gibson, James J.: *The senses considered as perceptual systems* (Boston. 1966); German ed.: *Die Sinne und der Prozess der Wahrnehmung* (Berne, 1973).
Giedion, Sigfried: *Space, time and architecture: the growth of a new tradition,* 5th ed. rev. and enl. (Cambridge, MA: Harvard UP, 1967).
Gill, Dominic: *The book of the violin* (Oxford, 1984).
Gishford, Anthony: *Grand Opera: the story of the world's leading opera houses and personalities* (London, 1972).
Glogau, Ulrich: *Der Konzertsaal: zur Struktur alter und neuer Konzerthäuser* (Hildesheim, 1989).
Göllerich, August and Max Auer: *Anton Bruckner,* 4 vols. (Regensburg, 1922–37).
Goldmann, Nikolaus: *Vollständige Anweisung zu der Civil-Baukunst* (Wolfenbüttel, 1698).
– *Vollständige Anweisung alle Arten von Kirchen wohl anzugeben* (Augsburg, 1718).
Gosset, Alphonse: *Traité de la construction des théâtres* (Paris, 1886).
Gottlob, D.: *Vergleich objektiver akustischer Parameter mit Ergebnissen subjektiver Untersuchungen an Konzertsälen.* Dissertation, Universität Göttingen (1973).
Granville-Delany, Mary: *Autobiography and correspondence of Mary Granville, Mrs. Delany,* ed. Lady Llanover, 6 vols. (London, 1861–62).
Gregor-Dellin, Martin: *Wagner* (Munich, 1980).
Gregor, Joseph: *Geschichte des österreichischen Theaters* (Vienna, 1948).
Gutknecht, Dieter: "Alte Musik heute, Geschichte und Stand der Aufführungspraxis der Musik des 18. Jahrhunderts", *Historische Aufführungspraxis* 2 (1989), 9–28.
Gysi, Barbara: *Die Entwicklung der Tontechnik und deren Einfluss auf musikaesthetische Veränderungen.* Lizentiatsarbeit, Universität Zürich (1992).
Haas, Helmut: "Über den Einfluss eines Einfachechos auf die Hörsamkeit von Sprache", *Acustica* 1 (1951), 49–58.
Habel, Heinrich: *Das Odeon in München und die Frühzeit des öffentlichen Konzertsaalbaus* (Berlin, 1967).
– "Die Konzertsäle", *Die Welt der Symphonie,* ed. Ursula Rauchhaupt (Hamburg and Braunschweig, 1972), 17–33.
– *Festspielhaus und Wahnfried. Geplante und ausgeführte Bauten Richard Wagners* (Munich, 1985).
Hahnloser, Hans Robert: *Das Bauhüttenbuch des Villard de Honnecourt,* 2nd ed. (Graz, 1972).
Haider, Edgar: *Verlorenes Wien: Adelspaläste vergangener Tage* (Vienna, 1984).
Hammitzsch, Martin: *Der moderne Theaterbau* (Berlin, 1906).
Hammond, Frederick: "More on music in Casa Barberini", *Studi Musicali* 15 (1986), 235–61.
Handbuch der Architektur
– IV/4/4 → Hilbersheimer, Ludwig
– IV/4/4/3 → Durm, Josef
– IV/4/6/5 → Semper, Manfred
– IV/6/3 → Schmitt, Eduard
Händel-Handbuch, ed. Walter and Margret Eisen (Supplement of Hallische Händel-Ausgabe): vol. 1–3: *Thematisches Werkverzeichnis,* ed. Bernd Baselt (Leipzig and Kassel. 1978–86); vol. 4: *Dokumente zu Leben und Schaffen,* based on Otto Erich Deutsch: Handel, a documentary biography, ed. Bernd Baselt (Leipzig and Kassel, 1985).
Händel, Georg Friedrich: Music editions → see 3.1.2
Handel, Stephen: *Listening: an introduction to the perception of auditory events* (Cambridge, MA, 1989; 2nd ed. 1991).
Hansell, Sven Hostrup: "Orchestral practice at the court of Cardinal Pietro Ottoboni", *JAMS* 19 (1966), 398–403.

Hanslick, Eduard: *Geschichte des Concertwesens in Wien,* 2 vols. (Vienna, 1869; repr. Hildesheim, 1979).
Hanson, Alice M.: *Musical life in Biedermeier Vienna* (London, 1985).
Harley, John: "Music at the English court in the 18th and 19th centuries", *ML* 50 (1969), 332–51.
Harrer, G. and H.: "Emotion and autonomic functions", *Music and the brain* (London, 1977), 212–13.
Harris, Ellen T.: "Handel's ghost: the composer's posthumous reputation in the eighteenth century", *Companion to contemporary musical thought* (London, 1992), vol. 1, 208–25.
Harris, John → *King's Arcadia: Inigo Jones*
Haskell, Harry: *Early music revival* (London, 1988).
Hawkins, Sir John: *A general history of science and practice of music,* 5 vols. (London, 1770–76; new ed. 2 vols. London, 1853; facs. ed. New York, NY, 1963).
Heartz, Daniel: "Nicolas Jadot and the building of the Burgtheater", *MQ* 68 (1982), 1–31.
Helmholtz, Hermann von: *Die Lehre von den Tonempfindungen als physiologische Grundlage für die Theorie der Musik* (Braunschweig, 1863; 3rd ed., Braunschweig, 1896).
Henry, Joseph: "On acoustics applied to public buildings", *Annual Report of the Smithsonian Institution* (Washington, 1856–57), 221–34.
– "Researches in sound", *Annual Report of the Smithsonian Institution* (Washington, 1878–79), 492, 527, 556–59.
Hensel, Sebastian: *Die Familie Mendelssohn* (Berlin, 1879).
Hermann, Wolfgang: *Gottfried Semper: theoretischer Nachlass* (Basel, 1981).
Heron of Alexander: *Katoptrika.* German translation W. Schmidt (Leipzig, 1906).
Herschel, Sir J.: "Sound", *Encyclopedia Metropolitana,* vol. 28, 813.
Heutschi, Kurt: *Computersimulation in der Raumakustik* (Dübendorf: EMPA, März 1995).
Hicks, Anthony: "G. F. Handel, work list", *NG*^{2}, vol. 8, 114–16.
– Booklet to the recording of *La Resurrezione,* (DECCA, 1982) → 3.2.5
HiFi, Ultraschall und Lärm. Die Welt des Schalls, ed. Jean Pütz, 1st ed. (Cologne: Verlagsgesellschaft Schulfernsehen, 1973)
Hilbersheimer, Ludwig: "Hallenbauten", *Handbuch der Architektur* IV/4/4 (Leipzig, 1931).
Hiller, Johann Adam: *Nachricht zur Aufführung des Händelschen Messias in der Domkirche zu Berlin* (Berlin, 1786).
– *Der Messias, nebst Betrachtungen über die Aufführung Händelscher Singkompositionen* [1787].
Hinrichsen, Hans-Joachim: " 'Urvater der Harmonie ?' Die Bach-Rezeption", *Bach-Handbuch,* ed. Konrad Küster, Kassel (1999), 31–67.
Hoeg, Wolfgang and Gerhard Steinke: *Stereofonie-Grundlagen,* 2nd rev. ed. (Berlin, 1972).
Hoffmann, Ernst Theodor Amadeus: "Johannes Kreislers, des Kapellmeisters, musikalisches Leiden", *Poetische Werke* in sechs Bänden, vol. 1 (Berlin, 1963), 80–88.
Hoffmann, Hans Christoph: *Die Theaterbauten von Fellner und Helmer* (Munich, 1966), (Studien zur Kunst des 19. Jahrhunderts, 2).
Höger, Rainer: *Lautheitsurteil und akustischer Kontext: experimentelle Untersuchungen zu einer mehrdimensionalen Bezugssystemtheorie* (Munich, 1987).
Hogwood, Christopher: *Handel* (London, 1984); rev. ed. (London 2007); paperback ed. (London, 1988 and 1995); German translation by Bettina Obrecht (Stuttgart, 1992).
– "Messiah: Notes on the music", Comment to *Handel, Messiah, a sacred oratorium,* directed by Chr. Hogwood, *L'Oiseau Lyre* D189D3 (Florilegium Series) → 3.4
Howard, Deborah and Laura Moretti: *Sound and space in Renaissance Venice: architecture, music, acoustics* (Yale and London: Yale UP, 2009).

Homes, William C.: "Venetian theatres during Vivaldi's era", *Opera and Vivaldi,* ed. Dennis Lynds (Austin, TX, 1984), 131–48.
Hume, Robert Douglas: *The London Theatre World* (Carbondale, 1980).
Hundert Jahre Goldener Saal: das Haus der Gesellschaft der Musikfreunde am Karlsplatz, ed. Gesellschaft der Musikfreunde in Wien (Vienna, 1970).
Hunt, Frederick V.: *Origins in acoustics: the science of sound from antiquity to the age of Newton* (New Haven: Yale University, 1978).
Inventory of historical monuments in London, ed. Royal Commission of Historical Monuments, 5 vols. (London, 1924–30).
Izenour, George C.: *Theater design* (New York, NY, 1977).
– "The ancient Roman roofed theater", *Perspecta,* Vol. 26: "Theater, Theatricality, and Architecture" (1990), 69-82.
Jackson, Allan S. → Moynet, M. J.
Jacobi, Erwin R.: "Notes inégales", *Riemann Lexikon,* 12th ed., *Sachteil* (Mainz, 1967), 642–43.
Jacquot, Jean: *Le lieu théâtrale à la Renaissance* (Paris: CNRS, 1964).
JAES: The Phonograph and Sound Recording after One-hundred Years, JAES, Centennial Issue, vol. 25 /10–11, ed. Audio Engineering Society (New York, NY, 1977).
Jäncke, Lutz: *Macht Musik schlau? Neue Erkenntnisse aus den Neurowissenschaften und der kognitiven Psychologie* (Berne, 2008).
Jakob, Friedrich: "Die Orgel des westlichen Kulturkreises im Mittelalter", *Orgel,* ed. Alfred Reichling (Kassel and Stuttgart, 2001), 53–55.
Jauk, Werner: "Lautheit: Ein kognitionstheoretisches Modell der Lautstärkewahrnehmung und seine Anwendung in der Musik", *AMl* 66 (1995), 67–77.
Jecklin, Jürg: *Musikaufnahmen: Grundlagen, Technik, Praxis* (Munich, 1980).
– "Mono – Stereo – Quadro: die Aufnahme und Wiedergabe von Musik", *Mono – Stereo – Quadro* (Basel, 1986), 81–178.
– "Die notwendige Signalaufbereitung: Der Transdyn-Process", *Mono – Stereo – Quadro,* (Basel, 1986), 156–59.
Johnson, H. Earle: "Handel and Haydn Society of Boston", *MGG,* vol. 5 (1956), col. 1291–94.
Johnson, Russell: *Newsletter ARTEC* (New York, NY, 1995).
Jordan, Vilhelm Lassen: "Einige Bemerkungen über Anhall und Anfangsnachhall in Musikräumen", *Applied Acoustics* 1 (1968), 29–36.
– *Acoustical design of concert halls and theatres* (London, 1980).
Jürgens, Karlheinz → Braunbehrens, Volkmar
Kaiser, Hermann: *Der Bühnenmeister Carl Brandt und Richard Wagner* (Darmstadt, 1968).
Kandel, Eric R. and Robert D. Hawkins: "The biological basis of learning and individuality", *Scientific American* 271 (1992), no. 9, 79–86.
Kelly, Michael: *Reminiscences* (London, 1826; reprint ed. R. Fiske, London 1968 and 1975).
Kenyon, Nicholas → *Authenticity*
Kiesel, Markus: "Construction History", *The Richard Wagner Festival Theatre Bayreuth* (Cologne, 2007), 56–87.
King, Alec Hyatt: "The London Tavern: a forgotten concert hall", *MT* 127 (1986), 382–85.
Kircher, Athanasius: *Ars magna lucis et umbrae* (Rome, 1645–46; 2nd ed. 1671).
– *Musurgia universalis, sive Ars magna consoni et dissoni* (Rome, 1650). German translation by A. Hirsch: *Philosophischer Extract und Auszug aus dess Welt-berühmten teutschen Jesuiten Athanasii Kircheri von Fulda Musurgia universali* (Schwäbisch Hall, 1662).
– *Phonurgia nova, sive Coniugium mechanico-physicum arti set naturae* (Campidonae-Kempten, 1673). German translation by Agathus Carion (Christoph Fischer): *Neue Hall- und Tonkunst* (Nördlingen, 1684).

– *Athanasius Kircher und seine Beziehungen zum gelehrten Europa seiner Zeit,* ed. John Fletcher (Wiesbaden, 1988) (Wolfenbütteler Arbeiten zur Barockforschung, 17).
Kirkendale, Ursula: "The Ruspoli documents on Handel", *JAMS* 20 (1967), 222–73 and 517-8. (KirkR)
– „The Ruspoli documents on Handel", *Music and meaning, studies in music history and the neighbouring disciplines,* ed. Warren and Ursula Kirkendale, Florence 2007, 287-346.
– *Antonio Caldara, sein Leben und seine venezianisch-römischen Oratorien* (Graz. 1966). (KirkC)
Kisch, Wilhelm: *Die alten Strassen und Plätze Vienna's und ihre historisch interessanten Häuser,* 3 Bde. (Vienna, 1885–88).
Klein, Rudolf: "Ein alt-Wiener Konzertsaal: das Etablissement Jahn an der Himmelpfortgasse", *ÖMZ* 18 (1963), 12–18.
Knapp, J. Merrill: "Handel's Roman church music", *Händel e gli Scarlatti,atti del convengno internazionale di studi, Roma 1985,* ed. N. Pirrotta, A. Ziino (Firenze, 1987), 15–27.
Knaus, Patrick: *Studie zur auditiven Wahrnehmung auf der Grundlage der Theorie von J. J. Gibson.* Lizentiatsarbeit, Universität Bern (1987).
Knudsen, Vern Oliver: *Architectural acoustics* (New York, NY, 1932).
– and Cyril M. Harris: *Architecture* (New York, NY, 1950).
– and Cyril M. Harris: *Acoustical design in architecture* (New York, NY, 1978).
Koch, Heinrich Christoph: *Muskalisches Lexikon, welches die theoretische und praktische Tonkunst, encyclopädisch bearbeitet, alle alten und neuen Kunstwörter erklärt, und die alten und neuen Instrumente beschrieben, enthält* (Frankfurt am Main, 1802; 2nd ed. 1817).
Koepf, Hans: *Baukunst in 5 Jahrtausenden,* 10th rev. ed. (Stuttgart, 1990).
Koury, Daniel J.: *Orchestral performance practices in the nineteenth century – size, proportions, and seating* (Ann Arbor: UMI, 1986).
Kowar, Helmut: "Liebhaber- und Amateurorchester in Wien: Versuch einer überblicksmässigen Darstellung", *Studien zur Musikwissenschaft, Beihefte DTÖ* 34 (1983), 131–40.
Kropfinger, Klaus: "Händels Oratorium: Aspekte der Gattungsprofilierung, Werkenstehung und Rezeption", *Gattungen der Musik* (Laaber, 1988), 87–108.
Kühl, Ole: *Musical semantics* (Berne. 2007).
Kuhl, W.: "Über Versuche zur Ermittlung der günstigsten Nachhallzeit grosser Musikstudios", *Acustica* 4 (1954), 618 and 8 (1958), 280.
Kuttruff, Heinrich: *Room acoustics* (London: Applied Science Publishers, 3rd. ed. 1991).
Lachez, Theodore: *L'acoustique et optique des salles de réunions publiques* (Paris, 1848, 2nd ed. 1879).
Lamoral, Roger: *Acoustique et architecture* (Paris: Masson, 1975), (G.A.L.F.).
Landels, John G.: "Assisted resonance in ancient theatres", *Greece and Rome* 14 (1967), 80.
Landon, H. C. Robbins: *Haydn,* 5 vols. (London, 1977).
– *Haydn in England* (London, 1976).
– *Handel and his world* (London, 1984).
Lang, Alfred, Roland Calmonte and Heinrich Zimmermann: *Klangkonstanz im Raum: Hören und Überhören von raumbedingter Schallvariation.* Appendix A: On timbre constancy in space. Preliminary Report. Nationalfonds Projekt, Psychologisches Institut Universität Bern, (1986–88).
Langhans, Carl F.: *Über Theater oder Bemerkungen über Katakustic in Beziehung auf Theater* (Berlin, 1810).
Langhans, Edward A.: "Theatre architecture", *New Grove Dictionary of Opera,* Vol. 4 (London, 1992).
Larsen, Jens Peter: *Handel's Messiah: origins, composition, sources* (London, 1957; 2nd ed.

New York, NY, 1972).
– "Wandel zur Auffassung von Händels 'Messias'", *Göttinger Händel-Beiträge* 1 (1984), 7–20.
– "Georg Friedrich Händel – Oper und Oratorium: Gattungstypologie und Aufführungspraxis", *Händel-Symposium, Karlsruhe 1986–*87 (Laaber, 1988), vol. 2, 17–28.
Lauber, Anselm → Furrer, Willi
Lavin, Irving: "Lettres de Parmes (1618, 1627–28) et débuts du théâtre baroque", *Le lieu théâtrale à la Renaissance* (Paris: CNRS, 1986), 105–157.
Le Huray, Peter: "Handel's 'Messiah'", *Authenticity in performance: eighteenth-century case studies* (Cambridge. 1990), 82–101.
Leacroft, Richard: *The development of the English playhouse* (London, 1973; 2nd ed. 1988).
Leacroft, Richard and Helen Leacroft: *Theatre and playhouse: an illustrated survey of theatre building from ancient Greece to the present day* (London and New York, NY, 1984).
Leclerc, Hélène: *Les origines italiennes de l'architecture théâtrale* (Paris, 1946).
Lehmann, Peter: „Über die Ermittlung raumakustischer Kriterien und deren Zusammenhang mit subjektiven Beurteilungen der Hörsamkeit", Dissertation, TU Berlin (1976).
– and H. Wilkens: "Zusammenhang subjektiver Beurteilungen von Konzertsälen mit raumakustischen Kriterien", *Acustica* 45 (1980), 256–68.
Leipp, Emile: *La machine à ecouter: essai de psycho-acoustique* (Paris: Masson, 1977).
– *Acoustique et musique,* 4e éd. (Paris: Masson, 1984).
– "A propos du Grand Orgue de la cathédrale de Strasbourg: Pourquoi des orgues en nids d'hirondelles", *Bulletin de la Cathédrale de Strasbourg,* 17 (1986), 17–30.
Lenz, Gunhild: "Zum Wandel des Konzertrepertoires in Dublin 1750–1800", *Mf* 46 (1989), 307–15.
Leonardo da Vinci: *Literary works,* ed. P. Richter (Oxford: OUP, 1939; 1st ed. 1883).
– *Leonardo da Vinci, das Lebensbild eines Genies,* ed. Sandro Piantanida et al. (Wiesbaden, 1972).
– *Il Codice Arundel 263 nella British Library (1508);* facs. ed. Carlo Pedretti; transcription and critical comment by Carlo Vecce (Florence, 1998).
Leopold, Silke: *Claudio Monteverdi* (Laaber, 1982).
– *Stefano Landi* (Hamburg, 1976).
Lesure, François: *L'Opéra Classique Français: XVIIe et XVIIIe siècles* (Geneva, 1972).
Levarie, Siegmund and Ernst Levy: *Tone: a study in musical acoustics,* 2nd ed. (Kent, Ohio, 1980).
Lewcock, Ronald: "Acoustics I: Rooms", *NG,* vol. 1, 43–67.
Lewcock, Ronald et al.: "Acoustics, § I, 8–9: Room acoustics: Medieval times; Renaissance and Baroque periods", *NG^2,* vol. 1, 82–87.
Lexikon der Weltarchitektur, ed. Nikolaus Pevsner, Hugh Honour, 3rd enlarged ed. (Munich, 1992).
Ley, Stephan: "Beethoven und Fürst Lobkowitz", *Atlantis* 9 (1937), 748–52.
– "Schloss Grätz bei Troppau: Beethoven und die fürstliche Familie Lichnowsky", *Atlantis* 9 (1937), 59–64.
Lichte, H. → Fischer, Fritz
Lichtenhahn, Ernst: "Musikalische Interpretation – ein romantisches Konzept", *Musikalische Interpretation: Reflexionen im Spannungsfeld von Notentext, Werkcharakter und Aufführung: Symposion zum 80. Geburtstag von Kurt von Fischer, Zürich 1993,* ed. Dorothea Baumann and Joseph Willimann (Berne, 1999), 107–14.

– Introduction to *Bearbeitung in der Musik, Colloquium Kurt von Fischer zum 70. Geburtstag, Bern 1983,* ed. Dorothea Baumann (Berne, 1986), (Schweizer Jb. für Musikwissenschaft, Neue Folge 3 (1983)).
– "*Gefährdete Innenräume:* Perspektiven des Musikdenkens im späten 18. und 19. Jahrhundert", *Schweizerische Musikpädagogische Blätter* 79 (1991), 139–46.
– "Das bürgerliche Musikfest im 19. Jahrhundert", *Stadt und Fest* (Stuttgart, 1987), 161–79.
– " Musik und Raum: Gesellschaftliche und ästhetische Perspektiven zur Situation um 1800", *Musik und Raum* (Mainz, 1989), 8–19.
Liebscher, Julia: "Händels Kammerduette: Höhepunkt und Wandlung einer Gattung", *Kongressbericht Stuttgart 1985,* I (Kassel. 1987), 59–65.
Le lieu théâtrale à la Renaissance, ed. Jean Jacquot (Paris: CNRS, 1986), (Colloque International, Royaumont 1963).
Lindsay, Robert Bruce: *The nature of Physics* (Providence, 1968).
– *Historical introduction to Lord Rayleigh: Theory of Sound,* (New York, NY: Reprint Dover, 1945), v–xlii.
– "The Story of acoustics", *JASA* 39 (1966), 629–44.
Lippmann, Walter: *Public Opinion* (New York, NY, 1922).
Loesch, Heinz von: "Musikwissenschaft nach 1945", *MGG*^{2}, vol. 6, col. 1807 ff.
The London Stage 1660–1800, ed. William Van Lenep, index, ed. B. R. Schneider, foreword by George Winchester Stone (Carbondale: Southern Illinois UP, 1960–79).
The London theatre world, ed. Robert D. Hume (London and Amsterdam, 1980).
Longsdale, Roger: *Dr. Charles Burney: a literary biography* (Oxford: Clarendon, 1986).
Lottermoser, Werner: "Nachhallzeiten in Barockkirchen", *Acustica* 2 (1952), 109.
Loukomski, Georgii Kreskent: *Les Théâtres anciens et modernes* (Paris: Firmin-Didot, 1935).
Lütolf, Max → Scarlatti, Alessandro
– Corelli, Arcangelo, music editions → see 3.1.2
Lysons, xxx:
MacClintock, Carol: *Readings in the history of music in performance* (Bloomington and London: Indiana UP, 1979).
McKinnon, John W. → Brown, Howard M.
Mackintosh, Iain: *Architecture, actor, and audience* (London and New York, NY, 1993).
Mahling, Christoph-Hellmut: "Zur Aufführung und Stellung der Oratorien Händels im 19. Jahrhundert", *Kongressbericht Stuttgart 1985,* I (Kassel. 1987), 264–73.
Mainwaring, John: *Memoirs of the life of the late George Frederic Handel* (London, 1760).
– *Georg Friedrich Händel, Biographie, Briefe und Schriften,* ed. H. and H. E. Müller von Asow (Lindau, 1949).
Makkreel, Rudolf A.: "Cassirer zwischen Kant und Dilthey", *Ernst Cassirers Werk und Wirkung,* ed. D. Frede and R. Schmücker (Darmstadt: WBG, 1997), 145–62.
– "Wilhelm Dilthey", *Stanford Encyclopedia of Philosophy*, electronic version (2008).
Malipiero, Gian Francesco: *Claudio Monteverdi* (Milan, 1929).
Mancini, Franco; Muraro, Maria Teresa; Povoledo, Elena: *I Teatri del Veneto:* 1. *Venezia, teatri effimeri e nobili imprenditori;* 2. *Venezia e il suo territorio*; 3. *Verona Vicenza Belluno* (Venice, 1985–1996).
Mander, Raymond: *The lost theaters of London* (London, 1968).
Marchesi Cappai, Carlo: *Acustica applicata all'architettura,* 2nd ed. (Milan: Hoepli, 1939).
Mariani, Carlo: *Progetto di un teatro moderno e nozioni generali intorno la costruzione di questo* (Milan, 1846).
Markowitsch, Hans. J.: *Neuropsychologie des Gedächtnisses* (Göttingen, 1992).

Marpurg, Friedrich W.: *Anleitung zur Musik überhaupt und zur Singekunst* (Berlin, 1763, repr. 1975).

Marshall, A. H.: "A note on the importance of room cross section in concert halls", *Journal of Sound Vibration* 5 (1967), 100–12.

Marshall, Robert L.: "Bach's 'Orchestre'", *EM* 13 (1985), 176–79.

Marx, Hans Joachim: "Die Musik am Hofe Pietro Kardinal Ottobonis unter Arcangelo Corelli", *Studien zur italienisch-deutschen Musikgeschichte* (Graz, 1968), vol. 5, 104–77.

– "Ein Beitrag Händels zur Accademia Ottoboniana in Rom", *Hamburger Jb. für Musikwissenschaft* 1 (1976), 69–86.

– *Die Überlieferung der Werke Arcangelo Corellis: Catalogue raisonné* (Köln, 1980), (A. Corelli, Historisch-krit. Gesamtausgabe, Suppl.)

– "Händel in Rom – seine Beziehung zu Benedetto Card. Pamphilj", *Händel-Jb.* 29 (1983), 107–18.

– "Die Musik am Hofe Pietro Kardinal Ottobonis unter Arcangelo Corelli", *Studi Musicali* 12 (1983), 121–87.

– "Zur Kompositionsgeschichte von Händels Pastoralkantate Apollo e Dafne HWV 122", *Göttinger Händel-Beiträge* 1 (1984), 77–92.

– "Die Instrumentation in Händels frühen italienischen Werken", *Göttinger Händel Beiträge* 2 (1986), 77–92.

– "Italienische Einflüsse in Händels früher instrumentalen Ensemblemusik", *Studi Musicali* 16 (1987), 381–93.

– "The instrumentation of Handel's early italian works", *EM* 16 (1988), 406–505.

– *Händel und seine Zeitgenossen. Eine biographische Enzyklopädie* (Laaber, 2008).

– Corelli, Arcangelo, music editions → see 3.1.2

Mattheson, Johann: *Der vollkommene Kapellmeister* (Hamburg, 1739).

Mattia, Roberto: "Feste musicali a palazzo Bonelli", *Palazzo Valentini* (Rome, 1985), 213–30.

Mazzola, Guerino: *Mathematische Musiktheorie* (Basel, 1989).

McVeigh, Simon: *Concert life in London from Mozart to Haydn* (Cambridge. 1993).

Mee, John H.: *The oldest music room in Europe* (London, 1911).

Meek, Harold Alan: *The Synagogue* (London, 1995).

Meinel, Rüdiger: *Das Odeion* (Frankfurt, 1980).

Mersenne, Marin: *Harmonie universelle,* 3 vols. (Paris, 1636; repr. Paris: CNRS, 1963).

Meyer, Erwin and Lothar Cremer: "Über die Hörsamkeit holzausgekleideter Räume", *Zeitschrift für Technische Physik* 14 (1933), 500.

Meyer, Erwin and Ernst Georg Neumann: *Physikalische und Technische Akustik* (Braunschweig, 1974).

Meyer, Jürgen: "Raumakustik und Orchesterklang in den Konzertsälen Joseph Haydns", *Acustica* 41, 3 (1978), 145–62.

– "Zum Hörerlebnis des Musikers im Konzertsaal", *Tiefenstruktur der Musik, Festschrift Fritz Winckel,* ed. Carl Dahlhaus (Berlin, 1982), 157–94.

– "Raumakustik und Orchesterklang: zur Aufführungspraxis der Sinfonien Joseph Haydns", *Musik und Raum* (Mainz. 1989), 20–39.

– *Akustik und musikalische Aufführungspraxis,* 4th enl. ed. (Frankfurt a. M., 1999; 5th ed. Bergkirchen, 2004).

– *Acoustics and the Performance of Music,* 5th ed. (New York, NY: Springer, 2009).

Meyer, Martin: "Functions of the left and right posterior temporal lobes during segmental and suprasegmental speech perception", *Zeitschrift für Neuropsychologie* 19 (2008), 101–14.

Milizia, Francesco: *Del teatro* (Rome: Casaletti, 1771).

– *Trattato completo formale e materiale del teatro* (Venice: Pasquali, 1794).

– *Principj di architettura civile,* 3 vols. (Bassano, 1875).
Milligan, Thomas B.: *The concerto and London's musical culture in the late eighteenth century* (Ann Arbor, 1983).
Miura, Tanetoshi → Nakayama, Takeshi
Mohrmann, K. "Lautheitskonstanz im Entfernungswechsel", *Zeitschrift für Psychologie* 145 (1939), 146–99.
Moir, James: *High Quality Sound Reproduction* (London, 1958).
Möller, Dirk: *Besetzung und Instrumentation in den Opern Georg Friedrich Händels* (Frankfurt, 1988; Europäische Hochschulschriften 36/38).
Mommertz, Eckard: "Einige Messungen zur streifenden Schallausbreitung über Publikum", *Acustica* 79 (1993), 42–52.
Mommertz, Eckard, Gerhard Müller and M. Dörle: "Messung der Einfallsrichtungen in Räumen mit Hilfe einer Korrelationsanalyse", *Fortschritte der Akustik – DAGA* 96, ed. T. Portele and W. Hess (Oldenburg, 1996), 174–75.
Mönck, Willi: *Holzbau,* 12th ed. (Berlin. 1995).
Mono – Stereo – Quadro: die Aufnahme und Wiedergabe von Musik, ed. Dorothea Baumann et al. (Basel: Radio DRS, 1986).
Monteverdi, Claudio: *Briefe,* transl. by Sabine Ehrmann and Hans-Horst Henschen, ed. Denis Stevens (Munich and Zurich, 1989); Engl. translation: *The letters of Claudio Monteverdi,* by Denis Stevens (Cambridge: Cambridge UP, 1980).
Moretti, Laura: "Architectural Spaces for Music: Jacopo Sansovino and Adrian Willaert at St Mark's", *EMH* 23 (2004), 153–84.
→ Howard, Deborah
Morris, Richard and Keith Spence: *Cathedrals and abbeys of England and Wales: the building church, 600–1540* (London, 1979).
Morrow, Mary Sue: *Concert life in Haydn's Vienna: aspects of developing and social institution* (New York, NY: Stuyvesant, 1988), (Sociology of Music Series 7).
Moss, F. and Kurt Wiesenfeld: "The benefits of background noise – stochastic resonance", *Scientific American* 273, 8 (1995), 50–53.
Mosse, George Lachmann: *Die Nationalisierung der Massen* (Frankfurt a. M., 1976).
Motta, Fabricio Carini: *Trattato sopra la struttura de theatri e scene* (Guastalla, 1676).
Moynet, M. Jules, Allan S. Jackson et al.: *French Theatrical Production in the Nineteenth Century. L'envers du théâtre* (New York, NY: Binghanton, 1976) (American Theatre Association).
Mozart, Wolfgang Amadeus: Music editions → 3.1.3
– *Briefe und Aufzeichnungen.* Gesamtausgabe ed. W. A. Bauer and Otto Erich Deutsch, 4 vols. (Kassel, 1962–63): Kommentarbände, 2 vols. ed. J. H. Eibl. (Kassel, 1975).
Mozart und seine Welt in zeitgenössischen Bildern, ed. Etto Erich Deutsch (Kassel, 1961) (W. A. Mozart, *Neue Ausgabe sämtlicher Werke* X, 32).
Müller, Helmut A. → Cremer, Lothar
Müller, Karlheinz: "The democratization of acoustics", *The Richard Wagner Festival Theatre Bayreuth* (Cologne, 2007), 174–94.
Müller, Volkmar: "Das akustische Geheimnis des alten Gewandhaussaales in Leipzig", *Zeitschrift für Instrumentenbau* (1894–95), 478 f.
Müller-Blattau, Joseph: "Händel", *MGG,* vol. 5 (1956), 1274–86.
Müller von Asow, H. and H. E. → Mainwaring, John
Music and the brain: studies in neurology of music, ed. Macdonald Critchley and R. A. Henson (London, 1977).
"Music in Large Buildings". *The Chamber's Journal* 1 (1854), 35.

Musical Life in Europe 1600–1900. Circulation, Institutions, Representation (Strassburg: European Science Foundation, 2003 f.); vol. 9: *Espaces et lieux de concert en Europe, 1700–1920: architecture, musique, société* (Berlin, 2008).
The Musical World (London, 1836 f.).
Musik – Gehirn – Spiel. Beiträge zum 4. Karajan Symposium Vienna 1988, ed. Hellmuth Petsche (Basel, 1989).
Die Musik in Geschichte und Gegenwart (MGG), ed. Friedrich Blume (Kassel. 1949–79).
Die Musik in Geschichte und Gegenwart (MGG²), ed. Ludwig Finscher, 2nd ed. (Kassel. 1992–2007).
Musik und Raum, eine Sammlung von Beiträgen aus historischer und künstlerischer Sicht zur Bedeutung des Begriffes "Raum" als Klangträger für die Musik, ed. Thüring Bräm (Basel, 1986).
Musik und Raum: vier Kongressbeiträge und ein Seminarbericht, ed. Marietta Morawska-Büngeler (Mainz. 1989), (Veröff. Inst. für Neue Musik und Musikerziehung Darmstadt 30).
Musikpsychologie, ein Handbuch, ed. Herbert Bruhn, Rolf Oerter (Reinbek b. Hamburg, 1993).
Musikpsychologie, das neue Handbuch, ed. Herbert Bruhn, Reinhard Kopiez, Andreas C. Lehmann (Reinbek b. Hamburg, 2008).
Myers, Howard Gregory: *Andrea Palladio: activities with the Accademia Olimpica.* Dissertation Kent State University (1978), Ann Arbor Microfilm No. 78-21088.
Myers, Robert Manson: *Early moral criticism of Handelian oratorio* (Williamsburg, VA, 1947).
– *Handel's Messiah: a touchstone of taste* (New York, NY, 1948).
Nagler, Alois Maria: *A source book in theatrical history* (New York, NY: Dover Paperback, 1952).
Nakayama, Takeshi and Tanetoshi Miura et al.: "Subjectiv assessment of multichannel reproduction", *JAES* 17 (1969), repr. *Mono – Stereo – Quadro* (Basel, 1986), 161–73.
Nasshan, Kimberly, G. Schupp and U. Stephenson: "Echtzeitauralisation als Element optischer und akustischer virtueller Realität", *Fortschritte der Akustik – DAGA 95* (Oldenburg, 1995), 307–10.
Naredi-Rainer, Paul von: *Architektur und Harmonie: Zahl, Mass und Proportion in der abendländischen Baukunst* (Köln, 1982).
– "Musiktheorie und Architektur", *Geschichte der Musiktheorie*, ed. Fred Zaminer (Darmstadt, 1985), vol. 1, 149–74.
Neuro- und Sinnesphysiologie, ed. Robert F. Schmidt (Berlin. 1993).
New Grove Dictionary of Musical Instruments, 3 vols., ed. Stanley Sadie (London, 1984).
New Grove Dictionary of Opera, 4 vols., ed. Stanley Sadie (London, 1992).
New Grove Dictionary of Music and Musicians (NG), 20 vols., ed. Stanley Sadie (London, 1980).
New Grove Dictionary of Music and Musicians (NG²), 29 vols., ed. Stanley Sadie and John Tyrrell (London, 2001).
Niccolini, Antonio: *Alcune idee sulla risonanza del teatro dell'architetto C. A. Nicolini* (Naples, 1816).
Nicoll, Allardyce: *The development of the theatre: a study of theatrical art from the beginnings to the present day,* 5th ed. (London, 1966).
Niederrheinische Musik-Zeitung für Kunstfreunde und Künstler, ed. L. Bischoff (Cologne, 1853–67).
Nottebohm, Gustav: "Aufführungen der Dritten und Neunten Sinfonie in Beethovens Anwesenheit", *AMZ* III, 11 (1876), 49 and 65.
Olson, Harry F.: *Music, Physics and Engineering,* 2nd rev. ed. (New York, NY, 1967).
– "Microphones for recording", *Journal of the AES* 25, 10/11 (1977), 676–84.

Onorati, F.: "La confraternità del Gonfalone", *Roma* 6 (1993), 24–25.
Oosting, Thomas: *The teatro Olimpico: an analysis of design sources.* Dissertation, Iowa (1970).
The orchestra: origins and transformations, ed. Joan Peyser (New York, NY: Scribner, 1986).
Orchestra → Westrup, Jack
Orel, Alfred → Wiener Beethoven-Buch
Orgel, ed. Alfred Reichling (Kassel etc. and Stuttgart, 2001), (181. Veröffentlichung der GdO).
Orrell, John: *The theatres of Inigo Jones and John Webb* (Cambridge: Cambridge UP, 1985).
Ouvard, René: *L'architecture harmonique ou application de la doctrine des proportions de la musique à l'architecture* (Paris: De la Caille, 1679).
Owen, Barbara and Peter Williams: "Organ, § IV, 6: The church organ, 1100–1450", *NG²,* vol. 18, 588.
Il palazzo dal Rinascimento a oggi, ed. Simonetta Valtieri (Rome, 1990).
Palazzo della Provincia, ed. Gennaro Farina (Rome: Palazzo Valentini, 1985).
Palazzo Valentini, ed. Giulio Carlo Argan (Rome: Editalia, 1984).
Palisca, Claude V.: "Interactions between acoustics and musical practice in the Renaissance", *Tiefenstruktur der Musik, Festschrift Fritz Winckel* (Berlin, 1982), 119–28.
Palladio, Andrea: *I Quattro Libri dell'Architettura* (Venice, 1570); German transl. Andreas Beyer and Ulrich Schütte (Zurich, 1983).
Patte, Pierre: *Essai sur l'architecture théâtrale ou l'ordonnace la plus avantageuse à une salle de spectacle relativement aux principes de l'optique et de l'acoustique [...]* (Paris, 1782).
→ Landriani, Paolo
Pelorsen, X. → Vian, J. P.
Pepys, Samuel: *Diary*: Commentary on the diary's musical references, ed. R. Latham, W. Matthews (London, 1970–83).
Performance practice, ed. Howard M. Brown and Stanley Sadie, 2 vols. (London, 1989).
Petsche, Hellmuth: "Zerebrale Verarbeitung", *Musikpsychologie* (Hamburg, 1993), 630–38.
Pevsner, Nicolaus: *A history of building types* (London, 1976).
Peyser, Joan → Orchestra
Pierce, John R.: *Klang: Musik mit den Ohren der Physik* (Heidelberg, 1985).
– *The science of musical sound* (New York, NY, 1992).
Pinthus, Gerhard: *Das Konzertleben in Deutschland. Ein Abriss seiner Entwicklung bis zum Beginn des 19. Jahrhunderts* (Hamburg, 1935).
Plenge, Georg: "Ausgewählte Aufsätze zum Thema Kunstkopfstereofonie, Vorwort", *25 Jahre RTM* (RTM Sept. 1981), 3–4.
Popper Karl, and John Eccles: *The self and its brain: an argument for interactionism* (Heidelberg, 1977); *Das Ich und sein Gehirn,* German transl. Angela Hartung and Willy Hochkeppel (Munich, 1982).
Poley, Arthur F. E.: *St. Paul's Cathedral,* London 1927
Porter, Roy: *English Society in the Eighteenth Century,* Allen Lane (London, 1982).
Portoghesi, Paolo: *Francesco Borromini: Baumeister des römischen Barock.* German transl. Julia Schlechta (Stuttgart, 1977).
Poultney, David: "Alessandro Scarlatti and the transformation of the oratorio", *MQ* 59 (1973), 584–601.
– *The oratorios of Alessandro Scarlatti: their lineage, milieu, and style,* Ph. D. dissertation, University of Michigan (1968).
Prante, Holger et al.: "Experimente zur binauralen Lautstärkeempfindung im diffusen Schallfeld", *Fortschritte der Akustik, DAGA '90* (Bad Honnef, 1990), 711–14.
Praschl-Bichler, Gabriele: *Wien speziell: Architektur des Barock* (Vienna: Brandstätter, 1990).
Preussner, Eberhard: *Die musikalischen Reisen des Herrn Uffenbach* (Kassel, 1949).

– *Die bürgerliche Musikkultur. Ein Beitrag zur deutschen Musikgeschichte des 18. Jahrhunderts,* 2nd ed. (Kassel, 1950).
Price, Curtis A.: "Italian opera and Arson in late eighteenth-century London", *JAMS* 42 (1989), 55–107.
– *Music in the Restoration Theatre* (Ann Arbor: UMI, 1979).
Pritchard, Brian W. → Caldara, Essays
Pütz, Jean.→ *HiFi, Ultraschall und Lärm. Die Welt des Schalls*
Purkis, Hubert John: "The reverberation time of some English cathedrals", *Bulletin of the Institute of Physics* 14 (1963), 8.
Quantz, Johann Joachim: *Versuch einer Anweisung die Flöte traversière zu spielen,* 3rd ed. (Breslau, 1789; repr. Kassel. 1953).
– *On Playing the Flute,* 2nd ed., transl. Edward R. Reilly (Northeastern UP, 2001).
Quellentexte zur Konzeption der europäischen Oper im 17. Jahrhundert, ed. Heinz Becker et al. (Kassel, 1981).
Racek, Jan: "Oratorien und Kantaten von Georg Friedrich Händel auf dem Mährischen Schlosse von Námest", *Händel-Jb.* 6 (1960), 175–93.
Rackwitz, Werner: "Bemerkungen zur theatralischen Darstellung oratorischer Kompositionen Händels", *Händel-Jb.* 35 (1989), 55–60.
– "Dramatische Aspekte in Händels Oratorium 'Messiah'", *Händel-Jb.* 37 (1991), 85–89.
Ramber, Bjorn and Kristin Gjesdal: "Hermeneutics", *Stanford Encyclopedia of Philosophy,* electronic version (2005).
Ravizza, Victor: "Musikalischer Satz und räumliche Dispostion: zur frühen venezianischen Doppelchörigkeit", *Festschrift Max Lütolf zum 60. Geburtstag* (Basel, 1994), 117–34.
Raynor, Henry: *A social history o music from the Middle Ages to Beethoven* (London, 1962).
Read, Oliver and Walter L. Welch: *From Tin Foil to Stereo: Evolution of the Phonograph,* (Indianapolis. 1959; 2nd ed. 1976).
Reallexikon der Akustik, ed. Michael M. Rieländer (Frankfurt a. M., 1982).
Rees, Abraham (ed.): *The Cyclopedia,* 39 vols. (London, 1802–20).
Reichardt, Walter: *Grundlagen der technischen Akustik* (Leipzig, 1968).
– *Grundlagen der Beschallungstechnik* (Stuttgart, 1981).
Reichling, Alfred → *Orgel*
Reid of Edinburgh, D. B.: "On the construction of Public Buildings in Reference to the communications of sound", *Transactions fo the British Association* (London, 1836).
Reinle, Adolf: "Architekturlehre", *Renaissance der Wissenschaften* (Zurich, 1981), 212.
Répertoire d'Art et d'Archéologie, (RAA), éd. Institut de l'Information Scientifique (Paris: CNRS, 1910–89).
Répertoire International de la Littérature d'Art, (RILA), ed. J. Paul Getty Trust, S. and F. Clark Art Institute (Williamstown, MA, 1975–89). Continuation → *Bibliography of the History of Art (BHA)*
Répertoire International de Littérature Musicale, (RILM), ed. RILM Center, CUNY (New York, NY, 1967 f.).
Répertoire International de la Presse Musicale; Retrospective Index to Music Periodicals, (RIPM), ed. H. Robert Cohen (College Park: University of Maryland, 1988 f.).
Richter, P.→ Leonardo da Vinci
Rieländer, Martin → Reallexikon
Riethmüller, Albrecht: "Wellingtons Sieg op. 91", *Beethoven, Interpretationen seiner Werke* (Darmstadt, 1994; 2nd ed. 1996), vol. 2, 39.
Rifkin, Joshua: "More (and less) on Bach's orchestra", *Performance Practice Review* 4 (1991), 5–13.

"Risuonanza", *Dizionario e bibliografia della musica,* ed. Pietro Lichtenthal (Milan, 1836), vol. 2, 158.
Roederer, Juan G.: *Introduction to the physics and psychophysics of Music* (Berlin, 1973; 2nd ed. 1975).
– *Physikalische und psychoakustische Grundlagen der Musik* (Berlin, 1977).
Rösing, Helmut: "Musikpsychologie", *MGG²,* vol. 6, col. 1551–1601.
Roper, E. S.; "Music at the English Chapel Royal c. 1135 – present day". *PRMA* 54 (1927/28), 19–33.
Roselli, Piero: *I teatri di Firenze* (Florence, 1978).
Rosselli, John: "La Mara, Getrud Elisabeth Schmeling", *New Grove of Opera* (London, 1984), vol. 3, 198.
Roubo, André-Jacob: *Traité de la construction des théâtres et des machines théâtrales* (Paris, 1780).
Rundfunktechnische Mitteilungen (RTM): Fünfundzwanzig Jahre, ausgewählte Aufsätze zum Thema Kunstkopfstereofonie. Funkausstellung Berlin, Tonmeistertagung (Munich and Hamburg, 1981).
Rudé, George: *Hanoverian London 1714–1808,* (Berkeley CA: UCL Press, 1971).
Russell, John Scott: "Treatise on sightlines and seating", *Edinburgh New Philosophical Journal* 27 (1838).
Russell, Terence: "Acoustics", *The built environment, a subject index 1800–1960* (Godstone, 1989), vol. 2, 344–68.
Saalman, Howard: *Filippo Brunelleschi, The Cupola of Santa Maria del Fiore* (London: Zwemmer, 1980).
Sabine, Wallace Clement W.: *Collected papers on acoustics,* (Cambridge MA, 1922; repr. New York, NY 1964; repr. Los Altos, 1992).
Sachs, Edwin O.: *Modern opera houses and theatres.* 3 vols. (London, 1896–98; repr. New York, NY, 1968).
Sadie, Stanley: "Concert life in 18th century England", *PRMA* 85 (1958/59), 17–30.
Saeltzer, Alexander: *A treatise on acoustics in connection with ventilation; and an account of the modern and ancient methodsof heating and ventilation* (New York, NY, 1873).
Saint, Andrew John: "The three Covent Gardens", *MT* 123 (1982), 826–31.
Salmen, Walter: *Musikleben im 16. Jahrhundert* (Leipzig, 1976), (Musikgeschichte in Bildern III/9).
– *Das Konzert, eine Kulturgeschichte* (Munich, 1988).
Saunders, George: *A treatise on theatres including some experiments in sound* (London, 1790; repr. New York, NY, 1968).
Savart, Félix: "Recherches sur les vibrations de l'air", *Annales de chimie et physique* 24 (1823), 56–88.
Scamozzi, Vincenzo: *Dell'idea della archittetura universale* (Venice, 1615; repr. Ridgewood, 1964).
[Scarlatti] *Alessandro Scarlatti und seine Zeit,* ed. Max Lütolf (Berne, 1995).
Schaal, Richard: "Konzertwesen", *MGG,* vol. 7 (Kassel, 1958), col. 1587–1605.
– "Feste und Festspiele", *MGG,* vol. 4 (Kassel, 1955) col. 104–28.
Schafer, R. Murray: *Klang und Krach: eine Kulturgeschichte des Hörens* (Frankfurt a.M., 1988).
Schaupert, Carl → Schmitt, Eduard
Schiavo, Armando: *Il palazzo della Cancelleria* (Rome, 1964).
Schiess, Ernst: *Nachlass,* Eidgenössisches Archiv für Denkmalschutz Berne.
Schimanek, H.: "Zur Frühgeschichte der Akustik", *Zeitschrift für technische Physik* 17 (1936), 500 f.

Schmidt, F.: *Festigkeitsnachweis für eine von Dreiecksbindern getragene Satteldachkonstruktion* (Köln, 1971–72); repr. Meinel, *Das Odeion* (Frankfurt, 1980), appendix.
Schmitt, Eduard, Carl Schaupert and Conradin Walther: *Künstler-Ateliers, Kunstakademien und Kunstgewerbeschulen, Konzerthäuser und Saalbauten* (Stuttgart, 1901), (Handbuch der Architektur IV/6/3).
Schneider, Otto: "Musik im Augarten", *ÖMZ* 21 (1966), 15–19.
Schoelcher, Victor: *The life of Handel* (London, 1857).
Scholes, Percy A.: *The mirror of music 1844–1944,* 2 vols. (Oxford, 1947).
Schreiber, Ottmar: *Orchester und Orchesterpraxis in Deutschland zwischen 1780 und 1850* (Berlin, 1938).
Schroeder, Manfred R., D. Gottlob and K. F. Siebrasse: "Comparative study of European concert halls: correlation of subjective preference with geometric and acoustic parameters", *JASA* 56 (1974), 1195–201.
Schroeder, Manfred R.: *Music perception in concert halls* (Stockholm, 1979).
Schünemann, Georg: *Geschichte des Dirigierens* (Leipzig, 1913).
Schwab, Heinrich W.: *Das Konzert: öffentliche Musikdarbietung vom 17. bis 19. Jahrhundert* (Leipzig, 1971) (Musikgeschichte in Bildern IV/2).
Scott, Hugh Arthur: "London's earliest public concerts", *MQ* 22 (1936), 194–209.
Seiffert, Max: "Die Mannheimer Messias-Aufführung 1777", *Peters-Jb.* (1916), 61–71.
Selfridge-Field, Eleanor → Westrup, Jack
– "Introduction": *Performance practice: music after 1600,* ed. Howard M. Brown and Stanley Sadie (London, 1989), 3–19.
Semper, Manfred: *Handbuch der Architektur IV/6/5: Theater,* ed. Josef Durm et al. (Stuttgart, 1904).
– *Das Münchner Festpielhaus: Gottfried Semper und Richard Wagner* (Hamburg, 1906).
Serauky, Walter: "Johann Adam Hiller als Erwecker der Händel-Tradition im 18. Jahrhundert", *Festschrift zum 175 jährigen Bestehen der Gewandhauskonzerte 1781–1956* (Leipzig, 1956), 36–39.
Serlio, Sebastiano: *Il Secondo libro di perspettiva = Le Second livre de perspective,* French translation Iehan Martin (Paris, 1545).
Shapiro, Alexander H.: "Drama of an infinitely superior nature: Handel's early English oratorios and the religious sublime", *ML* 74 (1993), 215–45.
Shankland, R. S. and H. K.: "Acoustics of St. Peter's and Patriarchal Basilicas in Rome", *JASA* 1 (1971), 2.
Shaw, George Bernard: *Music in London 1890–94,* rev. ed., 3 vols. (London, 1932).
Shaw, Watkins: *The story of Handel's Messiah 1741–84* (London, 1963).
– *A textual and historical companion to Handel's Messiah* (London, 1965).
→ Donald Burrows
Siegmund-Schultze, Dorothea: "Zur gesellschaftlichen Situation in London zur Zeit Händels", *Händel-Jb.* 32 (1986), 85–98.
Siegmund-Schultze, Walther: "Über die ersten Messias-Aufführungen in Deutschland", *Händel-Jb.* 6 (1960), 51–109.
– *G. F. Händel: sein Leben, sein Werk* (Munich, 1984).
Sheppard, F. H. W. → Survey of London
Skoda, Rudolf: *Das Gewandhaus Leipzig, Geschichte und Gegenwart* (Berlin, 1984); enl. ed.: *Die Leipziger Gewandhausbauten. Konzertgebäude im internationalen Vergleich* (Berlin 2001).
Skudrzyk, Eugen: *Die Grundlagen der Akustik* (Vienna, 1954); Engl. translation: *Foundations of Acoustics* (Vienna, 1971).

Smekal, Richard: *Das alte Burgtheater 1776–1888* (Vienna, 1916).
Smith, E. Baldwin: *The dome* (Princeton, 1974).
Smither, Howard E.: "'Messiah' and Progress in Victorian England", *EM* 13 (1985), 339–48.
– *A history of the oratorio in the Baroque era,* 2 vols. (Chapel Hill, 1977).
– *The oratorio in the Classical era* (Oxford, 1987).
Sonntag, Ernst → Fasold, Wolfgang
"Sound", *Encyclopedia Metropolitana,* re-issue (London, 1849), vol. 25, 752.
Sound recording practice, ed. John Borwick (London, 1976).
Specchi, A. → Falda, G. B.
Spohr, Louis: *Selbstbiographie* (Cassel and Göttingen, 1860–61).
Spelda, Antonín → Burghauser, Jarmil
Springer, Sally, and Georg Deutsch: *Left brain, right brain,* 4th ed. (San Francisco, 1993).
Stadlen, Peter: "Zu Schindlers Fälschungen in Beethovens Konversationsheften", *ÖMZ* 32 (1977), 246–52.
– "Schindler und die Konversationshefte", *ÖMZ* 34 (1979), 2–18.
Steinhauser, Monika: *Die Architektur der Pariser Oper* (Munich, 1978).
Steinke, Gerhard → Hoeg, Wolfgang
Stone, George Winchester: *The London Stage 1660–1800,* ed. William Van Lenep, foreword by George Winchester Stone (Carbondale: Southern Illinois UP, 1960–79).
Storia dell'Opera Italiana 5: La Spettacolarità, ed. Lorenzo Bianconi e Giorgio Pestelli (Turin, 1988).
Straub, Hans: *Die Geschichte der Bauingenieurkunst: ein Überblick von der Antike bis in die Neuzeit,* 4th ed. (Basel, 1992).
Strohm, Reinhard: "Haendel in Italia: nuovi contributi", *RIM* 9 (1974), 169.
– "Italienische Opernarien des frühen Settecento (1720–30)", *Analecta Musicologica* 14 (1976).
– "Il viaggio di Haendel come esperienzia europea", *Festival Vivaldi, Haendel e Italia* (Milan, 1981), 67.
– *Essays on Handel and Italian opera* (Cambridge, 1985).
– "Händel und Italien, ein intellektuelles Abenteuer", *Göttinger Händel-Beiträge* 5 (1993), 5–43.
– "A context for Griselda: the Teatro Capranica, 1711–24", *Alessandro Scarlatti und seine Zeit,* ed. Max Lütolf (Berne, 1995), 79–115.
Stromer, Wolfgang von: "Theophilus presbyter, *Diversarum artium schedula",* rev. ed., introduction with partial German transl. W. Theobald (Düsseldorf, 1984).
Stumpf, Carl: *Tonpsychologie* (Leipzig, 1883–90).
Sulzer, Johann Georg: "Kirchenmusik", *Allgemeine Theorie der schönen Künste,* 2nd ed. (Leipzig, 1779), vol. 3, 21 f.
Summers, William John: "George Frideric Handel in Rome as seen from the vantage point of the 'confraternità dei musici di Roma'", *Kongressbericht Stuttgart* (Kassel, 1987), vol. 2, 84–103.
Sundberg, Johan: *The science of musical sounds* (San Diego, 1991).
– "The Acoustics of the Singing Voice", *Scientific American* 236 (1977), 82.
Survey of London, ed. F. H. W. Sheppard (London, 1970).
Taccani, Francesco: *Sull'origine, la significatione e gli usi che si attribuiscono ai membri architettonici* (Milan, 1829).
– *Sulla forma della platea e del Proscenio di un teatro più propria alla propagazione del suono e sulla materia più atta a rinforzarlo e a sostenerlo* (Milan: Monti, 1840).
Talbot, Michael: "Corelli", *NG,* vol. 4, 769.

Tamburini, Elena: *Il luogo teatrale nella trattatistica italiana dell '800: dall'utopia alla prassi borghese* (Rome, 1984).
Taylor, Charles: *Der Ton macht die Musik: Die Wissenschaft von Klängen und Instrumenten* (Braunschweig and Wiesbaden, 1994).
– *Exploring music: the science and technologiy of tones and tunes* (Bristol, 1994).
Temperley, Nicholas: "Bach Revival", *NG,* vol. 1, 883–86.
– *Haydn: the Creation* (Cambridge, 1991; Cambridge Music Handbooks).
– "London: Charities and proprietary chapels (§ I, 5)", *NG,* vol. 11, 149.
Tevo, Zaccaria: *Il musico testore* (Venice, 1706).
Thayer, Alexander Wh.: *Ludwig van Beethovens Leben.* German ed. Hugo Riemann, 5 vols. (Leipzig, 1908).
Theatres in Europe: Restoration and Georgian England 1660–1788, ed. David Thomas (Cambridge, 1989).
Theile, Günther: "Zur Theorie der optimalen Wiedergabe stereofoner Signale über Lautsprecher und Kopfhörer", *25 Jahre RTM* Sept. (1981), 32–47.
Theophilus Presbyter: *Diversarum artium schedula. Technik des Kunsthandwerks im Zwölften Jahrhundert.* repr. of rev. ed. with partial German translation and comment by Wilhelm Theobald, introduction Wolfgang von Stromer (Düsseldorf: VDI Verlag, 1984).
Thomas, David → *Restoration and Georgian England*
Thompson, Homer A.: "American Excavations in the Athenian Agora: Thirty-Ninth Report (April - June, 1950)", *Hesperia*, vol. 19, 2, 31-141.
Thompson, Richard F.: *The brain, a neuroscience primer* (New York, NY, 1993), 2000.
– *Das Gehirn: von der Nervenzelle zur Verhaltenssteuerung,* 2nd ed. (Heidelberg, 1994).
Thorndyke, Lynn: *History of magic and experimental science.* 6 vols. (New York, NY, 1923).
Tiefenstruktur der Musik. Festschrift Fritz Winckel zum 75. Geburtstag, ed. Carl Dahlhaus et al. (Berlin, 1982).
Tiefenstruktur Musik und Baukunst. Festschrift Fritz Winckel zum 80. Geburtstag, ed. Carl Dahlhaus (Berlin, 1987).
Tilmouth, Michael: "Chamber Music", *NG,* vol. 4, 113–18.
Tolman, Edward C.: "Cognitive maps in rats and men", *Psychological Review* 55 (1948), 189–208.
Tontechnik, ed. Ausbildung Radio und Fernsehen DRS, 2 vols. (Zurich, 1978).
Trepel, Martin: *Neuroanatomie,* (Munich etc.: Urban & Schwarzenberg, 1995, 4th ed. Munich: Elsevier, 2008).
Torrington, Lord: *The Torrington diaries,* ed. Andres Bruyn (London, 1938).
Torselli, Giorgio: *Palazzi di Roma* (Milan, 1965).
Tryggvason, Tryg: "Classical Music", *Sound recording practice,* ed. John Borwick (London, 1976), 211–28.
Truesdell, Clifford → Dostrovsky, Sigalia
Tyndall, John: *Sound: a course of 8 lectures delivered at the Royal Institution of Great Britain* (London, 1868).
Uffenbach, Zacharias Conrad von: *Herrn Zacharias Conrad von Uffenbach merwürdige Reise* (Ulm and Memmingen, 1753–53) → Preussner, Eberhard.
Ullmann, Dieter: *Chladni und die Entwicklung der Akustik von 1750–1860* (Basel, 1996).
Ullrich, Hermann: "Beethovenpflege im Wiener Vormärz", *ÖMZ* 26 (1971), 18–36.
Unger, Th.: "Der akustische Musiksaal", *Die Musik* 8, 10 (1902) and 2 (1903).
Upham, J. B.: *Acoustic architecture* (New Haven, 1853).
Valesio, Francesco: *Diario.* Archivo Storico Capitolino di Roma, Diari di Roma 1708–11.
Valtieri, Simonetta → Palazzo dal Rinascimento a oggi

Van Lenep, William (ed.): *The London Stage 1660–1800* (Carbondale, 1960–79).
Vaterländische Blätter für den österreichischen Kaiserstaat (Vienna, 1808–20).
Vendrix, Philippe: "Proportions harmoniques et proportions architecturales dans la théorie française des XVIIe et XVIIIe siècles", *IRASM* 20 (1989), 3–10.
Venzke, G.: "Die Raumakustik der Kirchen verschiedener Baustilepochen", *Acustica* 9 (1959), 151–54.
Viale Ferrero, Mercedes: "Luogo teatrale e spazio scenico", *Storia dell'Opera Italiana,* (Turin, 1988), 1–122.
– *Juvarra* (Turin, 1970).
Vian, J. P. and X. Pelorsen: "Auditorium acoustics. What should we measure? What do we measure and what does it mean?" *JASA* 89 (1991), 1856.
Villard de Honnecourt: *Exempla. Ms. Paris, BN Ms. fr. 19093,* facs. ed. H. R. Hahnloser (Graz: ADEVA, 1972).
Vitruvius: *De Architectura Libri X,* Giovanni Sulpicio da Veroli, n.d., n.p. (orig. publ. 1487?); ed. Carl Fensterbusch (Darmstadt: WBG, 1964; 4th ed. 1987); *Vitruvius, the ten books on architecture,* English transl. by Morris Hicky Morgan (New York, NY, 1960).
Vogler, Georg Joseph A.: "Data zur Akustik, 4 parts", *AMZ* 3 (1800/01), col. 522 f.
Volek, Tomislav, and Jaroslav Macek: "Beethoven's rehearsals at the Lobkovitz's", *MT* 127 (1986), 75–80.
Vorländer, Michael: *Round robin on room acoustical computer simulations,* Labor für Raumakustik PTB (Braunschweig, 1994).
– "Virtual Acoustics – opportunities and limits of spatial sound reproduction", Archives of Acoustics, 33 (2008), S. 413–422.
Waddy, Patricia: *Seventeenth Century Roman palaces: use and the art of the plan* (New York, NY: Arch. Hist. Foundation, 1990).
Wagner, Heinrich → Durm, Josef
Wagner, Richard: *Mein Leben,* ed. Martin Gregor-Dellin (Munich, 1963; Zurich: Ex Libris, 1977); English transl. Andrew Gray and Mary Whittal: *My Life by Richard Wagner* (Cambridge: Cambridge UP, 1983).
– *Sämtliche Schriften und Dichtungen* (SS) (Leipzig, 1911–14).
– *Gesammelte Schriften und Dichtungen* (GS), in 10 vols. ed. Wolfgang Golther, 4th ed. (Leipzig, 1907).
Wallace, Robin: *Beethoven's critics: aesthetic dilemmas and resolutions during the composer's lifetime* (Cambridge, 1988).
Wallin, Nils: *Biomusicology: neurophysiological, neuropsychological, and evolutionary perspectives on the origins and purposes of music* (New York, NY, 1991).
Walter, C. → Schmitt, E.
Weber, William: "Intellectual bases of the Handelian tradition 1759–1800", *PRMA* 108 (1981/82), 100–14.
– *Music and the middle class* (New York, NY, 1975).
– *The rise of musical classics, performance practices in the Nineteenth century* (Ann Arbor: UMI, 1986).
– *The rise of musical classics in eighteenth-century England: a study in canon, ritual, and ideology* (New York, NY: Oxford UP, 1992).
Webers, Johannes: *Tonstudiotechnik: Schallaufnahme und -Wiedergabe bei Rundfunk, Fernsehen, Film und Schallplatte* (Munich, 1968).
– *Handbuch für Tonstudiotechnik,* 3rd ed. (Munich, 1979).
Weibel, Samuel: *Die Deutschen Musikfeste des 19. Jahrhunderts im Spiegel der zeitgenössischen musikalischen Fachpresse* (Berlin, 2006).

Weinzierl, Stephan: *Beethovens Konzerträume, Raumakustik und symphonische Aufführungspraxis an der Schwelle zum modernen Konzertwesen* (Frankfurt a. M.: Bochinsky, 2002).
– *Handbuch der Audiotechnik* (Berlin, 2008).
Welch, Walter L. → Read, Oliver
– "Edison and his contributions to the record industry", *JAES* 25, 10/11 (1977), 660–65.
Wellek, Albert: *Musikpsychologie und Musikästhetik* (Frankfurt a. M., 1963).
Welzl-Müller, Kunigunde → Böhme, Gerhard
Wenzinger, August: "Die Viola da gamba in Händels Oratorium *La Resurrezione*", *ÖMZ* 42 (1987), 80–83.
Werner, Eric: *Mendelssohn, Leben und Werk in neuer Sicht* (Zürich, 1980).
Westrup, Jack, Neal Zaslaw and Eleanor Selfridge-Field: "Orchestra", *New Grove Dictionary of Musical Instruments,* vol. 3 (London, 1984), 823–37.
Westrup, Jack, and Neal Zaslaw: "Orchestra", *NG,* vol. 13, 679–91.
Westrup, Jack, and J. Snitzer: "Orchestra" *New Grove Dictionary of Opera,* vol. 3 (London, 1992), 719–35.
Wever, Klaus: "K. F. Schinkels Position und Beitrag zur Reform des Theaterraums", *Schinkel, Ausstellungskatalog* (Berlin, 1981).
Vienna, Musikverein → *Hundert Jahre Goldener Saal*
Wiesenfeld, Kurt → Moss, F.
Wieser, Heinz-Gregor: "Musik und Gehirn", *Schweiz.Rundschau für Medizin* 76 (1986), 153–62.
Wild, Nicole: *Dictionnaire des théâtres Parisiens au XIXe siècle* (Paris, 1990).
Wilkens, H.: "Kopfbezügliche Stereophonie – ein Hilfsmittel für Vergleich und Beurteilung verschiedener Raumeindrücke", *Acustica* 26 (1972), 213–21.
→ Lehmann, Peter
Wilson, Robert: *The life and times of Queen Victoria* (London, 1900).
Winckel, Fritz: "Die besten Konzertsäle der Welt", *Baukunst und Werkform,* 8 (1955), 750–53.
– "Raumakustische Kriterien hervorragender alter und neuer Konzertsäle", *Frequenz* 12 (1958), 50–59.
– *Phänomene des musikalischen Hörens* (Berlin, 1960).
– "Optimum acoustic criteria of concert halls for the performance of classical music", *JASA* 34 (1962), 81–86.
– "Der beste Saal", *Hundert Jahre Goldener Saal* (Vienna: GdM, 1970), 51–55.
– *Music, sound, and sensation* (New York, NY, 1971).
– "How to measure the effectiveness of stage singer's voices", *Folia phoniatrica* 23 (1971), 228–33.
– "Das Stimmverhalten des Sängers in Abhängigkeit von den Musikgattungen und den Aufführungsstätten während der florentiner Renaissance", *Poesia e Musica nell'Estetica del XVI e XVII Secolo, Convegno Villa Medici* (Florence, 1976), 91–100.
Windelband, Wilhelm: *Geschichte und Naturwissenschaft* (Strassburg, 1894).
Winkler, Helgo → Fasold, Wolfgang
Winternitz, Emanuel: *Leonardo da Vinci and Music* (New Haven & London1982).
Wittkower, Rudolf: *Architectural principles in the age of Humanism* (London, 1949, 1971).
– *Grundlagen der Architektur im Zeitalter des Humanismus* (Munich, 1969; 1983).
Wolff, Christoph: "Mozart's 'Messiah' – 'The Spirit of Handel' from van Swieten's Hands", *Festschrift P. H. Lang* (New York, NY, 1984), 1–15.
Wolff, Helmuth Christian: *Oper: Szene und Darstellung von 1600 bis 1900* (Leipzig, 1968; Musikgeschichte in Bildern IV/1).
Wollenberg, Susan: "Music in 18th-century Oxford", *PRMA* 108 (1981/82), 69–99.
Wood, Alexander B.: *A Textbook of Sound* (London: Bell, 1944).

Wyatt, Benjamin: *Observations on the design for the Theatre Royal now building in Drury Lane* (London, 1811).
Young, Elizabeth and Wayland: *London's churches* (London, 1986).
Young, Percy M.: "Die Händel-Pflege in den Englischen Provinzen", *Händel-Jb.* 6 (1960), 31–49.
– *The concert tradition from the Middle Ages to the Twentieth century* (London, 1965).
Zarlino, Gioseffo: *Le istitutioni harmoniche* (Venice, 1558); Engl. transl. of the 3rd. rev. ed. 1573, ed. G. A. Marco and C. V. Palisca (New Haven. 1968).
Zaslaw, Neal: "Toward the revival of the classical orchestra", *PRMA* 103 (1976–77), 158–87.
– "The origins of the classical orchestra", *Basler Jb. für Hist. Musikpraxis* 17 (1993), 9–40.
→ Westrup, Jack and Neal Zaslaw
Zenner, Hans-Peter: "Hören", *Neuro- und Sinnesphysiologie* (Berlin, 1993), 305–28.
Zimmermann, Heinrich → Lang, Alfred
Zorkóczy, Levente: *Hörsamkeit in Kirchen* (Berlin, 1976).
Zorzi, Ludovico and Maria Teresa Muraro: *I teatri pubblici di Venezia dal secolo XVII–XVIII* (Turin, 1970).
Zwicker, Eberhard: *Psychoakustik* (Berlin, 1982).
Zwicker, Eberhard and Hugo Fastl: *Psychoacoustics, facts and models* (Berlin, 1990).
Zwicker, Eberhard and Feldtkeller Rudolf: *Das Ohr als Nachrichtenempfänger,* 2nd ed. (Stuttgart, 1976).

3.2 Discography

Georg Frideric Handel: A Collector's Messiah. Historic Handel Oratorio Recordings 1899–1930. Koch Historic 3-7703-2Y6x2: CD 1, no. 20: "Lift up your heads", Handel Festival Choir, Orchestra, Organ; Sir Henry Wood, conductor; English Columbia L 1768 (Matrix no. AX 1595 6270). Live recording Handel Festival 1926 in Crystal Palace, Sydenham: CD 2, No. 10: "Lift up your heads", Royal Choral Society, Royal Albert Hall Orchestra, R. Arnold Greir, organ, H. L. Balfour, conductor; HMV 1057 (Matrix no. CR 118 A 04861). Recorded 2 January 1926 in the Royal Albert Hall, London.

Handel Messiah: Live Recording Maryland Handel Festival in Washington Cathedral, commemorating the Westminster Abbey Performance of 1784. Edith Mathis, James Bowman, Claes Hahnsjö, Tom Krause, University of Maryland Chorus, Cathedral Choral Society, Smithsonian Concerto Grosso, Antal Dorati, conductor; © Pro Arte CDD 232. Recorded 9–10 October 1984 in Washington Cathedral, Washington D. C.

George Fredric Handel: Messiah (1752, version 1, ed. Donald Burrows). Lynne Dawson, Hilary Summers, John Mark Ainsley, Alastair Miles, David Goode, organ continuo, Alastair Ross, harpsichord, Cecelia Bruggemeyer, Judith Evans, bass, Crispian Steele-Perkins, trumpet, The Brandenburg Consort, The Choir of King's College, Cambridge, Stephen Cleobury, conductor; ARGO 440 672–2. Recorded 16–18 December 1992 and 14–16 March 1993 in King's College Chapel, Cambridge, © DECCA.

Handel: Messiah. Judith Nelson, Emma Kirkby, Carolyn Watkinson, Paul Elliott, David Thomas, Choir of Christ Church Cathedral, Oxford, directed by Simon Preston, The Academy of Ancient Music, Christopher Hogwood, conductor. Editions de l'Oiseau-Lyre 430 488–2, London ℗ 1980, © 1991. Recorded in St. Jude-on-the-Hill, London (1980), © DECCA.

Handel: Messiah. Lorraine Hunt, Janet Williams, Patricia Spence, Drew Minter, Jeffrey Thomas, William Parker, U. C. Berkeley Chamber Chorus, Philharmonia Baroque Orchestra, Nicholas McGegan, conductor. © Harmonia Mundi France HMU 908050.52, ℗ & ©: 1991. Recorded 4–7 January 1991 in Hertz Hall, U. C. Berkeley.

Handel: The Messiah. The Scholars Baroque Ensemble, Kym Amps, Angus Davidson, Robin Doveton, David van Asch, Crispian Steele-Perkins, trumpet, Terence Charlston, organ / harpsichord, David van Asch, artistic director. Naxos 8.550667-668, HNH International, ℗ & © Naxos Rights International Ltd. 1992. Recorded 6–9 July 1992 in Rosslyn Hill Chapel, London.

George Frideric Handel: Coronation Anthems (1727). John Butt, harpsichord, David Briggs, organs, Choir of King's College, Cambridge, English Chamber Orchestra, Philip Ledger, conductor. EMI CDM 7 69753 2, ℗ EMI 1988. Recorded 21–23 July 1982 in King's College Chapel, Cambridge.

Quel fior che all'alba ride HWV 192 (Georg Friedrich Händel). Emma Kirkby, Judith Nelson, soprano; Susan Sheppard, vc, Christopher Hogwood, harpsichord. L'Oiseau Lyre 430 282–2, © DECCA.

Table A: Orchestra size, chronological list

Year, month, day Place, *Event*, Source	vn1	vn2	va	vc	db str	fl	ob	cl	bn wd	hn	trbn	tpt brs	bc	instr.	soli, choir	total
1687.02.																
Rome, P. Riario																
Pasquini, Accademia (Corelli)	0	0	0	0	0	0	0	0	0	0	0	0	0	**150**		
NG, Corelli, 769					**150**				0			0				
1689.03.																
Rome, P. Pamphilij, Lulier																
S. Beatrice (Corelli)	20	20	10	17	7	0	0	0	0	0	1	2	1	**78**		
Marx, Corelli WV, 43, 208					**74**				0			**3**				
1690.02.																
Rome, SL:San Lorenzo in Damaso																
Concerto (Corelli)	18	0	7	7	7	0	0	0	0	0	0	2	5	**43**		
Marx, Ottoboni, 124					**39**				0			**2**				
1690.02.																
SL: *Corelli*	12	0	4	4	4	0	0	0	0	0	0	0	4	**28**		
Marx, Ottoboni, 124					**24**				0			0				
1691																
SL: *Corelli*	19	0	5	8	5	0	0	0	0	0	1	2	2	**42**	**16**	**58**
Marx, Ottoboni, 128					**37**				0			**3**				
1692																
SL: *40 hore (Corelli)*	4	0	0	1	1	0	0	0	0	0	0	0	1	**7**		
Marx, Ottoboni, 129					**6**				0			0				
1692.08.10.																
SL: *Corelli*	17	0	4	7	4	0	0	0	0	0	0	2	3	**37**	**19**	**56**
Marx, Ottoboni, 132					**32**				0			**2**				
1693																
SL: *Oratorio (Corelli)*	12	0	4	6	3	0	0	0	0	0	0	0	0	**25**		
Marx, Ottoboni, 139					**25**				0			0				
1693																
SL: *Corelli*	16	0	4	5	5	0	0	0	0	0	1	2	2	**33**		
Marx, Ottoboni, 139					**30**				0			**3**				
1694																
SL: *40 hore, Giovedì (Corelli)*	6	0	2	2	2	0	0	0	0	0	0	0	1	**13**		
Marx, Ottoboni, 139					**12**				0			0				
1694																
SL: *40 hore, Domenica (Corelli)*	15	0	4	3	4	0	0	0	0	0	0	0	1	**27**		
Marx, Ottoboni, 140					**26**				0			0				
1694																
SL: *Festa di S. Lorenzo (Corelli)*	4	0	0	2	0	0	0	0	0	0	0	0	1	**7**		
Marx, Ottoboni, 143					**6**				0			0				
1695																
SL: *40 hore*	9	16	4	6	4	0	0	0	0	0	0	0	1	**42**		
Marx, Ottoboni, 145					**41**				0			0				
1696.08.10.																
SL: *Festa di San Lorenzo*	4	0	0	2	2	0	0	0	0	0	0	0	1	**9**		
Marx, Ottoboni, 150					**8**				0			0				
1697																
SL: *40 hore*	14	0	4	5	5	0	0	0	0	0	0	0	1	**29**		
Marx, Ottoboni, 151					**28**				0			0				

Year, month, day Place, *Event*, Source	vn1	vn2	va	vc	db str	fl	ob	cl	bn wd	hn	trbn	tpt brs	bc	instr.	soli, choir	total
1698																
Rome, Oratorio della Chiesa nuova																
Oratorio (Corelli)	27	0	6	8	9	0	0	0	0	0	0	0	0	**50**	**38**	**88**
Marx, Ottoboni, 153: 15/9/8/6					**50**				0			0				
1699.12.25.																
SL: *La notte di Natale*	30	0	6	7	6	0	0	0	0	0	0	2	6	**57**		
Marx, Ottoboni, 155: 6 choirs					**49**				0			**2**				
1700																
SL: *40 hore*	29	0	7	6	8	0	0	0	0	0	0	0	1	**51**		
Marx, Ottoboni, 156					**50**				0			0				
1704.08.10.																
SL: *Festa di S. Lorenzo (Corelli)*	15	0	5	7	5	0	0	0	0	0	0	0	0	**32**		
Marx, Ottoboni, 160					**32**				0			0				
1705.08.10.																
SL: *Festa di S. Lorenzo (Corelli)*	21	0	3	8	4	0	0	0	0	0	0	0	0	**36**		
Marx, Ottoboni, 161					**36**				0			0				
1705.08.23.																
Rome, P. della Cancell., court yard																
Scarlatti, Regno di MV 2x	0	0	0	0	0	0	0	0	0	0	0	0	0	**100**	**4**	**104**
Smither I, 274; Morelli, 144					**100**				0			0				
1708.04.01., 02., 07.																
Rome, Palazzo Bonelli, Stanzione delle Accademie																
Handel Resurrezione rehearsal	18	0	4	4	4	2	4	0	1	0	1	2	2	**42**	**5**	**47**
KirkR, 256					**30**				**7**			**3**				
1708.04.08.																
Rome, Palazzo Bonelli, Salone																
Handel Resurrezione 2x	22	0	4	6	6	2	4	0	1	0	1	2	2	**52**	**5**	**57**
HHdb.2,38, 265; KirkR, 231					**40**				**7**			**3**				
1710																
Rome, P. Bonelli, Stanzione																
Oratorio per S. Francesca	8	0	2	1	1	0	0	0	0	0	0	0	1	**13**	**0**	
KirkC, 52					**12**				0			0				
1710.11.22.																
London, King's Theatre																
Mancini, Hydaspes	6	5	2	6	1	0	2	0	4	0	0	1	2	**29**	**0**	
Burrows (1985), 349–357					**20**				**6**			**1**				
1714																
London, Guildhall																
Royal Entertainment	8	8	4	0	2	0	6	0	10	0	0	2	0	**41**	**0**	
Burrows (1985), 349–357					**22**				**16**			**2**				
1718																
Cannons, P. Chandos																
Acis and Galatea	2	2	1	2	1	2	2	0	1	0	0	0	1	**15**	**5**	**20**
HHA 1/9,1; Gardiner					**8**				**5**			0				
1720																
London, King's Theatre																
Royal Academy	8	9	2	4	2	0	4	0	3	0	0	1	1	**34**	**0**	
Hogwood, 79					**25**				**7**			**1**				
1727																
London, Guildhall																
Royal Entertainment	9	9	5	2	2	0	6	0	7	4	0	2	0	**47**	**0**	
Burrows (1985), 349–357					**27**				**13**			**6**				

Year, month, day Place, *Event*, Source	vn1	vn2	va	vc	db str	fl	ob	cl	bn wd	hn	trbn	tpt brs	bc	instr.	soli, choir	total
1727																
Rome, P.della Cancell., Teatro																
Costanzi, Componimento sacro	18	17	0	0	2	0	0	0	2	2	0	4	2	**56**	**3**	**59**
Smither 1, 269, fig. VI-6					**37**				**11**			**6**				
1727.10.11.																
London, Westminster Abbey																
Coronation Anthems	0	0	0	0	0	0	0	0	0	0	0	0	1	**90**	**52**	**142**
HHdb.2,702; HHdb.4,154									0			0				
1728																
London, King's Theatre																
Royal Acad., Handel opera	11	11	2	3	2	2	2	0	3	2	0	0	3	**41**	**0**	
HHdb.4,167: Fougeroux					**30**				**7**			**2**				
1733.03.17.																
London, King's Theatre																
Deborah 6x	0	0	0	0	0	0	2	2	2	3	0	3	4	**75**		**107**
HHdb.2,117; HHdb.4,210									**6**			**6**				
1733.07.05.																
Oxford, Sheldonian Theatre																
Esther 2x	0	0	0	0	0	0	0	0	0	0	0	0	0	**0**	**0**	**70**
HHdb.4,218, 242									0			0				
Athalia 2x	0	0	0	0	0	0	0	0	0	0	0	0	0	**0**	**6**	**70**
HHdb.4,218: 3700 listeners									0			0				
1733.07.12.																
Oxford, Sheldonian Theatre																
Deborah	0	0	0	0	0	0	0	0	0	0	0	0	0	**0**	**6**	**70**
HHdb.4, 219 f.									0			0				
1737.12.14.																
London, Whitehall Palace, Banqueting Hall																
Funeral Anthem, rehearsal	0	0	0	0	0	0	0	0	0	0	0	0	0	**100**	**80**	**180**
HHdb.4,286,287,289									0			0				
1737.12.17.																
London, Westminster Abbey																
Funeral Anthem, chapel	0	0	0	0	0	0	0	0	0	0	0	0	0	**100**	**80**	**180**
HHdb.2, 735; HHdb.4, 286–9									0			0				
1742.04.13.																
Dublin, Neal's Music Hall																
Messiah, first performance	0	0	0	0	0	0	0	0	1	0	0	2	2	**25**	**6,20**	**48?**
HHdb.2,195; HHdb.4, 348					**20**				**1**			**2**				
1742.06.03.																
Dublin, Neal's Music Hall																
Messiah	0	0	0	0	0	0	0	0	0	0	0	0	0	**0**	**7,?**	**49?**
HHdb.4,350									0			0				
1749.04.21.																
London, Vauxhall Gardens																
Fireworks Music, rehearsal	0	0	0	0	0	0	24	0	12	9	0	9	0	**58**	**0**	**58**
HHdb.4,424: 12'000 listeners					0				**36**			**18**				
1749.04.27.																
London, Green Park																
Fireworks Music	0	0	0	0	0	0	24	0	12	9	0	9	0	**58**	**0**	**58**
HHdb.4, 423, 426									**37**			**18**				
1749.05.27.																
London, Foundling Hospital																
Anthem, Solomon excerpts	0	0	0	0	0	0	0	0	0	0	0	0	0	**50**	**5,50**	**>100**
HHdb.4,429: > 1000 listeners					**36**				**8**			**4**				

Year, month, day Place, *Event,* Source	vn1	vn2	va	vc	db str	fl	ob	cl	bn wd	hn	trbn	tpt brs	bc	instr.	soli, choir	total
1749.05.27.																
London, Foundling Hospital																
Fireworks Music	12	12	4	4	4	0	24	0	12	9	0	9	0	**94**	**0**	**94**
HHdb.4,429: no. of score					**36**				**37**			**18**				
1752.09.27.																
Salisbury, Cathedral																
Te Deum, Messiah exc., 2x	8	8	2	4	2	0	2	0	1	2	0	2	1	**32**	**18**	**51**
HHdb.4,461					**24**				**3**			**4**				
1754.05.15.																
London, Foundling Hospital																
Messiah (Handel last time)	8	6	6	3	2	0	4	0	4	2	0	2	1	**39**	**5,19**	**61**
HHdb.4,481;Hogwood,228; Landon,19					**25**				**8**			**4**				
1756.05.19.																
London, Foundling Hospital																
Messiah (Smith)	6	6	3	3	2	0	4	0	4	2	0	2	1	**33**	**5,17**	**53**
HHdb.4,498					**20**				**8**			**4**				
1758.04.27.																
London, Foundling Hospital																
Messiah (Smith)	6	6	3	3	2	0	4	0	4	2	0	2	1	**34**	**6,17**	**57**
HHdb.4,516: 6 Boys					**20**				**8**			**4**				
1759.05.03.																
London, Foundling Hospital																
Messiah (Smith)	6	6	3	3	2	0	4	0	4	2	0	2	1	**35**	**5,18**	**58**
HHdb.4,532					**20**				**8**			**4**				
1763																
Gloucester																
Messiah	8	8	4	4	2	0	4	2	4	0	0	3	1	**42**	**0**	**>50**
Dean, 104					**26**				**10**			**3**				
1767.10.23.																
Birmingham																
Messiah	8	8	4	5	2	0	4	0	4	2	0	2	1	**41**	**40**	**~85**
Dean, 104					**27**				**8**			**4**				
1780–1790																
Vienna, Tonkünstlersocietät																
Handel Samson (Salieri)	20	20	8	20	0	0	4	2	3	4	0	2	1	**85**	**0**	**~150**
Edelmann, 183: no. per parts					**68**				**9**			**6**				
1780–1790																
Vienna, Tonkünstlersocietät																
Handel Israel (Salieri)	20	20	8	17	0	2	4	2	3	4	3	2	1	**88**	**45****	**133**
Edelmann, 183: no. per parts					**66**				**11**			**9**				
1784.05.26.																
London, Westminster Abbey																
Messiah, Handel Commemoration	48	47	26	21	15	6	26	0	26	12	6	12	1	**250**	**283**	**533**
Burney, 8; Shaw, 69; Koury, 46					**157**				**59**			**30**				
1784.05.27.																
London, Pantheon																
Handel Commemoration	0	0	0	0	0	0	0	0	0	0	0	0	0	**0**	**0**	**200**
Burney, Account, 49					0				0			0				
1785																
London, Westminster Abbey																
Messiah	0	0	0	0	0	0	0	0	0	0	0	0	0	**0**	**0**	**616**
Hogwood, 240					0				0			0				

Year, month, day Place, *Event,* Source	vn1	vn2	va	vc	db str	fl	ob	cl	bn wd	hn	trbn	tpt brs	bc	instr.	soli, choir	total
1786																
London, Westminster Abbey																
Messiah	0	0	0	0	0	0	0	0	0	0	0	0	0	**0**	**0**	**640**
Hogwood, 240					0				0			0				
1787																
London, Westminster Abbey																
Messiah	0	0	0	0	0	0	0	0	0	0	0	0	0	**0**	**0**	**828**
Koury 27; Hogwood, 242					0				0			0				
1787–																
Dublin, church (yearly)																
Handel Commemoration, 2x	0	0	0	0	0	0	0	0	0	0	0	0	0	**0**	**0**	**>200**
Lenz, 309; Scholes,181					0				0			0				
1788.02.26.																
Wien, P. Johann Esterházy																
C.Ph.E.Bach, Auferstehung (Mozart	6	6	4	4	2	2	2	0	2	2	0	3	1	**36**	**30**	**66**
NMA X/28 I,2					**22**				**6**			**5**				
1788.05.30.																
Breslau, Maria-Magdalena Kirche																
Messias (Hiller) German	0	0	0	0	0	0	0	0	0	0	0	0	0	**0**	**0****	**259**
Koury, 27; HJb.1960, 66					**45**				**85**			**38**				
1789.03.06.																
Wien, Palais Johann Esterházy																
Messias (Mozart) 2x	6	6	8	4	1	2	2	2	2	2	3	2	3	**45**	**12**	**~50?**
NMA X/28 I, 2 ; Wolff 1984,12					**25**				**9**			**7**				
1791.06.01.																
London, Westminster Abbey																
Messiah	0	0	0	0	0	0	0	0	0	0	0	0	0	**0**	**0**	**1068**
Hogwood, 243: Haydn present					0				0			0				
1798.04.01.																
Wien, Burgtheater																
Haydn Sieben Worte 2x	21	20	9	8	8	2	6	2	6,1	4	0	4	1	**95**	**0**	**152**
Landon 4,316					**66**				**16**			**8**				
1799.03.19.																
Wien, Burgtheater																
Haydn Schöpfung, first public performance.	19	18	12	11	10	6	6	6	6,3	6,1	4	4	1	**115**	**60**	**~180**
Brown, 20; Landon IV, 455					**70**				**27**			**15**				
1801.01.16.																
Wien, Grosser Redoutensaal																
Haydn Schöpfung	0	0	0	0	0	0	0	0	0	0	0	0	0	**0**	**0**	**200**
Morrow, 307; Landon 5, 22					0				0			0				
1808.03.27.																
Wien, Alte Universität																
Haydn Schöpfung	13	12	7	6	4	2	2	2	2	2	0	2	1	**55**	**32**	**103**
Morrow,176; Biba,103					**42**				**6**			**4**				
1812																
Worcester Cathedral																
Three Choirs Festival	0	0	0	0	0	0	0	0	0	0	0	0	0	**0**	**0**	**174**
Lysons, 97					0				0			0				
1812.11.29.																
Wien, Winterreitschule																
Alexandersfest (Mozart)	60	60	37	33	21	12	12	12	12,2	12	9	12	1	**302**	**7,280**	**582**
AMZ 4, 851; Schwab, 90; EM (1988),11					**211**				**50**			**33**				

Year, month, day Place, *Event*, Source	vn1	vn2	va	vc	db str	fl	ob	cl	bn wd	hn	trbn	tpt brs	bc	instr.	soli, choir	total
1815.10.31.																
Edinburgh, Festival																
Messiah et al. 4x	12	12	7	6	5	3	2	2	2	2	2	2	1	**59**	**6,58**	**123**
AMZ 18 (1816), 633					**42**				**9**			**6**				
1823																
York, Cathedral																
Handel Te Deum (Festival)	0	0	0	0	0	0	0	0	0	0	0	0	0	**285**	**280**	**565**
Hogwood					0				0			0				
1823																
York, Assembly Room																
Handel Messiah (Festival)	0	0	0	0	0	0	0	0	0	0	0	0	0	**177**	**269**	**446**
Harmonicon 1(1823), 152 f.					0				0			0				
1823																
Gloucester																
Festival	0	0	0	0	0	0	0	0	0	0	0	0	0	**47**	**83**	**130**
Koury,170; Harmonicon (1823),148					0				0			0				
1828																
Paris, Conservatoire																
Beethoven 3rd symphony (Habeneck)	15	16	8	12	8	4	3	4	4	2	0	4	0	**86**	**0**	
Elwart, 98					**59**				**15**			**6**				
1834																
London, Westminster Abbey																
Messiah	40	40	32	18	18	10	12	8	12,4	10	8	8	0	**227**	**5,392**	**644**
Moscheles; Galkin; Hogwood, 249					**148**				**42**			**26**				
1836																
London, Exeter Hall																
Messiah	0	0	0	0	0	0	0	0	0	0	0	0	0	**0**	**0**	**>400**
Smither, EM (1985), 342					0				0			0				
1837																
Paris, Dôme des Invalides																
Berlioz Requiem	25	25	20	20	18	4	4	4	8	12	0	0	0	**207**	**210**	**420**
Berlioz Edition 1, 19					**108**				**22**			**53**				
1837.12.																
London, Exeter Hall																
Messiah	0	0	0	0	0	0	0	0	0	0	0	0	0	**0**	**400**	**500?**
Smither, EM (1985), 342					0				0			0				
1843.11.05.																
Vienna, Winterreitschule																
Haydn Schöpfung	59	59	40	41	25	13	12	12	12	12	9	8	1	**320**	**660**	**960**
Gassner, appendix; Schwab, 90					**224**				**49**			**29**				
1844																
Vienna																
Hofkapelle	6	6	2	3	3	2	2	2	2	2	2	2	1	**36**	**20**	
Gassner, appendix 12					**20**				**8**			**6**				
1857																
Sydenham, Crystal Palace																
Messiah (Costa)	75	75	50	50	50	9	9	9	9	12	9	12	1	**397**	**2'000**	**2400**
Scholes, 178; Hogwood, 256					**300**				**36**			**51**				
1859																
Sydenham, Crystal Palace																
Messiah (Costa)	92	90	60	60	61	10	10	10	10	12	9	6	1	**453**	**2'765**	**3218**
Niederrheinische Musikzeitung 7(1859), 226					**363**				**40**			**52**				

Year, month, day Place, *Event*, Source	vn1	vn2	va	vc	db str	fl	ob	cl	bn wd	hn	trbn	tpt brs	bc	instr.	soli, choir	total
1894.06.13.																
Cambridge, King's College Chapel																
Messiah (rev. version)	0	0	0	0	0	0	0	0	0	0	0	0	0	**63**	**200**	**>263**
MT (1894), 464					0				0			0				
1899.03.31.																
London, Albert Hall																
Messiah (Bridge, rev. version)	19	19	12	12	12	0	12	0	6	2	0	2	0	**100**	**4,842**	**946**
Scholes,68; MT (1899), 22 ,97					**75**				**18**			**4**				
1900.01.05.																
Scarborough, Grand Hotel																
Alexander Balus (rev. version)	5	5	3	3	2	2	4	0	4	2	0	2	4	**37**	**24**	**>61**
MT 1900, 116					**10**				**10**			**4**				
1906																
Cambridge, King's College Chapel																
Messiah (A. Manns)	0	0	0	0	0	0	0	0	0	0	0	0	0	**32**	**24**	
Scholes, 87					0				0			0				
1920																
Sydenham, Crystal Palace																
Messiah (Cowan)	0	0	0	0	0	0	0	0	0	0	0	0	0	**0**	**0**	**4000**
Scholes, 180; Hogwood, 265					0				0			0				
1926																
Sydenham, Crystal Palace																
Messiah recording (Wood)	0	0	0	0	0	0	0	0	0	0	0	0	0	**0**	**0**	**4000**
Scholes, 180					0				0			0				
1980																
London, St. Jude-on-the-Hill																
Messiah (Hogwood)	8	7	6	3	2	0	4	0	4	2	0	2	2	**41**	**4,31**	**76**
CD, choir 16 boys -5-5-5					**26**				**8**			**6**				
1984																
Washington, Cathedral																
Messiah (Dorati)	18	18	10	10	6	0	12	0	10	6	3	5	2	**103**	**4,405**	**512**
CD, choir: 100-100-100-100					**62**				**22**			**15**				
1991																
Berkeley, Hertz Hall																
Messiah (McGegan)	6	6	4	3	2	0	2	0	2	2	0	2	2	**32**	**6,42**	**80**
CD, choir 11-11-10-10					**21**				**4**			**4**				
1992																
London, Rosslyn Hill Chapel																
Messiah (van Asch)	4	3	2	1	1	0	0	0	0	0	0	2	1	15	4,14	33
CD, choir 4-3-4-3					**11**				0			**2**				
1992/93																
Cambridge, King's College Chapel																
Messiah (Cleobury)	6	6	4	4	2	0	2	0	1	0	0	2	2	**30**	**4,18?**	**52?**
CD					**22**				**3**			**2**				

Table B: Oratorios, orchestral and choir compositions, chronological list of cited examples

Year, month, day	Place	Event	Sources	Total v., or.
1694–	London, St. Bride's Church	Festival St. Cecilia: Purcell Te Deum	Shapiro, 219	
1698–	London, St. Paul's	Festival Sons of Clergy: Purcell Te Deum	Shapiro, 219	
1702–1707?	London, Stationer's Hall	Weedon Concerts: Blow's Coronation Anthem	Shapiro, 220	
1708.04.01–07.	Rome, P. Bonelli, Stanzione	Handel Resurrezione, rehearsal	Kirkpatrick Ruspoli, 256	5; 47
1708.04.08.	Rome, P. Bonelli, Salone	Handel Resurrezione 2x	HHdb.2, 38; Smither I, 265; Kirkpatrick Ruspoli, 231	5; 57
1710	Rome, P. Bonelli, Stanzione	Oratorio per S. Francesca	Kirkpatrick Corelli, 52	
1713.03.19.	London, Banqueting Hall	Utrecht Te Deum, rehearsal	HHdb.4, 61	
1713.07.07.	London, St. Paul's	Utrecht Te Deum, Jubilate	HHA III/6, HHdb.4, 64	
1714.09.26.	London, Chapel Royal, St.J.P.	Te Deum	Marx, Händel Enzykl., 1040	
1714.10.17.	London, Chapel Royal, St.J.P	Caroline Te Deum and Anthem	Marx, Händel Enzykl., 1040	
1718	Cannons, P. D. of Chandos	Acis and Galatea	HHA 1/9,1: Choir STTTB	20
1718	Hamburg, Kathedrale	Handel Brockes Passion	Smither, Oratorio II, 110	
1719.02.18.	London, Hickford's Room	Handel concert (Dubourg)	HHdb.4, 78; Deutsch, 83	
1720	London, King's Theatre	Royal Academy	Hogwood 1984, 79	
1720.08.29?	Cannons, P. D. of Chandos	Esther, Music Room?	HHA I/8; HHdb.2, 81	
1724.01.05.	London, Chapel Royal, St.J.P.	New Anthem	Marx, Händel Enzykl., 1047	
1724.09.	Hereford, Cathedral	Te Deum, Jubilate (3 choirs)	HHdb.4, 129	
1727	London, Guildhall	Royal Entertainment	Burrows (1985), 349	
1727.10.11.	London, Westminster Abbey	Coronation Anthems	HHdb.2, 702; HHdb.4, 154	142
1727.11.25.	London, Crown & Anchor	Dubourg Concert	HHdb.2, 156: 200 ladies	
1728	London, King's Theatre	Royal Academy, Handel opera	HHdb.4, 167: Fougeroux	
1731.03.26.	London, Lincoln's Inn	Acis and Galatea	HHdb.4, 190	
1732.02.23.	London, Crown & Anchor	Esther (Bernard Gates) 3x	HHdb.4, 198	
1732.04.20.	London, York Buildings	Esther (not Handel)	HHdb.4, 199; Dean, 205	
1732.05.02.	London, King's Theatre	Esther (Handel) 6x	HHdb.4, 200	
1732.05.17.	London, Little Theatre	Acis and Galatea (Arne)2x	HHdb.4, 201; Dean, 171 f.	
1732.06.10.	London, King's Theatre	Acis and Galatea Italian/English 4x	HHdb.4, 203; Dean, 173	
1732.12.05.	London, King's Theatre	Acis and Galatea Italian/English 4x	HHdb.4, 205	
1733.03.17.	London, King's Theatre	Deborah 6x	HHdb.2, 117; HHdb.4, 210	8,107
1733.04.14.	London, King's Theatre	Esther, organ concerto 2x	HHdb.4, 241	
1733.07.05.	Oxford, Sheldonian Th.	Esther 2x	HHdb.4, 218, 242	70
1733.07.08.	Oxford, St. Mary's Church	Te Deum, Anthems	HHdb.4, 218: 800 ladies	
1733.07.10.	Oxford, Sheldonian Th.	Athalia 2x	HHdb.4, 218: 3700 listeners	6, 70

Year, month, day	Place	Event	Sources	Total v., or.
1733.07.11.	Oxford, Christ Church Hall	Acis and Galatea	HHdb.4, 219 f.	
1733.07.12.	Oxford, Sheldonian Th.	Deborah	HHdb.4, 219 f.	70
1734.01.19.	Stockholm, Riddarhuset	Acis and Galatea (Romean)	HHdb.4, 229; Dean, 629	
1734.02.16.	London, Crown & Anchor	Te Deum, rehearsal	HHdb.4, 237: Earl of Egmont	
1734.02.19.	London, St. Paul's	Te Deum, Jubilate (Festival)	HHdb.4, 237: Earl of Egmont	
1734.03.13.	London, King's Theatre	Il Parnasso in festa, Conc.i op. 3?	HHdb.4, 238	
1734.03.14.	London, St. James's French Chapel	Wedding Anthem	HHdb.4, 239: great number	
1734.04.02.	London, King's Theatre	Deborah I/E 3x	HHdb.4, 239	
1734.05.01.	Dublin, Crow St. M. Hall	Acis and Galatea (Raffa)	HHdb.4, 241	
1734.05.07.	London, Covent Garden Th.	Acis und Galatea Italian/English	HHdb.4, 241	
1735	Dublin, Aungier St. Th.	Acis and Galatea	HHdb.4, 259	
1735.03.05.	London, Covent Garden Th.	Esther, 2 organ concertos 6x	HHdb.4, 250; Dean, 211	
1735.03.26.	London, Covent Garden Th.	Deborah, organ concerto 3x	HHdb.4, 251; Dean, 211	
1735.04.01.	London, Covent Garden Th.	Athalia, organ concerto 5x	HHdb.4, 252; Dean, 211	
1736.02.19.	London, Covent Garden Th.	Alexander's Feast, concert 5x	HHdb.4, 259–60: 1300 listen.	
1736.02.19.	London, St. Paul's	Jubilate, Coronation Anthem	HHdb.4, 259: vast number	
1736.03.24.	London, Covent Garden Th.	Acis and Galatea Italian/Engl. 2x	HHdb.4, 261	
1736.04.07.	London, Covent Garden Th.	Esther 2x	HHdb.4, 262	
1736.04.08.	Dublin, St. Andrew's	Te Deum, Jubilate, Coron. Anth.	HHdb.4, 261–62:	>70
1736.04.27.	London, Chapel Royal, St.J.P.	Wedding Anthem	Marx, Händel Enzykl., 1066	
1737.03.09.	London, King's Theatre	Parnasso in festa 2x	HHdb.4, 267f.	
1737.03.16.	London, Covent Garden Th.	Alexander's Feast 6x	HHdb.4, 276–77	
1737.03.23.	London, Covent Garden Th.	Il Trionfo del Tempo 4x	HHdb.4, 276, 278	
1737.03.23.	London, Crown & Anchor	Chandos, Te Deum (not Handel)	HHdb.4, 278	
1737.04.06.	London, Covent Garden Th.	Esther 2x	HHdb.4, 276, 279	
1737.06.25.	London, Covent Garden Th.	Alexander's Feast	HHdb.4, 279	
1737.12.14.	London, Whitehall B. Hall	Funeral Anthem, rehearsal	HHdb.4, 286, 287, 289	180
1737.12.17.	London, Westminster Abbey, King Henry VII' Chapel	Funeral Anthem	HHdb.2, 735; HHdb.4, 286f.	180
1738.03.10.	London, Swan Tavern	Coronation Anthem (not Handel)	HHdb.4, 292	
1738.03.28.	London, King's Theatre	An Oratorio, 500 listeners on stage	HHdb.4, 292–293	
1738.03.28.	Oxford	Handel Oratorio	Burrows, ML 61(1980), 180	
1738.07.13.	Oxford	Alexander's Feast (Hayes)	Burrows, ML 61(1980), 179	
1738.08.19.	London, Vauxhall Gardens	Coronation Anthem (not Handel)	HHdb.4, 298	
1738.09.12.	Worcester, Town Hall	Esther (Merrifield)	HHdb.4, 299	
1739.01.16.	London, King's Theatre	Saul, organ concertos 4x	HHdb.4, 304: carillon, timp.	
1739.01.18.	London, Crown & Anchor	Alexander's Feast (not Handel)	HHdb.4, 299	
1739.02.	London, Hickford's Room	Handel choirs (Snow)	HHdb.4, 306	
1739.02.17.	London, King's Theatre	Alexander's Feast 3x	HHdb.4, 305: 20.2. charity	
1739.03.03.	London, King's Theatre	Il Trionfo del Tempo	HHdb.4, 306	
1739.04.04.	London, King's Theatre	Israel in Egypt 3x	HHdb.4, 307	
1739.04.19.	London, King's Theatre	Saul 2x	HHdb.4, 309	
1739.05.10.	London, Crown & Anchor	Israel in Egypt (not Handel)	HHdb.4, 310	

Year, month, day	Place	Event	Sources	Total v., or.
1739.09.05.	Gloucester	Alexander's Feast (3 choirs)	HHdb.4, 311	
1739.11.22.	London, Lincoln's Inn	Alexander's Feast, Ode St. C. 2x	HHdb.4, 313–314	
1739.11.28.	Newcastle, Assembly Rooms	Handel Choirs (Avison)	HHdb.4, 314	<40
1739.12.13.	London, Lincoln's Inn	Acis, Ode for St. Cecilia 2x	HHdb.4, 315	
1740	London, Vauxhall Gardens	Hornpipe HWV 356	HHdb.4, 299	
1740.02.21.	London, Lincoln's Inn	Acis, Ode for St. Cecilia	HHdb.4, 319	
1740.02.27.	London, Lincoln's Inn	L'Allegro 5x	HHdb.4, 319	
1740.02.29.	Dublin, Music Hall	Arien aus Esther (charity)	HHdb.4, 319	
1740.03.06.	Dublin, St. Andrew's	Te Deum, Jubilate (charity)	HHdb.4, 318	
1740.03.21.	London, Lincoln's Inn	Saul, Concerto grosso	HHdb.4, 320	
1740.03.26.	London, Lincoln's Inn	Esther	HHdb.4, 320	
1740.03.28.	London, Lincoln's Inn	Acis, Ode for St. Cecilia	HHdb.4, 319, 320	
1740.04.01.	London, Lincoln's Inn	Israel in Egypt	HHdb.4, 320	
1740.04.02.	London, Hickford's Room	2 Anthems (not Handel)	HHdb.4, 320	
1740.04.23.	London, Lincoln's Inn	L'Allegro, Concerto grosso	HHdb.4, 321	
1740.04.24.	London, Crown & Anchor	Saul (not Handel)	HHdb.4, 321	
1740.05.08.	London, Chapel R. St.J.P.	Wedding Anthem	Marx, Händel Enzykl., 1070	
1740.11.08.	London, Lincoln's Inn	Il Parnasso in festa	HHdb.4, 323	
1741	Dublin, Dubourg	Arne: Comus	NG 5, 666: orch. enlarged	
1741.01.10.	London, Lincoln's Inn	Deidamia (Handel's last opera) 2x	HHdb.4, 329	
1741.01.31.	London, Lincoln's Inn	L'Allegro 3x	HHdb.4, 330	
1741.02.05.	London, Hickford's Room	Saul Ouverture & March	HHdb.4, 330	
1741.02.14.	Dublin, St. Andrew's	Te Deum, Jubilate, Coron. Anth.	HHdb.4, 330	
1741.02.28.	London, Lincoln's Inn	Acis and Galatea 2x	HHdb.4, 331	
1741.02.28.	London, Lincoln's Inn	Ode for St. Cecilia	HHdb.4, 331	
1741.03.14.	London, Lincoln's Inn	Il Parnasso in festa	HHdb.4, 331–332	
1741.03.18.	London, Lincoln's Inn	Saul	HHdb.4, 332	
1741.04.08.	London, Lincoln's Inn	L'Allegro, Ode for St. Cecilia	HHdb.4, 333	
1741.04.11.	London, Lincoln's Inn	"Handel's last concert"	HHdb.4, 333	
1741.12.10.	Dublin, St. Andrew's	Te Deum, Anthem, organ concerto	HHdb.4, 339,342: Handel	
1741.12.23.	Dublin, Neal's Music Hall	L'Allegro, ca. 600 listeners	HHdb.4, 340	
1742.01.13.	Dublin, Neals Music Hall	L'Allegro	HHdb.4, 342	
1742.01.20.	Dublin, Neal's Music Hall	Acis, Ode for St. Cecilia 2x	HHdb.4, 342	
1742.02.03.	Dublin, Neal's Music Hall	Esther 3x	HHdb.4, 345	
1742.02.08.	Dublin, St. Andrew's	Te Deum, Jubilate, Anthems	HHdb.4, 342–345: Handel	
1742.02.17.	Dublin, Neal's Music Hall	Alexander's Feast 2x	HHdb.4, 345	
1742.03.24.	Dublin, Neal's Music Hall	Hymen 2x	HHdb.4, 346–347	
1742.04.07.	Dublin, Neal's Music Hall	Esther	HHdb.4, 348	
1742.04.09.	Dublin, Neal's Music Hall	Messiah, publ. rehearsal	HHdb.4, 348	
1742.04.13.	Dublin, Neal's Music Hall	Messiah 1st perf., ca. 700 listen.	HHdb.2, 195; HHdb.4, 348	48?
1742.05.25.	Dublin, Neal's Music Hall	Saul	HHdb.4, 349	
1742.06.03.	Dublin, Neal's Music Hall	Messiah, organ concerts	HHdb.4, 350	48?
1742.11.25.	Salisbury, Cathedral	Te Deum, Jubilate, Anthems	HHdb.4, 354: St. Cecilia	
1743	Oxford, Christ Church Hall	Saul	HHdb.4, 356	

Year, month, day	Place	Event	Sources	Total v., or.
1743–1744	Dublin, Dubourg	Handel, oratorios 6x	NG 5, 666	
1743.02.08.	Dublin, St. Andrew's	Te Deum, Jubilate, Anthem	HHdb.4, 355	
1743.02.08.	Dublin, Great Music Room	Acis, Coronation Anthem (Arne)	HHdb.4, 356	
1743.02.18.	London, Covent Garden Th.	Samson 8x	HHdb.4, 356–357	
1743.02.24.	London, Crown & Anchor	Esther (not Handel)	HHdb.4, 358	
1743.03.18.	London, Covent Garden Th.	L'Allegro, Ode for St. Cecilia	HHdb.4, 359	
1743.03.23.	London, Covent Garden Th.	Messiah 3x (Sacred Oratorio)	HHdb.4, 359	
1743.04.08.	Dublin, Great Music Room	Alexander's Feast (charity)	HHdb.4, 361	
1743.09	Worcester, Town Hall ?	Athalia (3 choirs)	HHdb.4, 365	
1743.11.27.	London, Chapel Royal St.J.P.	Dettingen Te Deum	HHdb.4, 367	
1744/45	Dublin, Neal's Music Hall	Acis and Galatea	Dean, 630	
1744/45	Dublin, Neal's Music Hall	Athalia (Philharmonic Society)	Dean, 630	
1744.02	London, Crown & Anchor T.	Messiah (Needler)	Shaw, 36	
1744.02.07.	Dublin, Great Music Room	Messiah (charity)	HHdb.4, 367	
1744.02.09.	London, Stationers Hall	Acis, Coronation Anthem (Larken)	HHdb.4, 371	
1744.02.10.	London, Covent Garden Th.	Semele 4x	HHdb.4, 372	
1744.02.24.	London, King's Theatre	Samson 2x	HHdb.4, 378	
1744.03.02.	London, Covent Garden Th.	Joseph 4x	HHdb.4, 374	
1744.03.16.	London, Covent Garden Th.	Saul 2x	HHdb.4, 375	
1744.06.11.	Low-Layton, Ruckolt-House	Alexander's F. 3x (not Handel)	HHdb.4, 377	
1744.11.03.	London, King's Theatre	Deborah 2x	HHdb.4, 380	
1744.12.01.	London, King's Theatre	Semele 2x	HHdb.4, 381	
1745.01.05.	London, King's Theatre	Hercules 2x	HHdb.4, 383	
1745.01.17.	London, Swan Tavern	Coronation Anthem (charity)	HHdb.4, 384	
1745.02.14.	Dublin, St. Michan's	Dettingen Te D., Jubilate, Anth.	HHdb.4, 381, 386	
1745.02.20.	London, New Theatre Haym.	Coronation Anthem (charity)	HHdb.4, 386	
1745.03.01.	London, King's Theatre	Samson 2x	HHdb.4, 387	
1745.03.13.	London, King's Theatre	Saul	HHdb.4, 387	
1745.03.15.	London, King's Theatre	Joseph 2x	HHdb.4, 387	
1745.03.27.	London, King's Theatre	Belshazzar 3x	HHdb.4, 388	
1745.04.09.	London, King's Theatre	Messiah 2x (Sacred Oratorio)	HHdb.4, 388	
1745.04.10.	London, Covent Garden	Handel works (charity)	HHdb.4, 389	
1745.05.10.	London, Hickford's Room	Handel concert (charity)	HHdb.4, 392	
1745.09.05.	Gloucester, Booth Hall	Acis & Galatea (3 choirs)	HHdb.4, 394	
1745.09.09.	London, Vauxhall Gardens	Coronation Anthems	HHdb.4, 394	
1745.12.12.	Dublin, St. Michan's	Te Deum, Jubilate, Coron. Anth.	HHdb.4, 396	
1745.12.21.	Dublin, New Music Hall	Messiah	HHdb.4, 396	
1746.01.23.	Dublin, Neal's Music Hall	Deborah (Dubourg)	HHdb.4, 399	
1746.02.14.	London, Covent Garden Th.	Occasional Oratorio 3x	HHdb.4, 400	
1746.02.20.	Dublin, Music Hall	Esther	HHdb.4, 399	
1746.03.25.	London, King's Theatre	Handel arias (charity)	HHdb.4, 401	
1746.10.17.	Salisbury, Cathedral	Dettingen Te Deum	HHdb.4, 403: Gluck	
1747	London, Castle Tavern	Acis and Galatea	HHdb.4, 409	
1747	Dublin	Deborah	NG 5, 666	

Year, month, day	Place	Event	Sources	Total v., or.
1747.02.05.	Dublin, St. Michan's	Te Deum, Jubilate, Coron. Anth.	HHdb.4, 406	
1747.03.06.	London, Covent Garden Th.	Occasional Oratorio 3x	HHdb.4, 406	
1747.03.20.	London, Covent Garden Th.	Joseph 2x	HHdb.4, 406	
1747.04.01.	London, Covent Garden Th.	Judas Maccabaeus 6x	HHdb.4, 407	
1747.04.14.	London, King's Theatre	Handel works (charity)	HHdb.4, 407	
1747.04.30.	London, Crown & Anchor	Messiah	HHdb.4, 408	
1747.05.07.	London, St. Paul's	Te Deum, Jubilate, Coron. Anth.	HHdb.4, 408	
1747.12.03.	Dublin, St. Andrew's	Te Deum, Jubilate, Coron. Anth.	HHdb.4, 409	
1748.01.21.	Dublin, New Music Hall	L'Allegro	HHdb.4, 411	
1748.02.04.	Dublin, Neal's Music Room	Samson (Dubourg)	HHdb.4, 411; NG 5, 666	
1748.02.11.	Dublin, Neal's Music Hall	Judas Maccabaeus (Dubourg)	HHdb.4, 412; NG 5, 666	
1748.02.23.	Dublin, Neal's Music Hall	Acis and Galatea	HHdb.4, 412	
1748.02.26.	London, Covent Garden Th.	Judas Maccabaeus 6x	HHdb.4, 412	
1748.03.09.	London, Covent Garden Th.	Joshua 4x	HHdb.4, 412	
1748.03.23.	London, Covent Garden Th.	Alexander Balus 3x	HHdb.4, 413	
1748.04.05.	London, King's Theatre	Handel works (charity)	HHdb.4, 413	
1748.04.07.	London, Covent Garden	Judas Maccabaeus	HHdb.4, 414	
1748.04.27.	London, Hickford's Room	Aria from Messiah (Galli)	HHdb.4, 414	
1748.07.09.	Oxford, Holywell Music Room	Esther (not Handel)	HHdb.4, 416: opening!	
1748.09.14.	Gloucester, Cathedral	Te Deum, Jubilate (3 choirs)	HHdb.4, 416	
1748.09.14.	Gloucester, Booth Hall	Samson (3choirs)	HHdb.4, 417	
1748.09.15.	Gloucester, Cathedral	Coronation Anthem (3 choirs)	HHdb.4, 416	
1748.10.19.	Salisbury, Assembly Room	Alexander's Feast (charity)	HHdb.4, 417	
1748.10.20.	Salisbury, Assembly Room	Acis and Galatea (charity)	HHdb.4, 417	
1748.11.10.	Dublin, Philharmonic Room	Alexander's Feast (Dubourg)	HHdb.4, 417	
1748.11.11.	Dublin, Philharmonic Room	Esther (Dubourg)	HHdb.4, 417	
1748.11.17.	Dublin, Philharmonic Room	Samson (Oldmixon)	HHdb. 4, 417	
1748.11.18.	Dublin, Neal's Music Hall	Acis and Galatea 2x	HHdb. 4, 417	
1749	Dublin	Deborah	NG 5, 666	
1749.02.07.	Dublin, Neal's Music Hall	Acis and Galatea (Lampe)	HHdb.4, 419	
1749.02.10.	London, Covent Garden Th.	Susanna 4x	HHdb.4, 419	
1749.02.24.	London, Covent Garden Th.	Hercules 2x	HHdb.4, 420	
1749.03.03.	London, Covent Garden Th.	Samson 4x	HHdb.4, 420	
1749.03.15.	Dublin, Music Hall	Joshua	HHdb.4, 420	
1749.03.17.	London, Covent Garden Th.	Solomon 3x	HHdb.4, 420: double orch.	100?
1749.03.21.	London, King's Theatre	Handel works (charity)	HHdb.4, 422	
1749.03.23.	London, Covent Garden Th.	Messiah	HHdb.4, 422: double orch.	100?
1749.04.12.	Oxford	Esther	HHdb.4, 424	<50
1749.04.13.	Oxford, Sheldonian Theatre	Samson, Coronation Anthem	HHdb.4, 424	<50
1749.04.14.	Oxford, Sheldonian Theatre	Messiah (Hayes)	HHdb.4, 424: large choir	
1749.04.21.	London, Hickford's Room	Acis and Galatea (Dubourg)	HHdb.4, 425	
1749.04.21.	London, Vauxhall Gardens	Fireworks Music (Rehearsal)	HHdb.4, 424: 12'000 listen.	58
1749.04.25.	London, St. James's Chapel	Peace Anthem and Te Deum	HHdb.4, 425	
1749.04.25.	Dublin, St. Andrew's	Te Deum, Jubilate, Coron. Anthem	HHdb.4, 423	

Year, month, day	Place	Event	Sources	Total v., or.
1749.04.27.	London, Green Park	Fireworks Music	HHdb.4, 423, 426	58
1749.05.27.	London, Foundling Hospital	Foundling H. Anth., Solomon exc.	HHdb.4, 429: > 1000 listen.	50
1749.05.27.	London, Foundling Hospital	Fireworks Music	HHdb.4, 429: score	94
1749.09.19.	Salisbury, Assembly Room	Acis and Galatea (not Handel)	HHdb.4, 431	
1749.09.19.	Salisbury, Cathedral 2x	Te Deum, Coronation Anthems	HHdb.4, 431	
1749.09.20.	Salisbury, Assembly Room	Fireworks Music, Ode for S.C.	HHdb.4, 431	
1749.11.06.	Bath, Wiltshire's Room	Firemusic	HHdb.4, 431	
1749.11.18.	Durham, Assembly Room	Alexander's Feast	HHdb.4, 434	
1749.12.07.	Dublin, St. Andrew's	Te Deum, Jubilate, Coron. Anth.	HHdb.4, 431–432	
1750	London, Crown & Anchor	Acis (not Handel)	HHdb.4, 446	
1750.02.22.	Dublin, Neal's Music Hall	Judas Maccabaeus (charity)	HHdb.4, 431	
1750.03.02.	London, Covent Garden Th.	Saul 2x	HHdb.4, 435	
1750.03.09.	London, Covent Garden Th.	Judas Maccabaeus 4x	HHdb.4, 435	
1750.03.16.	London, Covent Garden Th.	Theodora 3x	HHdb.4, 436: empty theater	
1750.04.04.	London, Covent Garden Th.	Samson 2x	HHdb.4, 437	
1750.04.12.	London, Covent Garden Th.	Messiah	Larsen, 1984, 13	
1750.04.24.	London, St. Paul's	Te Deum, Jubilate, Coron. Anth.	HHdb.4, 438	
1750.05.01.	London, Foundling Hospital	Messiah 1st perf.	HHdb.4, 438: 1400 listeners	
1750.05.15.	London, Foundling Hospital	Messiah	HHdb.4, 440	
1750.09.	Hereford, College Hall	Messiah (Boyce, 3 choirs)	HHdb.4, 443	
1750.10.04.	Salisbury, Assembly Room	Messiah, organ consecration	HHdb.4, 443: 400 listeners	30
1750.10.04.	Salisbury, Cathedral	Te Deum und Anthem 2x	HHdb.4, 443	
1750.10.18.	London, King's Theatre	Handel works (charity)	HHdb.4, 443	
1750.11.30.	Dublin, St. Andrew's	Te Deum, Jubilate, Anthem	HHdb.4, 443	
1750.12.14.	Dublin, Music Hall	Messiah (Marella)	HHdb.4, 443–445	
1751.01.17.	Dublin, Neal's Music Hall	Judas Maccabaeus (Marella)	HHdb.4, 446	
1751.01.23.	Dublin, Philharmonic Room	Acis 4x (charity)	HHdb.4, 443–445	
1751.01.31.	Dublin, Neal's Music Hall	Joshua (charity)	HHdb.4, 447	
1751.02.14.	Dublin, Neal's Music Hall	Deborah (charity) 2x	HHdb.4, 447	
1751.02.18.	Dublin, Philharm. Room	Esther (charity)	HHdb.4, 447	
1751.02.11.	London, Covent Garden Th.	Deborah	HHdb.4, 447	
1751.02.22.	London, Covent Garden Th.	Belshazzar 2x	HHdb.4, 447–448	
1751.03.01.	London, Covent Garden Th.	Alexander's Feast 4x	HHdb.4, 448	
1751.03.01.	London, Covent Garden Th.	The Choice of Hercules 4x	HHdb.4, 448	
1751.03.12.	Dublin	Samson (not Handel)	HHdb.4, 449	
1751.03.15.	London, Covent Garden Th.	Esther	HHdb.4, 449	
1751.03.16.	London, Castle Tavern	Samson	HHdb.4, 449	
1751.03.20.	London, Covent Garden Th.	Judas Maccabaeus	HHdb.4, 449	
1751.03.22.	Dublin, Philharm. Room	Acis (Charitable Society)	HHdb.4, 449	
1751.03.26.	Edinburgh, New Concert Hall	Acis and Galatea (Lampe)	HHdb.4, 449	
1751.04.16.	London, King's Theatre	Handel works (charity)	HHdb.4, 451	
1751.04.18.	London, Foundling Hospital	Messiah, organ consecration 2x	HHdb.4, 450	
1751.04.23.	London, Haymarket	Coronation Anthem et al.	HHdb.4, 451	
1751.04.30.	London, St. Paul's	Te Deum, Jubilate, Coron. Anthem	HHdb.4, 450	

Year, month, day	Place	Event	Sources	Total v., or.
1751.04.30.	London, Hickford's Room	Handelwerke (Davies)	HHdb.4, 451	
1751.05.16.	London, Foundling Hospital	Messiah	HHdb.4, 452	
1751.05.23.	London, Hickford's Room	Handelarien (Cuzzoni)	HHdb.4, 452	
1751.08.28.	Gloucester, Cathedral	Te Deum, Coronation Anthem	HHdb.4, 452	
1751.08.29.	Gloucester, Booth Hall	Alexander's Feast, L'Allegro	HHdb.4, 453	
1751.10.23.	London, Castle Tavern	Acis & Galatea (not Handel)	HHdb.4, 453	
1751.10.23.	London, Castle Tavern	Acis and Galathea (not Handel)	HHdb.4, 454	
1751.10.26.	Salisbury, Cathedral	Te Deum, Coron. Anthem 2x	HHdb.4, 453	
1751.10.27.	Salisbury, Assembly Room	Samson (charity)	HHdb.4, 453	
1751.11.09.	Dublin, Philharmonic Room	Acis (Dubourg)	HHdb.4, 454	
1751.11.15.	Dublin, Philharmonic Room	Athalia (charity)	HHdb.4, 454	
1751.11.16.	Dublin	Deborah	HHdb.4, 454	
1751.11.21.	Dublin, Music Hall	Alexander's Feast, Coron. Anth.	HHdb.4, 454	
1751.11.29.	Dublin, Music Hall	Messiah (Marella)	HHdb.4, 454	
1751.12.05.	Dublin, St. Andrew's	Te Deum, 2 Anthems	HHdb.4, 454	
1751.12.27.	London, Haymarket	Water-Piece at the end	HHdb.4, 454	
1751.12.28.	London, Great R., Dean St.	Handel works (not Handel)	HHdb.4, 454	
1751.12.30.	London, Haymarket New Th.	Handelwerke (charity)	HHdb.4, 454	
1752	London, Castle Tavern	Esther (not Handel)	HHdb.4, 462	
1752	Oxford	Messiah	HHdb.4, 463	
1752.01.11.	London, Great R., Dean St.	Handel works (not Handel)	HHdb.4, 454	
1752.01.14.	London, Hickford's Gr. Room	Handel works (not Handel)	HHdb.4, 454	
1752.01.21.	London, Castle Tavern	Water Piece (not Handel)	HHdb.4, 456	
1752.01.27.	London, King's Arms	Handel works (not Handel)	HHdb.4, 456	
1752.02.05.	Dublin, Neal's Music Hall	Joshua (charity)	HHdb.4, 454	
1752.02.06.	London, Haymarket	Handel works (not Handel)	HHdb.4, 456	
1752.02.07.	Dublin, Crow St., Music Hall	Judas Maccabaeus (charity)	HHdb.4, 457	
1752.02.14.	London, Covent Garden Th.	Joshua 2x	HHdb.4, 457	
1752.02.21.	London, Covent Garden Th.	Hercules 2x	HHdb.4, 457	
1752.02.22.	London, Great R. Dean St.	Handel works (not Handel)	HHdb.4, 457	
1752.02.26.	London, Covent Garden Th.	Jephta 3x	HHdb.4, 457	
1752.03.02.	London, Haymarket	Handel works (not Handel)	HHdb.4, 457	
1752.03.06.	London, Covent Garden Th.	Samson 3x	HHdb.4, 4 58	
1752.03.14.	London, Christ Church	Te Deum, Jubilate	HHdb.4, 458	
1752.03.18.	London, Covent Garden Th.	Judas Maccabaeus 2x	HHdb.4, 459	
1752.03.24.	London, King's Theatre	Handel works (charity)	HHdb.4, 458	
1752.03.25.	London, Covent Garden Th.	Messiah 2x	HHdb.4, 460	
1752.03.31.	London, Foundling Hospital	Messiah 2x	HHdb.4, 459	
1752.04.11.	London, Great R., Dean St.	Handel works (not Handel)	HHdb.4, 461	
1752.04.16.	London, St. Paul's	Te Deum, Jubilate, Coron. Anth.	HHdb.4, 459	
1752.04.22.	London, King's Theatre	Handel works (charity)	HHdb.4, 460	
1752.04.24.	Dublin, Music Hall	Samson (Dubourg)	HHdb.4, 461	
1752.09.20.	Worcester, Cathedral	Te Deum, Coronation Anthem	HHdb.4, 461	
1752.09.21.	Worcester, Cathedral	Te Deum, Coronation Anthem	HHdb.4, 461	

Year, month, day	Place	Event	Sources	Total v., or.
1752.09.21.	Worcester, Town Hall	Samson	HHdb.4, 461	
1752.09.27.	Salisbury, Cathedral	Te Deum, Messiah parts 2x	HHdb.4, 461	51
1752.09.27.	Salisbury, Assembly Room	Samson (not Handel)	HHdb.4, 461	
1752.09.27.	Salisbury, Assembly Room	Judas Maccabaeus (not Handel)	HHdb.4, 461	
1752.12.19.	Dublin	Messiah (Marella)	HHdb.4, 462	
1753	Edinburgh	Acis and Galatea (not Handel)	HHdb.4, 462: first perform.	
1753	Edinburgh	Acis, Alexander's (not Handel)	HHdb.4, 478	
1753.01.23.	Dublin, Neal's Music Hall	Deborah (charity)	HHdb.4, 462	
1753.01.27.	London, Great R., Dean St.	Handel works (not Handel)	HHdb.4, 465	
1753.02.13.	Dublin, Neal's Music Hall	Joshua (charity)	HHdb.4, 462	
1753.02.20.	Dublin, Neal's Music Hall	Judas Maccabaeus (Marella)	HHdb.4, 467	
1753.02.22.	London	Esther (not Handel)	HHdb.4, 468	
1753.02.26.	Dublin, Neal's Music Hall	Esther (Marella)	HHdb.4, 467	
1753.03.02.	Lond, King's Theatre	Alexander's Feast (Stanley)	HHdb.4, 468	
1753.03.03.	London, Great R., Dean St.	Handel works (not Handel)	HHdb.4, 468	
1753.03.09.	London, Covent Garden Th.	Alexander's Feast 2x	HHdb.4, 469	
1753.03.16.	London, Covent Garden Th.	Jephta 2x	HHdb.4, 469	
1753.03.19.	London, Great R., Dean St.	Handel works (not Handel)	HHdb.4, 469	
1753.03.23.	London, Covent Garden Th.	Judas Maccabaeus 3x	HHdb.4, 469	
1753.04.02.	London, Little Theatre	Acis and Galatea (Frasi)	HHdb.4, 470	
1753.04.04.	London, Covent Garden Th.	Samson 3x	HHdb.4, 470	
1753.04.13.	London, Covent Garden Th.	Messiah	HHdb.4, 470	
1753.04.16.	London, Foundling Hospital	Foundling Hospital Anthem	HHdb.4, 470	
1753.04.30.	London, Haymarket	Handel works (not Handel)	HHdb.4, 471	
1753.05.01.	London, Foundling Hospital	Messiah, organ concerto	HHdb.4, 471: Handel blind	
1753.05.07.	London, King's Theatre	Judas Maccabaeus (charity)	HHdb.4, 472	
1753.05.08.	London, St. Paul's	Te Deum, Jubilate, Coron. Anth.	HHdb.4, 471: rehearsal	
1753.05.10.	London, St. Paul's	Te Deum, Jubilate, Coron. Anth.	HHdb.4, 471	
1753.07.07.	Oxford	Alexander's Feast	HHdb.4, 472	
1753.09.12.	Hereford, Cathedral Choir	Te Deum, Coronation Anthem 2x	HHdb.4, 472	
1753.09.13.	Hereford, College Hall	Samson (not Handel)	HHdb.4, 472	
1753.09.19.	Salisbury, Cathedral	Ouv., Te Deum, Coron. Anth. 2x	HHdb.4, 472	
1753.09.20.	Salisbury, Assembly Room	Judas Maccabaeus (not Handel)	HHdb.4, 472	
1753.11.19.	Dublin, Great Music Room	Samson	HHdb.4, 473	
1753.12.14.	Dublin, Great Music Room	Messiah (Marella)	HHdb.4, 473	
1754.02.13.	London, Little Theatre	Acis and Galatea (Galli)	HHdb.4, 478	
1754.03.01.	London, Covent Garden Th.	Alexander Balus	HHdb.4, 479	
1754.03.08.	London, Covent Garden Th.	Deborah	HHdb.4, 479	
1754.03.15.	London, Covent Garden Th.	Saul 2x	HHdb.4, 479	
1754.03.22.	London, Covent Garden Th.	Joshua	HHdb.4, 479	
1754.03.27.	London, Covent Garden Th.	Judas Maccabaeus 2x	HHdb.4, 480	
1754.03.29.	London, Covent Garden Th.	Samson 4x	HHdb.4, 480	
1754.04.05.	London, Covent Garden Th.	Messiah 2x	HHdb.4, 480	
1754.05.13.	Oxford, Holywell M. Room	Acis and Galatea (Oxford Mus.Soc.)	HHdb.4, 480	

Year, month, day	Place	Event	Sources	Total v., or.
1754.05.15.	London, Foundling Hospital	Messiah (Handel last time)	HHdb.4, 481; Hogwood, 228; Landon, 19	61
1754.05.23.	London, Covent Garden Th.	Ode for S. Cecilia, L'Allegro	HHdb.4, 481	
1754.07.03.	Oxford, Sheldonian Theatre	L'Allegro (Hayes)	HHdb.4, 483	
1754.07.04.	Oxford, Sheldonian Theatre	Judas Maccabaeus (Hayes)	HHdb.4, 483	
1754.07.05.	Oxford, Sheldonian Theatre	Messiah (Hayes)	HHdb.4, 483	~100
1754.09.11.	Gloucester, Booth Hall	L'Allegro (3 choirs)	HHdb.4, 483	
1754.09.12.	Gloucester, Booth Hall	Judas Maccabaeus (3 choirs) 2x	HHdb.4, 483	
1754.10.22.	King's Lynn St.Margaret's	Zadok (organ consecration)	HHdb.4, 484: Burney	
1754.11.15.	Oxford, Holywell Music Room	Alexander's Feast	HHdb.4, 484	
1754.12.07.	Dublin, St. Andrew's	Te Deum, 2 Anthems	HHdb.4, 484	
1755	Oxford, Holywell Music Room	Messiah (not Handel)	HHdb.4, 485	
1755	London, Haberdasher's Hall	Acis and Galatea (not Handel)	HHdb.4, 485	
1755	London	Joshua	HHdb.4, 485	
1755.02.14.	London, Covent Garden Th.	Alexander's Feast 2x	HHdb.4, 487	
1755.02.14.	London, Covent Garden Th.	The Choice of Hercules	HHdb.4, 487	
1755.02.14.	London, Covent Garden Th.	Alexander's Feast	HHdb.4, 487	
1755.02.21.	London, Covent Garden Th.	Ode St. Cecilia, L'Allegro	HHdb.4 ,487	
1755.02.26.	London, Covent Garden Th.	Samson 2x	HHdb.4, 488	
1755.02.28.	London, Covent Garden Th.	Joseph	HHdb.4, 488	
1755.03.05.	London, Covent Garden Th.	Theodora	HHdb.4, 489	
1755.03.11.	London, Great R. Dean St.	Esther (not Handel)	HHdb.4, 489	
1755.03.12.	London, Covent Garden Th.	Judas Maccabaeus 2x	HHdb.4, 489	
1755.03.17.	Oxford, Holywell M. Room	Judas Maccabaeus 2x (Oxf. Mus. Soc.)	HHdb.4,489	
1755.03.19.	London, Covent Garden Th.	Messiah 2x	HHdb.4, 489	
1755.03.31.	Oxford, Holywell M. Room	Messiah (last)	HHdb.4, 489	
1755.04.30.	Bath, Orchard Str.Theatre	Judas Maccabaeus (Chilcot)	HHdb.4, 489	
1755.05.01.	London, Foundling Hospital	Messiah (Smith)	HHdb.4, 489	
1755.05.03.	Bath, Orchard Str. Theatre	Samson (not Handel)	HHdb.4, 489	
1755.05.05.	Oxford, Holywell Music Room	Athalia (Oxf. Mus. Soc.)	HHdb.4, 490	
1755.05.14.	Bath, Wiltshire's Room	Alexander's Feast (Hayes)	HHdb.4, 490	
1755.05.17.	Bath, Wiltshire's Rooms	Messiah (not Handel)	HHdb.4, 490	
1755.06.23.	Oxford, Holywell M. Room	Acis and Galatea (Oxf. Mus. Soc.)	HHdb.4, 490	
1755.07.02.	Oxford, Holywell M. Room	Judas Maccabaeus. (Oxf. M.Soc.)	HHdb.4, 491	
1755.09.10.	Worcester, College Hall	Samson (3 choirs)	HHdb.4, 491	
1755.09.10.	Worcester, College Hall	Te Deum, Jubilate, Coron. Anthem	HHdb.4, 491	
1756.01.14.	Bristol, New Music Room	Messiah (not Handel)	HHdb.4, 492 : opening	
1756.02.02.	Oxford, Holywell M. Room	Esther	HHdb.4, 495	
1756.02.26.	Cambridge, Trinity C.Hall	Acis and Galatea (Randall)	HHdb.4, 496	
1756.03.01.	Oxford, Holywell Music Room	Samson in 2 parts	HHdb.4, 496	
1756.03.05.	London, Covent Garden Th.	Athalia 3x	HHdb.4, 496	
1756.03.17.	London, Covent Garden Th.	Israel in Egypt 2x	HHdb.4, 496	
1756.03.19.	London, Covent Garden Th.	Deborah	HHdb.4, 496	

Year, month, day	Place	Event	Sources	Total v., or.
1756.03.22.	Oxford	Te Deum	HHdb.4, 496	
1756.03.26.	London, Covent Garden Th.	Judas Maccabaeus 2x	HHdb.4, 497	
1756.04.02.	London, Covent Garden Th.	Jephta	HHdb.4, 497	
1756.04.07.	London, Covent Garden Th.	Messiah 2x	HHdb.4, 497	
1756.04.26.	Oxford	Messiah, 3rd part	HHdb.4, 498	
1756.05.04.	London, St. Paul's	Te Deum, Jubilate, Coron. Anth.	HHdb.4, 497	
1756.05.19.	London, Foundling Hospital	Messiah (Smith)	HHdb.4, 498	53
1756.05.24.	Oxford, Holywell Music Room	Acis and Galatea (Oxf. Mus. Soc.)	HHdb.4, 498	
1756.07.06.	Oxford, Sheldonian Theatre	Judas Maccabaeus (Hayes)	HHdb.4, 499	
1756.07.07.	Oxford, Sheldonian Theatre	Joshua (Hayes)	HHdb.4, 499	
1756.07.08.	Oxford, Sheldonian Theatre	Messiah	HHdb.4, 499: > 3000 listen.	
1756.09.15.	Hereford, Cathedral 2x	Te Deum, Jubil. Coronation Anth.	HHdb.4, 500	
1756.09.17.	Hereford, College Hall	L'Allegro, Ode St. Cecilia	HHdb.4, 500	
1756.10.06.	Salisbury, Assembly Room	Hercules (not Handel)	HHdb.4, 500	
1756.10.06.	Salisbury, Cathedral 2x	Te Deum, 2 Anthems	HHdb.4, 500	
1756.10.07.	Salisbury, Assembly Room	Esther (not Handel)	HHdb.4, 500	
1756.10.23.	Oxford, Holywell M. Room	Esther, Coronation Anthem	HHdb.4, 501	
1756.11.24.	Bath, Simpson's Great R.	Messiah (not Handel)	HHdb.4, 501	
1756.12.09.	London, Haberdasher's Hall	Samson (charity)	HHdb.4, 501	
1756.12.16.	Dublin, Great Music Room	Messiah (Lee)	HHdb.4, 501	
1757.07.16.	Gloucester, Cathedral	Te Deum, Anthem (3 choirs)	Jackson's Oxford Journal	
1757	Gloucester, Booth Hall	Messiah (Hayes, 3 choirs)	HHdb.4, 502: numerous	
1757,05.02.	London, Little Theatre	Acis and Galatea (Snow)	HHdb.4, 507	
1757.02.03.	Bristol, New Music Room	Judas Maccabaeus	HHdb.4, 504: new organ	
1757.02.07.	Oxford, Holywell Music Room	Messiah (selection)	HHdb.4, 505	
1757.02.21.	Oxford, Holywell Music Room	Acis and Galatea (Oxf. Mus. Soc.)	HHdb.4, 505	
1757.02.25.	London, Covent Garden Th.	Esther 2x	HHdb.4, 505	
1757.03.02.	Bristol, Prince's St., Music Room	Judas Maccab., organ consecr.	HHdb.4, 504	
1757.03.03.	Bristol, New Music Room	Messiah (not Handel)	HHdb.4, 504: large choir!	
1757.03.04.	London, Covent Garden Th.	Israel in Egypt	HHdb.4, 505	
1757.03.09.	London, Covent Garden Th.	Joseph	HHdb.4, 505	
1757.03.11.	London, Covent Garden Th.	Triumph of Time and Truth 4x	HHdb.4, 505	
1757.03.12.	Oxford, Holywell Music Room	Alexander's Feast	HHdb.4, 506	
1757.03.14.	London, Great Room	Samson (Frasi)	HHdb.4, 506	
1757.03.24.	London. King's Theatre	Coronation Anthem (charity)	HHdb.4, 506	
1757.03.25.	London, Covent Garden Th.	Judas Maccabaeus	HHdb.4, 506	
1757.03.28.	Oxford, Holywell Music Room	Messiah (end)	HHdb.4, 505	
1757.03.30.	London, Covent Garden Th.	Messiah 2x	HHdb.4, 506	
1757.04.18.	Bath, Orchard Str.,Theatre	Acis and Galatea (Linley)	HHdb.4, 506	
1757.04.20.	Bath, Orchard Str.,Theatre	Alexander's Feast	HHdb.4, 506	
1757.04.25.	Oxford, Holywell Music Room	Judas Maccabaeus, 1st part	HHdb.4, 507	
1757.04.25.	London, New Theatre	Acis and Galatea (charity)	HHdb.4, 507	
1757.05.05.	London, Foundling Hospital	Messiah	HHdb.4, 507	
1757.05.16.	Oxford, Holywell Music Room	Choice of Hercules (Oxf. Mus. S.)	HHdb.4, 507	

Year, month, day	Place	Event	Sources	Total v., or.
1757.06.09.	London, Ranelagh House	Acis and Galatea (Stanley)	HHdb.4, 508	
1757.06.27.	Oxford, Holywell Music Room	Saul (Oxf. Mus. Soc.)	HHdb.4, 508	
1757.07.06.	Oxford, Holywell Music Room	Messiah	HHdb.4, 508	
1757.07.07.	Oxford, Holywell Music Room	Esther	HHdb.4, 508	
1757.08.01.	Oxford	L'Allegro	HHdb.4, 508	
1757.09.07.	Bristol, New Assembly Room	Samson (Hayes)	HHdb.4, 509	
1757.09.14.	Gloucester, Booth Hall	Judas Maccabaeus (3 choirs)	HHdb.4, 508	
1757.09.14.	Gloucester, Cathedral 2x	Te Deum, Jubilate, Coron. Anth.	HHdb.4, 508	
1757.09.15.	Gloucester, Booth Hall	Acis & Galatea (3 choirs)	HHdb.4, 508	
1757.09.16.	Gloucester, Booth Hall	Messiah (3 choirs)	HHdb.4, 508	
1757.10.31.	Oxford, Holywell Music Room	Choice of Hercules, Ode S. C.	HHdb.4, 510	
1757.11.28.	Oxford, Holywell Music Room	Acis and Galatea (Oxf. Mus. Soc.)	HHdb.4, 510	
1757.12.01.	Dublin, St. Andrew's	Te Deum, Jubilate, Anthems	HHdb.4, 510	
1757.12.29.	Bath, Wiltshire's Rooms	Samson (Passerini)	HHdb.4, 511	
1758	Worcester, College Hall	Messiah (Hayes, 3 choirs)	HHdb.4, 512–513: Evening	
1758	Bristol	Saul (not Handel)	HHdb.4, 521	
1758	London, Castle Tavern	Samson (not Handel)	HHdb.4, 521	
1758.02.10.	London, Covent Garden Th.	Triumph of Time & Truth 2x	HHdb.4, 512	
1758.02.22.	London, Covent Garden Th.	Belshazzar	HHdb.4, 512	
1758.02.24.	London, Covent Garden Th.	Israel in Egypt	HHdb.4, 513	
1758.03.01.	London, Covent Garden Th.	Jephta	HHdb.4, 513	
1758.03.03.	London, Covent Garden Th.	Judas Maccabaeus 2x	HHdb.4, 513	
1758.03.06.	London, King's Theatre	Samson (Frasi)	HHdb.4, 514	
1758.03.10.	London, Covent Garden Th.	Messiah 4x	HHdb.4, 514	
1758.03.13.	Oxford	Messiah 2 parts	HHdb.4, 513	
1758.03.27.	Oxford, Holywell Music Room	Judas Maccbaeus (Oxf. Mus. S.)	HHdb.4, 514	
1758.03.31.	London, Great Room	Acis & Galatea 2x (charity)	HHdb.4, 514	
1758.04.04.	London, St. Andrew's	Te Deum, Anthem (charity)	HHdb.4, 515	
1758.04.06.	London, King's Theatre	Handel works (charity)	HHdb.4, 514	
1758.04.07.	London, St. Margaret's	Te Deum, Coronation Anthem, Hallelujah from Messiah	HHdb.4, 515: first time	
1758.04.18.	London, St. Paul's	Te Deum, Coronation Anthem, Hallelujah from Messiah	HHdb. 4, 515: first time	
1758.04.26.	Bath, Wiltshire's Rooms	Samson (charity)	HHdb.4, 516	
1758.04.26.	London, St. Anne	Te Deum, Coron. Anth., Hall.	HHdb.4, 516	
1758.04.27.	London, Foundling Hospital	Messiah (Smith)	HHdb.4, 516: 6 boys	57
1758.04.27.	Bath, Wilshire's Rooms	L'Allegro (charity)	HHdb.4, 515	
1758.05.11.	London	Messiah (A.A.M.)	HHdb.4, 517	
1758.06.12.	Oxford, Holywell Music Room	Acis & Galatea (Oxf. Mus. Soc.)	HHdb.4, 517	
1758.07.04.	Oxford, Holywell Music Room	Coronation Anthems et al.	HHdb.4, 518	
1758.08.16.	Bristol, Assembly Room	L'Allegro	HHdb.4, 518	
1758.08.17.	Bristol, Cathedral	Messiah	HHdb.4, 518: Evening	
1758.08.30.	Worcester, College Hall	Judas Macc. (3 choirs)	HHdb.4, 520	
1758.08.31.	Worcester, College Hall	Alexander's Feast	HHdb.4,520	
1758.09.01.	Worcester, College Hall	Messiah	HHdb.4, 520	

Year, month, day	Place	Event	Sources	Total v., or.
1758.10.30.	Dublin, Music Room	L'Allegro	HHdb.4, 521	
1758.12.14.	Dublin, Music Room	Messiah (Lee)	HHdb.4, 521	
1759	London	Joshua	HHdb.4, 521	
1759	London, Haberdasher's Hall	Esther (Castle Society)	HHdb.4, 541	
1759–1834	Hereford, Cathedral Choir	Messiah als Anthem	HHdb.4, 521, 539	
1759.02.06.	Dublin, Neal's Music Hall	Acis and Galatea (charity)	HHdb.4, 525	
1759.03.02.	London, Covent Garden Th.	Solomon 2x	HHdb.4, 525	
1759.03.09.	London, Covent Garden Th.	Susanna	HHdb.4, 526	
1759.03.14.	London, Covent Garden Th.	Samson 3x	HHdb.4, 526	
1759.03.23.	London, Covent Garden Th.	Judas Maccabaeus 2x	HHdb.4, 526	
1759.03.30.	London, Covent Garden Th.	Messiah 3x	HHdb.4, 527: Handel plays	
1759.04.07.	London, Foundling Hospital	Messiah (Smith, Handel last time)	HHdb.4, 527	
1759.04.26.	London, St. Margaret's	Te Deum, Messiah Hallelujah	HHdb.4, 528	
1759.04.28.	Bath, Assembl.Hall, Abbey Church	Messiah (Hayes)	HHdb.4, 533	
1759.05.03.	London, Foundling Hospital	Messiah (Smith)	HHdb.4, 532	5, 58
1759.05.10.	London, St. Paul's, Choir	Messiah	Deutsch, 825; Schoelcher, 337	
1759.05.17.	Bath, Assembly Hall	Messiah	HHdb.4, 536	
1759.05.17.	Cambridge, Senate Hall	Messiah	HHdb.4, 536	
1759.05.24.	London, Foundling Hospital	Foundling Hospital and Coronation Anthem (Smith)	HHdb.4, 536	
1759.05.28.	Oxford, Holywell Music Room	2 Coronation Anthems	HHdb.4, 536	
1759.07.03.	Oxford, Sheldonian Theatre Commemoration	Samson	HHdb.4, 537	
1759.07.04.	Oxford, Sheldonian Theatre Commemoration	Esther	HHdb.4, 537	
1759.07.05.	Oxford, Sheldonian Theatre Commemoration	Messiah	HHdb.4, 536	
1759.08.13.	London, Hampstead L. Room	Acis	HHdb.4, 535	
1759.09.	Salisbury, Cathedral	Samson (3 choirs)	HHdb.4, 540: in church!	
1759.09.12.	Hereford, Cathedral	Te Deum, Jubilate, Coron. Anth.	HHdb.4, 539	
1759.09.12.	Hereford, College Hall	Joshua	HHdb.4, 539	
1759.09.13.	Hereford, Cathedral	Te Deum, Jubilate, Coron. Anth.	HHdb.4, 539	
1759.09.14.	Hereford, Cathedral	Messiah	HHdb.4, 539: liturgy!	
1759.09.27.	Church-Langton	Messiah (Hayes)	HHdb.4, 539	
1759.10.29.	Oxford, Holywell Music Room	Acis	HHdb.4, 540	
1759.12.10.	Oxford, Holywell Music Room	L'Allegro	HHdb.4, 541	
1760.03.07.	Edinburgh	Messiah	HHdb.4, 161	
1763	Gloucester	Messiah	Dean, 104	>50
1767	London, Randall & Abell	Messiah, first printed score	HHA I/17	
1767.10.23.	Birmingham	Messiah	Dean, 104	~85
1770	New York	Messiah, selection	Hogwood, 254	
1770–1783	Winchester	Messiah	Dean, 104	
1771.09.23.	Hamburg	Alexander-Fest (Arne) German	Siegmund-Schulze, 1960, 52	
1772	New York	Messiah, Auswahl	Hogwood, 254	
1772.04.15.	Hamburg, Bosselhof, private	Messiah (Michael Arne) English	NMA X/28/1/2,41; HJb.1960, 52	

Year, month, day	Place	Event	Sources	Total v., or.
1772.05.21.	Hamburg, Drillhaus	Messiah (Michael Arne) English	NMA X/28/1/2,41; HJb.1960, 52	
1772.08.02.	Braunschweig, Carolinum	Judas Maccabaeus German	HJb.1960, 68	
1772.12.04.	Edinburgh	Messiah 2x	HHdb.4,161	
1773	Boston	Messiah, selection	Hogwood, 254	
1773.05.10.	Stockholm	Acis and Galatea, ballet	Dean, 104	
1773.05.27.	London, Marylebone Garden	Acis and Galatea, staged	Dean, 104	
1774	Berlin	Judas Maccabaeus	Bernhardt, ZfMw 7(1935),516	
1775.12.31.	Hamburg, Concertsaal auf der Kamp	Messias (Ph.E.Bach) German	HJb.1960,52–54	
1777	Berlin	Judas Maccabaeus	Bernhardt, ZfMw 7(1935),516	
1777.02.23.	Hamburg, Freimaurerloge	Messias (Ph.E.Bach) German	NMA X/28/1/2, 41; Sittard, 108	
1777.11.01.	Mannheim, Kirche	Messias (Abbé Vogler) Italian	Seiffert, Peters Jb. 1916, 61 f.	
1778.03.18.	Hamburg, Concertsaal auf der Kamp	Ph.E.Bach, Auferstehung	Hamburger Correspondent	
1778.04.12.	Hamburg, Freimaurerloge	Messias (Ph.E.Bach) German	NMA X/28/1/2, 41; Sittard, 108	
1778.12.20.	Vienna, Kärtnertortheater	Handel choirs (Tonkünstlersoc.)	Morrow, 246; Edelmann, 178	
1779.03.21.	Vienna, Kärtnertortheater	Judas Maccabaeus (Starzer) 2x	NMA X/28/1/1; Morrow, 246	
178.	Breslau, Maria-Magdalena-Kirche	Messias (Hiller)	Hiller	
1780	Betlehem, Pennsylvania	Messiah, selection	Hogwood, 254	
1780	Schwerin, Ludwigslust?	Messias German	HJb.1960,57	
1780–	New York	Messiah, selection	Hogwood, 254	
1780–1790	Vienna, Tonkünstlersocietät	Handel Samson (Salieri)	Edelmann, 183: no. of parts	~150
1780–1790	Vienna, Tonkünstlersocietät	Handel Israel (Salieri)	Edelmann, 183: no. of parts, double choir	133
1780.03.12.	Vienna, Kärntnertortheater	Handel choirs	Morrow, 247	
1780.05.13.	Weimar, Wittumspalais	Messias (Wolf), German 3x	HJb.1960, 56	
1782.03.	Edinburgh	Messiah	HHdb.4, 161	
1782.03.03.	Hamburg, Subskription	Messias, German	NMA X/28/1/2, 41; Sittard, 133	
1782.12.22.	Vienna, Kärntnertortheater	Handel choirs	Morrow, 252	~150
1783	Potsdam, Saal	Messias (Zelter)	Larsen, 1984, 14	
1783.05.06.	London, Covent Garden Th.	Acis and Galatea, staged	HHdb.4, 161	
1784	Paris	Messiah, aria (La Mara)	Hogwood, 252	
1784	Philadelphia	Messiah	Hogwood, 254	280
1784	Boston	Messiah	MMG 5, 1291 f.: Charity	
1784.03.28.	Vienna, Burgtheater	Haydn Ritorno di Tobia 2x	Morrow, 256; Biba, 94	153
1784.05.26.	London, Westminster Abbey	Messiah, Handel Commemoration	Burney, Account, 8; Shaw, 69; Koury, 46	533, 283
1784.05.27.	London, Pantheon	Handel Commemoration	Burney, Account, 49	200
1784.12.23.	Vienna, Burgtheater	Handel, choirs (Tonkünstlersocietät)	Morrow, 258	
1785	London, Westminster Abbey	Messiah	Hogwood, 240	616
1785.03.13.	Vienna, Burgtheater	Mozart Davidde penitente 2x	KV 469	
1786	London, Westminster Abbey	Messiah	Hogwood, 240	640
1786.03.	Kopenhagen	Messias	HJb.1960, 69	
1786.04.	Stockholm	Messias	Myers (1948), 269 f.	

Year, month, day	Place	Event	Sources	Total v., or.
1786.05.19.	Berlin, Dom	Messias (Hiller, Zelter), Italian	Hiller (1786), 28; Koury, 27, 48	
1786.11.03.	Leipzig, Paulinerkirche	Messias (Hiller), German	HJb.1960, 64; Schreiber, 204	
1787	London, Westminster Abbey	Messiah	Koury 27; Hogwood, 242	828
1787	Leipzig, Hotel de Saxe	Messias, selection (Hiller)	Hiller, 1787	
1787–	Dublin, Kirche jährlich	Handel Commemoration, 2 days	Lenz, 309; Scholes,181	>200
1787.05.11.	Leipzig, Paulinerkirche	Messias (Hiller), German	Hiller, 1787	
1788.02.26.	Vienna, P. Joh. Esterházy	Ph.E.Bach, Auferstehung (Mozart)	NMA X/28 I, 2; not 86!	66
1788.03.04.	Vienna, P. Joh. Esterházy	Ph.E.Bach, Auferstehung (Mozart)	Morrow, 11	
1788.03.07.	Vienna, Burgtheater	Ph.E.Bach, Auferstehung (Mozart)	Morrow, 11, 272; NMA X/28 I, 1; not 86!	66
1788.05.30.	Breslau, Maria-Magdalena Kirche	Messias (Hiller), German	Koury, 27 Anm.65; HJb.1960, 66; double choir	259
1788.11.	Vienna, Jahn Himmelspforte	Acis und Galatea (Mozart KV 566)	Windszus, HHA 1/9,1	
1788.12.22.	Vienna, Burgtheater	Handel, aria and choir (Salieri)	Morrow, 273	
1788.12.23.	Vienna, Burgtheater	Handel, aria and choir (Salieri)	Morrow, 273	
1788.12.30.	Vienna, P. Joh. Esterházy	Acis und Galatea (Mozart)	KV 566, 637; MJb. 1960/61, 173	
1789.03.06.	Vienna, P. Joh. Esterházy	Messias (Mozart) 2x	NMA X/28 I, 2; Wolff, Fs. Lang, 12	~50?
1789.04.07.	Vienna, P. Joh. Esterházy	Messias (Mozart)	Wolff, Fs.Lang, 12: Morrow, 180	
1790	Vienna	Alexandersfest / Cäcilienode (Mozart)	NMA X/28/1/1: probable	
1790	Vienna, v. Swieten	Messias (Mozart)	Morrow, 383	
1791	Vienna, v. Swieten	Alexandersfest (Mozart)	KV 591, 677; Morrow, 384	
1791	Vienna, v. Swieten	Caecilienode (Mozart)	KV 592, 678	
1791.04.16.	Vienna, Burgtheater	Mozart, various works 2x	Morrow, 278	
1791.06.01.	London, Westminster Abbey	Messiah	Hogwood, 243: Haydn	1068
1792.12.22.	Vienna, Burgtheater	Weigl, Venere e Adone	Morrow, 280	
1793.01.02.	Vienna, Jahn	Mozart, Requiem	Morrow, 281: v. Swieten	
1793.03.	Vienna, P. Dietrichstein	Alexandersfest (Mozart)	NMA X/28/1/3, S.VIII	
1793.03.23.	Vienna, Burgtheater	Weigl, Venere e Adone 2x	Morrow, 282	
1793.12.22.	Vienna, Burgtheater	Haydn, symphony, choir pieces 2x	Morrow, 283	
1793.12.24.	Vienna, P. Lichnowsky	Caecilienode, Herkules	Morrow, 386; NMA X/28/1/4	
1793.12.28.	Vienna, P. Paar	Caecilienode	Zinzendorf; NMA X/28/1/4	
1793.12.28.	Vienna, P. Schwarzenberg	Alexandersfest	Morrow, 386	
1794.04.12.	Vienna, Burgtheater	Haydn, symphony et al. 2x	Morrow, 285	
1794.04.15.	Vienna, P. Lichnowsky	Judas Maccabaeus	Zinzendorf	
1794.12.22.	Vienna, Burgtheater	Eybler, Die Hirten 2x	Morrow, 286	
1794.12.31.	Vienna, P. Paar	Athalia	Zinzendorf; Textbuch	
1795	Berlin, Singakademie	Judas Maccabaeus	Koury, 27	
1795.03.29.	Vienna, Burgtheater	Cartellieri, Gioas 2x	Morrow, 287: Beethoven piano concerto	
1795.04.05.	Vienna, P. Paar	Messias	NMA X/28/1/2, 33	
1795.12.22.	Vienna, Burgtheater	Handel, choir (Salieri)	Morrow, 289	

Year, month, day	Place	Event	Sources	Total v., or.
1795.12.23.	Vienna, Burgtheater	Handel, choir (Salieri)	Morrow, 290	
1796.03.20.	Vienna, Burgtheater	2 cantatas 2x	Morrow, 290	
1796.03.26.	Vienna, P. Schwarzenberg	Haydn Sieben Worte 2x	Landon 4, 97	
1797.03.24.	Vienna, P. Schwarzenberg	Acis und Galatea	NMA X/28/1/1	
1797.03.27.	Vienna, P. Schwarzenberg	Acis und Galatea	NMA X/28/1/3, VIII	
1797.12.22.	Vienna, Burgtheater	Handel, choir (Salieri)	Morrow, 295	
1797.12.23.	Vienna, Burgtheater	Handel, choir (Salieri)	Morrow, 295	
1798–	London, Vauxhall Gardens	Fireworks music	regularly since 1798	
1798.04.01.	Vienna, Burgtheater	Haydn Sieben Worte 2x	Landon 4, 316: not 180!	152
1798.04.27.	Vienna, P. Schwarzenberg	Haydn Schöpfung, 2 rehearsals	Landon 4, 321	
1798.04.29.	Vienna, P. Schwarzenberg	Haydn Schöpfung, 1st p. 2x	Landon 4, 321; Morrow, 391	
1798.05.07.	Vienna, P. Schwarzenberg	Haydn Schöpfung 2x	Landon Haydn 4, 322	
1799.03.02.	Vienna, P. Schwarzenberg	Haydn Schöpfung 2x	Brown, Tab. I; Morrow, 391	
1799.03.19.	Vienna, Burgtheater	Haydn Schöpfung 1st publ. perf.	Brown, 20; Landon 4, 455	180, 60
1799.03.23.	Vienna, P. Schwarzenberg	Messias 1. Teil	Morrow, 180: Küttner 3, 297	~150
1799.03.24.	Vienna, P. Schwarzenberg	Messias 2. Teil	Morrow, 180: Küttner 3, 297	~150
1799.12.22.	Vienna, Burgtheater	Haydn Schöpfung (Haydn) 2x	Landon 4, 498; Morrow, 303	>200
1800.03.08.	Ofen, für Erzherzog Joseph	Haydn Schöpfung (Haydn), dilettanti	Landon 4, 542	
1800.03.11.	Vienna, P. Trautmannsdorf	Haydn Schöpfung	Brown, Table I	
1800.03.28.	London, Covent Garden	Haydn Schöpfung, English 1st perf.	Landon 4, 573; Temperley, 111	120
1800.04.04.	Vienna, P. Fries	Haydn Schöpfung, reduced version	Landon Haydn 4, 546: soli + strings only.! Wranitzky	3, 12
1800.04.06.	Vienna, Burgtheater	Haydn Schöpfung (Haydn) 2x	Morrow, 304; EM 1988, 8	3, 200
1800.04.12.	Vienna, P. Schwarzenberg	Haydn Schöpfung (Weigl) 2x	Landon 4, 570	200?
1800.04.21.	London, Concert R. K.T.	Haydn Schöpfung (Salomon)	Landon 4, 573	
1800.08.19.	Salzburg, Aula Magna, Univ.	Haydn Schöpfung (M. Haydn)	Landon 4, 577	95
1800.09.06.	Eisenstadt, P. Esterhazy	Haydn Schöpfung	Brown, Table I	
1800.11.15.	Vienna, Theater Leopoldstadt	Haydn Schöpfung (Haydn?)	Brown, Table I	
1800.12.22.	Vienna, Burgtheater	Haydn Schöpfung (Wranitzki) 2x	Landon 4, 570; Morrow, 307	<200
1800.12.24.	Paris, Théâtre des arts	Haydn Création	Landon 4, 579: AMZ 3.1800, 269	309
1800.12.24.	Paris, Opéra	Haydn Création (Steibelt)	AMZ 5, 737	250
1801.01.05.	Berlin, Oper	Haydn Schöpfung	Landon 4, 586: AMZ 3.1801, 289	>150
1801.01.16.	Vienna, gr. Redoutensaal	Haydn Schöpfung	Morrow, 307; Landon 5, 22	200
1801.03.25.	Vienna, Theater Leopoldstadt	Haydn Schöpfung	Morrow, 308; Landon 5, 32	
1801.03.28.	Vienna, Burgtheater	Haydn Sieben Worte	AMZ 3, 498	
1801.04.04.	Vienna, P. Lobkowitz	Haydn Schöpfung Italian	Landon 5, 29	
1801.04.21.	Vienna, P. Schwarzenberg	Haydn Schöpfung, rehearsal 2x	Landon 5, 30	
1801.04.24.	Vienna, P. Schwarzenberg	Haydn Jahreszeiten, 1st perf. 3x	NG 8, 363; AMZ 2.5.1801	
1801.05.13.	Vienna, Salon Schmierer	Haydn Jahreszeiten	Brown, Table I, Rosenbaum	
1801.05.29.	Vienna, gr. Redoutensaal	Haydn Jahreszeiten	Morrow, 309	
1801.12.	Leipzig, Theater?	Haydn Jahreszeiten (Haydn) 2x	Landon 5, 183: AMZ 4.1801, 239f.	
1801.12.22.	Vienna, Burgtheater	Haydn Jahreszeiten (Haydn) 2x	Morrow, 311; Landon 5, 90	

Year, month, day	Place	Event	Sources	Total v., or.
1801.12.27.	Vienna, gr. Redoutensaal	Haydn Schöpfung (Haydn)	Morrow, 311; Landon 5, 91	200?
1802.01.22.	St.Petersburg, Saal	Haydn Schöpfung 2x	Landon, Haydn 5, 214: hall too small, 1000 listeners	250
1802.03.25.	Vienna, Th. an der Wien	Haydn Schöpfung (Haydn)	Brown, Table I	
1802.04.11.	Vienna, Burgtheater	Haydn Jahreszeiten (Haydn)	Morrow, 314	
1802.06.06.	Pressburg, Stadttheater	Haydn Schöpfung	Brown, Table I	
1802.09.30.	Vienna, Augarten	Haydn Schöpfung (Schuppanzigh)	Brown, Table I	
1802.12.22.	Vienna, Burgtheater	Haydn Jahreszeiten (Haydn) 2x	Morrow, 317; AMZ 4.1802, 263	200
1802.12.26.	Vienna, gr. Redoutensaal	Haydn Schöpfung (Haydn)	Morrow, 317	200?
1803	Halle	Messias etc. (Reichardt)	New Grove	
1803.02.18.	Vienna, gr. Redoutensaal	Messias, better than Th.a.d. Wien	AMZ 1803, 336	
1803.04.03.	Leipzig, Altes Gewandhaus	Messias (Mozart-version)	AMZ 4.1803, 482	
1803.04.05.	Vienna, Burgtheater	Haydn Schöpfung	Morrow, 319; Landon, 5, 258	~70
1803.05.13.	Vienna, Salon Schmierer	Haydn Jahreszeiten	Morrow, 32; Landon 5, 261	
1804.03.25.	Vienna, Burgtheater	Haydn Schöpfung 2x	Morrow, 323	~200
1804.09.	Eisenstadt, P. Esterhazy	Haydn Schöpfung (Hummel)	Brown, Table I	
1804.12.25.	Vienna, gr. Redoutensaal	Handel Saul (Weigl)	Morrow, 327: AMZ 13.2.1805	
1805	London, Covent Garden Th.	Messias (Mozart-version)	Hogwood 1984, 246	
1805.04.07.	Vienna, Burgtheater TK	Haydn Schöpfung (Salieri?) 2x	Morrow, 329; Brown, Table I	~200
1805.12?	Leipzig, Theater?	Haydn Schöpfung	AMZ 8 (1806), 224	
1805.12.24.	Vienna, gr. Redoutensaal	Haydn Schöpfung	Brown, Table I	
1806–	Berlin, Singakademie	Handel (Zelter), regularly	NG, Berlin	
1806.03.30.	Vienna, Th. an der Wien	Messias (Mozart-version)	AMZ 8 (1806), 461; Morrow, 337, ensemble too small!	>100
1806.12.	Vienna, Burgtheater	Judas Maccabaeus	AMZ 9 (1807), 233	
1806.12.	Leipzig	Messias (Mozart-version, Müller)	AMZ 9 (1807), 476	
1806.12.25.	Vienna, gr. Redoutensaal	Messias (Salieri)	AMZ 9 (1807), 336; Morrow, 341; hall too large!	
1807	Vienna, Theater an der Wien	Messias	AMZ 18.2.1807	
1807.03.22.	Vienna, Theater an der Wien	Alexandersfest (Mozart-version)	Morrow, 342: AMZ 22.4.1807	
1807.03.22.	Vienna, Burgtheater	Messias Alleluja	Morrow, 342	
1807.03.24.	Vienna, Burgtheater	Haydn Schöpfung	Morrow, 343	
1807.05.17.	Vienna, Burgtheater	Haydn Schöpfung	Morrow, 344	
1807.12.22.	Vienna, Burgtheater	Haydn Schöpfung 2x	Morrow, 346; Brown, Table I	~200
1807.12.22.	Vienna, Theater an der Wien	Alexandersfest (Mozart-version)	Morrow, 346	
1807.12.23.	Vienna, Burgtheater	Haydn Schöpfung 2x	Morrow, 346	
1808	Vienna, Burgtheater	Resonanzkuppel		
1808.03.27.	Vienna, Alte Universität	Haydn Schöpfung	Morrow, 176; Biba, 103	103
1808.04.17.	Vienna, Burgtheater	Haydn Schöpfung	Morrow, 350	200
1808.12.22.	Vienna, Burgtheater	Haydn Ritorno di Tobia 2x	Morrow, 352	
1809.03.26.	Vienna, Burgtheater	Haydn Schöpfung (Salieri) 2x	Brown, Table I, 652	200
1809.03.27.	Vienna, Theater an der Wien	Messias	Morrow, 354	
1809.11.15.	Vienna	Hofkapelle	AMZ 3.1.1810	
1809.12.22.	Vienna, Burgtheater	Haydn Jahreszeiten	Morrow, 358	
1809.12.23.	Vienna, Burgtheater	Haydn Jahreszeiten	Morrow, 358	

Year, month, day	Place	Event	Sources	Total v., or.
1810	Vienna, P. Freyin v. Sala	Haydn Schöpfung	Morrow, 180	
1810.12.22.	Vienna, Burgtheater	Haydn Schöpfung 2x	Morrow, 363	
1812.11.29.	Vienna, Winterreitschule	Alexandersfest (Mozart-version)	Schwab, 90; EM (1988),11	582
1813.11.11.	Vienna, Winterreitschule 2x	Alexandersfest (GdM)	Hanson, 93	704
1814	Vienna, Winterreitschule	Samson (Mosel)	Dean, 355	700
1814.10.16.	Vienna, Winterreitschule	Samson (GdM)	Bernhardt, 440; Hanson, 93	900
1814.12.22.	Vienna, Burgtheater	Haydn Schöpfung	AMZ 17, 46	200
1815	Boston, Handel & Haydn S.	Messiah, selection	Hogwood, 254	
1815–	Boston, Boylston Hall	Messiah, 1st compl. perf. USA	Hogwood, 254: yearly	
1815.05.20.	Vienna, Winterreitschule 2x	Messias (Mozart-version, GdM)	Hanslick I, 148; Hanson, 93	
1815.10.31.	Edinburgh, Festival	Messiah et al. 4x	AMZ 18 (1816), 633	123
1816.12.22.	Vienna	Haydn Schöpfung 2x	AMZ 19 (1816), 65	
1817	Vienna	Beethoven Christus	EM (1988), 8, Anm. 25	100
1818	Salzburg, Landständischer Saal	Timotheus (Alexandersfest)	AMZ 20,626	
1819	Elberfeld, Niederrhein. MF	Handel	NG, Festival: 200–600	
1819	Vienna, Tonkünstlersocietät	Stadler, Befreiung Jerusalems	Biba, Haydn, 94	
1820–1840	Namest, Schloss	Messias, other oratorios	Racek, HJb. 1959, 175	
1823	York, Cathedral	Handel Te Deum (Festival)	Hogwood, 245	565
1823	York, Assembly Room	Handel Messiah (Festival)	Harmonicon 1 (1823), 152 f.	446
1824	Wakefield	Festival	Harmonicon 3 (1825), 19	
1828	Paris, Conservatoire	Beethoven 3rd Symph. (Habeneck)	Elwart, 98	
1829.03.11.	Berlin, Singakademie	Bach, Matthäuspassion (Mendelssohn)	Geck, Wiederentdeckung, 34	158 v.
1829.06.22.	London, King's Theatre	Acis and Galatea, staged	Dean, 630	
1830	Vienna, Burgtheater	Messias Italian (Salieri-version)	Morrow, 188 f.; Edelmann, 177	
1830	Berlin, Singakademie	Bach, Matthäuspassion	Young, 116	
1831	Berlin, Singakademie	Bach, Matthäuspassion	Young, 116	
1831.02.03.	London, Queen's Theatre	Acis and Galatea, staged	Dean, 630	
1832	London, Drury Lane	Acis and Galatea, staged	Dean, 630	
1832	Berlin, Singakademie	Bach, Matthäuspassion	Young, 116	
1833	Berlin, Singakademie	Bach, Matthäuspassion	Young, 116	
1833.05.26.	Düsseldorf, Mendelssohn	Israel in Egypt	Hogwood, 252	
1833.11.22.	Düsseldorf, Mendelssohn	Alexander's Feast	Hogwood, 252	
1834	London, Westminster Abbey	Messiah	Hogwood, 249	644
1834	Vienna, Burgtheater	Messias Alleluja (Salieri-version)	Morrow, 188 f.	
1834.03.08.	Elberfeld, Mendelssohn	Messias	NG, Festival	
1834.08	Düsseldorf, Mendelssohn	Dettingen Te Deum	NG, Festival	
1834.12.16.	Düsseldorf, Mendelssohn	Judas Maccabaeus	NG, Festival	
1835	Hereford, Cathedral Choir	Messiah, as anthem	NG, Festival	
1835.05	Cologne, Mendelssohn	Solomon	Werner, Mendelssohn 254	
1836	Vienna, Burgtheater	Messias (Tonkünstlersocietät)	Morrow, 188 f.	
1836.05.22.	Niederrheinisches Musikfest	Mendelssohn Paulus	NG, Festival	
1836.11.07.	Leipzig, Paulinerkirche	Handel Israel (Mendelssohn)	Dörffel, 89	

Year, month, day	Place	Event	Sources	Total v., or.
1837	London, Exeter Hall	Mendelssohn Paulus	Young, 172	
1837.11.16.	Leipzig, Paulinerkirche	Messias (Mendelssohn)	NG, Festival	
1838	London, Westminster Abbey	Messiah	NG, Festival	
1838	Gloucester Cathedral	Three Choirs	Lysons, 133	301
1842.02.05.	London, Drury Lane	Acis and Galatea, staged	Dean, 630	
1842.11.21.	New York	Acis and Galatea, staged	Dean, 630	
1843	Hereford Cathedral	Three Choirs	Lysons, 146	110
1843.05.05.	London, Drury Lane	Acis and Galatea, staged	Dean, 630	
1843.11.05.	Vienna, Winterreitschule	Haydn Schöpfung	Gassner, app.I; Schwab, 90	980
1844	Paris, Opéra	Haydn Création	Blaze, Acad. II, 372	
1855	London, Exeter Hall	Messiah (Sacred Harmonic Society)	Hogwood, 250: Wagner	>700
1857	Sydenham, Crystal Palace	Messiah (Costa)	Scholes, 178; Hogwood, 256	2400
1859	Sydenham, Crystal Palace	Messiah (Costa)	Niederrh.MZ 7(1859), 226	3218
1859	Cologne, Niederrhein. Musikf.	Susanna	Dean, 535	
1869.08.02.	London, Princess's Theatre	Acis and Galatea, staged	Dean, 536	
1871.06.07.	Sydenham, Crystal Palace	Acis and Galatea, staged	Dean, 536	
1883	Sydenham, Crystal Palace	Messiah	Forsyth, 152: 87'769 listen.	4500
1884.12.12.	Leipzig, Neues Gewandhaus	Messias (opening ceremony)	Creuzburg, Gewandhaus, 118	
1888	Sydenham, Crystal Palace	Israel in Egypt (Manns)	Hogwood, 273; wax cylinder recording	4000
1888	Amsterdam, Concertgebouw	Messias Hallelujah (opening)	Forsyth, 1985	
1888.10.05.	Berlin, Philharmonie	Messias Hallelujah (opening)	Forsyth, 1985	
1895	Mainzer Händelfestspiele	Messias (Volbach)	NG, Festival: Chrysander	
1899.03.31.	London, Albert Hall	Messiah (Bridge, rev. version)	Scholes, 68; MT 1899, 22, 97	842, 100
1900	Paris, World Exposition	Messias (opening)	Hogwood, 254	
1902	London, Queen's Hall	Messiah	Scholes, 81	165
1902.05.10.	London, Great Queen Str., Theater	Acis and Galatea, staged 6x	Dean, 536	
1920	Sydenham, Crystal Palace	Messiah (Cowan)	Scholes, 180; Hogwood, 265	4000

Table C: Rooms, alphabetical list

Place, *Hall*, Year	Gr.	Ceiling shape	Length (m)	Width (m)	Height (m)	H/W	T sec [calcu-lated]	Vol. (m^3)	Seats	Sources (*measured by author)
Amsterdam, *Concertgebouw,* 1888	( ]	+++	43.15	20.50 28.15	16.40	0.58	2.0	18'700	2'206	Beranek
Athens, *Agora Odeion,* 15 B.C.	[))]	_+_+	24.60	24.75	22.50	0.90			1'000	Meinel
Basel, *Sängerfest Halle,* 1875	[ ]	∩	120.00	31.00	21.00	0.67		~75'000	5'300	Hdb. Arch.
Basel, *Stadt Casino,* *great hall,* 1876	[ ]	_	33.00	19.00	14.50	0.76	1.75	11'500	1'400	Furrer T Beranek
Bayreuth, *Festspielhaus,* 1876	[ ))]	_	28.20	30.00	15.90	0.43	1.6	10'300	1'345	Habel, 1985
Berlin, *Neue Philharmonie,* 1963	hept.	convex	70.00	50.00	21.00	0.42	1.95	24'500	2'218	Forsyth (ITA)
Berlin, *Philharmonie,* 1888–1944	[ ]	_	43.00	24.00	15.40	0.64	2.0	15'500	1'614	Skoda
Berlin, *Philharmonie,* *Beethovensaal,* 1888–1944	[ ]	_	37.90	19.30	12.50	0.65	1.7	6'830	1'103	Bagenal
Berlin, *Singakademie,* 1829–1945	( ]	+++	32.40	12.80	9.70	0.76	1.3 – 1.7	~3'300	1'200	Schiess, T Bagenal
Birmingham, *Symphony Hall,* 1991–	[ )	___	47.30	27.40	22.90	1.00	1.8	25'000	2'211	Beranek, 1997
Boston, *Music Hall,* 1863–	[ ]	_	45.50	22.80	15.80	0.69		18'400	2'400	Sabine
Boston, *Symphony Hall,* 1900	[ ]	+++	46.35	22.85	19.00	1.00	1.8	18'740	2'631	Levarie, T Beranek
Bristol, *Colston Hall,* 1876–1945	[ ]	_	43.65	24.40	17.50	0.72	1.6	14'200	2'560	Bagenal
Buenos Aires, *Teatro Colón,* 1908–	U	D	34.30	30.00 21.50	26.80	0.88	1.8	20'870	2'487	Beranek
Cambridge, *King's College Chapel,* (1446–1515) –	sN	gV	84.00	12.80 20.80	26.40	2.00	5.0	46'500		Great Buildings, T Bagenal
Constantinople, *Hagia Sophia,* (525–537) –	ell.	D	80.00	64.00	40.00 56.00	1.25				Great Buildings

Place, *Hall*, Year	Gr.	Ceiling shape	Length (m)	Width (m)	Height (m)	H/W	T sec [calculated]	Vol. (m^3)	Seats	Sources (*measured by author)
Dallas (TX), *Eugene-Mc-Dermott Hall*, 1990–	[ )	_	40.50	25.60	26.20	1.02	2.1	23'400	2'062	Beranek, 1997
Dresden, *Opernhaus*, 1878–1945 / *1985–*	horse shoe	_	20.25	17.25	18.50	1.07	1.5	7'400	2'000	Fasold
Düsseldorf, *Saal Niederrheinisches Musikfest* 1833	[ ]	_?	38.50	20.00	7.85	0.40		7'983	1'300	Hensel
Edinburgh, *St. Cecilia's Hall* 1762 / 1801–	ellipt.	flV	19.20	10.70	5.20	0.49	[0.8]	1'394	500	Forsyth, T estimate
Esterháza castle, *music hall*, 1766	[ ]	_	15.50	10.30	9.20	0.89	[1.2]	1'530	200	Meyer
Exeter, *Cathedral* (1107–1455)	--+-	gV	118.40	22.40	18.40	1.53				Great Buildings
Florence, *S. Maria del Fiore*, (1296–1436)	basilica oct.	gV D	78.40 153.00	19.00 40.80	40.80 86.40	1.00	10.0	250'000	10'000	Saalman
Florence, *Uffici, Teatro Mediceo*, 1585	[ ]	_	22.0? 44.00	20.40	14.00	0.69		11'424	2'000	NG 13
Geneva, *Victoria–Hall*, 1891–	( )	flV	36.00 48.00	11.50 19.50	17.00	0.87	1.3	11'500	1'754	Furrer
Granada, *P. Carlos V, Whispering gallery* (1538-42) –	oct.	ellD	13.80	13.80	4.70					Baumann, (*2010)
Leeds, *Town Hall*, 1858	[ ]	bV	49.10	21.90	22.90	1.05			1'800	Forsyth
Leipzig, *Old Gewandhaus*, 1781– 1842	[ ]	_	22.85	11.50	7.40	0.65	[1.9]	1'800	430	Skoda
Leipzig, *Old Gewandhaus*, with gallery, 1842–84	[ ]	_	22.85	11.50	7.40	0.65	[1.2]	2'396	570 -1000!	Skoda
Leipzig, *New Gewandhaus, great hall*, 1884–1944	[ ]	_	38.00	19.00	14.60	0.77	1.6	10'600	1'560	Skoda
Leipzig, *New Gewandhaus, small hall*, 1894–1944	[ ]	_	23.00	11.50	7.75	0.67		2'520	643	Skoda
Liverpool, *St. George's Hall*, 1854	[ ]	bV		25.30					3'000	Forsyth
London, *Chapel Royal, St. James's Palace*, 17th century	[ ]	+++	[20.00]	[10.00]	[10.00]	1.00		2'000		(*2006, outside)
London, *Covent Garden Th.*, 1732–82	[<]	_,-	13.20 22.70	10.00 12.50	9.90	1.00		3'200	1'400	Hume

Place, *Hall*, Year	Gr.	Ceiling shape	Length (m)	Width (m)	Height (m)	H/W	T sec [calcu-lated]	Vol. (m^3)	Seats	Sources (*measured by author)
London, *Covent Garden Th.,* 1858	U	D	23.50	18.90	19.25	1.01	1.1	12'240	2'180	Hume
London, *Crown & Anchor Tavern,* 1790–			24.70	11.00						Forsyth
London, *Drury Lane, Theatre Royal,* 1775–1791	<))	_	18.30 15.60	9.60 12.50					2'000	Hume
London, *Drury Lane, Theatre Royal,* 1794–1804	horse shoe	D velum	30.50 22.60	24.50	[14.00	0.57		9'000	3'611	Hume
London, *Exeter Hall in the Strand,* 1831–1880	[ ]	_ / V	45.00	27.00	14.60	0.54		17'739	3'000	Forsyth
London, *Hanover Square Rooms,* 1775–1874	[ ]	flV	24.10	9.80	[8.50]	0.78	[0.95]	1'875	800 -1500	Meyer
London, *Haymarket, King's Theatre,* 1709 –89	[<]	_,-	15.80 20.50	11.60 13.20	10.50	0.90		3'250	1'400	Hume
London, *Haymarket, King's Theatre, Concert Hall,* 1792–1843	[ ]	+++	29.60	14.60	11.90	0.82	[1.55]	4'550	1'000	Meyer
London, *Haymarket, Queen's Theatre, Concert Room,* 1705	[ ]	+++	17.30	10.00						Barlow
London, *Haymarket, Queen's Theatre,* 1705–09	[))]	D	16.65	13.95	16.00	1.15		3'716	1'000	Barlow
London, *Hickford's Rooms, Brewer Street,* 1729–1779	[ ]	_	15.20	9.10	6.70	0.74	[0.85]	934	300	Salmen
London, *Pantheon, Oxford Street,* 1772–92	( )	D++	36.10	20.15						Forsyth
London, *Regent Str., Queen's Hall,* 1893–1941	[ ))	_	32.10	26.45 17.45	17.45	1.00	1.5	12'000	2'026	Bagenal
London, *Ranelagh Garden Rotunda,* 1742–1805	ell.	_	45.70	32.00					3'000	Forsyth
London, *Royal Albert Hall,* 1871	ell.	ell.V	66.80	56.40	41.80	0.74	2.5	86'600	6'080	Beranek
London, *St. James's Hall,* 1858–1905	( ]	bV_	42.40	18.30	18.30	1.00		13'000	2'500	Forsyth

Place, *Hall*, Year	Gr.	Ceiling shape	Length (m)	Width (m)	Height (m)	H/W	T sec [calcu-lated]	Vol. (m^3)	Seats	Sources (*measured by author)
London, *St. Paul's Cathedral* *(1666)*,1675–1710	--+-- sN	bV	56.00 140.00	14.40 30.00	28.00 65.60	2.00	6.5 12.0	155'000		Great Buildings, T Bagenal
London, *Westminster Abbey,* (1275–1519)	--+- sN	gV	67.00 113.00	11.50 22.00	31.00	2.70	5.0	72'500		Hist. of Architecture
London, *Westminster Abbey,* *Chapel of Henry VII* *(1503-19)*	sN	gV	18.70 39.00	11.50 22.00	22.00	1.00				Hist. of Architecture
London, *Westminster Abbey,* *West Nave,* 1784	[ ] sN	gV	46.00	11.00	31.00	2.82		15'600	3'000	Burney
London, *Whitehall Palace,* *Banqueting Hall,* 1621–	[ ]	+-+-+	32.00	16.00	16.00	1.00		8'192	400	Fletcher
London, *Crystal Palace* →Sydenham										
Manchester, *Free Trade Hall,* 1856–1943	[ )		41.18	23.79	15.86	0.67	1.8	15'400	2'280	Forsyth
Milan, *Teatro alla Scala,* 1778–	horse shoe	flV	24.85	21.50	20.00	0.91	1.25	11'250	3'000	Garnier, T Beranek
Munich, *Cuvilliés-Theater,* 1753–	bell	_	14.00	10.00	10.85	1.07	[0.9]	1'512	436	Forsyth, T estimate
Naples, *Teatro* *S. Carlo,* 1737–1817	Huf	_vel	25.00	22.50	22.50	1.00		12'375	2'550	Diderot
New York, *Carneghie Hall,* 1891–	[ ))	,----	41.85	31.00 24.00	24.00	0.77 1.00	1.7	24'250	2'760	Beranek
New York, *Lincoln Center,* *Avery Fisher Hall* 1976–	[ ]	_	51.50	25.90	16.80	0.65	1.8	20'400	2'741	Beranek, 1997
New York, *Lincoln Center,* *Metropolitan Opera,* 1966–	/ \	_	52.00 31.00	22.0 33.0	25.00	0.75	1.8	30'500	3'816	Beranek, 1997
New York, *Lincoln Center,* *Philharmonic Hall,* 1962–76	( )	conc	46.20	32.33	20.00	0.61	2.0	24'479	2'658	Beranek
New York, *Metropolitan Opera,* 1883–1966	U	_	32.00	20.65 18.75	23.45	1.14	1.2	19'500	3'639	Beranek
Oxford, *Holywell Music Room* 1748–	( ]	_	19.85	9.80	9.15	0.93	[1.5]	1'660	300 –500	Bagenal
Oxford, *Sheldonian Theatre,* 1668–	U	_	24.10	21.90	14.60	0.66		7'035	2'000	Downes

Place, *Hall*, Year	Gr.	Ceiling shape	Length (m)	Width (m)	Height (m)	H/W	T sec [calculated]	Vol. (m^3)	Seats	Sources (*measured by author)
Paris, *Concerts rue Cléry,* 1791–1801	O	D	8.00	8.00	6.25	0.78			627	Habel, 1967
Paris, *Conservatoire,* 1811–	U	flV	24.45	11.10	10.40	0.94		2'800	1'021	Donnet
Paris, *Opéra Palais Garnier* 1875 / 1937–	horse shoe	flV/D	25.60	20.00	20.00	1.00	1.1	9'960	2'156	Garnier, T Forsyth
Paris, *Opéra, rue le Pelletier* 1821–73	horse shoe	flV/D	22.50	16.80	18.50	1.10		6'100	1'783	Garnier, T Forsyth
Paris, *St. Louis les Invalides,* 1675–1706–	centr	D	64.80 16.00	12.00 53.60	28.00 64.00	2.25				Great Buildings
Paris, *Trocadéro, Exposition Universelle,* 1878–1937	ellipt.	parab.	50.00	50.00	30.00	0.60	1.45	63'000	4'665	Favaro
Parma, *Teatro Farnese,* 1619–1732	U	_	42.25	30.00	20.70	0.70		32'000	2'500	Diderot
Pompeji, *Great Theatre.* -200 / -2 –	C	open	33.60	49.60					5'000	Great Buildings
Pompeji, *Theatrum tectum,* -080 – -063	[))]	+++	28.60	26.60					1'300	Meinel
Rome, *Chiesa nuova, Oratorio,* 1640–	[ ]	+++	25.50	17.00	25.50	1.50		11'000	515	Connors
Rome, *Oratorio del Gonfalone,* 1544–	[ ]	+++	18.00	9.00	9.00	1.00		1'458		Onorati , (*1996)
Rome, *Pantheon,* ca. -120–	O	D +++	43.00	43.00	43.00	1.00		52'037		Great Buildings
Rome, *Palazzo Barberini, Salone Grande,* 1639?–	[ ]	V	24.70	14.65	14.65	1.00		5'000		Waddy
Rome, *Palazzo Barberini, Salotto delle comedie* (1632–1637)	[ ]	V	17.15	12.45	9.35	0.75		2'550		Waddy
Rome, *Palazzo Bonelli, Salone grande, 1708*	[ ]	V	[31.00]	[12.5(	[19.00	[1.5(		6'750	[400]	Farina
Rome, *Palazzo Bonelli, Stanzione delle Accademie, 1708*	[ ]	V	[21.20]	[7.70]	[10.60	[1.3(		1'580	[200]	Farina
Rome, *Palazzo della Cancelleria, Sala Riaria,* 1485–	[ ]	+++	26.90	15.80	12.40	0.78		5'270		Frommel
Rome, *Palazzo della*	U	conc	9.20	8.95	10.60	1.18		856		Ferrero

Place, *Hall*, Year	Gr.	Ceiling shape	Length (m)	Width (m)	Height (m)	H/W	T sec [calculated]	Vol. (m³)	Seats	Sources (*measured by author)
Cancelleria, *Teatro Ottoboni,* 1690–1740										
Rome, *Palazzo della Cancelleria, S. Lorenzo in Damaso* 1640 - 1813	[ ]	+++	22.00 48.00	16.00 26.00	24.00	0.90		29'952		Hist. of Architecture
Rome, *Palazzo Doria-Pamphilij, Galeria,* 1684–	[ ]	V	51.00	15.00	20.40	1.36		12'675		Frommel
Rome, *St. Peter's,* *(1508–1626)* –	--+-	bV D	96.00 180.00	26.40 69.60	45.00 101.00	1.70	5.00	550'000		Great Buildings
Rome, *Teatro Argentina,* 1732– ~1832	horse shoe	flV	18.15	15.15	15.00	0.99		5'120	1'872	Diderot
Rome, *Teatro della Valle,* 1727–	U	_	14.50	12.00	12.00	1.00		1'740	1'100	Hdb. Arch.
Salisbury, *Cathedral* (1220–1333) –	-+--	gV	58.40 136.00	10.40 23.20	12.00 24.00	1.00				Hist. of Architecture
Sydenham, *Crystal Palace, Transept,* 1854–1937	T	V	117.00	66.00	51.00	0.77		847'855	12'000	Forsyth
Venice, *Teatro La Fenice,* 1792–	U	flV	22.00	18.00	14.00			5'420	2'000	Hdb. Arch.
Venice, *San Marco,* 1063–	--+--	5 D	48.00	13.00 26.00	29.60 21.60	1.66				Great Buildings
Venice, *Teatro SS. Giovanni e Paolo,*1639–1715	U	_	17.00	13.75	11.65	0.85		2'400	1'000	Forsyth
Vienna, *Burgtheater,* 1741–1889	bell	_	23.60	10.30	12.00	1.16	[0.7]	4'250	1'300	Hdb Arch. T Weinzierl
Vienna, *Deutsches Sängerfest, Halle,* 1875	oval	∩	116.00	56.00	23.00	0.41		115'000	5'867	Hdb. Arch.
Vienna, *Hofburg, great Redoutensaal,* 1748– / 1992–	[ ]	_	39.90	16.95	15.60	1.08	[2.3]	10'440	1'500	Burghauptmannschaft, T Weinzierl
Vienna, *Hofburg, small Redoutensaal,* 1748– / 1992–	[ ]	_	22.85	11.55	15.60	1.35	[2.1]	3'220	400	Burghauptmannschaft, T Weinzierl
Vienna, *old University, Aula,* 1753–	[ ]	_	28.00	14.00	10.00	0.71	[1.5]	4'800	750	(*2004), T Weinzierl
Vienna, *Hofburgtheater,* 1888–	bell	flV	21.40	14.60	17.50	1.20		5'500	1'474	Hdb. Arch.
Vienna, *Musikverein,*	[: :]	+++	32.80	10.60	10.60	1.00		3'500	625	GdMF

Place, *Hall*, Year	Gr.	Ceiling shape	Length (m)	Width (m)	Height (m)	H/W	T sec [calcu-lated]	Vol. (m^3)	Seats	Sources (*measured by author)
Brahmssaal, 1870–										
Vienna, *Musikverein, great hall,* 1870–	[ ]	+++	45.70	19.10	17.70	0.89	2.0	15'650	1'600	GdMF T Beranek
Vienna, *Palais Lobkowitz, hall,* 1804–	[]	_	15.20	7.15	8.25	1.16	[1.45]	900	120	(*1997), T Meyer
Vienna, *Theater an der Wien,* 1801–	U	_	18.00	20.00	15.00	0.75	1.15	5'200	1'060	(*1997), T Meyer
Vienna, *Winterreitschule,* 1735–1847	[ ]	_	55.00	18.00 21.00	17.00	0.94		16'830	1'500	Burghaupt-mannschaft
Washington, *Cathedral,* (1907–1990) –	-+--	gV	55.55 120.30	12.50 32.40	31.70	1.00				Guide
Winchester, *Cathedral,* (1079-93) –	-+---	gV	76.80 142.40	10.40 24.00	24.00	1.00				Hist. of Architecture
York, *Minster Saint Peter,* 1291–	-+--	gV	61.60 116.60	12.80 32.80	34.40	1.00	4.75 10.00			Great Buil-dings; T audio waves 2004 & Bagenal
Zürich, *Schützenfest Halle,* 1859	[ ]	^	98.00	28.00	16.00	0.57		2'300		Hensel
Zürich, *Eidgenössisches Schützenfest,* 1872	[ ]	^	115.00	45.00	21.00	0.47		5'300		Hdb.Arch.
Zürich, *Hallenstadion,* 1939–	oval	flV	112.00	62.00	24.00	0.38		240'000	11'000	Hdb.Arch.
Zürich, *Stadttheater / Opernhaus,* 1891–	U	flV	19.00	13.50	14.00	1.04		6'800	1'139	SBZ 1891
Zürich, *Tonhalle, great hall,* 1895–	[ ]	_	36.00	27.00	13.00	0.68	1.95	11'400	1'409	SBZ 1895, T Beranek
Zürich, *Tonhalle, small hall,* 1895–	( ]	flV	26.80	12.00	9.00	0.75		2'900	636	SBZ 1895

Index

E

F

M

P

natur, wissenschaft und die künste
nature, science et les arts
nature, science and the arts

Edited by

Julia Burbulla, Bernd Nicolai, Ana-Stanca Tabarasi-Hoffmann, Philip Ursprung & Wolf Wucherpfennig

Nature, Science and the Arts is an international scholarly series dealing with the history of cultural interplay between arts, humanities, natural sciences and technology, both on the level of theoretical reflection and in artistic enunciations. It is not restricted to any particular epoch, society, medium or region. By publishing contributions to this new interdisciplinary research area, the series illuminates the traditional connection between two ways of interpreting the world, a connection that has been largely marginalized since Wilhelm Dilthey's strict dissociation between humanities and natural sciences.

Vol. 1 Julia Burbulla
Allumfassende Ordnung. Gartenkunst und Wissenschaft in Gotha unter Ernst II. von Sachsen-Gotha-Altenburg (1772–1804).
ISBN 978-3-0343-0038-4. 2010.

Vol. 2 James King
Beyond *The Great Wave*. The Japanese Landscape Print, 1727–1960.
ISBN 978-3-0343-0317-0. 2010.

Vol. 3 Sabine Flach, Daniel Margulies & Jan Söffner (eds)
Habitus in Habitat I. Emotion and Motion.
ISBN 978-3-0343-0530-3. 2010.

Vol. 4 Sabine Flach & Jan Söffner (eds)
Habitus in Habitat II. Other Sides of Cognition.
ISBN 978-3-0343-0531-0. 2010.

Vol. 5 Glon Emmanuelle
Cinéma dans la Tête. L'esthétique du film à la lumière des neurosciences.
ISBN 978-3-0343-0603-4. 2011.

Vol. 6 Julia Burbulla & Ana-Stanca Tabarasi-Hoffmann (Hg.)
Gartenkunst und Wissenschaft. Diskurs, Repräsentation, Transformation seit dem Beginn der Frühmoderne.
ISBN 978-3-0343-0547-1. 2011.

Vol. 7 Dorothea Baumann
Music and Space. A systematic and historical investigation into the impact of architectural acoustics on performance practice, followed by a study of Handel's Messiah.
ISBN 978-3-0343-0615-7. 2011.

Vol. 8 Joerg Fingerhut, Sabine Flach & Jan Söffner (eds)
Habitus in Habitat III. Synaesthesia and Kinaesthetics.
ISBN 978-3-0343-0532-7. 2011.

Vol. 9 Sixto Castro & Alfredo Marcos (eds)
The Paths of Creation. Creativity in Science and Art.
ISBN 978-3-0343-0511-2. 2011.

Juridicum – Bochumer Schriften zur Friedenssicherung und zum Humanitären Völkerrecht | Bochum Studies on the International Law of Peace and Armed Conflict

Volume 73

In der Schriftenreihe des Bochumer Instituts für Friedensicherungsrecht und Humanitäres Völkerrecht werden regelmäßig Publikationen veröffentlicht, die sich mit aktuellen Fragen von Friedenssicherung und humanitärem Völkerrecht beschäftigen. Die Reihe erscheint seit 1988 unter dem Titel **Bochumer Schriften zur Friedenssicherung und zum Humanitären Völkerrecht**. Sie wurde 2017 nach einer Unterbrechung von zehn Jahren zunächst neu aufgelegt und wird seit 2025 bei Springer weitergeführt.
Die Bände behandeln vielfältige Themen der Friedens- und Konfliktforschung, etwa das Gewaltverbot und Selbstverteidigung, kollektive Sicherheit und das Recht der Vereinten Nationen, Friedenseinsätze, sowie die Menschenrechte, das Völkerstrafrecht, das Umweltrecht und – in all seinen Facetten – das humanitäre Völkerrecht und die humanitäre Hilfe.

This book series regularly publishes monographs and edited collections on current issues concerning the maintenance of international peace and security as well as international humanitarian law. The series has been published since 1988 under the title **Bochumer Schriften zur Friedenssicherung und zum Humanitären Völkerrecht**. It was reissued in 2017 after an interruption of ten years and has been continued by Springer since 2025.
The volumes cover a wide range of topics in peace and conflict studies, such as the use of force and self-defense, collective security and the law of the United Nations, peace missions, as well as human rights, international criminal law, environmental law and—in all its aspects—international humanitarian law as well as humanitarian action.

Laura Hofmann

Cultural Heritage through the Lens of Human Rights Bodies

Meaning, Rights, and Obligations

Springer

Laura Hofmann
Bonn, Germany

PhD thesis submitted at the Institute of Development Research and Development Policy, Ruhr-Universität Bochum
Date of oral examination (disputation): 7 May 2025
Professor Dr. Pierre Thielbörger (1st supervisor)
Professor Dr. Markus Kaltenborn (2nd supervisor)
Professor Dr. Wilhelm Löwenstein (chair of the oral exam)

ISSN 3091-4477 ISSN 3091-4485 (electronic)
Juridicum – Bochumer Schriften zur Friedenssicherung und zum Humanitären Völkerrecht | Bochum Studies on the International Law of Peace and Armed Conflict
ISBN 978-3-658-51202-6 ISBN 978-3-658-51203-3 (eBook)
https://doi.org/10.1007/978-3-658-51203-3

This Springer imprint is published by the registered company Springer Fachmedien Wiesbaden GmbH, part of Springer Nature.
The registered company address is: Abraham-Lincoln-Str. 46, 65189 Wiesbaden, Germany

Acknowledgments Completing this PhD thesis would not have been possible without the support of many wonderful individuals. First and foremost, I wish to thank my PhD supervisor, Professor Pierre Thielbörger, for his generous and invaluable support, guidance and patience during this process. I am very grateful to my second supervisor, Professor Markus Kaltenborn, and to Professor Wilhelm Löwenstein, who chaired the oral examination of my PhD, for their feedback and engagement with my doctoral research.

I benefitted from being affiliated with two institutions at Ruhr University Bochum: the Institute of Development Research and Development Policy (IEE) and the Institute for International Law of Peace and Armed Conflict (IFHV). I am incredibly grateful to my colleagues at both institutions for the inspiring and supportive work environment. The feedback and discussion resulting from the biannual doctoral colloquia at the IFHV were instrumental in advancing and completing my Ph.D. I would like to extend a special thanks to Professor Hans-Joachim Heintze for his support. Thanks are also due to Dr. Britta Niklas for all her support.

I am thankful to Ruhr University Bochum's Research School for providing the opportunity to attend conferences, summer schools, and complete a research visit at the University of Cambridge's Lauterpacht Centre for International Law.

I received incredible support from Theresa Ackermann and Tobias Ackermann. I am thankful for their unwavering encouragement, inspiration, and most importantly, their friendship on this journey. Thanks to Theresa for going above and beyond to support me then and now. I would also like to thank Benedikt Behlert, Maximilian Bertamini, Timeela Manandhar, Robin Ramsahye and Robin Pass for their support and friendship.

I am grateful to my family for their unwavering support throughout this endeavour. I owe my biggest thanks to my mum for always having confidence in me and for sparking my interest in the topic of cultural heritage protection many years ago, and to my dad, whose belief in me during this journey has meant more than I can express.

Competing Interests The author has no competing interests to declare that are relevant to the content of this manuscript.

Abstract

This PhD thesis asks, if and how international human rights law can offer a legal basis for the protection of cultural heritage. Against the backdrop of the Committee on Economic, Social and Cultural Rights' General Comment No. 21 in 2009 and the 2011 thematic report of the Special Rapporteur in the field of cultural rights on access and enjoyment of cultural heritage, this PhD thesis examines one possible answer—through international human rights bodies.

To this end, it reviews the pronouncements and work of human rights bodies, specifically the Human Rights Council, the Human Rights Committee and the Committee on Economic, Social and Cultural Rights. The research reviews the pronouncements and work of selected human rights bodies, focusing on the meaning of cultural heritage, the rights of individuals and persons belonging to minorities and Indigenous peoples vis-à-vis cultural heritage as well as on the corresponding state obligations.

Organised along the threefold structure of "meaning", "rights", and "obligations", the PhD thesis examines the meaning of cultural heritage, the relevant rights of individuals and groups, and the corresponding state obligations vis-à-vis cultural heritage. It further assesses to what extent rights, such as access to cultural heritage, and obligations serving the protection of (aspects of) cultural heritage, are (still) emerging in the international human rights framework.

It finds that although human rights law does not provide for an explicit right to cultural heritage, rights and obligations vis-à-vis cultural heritage form part of human right law. These rights and obligations form part of human rights provisions and are in the human rights bodies' views an important component in their proper implementation. By reviewing how human rights tools and mechanisms are used to protect rights relating to cultural heritage, the thesis suggests that the international human rights framework offers (soft) enforcement tools to strengthen rights and obligations vis-à-vis cultural heritage.

Contents

Abbreviations

AP I	First Additional Protocol to the Geneva Conventions
AP II	Second Additional Protocol to the Geneva Conventions
CCPR	Human Rights Committee (in footnotes)
CERD	Committee on the Elimination of Racial Discrimination
CESCR	Committee on Economic, Social and Cultural Rights
CRC	Committee on the Rights of the Child
ECOSOC	Economic and Social Council
ECtHR	European Court of Human Rights
fn	reference to footnote(s) cited in sources
HRC	Human Rights Committee
IACHR	Inter-American Commission on Human Rights
IACtHR	Inter-American Court of Human Rights
ICC	International Criminal Court
ICCPR	International Covenant on Civil and Political Rights
ICERD	International Convention on the Elimination of All Forms of Racial Discrimination
ICESCR	International Covenant on Economic, Social and Cultural Rights
ICJ	International Court of Justice
ICTY	International Criminal Tribunal for the Former Yugoslavia
ILA	International Law Association
ILC	International Law Commission
OHCHR	The Office of the High Commissioner for Human Rights
Res	Resolution
UDHR	Universal Declaration of Human Rights

UN	United Nations
UNDRIP	United Nations Declaration on the Rights of Indigenous Peoples
UNESCO	United Nations Educational, Scientific and Cultural Organization
UNGA	United Nations General Assembly
UNHRC	United Nations Human Rights Council
UPR	Universal Periodic Review
VCLT	Vienna Convention on the Law of Treaties

Introduction

1

> "There is an urgent need to protect the cultural heritage of individuals and groups; and such protection cannot take place without the involvement and implementation of a strong human rights system."[1]

The legal protection and accompanying discourse surrounding cultural heritage in international law has undergone several shifts: its protection has developed from armed conflict to peacetime, from cultural property to cultural and natural heritage, from the tangible to the intangible, and from the perception of its destruction as a "the destruction of stones" to a recognition that the direct victims are people(s). More recently, the state centricity of cultural heritage law has come under scrutiny. The discourse—and to a certain extent the legal framework—has acknowledged the importance and participation of communities[2], groups, and in some cases, individuals in international cultural heritage law.

Set against this backdrop, the thesis focuses on two developments: (1) international heritage law's increasing consideration of the role and relationship of people(s) and cultural heritage and (2) the mainstreaming of cultural heritage

[1] Xanthaki (2017), p. 2.

[2] Cognisant of the "colonial baggage tied to the term", communities in international heritage law can be defined as "groups who are not states, nor UNESCO, and at the same time have no claim to scholarly expertise with respect to their heritage, but have expertise based on their experience with it, by living in, with, or around heritage, or practicing it as part of their cultural lives." Jakubowski et al. (2022), para. 47–48.

L. Hofmann, *Cultural Heritage through the Lens of Human Rights Bodies*, Juridicum – Bochumer Schriften zur Friedenssicherung und zum Humanitären Völkerrecht | Bochum Studies on the International Law of Peace and Armed Conflict 73, https://doi.org/10.1007/978-3-658-51203-3_1

in the international human rights framework.[3] The protection of cultural heritage law and human rights law indeed share common goals. At the core, this includes the protection of the human dignity and the cultural identity of individuals and groups.[4] In 2007, Human Rights Council (UNHRC) resolution 6/11 entitled "Protection of cultural heritage as an important component of the promotion and protection of cultural rights" included the request to the UN High Commissioner for Human Rights "to encourage all relevant human rights bodies and mechanisms to pay due attention to the issue of […] protection of cultural heritage *as an important component of the promotion and protection of all human rights*".[5] The resolution not only recognises cultural heritage as "an important component of the cultural identity of communities, groups and individuals", it also understands that "its intentional destruction may have adverse consequences on human dignity and human rights".[6] Shortly after, General Comment No. 21 by the Committee on Economic, Social and Cultural Rights (CESCR) and the 2011 report on access to cultural heritage as a human right[7] by the then-Independent Expert in the field of cultural rights addressed and affirmed that Article 15 (1) (a) of the ICESCR, setting out the right to take part in cultural life, includes rights relating to cultural heritage and state obligations towards the protection of cultural heritage.

---

[3] UNHRC Res 6/11 (28 September 2007) UN Doc A/HRC/RES/6/11. UNHRC Res 49/7 (31 March 2022) UN Doc A/HRC/RES/49/7, preamble, 23rd recital (Adopted without a vote).

[4] Blake (2020a), p. 361.

[5] UNHRC Res 6/11 (28 September 2007) UN Doc A/HRC/RES/6/11, para. 8. (Emphasis added).

[6] UNHRC Res 6/11 (28 September 2007) UN Doc A/HRC/RES/6/11, preamble, 7th recital.

[7] UNHRC, Report of the independent expert in the field of cultural rights, Farida Shaheed (21 March 2011) UN Doc A/HRC/17/38.

The thesis asks if and how international human rights law can offer a legal basis for the protection of cultural heritage. Amid the absence of an explicit right to cultural heritage in international human rights treaties[8], the analysis focuses on the interpretation of human rights bodies. To this end, it reviews the pronouncements[9] and work of human rights bodies, specifically the UNHRC, the Human Rights Committee (HRC) and the CESCR. Organised along the threefold structure of "meaning", "rights", and "obligations", the thesis examines the meaning of cultural heritage in the human rights framework, the relevant rights of individuals and groups, and the corresponding state obligations vis-à-vis cultural heritage. It further assesses to what extent rights and obligations serving the protection of cultural heritage, are (still) emerging in the international human rights framework.

This thesis seeks to attain a better understanding of cultural heritage in the international human rights framework, including its meaning, rights and obligations towards heritage, as well as the use of soft enforcement tools by the human rights bodies. The latter refers to the monitoring and complaints tools of the treaty bodies, the UNHRC's Special Procedures which engages Special Rapporteurs and offers a complaints tool, and the Universal Periodic Review (UPR)—a peer review mechanism to which all states are subject. Particularly the establishment of treaty bodies by human rights treaties shifts "the centre of gravity in treaty interpretation from states to" the monitoring mechanisms.[10] This suggests a less state-centred framework and might contribute to strengthening the rights of individuals and members of groups. By analysing the work and pronouncements of selected human rights bodies as well as the use of the tools of the human rights framework, the thesis seeks to gauge if and how the international human rights framework can offer a complementary set of protection to cultural heritage and to what extent individuals and persons belonging to minorities or Indigenous[11] communities can rely on human rights provisions in matters concerning

[8] The Council of Europe's Faro Convention recognises a right to cultural heritage in its Art. 4 (c). Framework Convention on the Value of Cultural Heritage for Society, Signed in Faro on 27 October 2005. Council of Europe, ETS No. 199. Entry into force: 1 June 2011 (Faro Convention).

[9] The use of the term pronouncement stems from its designation in the Fourth Report by Special Rapporteur Nolte and covers the forms of action of expert bodies. ILC 'Fourth Report on Subsequent Agreements and Subsequent Practice in Relation to The Interpretation of Treaties by Georg Nolte, Special Rapporteur' (7 March 2016) UN Doc A/CN.4/694, para. 14.

[10] Scheinin (2017), pp. 25–26.

[11] In accordance with guidance by style guides and governments, Indigenous is capitalized as it refers to people, groups, culture, and society. See, e.g., The Chicago Manual of Style

cultural heritage. The reliance on the language of "rights and obligations" could strengthen the participation rights of local communities and Indigenous peoples or rights relating to access to cultural heritage. In addition to the shift in rhetoric, the tools of the international human rights framework can contribute to further strengthening these rights and to expounding on obligations in a meaningful way. Although international cultural heritage law has increasingly recognised the role and rights of individuals and members of communities, cultural heritage treaties were drafted "to create obligations for States parties to recognize, respect and protect cultural heritage".[12] Even in those cases, in which individuals and communities are given a particular role, provisions are phrased "as assignments to States parties to ensure involvement and participation".[13] They are not phrased as rights.[14]

The introduction proceeds with an overview of the development of the protection of cultural heritage, highlighting its origins in international humanitarian law and showing how the protection has evolved in peacetime. As the primary focus of the thesis lies on the World Heritage Convention and the Intangible Cultural Heritage Convention, Sect. 1.1 primarily focuses on the protection and safeguarding under these conventions. Subsequently, the chapter shows how the challenges vis-à-vis heritage have evolved. The analysis focuses on the one hand, on the form of challenges, and on the other hand on concrete challenges in implementing the World Heritage Convention and the Intangible Cultural Heritage Convention, addressing the practice of listing, state sovereignty, and by highlighting human rights issues. In the following, Sect. 1.3 introduces the intersection of cultural heritage protection and human rights bodies and outlines the research question and research methods (1.4). Finally, the introduction provides an overview of the thesis' structure and the subsequent chapters.

Online (2017): Capitalization (https://www.chicagomanualofstyle.org/qanda/data/faq/topics/Capitalization/faq0106.html) (accessed 31 January 2025), Government of British Columbia (2024): Capitalization and formatting of Indigenous terms (https://www2.gov.bc.ca/gov/content/governments/services-for-government/service-experience-digital-delivery/web-content-development-guides/web-style-guide/writing-guide-for-indigenous-content/capitalization-and-formatting-of-indigenous-terms) (accessed 31 January 2025).

[12] Donders (2018), p. 60.

[13] *ibid* p. 61.

[14] *ibid*.

1.1 From Armed Conflict to Peacetime Protection

The origins of cultural heritage protection at the international level have been ascribed to international humanitarian law.[15] Subsequently, the international legal protection of cultural heritage has evolved parallel but in a separate manner for times of peace and armed conflict.[16] This bifurcation has been subject to criticism, which questions the "historical division of the international law of war and peace."[17] Despite a certain trend that aims to dissolve the divide between protection in peacetime and armed conflict, such progress has been met with trepidation by states in the past.[18]

In the event of an armed conflict, international humanitarian law offers a specialised protection regime for cultural property.[19] The Convention for the Protection of Cultural Property in the Event of Armed Conflict (1954 Hague Convention) and its two Protocols from 1954 and 1999 form the centrepiece of the protection system.[20] The 1977 Additional Protocols I and II to the 1949 Geneva

---

[15] See rules pertaining to the protection of cultural property in early international (legal) instrument applicable in the event of an armed conflict. E.g., General Orders No 100: Instructions for the Government of Armies of the United States in the Field (prepared by Francis Lieber) (signed 24 Apr 1863) (Lieber Code), Arts. 35–36; Project of an International Declaration concerning the Laws and Customs of War. Brussels, 27 August 1874, Institute of International Law, he Laws of War on Land, Manual published by the Institute of International Law (Oxford Manual), 9 September 1880, Arts. 34, 53; Art. 8; Convention (II) with Respect to the Laws and Customs of War on Land and its annex: Regulations concerning the Laws and Customs of War on Land. The Hague, 29 July 1899, Arts. 27, 56; Regulations Respecting the Laws and Customs of War on Land, Annex to Convention (IV) Respecting the Laws and Customs of War on Land (adopted 18 Oct 1907, entered into force 26 Jan 1910) 205 CTS 277 (Hague Regulations), Arts. 27, 56.

Also see specialised international (legal) instruments Treaty on the Protection of Artistic and Scientific Institutions and Historic Monuments (Roerich Pact) (adopted 15 April 1935, entered into force 26 August 1935), and the Convention for the Protection of Cultural Property in the Event of Armed Conflict (adopted 14 October 1954, entered into force 7 August 1956) 249 UNTS 215 (*1954 Hague Convention*).

[16] Lenzerini (2003), p. 132.

[17] Bassiouni (1983), p. 287.

[18] Vrdoljak (2007).

[19] For a detailed (historical) overview see, e.g., Gerstenblith (2008), p. 681 f.; Keane (2004), p. 4 f.

[20] Despite the strict distinction, the 1954 Hague Convention also contains provisions that take effect in peacetime. See 1954 Hague Convention, Arts. 7(1) & (2), 8(5), 18(1), 25.

Conventions (AP I, AP II)[21] as well as customary international humanitarian law provide further protection for cultural heritage. The principal rule of the 1954 Hague Convention, referring to the prohibition of acts of violence against cultural property in an armed conflict, is interpreted to have reached the status of customary international law by leading scholars in the discipline.[22] Moreover, the case law of the International Criminal Tribunal for the Former Yugoslavia (ICTY) has continuously asserted that specific rules concerning the protection of cultural heritage have become customary international law. In 1995, *Tadić* confirmed that Article 19 of the 1954 Hague Convention has "gradually become part of customary international law".[23] In 2005, the Trial Chamber of the ICTY in the *Strugar* case confirmed Article 3 (d) of the Statute of the ICTY to be part of international customary law for both IACs and NIACs.[24] Article 3 (d) refers to the violation of laws or customs of war by the "seizure, destruction, or wilful damage done to institutions dedicated to religion, charity and education, the arts and sciences, historic monuments and works of art and science."[25]

Protection has been gradually extended to non-international armed conflicts with the help of inter alia the Second Protocol of 1999 to the 1954 Hague Convention as well as international criminal law.[26] The jurisprudence of the International Criminal Court (ICC)[27], and the ICTY have played a seminal role in developing the protection of cultural heritage in non-international armed conflicts.

At least five UNESCO Conventions can be highlighted in the evolution of cultural heritage protection and show how cultural heritage protection has been expanded to peacetime: the 1970 Convention on the Means of Prohibiting and Preventing the Illicit Import, Export and Transfer of Ownership of Cultural

[21] AP I and AP II provide provisions in their articles 53 and 16 respectively. Protocol Additional to the Geneva Conventions of 12 August 1949, and Relating to the Protection of Victims of International Armed Conflicts (adopted 8 June 1977, entered into force 7 Dec 1978) 1125 UNTS 3 (*AP I*); Protocol Additional to the Geneva Conventions of 12 August 1949, and Relating to the Protection of Victims of Non-International Armed Conflicts (adopted 8 June 1977, entered into force 7 Dec 1978) 1125 UNTS 609 (*AP II*).

[22] See e.g., Francioni (2004), p. 6; Lenzerini (2013), p. 256.

[23] See, *Prosecutor v Dusko Tadić aka 'Dule'* (Jurisdiction) ICTY 94 1 (2 Oct 1995), para. 98. Article 19 (1) of the 1954 Hague Convention applies to non-international armed conflicts and states that "each party to the conflict shall be bound to apply, as, a minimum, the provisions of the present Convention which relate to respect for cultural property."

[24] *Prosecutor v. Pavle Strugar* (Trial Judgment) IT-01–42 (31 Jan 2005), para. 230.

[25] UNSC Res 827 (25 May 1993) UN Doc S/RES/827, Art. 3 (d).

[26] See Lenzerini (2003), p. 138.

[27] *Prosecutor v. Ahmad Al Faqi Al Mahdi* (Judgment and Sentence) ICC-01/12-01/15 (27 September 2016).

Property; the 1972 Convention Concerning the Protection of the World Cultural and Natural Heritage (World Heritage Convention); the 2001 Convention on the Protection of the Underwater Cultural Heritage; the 2003 Convention for the Safeguarding of the Intangible Cultural Heritage (Intangible Cultural Heritage Convention); and the 2005 Convention for the Protection and Promotion of the Diversity of Cultural Expressions. In addition to the distinction between tangible, intangible and natural heritage emanating from these treaties, the vocabulary of material and immaterial heritage is also used at the national level[28] and intellectual property regimes are utilised for the protection of traditional cultural expressions and other forms of intangible cultural heritage[29].[30]

Beyond these UNESCO treaties, non-binding instruments, international organisations, and experts play an influential role in the field of international heritage law.[31] Notwithstanding the importance of the work of experts, the academic discourse has shed light on how the considerable input from experts in a top-down approach shapes international cultural heritage law and enforces a way of thinking about heritage that neglects its meaning for other stakeholders.[32] In this context, the academic discourse speaks of the "Authorized Heritage Discourse", which describes the reliance "on the power/knowledge claims of technical and aesthetic experts" and constitutes "a hegemonic discourse about heritage, which acts to constitute the way we think, talk and write about heritage".[33] The manner in which expertise is governed in international cultural heritage law can result in experts "losing sight of the fact that their role is 'to protect not only objects, but also the relationships between those objects and people'".[34]

In accordance with Article 4 of the World Heritage Convention, state parties have the primary duty to ensure "the identification, protection, conservation, presentation and transmission to future generations" of the cultural and natural heritage, situated in their respective territories, and referred to in the Convention. States commit to achieve this "to the utmost" of their resources and with

---

28 See, Lixinski (2013a), pp. 127, 134

29 See, *id.*, p. 120 f.

30 See Gilbert (2017), p. 20.

31 Lixinski (2019), pp. 17, 68; Jakubowski et al. (2022), para. 43 f.

Also see the 2003 UNESCO Declaration Concerning the Intentional Destruction of Cultural Heritage has been interpreted as indicative of an increasing perception of the intentional destruction of cultural heritage as a gross violation of the international obligation to respect cultural heritage in peacetime. Francioni (2004), p. 1219; Francioni (2020a), p. 536.

32 Lixinski (2019), p. 14. This is referred to as the "conservation paradigm".

33 Smith (2006), p. 11.

34 Watkins in Jakubowski et al. (2022), para. 46.

the help of "any international assistance and co-operation".[35] Article 5 lists several measures to ensure the protection, conservation, and presentation of said properties, which the state parties "shall endeavor, in so far as possible, and as appropriate for each country" to implement. State parties further commit to not undertake any deliberate measures that might damage heritage situated on the territory of another state party.[36] In accordance with the Intangible Cultural Heritage Convention, state parties "shall take the necessary measures to ensure the safeguarding of the intangible cultural heritage present in its territory".[37] Furthermore, state parties shall identify and define elements of the intangible cultural heritage "with the participation of communities, groups and relevant non-governmental organizations" and with the help of using inventories.[38] Articles 13 and 14 of the Intangible Cultural Heritage Convention stipulate further safeguarding and awareness-raising and capacity-building measures the state parties "shall endeavour" to adopt.

In addition to setting out state obligations to protect heritage, both conventions operate on the basis of lists. Listing is among the most visible protection mechanisms at the international level.[39] It mirrors the national obligation of inventorying[40] and contributes to awareness-raising of cultural heritage.[41] In the case of the World Heritage Convention, this refers to the "World Heritage List" and the "List of World Heritage in Danger".[42] The Intangible Cultural Heritage Convention sets up the "Representative List of the Intangible Cultural Heritage of Humanity" and the "List of Intangible Cultural Heritage in Need of Urgent Safeguarding"[43].

The World Heritage List "allows state parties to identify heritage sites of outstanding universal value and place them under the protection of the global

[35] Convention concerning the Protection of the World Cultural and Natural Heritage (adopted 16 November 1972, entered into force 17 December 1975) 1037 UNTS 151(*World Heritage Convention*), Art. 4. The principle of cooperation is further outlined in Arts. 6 and 7 of the Convention.

[36] World Heritage Convention, Art. 6 (3).

[37] Convention for the Safeguarding of the Intangible Cultural Heritage (adopted 17 October 2003, entered into force 20 April 2006) 2368 UNTS 3 (*Intangible Cultural Heritage Convention*), Art. 11 (a).

[38] Intangible Cultural Heritage Convention, Arts. 11 (b) and 12.

[39] Lixinski (2017), p. 227.

[40] Lenzerini (2020), p. 308.

[41] Lixinski (2013b), p. 139.

[42] World Heritage Convention, Art. 11.

[43] Intangible Cultural Heritage Convention, Arts. 16–17.

community".[44] Inscription of a site on the World Heritage List requires the site to meet at least one of ten criteria determining its outstanding universal value. Its outstanding universal value is subject to its integrity and/or authenticity and to an adequate protection and management system ensuring its safeguarding.[45] Inscription on the List is only possible if the state party (or state parties) in which the site is situated nominates the site to be placed on the List.[46] The World Heritage Committee, an intergovernmental committee of twenty-one members established in line with the Convention,[47] administers the List and takes the final decision in the inscription process of sites.[48] This process is supported by two advisory bodies of experts, namely the International Council on Monuments and Sites (ICOMOS) and the International Union for the Conservation of Nature (IUCN).

If one of the listed sites faces "serious and specific dangers", it can be moved on the List of the World Heritage in Danger.[49] The List of World Heritage in Danger is intended "to further galvanize international action to safeguard heritage threatened by major changes in human societies".[50] The inscription on this list alerts the international community to the situation and facilitates assistance by the World Heritage Committee, including financial assistance from the World Heritage Fund to the respective property in danger.[51]

The Intangible Cultural Heritage Convention employs the Representative List of Intangible Heritage of Humanity and the List of Intangible Cultural Heritage in Need of Urgent Safeguarding.[52] The prefix "representative" is important,[53] as in negotiations leading up to the Convention, concerns were addressed whether replicating the system of the World Heritage Convention would, as a by-product, create and perpetuate a hierarchy of forms of cultural heritage.[54] This criticism

---

44 Bertacchini, Liuzza, Meskell and Saccone (2016), pp. 167, 96.

45 UNESCO 'Operational Guidelines for the Implementation of the World Heritage Convention' (31 July 2024) WHC.24/01 (*Operational Guidelines*), paras. 77–78.

46 Lenzerini (2016a), p. 70. Although marine sites have been inscribed by extending the territorial jurisdiction of the state in the maritime area. Francioni (2020c), p. 259.

47 World Heritage Convention, Art. 8.

48 Francioni (2020c), pp. 258–259.

49 World Heritage Convention, Art. 11 (4).

50 Lixinski (2018a), p. 4.

51 UNESCO (2025) World Heritage in Danger. https://whc.unesco.org/en/158 (accessed 31 January 2025).

52 See Intangible Cultural Heritage Convention, Arts. 16–17.

53 Lixinski (2011), p. 82; Lixinski (2015), p. 394.

54 Lixinski (2013a), p. 43.

emanates from the risk of perceiving listed heritage to be of greater significance than the practices not listed[55] and ultimately resulting in the creation of an unconscious public perception that communities whose intangible cultural heritage is listed are of more value than others.[56] In spite of the List's prefix, it is unlikely that the public will not perceive the list as "creating a value-based classification" among forms of intangible cultural heritage.[57]

In contrast to the World Heritage List's requirement of outstanding universal value, the Representative List does not include such a requirement. Pursuant to the "Operational Directives for the Implementation of the Convention for the Safeguarding of the Intangible Heritage" (Operational Directives), inscribing intangible cultural heritage on the Representative List requires state parties to demonstrate five requirements. The element needs to fall within the definitional scope of Article 2 of the Convention and its inscription "will contribute to ensuring visibility and awareness of the [element's] significance" thereby "encouraging dialogue" and "reflecting cultural diversity worldwide".[58] Safeguarding measures are required that further the element's protection and promotion and the element needs to be included in an inventory and it should have been "nominated following the widest possible participation of the community, group or, if applicable, individuals concerned and with their free, prior and informed consent."[59] Analogously to the World Heritage List, many state parties continue to nominate elements for their unique value or their high culture character thereby misunderstanding the Representative List's "fundamental philosophy".[60]

In accordance with the criteria laid out in the Operational Directives, intangible cultural heritage inscribed on the Representative List can be placed on the "List of Intangible Cultural Heritage in Need of Urgent Safeguarding".[61] Elements are nominated by state parties[62] and need to show that the element is in "urgent need" or "extremely urgent need of safeguarding" because "its viability

---

55 Lenzerini (2015), pp. 145–46.

56 *ibid* p. 146.

57 *ibid*.

58 UNESCO 'Operational Directives for the implementation of the Convention for the Safeguarding of the Intangible Heritage' (7 July 2022) (*Operational Directives*), para. I. 2, R2.

59 Operational Directives, para. I. 2, R3–5.

60 Blake (2020a), p. 355.

61 Operational Directives, para. I.1 (para. I.1.1.U.1–6).

62 Although it is also possible for "communities, groups and, where appropriate, individuals concerned" to express the wish to move an element from one list to the other. Operational Directives, para. 382.

is at risk" or because "it is facing grave threats as a result of which it cannot be expected to survive without immediate safeguarding".[63] The nomination also requires the existence of a safeguarding plan enabling the continuity of the practice and its transmission; that the nomination was subject to the free, prior and informed consent and "widest possible participation" by the community, group, or individuals; that the heritage is part of the national inventory; and finally in the case of extreme urgency, that the state party has been duly consulted.[64]

A site or an element can also be removed from the World Heritage List or the Representative List. Removal from both lists can be prompted when the criteria for inscription are no longer satisfied.[65] Yet, it has been pointed out that deletion from the lists is a somewhat paradoxical remedy,[66] as the removal of a site from the list "punishes" the state party in question by withdrawing inter alia status and potential financial resources. However, delisting an item by extension also "punishes" the entire international community, which has demonstrated a shared interest in the preservation of the item.[67] Delisting can therefore be interpreted as a "double injury", the first occurring on the hands of the territorial state and the second through the delisting and the consequent deprivation of special protection of the respective site or practice.[68] This results, on the one hand, in withdrawing international protection, and on the other, in the World Heritage Committee losing any say it had in the management of the site[69], leaving the site or practice more vulnerable than before. Until today, three items have been removed from the list: Arabian Oryx Sanctuary in 2007, Dresden Elbe Valley in 2009, and Liverpool in 2021.[70]

---

[63] Operational Directives, para. I.1.

[64] *ibid.*

[65] *ibid* para. 40.1; Operational Guidelines, para. 192. It is interesting to note that in the case of the Operational Directives, removal can be requested by "concerned State Party, communities, groups or, where appropriate, individuals, or any other third party" Operational Directive (2022), para. 40.1.

[66] Lenzerini (2015), p. 148.

[67] Lenzerini (2013), p. 255.

[68] Lenzerini (2015), p. 148. Also characterised as a "double-edged sword" in Litton (2011), p. 229.

[69] Litton (2011), p. 229.

[70] UNESCO (2025): World Heritage List: Delisted, https://whc.unesco.org/en/list/?delisted=1, accessed 31 January 2025.

1.2 Evolving Challenges in Heritage Protection

Whilst armed conflicts continue to pose an ongoing threat to all forms of cultural heritage, including the specific threat of the intentional targeting and destruction of cultural heritage, the 2011 thematic report of the then-independent expert in the field of cultural rights, Farida Shaheed, points to further challenges:

> "Globalization, the exploitation of economic resources, the promotion of tourism, and development programmes may have a detrimental impact on the ability of individuals and communities to preserve/safeguard, develop and transmit their cultural heritage, including cultural practices, specific ways of life, and cultural sites and landscapes."[71]

Further challenges include natural catastrophes, urbanisation and industrialisation, pollution, social and economic upheaval, and, chiefly, the effects of climate change.[72] Climate change has been described as a "'threat multiplier" that magnifies existing threats.[73] Moreover, the Covid-19 pandemic posed significant challenges to the management, safeguarding and accessibility of cultural heritage.[74]

Negligence, lack of means, or competing interests by states also constitute a threat. In 2018, India's Supreme Court instructed the government to respond to the Taj Mahal's change in colour by inter alia seeking foreign help and by reprehending the government on its negligence.[75] The Taj Mahal was inscribed

---

[71] UNHRC 'Report of the independent expert in the field of cultural rights, Farida Shaheed' (21 March 2011) UN Doc A/HRC/17/38, para. 13.

[72] See Forrest (2010), p. 224; UNGA 'Report of the Special Rapporteur in the Field of Cultural Rights, Karima Bennoune' (10 August 2020) UN Doc A/75/298; Casini (2018), p. 179.

Also see Somers Cocks (2021): Seven sites, including Venice, may be added to UNESCO in Danger list, https://www.theartnewspaper.com/news/seven-sites-may-be-added-to-unesco-in-danger-list, accessed 31 January 2025; Somers Cocks (2021): Italy bans cruise ships from the Venice lagoon from 1 August, https://www.theartnewspaper.com/2021/07/14/italy-bans-cruise-ships-from-the-venice-lagoon-from-1-august, accessed 31 January 2025, Somers Cocks (2021): Venice is not in danger, according to UNESCO, https://www.theartnewspaper.com/2021/07/22/venice-is-not-in-danger-according-to-unesco, accessed 31 January 2025.

[73] UNHRC 'Cultural Rights: Tenth Anniversary Report' (17 January 2019) UN Doc A/HRC/40/53, para. 70.

[74] See UNHRC 'COVID-19, Culture and Cultural Rights' (17 February 2021) UN Doc A/HRC/46/34, para. 17.

[75] BBC (2018): Taj Mahal colour change worries India Supreme Court, https://www.bbc.com/news/world-asia-india-43967685, accessed 31 January 2025; Daley (2018): Indian

on the world heritage list in 1983.[76] Causes for the marble's loss of lustre stem from pollution, construction, and insect dung, among other factors.[77]

Furthermore, a state-driven reading and usage of cultural heritage, in an attempt to strengthen the national cultural identity, can result in exclusion of those that are not part of the majority culture.[78] States may also not recognise or consider the importance of a site or a space associated with cultural practices of a minority group or Indigenous community. In these situations, the dominant cultural group may impose "its own view of the shared cultural heritage on all others in society."[79] These power relationships within one state are important to consider, as more than one cultural or religious community "may lay claim to the same location as 'belonging' to their heritage", resulting in denial of access to that cultural heritage of the other group.[80] The holy sites in Jerusalem constitute one example, as they house holy sites for persons of the Christian, Jewish and Muslim faith.[81] Endemic structural issues, such as systemic racism and discrimination, can impact if and how heritage (of a minority group or of Indigenous peoples) is protected. For example, the art and traditional knowledge of Indigenous peoples is oftentimes ignored or used without the consent of the specific group.[82]

A closer look at the preambles of the UNESCO conventions shows cognisance of these challenges. For example, the preamble of the World Heritage Convention opens by noting that the threat of destruction does not only emanate from "traditional causes of decay", but also "by changing social and economic conditions".[83] The World Heritage Convention understands that cultural heritage is exposed to

Supreme Court Orders Government to Restore the Taj Mahal — or Demolish It, https://www.smithsonianmag.com/smart-news/indian-supreme-court-restore-taj-mahal-or-demolish-it-180969668/, accessed 31 January 2025.

[76] Decision 7 COM VIII Inscription Taj Mahal (India) in WHC 'Seventh Ordinary Session' SC83/CONF.009/2 ICOMOS (17 November 1983).

[77] BBC (2018): Taj Mahal colour change worries India Supreme Court, https://www.bbc.com/news/world-asia-india-43967685, accessed 31 January 2025; Daley (2018): Indian Supreme Court Orders Government to Restore the Taj Mahal — or Demolish It, https://www.smithsonianmag.com/smart-news/indian-supreme-court-restore-taj-mahal-or-demolish-it-180969668/, accessed 31 January 2025.

[78] Tünsmeyer (2020), p. 320; Donders (2017), p. 23.

[79] See Blake (2015), p. 285.

[80] *ibid* p. 284.

[81] *ibid.*

[82] Xanthaki (2017), p. 1.

[83] World Heritage Convention, preamble, first recital.

numerous and continuously evolving hazards.[84] The preamble of the Intangible Cultural Heritage Convention also refers to threats facing the intangible cultural heritage. It recognises

> "that the processes of globalization and social transformation, alongside the conditions they create for renewed dialogue among communities, also give rise, as does the phenomenon of intolerance, to grave threats of deterioration, disappearance and destruction of the intangible cultural heritage, in particular owing to a lack of resources for safeguarding such heritage."[85]

While these threats may not carry the same celerity as an attack on cultural heritage within or outside the context of an armed conflict, they contribute to its decay, degradation, destruction, and eventual loss. The effects of climate change, including sea level rise, soil erosion, forest fires, or other extreme weather events, are indiscriminate and affect all forms of heritage.[86] Therefore, they will impact "all of the values associated with heritage, including its intrinsic, touristic and economic values, as a marker of identity and attachment to place and as 'an embodiment of accumulated knowledge'".[87] International law's state-centricity results in the prevention of offences against heritage becoming virtually impossible "when a state considers an interest at odds with the protection of word heritage".[88]

---

[84] Art. 11 (4) addresses these hazards in the context of listing sites on the list of world heritage in danger and refers to: "the threat of disappearance caused by accelerated deterioration, large- scale public or private projects or rapid urban or tourist development projects; destruction caused by changes in the use or ownership of the land; major alterations due to unknown causes; abandonment for any reason whatsoever; the outbreak or the threat of an armed conflict; calamities and cataclysms; serious fires, earthquakes, landslides; volcanic eruptions; changes in water level, floods and tidal waves."

It is interesting to note that Art. 11 (4) of the World Heritage Convention not only refers to the outbreak of an armed conflict, but also the "threat thereof". This suggests not only the application of the World Heritage Convention in the event of an armed conflict. Confirmed by the ICJ's application of the World Heritage Convention in the 2013 judgement of the Temple of Preah Vihear case. *Request for Interpretation of the Judgement of 15 June 1962 in the Case Concerning the Temple of Preah Vihear* (Cambodia v. Thailand) (judgement) [2013] ICJ Rep 281.

[85] Intangible Cultural Heritage Convention, preamble.

[86] See UNGA 'Report of the Special Rapporteur in the Field of Cultural Rights, Karima Bennoune' (10 August 2020) UN Doc A/75/298, para. 30.

[87] *ibid.*

[88] Lenzerini (2013), p. 255.

These evolving challenges in the protection of cultural heritage (primarily in peacetime but such peacetime challenges can of course also coexist with armed conflicts) link to the challenges in the implementation of the World Heritage Convention and the Intangible Cultural Heritage Convention. Inscription on the lists is desired by state parties, as it is generally accompanied by international attention, an increase in tourism, and therefore economic incentives.[89] However, the mechanism of listing is not uncontested.

The over-representation of European and "Western" heritage has been subject to criticism.[90] In spite of the World Heritage Committee's Global Strategy for a Balanced, Representative and Credible World Heritage List, launched in 1994, the imbalance of geographic distribution of sites remains palpable.[91] Although the List of Intangible Cultural Heritage in Need of Urgent Safeguarding seeks to respond to "a generally accepted vulnerability of" intangible cultural heritage, a certain stigma remains associated with inscribing the elements on the list.[92]

The List of World Heritage in Danger is often associated with "a stronger sense of a failure" by the state parties.[93] Although UNESCO's World Heritage Centre underlines that inscription on the List of World Heritage in Danger should not be considered as a sanction, it recognises that such listing "is not perceived in the same way by all parties concerned".[94] This includes state parties wishing "to avoid an inscription, which they perceive as a dishonour"[95] or as a form of "naming and shaming".[96] The effectiveness of the List of World Heritage in Danger struggles with resistance by state parties. While these struggles stem from its perception as a tool of public reprimand, they can also be ascribed to the question of who can nominate a site to be placed on the List of World Heritage in Danger and whether the respective state must consent to the inscription.[97]

---

[89] Bertacchini, Liuzza, Meskell and Saccone (2016), pp. 167, 96.

[90] Lixinski (2013a), p. 23; Francioni (2020c), p. 254.

[91] See e.g., Steiner and Frey (2011); Hølleland, Hamman and Phelps (2019), p. 42.

[92] Blake (2020a), p. 354.

[93] *ibid.*

[94] UNESCO (2025): World Heritage in Danger. https://whc.unesco.org/en/158, accessed 31 January 2025.

[95] *ibid.*

[96] Affolder (2007), p. 354; Hølleland, Hamman and Phelps (2019), p. 38.

[97] See in the academic discourse: Buzzini and Condorelli (2008), p. 176; Francioni (2020c), p. 261; Affolder (2007), p. 341; See at the WHC: WHC 'Sixth extraordinary session' (3 December 2002) WHC-03/6 EXT.COM/INF.4A.

Placing a site on the List of World Heritage in Danger without the state's consent has, in part, elicited negative responses and resistance by state parties.[98] Whereas Article 11 (3) of the World Heritage Convention clearly sets out that including "a property in the World Heritage List requires the consent of the State concerned", Article 11 (4), which established the list of World Heritage in Danger, does not include a similar wording. *Buzzini* and *Condorelli* argue that once a property is inscribed on the World Heritage List and once the state has made a request for assistance under the Convention with respect to that property, no further state consent is required in placing a property on the List of World Heritage in Danger.[99] Moreover, as Article 11 (4) states that the World Heritage Committee "may at any time, in case of urgent need, make a new entry in the List of World Heritage in Danger", *Buzzini* and *Condorelli* conclude that the World Heritage Committee has discretionary power to inscribe a property on the List of World Heritage in Danger when a situation is of "urgent need", regardless of a prior request for assistance by the state party.[100] The inscription of Dubrovnik on the List of World Heritage in Danger in 1991 without the consent of Yugoslavia supports this argument.[101]

To what extent these "in danger" listings pose a challenge to state sovereignty has been subject to debate.[102] In the case of Australia's Kakadu National Park in 1998 the issue at stake was the opening of a uranium mine in the park.[103] In spite of protests by Indigenous groups and the recommendation of a fact-finding mission, organised by the World Heritage Committee, the Australian government refused to consent to moving Kakadu National Park on the List of the World Heritage in Danger.[104] *Francioni* explains:

> "This issue touches a raw nerve of the Convention—that is, the difficult balance between the principle of sovereignty of each State Party over the World Heritage Sites located in its territory, as recognized by article 6(1) of the Convention, and the concept of 'world heritage', which presupposes the recognition of a general interest of all

[98] Battini (2011), p. 354; Brown, Liuzza and Meskell (2019).

[99] Buzzini and Condorelli (2008), pp. 185–186.

[100] *ibid* p. 186.

[101] See "the landmark case of Dubrovnik clearly established that State Party consent was not a requirement for inscription on the List in Danger, a position that was reflected in the 1994 revision to the Operational Guidelines". Brown, Liuzza and Meskell (2019), p. 293.

[102] See, e.g., Buzzini and Condorelli (2008), p. 176; Affolder (2007), p. 354; Hølleland, Hamman and Phelps (2019), p. 39.

[103] Francioni (2020c), p. 262.

[104] *ibid.*

> States Parties, and of the international community as a whole, in the safeguarding of the outstanding universal value of listed properties under the system of international cooperation established by the Convention."[105]

The decision of the World Heritage Committee to not inscribe Kakadu National Park on the list is indicative of "giving overriding importance to sovereignty concerns as opposed to concerns for the proper conservation of World Heritage Sites under threat."[106] More recent cases, where the Committee either delayed or decided against an inscription on the List of World Heritage in Danger amid the scientific recommendations of its advisory bodies, include the Great Barrier Reef and Venice.[107] Journalistic coverage of the case of the Great Barrier Reef shed light on the politicisation of heritage protection[108] and lobbying efforts by the state to keep the site off the list.[109]

---

[105] *ibid* pp. 261–262.

[106] *ibid* p. 263.

[107] For coverage on the debates, see Somers Cocks (2023): For the second time, Venice escapes listing as at risk, but now Unesco is the enemy, https://www.theartnewspaper.com/2023/09/15/for-the-second-time-venice-escapes-listing-as-at-risk-but-now-unesco-is-the-enemy, accessed 31 January 2025); Readfearn (2023): Unesco recommends against Great Barrier Reef 'in danger' listing but Australia warned more action needed, https://www.theguardian.com/environment/2023/aug/01/climate-crisis-great-barrier-reef-unesco-australia, accessed 31 January 2025; Hofmann (2024): The Great Barrier Reef and the List of World Heritage in Danger, https://voelkerrechtsblog.org/the-great-barrier-reef-and-the-list-of-world-heritage-in-danger/, accessed 31 January 2025.

[108] See e.g., Readfearn (2021): World Heritage Committee agrees not to place Great Barrier Reef on 'in danger' list, https://www.theguardian.com/environment/2021/jul/23/world-heritage-committee-agrees-not-to-place-great-barrier-reef-on-in-danger-list, accessed 31 January 2025.

Also see, in the context of Venice: Somers Cocks (2021): The IPCC's predictions of sea-level rise are why Italy should beg UNESCO to put Venice on its at risk list, https://www.theartnewspaper.com/comment/the-ipcc-s-predictions-of-sea-level-rise-are-why-italy-should-beg-unesco-to-put-venice-on-its-at-risk-list, accessed 31 January 2025.

As well as the case of the Karen people: Readfearn (2021): 'Low point' in world heritage committee history as politics 'tramples' human rights of the Karen people, https://www.theguardian.com/environment/2021/aug/06/low-point-in-world-heritage-committee-history-as-politics-tramples-human-rights-of-the-karen-people, accessed 31 January 2025.

[109] The Guardian (2021): 'The lobbying push that killed off a fight to save the Great Barrier Reef', https://www.theguardian.com/news/audio/2021/jul/29/the-lobbying-push-that-killed-off-a-fight-to-save-the-great-barrier-reef-podcast, accessed 31 January 2025. Also see on lobbying: Brown, Liuzza and Meskell (2019); Lafrenz Samuels and Platts (2022), p. 422; Litton (2011), p. 220.

For groups and communities, heritage listing can serve as an important mechanism to recognise cultural identity at the international level as well as identity that transcends national borders.[110] Furthermore, it can contribute to a feeling of "an international presence against their state".[111] However, listing can run the risk of being ignorant to different perceptions of heritage, "assuming that heritage is to be made public in all instances, and given broad knowledge, and join in a universal pool of 'cultural commons'."[112] In other cases, heritage may be ignored or misappropriated. One such example is the Uluru-Kata Tjuta National Park on Australian territory, which was inscribed in 1987 as a natural property (formerly called Uluru (Ayers Rock—Mount Olga) National Park)[113].[114] Seven years later, the protection of the Uluru was extended by acknowledging its significant cultural criteria, specifically the site's importance as "an integral part of the traditional belief system of one of the oldest human societies in the world", the Anagu community.[115] This re-creation as world heritage[116] arguably contributed to the disenfranchisement of the Indigenous community from using the site as well as to a blatant disregard for the culture and beliefs of the Anagu community. This was long reflected by the inability to stop people from climbing Uluru[117] despite the community's efforts to explain that under their traditional law, climbing Uluru is not permitted.[118] Whilst on 26 October 2019, the climb to the top of Uluru was closed,[119] the example of the Anagu community and Uluru shows

[110] Lixinski (2017), pp. 235, 238.

[111] *ibid* pp. 235.

[112] Lixinski (2013b), pp. 147–148.

[113] UNESCO (2025): Uluru-Kata Tjuta National Park, https://whc.unesco.org/en/list/447/, accessed 31 January 2025.

[114] Advisory Body Evaluation—IUCN, 1987, available at: UNESCO (2025): Uluru-Kata Tjuta National Park: Documents, whc.unesco.org/en/list/447/documents/.

[115] Advisory Body Evaluation (ICOMOS), 1994, p. 105, UNESCO (2025): Uluru-Kata Tjuta National Park: Documents, whc.unesco.org/en/list/447/documents.

[116] Lixinski (2013b), p. 149.

[117] Francioni and Lixinski (2017), p. 21.

[118] A sign at Uluru read: "We, the Anangu traditional owners, have this to say. Uluru is sacred in our culture. It is a place of great knowledge. Under our traditional law climbing is not permitted. This is our home … Please don't climb. We invite you to walk around the base and discover a deeper understanding of this place." See Hobbs (2016): The Uluru decision highlights what's so wrong with Indigenous policy, https://www.theguardian.com/commentisfree/2016/apr/21/the-uluru-decision-highlights-whats-so-wrong-with-indigenous-policy, accessed 31 January 2025.

[119] Parks Australia (Commonwealth of Australia) (2019): Uluṟu climb closure, https://parksaustralia.gov.au/uluru/discover/culture/uluru-climb/, accessed 31 January 2025.

how essential the participation of Indigenous communities is in decision-making processes and for respecting cultural privacy. Furthermore, listing can become problematic when it becomes a political tool. China's inscription of the Tibetan Opera can be interpreted as an assertion of control "over the cultural manifestation, and the culture as a whole, subordinating its political caveats to tourism promotion and other economic interests, as well as to a larger national Chinese identity, ultimately diminishing the political strength of the Tibetan culture and all political claims of Tibetans."[120]

The Intangible Cultural Heritage Convention and the Operational Guidelines of the World Heritage Convention recognise participation of non-state actors in different processes, including in the identification, nomination and safeguarding. This includes "communities, groups and individuals" in the case of the Intangible Cultural Heritage Convention and local communities, Indigenous peoples, and other stakeholders in the case of the Operational Guidelines of the World Heritage Convention.[121] Yet, the leeway given to states to ensure this participation has been described as "more exhortatory than obligatory" resulting in participation remaining weak.[122]

1.3 Research Question

This brief, and by no means exhaustive, account of the challenges in protecting heritage points to at least two observations. The first stems from an observation that considers procedural aspects. Whilst the implementation of the World Heritage Convention and the Intangible Cultural Heritage Convention are supported by a monitoring mechanism,[123] the Conventions do not set up further enforcement tools. Although state-centricity is inherent in any branch of international law, the availability of an additional monitoring system and an individual complaint system could further strengthen protection. The second observation considers substantive aspects. The protection of cultural heritage and human rights intersects significantly. What is more, the protection of cultural heritage and continuously evolving threats can have a considerable impact on human rights, as access to heritage during the COVID-19 pandemic and the ongoing and accelerating threat

[120] Lixinski (2013a), p. 53.

[121] Intangible Cultural Heritage Convention, Art. 15. Also see Art. 11 (b). Operational Guidelines, para. 123.

[122] Blake (2020a), p. 356. Also see Disko (2017), p. 64; Gilbert (2018), p. 131.

[123] World Heritage Convention, Art. 29; Intangible Cultural Heritage Convention, Art. 29.

of climate change on persons' way of life have shown.[124] It is therefore worth considering cultural heritage in the human rights framework.

This thesis sets out to analyse if and how international human rights law can offer a legal basis for the protection of cultural heritage. As international human rights law does not offer an explicit right to cultural heritage, the analysis focuses on the interpretation of a selection of human rights bodies. The UN system is twofold and is comprised by the charter-based bodies and the treaty bodies. Charter-based bodies, established under the UN Charter, promote human rights in all UN member states. Treaty bodies, established under the individual treaties, are responsible for monitoring the implementation of the particular treaty.

The analysis examines the pronouncements of two treaty bodies: the Human Rights Committee[125] (HRC), established under the ICCPR, and the Committee on Economic, Social and Cultural Rights (CESCR), monitoring the implementation of the ICESCR.[126] Treaty bodies are composed of independent experts serving in their individual capacities, and are tasked with monitoring national implementation of international human rights treaties.[127] The pronouncements of these expert bodies can take several forms. For the purposes of this thesis, the principal activities of both the HRC and the CESCR will be grouped into three activities: authoring general comments, examining state reports, and offering views on individual communications.[128] The focus on these three activities excludes some of their functions, such as the capacity of both Committees to consider interstate communications or the "opt in clause"[129] of the Optional Protocol to the ICESCR for inquiry procedures.[130]

---

124 See UNHRC 'COVID-19, culture and cultural rights' (17 February 2021) UN Doc A/HRC/46/34, para. 15; CCPR, Communication No. 3624/2019, *Daniel Billy et al. v. Australia* (22 September 2022) UN Doc CCPR/C/135/D/3624/2019, paras. 3.1, 3.5, 8.10, 8.13, 8.14.

125 International Covenant on Civil and Political Rights (adopted 16 Dec 1966, entered into force 23 Mar 1976) 999 UNTS 171 (*ICCPR*), Articles 28–45.

126 Established under resolution 1985/17 of 28 May 1985 by the Economic and Social Council. ECOSOC, 28/05/1985, E/RES/1985/17.

127 Keller and Ulfstein (2012), p. 2.

128 See *ibid* p. 3; Neuman (2016), p. 2. Whilst Neuman only speaks of the Human Rights Committee, the Optional Protocol to the ICESCR, in the meantime, entered into force. The Committee on Economic, Social and Cultural Rights therefore also has assumed all three functions.

129 Pinto (2018a), para. 10.

130 ICCPR, Art. 41; Optional Protocol to the International Covenant on Economic, Social and Cultural Rights (adopted 10 December 2008, entered into force 5 May 2013) 2922 UNTS 29 (*Optional Protocol to the ICESCR*), Art. 10.

The thesis combines the analysis of the treaty bodies and the charter-based bodies, thereby allowing for a complementary approach. The primary focus for charter-based bodies in the framework of this thesis lies on the work of the UNHRC and its mechanisms, specifically the Universal Periodic Review (UPR) and the Special Procedures.[131] In addition, the analysis considers resolutions by the UN General Assembly and the Security Council that specifically address cultural heritage. The evident advantage of reviewing both treaty and charter-based body mechanisms is that *all* states are subject to the UPR, facilitating a "comprehensive and objective mapping of the human rights situation in the world".[132]

These mechanisms complement each other in the manner in which they consider the implementation of human rights. The reporting under the treaty bodies and within the framework of the UPR monitor the implementation at the state level. Country reports in the Special Procedures provide further insight into national human rights implementation. The thematic work of the Special Procedures or the treaty bodies' general comments, in contrast, consider substantive matters.[133]

Structured into "meaning", "rights", and "obligations", the thesis sets out to understand the meaning of cultural heritage in the human rights framework, the relevant rights of individuals and groups, and the corresponding state obligations vis-à-vis cultural heritage. It further assesses to what extent rights and obligations serving the protection of cultural heritage, are (still) emerging in the international human rights framework.

The thesis adopts the terminology of "rights" and "obligations" while being aware that it relies on formally non-binding sources. The legal significant and legal effect of the pronouncements and work of human rights bodies will be further explored in the following chapter. Whilst the thesis is structured around the notion of "protection", it is cognisant that "protection" has specific meanings in different international legal heritage instruments, including in the 1954 Convention and the World Heritage Convention. It is therefore important to underline

---

[131] Due to its confidentiality, the complaints procedure of the UNHRC was not considered. See, UNHRC, Human Rights Council Complaint Procedure, https://www.ohchr.org/en/hr-bodies/hrc/complaint-procedure/hrc-complaint-procedure-index, accessed 31 January 2025.

[132] Tistounet (2020), p. 12.

[133] See *ibid* p. 26.

that the use of "protection" and throughout the analysis is employed in accordance with its ordinary meaning[134] and detached from its meaning in specific Conventions.

1.4 Research Methods

The breadth of information offered by international human rights bodies requires a structured approach to data collection. The sample draws on reporting from approximately 15 years, taking 2009 as the starting year. This puts the adoption of General Comment No. 21 by the CESCR[135] in 2009 and the Special Rapporteur's report on access to cultural heritage as a human right[136] of 2011 at the beginning of the sampling timeline. The last rounds of reporting were considered up to January 2025. Earlier reporting was considered and consulted in individual cases to better understand the reporting history and wherever references to earlier reports or concluding observations rendered it necessary. Earlier reporting guidelines and complaints were similarly considered. The time frame allows the analysis to observe how the interpretation of human rights norms and standards has evolved vis-à-vis cultural heritage.

Rather than selecting a geographic region, the analysis draws on a sample. Considering the imbalance of the geographic distribution of world heritage sites, often subject to criticism,[137] the thesis relies on the regional division by the United Nation's standard country or area codes for statistical use, also known as the M49 standard.[138] Each geographic region—Africa, the Americas, Asia, Europe, and Oceania[139]—is divided into further geographic regions. Depending on the size of the region one or two states have been selected for the sample (see Table 1.1).

---

[134] Cambridge University Press & Assessment (2025): Dictionaries, https://dictionary.cambridge.org/dictionary/english/protection, accessed 31 January 2025.

[135] CESCR 'General Comment No. 21.' (21 December 2009) UN Doc E/C.12/GC/21.

[136] UNHRC 'Report of the independent expert in the field of cultural rights, Farida Shaheed' (21 March 2011) UN Doc A/HRC/17/38.

[137] See, e.g. Steiner and Frey (2011); Maurel, Chloé (2017): Whose World Heritage? The problem with UNESCO's famous list, https://www.equaltimes.org/whose-world-heritage-the-problem?lang=en#.WZmqN62B33B, accessed 31 January 2025.

[138] UN Statistics Division (2025): Methodology: Standard country or area codes for statistical use,, https://unstats.un.org/unsd/methodology/m49/, accessed 31 January 2025.

[139] For the purposes of this PhD thesis, Antarctica is omitted.

The selection of the sample considers several factors, including (1) the representation of minority groups and Indigenous peoples[140]; (2) conflict[141] and (3) non-conflict related challenges, including urban planning, development projects and tourism[142]; as well as (4) including states particularly vulnerable to natural catastrophes and climate change[143]. The factors are aimed to reflect or mirror the evolving challenges in heritage protection. The selection is further informed by the availability of concluding observations and state reports.

Table 1.1 Selection of States

Region	CESCR	HRC	UPR
Africa			
Northern Africa	Egypt	Egypt	Egypt
Eastern Africa	Tanzania	Tanzania	Tanzania
	Ethiopia	Ethiopia	Ethiopia
Middle Africa	Angola	Angola	Angola
Southern Africa	Namibia	Namibia	Namibia
Western Africa	Mali	Mali	Mali
Americas			
Caribbean	Dominican Republic	Dominican Republic	Dominican Republic
	n. a. (only signatory)	n. a. (only signatory)	Cuba
Central America	Costa Rica	Costa Rica	Costa Rica
	Mexico	Mexico	Mexico

(continued)

140 For example, Canada, the United States of America, Australia, New Zealand, Finland as well as Colombia, Costa Rica, and Brazil.

141 For example, Mali and Afghanistan.

142 For example, Italy, Viet Nam, Australia.

143 For example, small island developing states, such as Cuba, Dominican Republic, Federated States of Micronesia, Samoa, Tonga and Vanuatu.

Table 1.1 (continued)

Region	CESCR	HRC	UPR
South America	Brazil	Brazil	Brazil
	Colombia	Colombia	Colombia
Northern America	Canada	Canada	Canada
	n. a. (only signatory)	United States of America	United States of America
Asia			
Central Asia	Uzbekistan	Uzbekistan	Uzbekistan
Eastern Asia	Japan	Japan	Japan
South-eastern Asia	Viet Nam	Viet Nam	Viet Nam
Southern Asia	Afghanistan	Afghanistan	Afghanistan
Western Asia	Turkey	Turkey	Turkey
Europe			
Eastern Europe	Russian Federation	Russian Federation	Russian Federation
Northern Europe	Finland	Finland	Finland
	United Kingdom of Great Britain and Northern Ireland	United Kingdom of Great Britain and Northern Ireland	United Kingdom of Great Britain and Northern Ireland
Southern Europe	Italy	Italy	Italy
	Bosnia and Herzegovina	Bosnia and Herzegovina	Bosnia Herzegovina
Western Europe	France	France	France
Oceania			
Australia and New Zealand	Australia	Australia	Australia
	New Zealand	New Zealand	New Zealand
Melanesia	n. a. (no action)	no reports	Vanuatu
Micronesia	n. a (no action)	n. a. (no action)	Federated States of Micronesia
Polynesia	n. a (no action)	n. a. (no action)	Tonga
	n. a (no action)	no reports	Samoa

Documents were accessed through the official databases of the UN accessible to the public.[144]

The analysis of the pronouncements and output of the human rights bodies is limited in at least two ways. First, the breadth of information requires focusing on the most common and prevalent rights and obligations identified by the human rights bodies. Secondly, not all issues pertaining to heritage practices and protection could be considered in the same, detailed manner. It is for this reason the scope excludes the topic of religious clothing as a religious and cultural practice and only addresses sacred sites and places of worship. The Intangible Cultural Heritage Convention covers rituals that are associated with religions, however, religion in general terms is not included in the scope of the intangible cultural heritage.[145]

Although the primary focus of the analysis lies on cultural heritage, heritage can be both cultural and natural heritage. Especially when speaking of Indigenous heritage, legal definitions might fall short, as heritage can constitute a broader concept in this context that makes no distinction between natural, cultural, tangible, or intangible.[146]

1.5 Structure of the Thesis

The thesis is structured into seven chapters. After this introductory chapter offered a brief overview of the aims of this thesis, the protection of cultural heritage at the international level, and challenges in the implementation, Chap. 2 focuses on the legal significance of human rights bodies in international law. It offers a short introduction to the role and functions of human rights bodies in international law, discusses their interpretive authority, and considers how their work and pronouncements fit into the sources of international law. Arguing that pronouncements and the work of human rights bodies hold legal significance and should be understood on a continuum of legal effect, the chapter explores different ways in which they can contribute to the interpretation and development of human rights law.

---

[144] UNHRC (2025): Documentation by Country, https://www.ohchr.org/en/hr-bodies/upr/documentation, accessed 31 January 2025; UN Human Rights Treaty Body Database (20205): UN Treaty Body Database, https://tbinternet.ohchr.org/_layouts/15/TreatyBodyExternal/Home.aspx, accessed 31 January 2025.

[145] Ubertazzi (2020), p. 70; Francioni (2020b), p. 52.

[146] See Blake (2015), p. 280.

Chapter 3 analyses the human dimension of international heritage law in three key areas: (1) the evolution of the concept of cultural heritage, (2) the rationales for protecting cultural heritage and its protection mechanisms, and (3) the enforcement of crimes against cultural heritage. The chapter argues that increasingly, the role and rights of communities, groups and individuals are being considered as well as their relationship to cultural heritage. The chapter, thereby, lays the foundation and further establishes the impetus to focus on the international human rights framework.

Chapter 4, titled "Cultural Heritage through the Lens of the CESCR and HRC", is organised along the threefold structure of "meaning", "rights", and "obligations". The chapter examines the meaning of cultural heritage in the pronouncements of the CESCR and HRC, the relevant rights of individuals and persons belonging to groups, and the corresponding state obligations relating to cultural heritage. The aim of the chapter is to better understand if and how the committees address cultural heritage and which rights and obligations, if any, can be inferred from the pronouncements.

"Cultural Heritage through the Lens of the UNHRC and its Mechanisms", Chap. 5, is the counterpart to Chap. 4. It focuses on the UNHRC and, specifically on the Special Procedures and the UPR. The chapter's objective is to understand how the UNHRC and its mechanisms understand cultural heritage, the relevant rights of individuals and persons belonging to groups, and the corresponding state obligations relating to cultural heritage.

Chapter 6, entitled "Meaning, Rights, and Obligations in the Human Rights Bodies", appraises the findings of the human rights bodies with the aim of understanding if and how human rights law provides a legal basis for the protection of cultural heritage. It is structured into two parts: The first part juxtaposes and analyses the rights and obligations articulated by the human rights bodies, thereby including the state reports of the treaty bodies' periodic reporting with the aim of learning how these rights and obligations have been received. The second part looks at how the tools and mechanisms of the international human rights law framework can contribute to the protection of cultural heritage. It does so by delving further into how the mechanisms of human rights bodies can be used and have already been used to further the protection of cultural heritage.

The concluding Chap. 7 summarises and reflects on the main findings of the thesis. The thesis submits that although human rights law does not provide for an explicit right to cultural heritage, rights and obligations vis-à-vis cultural heritage form part of human right law. The chapter expounds on these rights and shows that these rights and obligations emanate from several legal bases and form part of the proper implementation of human rights provisions. It further argues that the human rights bodies equip individuals and persons belonging to a group with legal bases and tools to strengthen their claims. The chapter shows the ways in which they offer soft enforcement tools for the respective rights at the international level. The chapter concludes by addressing limitations and impetus for future research.

Reference List

Affolder, Natasha (2007): Democratising or Demonising the World Heritage Convention?. In: Victoria University of Wellington Law Review, Vol. 38, No. 2, S. 341–362.

Bassiouni, M. C. (1983). Reflections on Criminal Jurisdiction in International Protection of Cultural Property. *Syracuse Journal of International Law and Commerce, 10*(2), 281–322.

Battini, Stefano (2011): The Procedural Side of Legal Globalization: The Case of the World Heritage Convention. In: International Journal of Constitutional Law, Vol. 9, Issue 2, S. 340–368.

Bertacchini, E., Liuzza, C., Meskell, L., & Saccone, D. (2016). The Politicization of UNESCO World Heritage Decision Making. *Public Choice, 167*, 95–129.

Blake, Janet (2020a): Safeguarding Intangible Cultural Heritage. In: Francesco Francioni and Ana Filipa Vrdoljak (eds.): The Oxford Handbook of International Cultural Heritage Law. Place of Publication (Oxford University Press), S. 347–370.

Blake, Janet (2015): International Cultural Heritage Law. Oxford (Oxford University Press).

Brown, Nicholas E., Liuzza, Claudia and Meskell, Lynn (2019): The Politics of Peril: UNESCO's List of World Heritage in Danger. In: Journal of Field Archaeology, Vol. 44, No. 5, S. 287–303.

Buzzini, Gionata P. and Condorelli, Luigi (2008): Art.11: List of World Heritage in Danger and Deletion of a Property from the World Heritage List. In Francesco Francioni and Federico Lenzerini (eds.) The 1972 World Heritage Convention: A Commentary. 2nd edition, Oxford (Oxford University Press).

Casini, Lorenzo (2018): Cultural Sites Between Nationhood and Mankind. In: Eyal Benvenisti and Georg Nolte (eds.): Community Interests Across International Law. Oxford (Oxford University Press), S. 176–188.

Donders, Yvonne (2018): Protection and Promotion of Cultural Heritage and Human Rights through International Treaties: Two Worlds of Difference. In: Charlotte Waelde, Catherine Cummings, Mathilde Pavis and Helena Enright (eds.): Research Handbook on Contemporary Intangible Cultural Heritage. Cheltenham/Northampton (Edward Elgar), S. 54–77.

Donders, Yvonne (2017): Cultural Diversity and Cultural Identity in Human Rights. In: Andreas J. Wiesand, Kalliopi Chainoglou und Anna Sledzinska-Simon (eds.): Culture and Human Rights: The Wroclaw Commentaries. Berlin, Boston (De Gruyter), S. 23–32.

Disko, Stefan (2017): Indigenous Cultural Heritage in the Implementation of UNESCO's World Heritage Convention: Opportunities, Obstacles and Challenges. In: Indigenous Peoples' Cultural Heritage: Rights, Debates and Challenges. Leiden (Brill | Nijhoff), S. 39–77.

Forrest, Craig (2010): International Law and the Protection of Cultural Heritage. Abingdon/ New York, NY (Routledge).

Francioni, Francesco (2020a): Custom and General Principles of International Cultural Heritage Law. In: Francesco Francioni and Ana Filipa Vrdoljak (eds.): The Oxford Handbook of International Cultural Heritage Law. Oxford (Oxford University Press), S. 531–550.

Francioni, Francesco (2020b): Art.2(1) Defining Intangible Cultural Heritage. In: Janet Blake and Lucas Lixinski (eds.): The 2003 UNESCO Intangible Heritage Convention: A Commentary. Oxford (Oxford University Press), S. 48–57.

Francioni, Francesco (2020c): World Cultural Heritage. In: Francesco Francioni and Ana Filipa Vrdoljak (eds.): The Oxford Handbook of International Cultural Heritage Law. Oxford (Oxford University Press), S. 250–271.

Francioni, Francesco and Lixinski, Lucas (2017): Opening the Toolbox of International Human Rights Law in the Safeguarding of Cultural Heritage. In: Andrea Durbach and Lucas Lixinski (eds.): Heritage, Culture and Rights: Challenging Legal Discourses. Oxford (Hart Publishing.

Francioni, Francesco (2004): Beyond State Sovereignty: The Protection of Cultural Heritage as a Shared Interest in Humanity. In: Michigan Journal of International Law Vol. 25, No. 1, S. 1209–1228.

Gerstenblith, P. (2008). Protecting Cultural Heritage in Armed Conflict: Looking Back, Looking Forward. *Cardozo Public Law & Policy Ethics Journal, 7*, 677–708.

Gilbert, Jérémie (2018): Natural Resources and Human Rights: An Appraisal. Oxford (Oxford University Press).

Hølleland, Herdis, Hamman, Evan and Phelps, Jessica (2019): Naming, Shaming and Fire Alarms: The Compilation, Development and Use of the List of World Heritage in Danger. In: Transnational Environmental Law, Vol. 8, No. 1, S. 35–57.

Jakubowski et al. (2022): International Law Association Committee on Participation in Global Cultural Heritage Governance - Final Report (2022).

Keane, David (2004): The Failure to Protect Cultural Property in Wartime. In: DePaul Journal of Art, Technology & Intellectual Property Law, Vol. 14, Issue 1, S. 1–38.

Keller, H., & Ulfstein, G. (2012). *UN Human Rights Treaty Bodies*. Cambridge (Cambridge University Press).

Lafrenz Samuels, Kathryn and Platts, Ellen J. (2022): Global Climate Change and the UNESCO World Heritage Committee. In: International Journal of Cultural Property, Vol. 29, S. 409–432.

Lenzerini, Federico (2020): Art.16–17 Listing Intangible Cultural Heritage. In: Janet Blake and Lucas Lixinski (eds.): The 2003 UNESCO Intangible Heritage Convention: A Commentary, S. 306–328.

Lenzerini, Federico (2016a): Terrorism, Conflicts and the Responsibility to Protect Cultural Heritage. In: The International Spectator: Special Issue on the Responsibility to Protect, Vol. 51, No. 2, S. 70–85.

Lenzerini, Federico (2015); Protecting the Tangible, Safeguarding the Intangible: A Same Conventional Model for Different Needs. In: Sabine von Schorlemer and Sylvia Maus (eds.): Climate Change as a Threat to Peace: Impacts on Cultural Heritage and Cultural Diversity. Frankfurt am Main (Peter Lang Group AG), S. 141–160.

Lenzerini, Federico (2013): Suppressing and Remedying Offences against Culture. In: Ana Filipa Vrdoljak (ed.): The Cultural Dimension of Human Rights. Oxford (Oxford University Press), S. 240–272.
Lenzerini, Federico (2003): The UNESCO Declaration Concerning the Intentional Destruction of Cultural Heritage: One Step Forward and Two Steps Back. In: The Italian Yearbook of International Law, Vol. 3, Issue 1, S. 131–145.
Litton, Sam (2011): The World Heritage "In Danger" Listing as a Taking. In: New York University Journal of International Law & Politics, Vol. 44, No. 1, S. 219–265.
Lixinski, Lucas (2019): International Heritage Law for Communities: Exclusion and Re-Imagination. Oxford (Oxford University Press).
Lixinski, Lucas (2018a): Moving Cultures: Engaging Refugee and Migrant Cultural Rights in International Heritage Law. In: Indonesian Journal of International Law, Vol 16, No. 1, S. 1–27.
Lixinski, (2017): Heritage Listing as Self-determination. In: Andrea Durbach and Lucas Lixinski (eds.) Heritage, Culture and Rights: Challenging Legal Discourses. Oxford (Hart Publishing), S. 227–249.
Lixinski, Lucas (2015): Heritage Listing as a Tool for Advocacy: The Possibilities for Dissent, Contestation, and Emancipation in International Law Through International Cultural Heritage Law. In: Asian Journal of International Law, Vol. 5, No. 2, S. 387–409.
Lixinski, Lucas (2013b): Luci E Ombre: The Bright and Dark Sides of International Heritage Law. In: The Italian Yearbook of International Law Online, Vol. 22, Issue 1, S. 133–153.
Lixinski, Lucas (2013a): Intangible Cultural Heritage in International Law. Oxford (Oxford University Press).
Linxinski, Lucas (2011): Selecting Heritage: The Interplay of Art, Politics and Identity. In: European Journal of International Law, Vol. 22, No. 1, S. 81–100.
Neuman, Gerald L. (2016): Giving Meaning and Effect to Human Rights: The Contributions of Human Rights Committee Members. In: Human Rights Program Research Working Paper Series.
Pinto, Mónica (2018a): Inquiry Procedures: United Nations Human Rights Bodies. In: Max Planck Encyclopedias of International Law.
Scheinin, Martin (2017): The Art and Science of Interpretation in Human Rights Law. In: Bård A. Andreassen, Claire Methven O'Brien and Hans-Otto Sano (eds.): Research Methods in Human Rights: A Handbook. Cheltenham/Northampton (Edward Elgar), S. 17–37.
Smith, Laurajane (2006): Uses of Heritage. (Routledge).
Steiner, Lasse and Frey, Bruno S. (2011): Imbalance of World Heritage List: Did the UNESCO Strategy Work? University of Zurich, Economics Working Paper No. 14, https://ssrn.com/abstract=1807889 or https://doi.org/10.2139/ssrn.1807889.
Tistounet, Eric (2020): The UN Human Rights Council – A Practical Anatomy. Cheltenham/Northampton (Edward Elgar).
Tünsmeyer, Vanessa (2020): Bridging the Gap Between International Human Rights and International Cultural Heritage Law Instruments: A Functions Approach. In: Anne-Marie Carstens and Elizabeth Varner (eds.): Intersections in International Cultural Heritage Law. Oxford (Oxford University Press), S. 319–342.

Ubertazzi, Benedetta (2020). Art. 2(2) Manifesting Intangible Cultural Heritage. In J. Blake & L. Lixinski (Eds.), *The 2003 UNESCO Intangible Heritage Convention: A Commentary* (pp. 58–80). Oxford University Press.

Vrdoljak, Ana Filipa (2007): Intentional Destruction of Cultural Heritage and International Law. In: Kalliopi Koufa (ed.): Multiculturalism and International Law, XXXV Thesaurus Acroasium. Thessaloniki (Sakkoulas Publications), S. 377–396.

Xanthaki, Alexandra (2017): International Instruments on Cultural Heritage: Tales of Fragmentation. In: Alexandra Xanthaki, Sanna Valkonen, Leena Heinämäki and Piia Kristiina Nuorgam (eds.): Indigenous Peoples' Cultural Heritage: Rights, Debates, Challenges. Leiden (Brill | Nijhoff.), S. 1–19.

2 The Legal Significance of Human Rights Bodies in International Law

This chapter offers an introduction to different human rights bodies, presenting a brief overview of their work, pronouncements, and how they fit into the sources of international law. The chapter seeks to provide a better understanding of the role of human rights bodies in the international human rights framework, specifically focusing on the legal significance of their work. The chapter first offers a brief introduction to human rights bodies, which highlights their role and work in the human rights framework (Sect. 2.1). This is followed by a discussion of the legal effect of their pronouncements and output (Sect. 2.2). It discusses the sources of international law, the interpretive authority of the treaty bodies and considers how the work and the pronouncements of human rights bodies can contribute to the crystallisation of customary international law and subsequent practice in the application of a treaty. The chapter argues that each pronouncement and output of the human rights bodies holds legal significance. Rather than focusing on the binary of whether a source is legally binding or not, the chapter suggests that the work and pronouncements of human rights bodies should be seen on a continuum of legal effect.

L. Hofmann, *Cultural Heritage through the Lens of Human Rights Bodies*, Juridicum – Bochumer Schriften zur Friedenssicherung und zum Humanitären Völkerrecht | Bochum Studies on the International Law of Peace and Armed Conflict 73, https://doi.org/10.1007/978-3-658-51203-3_2

2.1 An Introduction to the Pronouncements and Work of Human Rights Bodies

The thesis examines the pronouncements of the Human Rights Committee[1] (HRC) and the Committee on Economic, Social and Cultural Rights (CESCR).[2] Both the HRC and the CESCR are composed of eighteen independent experts serving in their individual capacities.[3] These independent experts should be competent in the field of human rights, but are not required to have a legal background.[4] Membership in the committees must consider "equitable geographic distribution".[5] The resulting diversity in membership is aimed at producing linguistic, religious, ideological, political and cultural variety but also diversity in terms of gender or legal orders.[6] Whilst this is likely to result in a more holistic and considerate approach, it could also present challenges in finding common ground.

The analysis of pronouncements by the HRC and the CESCR considers their general comments, their examination of state reports, and their views on communications.[7] The analysis of the charter-based bodies primarily considers the Human Rights Council (UNHRC); its Special Procedures, specifically the mandate of the Special Rapporteur in the field of cultural rights; and the Universal Periodic Review (UPR).[8] The analysis similarly considers resolutions by the UN General Assembly and the Security Council that attend to cultural heritage. These

[1] Established under the International Covenant on Civil and Political Rights (adopted 16 Dec 1966, entered into force 23 Mar 1976) 999 UNTS 171 (*ICCPR*). ICCPR, Arts. 28–45.

[2] Established under resolution 1985/17 of 28 May 1985 by the Economic and Social Council. ECOSOC Res 1985/17 (28 May 1985) UN Doc E/RES/1985/17.

[3] ICCPR, Art. 28 (1); ECOSOC Res 1985/17 (28 May 1985) UN Doc E/RES/1985/17, para. (b).

[4] ICCPR, Art. 28 (2); ECOSOC Res 1985/17 (28 May 1985) UN Doc E/RES/1985/17, para. (b).

[5] ICCPR, Art. 31 (2); ECOSOC Res 1985/17 (28 May 1985) UN Doc E/RES/1985/17, para. (b).

[6] Medina (2013), p. 656.

[7] See Keller and Ulfstein (2012), p. 3; Neuman 2016, p. 2. Whilst Neuman only speaks of the Human Rights Committee, the Optional Protocol to the ICESCR, in the meantime, entered into force. The Committee on Economic, Social and Cultural Rights therefore also has assumed all three functions.

[8] Due to its confidentiality, the complaints procedure of the UNHRC was not considered. Further information on this procedure can be found at: UNHRC (2025): Human Rights Council Complaint Procedure, https://www.ohchr.org/en/hr-bodies/hrc/complaint-procedure/hrc-complaint-procedure-index, accessed 31 January 2025.

mechanisms complement each other in the way they consider the implementation of human rights. Yet, different human rights bodies exercise different functions and, as a consequence, their legal significance differs. Therefore, even the legal significance of different pronouncements within one body or mechanism will vary. The brief introduction to the human rights bodies shows the diversity of bodies, mechanisms, and sources.

2.1.1 Treaty Bodies

Treaty bodies play an important role in human rights law inter alia by concretising individual rights and state obligations.[9] The thesis focuses on three principal activities of both the HRC and the CESCR: authoring general comments, examining state reports, and offering views on communications.

2.1.1.1 General Comments

General comments are prepared and adopted by the Committees and focus on different aspects of the Covenants or their Protocols.[10] The Rules of Procedure of the CESCR, for example, make clear that general comments assist states "in fulfilling their reporting obligations".[11] General comments adopt an expository style[12] and are based on the findings of the treaty bodies in the concluding observations and individual communications, thereby representing the "cumulative experience acquired by a committee"[13]. In other words, general comments can be seen as a synthesis of the findings of the committees in their concluding observations

---

[9] OHCHR (2015): Statement by Mr. Zeid Ra'ad Al Hussein, United Nations High Commissioner for Human Rights, at the International Law Commission, https://www.ohchr.org/EN/NewsEvents/Pages/DisplayNews.aspx?NewsID=16254&LangID=E, accessed 31 January 2025.

[10] CCPR 'Rules of Procedure of the Human Rights Committee' (9 January 2019) UN Doc CCPR/C/3/Rev.11, Rule 76 (1). Also see ICCPR, Art. 40 (4); CESCR 'Rules of Procedure of The Committee: Provisional rules of procedure adopted by the Committee at its third session (1989)' (1 September 1993) UN Doc E/C.12/1990/4/Rev.1, Rule 65.

[11] CESCR 'Rules of Procedure of The Committee: Provisional rules of procedure adopted by the Committee at its third session (1989)' (1 September 1993) UN Doc E/C.12/1990/4/Rev.1, Rule 65.

[12] Rodley (2009), p. 57.

[13] Kälin and Künzli (2019), pp. 213–214. Also see, Medina (2013), p. 656; Rodley (2013), p. 631.

and views.[14] They serve as a representation of the committee's "understanding and interpretation of a specific guarantee", guiding the state parties "on the content and scope of [...] [their] human rights obligations in respect of a particular right".[15] Yet, the adoption procedure of a general comment also allows states to express their views and comments on the given topic, as the preliminary drafts are circulated among the state parties prior to the adoption of the document.[16] This means that a general comment is not a pure reflection of the committee's position, but may be influenced by the position of state parties too. Taking all this into consideration, they arguably carry the most weight of the three instruments of the treaty bodies.[17] General comments do not solely have a descriptive function but also contain normative elements.[18] This has been underlined as a prescriptive function that influences the normative development of human rights law and ventures into norm-creation.[19]

2.1.1.2 Concluding Observations

A treaty body's concluding observations are observations and recommendations on a state party's report. Compulsory state reporting under the ICCPR and the ICESCR requires state parties[20] to periodically report on "the measures they have adopted which give effect to the rights recognized" in the Covenants and "on the progress made in the enjoyment of those rights".[21]

State parties to the ICESCR have agreed to report on the progressive realisation of the Covenant rights.[22] The progressive realisation of the ICESCR rights is a key textual difference to the wording of the ICCPR: Whereas the ICCPR requires each state party to "respect and to ensure" Covenant rights[23],

---

[14] Kälin and Künzli (2019), pp. 213–214. Also see, Medina (2013), p. 656; Rodley (2013), p. 631; Shelton (2012), p. 563.

[15] Kälin and Künzli (2019), p. 214.

[16] CCPR 'Rules of Procedure of the Human Rights Committee' (9 January 2019) UN Doc CCPR/C/3/Rev.11, Rule 76 (4).

[17] See Keller and Grover (2012), pp. 117–118. Shelton (2012), p. 564.

[18] Bódig (2016), pp. 73–74.

[19] *ibid*, p. 74.

[20] ICCPR, Art. 40, International Covenant on Economic, Social and Cultural Rights (adopted 16 Dec 1966, entered into force 3 Jan 1976) 993 UNTS 3 (*ICESCR*), Art. 16.

[21] ICCPR, Art. 40 (1); also see CESCR 'Rules of Procedure of The Committee: Provisional rules of procedure adopted by the Committee at its third session (1989)' (1 September 1993) UN Doc E/C.12/1990/4/Rev.1, Rule 58 (1).

[22] See ICESCR, Part IV f.

[23] ICCPR, Art. 2.

a state party under the ICESCR undertakes steps "to the maximum of is available resources, with a view to achieving progressively the full realization of the" Covenant rights[24]. The wording was rooted in the belief that the implementation of economic, social, and cultural rights were fundamentally different from civil and political rights.[25] Although it has been pointed out that even at the time of drafting, the permissible wording of the article was critiqued.[26] The CESCR has acknowledged that although the full realisation of rights may be achieved progressively, "deliberate, concrete and targeted" steps need to be taken almost immediately after the Covenant's entry into force.[27] The CESCR therefore asks state parties to specify in their reports why the adopted measures are considered the most appropriate.[28] Reporting can thereby serve as a benchmark, for both the committee and the state, to evaluate the progress achieved in the realisation of the rights set out in the Covenant.[29]

At the core of the reporting system lies the facilitation of "constructive dialogue" between the treaty bodies and the state party.[30] Upon reviewing the state party's report, a list of issues is compiled by the committee (or a working group of the committee) to be returned to the state party.[31] The state's replies to this list of issues are shared with the committee to facilitate the "constructive dialogue" between committee and state party.[32] After the conclusion of the public dialogue between the committee and the state party, the committee drafts and adopts the concluding observations in closed session.[33] The concluding observations provide for a follow-up procedure, under which the committee can request the state party

---

[24] ICESCR, Art. 2.

[25] See UNGA 'Draft International Covenants on Human Rights' (1 July 1995) UN Doc A/2929, Chapter I. Also see Forman (2016), pp. 563–564.

[26] Forman (2016), pp. 563–564.

[27] CESCR 'General Comment No. 3' (14 December 1990) UN Doc E/1991/23, paras. 2–3.

[28] *ibid* para. 4.

[29] CESCR 'General Comment No. 1' (27 July 1981) UN Doc E/1989/22, para. 6.

[30] See CCPR 'Rules of Procedure of the Human Rights Committee' (9 January 2019) UN Doc CCPR/C/3/Rev.11, Rule 73 (1); Kälin and Künzli (2019), pp. 210–211.

[31] CESCR 'Report on the Forty-Fourth and Forty-Fifth Sessions' (3–21 May 2010, 1–19 November 2010) UN Doc E/2011/22—E/C.12/2010/3, para. 22; CCPR 'Rules of Procedure of the Human Rights Committee' (9 January 2019) UN Doc CCPR/C/3/Rev.11, Rule 73.

[32] See CCPR 'Rules of Procedure of the Human Rights Committee' (9 January 2019) UN Doc CCPR/C/3/Rev.11, Rule 73 (1); Kälin and Künzli (2019), pp. 210–211.

[33] Kälin and Künzli (2019), p. 211.

to submit information on particularly pressing topics in its next report or prior to the submission of its next report.[34]

General comment No. 1 of the CESCR, entitled "Reporting by States Parties", identifies seven objectives of state reporting. Although adopted by the CESCR, the objectives reflect the more general idea behind state reporting and can be extrapolated to the HRC.[35] The objectives refer to reviewing the state's "national legislation, administrative rules and procedures"; assessing "the actual situation" vis-à-vis human rights obligations; demonstrating the progress made by the state party; facilitating public scrutiny; evaluating the progress made in the implementation of the Covenant as well as understanding shortcomings and problems; facilitating exchange of information and understanding of common problems in the realisation of the rights of all state parties.[36]

The systematic and comprehensive nature of the reporting mechanism allows the treaty bodies to address positive and negative state obligations and therefore "does not suffer from the somewhat coincidental nature of complaints-driven jurisprudence".[37] Accordingly, the process has the capacity to "enrich the understanding and the scope and substance of treaty provisions".[38] Considering the importance of the progressive implementation of rights enshrined in the ICESCR, reflecting on progress as well as on obstacles is essential.[39] Concluding observations at times address matters extraneous to the specific treaty obligations, including, for example, the ratification of other treaties.[40] Yet, recommendations extraneous to a specific state obligation cannot be construed as indicative of how a provision should be interpreted.

It is important to note that time restraints, among other factors, can cause the concluding observations to remain at a general level.[41] Concluding observations are sometimes delayed by states submitting their reports late or even failing to submit their reports altogether.[42] Moreover, caution needs to be taken vis-à-vis the at times "self-congratulatory" nature of state reports, as these run the risk

[34] *ibid.*

[35] See *ibid* p. 209.

[36] CESCR 'General Comment No. 1' (27 July 1981) UN Doc E/1989/22, paras. 2–9.

[37] Scheinin (2007), p. 136.

[38] *ibid.*

[39] CESCR 'General Comment No. 1' (27 July 1981) UN Doc E/1989/22, paras. 7–8.

[40] O'Flaherty (2006), pp. 33, 42.

[41] Mechelm (2009), p. 923.

[42] Coomans (2018), p. 5.

of offering "little information on how the rights are enjoyed in practice".[43] This dependency on if and what kind of information states offer in their reports directly impacts how well the respective Committee can monitor the implementation of human rights in the first place. In the absence of a state report, the Committee can consider the state party in light of information provided from inter alia UN human rights monitoring procedures, NGOs and specialized agencies.[44] Notwithstanding this critique, the dialogue between the committees and the state parties can offer insights into the scope and interpretation of treaty provisions and can be framed as "a consensus on how the provisions [...] should be interpreted with regard to the particular situation in a country."[45].

2.1.1.3 Communications and Views

Communications are complaints by individuals concerning an alleged breach or breaches of one or more rights under the respective Covenants. Upon ratifying the First Optional Protocols to the ICCPR and the ICESCR, state parties recognise the Committees' competence to receive and consider communications.[46] This means that a state party's decision to allow communications is optional.

Whereas the Optional Protocol to the ICESCR allows both individuals and groups of individuals to submit a communication, under the Optional Protocol to the ICCPR only individuals can bring communications.[47] Communications may also be submitted on behalf of one or several individuals.[48] The HRC's views in *Chief Bernard Ominayak and the Lubicon Lake Band v Canada* have been interpreted to imply that "a community leader, acting on behalf of their community, may have standing to represent all member of the community without separate

---

[43] Rodley (2013), pp. 627–628.

[44] Coomans (2018), p. 5.

[45] Mechelm (2009), p. 923.

[46] Optional Protocol to the International Covenant on Economic, Social and Cultural Rights (adopted 10 December 2008, entered into force 5 May 2013) 2922 UNTS 29 (*Optional Protocol to the ICESCR*), Art. 1 (1); Optional Protocol to the International Covenant on Civil and Political Rights (adopted 16 December 1966, entered into force 23 March 1976) 999 UNTS 171 (*Optional Protocol to the ICCPR*), Art. 1.

[47] Optional Protocol to the ICESCR, Art. 2; Optional Protocol to the ICCPR, Art. 1.

[48] CCPR 'Rules of Procedure of the Human Rights Committee' (9 January 2019) UN Doc CCPR/C/3/Rev.11, Rule 91; Optional Protocol to the ICESCR, Art. 2. As established in CCPR, Communication No 167/1984, *Chief Bernard Ominayak and the Lubicon Lake Band v Canada* (26 March 1990) UN Doc CCPR/C/38/D/167/1984. See, Scheinin (2007).

written authorization from the individuals concerned".[49] Whilst the collective right to self-determination in Article 1 of the ICCPR cannot be subject of an individual complaint,[50] subsequent views of the HRC have conceded that Article 1 may indeed "be relevant in the interpretation of other rights protected by the Covenant, in particular article 27",[51] that is, the right of persons belonging to an ethnic, religious or linguistic minority "to enjoy their own culture, to profess and practice their own religion, or to use their own language."

The examination of communications takes place in closed sessions and the committees only release information once they have completed their deliberations.[52] The examination considers two aspects: first the admissibility and subsequently the merits.[53] The committee will further assess if the author of the communication is a victim, if the allegations concern the violation of a right guaranteed by the Covenant, if these are sufficiently substantiated, and if all available domestic remedies have been exhausted.[54] This inter alia means that "abstract" or "actio popularis" communications are inadmissible.[55] A communication will also be declared inadmissible if it is submitted anonymously, if the committee considers it abusive, or if the matter is already under examination by another procedure.[56] Furthermore, the alleged violation needs to have occurred after the

---

[49] Relying on paras. 2.2 and 14 of the views in *Chief Bernard Ominayak and the Lubicon Lake Band v Canada*. Scheinin (2007), p. 143.

[50] CCPR, Communication No 167/1984, *Chief Bernard Ominayak and the Lubicon Lake Band v Canada* (26 March 1990) UN Doc CCPR/C/38/D/167/1984, para. 13.3.

[51] E.g., CCPR, Communication No 547/1993, *Apirana Mahuika and Ors v New Zealand* (27 October 2000) UN Doc CCPR/C/70/D/547/1993, para 9.2. Also see Scheinin (2007).

[52] CCPR 'Rules of Procedure of the Human Rights Committee' (9 January 2019) UN Doc CCPR/C/3/Rev.11, Rule 110; CESCR 'Provisional Rules of Procedure under the Optional Protocol to the International Covenant on Economic, Social and Cultural Rights, Adopted by the Committee at its Forty-Ninth Session (12–30 November 2012)' (15 January 2013) UN Doc E/C.12/49/3, Rule 19 (1).

[53] When assessing the admissibility of a communication, the committee will consider its own competence. Furthermore, issues may arise if the state party has submitted reservations. Optional Protocol to the ICCPR, Art. 1. CESCR 'Rules of procedure under the Optional Protocol to the International Covenant on Economic, Social and Cultural Rights' (3 May 2022) UN Doc E/C.12/5, Rules 6, 10.

[54] Optional Protocol to the ICCPR, Art. 2. For the ICESCR, see Optional Protocol to the ICESCR, Art. 3.

[55] CCPR 'Rules of Procedure of the Human Rights Committee' (9 January 2019) UN Doc CCPR/C/3/Rev.11, Rule 96 (b).

[56] Optional Protocol to the ICCPR, Art. 3, 5 (2) (a). CESCR 'Rules of procedure under the Optional Protocol to the International Covenant on Economic, Social and Cultural Rights' (3 May 2022) UN Doc E/C.12/5, Rule 12.

entry into force of the Optional Protocols. Committees will distinguish between "ongoing violations" and "ongoing effects of past violations".[57] Ongoing effects of past violations, prior to the entry into force of the Optional Protocol, cannot be considered.[58] For the CESCR, the rule reads that the communication ought to be "submitted within one year after the exhaustion of domestic remedies, or the author has demonstrated that it has not been possible to submit the communication within this time limit".[59] Finally, committees can find a non-violation, a violation or a mixture of both.[60] If, after the examination of the merits, the committee finds a violation, it can also determine a remedy.[61] The guidelines of the HRC on measures of reparation under the Optional Protocol to the ICCPR specify that these remedies might entail "restitution, compensation, rehabilitation and measures of satisfaction" and measures that aim to prevent the reoccurrence of violations of a similar kind.[62]

Whilst the rules of procedure for the HRC state that views are adopted by a simple majority, in practice they are often adopted unanimously.[63] For both the HRC and the CESCR, separate opinions can be attached to the views.[64] The procedure takes place in writing and the alleged victim and the state are neither seen nor heard in person.[65] The communications procedure also allows the

---

[57] See Kälin and Künzli (2019), p. 216; Joseph (2018), p. 57.

[58] Kälin and Künzli (2019), p. 216.

[59] CESCR 'Rules of procedure under the Optional Protocol to the International Covenant on Economic, Social and Cultural Rights' (3 May 2022) UN Doc E/C.12/5, Rule 12 (e).

[60] Joseph (2018), p. 68.

[61] See ICCPR, Art. 2 (3) (a).

[62] CCPR 'Guidelines on measures of reparation under the Optional Protocol to the International Covenant on Civil and Political Rights' (30 November 2016) UN Doc CCPR/C/158.

[63] CCPR 'Rules of Procedure of the Human Rights Committee' (9 January 2019) UN Doc CCPR/C/3/Rev.11, Rule 105 (4); There is a paucity on measures of reparation in the guidelines of the CESCR. Also see Kälin and Künzli (2019), p. 217, fn. 78.

[64] CCPR 'Rules of Procedure of the Human Rights Committee' (9 January 2019) UN Doc CCPR/C/3/Rev.11, Rule 103; CESCR 'Rules of procedure under the Optional Protocol to the International Covenant on Economic, Social and Cultural Rights' (3 May 2022) UN Doc E/C.12/5, Rule 17.

[65] Medina (2013), p. 657; Optional Protocol to the ICCPR, Arts. 2, 4 (2), 5 (1). Although rare, in a few exceptional cases characterised by their complexity, the HRC has allowed oral evidence. Joseph (2018), p. 43.

The widely discussed communication *Sacchi et al v. Argentina et al* contained oral hearings. CRC, Communication No. 104/2019, *Sacchi et al v. Argentina et al.* (8 October 2021) UN Doc CRC/C/88/D/104/2019, para. 8.1.

committees to request, in exceptional circumstances and before it has considered the merits, interim measures after the receipt of a communication.[66] Such interim orders arise in situations of imminent danger of irreparable harm suffered by the alleged victim. They constitute an important feature of the complaint mechanism, as they perform a pre-emptive function so that harm can be stopped before it occurs.[67] The views of the committees are specific to the communications at hand and therefore restricted to the case at hand. Yet, the committees often refer to previous cases in their reasoning.

Joseph has raised the question if interim orders constitute an exception to the non-binding nature of the pronouncements. This question stems from the Committees' strong condemnations of non-compliance with these orders.[68] Arguing that non-compliance with interim measures "is incompatible with the obligation to respect in good faith the procedure of individual communication established under the Optional Protocol"[69] the binding nature would emanate from Article 26 of the VCLT's good faith requirement. Interim orders would thereby enjoy a status that the final decision on the merits of the communication does not.[70]

2.1.2 The Human Rights Council

Charter-based bodies are established under the UN Charter. The UNHRC, a subsidiary organ of the UN General Assembly, is composed of 47 member states with representation from the African, Asian, Eastern European, Latin American and Caribbean, and Western European regions.[71] Members serve for three years and are "elected directly and individually by secret ballot by the majority of the

---

[66] CCPR 'Rules of Procedure of the Human Rights Committee' (9 January 2019) UN Doc CCPR/C/3/Rev.11, Rule 94 (1); CESCR 'Provisional Rules of Procedure under the Optional Protocol to the International Covenant on Economic, Social and Cultural Rights, Adopted by the Committee at its Forty-Ninth Session (12–30 November 2012)' (15 January 2013) UN Doc E/C.12/49/3, Rule 7 (1).

[67] Chenwi (2009), p. 37.

[68] Joseph (2018), pp. 47–50.

[69] CCPR 'General Comment No. 33' (25 June 2009) UN Doc CCPR/C/GC/33, 19.

[70] Joseph (2018), p. 50.

[71] 'Group of African States, thirteen; Group of Asian States, thirteen; Group of Eastern European States, six; Group of Latin American and Caribbean States, eight; and Group of Western European and other States, seven', UNGA Res 60/251 (3 April 2006) UN Doc A/RES/60/251, para. 7.

members of the General Assembly".[72] Emanating from a reform of the Commission on Human Rights, the UNHRC, broadly speaking, is entrusted with the promotion of "universal respect for the protection of all human rights and fundamental freedoms for all".[73] This includes addressing "situations of violations of human rights",[74] issuing recommendations on the promotion and protection of human rights,[75] and promoting states' implementation of human rights[76]. UN General Assembly resolution 60/251 envisions the role of the UNHRC to be one of a forum, fostering dialogue on human rights.[77] Dialogue and cooperation are central elements in the Council's work in contributing to the prevention of human rights violations and in responding swiftly to ongoing violations.[78] Several mechanisms facilitate carrying out these tasks.[79] The UNHRC "may adopt resolutions, decisions, recommendations, conclusions, summaries of discussions, Presidential Statements, and its annual report to the General Assembly".[80] Decisions by the UNHRC are adopted by a majority of states that are present and voting.[81]

2.1.2.1 The Special Procedures

The UNHRC's Special Procedures engages independent experts that act as Special Rapporteurs or as members of working groups.[82] Individual mandate holders either work on a country mandate or on a thematic mandate. The procedure seeks to contribute to "standard-setting for the further development of human rights law"[83] and the role of the Special Rapporteurs is to develop "the international understanding of thematic or country situations".[84] Expertise, experience, independence, impartiality, personal integrity, and objectivity are the key criteria

[72] *Ibid*

[73] *Ibid* para. 2.

[74] *ibid* para. 3.

[75] *ibid* para. 5 ((i).

[76] *Ibid* para. 5 (d).

[77] *Ibid* para. 5 (b).

[78] *Ibid* para. 5 (f).

[79] Other mechanisms not considered include the advisory committee, the complaints procedure, and the commission of inquiry. For an exhaustive description of the mechanisms, see e.g., Kälin and Künzli (2019); Tistounet (2020).

[80] Kothari (2013), p. 594.

[81] UNHRC, Res 5/1 (18 June 2007) UN Doc A/HRC/RES/5/1, Rule 20.

[82] See Kälin and Künzli (2019), p. 240.

[83] Kothari (2013), p. 610.

[84] Tistounet (2020), p. 13.

considered in the selection of mandate-holders.[85] The expert input resulting from the Special Procedures is highly relevant to the work of the UNHRC.[86] Special mandate holders generally do not take decisions themselves, but rather "report and submit recommendations to the Human Rights Council, the General Assembly and occasionally to the Security Council, as well as to individual members of the United Nations".[87]

The work of Special Rapporteurs can be described as operating on three levels,[88] two of which concern official country missions and thematic questions. Special Rapporteurs present the outcomes of these investigations in the form of reports to the UNHRC and the collected information, in turn, contributes to the UPR. These reports contribute to the general body of knowledge by assessing "the conceptual framework, obligations, and rights underpinning any human rights issue".[89] The third layer engages a Special Rapporteur's diplomatic abilities, acting on, either individual cases or specific issues through letters (or communications) containing allegations or through urgent appeals.[90]

The Special Procedures, thereby, allow everyone with "reliable information" to approach the UN High Commissioner for Human Rights, providing for a more informal quasi-complaint mechanism.[91] Since the individual(s) bringing the allegation neither need to be personally affected nor exhaust local remedies, the Special Procedures are flexible and can move speedily.[92] At the same time, the informal nature of this mechanism carries risks, such as the credibility of the source.[93] Any "ill-founded or premature interventions" can result in justified criticism.[94] Although the communications cannot be compared to a judicial mechanism, this particular aspect of the mechanism is often described as "quasi-judicial"[95]. The issuance of joint communication reports by Special Rapporteurs, since September 2011, exposes governments to public scrutiny in how

[85] UNHRC, Res 5/1 (18 June 2007) UN Doc A/HRC/RES/5/1, para. 39.

[86] Subedi (2014), p. 242.

[87] *ibid* p. 249.

[88] See Bielefeldt (2011), pp. 230–232.

[89] Domínguez-Redondo (2020), p. 144.

[90] Bielefeldt (2011), p. 231; Pinto (2018b), paras. 50 ff.

[91] Bielefeldt (2011), p. 231.

[92] *ibid*.

[93] Domínguez-Redondo (2020), p. 136.

[94] *ibid*.

[95] Kothari (2013), p. 614.

they respond—or fail to respond—to the allegations.[96] In contrast to the treaty bodies, however, the Special Procedures do not have a monitoring mechanism to follow up on the recommendations.[97]

Special Rapporteurs have an important "public advocacy role" and their work often receives media attention.[98] This publicity, in turn, can be wielded as a "source of pressure on the governments concerned".[99] The Code of Conduct for Special Procedures Mandate-holders provides direction for the interaction with the public in certain respects.[100] Whilst the thematic reports are publicly presented and receive much attention, written communications, for example, have been found to have a somewhat "limited impact on influencing state behaviour".[101] Although they are an important tool in bringing issues to international attention, written communications are often rejected by states or receive immaterial responses.[102]

2.1.2.2 The Universal Periodic Review

The aim of the UPR is to establish whether and to what extent each state fulfils its human rights obligations. Cognisant of the monitoring system under the treaty bodies, this mechanism intends to complement the work of treaty bodies.[103] Therefore, the review of each state is based on three documents. This includes the state's own report and two reports presented by the Office of the High Commissioner for Human Rights.[104] The latter two are, first, a compilation of information provided by the treaty bodies, the Special Procedures and other UN documents,[105] and, second, reliable and credible information by national human rights institutions as well as by non-governmental organisations.[106] The Charter of the United Nations, the Universal Declaration of Human Rights (UDHR), and the human

---

[96] Bielefeldt (2011), p. 231.

[97] See Subedi (2014), p. 257.

[98] Subedi (2014), p. 249. Also see: Domínguez-Redondo (2020), p. 57.

[99] Subedi (2014), p. 249.

[100] UNHRC Res 5/2 (18 June 2007) UN Doc A/HRC/RES/5/2, Arts. 12, 13. Also see: Domínguez-Redondo (2020), p. 57.

[101] Subedi (2014), p. 260. Also see Piccone (2010), p. ix.

[102] Subedi (2014), p. 260.

[103] UNGA Res 60/251 (3 April 2006) UN Doc A/RES/60/251, para. 5 (e).

[104] UNHRC Res 5/1 (18 June 2007) UN Doc A/HRC/RES/5/1, para. 15.

[105] Tistounet (2020), p. 9.

[106] *ibid.*

rights instruments that a state has ratified form the yardstick of the UPR.[107] The outcome of the review is documented in a report and does not constitute a binding result.[108] Rather, at the end of each review a state identifies the recommendations it supports.[109]

The UPR can be described as a "peer review procedure", constituting what essentially is a political process, in which member states review each other every four and a half years.[110] The review takes place in an open session of three and a half hours and is subject to strict time limits.[111] As the allocated time does not allow for a "genuine exchange of views", the idea of a dialogue can be hampered.[112] This means that after presenting its position and listening to the recommendations by the participant states, the state under review frequently reverts to reaffirming its position.[113] Recommendations that are merely "noted" are effectively rejected.[114]

2.2 Human Rights Bodies and the Sources of International Law

The legal positivist approach is based on the core assumption of international law as a system of objective principles. These emanate from the primary actors of international law—states—and are expressed through treaty law or customary law.[115] In accordance with positivist legal thinking, "extra-legal considerations",[116] such as incorporating social, political, economic, or moral factors are outside its scope. This traditional approach to international law has helped transform Article 38 of the Statute of the International Court of Justice (ICJ) into

[107] UNHRC Res 5/1 (18 June 2007) UN Doc A/HRC/RES/5/1, para. 1.

[108] Tomuschat (2019), para. 23.

[109] UNHRC Res 5/1 (18 June 2007) UN Doc A/HRC/RES/5/1, para. 32.

[110] Kälin and Künzli (2019), p. 238; Kothari (2013), p. 595.

[111] Tomuschat (2019), paras. 19, 22.

[112] *ibid* para. 44.

[113] *ibid*.

[114] Tistounet (2020), p. 11.

[115] Bianchi (2016), p. 21.

[116] *ibid*.

a list of the sources of international law.[117] It follows that legal norms and concepts are subject to analysis and interpretation in line with Articles 31 and 32 of the Vienna Convention on the Law of Treaties (VCLT).[118]

The reliance on formally non-binding sources by human rights bodies throughout this thesis necessitates their contextualization and classification. This is why this section examines how the pronouncements and the output of human rights bodies fit into the sources of international law and into the dichotomy of defining something as legally binding or non-binding. The thesis draws on and consults inter alia UN General Assembly resolutions, UNHRC resolutions, the documentation of the UPR, the Special Procedures as well as the pronouncements of the HRC and the CESCR. It is evident from Article 38 of the Statute of the ICJ that these sources do not fall within the list of primary sources of international law. As they cannot in themselves constitute a binding source of international law, how can these "sources" nonetheless be used in attaining a better understanding of the scope of rights and obligations in the Covenants? This section suggests that in spite of not fitting within the sources of law laid down in Article 38 of the Statute of the ICJ, they can have legal significance and can be considered on a continuum of legal effect for sources of international law.

Article 38 (1) of the Statute of the ICJ specifies (a) international conventions, (b) international custom, (c) general principles of law, and, (d) as subsidiary means in determining rules of law, "judicial decisions and the teachings of the most highly qualified publicists of the various nations". Whilst the work of the UNHRC and the pronouncements of treaty bodies are neither by themselves conventions, custom, nor principles of law, one could ask the question whether they could possibly fall under "subsidiary means for the determination of rules of law".

Article 38 (1) (d) considers the interpretive authority of individual experts by allowing for the application of "judicial decisions and the *teachings of the most highly qualified publicists* of the various nations, as subsidiary means for the determination of rules of law" (emphasis added). The article itself does not further define who the "most highly qualified publicists" are. Research on the subject has approached the concept from different angles and has, for example, considered the most-cited writers in the jurisprudence of the ICJ. This approach shows "that the top 10 per cent most-cited writers have 2,077 citations, which

[117] *ibid* p. 24.

[118] Vienna Convention on the Law of Treaties (adopted 23 May 1969, entered into force 23 Jan 1980) 1155 UNTS 331 (*VCLT*), Art. 31.

is just over 50 per cent of the total"[119]. This suggests that reference to "the teachings" by the judges draws on a small pool of experts and suggests a very narrow interpretation. *Sivakumaran* has pointed out that although citation is "a useful measure of influence [...] citation is not the same as influence; and lack of citation does not necessarily mean lack of influence".[120] He takes a broader approach by identifying three types of publicists: entities empowered by states (such as the International Law Commission), expert groups (such as the International Law Association or non-permanent ones carrying out a specific project), and "ordinary" publicists.[121] Whilst the treaty bodies, as expert committees, do not fall within the category of expert groups, a "most highly qualified publicist" can clearly be part of a treaty bodies or of the Special Procedures. The presence of several "most highly qualified publicist" in a treaty body can further increase that the views or comments could be considered a subsidiary means of interpretation. However, considering the changing composition of the treaty bodies would render it impractical and inconsistent to elevate one pronouncement over another due to its authorship. In pursuing an approach that considers the treaty body as a whole, the subsequent analysis will further look into the interpretive authority of the human rights bodies, focusing specifically on the treaty bodies.

2.2.1 The Interpretive Authority of Human Rights Bodies

As there is no indication that human rights treaties confer the power to adopt binding interpretations of the treaty on their treaty bodies,[122] the authority of human rights treaty bodies is contested. The rules of treaty interpretation are laid down in Articles 31–33 of the VCLT. These rules, however, show insufficiencies in responding to questions that arise in the context of interpreting human rights treaties.[123] Contrary to treaties governing the reciprocal relationship between two or more states, human rights treaties create rights for individuals and in specific cases for persons belonging to groups.[124] As human rights treaties often establish

---

[119] Helmersen (2019), p. 514.

[120] Sivakumaran (2017), p. 3.

[121] *ibid.*

[122] See CESCR 'Rules of Procedure of The Committee: Provisional rules of procedure adopted by the Committee at its third session (1989)' (1 September 1993) UN Doc E/C.12/1990/4/Rev.1; CCPR 'Rules of Procedure of the Human Rights Committee' (11 January 2012) UN Doc CCPR/C/3/Rev.10; ILA, 2004, para. 18.

[123] Scheinin (2017), p. 22.

[124] *See ibid* p. 25.

monitoring mechanisms, like treaty bodies or courts,[125] they shift "the centre of gravity in treaty interpretation" from states to monitoring mechanisms, giving more room to the interpretive practice of treaty bodies and courts.[126] The VCLT, however, was not drafted with the idea of third-party beneficiaries (i.e. individuals and individuals belonging to a group) or international monitoring mechanisms in mind. Thus, the treaty is silent on the interpretive authority of courts, tribunals, or treaty bodies.[127] However, it has also been argued that

> "[t]he legal capacity of actors such as the General Assembly to interpret international law can be necessarily implied by the functions which are explicitly granted to such bodies."[128]

The ICJ prominently addressed the interpretive authority of the treaty bodies' pronouncement in the *Case concerning Ahmadou Sadio Diallo*, invoking specifically a communication and General Comments Nos. 15 and 8 by the HRC.[129] The court explained that while it "is in no way obliged [...] it believes that it should ascribe great weight to the interpretation" of the HRC.[130] This was argued with a view to achieving "the necessary clarity and the essential consistency of international law, as well as legal security".[131] The Inter-American Court referred to and cited general comments and concluding observations of the CESCR, decisions of the HRC, and concluding observations of the Committee on the Elimination of Racial Discrimination in its Advisory Opinion OC-23/17 on The Environment and Human Rights.[132]

The International Law Association (ILA) found it "to be well accepted that the findings of the treaty bodies do not themselves constitute binding interpretations

[125] *ibid.*

[126] *ibid* pp. 25–26.

[127] *ibid* p. 22.

[128] Provost (2015), pp. 295–6.

[129] *Ahmadou Sadio Diallo (Guinea v DRC) (Merits, Judgment) [2010] ICJ Rep 639*, paras. 66, 77. In the Advisory Opinion *Legal Consequences of the Construction of a Wall in the Occupied Palestinian Territory*, the ICJ relied on the concluding observations of the HRC's and the CESCR. *Legal Consequences of the Construction of a Wall in the Occupied Palestinian Territory (Advisory Opinion) [2004] ICJ Rep 136*, paras. 110–112.

[130] *Ahmadou Sadio Diallo (Guinea v DRC) (Merits, Judgment) [2010] ICJ Rep 639*, para. 66.

[131] *ibid.*

[132] *The Environment and Human Rights*, IACtHR, Advisory Opinion OC-23/17, 15 November 2017, fn 109–113, 140, 185, 194, 210, 213, 215–16, 226, 232–234, 239, 298–299, 344, 353.

of the treaties" in their Report on the Impact of Findings of the United Nations Human Rights Treaty Bodies.[133] For the practical purpose of exercising their functions, however, treaty bodies require the power to adopt interpretations of the respective human rights treaty.[134]

Legal scholarship generally attributes an important role to the interpretation of the committees, especially the HRC. This is why the role of treaty bodies has been conceived of as one of "establishing the normative content of human rights and [...] giving concrete meaning to individual rights and state obligations".[135] *Chinkin* suggests that the work of the treaty bodies "might be described as secondary treaty law, in that state consent can be implied from their acceptance of the treaties and the authority of the expert committees".[136] Accordingly, state's compliance and responsiveness to reporting and concluding observations can reinforce this view.[137]

Whereas there appears to be general agreement that the treaty bodies constitute *an* authoritative interpreter, a major point of contention concerns the question whether the committees are "*the* most authoritative interpreter of the treaty in question"[138]. In other words, should the states or the committees have the last word in treaty interpretation?[139] Amid the absence of a uniform approach,[140] one of the most compelling arguments for recognising the interpretive authority

[133] ILA (2004), para. 15.

[134] See ILA (2004), para. 18.

[135] Mechelm (2009), p. 908.

[136] Chinkin (2022), p. 81.

[137] *ibid.*

[138] Scheinin (1997), p. 369. Also, see '*the* most authoritative interpreter of the treaty it monitors' in Mechelm (2009), p. 929. Further, see 'No better expertise as to the scope and meaning of any of the human rights treaties can be found than in the expert bodies set up to monitor their observance by states.' Tomuschat (2003), p. 183.

[139] See Harrington (2015): The Human Rights Committee, Treaty Interpretation, and the Last Word, https://www.ejiltalk.org/the-human-rights-committee-treaty-interpretation-and-the-last-word/, accessed 31 January 2025; Citroni (2015): The Human Rights Committee and its Role in Interpreting the International Covenant on Civil and Political Rights vis-à-vis States Parties, https://www.ejiltalk.org/the-human-rights-committee-and-its-role-in-interpreting-the-international-covenant-on-civil-and-political-rights-vis-a-vis-states-parties/ accessed 31 January 2025; ILA, 2004; para. 8.

[140] Whilst there is no uniform approach by national courts, it is interesting to note that the Spanish Supreme Court decided that human rights treaty bodies' views in individual communications are binding on the state. Tribunal Supremo, Sala de lo Contencioso-Administrativo Sección Cuarta, Sentencia núm. 1263/2018, Madrid, a 17 de julio de 2018. Also see, Casla (2018): Supreme Court of Spain: UN Treaty Body Individual Decisions

of treaty bodies is the need for a consistent interpretation of covenant rights. In other words, the rights "cannot be left to the chaos of self-preserving interpretations by each indifferent or complacent state."[141] Furthermore, considering the pronouncements of treaty bodies as no more than opinions lacking any legal consequences would result in the fragmentation of human rights law.[142] *Scheinin* points out that the resulting variety in legal obligations by state parties would ultimately culminate in a denial of the covenant rights.[143]

Treaty bodies themselves have also brought forward that their pronouncements, especially their views and requests for interim measures, emanate from legal norms that are binding obligations and therefore their pronouncements cannot be treated as mere recommendations.[144] In the case of the HRC, state parties may voice their discontent with the observations of the committee in accordance with Article 40 (5) of the ICCPR. The antithetical nature of assuming that states can simply "impose their own interpretation" and declaring it better than that of the committee has been underlined.[145] In accordance with this view, the sheer impartiality and experience of the HRC may let its interpretation prevail over those of the state parties.[146] *Seibert-Fohr* and *Weniger* add that a state's "mere denial without argumentative rebuttable would be irreconcilable with the object and purpose of the Covenant's monitoring system."[147]

In draft General Comment No. 33 on the obligations of state parties under the ICCPR, the HRC contended that its

> "general body of jurisprudence [...] constitutes 'subsequent practice in the application of the treaty which establishes the agreements of the parties regarding its interpretation' within the sense of article 31 (3) (b) of the [VLCT]."[148]

Are Legally Binding, https://www.ejiltalk.org/supreme-court-of-spain-un-treaty-body-individual-decisions-are-legally-binding/, accessed 31 January 2025; Borlini and Crema (2020).

[141] Neuman (2016), p. 5.

[142] Scheinin (2017), p. 28.

[143] *ibid.*

[144] ILA (2004), para. 15.

[145] Citroni (2015): The Human Rights Committee and its Role in Interpreting the International Covenant on Civil and Political Rights vis-à-vis States Parties, https://www.ejiltalk.org/the-human-rights-committee-and-its-role-in-interpreting-the-international-covenant-on-civil-and-political-rights-vis-a-vis-states-parties/ accessed 31 January 2025.

[146] *ibid.*

[147] Seibert-Fohr and Weniger (2019), p. 9.

[148] CCPR 'Draft General Comment No. 33' (25 August 2008) UN Doc CCPR/C/GC/33/CRP.3, para. 18.

Calling this an "extraordinary assertion", the draft was heavily criticised by inter alia the United States of America,[149] resulting in the omission of this idea in the version eventually adopted by the committee.[150] Subsequent practice in the application of the treaty is further explored below (see Sect. 2.2.3). Amidst the US' submission that the committee's views do not constitute "a source of international legal obligation",[151] General Comment No. 33 now ascribes the committee's views "some of the principal characteristics of a judicial decision".[152] It contends the authoritative nature of the committee's views, emphasising that the committee is established under the ICCPR and tasked with its legal interpretation.[153] General Comment No. 33 further argues that the states' duty to cooperate with the HRC stems from their duty to perform their treaty obligations in good faith, as laid down in Article 26 of the VCLT.[154] Whilst this line of argumentation specifically concerns the ICCPR and the HRC, the reasoning can be extrapolated to the other treaty bodies as well.[155] The failure to implement the committee's views is criticised by the committee and addressed to the respective state party

[149] Government of the United States of America (2008): Comments of the United States on the Human Rights Committee's 'Draft General Comment 33: The Obligations of States Parties under the Optional Protocol to the International Covenant on Civil and Political Rights, https://2009-2017.state.gov/documents/organization/138851.pdf, accessed 31 January 2025, para. 9.

[150] In addition to the United States of America, several other states objected, including Australia, Japan, New Zealand, United Kingdom, Belgium, and Poland, as pointed out in Azaria (2020), p. 36, fn. 8.

[151] Government of the United States of America (2008): Comments of the United States on the Human Rights Committee's 'Draft General Comment 33: The Obligations of States Parties under the Optional Protocol to the International Covenant on Civil and Political Rights, https://2009-2017.state.gov/documents/organization/138851.pdf, accessed 31 January 2025, para 15. In full, the US concedes that 'the views of the Committee are entitled to respect and should be considered carefully by States Parties.' Yet, they emphasise that '[s]uch views are not, however, a source of international legal obligation, nor do they have a 'determinative,', 'authoritative,' or 'judicial' character.'

[152] CCPR 'General Comment No. 33 The Obligations of State Parties under the Optional Protocol to the International Covenant on Civil and Political Rights' (5 November 2008) UN Doc CCPR/C/GC/33, para. 11.

[153] *ibid* para. 13.

[154] However, a duty to cooperate appears to "fall short of a requirement of complete compliance". *Joseph* observes that "the routine failure of a State to comply with the determinations of a human rights body" can be read as "evidence of that State's bad faith towards the relevant binding international treaty obligations." Joseph (2018), para. 9.

[155] Joseph (2018), para. 9.

in their concluding observations along with questioning their commitment to the Optional Protocol.[156]

To conclude, while it is generally agreed that the treaty bodies have the authority to interpret human rights provisions, there is disagreement whether the committees should have the last say in treaty interpretation. Arguably, the most convincing argument brought forward in both jurisprudence and scholarship is the need to achieve clarity, consistency and legal security.[157] After having considered the interpretive authority, the following section will further analyse the potential legal effect of the work and pronouncements of human rights bodies.

2.2.2 Legal Effect as a Continuum

Considering UN General Assembly resolutions, *Lauterpacht* affirmed the "somewhat obvious principle" that these resolutions are not legally binding, yet also highlighted that

> "It is another thing to give currency to the view that they have no force at all whether legal or other and that therefore they cannot be regarded as forming in any sense part of a legal system of supervision. […] A Resolution recommending to an Administering State a specific course of action creates *some* legal obligation which, however rudimentary, elastic and imperfect, is nevertheless a legal obligation and constitutes a measure of supervision. The State in question, while not bound to accept the recommendation, is bound to give it due consideration in good faith."[158]

Further considering UN General Assembly resolutions, *Asamoah* distinguishes between the binding nature and legal effect: the legal effect can be seen as a broader concept that encompasses the notion of something being "legally binding".[159]

Building on these considerations, this thesis proposes that it is indeed more constructive to consider the "legal effect" and the creation of "*some* legal obligation" than to adhere to the dichotomy of binding or non-binding. The thesis

[156] CCPR 'Concluding Observations: Russian Federation' (2009) UN Doc CCPR/C/RUS/CO/6, para. 5. Also see, CCPR 'Concluding Observations: France' (2015) UN Doc CCPR/C/FRA/CO/5, para. 7.

[157] See *Ahmadou Sadio Diallo (Guinea v DRC) (Merits, Judgment) [2010] ICJ Rep 639*, paras. 66, 77. Neuman, 2016, p. 5; Scheinin, 2017, p. 28.

[158] *South-West Africa—Voting Procedure (Advisory Opinion) [1955] ICJ Rep 67 (Separate Opinion of Judge Lauterpacht)*, p. 55.

[159] Asamoah (1964), p. 214.

proposes to extrapolate this thought beyond the legal effect of a UN General Assembly resolution to resolutions of the UNHRC, the work of the Special Procedures, and the pronouncements of treaty bodies. All these formally non-binding sources can be considered on a continuum of legal effect (see Fig. 2.1). This spectrum can be thought of as containing "no legal effect whatsoever" on one side and a "formally binding legal source" on the other. Accordingly, sources are not dismissed if they are not formally legally binding. They are understood as being able to exert a legal effect that can contribute to or even eventually result in a rule, right or obligation becoming legally binding (but do not have to).

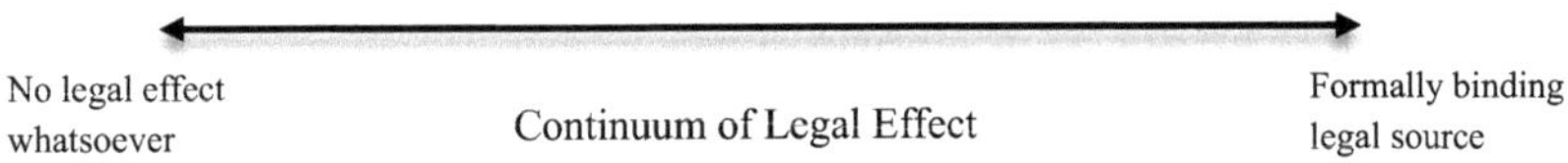

Fig. 2.1 Continuum of Legal Effect

One such legal effect can be the contribution to customary international law. In its Advisory Opinion on *Legality of the Threat or Use of Nuclear Weapons,* the ICJ addressed the legal effect of a series of resolutions adopted by the UN General Assembly. The Court showed that conduct in relation to non-binding resolutions, decisions or acts by the UN General Assembly can go beyond their declaratory or recommendatory effect and contribute to the development of customary international law. The ICJ ascribed "normative value" to such resolutions in spite of their non-binding nature[160] and continued to affirm that "in certain circumstances, [these resolutions] provide[d] evidence important for establishing the existence of a rule or the emergence of an *opinio juris*."[161] Establishing whether this is the case, firstly, requires examining the content and, secondly, the conditions of its adoption.[162] Similarly, the Court stated that it was "also necessary to see whether an *opinio juris* exists as to its normative character", although a series of resolutions may also demonstrate "the gradual evolution of the *opinio juris* required for the establishment of a new rule."[163] It is well established that determining a rule of customary international law requires identifying and ascertaining two constituent elements: "whether there is a general practice that is accepted as law

[160] *Legality of the Threat or Use of Nuclear Weapons (Advisory Opinion) [1996] ICJ Rep 226*, para. 70.

[161] *ibid.*

[162] *ibid.*

[163] *ibid.*

(*opinio juris*)".[164] This "two-element approach" requires establishing practice and determining whether such practice is exercised with a sense of "obligation or a right to act in that way".[165] This sense of obligation needs to be distinguished from habit.[166]

Although a rule set out in a resolution, decision or any other act adopted by an international organisation cannot, of itself, create customary international law, it may assist in "determining the existence and content of a rule of customary international law, or contribute to its development."[167] Resolutions can provide insights into the member states' attitudes on legal matters.[168] In this context, the role of the UN General Assembly can offer insights into the collective expression of views of the member states, as it enjoys near universal participation.[169] *Schachter* highlights that the UN General Assembly serves as an important forum for states to articulate interests and, as a consequence, resolutions emanating from this forum may be attached with expressions of "'general will' of the international community".[170] Thus, unanimous or overwhelming approval in the adoption of a UN General Assembly resolution can be indicative of *opinio juris*, but it will still need to be accompanied by concomitant state practice.[171] This makes clear that the adoption of a resolution or a series of resolutions "purporting to lay down a rule of customary international law does not create such law."[172]

The emphasis on near universal membership also implies that less weight is to be given to resolutions, decisions or any other acts adopted by organs with a more limited membership,[173] like the UNHRC. This does not exclude the Council's

---

[164] ILC 'Draft Conclusions on Identification of Customary International Law, with Commentaries (2018)' UN Doc A/73/10, para. 124. Statute of the International Court of Justice (18th April 1946) 33 UNTS 993, Art. 38 (1) b).

[165] ILC, 'Draft Conclusions on Identification of Customary International Law, with Commentaries' (2018) UN Doc A/73/10, para. 125.

[166] *Ibid* conclusion 9 (2).

[167] *ibid* para.141, conclusion 12 (1) (2).

[168] *Ibid* para.147.

[169] *ibid.*

[170] Schachter (1991), p. 85.

[171] ILC 'Third Report on Identification of Customary International Law, by Michael Wood, Special Rapporteur' (27 March 2015) UN Doc A/CN.4/682, para. 53.

[172] ILC 'Draft Conclusions on Identification of Customary International Law, with Commentaries (2018)' UN Doc A/73/10, 147.

[173] *ibid.*

resolutions, decisions or any other acts[174] but makes clear that they cannot be as determinative as a UN General Assembly resolution. Decisions by the UNHRC are adopted by a majority of states that are present and voting.[175] Although not formally tasked with the interpretation of human rights, its role in promoting "the full implementation of human rights obligations"[176] can be interpreted as contributing to defining what the full implementation of each specific right entails and therefore, by extension, to the interpretation of human rights.

Moreover, and very clearly, not all resolutions may contribute to the determination of customary international law. Therefore, negative votes, abstentions and disassociations from a consensus may constitute critical evidence that a rule is not accepted as law.[177] The choice as well as the avoidance of particular terms can, similarly, be decisive.[178]

Moreover, it is reasonable to question if the legal effect of a resolution *should* go beyond its recommendatory effect. *Schachter* points out that even if a state votes in favour of adopting a resolution, such a vote "may not be intended to signify agreement on the legal validity of the asserted norm."[179] Considering the political nature of the vote, it would be problematic to insinuate conviction and legal effect to these political expressions.[180]

Could statements and opinions asserted in the UPR be indicative of a state's legal conviction? At least one problem that emerges in seeking to discern a sense of legal obligation from the UPR not only concerns the political nature of the forum, but also the serious shortages in the mechanism that can result from the strict enforcement of time limits. It has been noted that the time limit makes it virtually impossible to develop a nuanced argument, going so far as threatening to "expose the entire procedure to ridicule".[181] Hence, it will be unlikely that states will present legal reasoning or conviction in these statements. Yet there

---

[174] The UNHRC "may adopt resolutions, decisions, recommendations, conclusions, summaries of discussions, Presidential Statements, and its annual report to the General Assembly". Kothari (2013), p. 594.

[175] UNHRC Res 5/1 (18 June 2007) UN Doc A/HRC/RES/5/1, Rule 20.

[176] UNGA Res 60/251 (3 April 2006) UN Doc A/RES/60/251, para. 5 (d).

[177] ILC 'Draft Conclusions on Identification of Customary International Law, with Commentaries (2018)' UN Doc A/73/10, para. 148.

[178] ILC 'Third Report on Identification of Customary International Law, by Michael Wood, Special Rapporteur' (27 March 2015) UN Doc A/CN.4/682, pp. 39, 35, para. 48.

[179] Schachter (1991), p. 88.

[180] *ibid*, pp. 87, 88. Also see, ILC 'Draft Conclusions on Identification of Customary International Law, with Commentaries (2018)' UN Doc A/73/10, para. 148.

[181] Tomuschat (2019), para. 22.

still might be room for states to express a sense of legal obligations through brief statements. Indeed, the "diktat of brevity" is not strictly disadvantageous.[182] Short, sober statements on a crucial issue repeated by several delegations can also be used strategically and have the effect that the issue cannot be ignored,[183] thereby expressing their (legal) conviction and exerting political pressure.

The pronouncements of treaty bodies can be thought of on the spectrum of legal effect. While states are not directly involved in the pronouncements, the dialogue between the committees and the state parties through, for example, the adoption process of a general comment, the views, and the monitoring process, may also be indicative. It is important to distinguish between the different types of pronouncements and to not give the same weight to general comments, views and concluding observations. Whilst general comments and views surely carry more weight than concluding observations, general comments are often the synthesis of the Committee's work on a specific subject matter and therefore arguably carry the most weight.

2.2.3 Subsequent Practice in the Application of Human Rights Treaties?

When considering the pronouncement of the treaty bodies, a further legal effect may emanate from subsequent practice in the application of human rights treaties. Taking a closer look at subsequent practice in the application of human rights treaties requires recourse to Article 31 (3) of the VCLT. Article 31 (3) allows taking into account

> "(a) any subsequent *agreement* between the parties regarding the interpretation of the treaty or the application of its provisions;
>
> (b) any subsequent *practice* in the application of the treaty which establishes the agreement of the parties regarding its interpretation;
>
> (c) any relevant *rules* of international law applicable in the relations between the parties."[184]

[182] *ibid.*

[183] *ibid.*

[184] Emphasis added. VLCT, Art. 31 (3).

As the pronouncement of the treaty bodies cannot be considered a formal agreement between the states and neither be seen as formal rules of international law, the analysis will examine subsequent practice.

Subsequent practice refers to any conduct in the application of a treaty after its conclusion[185] and covers the exercise of a state's "executive, legislative, judicial or other functions".[186] Article 31 (3) (b) of the VCLT, as an authentic means of interpretation, requires conduct that establishes the agreement of the parties. It sets out that "[a]ny subsequent practice in the application of a treaty which establishes the agreement of the parties regarding its interpretation" shall be taken into account, alongside its context.[187] From the wording of the article it is clear that relevant subsequent practice needs to stem from the parties to the treaties. In other words, it needs to be attributable to them.[188] Such practice cannot be established by reference to isolated acts but needs to be predicated on "a sequence of acts or pronouncements".[189]

Any practice that falls short of the requirements set out in Article 31(3) (b) may still be considered as a supplementary means of interpretation under Article 32 of the VCLT. This provision provides that "recourse may be had to supplementary means of interpretation" to

> "confirm the meaning resulting from the application of article 31, or to determine the meaning when the interpretation according to article 31: (a) leaves the meaning ambiguous or obscure; or (b) leads to a result which is manifestly absurd or unreasonable."

In contrast, agreement by the state parties is not required by Article 32 of the VLCT.[190] This, however, also implies that the means of interpretation of Article 32 of the VCLT are "considered to be considerably less reliable".[191]

In considering the pronouncements of the treaty bodies as subsequent practice two approaches can be identified: a traditional approach and a progressive

[185] ILC 'Draft Conclusions on Subsequent Agreements and Subsequent Practice in Relation to The Interpretation of Treaties, With Commentaries' (2018) UN Doc A/73/10, conclusion 4 (2).

[186] *ibid* conclusion 5 (1).

[187] VCLT, Art. 31 (3) (b).

[188] Dörr and Schmalenbach (2018), p. 599, para. 83.

[189] *ibid.*

[190] *Ibid* p. 618, para. 2.

[191] *ibid.*

one.[192] The traditional approach, does not view pronouncements of treaty bodies as amounting to state practice, but concedes that the responses of individual states or states as a whole in response to the pronouncements would constitute state practice.[193] In other words, states' responses to the pronouncements of the treaty bodies establish subsequent practice within the meaning of Article 31 (3) (b) of the VCLT. In comparison, the progressive approach builds on the special nature of human rights treaties, acknowledging inter alia third party-interests and the monitoring mechanisms established by the treaties.[194] It suggests that relevant subsequent practice "might be broader" in the case of human rights treaties than subsequent state practice.[195] This allows considering the pronouncements of treaty bodies "adopted in the performance of the functions conferred on them by States parties."[196] This, similarly, emphasizes the argument that state parties indeed have delegated and conferred powers to the treaty bodies and that as a practical matter treaty bodies need to adopt interpretations of the treaties in monitoring the covenants.[197]

Whilst the nature of human rights treaties *should* render subsequent practice broader and consider third party interests and the role and function of treaty bodies, the reading of Article 31 (3) (b) of the VCLT, which explicitly refers to establishing "the agreement of the parties", is clear and limits subsequent practice to state parties. Both the preparatory work of Special Rapporteur *Nolte* and the ILC's Draft conclusions on subsequent agreements and subsequent practice in relation to the interpretation of treaties reject the idea that the pronouncement of a treaty body can constitute subsequent practice under Article 31 (3) (b) of the VCLT by reference to its wording.[198]

This, however, does not have to mean that human rights bodies do not play a role in establishing subsequent practice. The draft conclusions give a more indirect role to pronouncements, one that "may give rise to, or refer to, a subsequent agreement or subsequent practice by parties" within the meaning of Article

[192] ILA (2004), para. 21 f.

[193] See *ibid*. Also see Mechelm (2009), p. 920.

[194] ILA (2004), para. 22. Also see, Tobin (2010), p. 20.

[195] ILA (2004), para. 22.

[196] *ibid*.

[197] See *ibid* para. 18.

[198] ILC 'Fourth Report on Subsequent Agreements and Subsequent Practice in Relation to The Interpretation of Treaties by Georg Nolte, Special Rapporteur' (7 March 2016) UN Doc A/CN.4/694, 19, para. 42; ILC 'Draft Conclusions on Subsequent Agreements and Subsequent Practice in Relation to The Interpretation of Treaties, With Commentaries' (2018) UN Doc A/73/10, para. 110.

31 (3) of the VCLT and Article 32 of the VCLT.[199] The legal effect thereby emanates from the agreement and conduct by state parties, not the treaty body.[200] Whilst the legal effect stems from the agreement and conduct by state parties, it also acknowledges the power and normative force human rights bodies can exert in seeking to contribute to developments in international human rights law.

Yet, establishing uniform agreement or practice by state parties vis-à-vis a particular pronouncement of an expert body that concerns the interpretation of a treaty is still a difficult criterion to fulfil.[201] Particularly, when recalling the draft conclusions' assertion that one cannot infer acceptance of the expert treaty body's interpretation from a state party's silence.[202] One potential avenue to identify the agreement of the parties on the interpretation of a treaty within the meaning of Article 31 (3) (b) of the VCLT is by means of UN General Assembly resolutions.[203] These have on occasion cited the pronouncements of treaty bodies in affirmation of their views.[204] As pronouncements by expert treaty bodies can only indirectly contribute to subsequent practice under Article 31 (3) (b) of the VCLT, the Fourth Report suggests that their more direct effect could be as a supplementary means of interpretation under Article 32 of the VCLT.[205]

When considering the more direct effect under Article 32 of the VCLT, an important distinction must be drawn between *shall* and *may*.[206] *Shall* within the meaning of Article 31 of the VCLT implies that something *needs* to be taken

---

199 ILC 'Draft Conclusions on Subsequent Agreements and Subsequent Practice in Relation to The Interpretation of Treaties, With Commentaries' UN Doc A/73/10 (2018), conclusion 13.

200 ILC 'Fourth Report on Subsequent Agreements and Subsequent Practice in Relation to The Interpretation of Treaties by Georg Nolte, Special Rapporteur' (7 March 2016) UN Doc A/CN.4/694, para. 43.

201 See ibid para. 44.

202 ILC 'Draft Conclusions on Subsequent Agreements and Subsequent Practice in Relation to The Interpretation of Treaties, With Commentaries' UN Doc A/73/10 (2018), conclusion 13 (3).

203 *ibid* p. 111.

204 *ibid* pp. 111–112.

205 ILC 'Fourth Report on Subsequent Agreements and Subsequent Practice in Relation to The Interpretation of Treaties by Georg Nolte, Special Rapporteur' (7 March 2016) UN Doc A/CN.4/694, para. 62 f.

206 See VCLT, Arts. 31 and 32; ILC 'Fourth Report on Subsequent Agreements and Subsequent Practice in Relation to The Interpretation of Treaties by Georg Nolte, Special Rapporteur' (7 March 2016) UN Doc A/CN.4/694, para. 50.

into account; *may* within the meaning of Article 32 of the VCLT signifies a supplementary means of interpretation.[207] It is therefore evident that the latter also may *not* be taken into account. In other words, the decision to have recourse to Article 32 of the VCLT lies with the interpreter.[208] This opens the door to a wide array of interpretations, depending on whether the interpreter deems it necessary to consider a pronouncement as a supplementary means of interpretation, leaving the legal effect and significance of pronouncements uncertain.

However, the draft conclusions by the ILC limit "subsequent practice as a supplementary means of interpretation under article 32" to conduct by "one or more parties in the application of the treaty".[209] This excludes the possibility for pronouncements to form part of "other supplementary means".[210] In accordance with draft conclusion 13 (3), pronouncements may only "give rise to, or refer to, [...] subsequent practice under article 32." In the view of the ILC, their effect under Article 32 of the VCLT remains indirect. Yet, Article 32 of the VCLT does not explicitly refer to the parties, but more ambiguously to "the preparatory work of the treaty and the circumstances of its conclusion" leaving the door open to an interpretation that includes treaty bodies. Moreover, the Fourth Report emphasises that domestic and international courts generally use the pronouncements "in a supplementary fashion" akin to the meaning of Article 32 of the VCLT, not out of a sense of obligation.[211]

Draft conclusion 13 on subsequent agreements and subsequent practice in relation to the interpretation of treaties excludes organs of international organisations, like the UN General Assembly or the UNHRC, from their definition of "expert body".[212] The draft conclusions, however, underline that in these cases it is the

---

207 See ILC 'Fourth Report on Subsequent Agreements and Subsequent Practice in Relation to The Interpretation of Treaties by Georg Nolte, Special Rapporteur' (7 March 2016) UN Doc A/CN.4/694, para. 50.

208 Van Alebeek and Nollenkaemper (2012), p. 411; Dörr and Schmalenbach (2018), p. 627, para.27.

209 ILC 'Draft Conclusions on Subsequent Agreements and Subsequent Practice in Relation to The Interpretation of Treaties, With Commentaries' UN Doc A/73/10, conclusion 4 (3).

210 *ibid.*

211 ILC 'Fourth Report on Subsequent Agreements and Subsequent Practice in Relation to The Interpretation of Treaties by Georg Nolte, Special Rapporteur' (7 March 2016) UN Doc A/CN.4/694, para. 53.

212 '[...] which is established under a treaty and is not an organ of an international organization.' ILC 'Draft Conclusions on Subsequent Agreements and Subsequent Practice in Relation to The Interpretation of Treaties, With Commentaries' (2018) UN Doc A/73/10, conclusion 13 (1).

state parties themselves engaging in practice through their organs.[213] In other words, the conduct is attributable to the state parties, not the international organs. *Dörr* and *Schmalenbach* describe that practice of an organisation can constitute subsequent practice within the meaning of Article 31 (3) (b) of the VCLT as "a permanent feature of international jurisprudence", referring to the ICJ's frequent references to UN organs in e.g., its opinion in *Namibia* or in *Legal Consequences of the Construction of a Wall in the Occupied Palestinian Territory*.[214]

What appears to be more difficult is to establish the agreement of the parties. *Dörr* and *Schmalenbach* explain that whilst it suffices that only some parties participated in the practice, it must be accepted by all parties. This requires "at least good evidence" that the inactive parties endorse such practice.[215] In the case of organs of international organisations, they add, "it is only relevant if it is not counteracted by acts or representations of the parties to the treaty in question".[216] This means that the resolutions adopted by the UNHRC, which in contrast to the UN General Assembly does not represent all member states, could still contribute to this form of interpretation. It is, however, likely that establishing agreement of the state parties would require the resolutions to be adopted by consensus or unanimously to ensure that they enjoy the support of all state parties.[217] Furthermore, establishing agreement is premised on all parties having knowledge or awareness of such practice.[218] Responses by states to the Draft Conclusions reflect differing views on the interpretive authority and value of pronouncements of treaty bodies.[219] This ranges from unconditional endorsements to an implicit suggestion that pronouncements cannot be attributed legal value or effect.[220]

---

213 *ibid*, 33. Draft conclusion 12 focuses exclusively on the constituent instruments of international organisations, excluding 'the interpretation of decisions by organs of international organizations as such'.

214 Dörr and Schmalenbach (2018), p. 600, para. 86.

215 *Ibid* para. 87.

216 *ibid.*

217 See *ibid* para 88.

218 *ibid* para 89.

219 See Azaria (2020), p. 53 f.

220 *ibid* p. 54.

2.3 Conclusions

This chapter has offered a brief introduction to the role and function of human rights bodies. It has positioned human rights bodies and their pronouncements and work in the interpretation and development of international law. By reference to their interpretive authority, sources of public international law, contribution to customary international law, and subsequent practice in the application of a treaty, the chapter has explored, shown, and dismissed different ways in which the work and pronouncement of human rights bodies can be legally significant. Rather than adhering to the binary of legally binding and non-binding, the chapter suggests considering the legal effect of a source. It proposes considering sources, including the work and pronouncements of human rights bodies, on a continuum of legal effect.

Reference List

van Alebeek, Rosanne and Nollenkaemper, André (2012): The Legal Status of Decisions by Human Rights Treaty Bodies In National Law. In: Helen Keller and Geir Ulfstein (eds.): UN Human Rights Treaty Bodies. Law and Legitimacy. Cambridge (Cambridge University Press), pp. 356–413.

Asamoah, Obed (1964): The Legal Effect of Resolutions of the General Assembly. In: Columbia Journal of Transnational Law, Vol. 3, No. 2, pp. 210–230.

Azaria, Danae (2020): The Legal Significance of Expert Treaty Bodies Pronouncements for the Purpose of the Interpretation of Treaties. In: International Community Law Review, Vol. 22, No.1, pp. 33–60.

Bianchi, Andrea (2016): International Law Theories: An Inquiry into Different Ways of Thinking. Oxford (Oxford University Press).

Bielefeldt, Heiner (2011): Protecting and Implementing the Right to Freedom of Religion or Belief: Interview with Heiner Bielefeldt. In: Journal of Human Rights Practice, Vol. 3, Issue 3, pp. 229–244.

Bódig, Mátyás (2016): Soft Law, Doctrinal Development, and the General Comments of the UN Committee on Economic, Social and Cultural Rights. In: Stéphanie Lagoutte, Thomas Gammeltoft-Hansen and John Cerone (eds.): Tracing the Roles of Soft Law in Human Rights. Oxford (Oxford University Press), pp. 69–88.

Borlini, Leonardo and Crema, Luigi (2020): The Legal Status of Decisions by Human Rights Treaty Bodies: Authoritative Interpretation or mission éducatrice?. In: Giuliana Ziccardi Capaldo (ed.): Global Community Yearbook of International Law and Jurisprudence. Oxford (Oxford University Press).

Casla (2018): Supreme Court of Spain: UN Treaty Body Individual Decisions Are Legally Binding, https://www.ejiltalk.org/supreme-court-of-spain-un-treaty-body-individual-decisions-are-legally-binding/, accessed 31 January 2025.

Chenwi, Lilian (2009): Correcting the Historical Asymmetry between Rights: The Optional Protocol to the International Covenant on Economic, Social and Cultural Rights. In: African Human Rights Law Journal, Vol. 9, No.1, pp. 23–51.

Chinkin, Christine (2022): Sources. In: Daniel Moeckli, Sangeeta Shah, Sandesh Sivakumaran and David Harris (eds.): International Human Rights Law. 4th ed., Oxford, Oxford University Press, p. 65–88.

Citroni (2015): The Human Rights Committee and its Role in Interpreting the International Covenant on Civil and Political Rights vis-à-vis States Parties, https://www.ejiltalk.org/the-human-rights-committee-and-its-role-in-interpreting-the-international-covenant-on-civil-and-political-rights-vis-a-vis-states-parties/ accessed 31 January 2025.

Coomans, Fons (2018): The UN Committee on Economic, Social, and Cultural Rights. In: Oberleitner, Gerd (ed): International Human Rights Institutions, Tribunals, and Courts: International Human Rights. Singapore (Springer).

Dörr, Oliver and Schmalenbach, Kirsten (2018): Vienna Convention on the Law of Treaties. A Commentary. 2nd edition, Berlin/Heidelberg (Springer).

Domínguez-Redondo, Elvira (2020): In Defense of Politicization of Human Rights: The UN Special Procedures. Oxford (Oxford University Press).

Forman, Lisa (2016): Can Minimum Core Obligations Survive a Reasonableness Standard of Review under the Optional Protocol to the International Covenant on Economic, Social and Cultural Rights. In: Ottawa L. Rev., Vol. 47, No. 2, pp. 561–573.

Harrington (2015): The Human Rights Committee, Treaty Interpretation, and the Last Word, https://www.ejiltalk.org/the-human-rights-committee-treaty-interpretation-and-the-last-word/, accessed 31 January 2025.

Helmersen, Sondre Torp (2019): Finding 'the Most Highly Qualified Publicists': Lessons from the International Court of Justice. In: European Journal of International Law, Vol. 30, Issue 2, pp. 509–535.

Joseph, Sarah (2018): Committees: Human Rights Bodies. In: Max Planck Encyclopedias of International Law.

Kälin, Walter and Künzli, Jörg (2019). The Law of International Human Rights Protection. 2nd edition, Oxford (Oxford University Press).

Keller, Helen and Grover, Leena (2012): General Comments of the Human Rights Committee and their Legitimacy. In: Helen Keller and Geir Ulfstein (eds.): UN Human Rights Treaty Bodies: Law and Legitimacy. Cambridge (Cambridge University Press), pp. 116–198.

Keller, Helen and Ulfstein, Geir (2012): UN Human Rights Treaty Bodies. Law and Legitimacy. Cambridge (Cambridge University Press).

Kothari, Miloon (2013): From Commission to the Council: Evolution of UN Charter Bodies. In: Dinah Shelton (ed.): The Oxford Handbook of International Human Rights Law. Oxford (Oxford University Press), pp. 587–620.

Mechelm, Kerstin (2009): Treaty Bodies and the Interpretation of Human Rights. In: Vanderbilt Journal of Transnational Law, Vol. 42, Issue 3, pp. 905–947.

Medina Quiroga, Cecilia (2013): The Role of International Tribunals: Law-Making or Creative Interpretation?. In: Dinah Shelton (ed.): The Oxford Handbook of International Human Rights Law. Oxford (Oxford University Press), pp. 649–669.

Neuman, Gerald L. (2016): Giving Meaning and Effect to Human Rights: The Contributions of Human Rights Committee Members. In: Human Rights Program Research Working Paper Series.

O'Flaherty, Michael (2006): Concluding Observations of United Nations Human Rights Treaty Bodies. In: Human Rights Law Review, Vol. 6, No. 1, pp. 27–52.
Piccone, Ted (2010): Catalysts for Rights: The Unique Contribution of the UN's Independent Experts on Human Rights. Final Report of the Brookings Research Project on Strengthening U.N. Special Procedures. The Brookings Institution, pp. 1–90.
Pinto, Mónica (2018b): Special Procedures: Human Rights Council. Max Planck Encyclopedias of International Law.
Provost, René (2015): Interpretation in International Law as a Transcultural Project. In: Andrea Bianchi, Daniel Peat and Matthew Windsor (eds.): Interpretation in International Law. Cambridge (Cambridge University Press), pp. 290–308.
Rodley, Sir Nigel (2013): The Role and Impact of Treaty Bodies. In: Shelton, Dinah (ed.): The Oxford Handbook of International Human Rights Law. Oxford (Oxford University Press), pp. 621–648.
Rodley, Sir Nigel (2009): The United Nations Human Rights Council, Its Special Procedures, and Its Relationship with the Treaty Bodies: Complementarity or Competition?. In: Kevin Boyle (ed.): New Institutions for Human Rights Protection, Collected Courses of the Academy of European Law. Oxford (Oxford University Press), pp. 49–74.
Schachter, Oscar (1991): International Law in Theory and Practice. Dordrecht (Martinus Nijhoff).
Scheinin, Martin (2017): The Art and Science of Interpretation in Human Rights Law. In: Bård A. Andreassen, Claire Methven O'Brien and Hans-Otto Sano (eds.): Research Methods in Human Rights: A Handbook. Cheltenham/Northampton (Edward Elgar), pp. 17–37.
Scheinin, Martin (2007): Access to Justice before International Human Rights Bodies: Reflections on the Practice of the UN Human Rights Committee and the European Court of Human Rights. In: Francesco Francioni (ed.): Access to Justice as a Human Right. Oxford (Oxford University Press), pp. 135–152.
Scheinin, Martin (1997): International Mechanisms and Procedures for Implementation. In: Raija Hanski Markku Suks (ed.): An Introduction to the International Protection of Human Rights: A Textbook. Turku (Institute for Human Rights).
Seibert-Fohr, Anja and Weniger, Christine (2019): Compliance Monitoring under the International Covenant on Civil and Political Rights: To Comply, or not to Comply, that is the Question. https://papers.ssrn.com/sol3/papers.cfm?abstract_id=3337435.
Shelton, Dinah (2012): The Legal Status of Normative Pronouncements of Human Rights Treaty Bodies. In: Hestermeyer, Holger Paul, König, Doris, Matz-Lück, Nele, Röben, Volker, Seibert-Fohr, Anja, Stoll, Peter-Tobias, Vöneky, Silja (eds.): Coexistence, Cooperation and Solidarity – Liber Amicorum Rüdiger Wolfrum (vol. I). Martinus Nijhoff Publishers, pp. 553–575.
Sivakumaran, Sandesh (2017): The Influence of Teachings of Publicists on the Development of International Law. In: International and Comparative Law Quarterly, Vol. 66, pp. 1–37.
Subedi, Surya P. (2014): Human Rights Experts in the United Nations: A Review of the Role of United Nations Special Procedures. In: Monika Ambrus, Karin Arts, Ellen Hey and Helena Raulus (eds.): The Role of 'Experts' in International and European Decision-Making Processes: Advisors, Decision Makers or Irrelevant Actors?. Cambridge (Cambridge University Press), pp. 241–262.

Tistounet, Eric (2020): The UN Human Rights Council – A Practical Anatomy. Cheltenham/ Northampton (Edward Elgar).

Tobin, John (2010): Seeking to Persuade: A Constructive Approach to Human Rights Treaty Interpretation. In: Harvard Human Rights Journal, Vol 23, pp. 201–250.

Tomuschat, Christian (2019). Universal Periodic Review Procedure: Human Rights Council. In: Max Planck Encyclopedias of International Law.

Tomuschat, Christian (2003): Human Rights: Between Idealism and Realism. Oxford (Oxford University Press).

3 The Human Dimension of International Cultural Heritage Law

3.1 Introduction

The protection of cultural heritage has increasingly attracted attention from an international human rights perspective.[1] Similarly, the "human dimension" of international cultural heritage law has made continuous developments—both in academic scholarship and in the law. The human dimension of cultural heritage protection is multifaceted and, broadly speaking, refers to the unifying narrative that cultural heritage is protected for itself *and* for its relationship to people, as individuals or groups.[2] This is reflected inter alia in the importance accorded to cultural heritage as a vehicle for the cultural identity of people. The "increasing cross-fertilization" between cultural heritage law and human rights is a particularly pertinent example of this human dimension.[3]

The objectives of the chapter go hand in hand with further examining the protection and governance of cultural heritage at the international level. Rather than providing a comprehensive overview of the international legal regulation of cultural heritage in international law, the chapter primarily focuses on the

[1] See e.g., UNHRC Res 6/11 (28 September 2007) UN Doc A/HRC/RES/6/; UNHRC Res 33/20 (6 October 2016) A/HRC/RES/33/20. Also see, Xanthaki (2017), p. 1.

[2] See Vrdoljak (2013a), p. 1.

[3] Francioni (2011), p. 14.

L. Hofmann, *Cultural Heritage through the Lens of Human Rights Bodies*, Juridicum – Bochumer Schriften zur Friedenssicherung und zum Humanitären Völkerrecht | Bochum Studies on the International Law of Peace and Armed Conflict 73, https://doi.org/10.1007/978-3-658-51203-3_3

1972 Convention concerning the Protection of the World Cultural and Natural Heritage[4] (World Heritage Convention) and the 2003 Convention for the Safeguarding of the Intangible Cultural Heritage[5] (Intangible Cultural Heritage Convention). This chapter analyses the human dimension of international cultural heritage law in three key areas: (1) the evolution of the concept of cultural heritage; (2) the rationales for protecting cultural heritage and its protection mechanisms; and (3) in the enforcement of cultural heritage protection[6]. The chapter argues that increasingly, the role and rights of communities, groups, and individuals as well as their relationship with heritage are considered. This provides the impetus to closely analyse the protection of rights of individuals and members of groups relating to cultural heritage in the international human rights framework.

3.2 The Evolving Meaning of Cultural Heritage

To illustrate the human dimension in international cultural heritage law, the chapter first turns to the concept of cultural heritage itself. Taking the 1954 Hague Convention, the World Heritage Convention, and the Intangible Cultural Heritage Convention as important markers, the analysis shows by means of the Conventions how the meaning of cultural heritage has evolved and integrated a human dimension.

3.2.1 Cultural Property in the Hague Convention

As an object of international legal protection, the term "cultural property" was introduced in the 1954 Convention for the Protection of Cultural Property in the

---

[4] Convention concerning the Protection of the World Cultural and Natural Heritage (adopted 16 November 1972, entered into force 17 December 1975) 1037 UNTS 151(*World Heritage Convention*).

[5] Convention for the Safeguarding of the Intangible Cultural Heritage (adopted 17 October 2003, entered into force 20 April 2006) 2368 UNTS 3 (*Intangible Cultural Heritage Convention*).

[6] International criminal law can not only be regarded as enforcing international humanitarian law but can also be thought of as "the internationalization of individual criminal responsibility for serious human rights abuses". Francioni (2008), p. 11. As the flip side of the human rights coin, international criminal law evidences important recognitions of human rights implications of heritage destruction.

Event of Armed Conflict[7] (1954 Hague Convention). Article 1 of the 1954 Hague Convention defines cultural property as "movable or immovable property of great importance to the cultural heritage of every people".[8] The article further continues with a non-exhaustive list of examples of such property, specifying:

> "[…] monuments of architecture, art or history, whether religious or secular; archaeological sites; groups of buildings which, as a whole, are of historical or artistic interest; works of art; manuscripts, books and other objects of artistic, historical or archaeological interest; as well as scientific collections and important collections of books or archives or of reproductions of the property defined above; […]."[9]

The definition further includes buildings that preserve or exhibit cultural property, referring to "museums, large libraries and depositories or archives" as well as those refuges that shelter movable cultural property during an armed conflict.[10] The definition extends to centres, which contain "a large amount of cultural property."[11] It is important to note that the protection extends to these two groups of buildings not for their inherent value, but for the cultural objects they host. Consequently, protection can cease if the cultural properties are removed from the building.[12] It follows that the definition of cultural property is premised on two criteria:[13] the importance of the cultural object and its typology. The use of these two criteria to define cultural property and cultural heritage can be found throughout subsequent conventions, although the criteria themselves are frequently amended or added to.

Since then, the use of the term has been liberal, "but elusive of definition."[14] International law has produced several definitions, showcasing that each multilateral treaty adopts its own definition.[15] For example, the 2001 Convention on the Protection of the Underwater Cultural Heritage defines underwater cultural

---

[7] Convention for the Protection of Cultural Property in the Event of Armed Conflict (adopted 14 October 1954, entered into force 7 August 1956) 249 UNTS 215 (*1954 Hague Convention*).

[8] *ibid* Art. 1 (a).

[9] *ibid.*

[10] *ibid* Art. 1 (b).

[11] *ibid* Art. 1 (c).

[12] O'Keefe (2007), p. 103.

[13] Francioni (2013a), p. 4.

[14] Forrest (2010), p. 1.

[15] *Ibid* p. 20.

heritage as "all traces of human existence having a cultural, historical or archaeological character which have been partially or totally under water, periodically or continuously, for at least 100 years".[16] The 1970 Convention on the Means of Prohibiting and Preventing the Illicit Import, Export and Transfer of Ownership of Cultural Property defines cultural property as "property which, on religious or secular grounds, is specifically designated by each State as being of importance for archaeology, prehistory, history, literature, art or science" and falls into one of the typologies identified in the treaty.[17] *Forrest* explains this evolution by arguing that the use of the concept in each convention mirrors the economic, political, and social context as well as value negotiated at the time. Accordingly, the use of the concept in the individual conventions constitutes a broader reflection of the perception of the notion at that time.[18] Cultural heritage also has an economic dimension, reflected in tourism, the creative industries, its relationship to the concept of sustainable development[19] and heritage practices that can be an important aspect of a person's livelihood, such as in the case of reindeer herding of Indigenous peoples.

3.2.2 Cultural and Natural Heritage in the 1972 World Heritage Convention

The object and purpose of the World Heritage Convention lie in recognising the collective interest of the international community in the preservation and protection of "world heritage", a concept referring to both cultural and natural heritage.[20] Cultural heritage within the meaning of Article 1 of the World Heritage Convention includes material and immovable properties of outstanding universal character. Pursuant to this definition, cultural heritage encompasses:

---

[16] Convention on the Protection of the Underwater Cultural Heritage (adopted 2 November 2001, entered into force 2 January 2009) 2562 UNTS (*Underwater Cultural Heritage Convention*), Art. 1.

[17] 1970 Convention on the Means of Prohibiting and Preventing the Illicit Import, Export and Transfer of Ownership of Cultural Property (adopted 14 November 1970, entered into force 24 April 1972) 823 UNTS 231 (*1970 Convention*), Art. 1.

[18] Forrest (2010), p. 23 f.

[19] Scovazzi (2020), p. 746.

[20] As defined in World Heritage Convention, Arts. 1 and 2. Also see Francioni (1993), p. 195. This chapter focuses only on cultural heritage but also addresses mixed heritage sites.

> "monuments: architectural works, works of monumental sculpture and painting, elements or structures of an archaeological nature, inscriptions, cave dwellings and combinations of features, which are of outstanding universal value from the point of view of history, art or science;
>
> groups of buildings: groups of separate or connected buildings which, because of their architecture, their homogeneity or their place in the landscape, are of outstanding universal value from the point of view of history, art or science;
>
> sites: works of man or the combined works of nature and man, and areas including archaeological sites which are of outstanding universal value from the historical, aesthetic, ethnological or anthropological point of view."

Natural heritage is defined in Article 2 and covers

> "natural features consisting of physical and biological formations or groups of such formations, which are of outstanding universal value from the aesthetic or scientific point of view;
>
> geological and physiographical formations and precisely delineated areas which constitute the habitat of threatened species of animals and plants of outstanding universal value from the point of view of science or conservation;
>
> natural sites or precisely delineated natural areas of outstanding universal value from the point of view of science, conservation or natural beauty."

The articles operate on the concept of "outstanding universal value", which means that the "world heritage status" is only available to "very special and exceptional items".[21] Whereas this thesis primarily focuses on cultural heritage, sites listed on the World Heritage List can be of a mixed character, containing both an outstanding universal cultural and natural value.[22] Awarding the status of world cultural heritage under the Convention requires an "extraordinary level of selectivity"[23] at the international level.

The shift from "a perception of culture simply as property to one of culture as heritage" can be traced to the World Heritage Convention.[24] This has been ascribed to the need of bringing both cultural properties and natural heritage under

---

[21] Francioni (2023), p. 26–27.

[22] Currently, 40 mixed sites are inscribed on the World Heritage List and on the List of World Heritage in Danger. See UNESCO (2025a, 2025b): World Heritage List: Mixed Properties, https://whc.unesco.org/en/list/?search=&type=mixed&components=0, accessed 31 January 2025.

[23] O'Keefe (1999), p. 44.

[24] Lenzerini (2016a), p. 70.

a "single conceptual umbrella" in the World Heritage Convention.[25] The use of "property" over "heritage" is often employed in those legal instruments that seek to regulate the illicit flow of cultural heritage.[26] For *Forrest*, the use of "property" underlines the object's commercial value.[27] Since this may result in the object's cultural value being perceived as secondary, he argues that employing the term "heritage" is favourable and conducive to its protection.[28] *Lixinski* highlights the positive effects of "the move from 'cultural property' to 'cultural heritage'"[29], but also alerts to the unintended by-product "of disassociating communities from heritage".[30] This is brought about by inter alia the "narrative of heritage as an interest of humankind", which "helps drive cosmopolitanism" and connects "heritage to a universal discourse of commonality".[31] However, it also "reinforce[s] the interests of the scientific community (at the expense of communities)".[32] Moreover, the use of "heritage" over "property" can be disadvantageous and lead to a disconnect, because while the international legal discourse has embraced the concept of "heritage", the implementation of cultural heritage law into domestic law often continues to rely on the term "property".[33]

3.2.3 Intangible Cultural Heritage in the 2003 Intangible Cultural Heritage Convention

The adoption of the Intangible Cultural Heritage Convention in 2003 has continued advancing[34] the notion by introducing the intangible constituent.[35] The Convention was adopted against the backdrop of unprecedented globalisation and

[25] Francioni (2008), p. 6. Also see, Francioni (2020c), p. 250.

[26] E.g., The 1970 Convention. Also see Forrest (2010), p. 22.

[27] Forrest (2010), p. 24.

[28] *ibid.*

[29] Lixinski (2019), p. 23.

[30] *ibid* p. 27.

[31] *ibid* p. 52.

[32] *ibid.*

[33] *ibid* p. 27. *Lixinski* further explores the idea of property as rights in the chapter "Definitions: From Cultural Property to Cultural Heritage (and Back?)" of the same book.

[34] For a critical appraisal of this shift, see *ibid* p. 28.

[35] However, also see the 1989 Recommendation on the Safeguarding of Traditional Culture and Folklore closely connected to the intangible cultural heritage and safeguarding cultural identity, as pointed out in Donders (2020), pp. 386–387.

the growing concern for cultural diversity, environmental protection, and sustainable development.[36] Including intangible elements of cultural heritage was originally envisaged in the World Heritage Convention but abandoned before the final adoption of the text of the convention.[37] The objectives of the Intangible Cultural Heritage Convention include to (a) safeguard intangible cultural heritage, (b) ensure its respect, (c) raise awareness, and (d) provide a system of international cooperation and assistance.[38]

Article 2(1) of the Intangible Cultural Heritage Convention defines intangible cultural heritage as

> "the practices, representations, expressions, knowledge, skills—as well as the instruments, objects, artefacts and cultural spaces associated therewith—that communities, groups and, in some cases, individuals recognize as part of their cultural heritage."

The protection afforded to tangible objects derives from the cultural practice, not the other way around.[39] Moreover, the definition recognises "communities, groups, and individuals" as holders of cultural heritage.[40] This can be interpreted as a step away from the state-centric nature of the World Heritage Convention,[41] and a step towards recognising the role of "communities, groups, and individuals" in the safeguarding of cultural heritage. However, it is important to recall that states still primarily "hold the power to select and legally recognize cultural heritage".[42] In accordance with this recognition, state parties are obligated to include communities, groups, and individuals in decision-making processes.[43] Although "communities", "groups", and "individuals" are not further defined, the repetition of the three in the same order has been interpreted to imply a hierarchy.[44]

---

[36] Forrest (2010), p. 28.

[37] Blake (2006), pp. 4–5. For an account of the historical developments leading up to the adoption of the Intangible Cultural Heritage Convention, see Lixinski (2013a), pp. 30–33.

[38] Intangible Cultural Heritage Convention, Art. 1.

[39] Lixinski (2013a), p. 9.

[40] The term "communities" is not defined as a legal term in the Convention and state parties adopt their own definition in the implementation of the Intangible Cultural Heritage Convention. See, Lixinski (2011), pp. 84–85.

[41] Francioni (2008), p. 15; Lixinski (2013a), pp. 2, 3.

[42] Donders (2020), p. 385.

[43] Cf. Kraak and Aykan (2018), p. 4.

[44] UNESCO and Asia/Pacific Cultural Centre for UNESCO 'Expert Meeting on Community Involvement in Safeguarding Intangible Cultural Heritage: Towards the Implementation of the 2003 Convention' (13–15 March 2006) CLT/CH/ITH/DOCEM0306 REV.1, p. 8.

In the absence of a definition, experts, at a UNESCO sponsored expert meeting, suggested that communities refer to "networks of people whose sense of identity or connectedness emerges from a shared historical relationship that is rooted in the practice and transmission of, or engagement with, their [intangible cultural heritage]".[45] Groups refer to "people within or across communities who share characteristics such as skills, experience and special knowledge, and thus perform specific roles in the present and future practice, re-creation and/or transmission of their intangible cultural heritage as, for example, cultural custodians, practitioners or apprentices."[46] Finally, individuals are persons "within or across communities who have distinct skills, knowledge, experience or other characteristics, and thus perform specific roles in the present and future practice, re-creation and/or transmission of their intangible cultural heritage as, for example, cultural custodians, practitioners and, where appropriate, apprentices."[47]

The Intangible Cultural Heritage Convention protects cultural heritage not as an object that has an intrinsic historical or archaeological value.[48] Intangible cultural heritage is protected because it is part of the living culture of a cultural community seeking to maintain and develop their identity as a human rights claim.[49] The Intangible Cultural Heritage Convention has brought the fields of cultural heritage law and international human rights law closer together, specifically the protection of minority groups and Indigenous peoples.[50]

Article 2 (1) of the Intangible Cultural Heritage Convention also shows a greater understanding that (intangible) cultural heritage is not static, but "transmitted from generation to generation, [...] constantly recreated by communities and groups in response to their environment, [and] their interaction with nature and their history".[51]

In defining intangible cultural heritage, Article 2 (2) lists:

"(a) oral traditions and expressions, including language as a vehicle of the intangible cultural heritage;

(b) performing arts;

(c) social practices, rituals and festive events;

---

[45] *ibid* p. 9.

[46] *ibid.*

[47] *ibid.*

[48] Francioni (2013b), p. 16.

[49] *ibid.*

[50] See Francioni (2011), p. 14; Francioni (2012), p. 23.

[51] Intangible Cultural Heritage Convention, Art. 2 (1).

(d) knowledge and practices concerning nature and the universe;

(e) traditional craftsmanship."

The definition of "intangible cultural heritage" itself integrates human rights as a limit to the identification and safeguarding of intangible cultural heritage by only considering practices that are "compatible with existing international human rights instruments".[52] The Intangible Cultural Heritage Convention thus refers to international human rights instruments as a yardstick, as the drafting of this provision was influenced by certain traditional cultural practices that are incompatible with the enjoyment of human rights.[53] Quite contrary to a recognition of the mutually reinforcing relationship of cultural heritage and human rights law, this definitional restriction can also be interpreted to place culture "in an antagonistic position with respect to human rights, in that it showcases culture as potentially primitive and uncivilised, and thus needing erasure by the imperatives of modernity."[54]

3.2.4 Interim Conclusions

Continuing to think of tangible and intangible cultural heritage in a binary way is somewhat misleading and can be seen as a historical byproduct of the cultural heritage treaty-making process.[55] This distinction fails to recognise a more holistic view of heritage and is increasingly viewed as arbitrary.[56] The intrinsic link between both legal concepts has been described as a symbiotic relationship in academia.[57] Tangible cultural heritage, for instance, requires to be claimed as heritage and accorded meaning, which is why there is an inherent intangible aspect to it.[58] Moreover, the destruction of material heritage has often been

---

[52] *ibid* Art. 2.

See Lixinski (2013a), p. 36. This "invisible asterisk" has been subject to criticism on the grounds of its "colonial imprint" but can also be viewed less critically by arguing that it promotes "the non-negotiable core of human rights". Lixinski (2013b), p. 150.

[53] Francioni and Lixinski (2017), p. 12.

[54] *ibid.*

[55] Blake (2011), p. 203; Blake (2020a), p. 353. Also see, Donders (2020), p. 385.

[56] Blake (2020a), p. 353.

[57] Lixinski (2013a), p. 19.

[58] Silverman and Fairchild Ruggles (2007), p. 12.

linked to the destruction of immaterial heritage.[59] An example of this is the displacement of populations in Northern Iraq and Syria, accompanied by the destruction of objects, text, and historic structures. This destruction of tangible heritage impedes the transmission of inter alia ancient languages and religious practices that are tied to such sites.[60] The destruction of religious heritage in Timbuktu at the hands of *Ansar Dine* in 2012 and across Syria and Iraq at the hands of the so-called Islamic State (IS) since 2014 are among the most infamous examples of the recent past.[61]

The Intangible Cultural Heritage Convention brought about a "dematerialisation"[62] of heritage. This incremental expansion is reflective of an evolving understanding from seeing cultural heritage as high culture, typically manifesting itself in tangible masterpieces of art and high culture,[63] to recognising the universal value of national cultural diversity in both material and immaterial forms.[64] The definition in the Intangible Cultural Heritage Convention also reflects an important difference as it moves away from outstanding universal value and adds "the self-recognition of its manifestations as part of the cultural heritage by the communities concerned".[65]

The shift from "cultural property" to "cultural heritage" can also be seen as a reflection of a less "Western" conceptualisation of heritage, which views it "in less transactional terms, with an emphasis on custodianship in communal and intergenerational terms", reflecting the views of "[s]tates in Africa, Asia, and the Global South".[66] This shift further implies

---

59 UNGA 'Report of the Special Rapporteur in the field of cultural rights' (9 August 2016) UN Doc A/71/317, para. 7.

60 See *ibid*.

61 See Clapperton, Jones and Smith 2017; U.S. Department of State (2014): Threats to Cultural Heritage in Iraq and Syria: Fact Sheet, https://2009-2017.state.gov/r/pa/prs/ps/2014/09/232028.htm, accessed 31 January 2025; Fletcher (2012): Timbuktu tomb destroyers pulverise Islam's history, https://www.reuters.com/article/business/timbuktu-tomb-destroyers-pulverise-islams-history-idUSLNE86202H/, 31 January 2025.

62 Francioni and Lixinski (2017), p. 17.

63 High culture generally refers to museums, libraries, and theatres. See Vrdoljak (2013b), p. 36.

64 See Vrdoljak (2013a), p. 1; Uerpmann-Wittzack (2018), p. 2.

65 Dorough and Wiessner (2020), p. 419.

66 Vrdoljak and Francioni (2020), p. 4.

> "acceptance that the protection of heritage goes beyond the protection of the actual sites, object, and artifacts; instead, what is to be protected is the relationship between these sites, objects, and artifacts and human beings."[67]

The recognition that protection goes beyond the physical object and extends to the relationship between cultural heritage and human beings is key. It can be interpreted as recognition that built heritage serves a function in society "as indispensable tools for the exercise of certain fundamental rights and freedoms".[68] The Intangible Cultural Heritage Convention "gives more value to the cultural heritage of 'communities, groups and … individuals'", a view that is similarly shared in the Convention on the Value of Cultural Heritage for Society (Faro Convention).[69]

The Council of Europe's Faro Convention[70] contains a strong recognition of the relationship between cultural heritage protection and human rights. The Convention emphasises participatory rights,[71] and is inter alia structured around the recognition that "rights relating to cultural heritage" constitute an inherent part of the right to take part in cultural life, as laid down in the Universal Declaration of Human Rights.[72] The definition of cultural heritage in the Convention emphasises the role of people in the identification of heritage, as its Article 2 understands cultural heritage as "a group of resources inherited from the past which people identify, independently of ownership, as a reflection and expression of their constantly evolving values, beliefs, knowledge and traditions."[73] Although the Faro Convention contains explicit recognitions of rights relating to cultural heritage, the Convention itself makes clear that it does not create enforceable rights.[74] This means that these references to rights "are not meant to create substantive rights, but serve as a reminder of the link between cultural heritage and human rights".[75] As a "Framework Convention", its aims lie in defining broad objective and actionable areas for progress, meaning that "[i]nstruments of this type

---

67 Lixinski (2013a), p. 7.

68 See Francioni (2008), p. 7.

69 Blake (2015), p. 273.

70 Framework Convention on the Value of Cultural Heritage for Society, Signed in Faro on 27 October 2005. Council of Europe, ETS No. 199. Entry into force: 1 June 2011 (*Faro Convention*).

71 *ibid* Art. (12) (a).

72 *ibid* Art. 1 (a).

73 *ibid* Art. 2 (a).

74 *ibid* Art. 6 (c).

75 Donders (2018), p. 61.

may identify generic activities, but unlike ordinary Conventions, do not create obligations to specific actions."[76]

3.3 Protection of Cultural Heritage

This section addresses the human dimension of cultural heritage in its protection. The analysis reviews the preambles of international heritage treaties with a view to analysing how the rationales to protect cultural heritage have evolved. Subsequently, it examines the protection and safeguarding of cultural heritage within the meaning of the World Heritage Convention and the Intangible Cultural Heritage Convention with a view to showing how the protection has come to integrate communities, groups, and individuals.

3.3.1 The Evolving Rationale to Protect Cultural Heritage

In examining the rationales for protecting cultural heritage and its protection mechanisms, the analysis starts with preambles. Although they do not set out legally binding obligations on the state parties, preambles are indicative of the treaty's objectives and are "exhortatory in character".[77]

The 1954 Hague Convention focuses on cultural property as an object of physical protection and preservation and cites the grave losses suffered in the First and Second World War in its preamble. Nevertheless, the Convention underscores the importance of preserving cultural heritage *for* the peoples.[78] The preamble conveys this by means of the well-known proclamation that "damage to cultural property belonging to any people whatsoever means damage to the cultural heritage of all mankind".[79] Rather than focusing on protection for local communities, groups, and individuals though, the preamble refers to an international idea of heritage and its importance to the international community. The World Heritage Convention also conveys this sentiment in its preamble by declaring

[76] Council of Europe 'Explanatory Report to the Council of Europe Framework Convention on the Value of Cultural Heritage for Society' (27 October 2005) ETS No. 199, section C.

[77] See, Blake (2020b), p. 20. See also, more generally, Hulme (2016), pp. 1281–1343.

[78] 1954 Hague Convention, preamble, 3rd recital.

[79] 1954 Hague Convention, preamble, 2nd recital. Also referred to as "cultural property internationalism".

that the "deterioration or disappearance of any item of the cultural or natural heritage constitute a harmful impoverishment of the heritage of *all the nations of the world*".[80] This notion of universality can also be seen in the 1970 Convention on the Means of Prohibiting and Preventing the Illicit Import, Export and Transfer of Ownership of Cultural Property, which refers to "the interchange of cultural property among nations for scientific, cultural and educational purposes" and the corresponding benefit for the cultural life "of all peoples".[81] The preamble and certain provisions of the 2001 Convention on the Protection of Underwater Cultural Heritage are structured around the notion of "humanity" and relate to inter alia scientific and educational benefits of preserving underwater cultural heritage for humanity.[82] These conventions exhibit the idea of cultural property internationalism and, accordingly, of global heritage protection for all peoples.[83]

The rationale of the 2003 Intangible Cultural Heritage Convention similarly contains notions of this idea by referring to "the value of such heritage 'as a mainspring of cultural diversity', corresponding to a 'common concern' of the international community as a whole".[84] Along the same lines, Article 19 (2) recognises "that the safeguarding of intangible cultural heritage is of general interest to humanity".[85] The preamble, furthermore, highlights the role of intangible cultural heritage in connecting human beings, fostering exchange and mutual

[80] Emphasis added. World Heritage Convention, preamble, 2nd recital. The fifth recital further states that the existing legal construct of binding and non-binding documents manifests the importance "of safeguarding this unique and irreplaceable property, to whatever people it may belong".

[81] The second recital reads "Considering that the interchange of cultural property among nations for scientific, cultural and educational purposes increases the knowledge of the civilization of Man, enriches the cultural life of all peoples and inspires mutual respect and appreciation among nations".

[82] Underwater Cultural Heritage Convention, preamble, 1st recital, 4th and 5th recital; Arts. 2, 12 (6).

[83] See Merryman (1986); Merryman (2005).

[84] Referring to the second recital of the Convention's preamble. In: Francioni (2004), p. 1225.

[85] The wording here is deliberate and noteworthy, as it implies that the interest lies in the safeguarding, not in the intangible heritage. The phrasing stems from concerns over the internationalisation of intangible cultural heritage, i.e., destroying local culture by elevating it to the international level, as well as from states' unease that elevating local culture to the international level could result in claims against the state, including in claims for statehood. Lixinski (2013a), p. 52; Lixinski (2011), p. 82.

understanding.[86] However, it also recognises the role of communities, groups, and individuals "in the production, safeguarding, maintenance and re-creation of the intangible cultural heritage".[87] Article 2 of the Intangible Cultural Heritage Convention explains that intangible cultural heritage "provides them [communities and groups] with a sense of identity and continuity, thus promoting respect for cultural diversity and human creativity." Underlining the protection of identity as the central protection objective of the Convention,[88] cultural heritage is conceived of as a vehicle for the promotion of cultural identity.[89]

Cultural identity and human dignity are also underlined in the preamble of the UNESCO Declaration Concerning the Intentional Destruction of Cultural Heritage. The Declaration points directly to the adverse effects of the (intentional) destruction of cultural heritage on human dignity and human rights. The fifth recital of the preamble reads in full:

> "Mindful that cultural heritage is an important component of the cultural identity of communities, groups and individuals, and of social cohesion, so that its intentional destruction may have adverse consequences on human dignity and human rights."[90]

The recital thereby highlights "communities, groups and individuals" as holders of the rights.[91] The significance of the recognition of communities backslides in the Declaration, as Article VI and Article VII on state responsibility and individual criminal responsibility, respectively, were, in the final version, deprived of a clause that emphasised the "special interest for the community directly affected by such destruction".[92] Section IX on human rights and international humanitarian law emphasises respect for international rules related to inter alia human

---

[86] Reading in full: "Considering the invaluable role of the intangible cultural heritage as a factor in bringing human beings closer together and ensuring exchange and understanding among them".

[87] Intangible Cultural Heritage Convention, preamble, 6th recital.

[88] "It is precisely *because* intangible cultural heritage provides a sense of identity that it is worthy of protection." Lixinski (2013a), p. 9.

[89] *ibid* p. 4.

[90] Reading in full: "Mindful that cultural heritage is an important component of the cultural identity of communities, groups and individuals, and of social cohesion, so that its intentional destruction may have adverse consequences on human dignity and human rights". UNESCO General Conference 'Declaration concerning the Intentional Destruction of Cultural Heritage' (17 October 2003), 5th recital.

[91] Vrdoljak (2007), p. 19.

[92] UNESCO 'Draft declaration concerning the intentional Destruction of Cultural Heritage' (17 July 2003) UNESCO Doc. 32 C/25. Also see: Lenzerini (2003), p. 142.

rights especially "when intentional destruction of cultural heritage is linked to those violations."[93]

Whilst the preambles of earlier international cultural heritage instruments highlight the physical protection, the scientific and educational advancements, and the international community or humanity as their driving force, a shift, or rather, an addition to the impetus to protect cultural heritage for its importance to peoples, groups, communities, and individuals is visible.[94] Whereas the rationale to protect cultural heritage stemmed from the physical preservation of an outstanding value for the international community or humanity, it evolved to recognising heritage's value for the cultural identity of communities, groups and individuals as well as their relationship to it. It is of further interest to note that *Vrdoljak* has pointed out that a draft version of the preamble of the 1970 Convention referred to cultural participation in Article 27 of the Universal Declaration on Human Rights and read "that it is incumbent upon States to protect the cultural property existing within their territory against the dangers from the illicit export and transfer of such property".[95]

3.3.2 The Evolving Recognition of the Role of Communities, Groups, and Individuals in Protecting and Safeguarding Cultural Heritage

Early cultural heritage conventions have a state-centric understanding of the protection of cultural heritage. This means that they focus on the role of the state in protecting cultural heritage. Article 4 of the World Heritage Convention sets out the duty of state parties to ensure heritage's proper "identification, protection, conservation, presentation and transmission to future generations."[96] The state duty extends to those cultural and natural heritage sites situated on its territory and belonging "primarily to that State."[97] Article 4 highlights the necessity, where appropriate, to rely on international assistance and cooperation of a financial, artistic, scientific, and technical nature to obtain the article's goal.[98] The

---

[93] UNESCO General Conference 'Declaration concerning the Intentional Destruction of Cultural Heritage' (17 October 2003), section IX.

[94] Also see Vrdoljak (2007), p. 12.

[95] UNESCO Doc.SHC/MD/3, paras.9 and 10 in Vrdoljak (2014), pp. 146–147.

[96] World Heritage Convention, Art. 4.

[97] *ibid.*

[98] World Heritage Convention, Art. 4.

duty of cooperation is subject to Article 6. Judge Cançado Trindade highlighted in his separate opinion to the ICJ order on provisional measures in *Request for Interpretation*, that

> "[i]t is beyond doubt that the States, as promoters of the *common good*, are under the duty of co-operation between themselves to the end of the safeguard and preservation of the cultural and spiritual heritage."[99]

The ICJ's judgement confirmed this opinion.[100]

In spite of the narrow and selective nature of the World Heritage Convention, the state duty of Article 4 has been interpreted to extend to all cultural heritage (within the meaning of Article 1 of the Convention) situated on the territory of a state, not merely to the listed sites.[101] The reason for this is that Article 4 refers to the cultural heritage in Article 1, which is the provision setting out the definition of cultural heritage, and not to "the cultural [...] heritage referred to in paragraphs 2 and 4 of Article 11", which is the provision establishing the lists.[102] In other words, rather than speaking of the cultural heritage listed pursuant to Articles 11 (2) and (4), the reference to Article 1 reverts to the definition of cultural heritage. Article 12 further assures that cultural heritage not included on the lists is not indicative of a lack of outstanding universal value and should be the object of protection too.

In the context of the demolition of the Buddhas of Bamiyan, *Francioni* and *Lenzerini* argue that mere membership in the World Heritage Convention "obliges state parties to conserve and protect their own cultural properties even if these are not inscribed in the World Heritage List."[103] The Buddhas of Bamiyan were at the time of their destruction not part of the World Heritage List, although Afghanistan had ratified the Convention and nominated the monument of the Bamiyan valley

---

[99] *Request for Interpretation of the Judgment of 15 June 1962 in the Case Concerning the Temple of Preah Vihear (Cambodia v. Thailand) (provisional measures) [2011] ICJ Rep 537 (separate opinion of Judge Cançado Trindade)*, para. 115.

[100] *Request for Interpretation of the Judgement of 15 June 1962 in the Case Concerning the Temple of Preah Vihear (Cambodia v. Thailand) (judgement) [2013] ICJ Rep 281*, para 106.

[101] Francioni and Lenzerini (2003), p. 631; O'Keefe (2004), p. 189, fn 7; Lenzerini (2015), pp. 142–43.

[102] World Heritage Convention, Art. 6 (2).

[103] Francioni and Lenzerini (2003), p. 631.

on 21 December 1981. As a result, *O'Keefe* argues in favour of both the application of the World Heritage Convention and a breach of Afghanistan's obligations under Article 4 of that Convention.[104]

The Intangible Cultural Heritage Convention sets up a system of "safeguarding" at both the national and international level.[105] Article 2 (3) of the Convention explains that "safeguarding" refers to all measures pertaining to the "identification, documentation, research, preservation, protection, promotion, enhancement, transmission, [...] [and] revitalization of the various aspects of such heritage."[106] In addition, the state parties' obligation to safeguard includes ensuring "the participation of communities, groups and relevant non-governmental organizations."[107] The practice under the Intangible Cultural Heritage Convention has shown that community involvement is by no means limited to a single community, but can include different communities, including non-governmental authorities and natural persons.[108] "Safeguarding" recognises the expertise of communities and groups by acknowledging the state's lack of expertise as an obstacle to the safeguarding of the intangible cultural heritage.[109] *Lixinski* argues that "as it is impossible to dissociate communities from their heritage", communities themselves have become "entitled to protection" under the Intangible Cultural Heritage Convention.[110] Yet, the requirement to involve communities[111] has been described as "more exhortatory than obligatory", meaning that state parties have leeway in choosing the participatory approach under the Convention.[112] "Real and effective community participation" has thereby remained weak in the application of the Intangible Cultural Heritage Convention.[113]

Community participation is also important for the World Heritage Convention and the World Heritage Committee has been criticised for ignoring "ongoing, significant human rights violations related to properties inscribed on the World

---

[104] O'Keefe (2004), p. 195 fn 44.

[105] Intangible Cultural Heritage Convention, Arts. 11–18.

[106] Intangible Cultural Heritage Convention, Art. 2 (3).

[107] Intangible Cultural Heritage Convention, Art. 11 (b), Art. 15.

[108] Lixinski (2011), p. 90.

[109] Lixinski (2013a), p. 42.

[110] *ibid* p. 3.

[111] See Intangible Cultural Heritage Convention, Art. 15: "each State Party shall endeavour to ensure the widest possible participation of communities, groups and, where appropriate, individuals".

[112] Blake (2020a), p. 356. Also see Disko (2017), p. 64.

[113] Blake (2020a), p. 356.

Heritage List" as well as for inscribing "such properties [...] despite clear evidence that the Indigenous peoples had not been consulted".[114] Whilst the preamble of the World Heritage Convention underlines participation in light of the international community's role of protecting heritage and providing collective assistance, the Operational Guidelines have come to include guidance on participation by inter alia local communities and Indigenous peoples[115] and the free, prior and informed consent of Indigenous communities in the nomination process of sites.[116]

The primary tool of safeguarding cultural heritage is through the use of inventories, set out in Article 12 of the Intangible Cultural Heritage Convention. Although Articles 13 to 16 specify other measures of safeguarding; emphasise the role of education, awareness-raising, and capacity-building; and seek to ensure the participation of communities, groups, and individuals, *Lixinski* argues that the language employed by the Convention ("each State Party shall endeavour to") results in making these provisions "subsidiary to the obligation of creating national inventories"[117]. Since international law remains inherently state centric, it is unsurprising that the conventions give state parties control over what constitutes cultural heritage in their territory and, accordingly, what is worthy of protection under the conventions.[118]

This bears at least two risks: on the one hand, the heritage of minority groups or Indigenous peoples might not be perceived as reflecting national heritage or not deemed worthy of international (or national) protection. As a consequence, such heritage might not be sufficiently protected from potential harm, including by the state in which it is situated or originates from.[119] On the other hand, even if heritage is recognised and protected, it has been argued that the design and

---

[114] Vrdoljak and Meskell (2020), p. 38. Also see the case of the indigenous Karen in Thailand, Readfearn, Graham (2021): 'Low point' in world heritage committee history as politics 'tramples' human rights of the Karen people, https://www.theguardian.com/environment/2021/aug/06/low-point-in-world-heritage-committee-history-as-politics-tramples-human-rights-of-the-karen-people, accessed 31 January 2025.

[115] UNESCO 'Operational Guidelines for the Implementation of the World Heritage Convention' (31 July 2024) WHC.24/01, I.C. 12.

[116] *ibid* para. 64.

[117] Lixinski (2013a), p. 42.

[118] See *ibid* p. 52; Lixinski (2011), pp. 94, 100. For a concise juxtaposition of the prevailing opinion in the literature and the more critical view of this, see Lixinski (2016), p. 190.

[119] See Lixinski (2013a), p. 52.

implementation of the key UNESCO treaties impedes local communities "from having a say in how their heritage is managed".[120]

3.3.3 Interim Conclusions

Francioni and *Vrdoljak* note that a "cursory examination of the culture conventions lays bare that it is the State that defines what, whether, how, and when heritage will be protected".[121] A state's interest can come into conflict with how cultural heritage is protected and such conflicts are most likely to concern the heritage of minority groups or Indigenous peoples. The academic discourse has pointed out that local communities often are excluded from the identification of heritage, from decision-making processes concerning their heritage and in some cases from the heritage itself.[122]

The Intangible Cultural Heritage Convention signifies a shift in integrating the obligation to ensure participation of communities and groups, albeit couched in soft terms ("shall endeavour to ensure")[123]. The preambles of international cultural heritage law have undergone a shift in the impetus to protect cultural heritage too. Earlier instruments tend to exclusively focus on international aspects of heritage protection, including its outstanding value for humanity and the international community or the international interchange of cultural property. Whilst these ideas are still present in the Intangible Cultural Heritage Convention, it also recognises the role and relationship of communities, groups, and individuals to cultural heritage.

Beyond the conventions' preambles, the evolving objectives and the shift to the importance of protecting cultural heritage for individuals and groups and their cultural identities can also be seen in the application of the World Heritage Convention and the World Heritage Committee's Operational Guidelines for the Implementation of the World Heritage Convention. The case of the Uluru-Kata Tjuta National Park in Australia (formerly called Uluru (Ayers Rock—Mount Olga) National Park)[124], which was inscribed in 1987 as a natural site, serves as

---

[120] This is the key argument explored in: Lixinski (2019), p. 1. Lixinski further argues that excluding and alienating communities from their heritage does not only affect decision-making process but contributes to the disappearance of heritage in the long term (p. 14).

[121] Francioni and Vrdoljak (2020), p. 5.

[122] Lixinski (2019).

[123] Intangible Cultural Heritage Convention, Art. 15.

[124] UNESCO (2025a, 2025b): Uluru-Kata Tjuta National Park, https://whc.unesco.org/en/list/447/, accessed 31 January 2025.

one example.[125] After its original inscription as natural heritage, the protection of Uluru was extended by recognising cultural criteria in 1994. This extension acknowledged "successful human adaption over many millennia", the "present management system based on the practices of its traditional owners", and, most importantly, the site forming "an integral part of the traditional belief system of one of the oldest human societies in the world."[126] The evaluation, moreover, referred to the religious, artistic, and cultural qualities of the park.[127] This can be interpreted as marking

> "an important change in World Heritage policies, reflected [...] in the evaluation of natural properties associated with the significance of sacred sites, and thus indirectly also with the rights of members of local communities."[128]

Extending the inscription to include cultural elements reflects an evolving understanding of the site. However, it also reveals the expert group's oversight in the representation and creation of the Uluru-Kata Tjuta Park as world heritage abroad.[129] The initial omission is significant considering the disenfranchisement of the Anangu community from using the sacred site, long reflected by the inability to stop tourists from continuing to climb it.[130] The case illustrates the importance of including local communities in the identification, management and protection process of heritage.

---

[125] See Advisory Body Evaluation—IUCN, 1987 in UNESCO (2025a, 2025b): Uluru-Kata Tjuta National Park: Documents, whc.unesco.org/en/list/447/documents/, accessed 31 January 2025.

[126] See Advisory Body Evaluation (ICOMOS), 1994, para. 105 in *ibid.*

[127] Advisory Body Evaluation (ICOMOS), 1994, para. 99 in *ibid.*

[128] Jokilehto (2012), pp. 228–29.

[129] Lixinski (2013b), p. 149.

[130] Francioni and Lixinski (2017), p. 21.

The climb to the top of Uluru was closed on 26 October 2019. Parks Australia (Commonwealth of Australia) (2019): Uluṟu climb closure, https://parksaustralia.gov.au/uluru/discover/culture/uluru-climb/, accessed 31 January 2025.

3.4 The Human Dimension of Enforcing Cultural Heritage Protection

The third and final section focuses on the human dimension in the enforcement of cultural heritage protection. For these purposes, the chapter attends to international criminal law, specifically cases of the International Criminal Court (ICC) and the International Criminal Tribunal for the Former Yugoslavia (ICTY). The section shows how the ICC and the ICTY have addressed heritage destruction. This includes their emphasis on communities as the main target of heritage destruction and their understanding that heritage destruction could be considered as evidence of discriminatory and genocidal intent.

3.4.1 Human Communities as the Target of Cultural Heritage Destruction

The evolving jurisprudence stemming from international criminal law, as enforcing cultural heritage protection, has contributed to international cultural heritage law. It is important to recognise the discriminatory and persecutory intent behind the destruction of cultural heritage and to understand that the destruction of cultural heritage itself is often not the goal.[131] Instead, the goal is the annihilation of communities for which the cultural heritage "is perceived as an essential element of their own life, cultural identity, and distinctiveness."[132] This makes the target "not the heritage in itself but the human communities for which such a heritage is of special significance."[133]

The close connection to human communities renders human rights implications for the (intentional) destruction of cultural heritage inevitable.[134] Considering the seriousness of the offence of directing an attack against cultural heritage, the ICTY Trial Chamber in *Strugar* understood the victim of such an offence to be "a 'people', rather than any particular individual" and noted the grave consequences for the victim.[135] In *The Prosecutor v. Al Mahdi,* the ICC had the

---

[131] Lenzerini (2020), pp. 76, 82.

[132] *Ibid* p. 76.

[133] *ibid* p. 77.

[134] *ibid* p. 86. Also see, "Today, international law provides for the criminalization of serious offences against cultural heritage. This is another manifestation of the increasingly important connection between cultural heritage and human rights." Francioni (2008), p. 11.

[135] *Prosecutor v. Pavle Strugar* (Trial Judgment) IT-01–42 (31 Jan 2005), para. 232.

opportunity to address the destruction of cultural heritage as the principal charge of a war crime in a non-international armed conflict.[136] The charge against Al Mahdi concerned the war crime of intentionally attacking protected buildings dedicated to religion and historic monuments in Timbuktu by armed forces under his control. The ICC acknowledged the local population as the "direct victims of the crimes".[137] Similarly, the ICC's Reparations Order emphasised the community of Timbuktu as the main victim.[138] In addition, it recognised that "people throughout Mali and the international community" were also victims of the destruction.[139]

Noting the religious motives to attack the sites, namely *Ansar Dine's* intent to stop the religious practices of the community of Timbuktu,[140] the ICC pointed to the role of the mausoleums in the cultural life of the community.[141] The ICC's Reparations Order further referred to the international community's recognition of the "importance of the human right to cultural life and its physical embodiments."[142] The mausoleums were emblematic of the locally practiced Islam in Timbuktu.[143] The destruction of the mausoleums and manuscripts, the physical sites and objects, is therefore linked to the attacks on cultural and religious practices[144] and has been regarded to have "gravely harmed" the intangible heritage of the local population.[145] In addition to seeing the attacks on cultural and religious practices as attacks against intangible cultural heritage,[146] they can also be

---

136 Although the ICC's recognition of the discriminatory intent has led to surprise that crimes against humanity were not considered in the case. See, Lenzerini (2020), p. 84.

137 *Prosecutor v. Ahmad Al Faqi Al Mahdi* (Judgment and Sentence) ICC-01/12–01/15 (27 September 2016), para. 80.

138 *Prosecutor v. Ahmad Al Faqi Al Mahdi* (Reparations Order) ICC-01/12–01/15 (17 August 2017), para. 56.

139 *Prosecutor v. Ahmad Al Faqi Al Mahdi* (Judgment and Sentence) ICC-01/12–01/15 (27 September 2016), para. 80.

140 *ibid* para. 81.

141 *ibid* para. 46.

142 *Prosecutor v. Ahmad Al Faqi Al Mahdi* (Reparations Order) ICC-01/12–01/15 (17 August 2017), para. 14.

143 ICC (2017): 'Brief by Ms. Karima Bennoune: Expert Appointed by the ICC in the Case of The Prosecutor v. Ahmad Al Faqi Al Mahdi, ICC-01/12–01/15, Reparations Phase', p. 19.

144 UNHRC, Report of the Special Rapporteur in the field of cultural rights (3 February 2016) UN Doc A/HRC/31/59, para. 77.

145 UNHRC, Intersessional seminar on cultural rights and the protection of cultural heritage, Report of the United Nations High Commissioner for Human Rights (27 December 2017) UN Doc A/HRC/37/29, para. 27.

146 See "…carried out destructions of tangible cultural heritage while also attacking intangible cultural heritage, including local music, religious practice and forms of dress". ICC

considered attacks targeting the exercise of human rights, specifically the right to take part in cultural life and the freedom of thought, religion and conscience. Beyond the religious component of the sites, the mausoleums were also frequently visited by the non-Muslim population as inter alia part of Friday blessings, making them an integral part of their cultural life.[147] These decisions make clear that the victim of such destruction are "humans whose rights have been violated"[148].

3.4.2 Cultural Heritage Destruction as Evidence of Discriminatory and Genocidal Intent

Prior to the *Al Mahdi* case at the ICC, judicial pronouncements of primarily the ICTY played a key role in developing and advancing cultural heritage protection.[149] The interpretation of war crimes, crimes against humanity, and genocide by the ICTY has resulted in a widening of the protection afforded to cultural heritage.[150] This has been accompanied by a growing acknowledgement that although destruction of cultural heritage frequently occurs "under the cloak of armed conflict, it is neither confined to nor necessarily related to the hostilities."[151] The elevated nature of crimes against cultural heritage was, for example, subject of the *Jokić* and *Strugar* judgments of the ICTY.[152] The Trial Chamber argued that "it is a crime of *even greater seriousness* to direct an attack on an especially protected site", comparing it to the attack on a civilian building.[153]

(2017): 'Brief by Ms. Karima Bennoune: Expert Appointed by the ICC in the Case of The Prosecutor v. Ahmad Al Faqi Al Mahdi, ICC-01/12–01/15, Reparations Phase', p. 19.

[147] *ibid* p. 20.

[148] Dijkstal (2019), p. 397.

[149] For a comprehensive analysis of international tribunals and courts' interpretation of the crime of genocide in the context of cultural heritage, see specifically Novic (2016), pp. 50–95 (Chap. 3 'The Cultural Dimension of the Crime of Genocide').

[150] Vrdoljak (2007), p. 9.

[151] *ibid.*

[152] Also see on cultural heritage in the *Strugar* case: Kalshoven (2005), p. 65.

[153] Employing identical wording in both: *Prosecutor v. Miodrag Jokić* (Sentencing Judgement) IT-01–42/1-S (18 March 2004), para. 53. *Prosecutor v. Pavle Strugar* (Trial Judgment) IT-01–42 (31 Jan 2005), para. 232. Emphasis added by the author.

Similarly, the Trial Chamber in *Jokić* confirmed the immunity of cultural property in international humanitarian law from direct attacks to be "clearly additional to the protection attached to civilian objects."[154]

In *Kupreškić*, the Trial Chamber explained that the core elements of crimes against humanity comprise firstly the existence of an armed conflict, secondly the occurrence of widespread or systematic crimes targeting the civilian population, and thirdly the knowledge of the perpetrator of the context of their crime.[155] Whilst addressing the deliberate destruction of ancient mosques, the Trial Chamber in *Kordic & Cerkez* found that:

> "when perpetrated with the requisite discriminatory intent, [the destruction] amounts to an attack on the very religious identity of the people. As such it manifests a nearly pure expression of the notion of 'crimes against humanity' for all of humanity is indeed injured by the destruction of a unique religious culture and its concomitant cultural objects."[156]

Kordic & Cerkez linked "crimes against humanity" to the destruction of cultural and religious heritage. The *Blaškić* case concerned persecution taking the form of confiscation or destruction of "symbolic buildings [...] belonging to the Muslim population of Bosnia-Herzegovina" by Bosnian Croat forces.[157] The Trial Chamber further noted that

> "[t]he methods of attack and the scale of the crimes committed against the Muslim population or the edifices symbolising their culture sufficed to establish beyond reasonable doubt that the attack was aimed at the Muslim civilian population."[158]

In spite of culture being at the heart of *Lemkin's* concept of genocide,[159] in 1948, the UN General Assembly rejected the notion of "cultural genocide".[160] *Novic* points out that the academic discourse on the concept of cultural genocide

---

[154] *Prosecutor v. Miodrag Jokić* (Sentencing Judgement) IT-01–42/1-S (18 March 2004), para. 50.

[155] *Prosecutor v. Kupreškić et al.* (Trial Judgement) IT- -95–16 (14 January 2000), para. 544.

[156] *Prosecutor v. Kordic & Cerkez* (Judgement) IT-95–14/2 (26 February 2011), para. 207.

[157] *Prosecutor v. Blaškić* (Trial Judgement) IT-95–14 (3 March 2000), para. 227.

[158] *ibid* para. 425.

[159] Lemkin (2008), pp. 84–85; also see: Francioni and Lixinski (2017), p. 12.

[160] UNGA Sixth Committee 'Genocide: Draft Convention and Report of the Economic and Social Council' (3 December 1948) UN Doc A/760, para. 11.

has persisted.[161] She explains that whilst *Lemkin* conceived of cultural genocide "as a technique of implementation", the contemporary understanding revolves around "a process of its own, which would result in the extinction of some ethnic groups' culture, either through its targeted destruction or its forced assimilation into the dominant culture."[162] During the negotiations of the UN Declaration on the Rights of Indigenous Peoples (UNDRIP) in 2007, the concept of "cultural genocide", which appeared in in draft article 7, was again rejected in the final declaration.[163]

In the *Krstić* case, the ICTY had the opportunity to consider the destruction of cultural heritage in the context of the atrocities related to the fall of Srebrenica. Considering the manner "in which the destruction of a group may be implemented so as to qualify as a genocide", the Trial Chamber noted that "one may also conceive of destroying a group through purposeful eradication of its culture and identity resulting in the eventual extinction".[164] Recognising that international customary law limits genocide to acts seeking biological or physical destruction, the Trial Chamber conceded that simultaneous attacks on "cultural or religious property and symbols of the targeted group" accompanying physical or biological destruction, could be evidence of the intent.[165]

3.4.3 Interim Conclusions

The case law of the ICC and the ICTY have underlined at least two important aspects of the human dimension of cultural heritage. First, they have highlighted the impact of cultural heritage destruction on people and have identified a people or community as the main victims of attacks. Furthermore, they have shown that beyond war crimes the destruction of cultural heritage can play into determining evidence of the intent of crimes against humanity and genocide.

The findings of the *Krstić* case were repeated and endorsed in the ICJ's case concerning the *Application of the Convention on the Prevention and Punishment of*

[161] Novic (2016), p. 4.

[162] *ibid.*

[163] UNCHR 'Report of the Sub-Commission on Prevention of Discrimination and Protection of Minorities on its Forty-Sixth Session' (28 October 1994) UN Doc E/CN.4/Sub.2/1994/56, para. 107. Also see Novic (2016), pp. 9–10.

[164] *Prosecutor v. Krstić* (Trial Judgement) IT-98–33 (2 August 2001), para. 574.

[165] *ibid* para. 580.

the Crime of Genocide (Bosnia and Herzegovina v. Serbia and Montenegro).[166] The ICJ confirmed the destruction of cultural, religious, and historical heritage not to constitute "the deliberate infliction of conditions of life calculated to bring about the physical destruction of the group"[167] but endorsed the finding "that 'where there is physical or biological destruction there are often simultaneous attacks on the cultural and religious property [...] which may legitimately be considered as evidence of an intent to physically destroy the group'".[168]

3.5 Conclusions: Turning to Human Rights Law

International cultural heritage law is undergoing changes by

> "moving away from physical remnants and towards living cultures; from States toward communities of practice; from preservation to safeguarding; and from Western-centric appreciations of "Outstanding Universal Value" towards culturally-sensitive engagements which represent all peoples."[169]

In other words, it is shifting from preserving the past to "safeguarding heritage (of the present)".[170] This has also been observed by *Donders*, who argues that

> "a shift appears to be visible from the protection of cultural heritage as such, based on its historical value for humanity, its uniqueness and non-renewable character, to the protection of cultural heritage as a crucial value for peoples and individuals in relation to the (re)construction of their cultural identity."[171]

This chapter has traced the evolving meaning of cultural heritage; the evolving rationale to protect cultural heritage and the role of communities, groups, and

166 *Application of the Convention on the Prevention and Punishment of the Crime of Genocide (Bosnia and Herzegovina v Serbia and Montenegro) [2007] ICJ Rep 43*, para. 344.

167 *ibid.*

168 *ibid*, 344. Also see *Application of the Convention on the Prevention and Punishment of the Crime of Genocide (Croatia v Serbia) [2015] ICJ Rep 3*, para. 390.

169 Lixinski and Schreiber (2017), p. 18.

170 *ibid* p. 19.

171 Donders (2020), p. 381. Also see, Jakubowski (2016), p. 161. "The legal notion of 'cultural heritage' has evolved into a more human-oriented, holistic concept, comprising both tangible as well as intangible manifestations of hu- man intellectual and spiritual achievements."

individuals in protection mechanisms; and finally, the enforcement of the protection of cultural heritage in international criminal law. These three broad examples serve to illustrate that the relationship and the role and rights of communities, groups and, in some cases, individuals to cultural heritage have been increasingly considered. The adoption of the Intangible Cultural Heritage Convention was pivotal in underlining cultural heritage as living heritage and a vehicle for the cultural identity of groups and individuals. In addition, it has given "communities, groups and, in some cases, individuals" an active role in the identification of intangible cultural heritage by including them in its definition.[172] The integration of "communities, groups and, in some cases, individuals" in the definition of intangible cultural heritage can be interpreted as challenging the preponderant state centric view of cultural heritage protection in international law.[173]

The shift to the role and rights of communities, groups, and individuals is also visible in the rationales for protecting cultural heritage. Although the preambles continuously made clear that cultural heritage is protected *for the people,* there was a greater focus on the international community or the notion of humanity. The protection of cultural identity in the Intangible Cultural Heritage Convention (and in the preamble of the 2003 UNESCO Declaration) has shifted the focus to the community at the local level and to Indigenous communities. The understanding of the relationship between people(s) and cultural heritage is not confined to the law but also evident in world heritage policies.

International cultural heritage law increasingly makes use of a language of "rights" through, for example, participation without creating rights.[174] The Intangible Cultural Heritage Convention has included the participation of communities, groups, and relevant non-governmental organisations in its notion of safeguarding. Yet, effective community participation under the Convention has remained weak in practice.[175] Although the World Heritage Convention does not contain a similar provision, the Operational Guidelines now "encourages" participation of inter alia local communities and Indigenous peoples in the process of identifying, nominating, managing and protecting World Heritage properties.[176]

---

[172] Intangible Cultural Heritage Convention, Art. 2 (1). The relevant sentence reads "The "intangible cultural heritage" means the practices, representations, expressions, knowledge, skills—as well as the instruments, objects, artefacts and cultural spaces associated therewith—that communities, groups and, in some cases, individuals recognize as part of their cultural heritage."

[173] Lixinski (2013a), pp. 2–3.

[174] Donders (2020), p. 388.

[175] Blake (2020a), p. 356.

[176] Operational Guidelines, I.C. 12.

In enforcing the protection of cultural heritage, the jurisprudence of the ICTY and ICC has significantly contributed to the understanding of cultural heritage for groups and individuals. Case law has demonstrated the discriminatory and persecutory element of the destruction of cultural heritage and has shown that such destruction can be linked to the suppression of religious freedom, religious practices, and participation in cultural life. Especially the focus on victims of the destruction of cultural heritage has highlighted the local community or a people as the direct victims.

The discourse surrounding international cultural heritage law has been critical towards the lack of enforcement.[177] This criticism includes the language of the conventions, including the ensuing lack of "legal bite".[178] Specifically the Intangible Cultural Heritage Convention and the Convention on the Protection and Promotion of the Diversity of Cultural Expressions have been described provokingly as resembling declarations of principles, instead of real treaties.[179] Furthermore, the conventions have been criticised for falling short of suppressing and remedying offences and not contributing significantly to the prevention of such offences.[180] Similarly, the Intangible Cultural Heritage Convention has been described as resembling more of a proclamation of states' rights than as imposing any state obligations.[181] Nevertheless, it is important to concede that the use of lists can play a role in the prevention of offences.[182]

The "interstate focus" of the UNESCO treaties concerning cultural heritage means that they are traditional treaties with a horizontal character.[183] The subsequent analysis explores the international human rights treaties for their vertical character in providing rights for individuals and groups. The absence of formal participation mechanisms in international cultural heritage law raises the question if individuals and members of groups could rely on human rights mechanisms to enforce inter alia their participation in the identification and management of cultural heritage or access to cultural heritage. Beyond the specific example of participation, the increasing recognition of the role and rights of groups and individuals warrants the question to what extent rights vis-à-vis cultural heritage

[177] Novic (2020), p. 642.

[178] Lixinski (2011), p. 81; Nafziger (2021), p. 386.

[179] Lenzerini (2013), p. 254; Also see Francioni and Lenzerini (2023), p. 426. Describing provisions of the World Heritage Convention as too soft.

[180] Lenzerini (2013), p. 254.

[181] Lenzerini (2015), p. 144.

[182] Lenzerini (2013), p. 254.

[183] Donders (2020), p. 384.

have emerged in the interpretation of human rights law. It is therefore important to closely analyse the protection of the rights of individuals and groups relating to cultural heritage in the international human rights framework. The international human rights framework could thereby complement the protection of cultural heritage that international cultural heritage law provides by extending it to the rights of communities, groups and individuals vis-à-vis their heritage offering a more holistic protection of cultural heritage overall.

Reference List

Blake, Janet (2020b): The Preamble. In: Janet Blake and Lucas Lixinski (eds.): The 2003 UNESCO Intangible Heritage Convention: A Commentary. Oxford (Oxford University Press), pp. 19–35.

Blake, Janet (2020a): Safeguarding Intangible Cultural Heritage. In: Francesco Francioni and Ana Filipa Vrdoljak (eds.): The Oxford Handbook of International Cultural Heritage Law. Place of Publication (Oxford University Press), pp. 347–370.

Blake, Janet (2015): International Cultural Heritage Law. Oxford (Oxford University Press).

Blake, Janet (2011): Taking a Human Rights Approach to Cultural Heritage Protection. In: Heritage & Society, Vol. 4 No. 2, pp. 199–238.

Blake, Janet (2006): Commentary on the UNESCO 2003 Convention on the Safeguarding of the Intangible Cultural Heritage. Leicester (Institute of Art and Law).

Clapperton, Matthew, Jones, David Martin and Smith, M. L. R. (2017): Iconoclasm and Strategic Thought: Islamic State and Cultural Heritage in Iraq and Syria. In: International Affairs, Vol. 93, No. 5, pp. 1205–1231.

Dijkstal, Haydee J (2019): Destruction of Cultural Heritage before the ICC: The Influence of Human Rights on Reparations Proceedings for Victims and the Accused. In: Journal of International Criminal Justice, Vol. 17, Issue 2, pp. 391–412.

Disko, Stefan (2017): Indigenous Cultural Heritage in the Implementation of UNESCO's World Heritage Convention: Opportunities, Obstacles and Challenges. In: Indigenous Peoples' Cultural Heritage: Rights, Debates and Challenges. Leiden (Brill | Nijhoff), pp. 39–77.

Donders, Yvonne (2020): Cultural Heritage and Human Rights. In: Francesco Francioni and Ana Filipa Vrdoljak (eds.): The Oxford Handbook of International Cultural Heritage Law. Oxford (Oxford University Press), pp. 379–406.

Donders, Yvonne (2018): Protection and Promotion of Cultural Heritage and Human Rights through International Treaties: Two Worlds of Difference. In: Charlotte Waelde, Catherine Cummings, Mathilde Pavis and Helena Enright (eds.): Research Handbook on Contemporary Intangible Cultural Heritage. Cheltenham/Northampton (Edward Elgar), pp. 54–77.

Dorough, Dalee Sambo and Wiessner, Siegfried (2020): Indigenous Peoples and Cultural Heritage. In: Francesco Francioni and Ana Filipa Vrdoljak (eds.): The Oxford Handbook of International Cultural Heritage Law. Oxford (Oxford University Press), pp. 407–430.

Fletcher (2012): Timbuktu tomb destroyers pulverise Islam's history, https://www.reuters.com/article/business/timbuktu-tomb-destroyers-pulverise-islams-history-idUSLNE86202H/, 31 January 2025.

Francioni, Francesco (2023): The Preamble. In: Francesco Francioni (ed.): The 1972 World Heritage Convention—A Commentary. 2nd edition, Oxford (Oxford University Press), pp. 21–29.

Francioni, Francesco and Lenzerini, Federico (2023): The Future of the World Heritage Convention: Problems and Prospects. In: Francesco Francioni (ed.): The 1972 World Heritage Convention: A Commentary. 2nd edition, Oxford (Oxford University Press), pp. 421–428.

Francioni, Francesco (2020c): World Cultural Heritage. In: Francesco Francioni and Ana Filipa Vrdoljak (eds.): The Oxford Handbook of International Cultural Heritage Law. Oxford (Oxford University Press), pp. 250–271.

Francioni, Francesco and Lixinski, Lucas (2017): Opening the Toolbox of International Human Rights Law in the Safeguarding of Cultural Heritage. In: Andrea Durbach and Lucas Lixinski (eds.): Heritage, Culture and Rights: Challenging Legal Discourses. Oxford (Hart Publishing.

Francioni, Francesco (2013a): Cultural Heritage. In: Max Planck Encyclopaedia of International Law.

Francioni, Francesco (2013b): Plurality and Interaction of Legal Orders in the Enforcement of Cultural Heritage Law. In: Francesco Francioni and James Gordley (eds.): Enforcing International Cultural Heritage Law. Oxford (Oxford University Press), pp. 9–21.

Francioni, Francesco (2012): The Evolving Framework for the Protection of Cultural Heritage in International Law. In: Silvia Borelli and Federico Lenzerini (eds.): In Cultural Heritage, Cultural Rights, Cultural Diversity. Leiden (Brill | Nijhoff).

Francioni, Francesco (2011): The Human Dimension of International Cultural Heritage Law: An Introduction. In: European Journal of International Law, Vol. 22, Issue 1, pp. 9–16.

Francioni, Francesco (2008): Culture, Heritage and Human Rights: An Introduction. In: Francesco Francioni and Martin Scheinin (eds.): Cultural Human Rights. Leiden (Brill / Nijhoff, pp. 1–15.

Francioni, Francesco (2004): Beyond State Sovereignty: The Protection of Cultural Heritage as a Shared Interest in Humanity. In: Michigan Journal of International Law Vol. 25, No. 1, pp. 1209–1228.

Francioni, Francesco and Lenzerini, Federico (2003): The Destruction of the Buddhas of Bamiyan in International Law. In: European Journal of International Law, Vol. 14, No. 4, pp. 619–651.

Francioni, Francesco (1993): World Cultural Heritage List and National Sovereignty. In: Humanitäres Völkerrecht—Informationsschriften, Vol. 4, Issue 4, pp. 195–198.

Forrest, Craig (2010): International Law and the Protection of Cultural Heritage. Abingdon/New York, NY (Routledge).

Hulme, Max H. (2016): Preambles in Treaty Interpretation. In: University of Pennsylvania Law Review, Vol. 164, Issue 5, pp. 1281–1343.

Jakubowski, Andrzej (2016): Cultural Heritage and the Collective Dimension of Cultural Rights in the Jurisprudence of the European Court of Human Rights. In: Jakubowski, Andrzej (ed.): Cultural Rights as Collective Rights. Leiden (Brill | Nijhoff), pp. 155–179.

Jokilehto, Jukka (2012): Human Rights and Cultural Heritage. Observations on the Recognition of Human Rights in the International Doctrine. In: International Journal of Heritage Studies, Vol. 18, No. 3, pp. 1–5.

Kalshoven, Frits (2005): The Protection of Cultural Property in the Event of Armed Conflict within the Framework of International Humanitarian Law. In: Museum International, Vol. 57, No. 4, pp. 61–70.

Kraak, Anne-Laura and Aykan, Bahar (2018): Editorial: The Possibilities and Limitations of Rights-Based Approaches to Heritage Practice. In: International Journal of Cultural Property, Vol. 25, No. 1, pp. 1–10.

Lemkin, Raphael (2008): Axis Rule in Occupied Europe: Laws of Occupation, Analysis of Government, Proposals for Redress. 2nd edition (first published 1944), Clark, NJ (Lawbook Exchange).

Lenzerini, Federico (2020): Intentional Destruction of Cultural Heritage. In: Francesco Francioni and Ana Filipa Vrdoljak (eds.) The Oxford Handbook of International Cultural Heritage Law. Oxford (Oxford University Press), pp. 75–99.

Lenzerini, Federico (2016a): Terrorism, Conflicts and the Responsibility to Protect Cultural Heritage. In: The International Spectator: Special Issue on the Responsibility to Protect, Vol. 51, No. 2, pp. 70–85.

Lenzerini, Federico (2015); Protecting the Tangible, Safeguarding the Intangible: A Same Conventional Model for Different Needs. In: Sabine von Schorlemer and Sylvia Maus (eds.): Climate Change as a Threat to Peace: Impacts on Cultural Heritage and Cultural Diversity. Frankfurt am Main (Peter Lang Group AG), pp. 141–160.

Lenzerini, Federico (2013): Suppressing and Remedying Offences against Culture. In: Ana Filipa Vrdoljak (ed.): The Cultural Dimension of Human Rights. Oxford (Oxford University Press), pp. 240–272.

Lenzerini, Federico (2003): The UNESCO Declaration Concerning the Intentional Destruction of Cultural Heritage: One Step Forward and Two Steps Back. In: The Italian Yearbook of International Law, Vol. 3, Issue 1, pp. 131–145.

Lixinski, Lucas (2019): International Heritage Law for Communities: Exclusion and Re-Imagination. Oxford (Oxford University Press).

Lixinski, Lucas and Schreiber, Hanna (2017): Guest Editorial, Intangible Cultural Heritage: Successes, Disappointments, and Challenges. In: Santander Art and Culture Law Review, Vol. 2, No. 4, pp. 17–20.

Lixinski, Lucas (2016): Intangible Cultural Heritage. In: Anna Sledzinska-Simon, Kalliopi Chainoglou and Andreas J Wiesand (eds.): Culture and Human Rights: The Wroclaw Commentaries. Berlin/Boston (De Gruyter), pp. 189–190.

Lixinski, Lucas (2013a): Intangible Cultural Heritage in International Law. Oxford (Oxford University Press).

Lixinski, Lucas (2013b): Luci E Ombre: The Bright and Dark Sides of International Heritage Law. In: The Italian Yearbook of International Law Online, Vol. 22, Issue 1, pp. 133–153.

Linxinski, Lucas (2011): Selecting Heritage: The Interplay of Art, Politics and Identity. In: European Journal of International Law, Vol. 22, No. 1, pp. 81–100.

Merryman, John Henry (2005): Cultural Property Internationalism. In: International Journal of Cultural Property, Vol. 12, Issue 1, pp. 11–39.

Merryman, John Henry (1986): Two Ways of Thinking About Cultural Property. In: The American Journal of International Law, Vol. 80, Issue 4, pp. 831–853.

Nafziger, James A. R. (2021): Reading: The World Court, the World Heritage, and the World of Non-state Actors. In: James A. R. Nafziger: Frontiers of Cultural Heritage Law. Leiden (Brill | Nijhoff), pp. 378–400.
Novic, Elisa (2020): Remedies. In: Francesco Francioni and Ana Filipa Vrdoljak (eds.): The Oxford Handbook of International Cultural Heritage Law. Oxford (Oxford University Press), pp. 642–662.
Novic, Elisa (2016): The Concept of Cultural Genocide. Oxford (Oxford University Press).
O'Flaherty, Michael (2006): Concluding Observations of United Nations Human Rights Treaty Bodies. In: Human Rights Law Review, Vol. 6, No. 1, pp. 27–52.
O'Keefe, Roger (2007): The Protection of Cultural Property in Armed Conflict. Cambridge (Cambridge University Press).
O'Keefe, R. (2004). World Cultural Heritage: Obligations to The International Community as A Whole? *International and Comparative Law Quarterly, 53*, 189–209.
O'Keefe, Roger (1999): The Meaning of 'Cultural Property' Under the 1954 Hague Convention. In: Netherlands International Law Review, Vol. 46, No. 1, pp. 26–56.
Parks Australia (Commonwealth of Australia) (2019): Uluṟu climb closure, https://parksaustralia.gov.au/uluru/discover/culture/uluru-climb/, accessed 31 January 2025.
Readfearn, Graham (2021): 'Low point' in world heritage committee history as politics 'tramples' human rights of the Karen people, https://www.theguardian.com/environment/2021/aug/06/low-point-in-world-heritage-committee-history-as-politics-tramples-human-rights-of-the-karen-people, accessed 31 January 2025.
Scovazzi, Tullio (2020): International Cultural Heritage Law: The Institutional Aspects. In: Francesco Francioni and Ana Filipa Vrdoljak (eds.): The Oxford Handbook of International Cultural Heritage Law, Oxford (Oxford University Press), pp. 737–759.
Silverman, Helaine and Fairchild Ruggles, D. (2007): Cultural Heritage and Human Rights. In. Helaine Silverman and D. Fairchild Ruggles (eds.): Cultural Heritage and Human Rights, New York, NY (Springer), pp. 3–29.
Uerpmann-Wittzack, Robert (2018): Introduction: Cultural Heritage Law and the Quest for Human Identities. In: Evelyne Lagrange, Stefan Oeter and Robert Uerpmann-Wittzack (eds.): Cultural Heritage and International Law. Cham (Springer), pp. 1–11.
UNESCO (2025): Uluru-Kata Tjuta National Park, https://whc.unesco.org/en/list/447/, accessed 31 January 2025.
UNESCO (2025): World Heritage List: Mixed Properties, https://whc.unesco.org/en/list/?search=&type=mixed&components=0, accessed 31 January 2025.
Vrdoljak, Ana Filipa and Francioni, Francesco (2020): Introduction. In: Ana Filipa Vrdoljak and Francesco Francioni (eds.): The Oxford Handbook of International Cultural Heritage Law. Oxford (Oxford University Press), pp. 1–10.
Vrdoljak, Ana Filipa and Meskell, Lynn (2020): Intellectual Cooperation Organisation, UNESCO, and the Culture Conventions. In: Francesco Francioni and Ana Filipa Vrdoljak (eds.): The Oxford Handbook of International Cultural Heritage Law, Oxford (Oxford University Press), pp. 13–40.
Vrdoljak, Ana Filipa (2014): Human Rights and Cultural Heritage in International Law. In: Federico Lenzerini and Ana Filipa Vrdoljak (eds.): International Law for Common Goods: Normative Perspectives on Human Rights, Culture and Nature. Oxford (Hart Publishing), Available at SSRN: https://ssrn.com/abstract=2200186, pp. 1–39.

Vrdoljak, Ana Filipa (2013b): Liberty, Equality, Diversity: States, Cultures, and International Law. In: Ana Filipa Vrdoljak (ed.) The Cultural Dimension of Human Rights. Oxford (Oxford University Press), pp. 26–70.
Vrdoljak, Ana Filipa (2013a): Introduction. In: Ana Filipa Vrdoljak (ed.): The Cultural Dimension of Human Rights. Oxford (Oxford University Press), pp. 1–10.
Vrdoljak, Ana Filipa (2007): Intentional Destruction of Cultural Heritage and International Law. In: Kalliopi Koufa (ed.): Multiculturalism and International Law, XXXV Thesaurus Acroasium. Thessaloniki (Sakkoulas Publications), pp. 377–396.
Xanthaki, Alexandra (2017): International Instruments on Cultural Heritage: Tales of Fragmentation. In: Alexandra Xanthaki, Sanna Valkonen, Leena Heinämäki and Piia Kristiina Nuorgam (eds.): Indigenous Peoples' Cultural Heritage: Rights, Debates, Challenges. Leiden (Brill | Nijhoff.), pp. 1–19.

Cultural Heritage through the Lens of the Human Rights Committee and the Committee on Economic, Social and Cultural Rights

4

4.1 Introduction

This chapter reviews the pronouncements of two human rights treaty bodies: the Human Rights Committee (HRC) and the Committee on Economic, Social and Cultural Rights (CESCR). It considers their pronouncements, which chiefly include their general comments, concluding observations, and views. The analysis pursues three goals: identifying (1) the meaning of cultural heritage in these pronouncements, (2) the rights asserted vis-à-vis cultural heritage, and (3) the corresponding state obligations that serve these rights and, more generally, cultural heritage. The chapter seeks to attain a better understanding of how the committees address cultural heritage, which legal bases they rely on, and which rights and obligations, if any, can be inferred from the pronouncements of the treaty bodies.

Cultural rights serve as a starting point for the chapter. Whilst there is a paucity of an international legal definition of "cultural rights", they can be understood as protecting the rights of individuals and groups of people "to develop and express their humanity, their world view and the meanings they give to their existence and their development through, inter alia, values, beliefs, convictions, languages, knowledge and the arts, institutions and ways of life".[1]

[1] OHCHR (2025b): International standards: Special Rapporteur in the field of cultural rights, https://www.ohchr.org/en/special-procedures/sr-cultural-rights/international-standards, accessed 31 January 2025.

L. Hofmann, *Cultural Heritage through the Lens of Human Rights Bodies*, Juridicum – Bochumer Schriften zur Friedenssicherung und zum Humanitären Völkerrecht | Bochum Studies on the International Law of Peace and Armed Conflict 73, https://doi.org/10.1007/978-3-658-51203-3_4

Furthermore, cultural rights can be categorised by distinguishing between rights that explicitly refer to culture and those that have a link to culture.[2] Whereas the former category includes the right to take part in cultural life and the rights of persons belonging to minorities (Articles 15 (1) (a) of the ICESCR and Article 27 of the ICCPR), the most apparent examples of the latter category are the right to self-determination (Article 1 of both the ICCPR and ICESCR), the right to freedom of thought, conscience and religion (Article 18 of the ICCPR) or the right to education (Articles 13 and 14 of the ICESCR).[3] Other human rights can also have a cultural dimension, such as the right to adequate food and the cultural component of the preparation of food or of a traditional cuisine (as part of Article 11 of the ICESCR).[4]

As several cultural rights are not only exercised individually but also in groups or communities, it is helpful to distinguish between those rights. The rights holders of community rights or groups rights are "a collective entity", including "a people" or "a community".[5] Articles 1 of the ICCPR and ICESCR serve as examples of such rights. In contrast, the rights of "persons belonging to [ethnic, religious or linguistic] minorities [...] to enjoy their own culture, to profess and practise their own religion, or to use their own language" emanating from Article 27 of the ICCPR are referred to as communal rights. The rights-holder of communal rights "is an individual recognized as a member of a collective entity, and in which this membership is often explicitly referred to."[6] Finally, individual rights can have a collective dimension, such as the rights to take part in cultural life or the right to education in the ICESCR.[7]

The analysis is structured into three parts: meaning, rights, and obligations. In the absence of the terms "cultural heritage" or "cultural property" in the ICESCR and the ICCPR, the first section shows that cultural heritage forms part of Article 27 of the ICCPR and Article 15 of the ICESCR. The following section addresses the interpretation of human rights provisions. Which provisions can be used as a legal basis for claims concerning cultural heritage? Which rights do the committees assert vis-à-vis cultural heritage? The analysis identifies three rights: (1) access to cultural heritage; (2) conserving, promoting, and developing one's culture, including cultural heritage; and (3) effective consultations with

---

[2] See Donders (2016), p. 91.

[3] *ibid.*

[4] *ibid.*

[5] Art. 1 ICCPR and ICESCR present examples of community rights. Donders (2016), p. 89.

[6] Donders (2016), p. 89.

[7] *ibid* p. 90.

a view to obtaining the free, prior and informed consent and participation in decision-making affecting Indigenous and local communities' rights. The subsequent section highlights the corresponding obligations vis-à-vis heritage and identifies five obligations: the two underlying obligations to (1) promote and protect cultural diversity and (2) respect and protect cultural heritage at all times; as well as the obligations to (3) guarantee access to cultural heritage; (4) respect, protect, and recognise the heritage of persons belonging to minorities and Indigenous communities and create favourable conditions; and (5) ensure effective participation in decision-making concerning cultural heritage. The analysis highlights the legal basis of these rights and obligations, the rights holders of these rights, and illustrates their scope by reference to the pronouncements of the committees.

4.2 Meaning

Neither the HRC nor the CESCR offer their own definition of cultural heritage. The section shows that cultural heritage forms part of the right to take part in cultural life, Article 15 (1) (a) of the ICESCR) (Sect. 4.2.1) as well as of the rights of minorities, as laid out in Article 27 of the ICCPR (Sect. 4.2.2). Finally, this section addresses the presence of manifestations of cultural heritage in the notions of "home", "family" and "privacy within the meaning of Article 17 of the ICCPR (Sect. 4.2.3).

4.2.1 Cultural Heritage Forms Part of Article 15 (1) (a) of the ICESCR

Article 15 (1) (a) of the ICESCR is a salient provision that recognises the right of everyone "[t]o take part in cultural life". Yet, the meaning of "cultural life" is not further defined in the provision. Before turning to the interpretation of the committees, the *travaux préparatoires* show that a more detailed draft of the article had contained the wording "access to manifestations of national and international cultural life", which was subsequently shortened to "cultural life".[8] The reference to manifestations of national and international cultural life can be interpreted to

[8] UNGA 'Third Committee' (31 October 1957) UN Doc A/C.3/SR.797, para.13.

include cultural products.[9] Whilst the wording, i.e. cultural products, appears strongly indicative of tangible cultural heritage, it is unclear whether the drafters of the ICESCR had the protection of intangible aspects of cultural heritage in mind[10].

The conceptualisation of "culture" within the work of the CESCR has gradually widened and extended to reading the notion of "culture" as encompassing different ways of life.[11] Accordingly, the CESCR considers (at least) two forms of culture: culture as cultural products and culture as a way of life.[12] At the General Discussion days of the CESCR in 1992, it was agreed that the definition of cultural life goes well beyond high culture and the arts and embraces immaterial aspects, including "food, languages, religion, music, dance and traditional activities and rituals".[13] The discussion also affirmed that culture referred to "a way of life", including:

> "language, non-verbal communication, oral and written literature, song, religion or belief systems which included rites and ceremonies, material culture, including methods of production or technology, livelihood, the natural and man-made environment, food, clothing, shelter, the arts, customs and traditions, plus a world view representing the totality of a person's encounter with the external forces affecting his life and that of his community."[14]

This broadening of "cultural life" by interpretation of the CESCR also reinforced that Article 15 (1) (a) of the ICESCR is closely connected to other human rights,

---

[9] "It seems clear that the drafter of Article 15 generally understood culture in terms of its external manifestations." Saul, Kinley and Mowbray (2014), p. 1180.

[10] Considering that the Intangible Cultural Heritage Convention was only adopted in 2003. Convention for the Safeguarding of the Intangible Cultural Heritage (adopted 17 October 2003, entered into force 20 April 2006) 2368 UNTS 3 (*Intangible Cultural Heritage Convention*).

[11] Saul, Kinley and Mowbray (2014), p. 1188.

[12] *ibid.*

[13] CESCR 'General Discussions on the Right to Take Part in Cultural Life as Recognized in Article 15 of the Covenant' (12 December 1992) UN Doc E/C.12/1992/SR.17, para. 23; CESCR 'Excerpt from the Report on the Seventeenth Session, General Discussion on the Right to Take Part in Cultural Life as Recognized in Article 15 of the International Covenant on Economic, Social and Cultural Rights' (1993) UN Doc E/1993/22 – E/C.12/1992/2, para. 210.

[14] CESCR 'Chapter VII, General Discussion on the Right to Take Part in Cultural Life as Recognized in Article 15 of the International Covenant on Economic, Social and Cultural Rights' (1993) UN Doc E/1993/22, para. 213.

including the rights to education, freedom of thought, religion, expression, and assembly.[15]

At the CESCR's Day of General Discussion in 2008, the affirmation that "cultural life" is inclusive of "ways of life" was picked up, arguing inter alia that it involves the protection of cultural heritage.[16] Days of general discussion, within the framework of the treaty bodies, seek to deepen the understanding of topics and specific provisions of the Conventions.[17] The days allow for input from interested parties and contribute to future general comments.[18]

A year later, General Comment No. 21 put into writing that "cultural life" constitutes "an explicit reference to culture as a living process, historical, dynamic and evolving, with a past, a present and a future."[19] The right to take part in cultural life includes following "a way of life associated with the use of cultural goods and resources such as land, water, biodiversity, language or specific institutions, and to benefit from the cultural heritage and the creation of other individuals and communities".[20] For the purpose of implementing Article 15 (1) (a), the CESCR asserted the following understanding of culture in General Comment No. 21:

> "[w]ays of life, language, oral and written literature, music and song, non-verbal communication, religion or belief system, rites and ceremonies, sport and games, methods of production or technology, natural and man-made environments, food, clothing and shelter and the arts, customs and traditions [...]."[21]

The interpretation of the CESCR addresses cultural heritage explicitly and covers elements of heritage under the right to take part in cultural life. The evolving interpretation of the committee also makes clear that the implementation of Article 15 (1) (a) requires state parties to go beyond "the material aspects of culture"

---

[15] CESCR 'Summary Record of the Seventeenth Meeting, Substantive Issues Arising on the Implementation of the International Covenant on Economic, Social and Cultural Rights, Day of General Discussion on Paragraph 1 (a) of Article 15 of the Covenant on the Right to Take Part in Cultural Life' (19 May 2008) UN Doc E/C.12/2008/SR.17, para. 3.

[16] *ibid.* para. 3.

[17] OHCHR (2025a): Days of general discussion, https://www.ohchr.org/en/treaty-bodies/cescr/days-general-discussion, accessed 31 January 2025.

[18] *ibid.*

[19] CESCR 'General Comment No. 21' (21 December 2009) UN Doc E/C.12/GC/21, para. 11.

[20] *ibid* para. 15 (b).

[21] *ibid* para. 13.

and ensure "effective access by all to intangible cultural goods", such as language, knowledge and traditions.[22]

4.2.2 Cultural Heritage Forms Part of Article 27 of the ICCPR

Article 27 of the ICCPR sets out the rights of persons belonging to ethnic, religious or linguistic minorities to "enjoy their own culture, to profess and practise their own religion, or to use their own language". Although the HRC's General Comment No. 23 on the rights of minorities makes no explicit mention of cultural heritage, it explains

> "that culture manifests itself in many forms, including a particular way of life associated with the use of land resources, especially in the case of indigenous peoples."[23]

Maintaining a particular way of life is closely associated with the use of land, especially but not exclusively in the case of Indigenous peoples.[24] This also may include "traditional activities", such as fishing or hunting.[25]

Although pronouncements will often refer to Indigenous peoples when speaking of minority groups, Indigenous peoples are distinct from minorities and therefore hold distinct rights. Indigenous peoples have rejected "their categorisation as 'minorities'", pointing to differences in historical circumstances.[26] While cognisant of the importance to distinguish the rights of Indigenous peoples from the rights of minorities, Article 27 of the ICCPR also includes Indigenous peoples when they constitute a minority group.[27]

---

[22] *ibid* para. 70.

[23] CCPR 'General Comment No. 23' (8 April 1994) UN Doc CCPR/C/21/Rev.1/Add.5, para. 7. This finding is repeated in the views of the HRC. Also see, CCPR, Communication No. CCPR/ C/124/D/2668/2015, *Tiina Sanila-Aikio v. Finland* (20 March 2019) UN Doc CCPR/ C/124/D/2668/2015, para. 6.8.

[24] CCPR 'General Comment No. 23' (8 April 1994) UN Doc CCPR/C/21/Rev.1/Add.5, para. 3.2.

[25] *ibid* para. 7.

[26] Vrdoljak (2014), p. 27.

[27] CCPR 'Guidelines for the treaty-specific document to be submitted by States parties under article 40 of the International Covenant on Civil and Political Rights' (22 November 2010) UN Doc CCPR/C/2009/1, para. 104.

The notion of "particular way(s) of life" is commonplace in the context of Article 27 of the ICCPR.[28] The HRC has emphasised the importance of Article 27 for the "survival and continued development of cultural identity" of minorities, including Indigenous peoples in its views.[29] In *Länsman et al. v. Finland*, the Committee explained that the protection of Article 27 of the ICCPR does not only extend to "traditional means of livelihoods"[30] but also covers its adaption with the help of modern technology.[31] The committee, furthermore, established that if economic activities are "an essential element" in the culture of a community, they may fall within the ambit of Article 27.[32]

The views of the HRC have contributed to further understanding which forms of cultural heritage can be (and have been) considered under Article 27. The views show that this includes, in the words of the Intangible Cultural Heritage Convention, "knowledge and practices concerning nature", such as traditional fishing, reindeer herding or llama-raising.[33] For example, *Äärelä and Näkkäläjärvi v. Finland* concerned reindeer breeders of Sami ethnic origin, who alleged that logging in the herding land and a reduction in the permissible number of reindeers resulted in a denial of their right to enjoy their culture.[34] *Howard v. Canada* addressed restrictions in aboriginal fishing rights, threatening the cultural, spiritual, and social survival of an individual belonging to an Indigenous community

---

[28] E.g., CCPR 'Concluding Observations: Tanzania' (2009) UN Doc CCPR/C/TZA/CO/4, para. 26; CCPR 'Concluding Observations: Angola' (2019) UN Doc CCPR/C/AGO/CO/2, para. 49; CCPR 'Concluding Observations: USA' (2023) UN Doc CCPR/C/USA/CO/5, para. 67; CCPR 'Concluding Observations: Mexico' (2019) UN Doc CCPR/C/MEX/CO/6, para. 45; CCPR 'Concluding Observations: Costa Rica' (2016) UN Doc CCPR/C/CRI/CO/6, para. 42; CCPR 'Concluding Observations: Colombia' (2016) UN Doc CCPR/C/COL/CO/7, para. 42; CCPR 'Concluding Observations: Dominican Republic' (2020) UN Doc CCPR/C/DMA/COAR/1, para. 48; CCPR 'Concluding Observations: Finland' (2013) UN Doc CCPR/C/FIN/CO/6, para. 16.

[29] CCPR, Communication No. 1457/2006, *Ángela Poma Poma v. Peru* (24 April 2009) UN Doc CCPR/C/95/D/1457/2006, para. 7.2.

[30] CCPR, Communication No. 511/1992, *Länsman v. Finland* (14 October 1993) UN Doc CCPR/C/49/D/511/1992, para. 9.3.

[31] *bid*

[32] See, e.g., CCPR, Communication No. 197/1985, *Kitok v. Sweden* (27 July 1988) UN Doc CCPR/C/33/D/197/1985, para. 9.2 or CCPR, Communication No 547/1993, *Apirana Mahuika and Ors v New Zealand* (27 October 2000) UN Doc CCPR/C/70/D/547/1993, para. 9.3.

[33] See Intangible Cultural Heritage Convention, Art. 2 (2) (d).

[34] CCPR, Communication No. 779/1997, *Äärelä and Näkkäläjärvi v Finland* (24 October 2001) UN Doc CCPR/C/73/D/779/1997, para. 3.1.

in Canada.[35] The communication *Poma Poma v. Peru* dealt with the consequences of state-authorised water diversion resulting in negative impacts on the author's rights as well as on llama-raising, an important activity the community's livelihood and survival relies on.[36]

The more recent communication of *Daniel Billy et al. v. Australia* alleged insufficient action over the reduction of emissions as well as the implementation of adaptation measures on the Torres Strait Islands.[37] The communication concerned inter alia the impacts of rising sea levels on traditional burial grounds as well as on the Torres Strait Islanders' way of life and culture,[38] claiming violations of their rights under Articles 2, 6, 17 and 27 of the ICCPR (read alone and in conjunction).[39] Considering the

> "failure to adopt timely adequate adaptation measures to protect the authors' collective ability to maintain their traditional way of life, to transmit to their children and future generations their culture and traditions and use of land and sea resources"

the HRC found a violation of Article 27.[40]

Benito Oliveira Pereira v. Paraguay considered the state's failure to provide protection against "large-scale, illegal fumigation [...] causing health problems among community members" of the Campo Agua'ẽ indigenous community and "contaminating the community's waterways, destroying its subsistence crops, killing its livestock and triggering the mass extinction of fish and bees"."[41]

---

[35] CCPR, Communication No. 879/1999, *Howard v. Canada* (4 August 2005) UN Doc CCPR/C/84/D/879/1999.

[36] CCPR, Communication No. 1457/2006, *Ángela Poma Poma v. Peru* (24 April 2009) UN Doc CCPR/C/95/D/1457/2006.

[37] CCPR, Communication No. 3624/2019, *Daniel Billy et al. v. Australia* (22 September 2022) UN Doc CCPR/C/135/D/3624/2019. Also see, Murphy (2019): Torres Strait Islanders take climate change complaint to the United Nations, https://www.theguardian.com/australia-news/2019/may/13/torres-strait-islanders-take-climate-change-complaint-to-the-united-nations, accessed 31 January 2025.

[38] Mosby (2020): Here is why we are taking the Australian government to the UN over its inaction on climate change, https://www.theguardian.com/commentisfree/2020/aug/14/here-is-why-we-are-taking-the-australian-government-to-the-un-over-its-inaction-on-climate-change, accessed 31 January 2025.

[39] CCPR, Communication No. 3624/2019, *Daniel Billy et al. v. Australia* (22 September 2022) UN Doc CCPR/C/135/D/3624/2019, para. 1.1.

[40] *ibid* para. 8.14.

[41] CCPR, Communication No. 2552/2015, *Benito Oliveira Pereira et al. v. Paraguay* (12 October 2021) UN Doc CCPR/C/132/D/2552/2015, para. 8.4.

Recalling that Indigenous peoples' right to enjoyment of their culture is "directed towards the survival and continued development of the cultural identity",[42] the Committee noted the contamination's impact on the community member's cultural life[43].

Article 27 of the ICCPR offers a provision for the protection of the culture of persons belonging to minorities, including Indigenous peoples. The pronouncements of the HRC show that the committee interprets culture broadly and understands that culture can imply a particular or traditional way of life. Yet, the focus on the notion of a particular way of life is not to imply that the HRC does not refer to cultural heritage explicitly in its pronouncements, for example, in the framework of the concluding observations.[44]

4.2.3 Cultural Heritage in the Notions of "Home", "Family" and "Privacy"

Cultural heritage also forms part of other human rights provisions. In *Mathieu Vakoumé et al. v. France* the authors sought interim measures in the face of the creation of a hotel complex citing Articles 17 (1), 23 (1) and 18 of the ICCPR,[45] referring to the rights to privacy, family, and home as well as to freedom of thought, conscience and religion. The complaint claimed that the site held special significance for their history, culture, and life[46] and declared the site on which the complex was built "an important part of their natural, historic and cultural heritage" including ancestral burial grounds[47]. The communication, however, was found inadmissible by the committee.

In *Hopu and Bessert v France*, the authors, both ethnic Polynesians, sought interim measures against France by contending that the construction of a luxury

---

[42] *ibid* para. 8.6.

[43] *ibid* para. 8.8. The Committee found violations of Articles 17 and 27 of the ICCPR read alone and in conjunction with Article 2 (3).

[44] E.g., CCPR 'Concluding Observations: Australia' (2017) UN Doc CCPR/C/AUS/CO/6, para. 50; CCPR 'Concluding Observations: Tanzania' (2009) UN Doc CCPR/C/TZA/CO/4, para. 26.

[45] CCPR, Communication No. 822/1998, *Mathieu Vakoumé et al v. France* (2000) UN Doc CCPR/C/70/D/822/1998, paras. 3.1, 7.

[46] *ibid* para. 3.2.

[47] *ibid* para. 3.4.

hotel complex in Tahiti, French Polynesia would result in the destruction of traditional burial grounds and adversely impact their fishing activities.[48] According to the authors, the burial grounds form part of "their history, culture and life" and the construction would be an arbitrary interference with their family life and privacy.[49] The authors, furthermore, contended that family members were buried on the site. The communication allowed the HRC to expound on the term "family" and the committee found that cultural traditions should be considered.[50] It further acknowledged the authors' relationship to their ancestors as "an essential element of their identity and to play an important role in their family life".[51] This led the committee to find that the construction of the hotel complex did interfere with the right to family and privacy, as laid down in Articles 17 (1) and 23 (1) of the ICCPR.[52]

The committee's interpretation was far from unanimous, as several committee members attached dissenting opinions to the views. The perhaps most important point of contention, for the purpose of this chapter, is the interpretation of the term "family". Whilst expressing concern at the state's failure to respect a site of "obvious importance in the cultural heritage of the indigenous peoples of French Polynesia", four members of the HRC found the views' interpretation of the term "family" and "privacy" distorted.[53] They argued that the term "family" should not be expanded to include "all members of one's ethnic or cultural group" or "all one's ancestors, going back to time immemorial".[54] Whereas the reasoning by the committee, referring to the protection of the "history, culture and life" of the authors was recognised worthy of protection, the dissenting opinion also mentioned that these are values protected under Article 27 of the ICCPR. The reason why the committee could not consider the issue under this provision can most likely be ascribed to the French declaration to Article 27 of the ICCPR, which has been interpreted as a reservation by the Committee[55]. The dissenting

[48] CCPR, Communication No. 549/1993, *Hopu and Bessert v. France* (29 December 1997) UN Doc CCPR/C/60/D/549/1993/Rev.1, para. 2.5.

[49] *ibid* para. 10.3.

[50] *ibid.*

[51] *ibid.*

[52] *ibid.*

[53] *ibid* para. 7.

[54] *ibid* para. 4.

[55] CCPR, Communication No. 220/1987, *T. K. v. France* (8 November 1989) UN Doc CCPR/C/37/D/220/1987; CCPR, Communication No. 222/1987, *H. K. v. France* (8 November 1989) UN Doc CCPR/C/37/D/222/1987.

opinion was clear in that "the mere fact that visits to a certain site play an important role in one's identity, does not transform such visits into part of one's right to privacy".[56] Instead, it argued, such interferences could constitute violations of Articles 18 or 27 of the ICCPR.[57]

More recent communications, such as *Portillo Cáceres v. Peru, Daniel Billy et al. v. Australia* and *Benito Oliveira Pereira v. Paraguay* have picked up the notions of home, family and privacy in the context of culture and heritage again, albeit in a slightly different manner. *Portillo Cáceres v. Peru* concerned the use of toxic agrochemicals by public officials on industrial farms adjacent to the homes of the authors resulting in serious health problems and the death of one person. The Committee found that the elements constituting "components of the way of life of the authors," who are *campesinos* and share "a special attachment to and dependency on the land [...] can be considered to fall under the scope of protection of article 17 of the Covenant."[58]

The views in *Daniel Billy et al. v. Australia* do not only highlight the impacts of climate change on the traditional ways of life of Indigenous communities. They also found that "components of the traditional indigenous way of life" fall under Article 17 of the ICCPR.[59] In the communication at hand these components refer to the authors dependence on "marine resources, land crops, and trees for their subsistence and livelihoods, and [...] their surrounding ecosystems for their own wellbeing".[60] Furthermore, the Committee also concluded that Australia failed "to discharge its positive obligation to implement adequate adaptation measures to protect the authors' home, private life and family" and found a violation of article 17 of the ICCPR.

In *Benito Oliveira Pereira v. Paraguay*, outlined above, the authors of the communication inter alia submitted that the contamination threatened ancestral practices, loss of traditional knowledge and loss of fundamental sacred rituals

---

This is also reflected in France's state reports, which emphasize that "France does not recognize the collective rights of any group defined by a common origin, culture, language or belief". See CESCR 'Periodic report submitted by France' (2021) UN Doc E/C.12/FRA/5, para. 200. Also see, in the academic discourse: Lixinski (2013a), p. 155.

[56] CCPR, Communication No. 549/1993, *Hopu and Bessert v. France* (29 December 1997) UN Doc CCPR/C/60/D/549/1993/Rev.1, para. 6.

[57] *ibid.*

[58] CCPR, Communication No. 2751/2016, *Norma Portillo Cáceres et al. v. Paraguay* (20 September 2019) UN Doc CCPR/C/126/D/2751/2016, para. 7.8.

[59] CCPR, Communication No. 3624/2019, *Daniel Billy et al. v. Australia* (22 September 2022) UN Doc CCPR/C/135/D/3624/2019, para. 8.10.

[60] *ibid.*

and ceremonies, including children's rite, key to their cultural identity.[61] Against the backdrop of the environmental pollution, the HRC also found a violation of privacy, family life and the home, noting the contamination's direct and serious repercussions and adverse effects on the right to one's privacy, family life and home.[62]

4.2.4 Interim Conclusions

The pronouncements from both treaty bodies show that cultural heritage forms part and falls within the scope of Articles 15 (1) (a) of the ICESCR and 27 of the ICCPR. The CESCR's predominant focus on "cultural heritage" and the HRC's focus on "way of life" can likely be ascribed to the CESCR's primary focus on popular cultural participation within the meaning of Article 15 of the ICESCR and the HRC's exclusive focus on the protection of the rights of persons belonging to minorities, including Indigenous peoples, in Article 27 of the ICCPR.

Whilst Article 15 (1) (a) of the ICESCR and Article 27 of the ICCPR appear to be the primary provisions the committees rely on, the communications *Hopu and Bessert v France, Portillo Cáceres v. Peru, Daniel Billy et al. v. Australia,* and *Benito Oliveira Pereira v. Paraguay* illustrate, in different ways, how the right to privacy, family and home can be utilised. The views of the HRC show that the protection of traditional ways of life also falls under the scope of the protection of Article 17 of the ICCPR. In the communications of *Portillo Cáceres v. Peru, Daniel Billy et al. v. Australia,* and *Benito Oliveira Pereira v. Paraguay,* the committee also highlighted the authors' dependence on land and marine resources for their subsistence and livelihoods. The committee considered these elements parts of "the traditional indigenous way of life"[63] or components of the author's way of life "who have a special attachment to and dependency on the land"[64] and confirmed that they fall under the scope of Article 17 of the ICCPR.

---

[61] CCPR, Communication No. 2552/2015, *Benito Oliveira Pereira et al. v. Paraguay* (12 October 2021) UN Doc CCPR/C/132/D/2552/2015, para. 8.5.

[62] *ibid* para. 8.4. The Committee also cited CCPR, Communication No. 2751/2016, *Norma Portillo Cáceres et al. v. Paraguay* (20 September 2019) UN Doc CCPR/C/126/D/2751/2016, para. 7.8.

[63] CCPR, Communication No. 3624/2019, *Daniel Billy et al. v. Australia* (22 September 2022) UN Doc CCPR/C/135/D/3624/2019, para. 8.10.

[64] CCPR, Communication No. 2751/2016, *Norma Portillo Cáceres et al. v. Paraguay* (20 September 2019) UN Doc CCPR/C/126/D/2751/2016, para. 7.8.

4.3 Rights

The HRC and the CESCR both identify rights vis-à-vis cultural heritage in their pronouncements. The committees mainly rely on Article 15 (1) (a) of the ICESCR and Article 27 of the ICCPR, although other important legal bases are also highlighted.

4.3.1 Access to Cultural Heritage

Access to cultural heritage takes different forms and legal bases. First, this includes access to cultural heritage through the concept of cultural participation and culture in Article 15 (1) (a) of the ICESCR and Article 27 of the ICCPR. Second, it covers access to places of worship and sacred sites through Article 27 of the ICCPR and, less commonly, Article 18 of the ICCPR. And, thirdly, it encompasses access to linguistic heritage and education through Article 27 of the ICCPR and Articles 13–14 and 15 (1) (a) of the ICESCR.

4.3.1.1 Access to Cultural Heritage through Cultural Participation

"Access" is a central component of the right to take part in cultural life in General Comment No. 21 and covers multiple rights, specifically the rights

> "to know and understand his or her own culture and that of others through education and information, and to receive quality education and training with due regard for cultural identity. Everyone has also the right to learn about forms of expression and dissemination through any technical medium of information or communication, to follow a way of life associated with the use of cultural goods and resources such as land, water, biodiversity, language or specific institutions, and to benefit from the cultural heritage and the creation of other individuals and communities."[65]

Thus, "access" itself includes different forms of cultural heritage and different rights vis-à-vis cultural heritage. This can be inferred from the assertion that everyone has the right to know and learn about one's culture and to "follow a way of life associated with the use of cultural goods and resources such as land,

---

[65] CESCR 'General Comment No. 21' (21 December 2009) UN Doc E/C.12/GC/21, para. 15.

water, biodiversity, language or specific institutions"[66] and even more explicitly from the right to "benefit from the cultural heritage and the creation of other individuals and communities". Access not only signifies physical access. It encompasses receiving information, education, following a way of life associated with one's culture and cultural heritage, to benefitting from cultural heritage and the creation of others. Reporting on the implementation of Article 15 of the ICESCR should contain information on inter alia participation in and access to cultural goods, institutions, activities, and the cultural heritage of mankind.[67]

"Availability" is a precondition of the right to take part in cultural life and requires "the presence of cultural goods", including libraries, museums, literature, the arts in all forms, natural heritage as well as intangible cultural goods.[68] Access further presupposes the (physical and financial) accessibility of culture, the acceptability of inter alia laws policies and measures adopted by the state, their adaptability as well as the (cultural and contextual) appropriateness of the realisation of the right.[69] The presence and availability of cultural heritage is thus intrinsically linked to participation in cultural life.[70]

Whereas the reporting guidelines of the HRC do not explicitly consider access to cultural heritage, the HRC has also addressed access in its views in the context of cultural participation in one's community, specifically in *Sandra Lovelace v. Canada*, in which the Committee found a violation of Article 27 of the ICCPR.[71]

---

[66] Also see, Donders (2020), p. 393.

[67] See CESCR 'Guidelines on Treaty-Specific Documents to be Submitted by State Parties under Articles 16 and 17 of the International Covenant on Economic, Social and Cultural Rights' (24 March 2009) UN Doc E/C.12/2008/2, para. 67 (b). See CESCR 'Concluding Observations: Angola' (2008) UN Doc E/C.12/AGO/CO/3, para. 40 where the CESCR requests information concerning "access to cultural activities and cultural property, especially by disadvantaged and marginalized groups in society".

[68] The list reading in full refers to "libraries, museums, theatres, cinemas and sports stadiums; literature, including folklore, and the arts in all forms; the shared open spaces essential to cultural interaction, such as parks, squares, avenues and streets; nature's gifts, such as seas, lakes, rivers, mountains, forests and nature reserves, including the flora and fauna found there, which give nations their character and biodiversity; intangible cultural goods, such as languages, customs, traditions, beliefs, knowledge and history as well as values, which make up identity and contribute to the cultural diversity of individuals and communities." *ibid* para. 16 (a).

[69] *ibid* para. 16.

[70] *ibid* para 16 (a).

[71] CCPR, Communication No. 24/1977, *Sandra Lovelace v. Canada* UN Doc CCPR/C/OP/1 at 10 (1984). Also see, CCPR, Communication No. 197/1985, *Kitok v. Sweden* (27 July 1988) UN Doc CCPR/C/33/D/197/1985.

The communication dealt with the exclusion from cultural practices and loss of indigenous status for women after marriage to a non-indigenous man. It found that the applicant could not exercise her right "to access to her native culture and language 'in community with the other members' of her group (…) because there is no place outside the Tobqiue Reserve where such a community exists".[72] The *Lovelace* case established that restrictions upon the right of a single members must have a reasonable and objective justification.[73] Furthermore, states need to show the necessity "for the continued viability and welfare of the minority as a whole".[74]

Finally, it is important to note that access to land, which is of fundamental importance to Indigenous peoples' culture, is addressed by the Committee in the reporting.[75]

4.3.1.2 Access to Cultural Heritage through Freedom of Worship and Professing and Practicing One's Religion

Article 18 of the ICCPR covers access to places of worship.[76] In the concluding observations, the HRC criticises restrictions on access to places of worship of religious minorities, albeit not necessarily by framing them as interferences with access to heritage.[77] Article 27 of the ICCPR addresses access to cultural and sacred sites in the context of Indigenous peoples. For example, the HRC reprehends the United States of America for restrictions on access to sacred areas for Indigenous communities "essential for the preservation of their religious, cultural and spiritual practices".[78] In the case of New Zealand, the HRC recommends

[72] CCPR, Communication No. 24/1977, *Sandra Lovelace v. Canada* UN Doc CCPR/C/OP/1 at 10 (1984), para. 15.

[73] *ibid* para. 16.

[74] CCPR, Communication No. 197/1985, *Kitok v. Sweden* (27 July 1988) UN Doc CCPR/C/33/D/197/1985, p 230, para. 9.8.

[75] CCPR 'List of issues: Namibia' (2023) UN Doc CCPR/C/NAM/Q/3, para. 23.

[76] CCPR 'Guidelines for the treaty-specific document to be submitted by States parties under article 40 of the International Covenant on Civil and Political Rights' (22 November 2010) UN Doc CCPR/C/2009/1, p. 21, para. 104.

[77] See CCPR 'Concluding Observations: Australia' (2017) UN Doc CCPR/C/AUS/CO/6, para. 19.

[78] CCPRR 'Concluding Observations: USA' (2014) UN Doc CCPR/C/USA/CO/4, para. 25.

paying special attention to "the cultural and religious significance of access to the foreshore and seabed for the Māori".[79]

In the pronouncements of the CESCR, the committee reiterates that freedom of worship forms part of the right to take part in cultural life.[80] In this context, however, access is only sporadically addressed. For instance, the concluding observations on Egypt's state report emphasise that the destruction of churches results in a hindrance to the full enjoyment of Article 15 of the ICESCR.[81]

4.3.1.3 Access to Cultural Heritage through Linguistic Rights and the Right to Education

The Intangible Cultural Heritage Convention views language as a vehicle for the transmission of the intangible cultural heritage.[82] As language is linked to the expression of cultural identity,[83] the preservation of traditional languages is of importance to the cultural identity of a community.[84]

In the framework of the ICCPR, the rights of linguistic minorities are protected in Article 27. The HRC has made a point of demarcating the rights of linguistic minorities "to use their language among themselves, in private or in public", protected under Article 27 of the ICCPR, from other rights in the Covenant, specifically from the general right to freedom of expression under Article 19 and the entitlement to interpretation where the language of the court is not understood under Article 14 (3) (f).[85]

The concluding observations of the HRC do not appear to link linguistic rights of persons belonging to minorities and Indigenous peoples to cultural identity or

---

[79] Articles 2, 26, and 27 ICCPR form the legal basis for this recommendation. CCPR 'Concluding Observations: New Zealand' (2010) UN Doc CCPR/C/NZL/CO/5, para. 19.

[80] CESCR 'Concluding Observations: Egypt' (2013) UN Doc E/C.12/EGY/CO/2-4, para. 23. See CESCR 'General Comment No. 21' (21 December 2009) UN Doc E/C.12/GC/21, para. 10.

[81] CESCR 'Concluding Observations: Egypt' (2013) UN Doc E/C.12/EGY/CO/2-4, para. 23.

[82] Intangible Cultural Heritage Convention, Art. 2 (2) (a).

[83] De Witte (2020), p. 371.

[84] CESCR 'Concluding Observations: Australia' (2009) UN Doc E/C.12/AUS/CO/4, para. 33 (a).

[85] The HRC has made a point of demarcating the rights of linguistic minorities "to use their language among themselves, in private or in public", protected under Article 27 ICCPR, from other rights in the Covenant, specifically from the general right to freedom of expression under Article 19 ICCPR and the entitlement to interpretation where the language of the court is not understood under Article 14 (3) (f) ICCPR. CCPR 'General Comment No. 23: Article 27 (Rights of Minorities)' (8 April 1994) UN Doc CCPR/C/21/Rev.1/Add.5, para. 5.3.

heritage.[86] The CESCR, on the other hand, underlines that "language, oral and written literature, music and song, [and] non-verbal communication" forms part of its definition of culture and are thereby included in Article 15 (1) (a) of the ICESCR.[87] Access to one's "own cultural and linguistic heritage and to that of others" is an important part of access to cultural heritage in General Comment No. 21 of the CESCR.[88] General comment No. 21 and the concluding observations show that "participation" covers the right to "express oneself in the language of one's choice".[89]

The committee highlights "the right to their [...] languages, communication media [...] and other manifestations of their cultural identity and membership", as part of "the right of minorities and of persons belonging to minorities to take part in the cultural life of society, and also to conserve, promote and develop their own culture".[90] This is further specified in the concluding observations[91], in which the CESCR refers to the right of persons belonging to a minority group to inter alia "enjoy [...] their own languages, as well as all other manifestations of their identities and cultural ties."[92] Moreover, the committee refers to the right

---

[86] CCPR 'Concluding Observations: Turkey' (2012) UN Doc CCPR/C/TUR/CO/1, para. 9; CCPR 'Concluding Observations: Bosnia and Herzegovina' (2017) UN Doc CCPR/C/BIH/CO/3, paras. 39–40.

The link between linguistic rights and Article 17 ICCPR, the right to privacy, was highlighted in *Raihman v. Latvia*, a communication concerning the spelling of a name. CCPR, Communication No. 1621/2007, *Raihman v. Latvia* (30 November 2010) UN Doc CCPR/C/100/D/1621/2007. On this case, also see Taylor (2020), p. 801.

[87] CESCR 'General Comment No. 21' (21 December 2009) UN Doc E/C.12/GC/21, para. 13.

[88] *ibid*. para. 49 (d).

[89] *ibid* para. 15 (a). Also see *ibid* p. 12, para. 49 (b). E.g., CESCR 'Concluding Observations: France' (2016) UN Doc E/C.12/FRA/CO/4, para. 57.

[90] CESCR 'General Comment No. 21' (21 December 2009) UN Doc E/C.12/GC/21, para. 32.

[91] CESCR 'Concluding Observations: New Zealand' (2016) UN Doc E/C.12/NZL/CO/3, para. 26; CESCR 'Concluding Observations: France' (2016) UN Doc E/C.12/FRA/CO/4, para. 57. See CESCR 'Concluding Observations: Angola' (2016) UN Doc E/C.12/AGO/CO-4-5, paras. 55–56; CESCR 'Concluding Observations: Russian Federation' (2017) UN Doc E/C.12/RUS/CO/6, paras. 58–59.

[92] CESCR 'Concluding Observations: Uzbekistan' (2014) UN Doc E/C.12/UZB/CO/2, para. 27. This also includes references to mother tongue instruction under Art. 15. See CESCR 'Concluding Observations: Italy' (2015) UN Doc E/C.12/ITA/CO/5, para. 56–57; CCPR 'Concluding Observations: Bosnia and Herzegovina' (2012) UN Doc CCPR/C/BIH/CO/2, para. 21.

of Indigenous peoples "under articles 1 and 15 to enjoy their identity and culture, including through the preservation of their traditional languages".[93]

The CESCR also addresses access to languages and linguistic heritage from the perspective of the right to education and to receive information.[94] The educational aspect of access to heritage is underlined by the CESCR's assertion that everyone has "the right to be taught about one's own culture as well as those of others".[95] The right to education is an important means for individuals and communities to transmit "values, religion, customs, language and other cultural references".[96] During a General Discussion Day on the Right to Education, the interdependency of the right to education and other cultural rights was highlighted, referring to "concomitant rights, such as linguistic freedom, minority rights, cultural identity and access to cultural properties" and heritage.[97]

The concluding observations of the CESCR primarily refer to Articles 13–14 and Article 15 of the ICESCR when addressing language and linguistic rights.[98] Access to and the use of Indigenous or minority languages–and mother-tongue instruction–is mainly dealt with from a general point of view that expresses concern of the threat of their disappearance and extinction, their absence from cultural life or from education.[99] Although the transmission of cultural heritage is implicit in these concerns, it is generally not explicitly addressed. A notable

[93] CESCR 'Concluding Observations: Australia' (2009) UN Doc E/C.12/AUS/CO/4, para. 33.

[94] CESCR 'General Comment No. 21' (21 December 2009) UN Doc E/C.12/GC/21, para. 15.

[95] *ibid* para. 49 (d).

[96] *ibid* para. 2.

[97] CESCR 'Excerpt from the Report on the Eighteenth and Nineteenth Sessions' (30 November 1998) UN Doc E/1999/22 -E/C.12/1998/26, para. 482. CESCR 'Summary Record of the 49th Meeting: General Discussion: "Right to education (Articles 13 and 14 of the Covenant)' (2 December 1998) UN Doc E/C.12/1998/SR.49, para. 54.

[98] See CESCR 'Concluding Observations: New Zealand' (2018) UN Doc E/C.12/NZL/CO/4, paras. 48-49; CESCR 'Concluding Observation: Canada' (2016) UN Doc E/C.12/CAN/CO/6, paras. 59-60; CESCR 'Concluding Observations: Japan' (2013) UN Doc E/C.12/JPN/CO/3, para. 30; CESCR 'Concluding Observations: Finland' (2014) UN Doc E/C.12/FIN/CO/6, para. 29; CESCR 'Concluding Observations: Uzbekistan' (2014) UN Doc E/C.12/UZB/CO/2, para. 26.

[99] See e.g., CESCR 'Concluding Observations: Mexico' (2018) UN Doc E/C.12/MEX/CO/5-6, para. 67; CESCR 'Concluding Observations: Russian Federation' (2017) UN Doc E/C.12/RUS/CO/6, para. 58; CESCR 'Concluding Observations: Canada' (2016) UN Doc E/C.12/CAN/CO/6, para. 59; CESCR 'Concluding Observations: Finland' (2014) UN Doc E/C.12/FIN/CO/6, para. 29; CESCR 'Concluding Observations: Australia' (2009) UN Doc E/C.12/AUS/CO/4, para. 33. Also see CESCR's recommendation vis-à-vis migrant children

exception are the concluding observations on Australia's state report, in which the committee, by reference to Articles 1 and 15 of the ICESCR, recommends strengthening the rights of Indigenous peoples to enjoy their identity and culture, for which the preservation of traditional languages forms part of.[100]

4.3.2 Conserve, Promote, and Develop One's Culture, Including Cultural Heritage

Whilst Article 27 of the ICCPR provides a legal basis for the rights of persons belonging to minorities, the CESCR's interpretation of Article 15 (1) (a) also "includes the right of minorities and of persons belonging to minorities to [...] conserve, promote and develop their own culture."[101] This entails the right to the enjoyment of their cultural diversity, traditions, customs, languages and manifestations of identities and cultural ties, religion, forms of education, languages, communication media as well as the preservation, development, expression, maintaining and sharing of culture, history, identity, languages, and traditions.[102]

The concluding observations reflect that Indigenous peoples have the right to "conserve, promote and develop their own culture, language and cultural heritage, traditional knowledge and traditional cultural expressions, and the manifestations of their sciences and cultures".[103] In one instance, the CESCR explicitly requests information on the measures adopted serving the implementation of the right of Angolan communities to preserve, protect, and develop their cultural heritage.[104]

---

and youth, see CESCR 'Concluding Observations: Italy' (2015) UN Doc E/C.12/ITA/CO/5, paras. 56-57.

[100] CESCR 'Concluding Observations: Australia' (2009) UN Doc E/C.12/AUS/CO/4, para. 33 (a).

[101] CESCR 'General Comment No. 21' (21 December 2009) UN Doc E/C.12/GC/21, para 32.

[102] *ibid* para. 32. Also see para. 33.

Also see CESCR 'Concluding Observations: Uzbekistan' (2014) UN Doc E/C.12/UZB/CO/2, para. 27; CESCR 'Concluding Observations: Brazil' (2023) UN Doc E/C.12/BRA/CO/3, para. 70 (b); CESCR 'Concluding Observations: France' (2016) UN Doc E/C.12/FRA/CO/4, para. 57. Also see, CCPR 'Concluding observations: Viet Nam' (2019) UN Doc CCPR/C/VNM/CO/3, para. 55.

[103] CESCR 'Concluding Observations: New Zealand' (2016) UN Doc E/C.12/NZL/CO/3, para. 26.

[104] CESCR 'Concluding Observations: Angola' (2008) UN Doc E/C.12/AGO/CO/3, para. 40.

The HRC's concluding observations repeatedly highlights the Māori's right to cultural development,[105] but general recommendations vis-à-vis cultural development in other geographical contexts are absent in the concluding observations. In *Benito Oliveira Pereira v Paraguay*, the HRC found that interpreting Article 27 of the ICCPR in light of the United Nations Declaration on the Rights of Indigenous Peoples (UNDRIP) "enshrines the inalienable right of indigenous peoples to enjoy the territories and natural resources that they have traditionally used for their subsistence and cultural identity."[106]

The committees' interpretation, however, also identifies more specific ways to conserve, promote, and develop one's cultural heritage (Sects. 4.3.2.1 and 4.3.2.2). For example, recommendations vis-à-vis the protection of cultural diversity are frequent and sometimes framed as a right.[107] This right not only includes cultural diversity but establishes the link to cultural heritage, as the Committee refers to "the right to enjoy their cultural diversity, their traditions, their customs and their own languages, as well as all other manifestations of their identities and cultural ties".[108]

4.3.2.1 Leading a Way of Life Associated with the Use of Land Resources and Traditional Activities

As previously outlined, the rights set out in Article 27 of the ICCPR entail leading "a way of life […] closely associated with territory and use of its resources", including "traditional activities".[109] Leading a particular or traditional way of life may be particularly important for Indigenous peoples,[110] but also covers persons belonging to minorities. The HRC has expounded on the right in its

---

105 CCPR 'Concluding Observations: New Zealand' (2016) UN Doc CCPR/C/NZL/CO/6, paras. 43–44.

106 CCPR, Communication No. 2552/2015, *Benito Oliveira Pereira et al. v. Paraguay* (12 October 2021) UN Doc CCPR/C/132/D/2552/2015, para. 8.6.

107 See CESCR 'Concluding Observations: Uzbekistan' (2014) UN Doc E/C.12/UZB/CO/2, para. 27.

108 *ibid.*

109 CCPR 'General Comment No. 23: Article 27 (Rights of Minorities)' (8 April 1994) UN Doc CCPR/C/21/Rev.1/Add.5, paras. 3.2, 7.

In the concluding observations, the Committee addresses the cultural heritage and traditional way of life of Indigenous communities under Articles 2 and 27 ICCPR, read together, as well as under Articles 26 and 27 ICCPR, read together. See, e.g., CCPR 'Concluding Observations: Japan' (2008) UN Doc CCPR/C/JPN/CO/5, para. 32; CCPR 'Concluding Observations: Tanzania' (2009) UN Doc CCPR/C/TZA/CO/4, para. 26.

110 CCPR 'General Comment No. 23: Article 27 (Rights of Minorities)' (8 April 1994) UN Doc CCPR/C/21/Rev.1/Add.5.

views, including in *Daniel Billy et al. v. Australia.* Here, the HRC inter alia found a violation of Article 27 in the state's

> "failure to adopt timely adequate adaptation measures to protect the authors' collective ability to maintain their traditional way of life, to transmit to their children and future generations their culture and traditions and use of land and sea resources."[111]

Following "a way of life associated with the use of cultural goods and resources such as land, water, biodiversity, language or specific institutions" also forms part of Article 15 (1) (a) of the ICESCR.[112] In the concluding observations on Finland's report, the CESCR frames maintaining culture, a way of life and traditional livelihoods as a right.[113] However, whenever the concluding observations address the protection of particular or traditional ways of life, it is more often put as an obligation to consider these aspects.[114]

4.3.2.2 Maintaining, Controlling, Protecting, and Developing Cultural Heritage, Traditional Knowledge, and Expressions

A further way of conserving, promoting, and developing cultural heritage is shown in the pronouncements of the CESCR and is the right of Indigenous peoples to

> "act collectively to ensure respect for their right to maintain, control, protect and develop their cultural heritage, traditional knowledge and traditional cultural expressions, as well as the manifestations of their sciences, technologies and cultures."[115]

---

[111] CCPR, Communication No. 3624/2019, *Daniel Billy et al. v. Australia* (22 September 2022) UN Doc CCPR/C/135/D/3624/2019, para. 6.

[112] CESCR 'General Comment No. 21' (21 December 2009) UN Doc E/C.12/GC/21, para. 15.

[113] CESCR 'Concluding Observations: Finland' (2021) UN Doc E/C.12/FIN/CO/7, paras. 50-51.

[114] See, e.g., CESCR 'Concluding Observations: Tanzania' (2012) UN Doc E/C.12/TZA/CO/1-3, para. 29; CESCR 'Concluding Observations: Namibia' (2016) UN Doc E/C.12/NAM/CO/1, para. 74; CESCR 'Concluding Observations: Brazil' (2023) UN Doc E/C.12/BRA/CO/3, para. 20 (a).

[115] CESCR 'General Comment No. 21' (21 December 2009) UN Doc E/C.12/GC/21, para. 37.

This right acknowledges acting collectively and goes beyond protection, preservation, and development by adding, inter alia, the element of controlling their cultural heritage, traditional knowledge, and other manifestations.

In the concluding observations, the CESCR highlights the negative effects of unresolved questions pertaining to ownership and the use of land on Indigenous peoples' right "to maintain and develop their traditional culture and way of life", referring specifically to reindeer herding.[116] Furthermore, the Committee "urges" "the protection of the right of Indigenous peoples to own, use, develop and control their lands, territories and natural resources with complete security, including through legal recognition".[117]

A further element of the right of Indigenous peoples to maintain, control, protect and develop important elements of their culture, traditional knowledge, heritage, and sciences links to the enjoyment of the benefits resulting from one's scientific, literary, or artistic productions. Set out in Article 15 (1) (c) of the ICESCR, General Comment No. 17 clarifies that this right seeks to safeguard "the personal link between authors and their creations and between peoples, communities, or other groups and their collective cultural heritage, as well as their basic material interest", and that this right is a necessity to ensure that authors "enjoy an adequate standard of living".[118] General comment No. 17 explains that the right concerns inter alia "knowledge, innovations and practices of indigenous and local communities" as well as art, performances, and oral traditions.[119]

---

[116] CESCR 'Concluding Observations: Finland' (2008) UN Doc E/C.12/FIN/CO/5, para. 11.

[117] CESCR 'Concluding Observations: Mexico' (2018) UN Doc E/C.12/MEX/CO/5-6, para. 68. Also see CESCR 'Concluding Observations: Russian Federation' (2011) (UN Doc E/C.12/RUS/CO/5), para. 34.

[118] CESCR 'General Comment No. 17' (12 January 2006) UN Doc E/C.12/GC/17, para. 2. This right needs to be distinguished from intellectual property rights.

[119] *ibid* para. 9.

4.3.3 Effective Consultations, the Principle of Free, Prior and Informed Consent and Participation in Decision-Making Affecting Indigenous and Local Communities' Rights

Effective participation in decision-making affecting minority groups is addressed in the framework of Article 27 of the ICCPR[120] as well as under Article 1 of the ICCPR. Reporting under the latter provision sets out duly consultations with Indigenous and local communities as well as the extent to which "their prior informed consent is sought in any decision-making processes affecting their rights and interests under the Covenant".[121]

When addressing persons and communities requiring special protection under Article 15 (1) (a) of the ICESCR, the CESCR states that the principle of free, prior and informed consent of Indigenous peoples should be respected "in all matters covered by their specific rights".[122] Moreover, the CESCR's General Comment No. 17 on Article 15 (1) (c) underlines the importance of free, prior and informed consent. It states that adopting measures that protect the interest of Indigenous communities and their productions should be subject to the "principle of free, prior and informed consent of the indigenous authors concerned".[123]

How do effective consultations, participation in decision-making, and free, prior and informed consent link to cultural heritage in the committees' pronouncements? Prevalent issues in the concluding observations[124] and the list of

---

[120] See CCPR 'Guidelines for the treaty-specific document to be submitted by States parties under article 40 of the International Covenant on Civil and Political Rights' (22 November 2010) UN Doc CCPR/C/2009/1, para. 104.

[121] *ibid* para. 28.

[122] CESCR 'General Comment No. 21' (21 December 2009) UN Doc E/C.12/GC/21, para. 37.

[123] CESCR 'General Comment No. 17' (12 January 2006) UN Doc E/C.12/GC/17, para. 32.

[124] For example: CESCR 'Concluding Observations: Mexico' (2018) UN Doc E/C.12/MEX/CO/5-6, para. 12; CESCR 'Concluding Observations: Tanzania' (2012) UN Doc E/C.12/TZA/CO/1-3, pp. 3-4; CESCR 'Concluding Observations: New Zealand' (2016) UN Doc E/C.12/NZL/CO/3, para. 11.

CCPR 'Concluding Observations: Angola' (2019) UN Doc CCPR/C/AGO/CO/2, para. 50 (c); CCPR 'Concluding Observations: Ethiopia' (2022) UN Doc CCPR/C/ETH/CO/2, para. 48 (b); CCPR 'Concluding Observations: Namibia' (2016) UN Doc CCPR/C/NAM/CO/2, para. 44; CCPR 'Concluding Observations: Mexico' (2019) UN Doc CCPR/C/MEX/CO/6, para. 44.

issues[125] of both committees are the adverse effects of operations, including those of extractive industries and development projects, on ancestral or traditional lands, including access to resources, that lack prior consultation and the free, prior and informed consent of the respective Indigenous communities. In this context, the CESCR refers to "the right to prior consultation with a view to obtaining their free, prior and informed consent"[126] or to the right to free, prior and informed consent[127].

Yet, even when the CESCR explicitly notes projects that are carried out on Indigenous lands or territories and impede access to land and natural resources or more generally the ability to exercise economic, social and cultural rights, there is a paucity of references to the way of life and heritage of Indigenous peoples.[128] Ensuring adequate redress for the Indigenous communities is emphasised in the CESCR's concluding observations on New Zealand's report, but is not commonly addressed in other concluding observations.[129]

Most comments by both committees concerning participation in decision-making processes concern Indigenous peoples.[130] The phrasing in the concluding

[125] See, e.g., CESCR 'List of Issues: Colombia' (2009) UN Doc E/C.12/COL/Q/5, para. 6; CESCR 'List of Issues: Costa Rica' (2016) UN Doc E/C.12/CRI/Q/5, para. 4; CESCR 'List of Issues: Australia' (2016) UN Doc E/C.12/AUS/Q/5, para. 4.

[126] CESCR 'Concluding Observations: Mexico' (2018) UN Doc E/C.12/MEX/CO/5-6, para. 12.

[127] CESCR 'Concluding Observations: Canada' (2016) UN Doc E/C.12/CAN/CO/6, para. 13; CESCR 'Concluding Observations: Colombia' (2010) UN Doc E/C.12/COL/CO/5, para. 9. Also see, CESCR 'Concluding Observations: Brazil' (2023) UN Doc E/C.12/BRA/CO/3, para. 20 (a).

[128] See e.g., CESCR 'Concluding Observations: Angola' (2016) UN Doc E/C.12/AGO/CO/4-5, para. 19; CESCR 'Concluding Observations: New Zealand' (2018) UN Doc E/C.12/NZL/CO/4, paras. 8, 10; CESCR 'Concluding Observations: Australia' (2017) UN Doc E/C.12/AUS/CO/5, paras. 15-16; CESCR 'Concluding Observations: Colombia' (2017) UN Doc E/C.12/COL/CO/6, para. 18; CESCR 'Concluding Observations: Costa Rica' (2016) UN Doc E/C.12/CRI/CO/5, paras. 8-9; CESCR 'Concluding Observations: Canada' (2016) UN Doc E/C.12/CAN/CO/6, paras. 13-14; paras. 19-20; CESCR 'Concluding Observations: Mexico' (2018) UN Doc E/C.12/MEX/CO/5-6, para. 12; CESCR 'Concluding Observations: Russian Federation' (2011) UN Doc E/C.12/RUS/CO/5, para. 7. The legal basis for such cases is predominantly Articles 1, 1(2), 2 ICESCR, read together as well as separately.

[129] CESCR 'Concluding Observations: New Zealand' (2016) UN Doc E/C.12/NZL/CO/3, para. 11.

[130] Also addressing (traditional) communities more generally: CCPR 'Concluding Observations: Viet Nam' (2019) UN Doc CCPR/C/VNM/CO/R.3, para. 55; CCPR 'Concluding

observations of both committees varies, but addresses effectively consulting communities,[131] speaks of their effective participation,[132] seeking or attempting to obtain their free and informed consent,[133] respecting the free, prior and informed consent,[134] and, in a few cases, recommends ensuring their *right* to free, prior, and informed consent[135].

4.3.4 Interim Conclusions

The analysis has identified three broad rights in the pronouncements of the treaty bodies: access to cultural heritage, the right of persons belonging to minorities to conserve, promote, and develop their culture, including cultural heritage, and the right of Indigenous and local communities to effective consultations with

---

Observations: Australia' (2017) UN Doc CCPR/C/AUS/CO/6, para. 50; CESCR 'Concluding Observations: Colombia' (2017) UN Doc E/C.12/COL/CO/6, para. 18 (b); CESCR 'Concluding Observations: Brazil' (2023) UN Doc E/C.12/BRA/CO/3, para. 20 (a).

[131] CCPR 'Concluding Observations: Costa Rica' (2016) UN Doc CCPR/C/CRI/CO/6, paras. 41-42; CCPR 'Concluding Observations: Mexico' (2010) UN Doc CCPR/C/MEX/CO/5, para. 22; CCPR 'Concluding Observations: USA' (2014) UN Doc CCPR/C/USA/CO/4, para. 25.

[132] CCPR 'Concluding Observations: New Zealand' (2016) UN Doc CCPR/C/NZL/CO/6, para. 46 (b) and (c); CCPR 'List of Issues: Australia' (2012) UN Doc CCPR/C/AUS/Q/6, para. 23.

[133] See CCPR 'Concluding Observations: Namibia' (2016) UN Doc CCPR/C/NAM/CO/2, para. 44; CCPR 'Concluding Observations: Colombia' (2016) UN Doc CCPR/C/COL/CO/7, para. 43; CCPR 'Concluding Observations: USA (2014) UN Doc CCPR/C/USA/CO/4, para. 25, CCPR 'Concluding Observations: Russian Federation' (2015) UN Doc CCPR/C/RUS/CO/7, para. 24.-Also see, CCPR 'Concluding Observations: Dominican Republic' (2020) UN Doc CCPR/C/DMA/COAR/1, para. 48 (c); CCPR 'Concluding observations: Mexico' (2019) UN Doc CCPR/C/MEX/CO/6, para. 45.

CESCR 'Concluding Observations: Russian Federation' (2011) UN Doc E/C.12/RUS/CO/5, para. 7 (b); CESCR 'Concluding Observations: Namibia' (2016) UN Doc E/C.12/NAM/CO/1, para. 16; CESCR 'Concluding Observations: Finland' (2014) UN Doc E/C.12/FIN/CO/6, para. 9 (b); CESCR 'Concluding Observation: Angola' (2016) UN Doc E/C.12/AGO/CO/4-5, para. 20 (c); CESCR 'Concluding Observations: Costa Rica' (2016) UN Doc E/C.12/CRI/CO/5, para. 9 (a); CESCR 'Concluding Observations: Colombia' (2017) UN Doc E/C.12/COL/CO/6, para. 18 (b).

[134] CESCR 'Concluding Observations: New Zealand' (2016) UN Doc E/C.12/NZL/CO/3, para. 11.

[135] CCPR 'Concluding Observations: Canada' (2015) UN Doc CCPR/C/CAN/CO/6, para. 16; CCPR 'Concluding Observations: Viet Nam' (2019) UN Doc CCPR/C/VNM/CO/3, para. 56 (b).

a view to obtaining the free, prior and informed consent and to participate in decision-making that affects their rights.

Access to cultural heritage is enjoyed by everyone and does not only cover cultural heritage, but also religious heritage or places of worship, and linguistic heritage. The committees rely on several legal bases in their pronouncements, including most prominently, Article 15 (1) (a) of the ICESCR and Article 27 of the ICCPR. In addition, the pronouncements underscore inter alia Articles 13–14 of the ICESCR in the context of access and transmission of linguistic heritage. While Article 18 of the ICCPR would appear to be a prominent provision to safeguard heritage of a religious or spiritual nature, the pronouncements of the HRC rather rely on Article 27 in this matter. This could be ascribed to the occurrences in the concluding observations concerning the protection of sacred sites of Indigenous peoples, which aligns with the primary objective of Article 27 being the protection of persons belonging to minority groups to practise their religion.

The committees enshrine the rights of persons belonging to minorities to conserve, promote, and develop their culture and cultural heritage in Article 15 (1) (a) of the ICESCR and Article 27 of the ICCPR. The right to lead a particular, traditional or Indigenous way of life has the same legal basis and is related to the right. For Indigenous peoples, "their right to maintain, control, protect and develop their cultural heritage, traditional knowledge and traditional cultural expressions, as well as the manifestations of their sciences, technologies and cultures" can be highlighted in this context[136]. The CESCR infers this right from its interpretation of the right to take part in cultural life in General Comment No. 21. The right of Indigenous peoples to maintain, control, protect and develop important elements of their culture, traditional knowledge, heritage, and sciences can be linked to another right under Article 15 of the ICESCR, namely the enjoyment of the benefits resulting from one's scientific, literary, or artistic productions (Article 15 (1) (c)).

Finally, the committees address the right of Indigenous and local communities to effective consultations, the principle of free, prior and informed consent and to participate in decision-making processes that affect their rights under Articles 1 common to the ICESCR and ICCPR, Article 15 (1) (a) of the ICESCR and Article 27 of the ICCPR. This typically concerns development projects that impact lands and environment, yet there is a paucity in the concluding observations in articulating the link to potential impacts on culture, way of life and elements of cultural heritage.

---

[136] CESCR 'General Comment No. 21' (21 December 2009) UN Doc E/C.12/GC/21, para. 37.

4.4 Obligations

The HRC and the CESCR both refer to obligations vis-à-vis cultural heritage. The analysis identifies more obligations than corresponding rights. This can be ascribed to more broadly framed obligations that ensure the underlying conditions for the existence and transmission of culture, cultural diversity, and cultural heritage.

4.4.1 Promote and Protect Cultural Diversity

Promoting and protecting cultural diversity is an underlying obligation that facilitates other rights and obligations vis-à-vis heritage. The CESCR anchors an obligation to protect and promote cultural diversity in the interpretation of "the obligation to fulfil" under Article 15 (1) (a) of the ICESCR.[137] Accordingly, the obligation corresponds to the right to take part in cultural life. On a more general level, respect for participation in cultural life and cultural diversity allows everyone to freely choose their own cultural identity.[138] In facilitating cultural participation within the meaning of Article 15 (1) (a), states ought to adopt policies that protect, promote, and raise awareness for cultural diversity through inter alia ensuring access to cultural expressions and infrastructure, measures that promote linguistic heritage, and enabling persons to choose freely their way of life.[139]

The obligation contributes to ensuring the existence and diversity of cultural heritage and forms a necessary basis for other obligations vis-à-vis cultural heritage. For example, ensuring the promotion and protection of cultural diversity facilitates the availability of and access to a breadth of cultural expressions.[140] Cultural diversity is essential to the protection of minority cultures, including to the preservation of minority languages, establishing a link to the right to education.[141]

---

[137] See *ibid* para. 52 (a). Also see CESCR 'Guidelines on Treaty-Specific Documents to be Submitted by State Parties under Articles 16 and 17 of the International Covenant on Economic, Social and Cultural Rights' (24 March 2009) UN Doc E/C.12/2008/2, 68.

[138] CESCR 'General Comment No. 21' (21 December 2009) UN Doc E/C.12/GC/2, para. 49 (a).

[139] *ibid* para. 52 (a) and (b) and 53.

[140] See *ibid* para. 52 (a).

[141] CESCR 'Concluding Observations: Costa Rica' (2016) UN Doc E/C.12/CRI/CO/5, para. 63 (a).

Recommendations in the concluding observations of both committees frequently address ensuring respect for and the protection of (cultural) diversity.[142] The concluding observations of the CESCR often express cultural diversity in the context of the cultural diversity of minority groups[143] and, occasionally, frame it as a right[144]. In contrast, the concluding observations of the HRC appear to focus on diversity in the context of incidents of (racial) discrimination and hate speech.[145] The latter relies on articles 2, 7, 18, 20, and 27 of the ICCPR[146], which do not all focus on culture.[147]

---

[142] For the CESCR under the implementation of Article 15. See, e.g., CESCR 'Concluding Observations: Mexico' (2018) UN Doc E/C.12/MEX/CO/5-6, para. 62; CESCR 'Concluding Observations: Mali' (2018) UN Doc E/C.12/MLI/CO/1, para. 53; CESCR 'Concluding Observations: Colombia' (2017) UN Doc E/C.12/COL/CO/6, para. 66; CESCR 'Concluding Observations: Costa Rica' (2016) UN Doc E/C.12/CRI/CO/5, para. 63 (a); CESCR 'Concluding Observations: Namibia' (2016) UN Doc E/C.12/NAM/CO/1, para. 73; CESCR 'Concluding Observations: France' (2008) UN Doc E/C.12/CO/FRA/CO/3, para. 50.

For the HRC see, e.g., CCPR 'Concluding Observations: Italy' (2017) UN Doc CCPR/C/ITA/CO/6, para. 13; CCPR 'Concluding Observations: Australia' (2017) UN Doc CCPR/C/AUS/CO/6, para. 20; CCPR 'Concluding Observations: Costa Rica' (2016) UN Doc CCPR/C/CRI/CO/6, para. 10; CCPR 'Concluding Observations: Russian Federation' (2015) UN Doc CCPR/C/RUS/CO/7, para. 8 (a); CCPR 'Concluding Observations: France' (2015) UN Doc CCPR/C/FRA/CO/5, para. 6; CCPR 'Concluding Observations: Bosnia and Herzegovina' (2012) UN Doc CCPR/C/BIH/CO/2; para. 20.

[143] See, for example, CESCR 'Concluding Observations: Mexico' (2018) UN Doc E/C.12/MEX/CO/5-6, para. 68; CESCR 'Concluding Observations: Colombia' (2017) UN Doc E/C.12/COL/CO/6, para. 65; CESCR 'Concluding Observations: Costa Rica' (2016) UN Doc E/C.12/CRI/CO/5, para. 62; CESCR 'Concluding Observations: France' (2008) UN Doc E/C.12/CO/FRA/CO/3, para. 50.

[144] CESCR 'Concluding Observations: Uzbekistan' (2014) UN Doc E/C.12/UZB/CO/2, para. 27

[145] Often under a heading referring to hate speech, racism and racial discrimination and concerning Articles 2, 7, 18, 20, 26, E.g., CCPR 'Concluding Observations: Italy' (2017) UN Doc CCPR/C/ITA/CO/6, para. 13; CCPR 'Concluding Observations: Australia' (2017) UN Doc CCPR/C/AUS/CO/6, para. 20; CCPR 'Concluding Observations: Costa Rica' (2016) UN Doc CCPR/C/CRI/CO/6, para. 10.

[146] *ibid.*

[147] Although it can be argued that Art. 26 includes elements that can have a link to a broad reading of cultural elements (e.g., race and language).

4.4.2 Respect and Protect Cultural Heritage at all Times

Respecting and protecting cultural heritage at all times serves as an underlying obligation that facilitates other rights and obligations vis-à-vis heritage. Prior to the adoption of General Comment No. 21 in 2009, the protection of cultural heritage was the subject of discussion at the 1992 General Discussion Days of the CESCR. An analysis of Article 15, prepared in advance of the sessions, argues that the right of access to culture includes the protection of the cultural and artistic heritage.[148]

The CESCR asserts that states have the obligation to "[r]espect and protect cultural heritage in all forms, in times of war and peace, and natural disasters".[149] The Committee specifies that respecting and protecting cultural heritage entails preserving, developing, enriching, and transmitting cultural heritage to future generations. This includes the "care, preservation and restoration of historical sites, monuments, works of art and literary works".[150] Accordingly, states have an obligation to respect and protect the cultural heritage and cultural productions of all groups and communities, including the traditional knowledge in the case of Indigenous communities.[151]

The CESCR's concluding observations address the protection of cultural heritage[152] as well as its destruction[153]. In addition to targeted attacks, the CESCR expresses concern over (urban) development projects near world cultural heritage sites.[154]

In comparison, the wording in the concluding observations is couched in softer terms. Even in the case of the highly publicised destruction of cultural heritage

---

[148] CESCR 'Excerpt from the Report on the Seventeenth Session, General Discussion on the Right to Take Part in Cultural Life as Recognized in Article 15 of the International Covenant on Economic, Social and Cultural Rights' Date UN Doc, E/1993/22 – E/C.12/1992/2, para. 205. Cultural heritage was, furthermore, subject in other contexts, raising the issues of *inter alia* cultural freedoms and the fact that cultural heritage is often on display outside its country of origin. See CESCR 'General discussion on article 15, paragraph 1 (a), of the Covenant, on the right to take part in cultural life (continued)' (16 May 2018) UN Doc E/C.12/2008/SR.18, para. 50.

[149] *ibid* para. 50 (a).

[150] *ibid.*

[151] *ibid* para. 50 (b) and (c).

[152] CESCR 'Concluding Observations: Brazil' (2023) UN Doc E/C.12/BRA/CO/3, para. 69.

[153] See, e.g., CESCR 'Concluding Observations: Afghanistan' (2010) UN Doc E/C.12/AFG/CO/2-4, para. 44.

[154] CESCR 'Concluding Observations: Uzbekistan' (2022) UN Doc E/C.12/UZB/CO/3, para. 56.

in Timbuktu, the committee merely "encourages" Mali "to strengthen its legal framework with a view to providing greater protection to victims of attacks targeting their cultural heritage and modes of cultural expression."[155] The committee's choice of not employing stronger language in the concluding observations by e.g., *urging* state parties to protect such sites or recalling the words of the general comment by referring to the state parties' *obligation*, could be interpreted as a reluctance to articulating a strong expectation—or even obligation—in the concluding observations towards the protection of cultural heritage.

The concluding observations provide further insight into measures that need to be adopted: the Committee recommends that state parties "ensure respect for cultural and linguistic cultural heritage and diversity"[156], "enhance its current practice" of registering and protecting heritage sites[157], "ensure that the management and development of world cultural heritage sites are in line with the best practice"[158], or "further strengthen its measures [...] to stop destruction and looting of sites and objects with cultural heritage significance, and take steps to bring perpetrators to justice".[159]

The quasi-absence of any recommendations by the HRC vis-à-vis the destruction of places of worship under Article 18 of the ICCPR is notable. Exceptions include the concluding observations on the implementation of Article 27 of the ICCPR in the United States of America and the Russian Federation, in which the Committee expresses concern at the insufficient protection of sacred places of Indigenous peoples against destruction.[160] Moreover, a closer look at the list of issues reveals that the Committee requests clarification on measures combatting the destruction of places of worship in an individual case.[161]

---

155 CESCR 'Concluding Observations: Mali' (2018) UN Doc E/C.12/MLI/CO/1, para. 54.

156 CESCR 'Concluding Observations: Afghanistan' (2010) UN Doc E/C.12/AFG/CO/2-4, para. 44.

157 *ibid.*

158 CESCR 'Concluding Observations: Uzbekistan' (2022) UN Doc E/C.12/UZB/CO/3, para. 57.

159 CESCR 'Concluding Observations: Afghanistan' (2010) UN Doc E/C.12/AFG/CO/2-4, para. 44.

160 CCPR 'Concluding Observations: United States of America' (2023) UN Doc CCPR/C/USA/CO/5, para. 66 and CCPR 'Concluding Observations: United States of America' (2014) UN Doc CCPR/C/USA/CO/4, para. 25 as well as CCPR 'Concluding Observations: Russian Federation' (2015) UN Doc CCPR/C/RUS/CO/7, para. 24. All recommendations concern ICCPR, Art. 27.

161 CCPR 'Concluding Observations: Bosnia and Herzegovina' (2006) UN Doc CCPR/C/BIH/Q/1, para. 20.

The few recommendations concerning the destruction of places of worship in this area could be ascribed to at least two factors: first, states generally suspend reporting during the occurrence of an armed conflict. This is reflected in gaps in the periodic reporting under the ICCPR: Mali's state report dates to 2003, while Syria and Yemen both show bigger gaps in reporting.[162] As Russia submitted its report in 2019 and the concluding observations by the HRC were issued on 1 December 2022, the committee directly addresses the current and ongoing armed conflict in Ukraine, which was initiated by the Russian Federation.[163] Second, although the notion that states are solely responsible for identifying heritage situated or finding expression on their territory—to a certain extent—has been challenged through inter alia the Intangible Cultural Heritage Convention, the places of worship referred to in the concluding observations (almost) exclusively belong to religious minorities. The committee may be more likely to address sacred places of minorities and Indigenous communities under Article 27 of the ICCPR, as seen in the cases of the United States of America and the Russian Federation.

4.4.3 Guarantee Access to Cultural, Linguistic and Religious or Sacred Heritage

4.4.3.1 Guarantee Access to Cultural Heritage

General Comment No. 21 addresses measures to ensure and enhance access to cultural heritage. The CESCR characterises Article 15 (1) (a) as a freedom, requiring, on the one hand, non-interference from the states "with the exercise of cultural practices and with access to cultural goods" and, on the other, providing "access to and preservation of cultural goods".[164] The interpretation of the CESCR is informed by equal treatment and the prohibition of discrimination.[165] This means that an immediate obligation under Article 15 (1) (a) is to guarantee the exercise of the right without discrimination.[166] This implies that "no one shall

[162] UN Human Rights Treaty Body Database (20205): UN Treaty Body Database, https://tbinternet.ohchr.org/_layouts/15/TreatyBodyExternal/Home.aspx.

[163] CCPR 'Concluding Observations: Russian Federation' (2022) UN Doc CCPR/C/RUS/CO/8, para 6 f.

[164] CESCR 'General Comment No. 21.' (21 December 2009) UN Doc E/C.12/GC/21, para. 6.

[165] *ibid* para. 21.

[166] *ibid* para. 44.

be excluded from […] cultural practices, goods and services".[167] The Committee states in clear terms that "(s)tates must respect free access by minorities to their own culture, heritage and other forms of expression".[168] Moreover, a core obligation requires the elimination of barriers or obstacles that inhibit or restrict an individual's access to their own culture or other cultures.[169]

The committee's interpretation shows that obligations pertaining to access can take different forms. A negative obligation is that states need to refrain from interfering, directly or indirectly, with the enjoyment of the right to take part in cultural life.[170] Positive obligations to guarantee access include adopting specific measures, which respect and facilitate having "access to cultural goods and services, understood as vectors of identity, values and meaning" as well as to cultural and linguistic heritage.[171] The latter implies that states "must respect" and guarantee "free access by minorities to their own culture, heritage and other forms of expression, as well as the free exercise of their cultural identity and practices".[172] Similarly, having access to cultural and linguistic heritage means that states must respect Indigenous peoples' rights "to their culture and heritage and to maintain and strengthen their spiritual relationship with their ancestral lands and other natural resources traditionally owned, occupied or used by them, and indispensable to their cultural life".[173]

The implementation and fulfilment of this obligation requires states to adopt policies that facilitate access to cultural expressions. This includes support of inter alia the cultural infrastructure and diversity by promoting regional and minority language.[174] Equally, fulfilling the obligation requires states to step in when individuals or communities cannot realise the right for themselves for whatever reason. This contains guaranteeing access to museums, libraries and other cultural institutions, events, and activities without discrimination.[175]

[167] *ibid* para. 22.

[168] *ibid* para. 49 (d).

[169] *ibid* para. 55 (d) and (e).

[170] *ibid* para. 48.

[171] *ibid* para. 49 (b) and (d).

[172] *ibid* para. 49 (d).

[173] *ibid.*

[174] *ibid* para. 52 (a).

[175] *ibid* para. 54 (d).

The prominent focus on access to cultural heritage in the CESCR's General Comment No. 21 is not equally reflected in the concluding observations. Recommendations that concern access are not frequent,[176] save for the concluding observations on Finland's report, in which the CESCR "encourages the State party to guarantee access to education in Saami languages" with the aim of preserving, protecting and promoting the culture of the Saami culture "as part of cultural diversity and heritage" under the implementation of Articles 13 and 15 of the ICESCR.[177] Furthermore, in the concluding observations on Tanzania's report the committee highlights restrictions to land and resources that have an impact on the cultural heritage of communities.[178] A more implicit recommendation can be inferred from the concluding observations on Canada's report, in which the Committee recommends taking "effective measures to protect access to traditional food" under Article 11 of the ICESCR.[179]

4.4.3.2 Guarantee Access to Religious and Sacred Heritage

The reporting guidelines of the HRC specify that state parties are to provide information on access to and the protection of places of worship, particularly with regard to individuals belonging to religious minorities under Article 18 of the ICCPR.[180] The HRC condemns restrictions on places of worship that take the form of closures or the denial of licenses to construct places of worship in the first place.[181] In the concluding observations on Brazil and Australia's reports, it addresses violence committed against sacred places of worship.[182] Yet,

---

[176] See, CESCR 'Concluding Observations: Russian Federation' (2011) UN Doc E/C.12/RUS/CO/5, para. 34. Here the Committee criticises obstacles in the accessibility of cultural facilities for persons with disabilities.

[177] CESCR 'Concluding Observations: Finland' (2014) UN Doc E/C.12/FIN/CO/6, para. 29.

[178] CESCR 'Concluding Observations: Tanzania' (2012) UN Doc E/C.12/TZA/CO/1-3, para. 29.

[179] CESCR 'Concluding Observations: Canada' (2016) UN Doc E/C.12/CAN/CO/6, para. 48.

[180] CCPR 'Guidelines for the treaty-specific document to be submitted by States parties under article 40 of the International Covenant on Civil and Political Rights' (22 November 2010) UN Doc CCPR/C/2009/1, para. 83.

[181] CCPR 'Concluding Observations: Egypt' (2023) UN Doc CCPR/C/EGY/CO/5, para. 43; CCPR 'Concluding Observations: Uzbekistan' (2015) UN Doc CCPR/C/UZB/CO/4, para. 22.

Concerning Art. 27: CCPR 'Concluding Observations: USA' (2014) UN Doc CCPR/C/USA/CO/4, para. 25.

[182] CCPR 'Concluding Observations: Brazil' (2023) UN Doc CCPR/C/BRA/CO/3, para. 55; CCPR 'Concluding Observations: Australia' (2017) UN Doc CCPR/C/AUS/CO/6, para. 19.

the recommendations in response to said restrictions or attacks do not frame them in the language of "heritage" but state that these (hate) crimes should be investigated.[183]

4.4.3.3 Guarantee Access to Linguistic Heritage

According to General Comment No. 23, protecting the rights enshrined in Article 27 of the ICCPR against denial or violation does not only entail negative obligations but also requires positive actions by the state parties against the acts of the state as well as those of other persons within the state.[184] The wording of Article 27 clarifies that a state shall not deny a person belonging to a minority the right "to use their own language". This obligation is reflected in the concluding observations in the context of adopting laws and measures promoting and protecting the right of minority groups and Indigenous communities, non-recognition of minorities, structural discriminatory practices, or the risk of disappearance of Indigenous languages.[185]

Positive obligations emanating from Article 27 in this context can further be discerned from the views and concluding observations. Chief among those is the committee's recommendation to withdraw reservations to Article 27 of the ICCPR[186] and addresses the recognition of linguistic minorities[187]. In addition to the request to increase efforts to promote and protect Indigenous languages and cultures,[188] the HRC emphasises the importance of offering adequate opportunities to receive "instruction in or of their language and about their culture"[189].

---

[183] CCPR 'Concluding Observations: Brazil' (2023) UN Doc CCPR/C/BRA/CO/3, para. 56; CCPR 'Concluding Observations: Australia' (2017) UN Doc CCPR/C/AUS/CO/6, para. 20.

[184] CCPR 'General Comment No. 23: Article 27 (Rights of Minorities)' (8 April 1994) UN Doc CCPR/C/21/Rev.1/Add.5, para. 6.1; CCPR 'Guidelines for the treaty-specific document to be submitted by States parties under article 40 of the International Covenant on Civil and Political Rights' (22 November 2010) UN Doc CCPR/C/2009/1, para. 104.

[185] See e.g., CCPR 'Concluding Observations: Viet Nam' (2019) UN Doc CCPR/C/VNM/CO/R.3, para. 56 (a); CCPR 'Concluding Observations: Canada' (2015) UN Doc CCPR/C/CAN/CO/6, para. 19; CCPR 'Concluding Observations: Turkey' (2012) UN Doc CCPR/C/TUR/CO/1, para. 9.

[186] CCPR 'Concluding Observations: Turkey' (2012) UN Doc CCPR/C/TUR/CO/1, para. 9.

[187] CCPR 'Concluding Observations: France' (2015) UN Doc CCPR/C/FRA/CO/5, para. 6; CCPR 'Concluding Observations: Ethiopia' (2011) UN Doc CCPR/C/ETH/CO/1, para. 26.

[188] E.g., CCPR 'Concluding Observations: Canada' (2015) UN Doc CCPR/C/CAN/CO/6, para. 19.

[189] CCPR 'Concluding Observations: Japan' (2008) UN Doc CCPR/C/JPN/CO/5, para. 32. Also see, CCPR 'Concluding Observations: Japan' (2014) UN Doc CCPR/C/JPN/CO/6,

The views of the HRC are instructive in understanding how the Committee defines the scope of this obligations under Article 27 of the ICCPR. In the views on *Cadoret and Bihan v. France*, a communication concerning the use of a minority language in court proceedings and an alleged violation of Article 14 (1) and (3) (e) and (f),[190] the HRC made clear that France was under no obligation to make available interpreters in court proceedings so members of the linguistic minority of Breton speakers could express themselves in the Breton language when they can express themselves adequately in French.[191] This finding was subsequently addressed in General Comment No. 23, which clarified that

> "the right protected under article 27 should be distinguished from the particular right which article 14.3 (f) of the Covenant confers [...] Article 14.3 (f) does not [...] confer on accused persons the right to use or speak the language of their choice in court proceedings."[192]

In the framework of the CESCR, the Committee emphasizes respect for the "freedom of expression in the language or languages" of one's choice.[193] In fulfilling the implementation of cultural participation, states ought to promote "the right of association for cultural and linguistic minorities for the development of their cultural and linguistic rights".[194] The CESCR underlines recognising and promoting the right of members of regional or minority linguistic communities "to use their own language as one aspect of their right to take part in cultural life" in public and in private.[195] Along the same lines, they recommend the adoption

---

para. 26; CCPR 'Concluding Observations: Finland' (2013) UN Doc CCPR/C/FIN/CO/6, para. 16; CCPR 'Concluding Observations: Bosnia and Herzegovina' (2012) UN Doc CCPR/C/BIH/CO/2, para. 21.

190 CCPR, Communication Nos. 221/1987 and 323/1988, *Yves Cadoret and Hervé Le Bihan v. France*, (11 April 1991) UN Doc CCPR/C/41/D/221/1987.

191 *ibid* paras. 5.6-5.7. Also see CCPR, Communication No. 219/1986, *Dominique Guesdon v. France*, (25 July 1990) UN Doc CCPR/C/39/D/219/1986, para. 10.2–10.3; CCPR, Communication No. 327/1988, *Hervé Barzhig v. France* (11 April 1991) UN Doc CCPR/C/41/D/327/1988, paras. 5.5-5.6.

192 CCPR 'General Comment No. 23: Article 27 (Rights of Minorities)' (8 April 1994) UN Doc CCPR/C/21/Rev.1/Add.5, para. 5.3.

193 CESCR 'General Comment No. 21' (21 December 2009) UN Doc E/C.12/GC/21, para. 49 (b).

194 *ibid* para. 52 (c).

195 CESCR 'Concluding Observations: France' (2016) UN Doc E/C.12/FRA/CO/4, para. 57.

of language acts with the aim of contributing to the promotion and preservation of minority languages and cultural heritage.[196]

As for the legal basis, obligations towards the use of minority languages are mainly anchored in Articles 13–15 of the ICESCR. With respect to the right to education, the current reporting guidelines request that state parties report on the opportunities of children belonging to minority groups and Indigenous peoples to "receive instruction in or of their native language".[197] Prior to this, the 1991 reporting guidelines did not contain any references to native languages of minority groups or Indigenous peoples.[198] In reporting on the implementation of Article 15 of the ICESCR, the guidelines now specify for states to "promote awareness of the cultural heritage of [...] linguistic minorities" and "create favourable conditions for them to preserve, develop, express and disseminate their [...] language [...]."[199] Educational programmes for linguistic minorities should be conducted on or in their own language considering "the wishes expressed by communities".[200]

Altogether, the CESCR's concluding observations highlight three aspects: (1) the official recognition of minority languages, (2) their promotion, and (3) their preservation—the latter two are often mentioned together. The Committee underlines the formal recognition of regional and minority languages, as the absence of such recognition contributes to their decline.[201] This is continuously addressed in the concluding observations on France's state reports.[202] Recommendations and

---

[196] CESCR 'Concluding Observations: UK' (2009) UN Doc E/C.12/GBR/CO/5, para. 37.

[197] CESCR 'Guidelines on Treaty-Specific Documents to be Submitted by State Parties under Articles 16 and 17 of the International Covenant on Economic, Social and Cultural Rights' (24 March 2009) UN Doc E/C.12/2008/2, para. 63.

[198] Cf. CESCR 'Revised General Guidelines Regarding the Form and Contents of Reports to be Submitted by States Parties under Articles 16 and 17 of the International Covenant on Economic, Social and Cultural Rights' (17 June 1991) UN Doc E/C.12/1991/1, pp. 17-19.

[199] CESCR 'Guidelines on Treaty-Specific Documents to be Submitted by State Parties under Articles 16 and 17 of the International Covenant on Economic, Social and Cultural Rights' (24 March 2009) UN Doc E/C.12/2008/2, para. 68.

[200] CESCR 'General Comment No. 21' (21 December 2009) UN Doc E/C.12/GC/21, para. 27.

[201] CESCR 'Concluding Observations: France' (2008) UN Doc E/C.12/FRA/CO/3, para. 30. CESCR 'Concluding Observations: Costa Rica' (2016) UN Doc E/C.12/CRI/CO/5, paras. 62-63.

[202] CESCR 'Concluding Observations: France' (2023) UN Doc E/C.12/FRA/CO/5, para. 58; CESCR 'Concluding Observations: France' (2016) UN Doc E/C.12/FRA/CO/4, para. 56; CESCR 'Concluding Observations: France' (2008) UN Doc E/C.12/FRA/CO/3, paras. 29-30.

requests to promote and preserve minority languages[203] are often linked to education. Realising the promotion and preservation of minority languages entails ensuring that persons belonging to minority groups and Indigenous communities may learn their languages and that they have access to education in Indigenous languages.[204]

In addition to viewing the right to education as a concomitant right whose realisation aids access to cultural heritage, the right serves the protection of cultural heritage more directly. The CESCR has interpreted the right to education to encompass "the right of national, ethnic and linguistic minorities to be taught in the language of their choice, in institutions outside the official system of public education".[205] This means that it can serve as an important resource for the heritage of minority groups or Indigenous peoples relating to inter alia the transmission of native languages and traditional knowledge. General Comment No. 13 on the right to education is not as explicit on this interdependence but explains that states have the specific legal obligation to take positive measures that ensure the cultural appropriateness of education for minorities and Indigenous communities.[206]

The importance of the protection and use of language for cultural heritage or as a vehicle for cultural heritage is not often explicitly addressed. One exception is the committee encouraging Finland "to guarantee access to education in Sami languages [...], to preserve, protect and promote the Saami culture as part of cultural diversity and heritage".[207] The concluding observations on Afghanistan's report recommend the adoption of "a comprehensive national cultural policy that ensures respect for cultural and linguistic cultural heritage and diversity."[208]

---

[203] See CESCR 'Concluding Observations: Russian Federation' (2017) UN Doc E/C.12/RUS/CO/6, para. 59; CESCR 'Concluding Observations: UK' (2016) UN Doc E/C.12/GBR/CO/6, paras. 67-68; CESCR 'Concluding Observations: Canada' (2016) UN Doc E/C.12/CAN/CO/6, paras. 59-60; CESCR 'Concluding Observations: Uzbekistan' (2014) UN Doc E/C.12/UZB/CO/2, para. 26; CESCR 'Concluding Observations: Japan' (2013) UN Doc E/C.12/JPN/CO/3, para. 30.

[204] CESCR 'Concluding Observations: Angola' (2016) UN Doc E/C.12/AGO/CO/4-5, para. 56; CESCR 'Concluding Observations: Finland' (2014) UN Doc E/C.12/FIN/CO/6, para. 29.

[205] CESCR 'Excerpt from the Report on the Eighteenth and Nineteenth Sessions' (30 November 1998) UN Doc E/1999/22 -E/C.12/1998/26, para. 494.

[206] CESCR 'General comment No. 13' (8 December 1999) UN Doc E/C.12/1999/10, para. 50.

[207] CESCR 'Concluding Observations: Finland' (2014) UN Doc E/C.12/FIN/CO/6, para. 29.

[208] CESCR 'Concluding Observations: Afghanistan' (2010) UN Doc E/C.12/AFG/CO/2-4, para. 44.

4.4.4 Respect, Protect, and Recognise the Heritage of Persons Belonging to Minorities and Indigenous Communities and Create Favourable Conditions

The committees recognise that persons belonging to minorities and Indigenous communities have the right to preserve and develop their culture, including their cultural heritage. The CESCR asserts that states have the obligations to "recognize, respect and protect minority cultures as an essential component of the identity of the States themselves" in General Comment No. 21.[209] Along the same lines,[210] the HRC underlines the legal recognition and protection of minorities[211] and Indigenous communities[212] in its concluding observations under inter alia Article 27 of the ICCPR.

4.4.4.1 General Obligations towards Persons Belonging to Minorities

In respecting cultural participation within the meaning of Article 15 (1) (a) of the ICESCR,[213] states ought to adopt measures that guarantee and not interfere with the free expression of cultural identity, practices, and one's way of life.[214] Positive obligations entail the promotion of "awareness of the cultural heritage of [...] minorities and of indigenous communities", and to "create favourable conditions for them to preserve, develop, express and disseminate their identity,

---

[209] CESCR 'General Comment No. 21' (21 December 2009) UN Doc E/C.12/GC/21, para. 32.

[210] CESCR 'Concluding Observations: Finland' (2021) UN Doc E/C.12/FIN/CO/7, paras. 50-1; CESCR 'Concluding Observations: Namibia' (2016) UN Doc E/C.12/NAM/CO/1, para. 15; CESCR 'Concluding Observations: Viet Nam' (2014) UN Doc E/C.12/VNM/CO/2-4, para. 33.

[211] E.g., CCPR 'Concluding Observations: Tanzania' (2009) UN Doc CCPR/C/TZA/CO/4, para. 26; CCPR 'Concluding Observations: Italy' (2017) UN Doc CCPR/C/ITA/CO/6, para. 15 (e).

[212] E.g., CCPR 'Concluding Observations: Japan' (2022) UN Doc CCPR/C/JPN/CO/7, para. 42; CCPR 'Concluding Observations: Russian Federation' (2015) UN Doc CCPR/C/RUS/CO/7, para. 24; CCPR 'Concluding Observations: Ethiopia' (2022) UN Doc CCPR/C/ETH/CO/2, para. 47.

[213] Whilst respect is also mentioned in the concluding observations of the HRC, recommendations are not as frequent and framed in more general terms. CCPR 'Concluding Observations: Russian Federation' (2022) UN Doc CCPR/C/RUS/CO/8, para. 39 (d); Also see, CCPR 'Concluding Observations: Japan' (2022) UN Doc CCPR/C/JPN/CO/7, para. 43.

[214] CESCR 'General Comment No. 21' (21 December 2009) UN Doc E/C.12/GC/21, para. 49 (a).

history, culture, language, traditions and customs".[215] This was already reflected in the 1991 reporting guidelines of the CESCR, which requested states to report on measures taken to promote "awareness and enjoyment of cultural heritage of national ethnic groups and minorities and of indigenous peoples".[216]

The addition, namely that states ought to create favourable conditions for minorities, which allow "them to preserve, develop, express and disseminate their identity, history, culture, language, traditions and customs"[217]—all of which form part of their heritage—is important. The concluding observations further elaborate on measures to promote, preserve, and protect cultural heritage and traditional ways of life of communities.[218] These may include "concrete measures to preserve, promote and develop the culture, language, religion and traditions of" minority groups as well as administrative and financial support by the state.[219] The HRC recommends carrying out studies, adopting specific legislation as well as special measures aimed at the protection, preservation, and promotion of their cultural heritage and traditional way of life.[220]

4.4.4.2 Specific Obligations Towards Indigenous Peoples

Recognizing, respecting, and protecting the cultural heritage of minorities entails more specific obligations towards Indigenous peoples. In the case of Indigenous peoples, the matter of heritage and land intersects. Considering Indigenous

---

[215] CESCR 'Guidelines on Treaty-Specific Documents to be Submitted by State Parties under Articles 16 and 17 of the International Covenant on Economic, Social and Cultural Rights' (24 March 2009) UN Doc E/C.12/2008/2, para. 68.

The CESCR recommends promoting awareness of the cultural heritage of minorities. See CESCR 'Concluding Observations: Mali' (2018) UN Doc E/C.12/MLI/CO/1, para. 54; CESCR 'Concluding Observations: UK' (2009) UN Doc E/C.12/GBR/CO/5, para. 37.

[216] CESCR 'Revised General Guidelines Regarding the Form and Contents of Reports to be Submitted by States Parties under Articles 16 and 17 of the International Covenant on Economic, Social and Cultural Rights' (17 June 1991) UN Doc E/C.12/1991/1, Art. 15, 1 (d).

[217] CESCR 'Guidelines on Treaty-Specific Documents to be Submitted by State Parties under Articles 16 and 17 of the International Covenant on Economic, Social and Cultural Rights' (24 March 2009) UN Doc E/C.12/2008/2, 68.

[218] CESCR 'Concluding Observations: Tanzania' (2012) UN Doc E/C.12/TZA/CO/1-3, para. 29. Also see, CESCR 'Concluding Observations: France' (2008) UN Doc E/C.12/FRA/CO/3, para. 51, where the Committee recommends to "increase (…) efforts to preserve and promote regional and minority languages and cultural heritage".

[219] CESCR 'Concluding Observations: Bosnia and Herzegovina' (2021) UN Doc E/C.12/BIH/CO/3, para. 53 (b).

[220] CCPR 'Concluding Observations: Tanzania' (2009) UN Doc CCPR/C/TZA/CO/4, para. 26.

peoples' way of life, ancestral lands, and traditional cultural knowledge, the committees tailor the obligations to respect, protect, and recognise Indigenous heritage and to create favourable conditions to Indigenous communities.

The concluding observations of the HRC address the insufficient protection of the cultural heritage of communities or of sacred areas of Indigenous communities[221] as well as the ramifications of the limited protection of the culture of Indigenous communities[222]. The HRC underlines the recognition and effective legal protection of the rights of Indigenous peoples to their lands, territories, and sacred areas that they traditionally occupy or use as sources of livelihood both under Articles 1 and 27 of the ICCPR.[223] For these purposes, the HRC's reporting guidelines request states to outline the measures taken to ensure that Indigenous communities can "lead their particular way of life which may be associated with the use of land resources and traditional activities".[224] Although the legal recognition of Indigenous peoples' right to their traditional lands and territories is addressed several times, recommendations vis-à-vis their lands and resources are usually not directly linked to heritage in the concluding observations of the HRC.[225]

---

221 *ibid*; CCPR 'Concluding Observations: Japan' (2008) CCPR/C/JPN/CO/5, para. 32; CCPR 'Concluding Observations: USA' (2014) UN Doc CCPR/C/USA/CO/4, para. 25.

222 CCPR 'Concluding Observations: Finland' (2013) UN Doc CCPR/C/FIN/CO/6, para. 16; CCPR 'Concluding Observations: Costa Rica' (2016) UN Doc CCPR/C/CRI/CO/6, para. 42.

223 CCPR 'Guidelines for the treaty-specific document to be submitted by States parties under article 40 of the International Covenant on Civil and Political Rights' (22 November 2010) UN Doc CCPR/C/2009/1, para. 28 (article 1 of the ICCPR); CCPR 'Concluding Observations: Russian Federation' (2015) UN Doc CCPR/C/RUS/CO/7, para. 24 (articles 2 and 27 of the ICCPR); CCPR 'Concluding Observations: Ethiopia' (2022) UN Doc CCPR/C/ETH/CO/2, para. 48 (a) (articles 2, 14 and 27 of the ICCPR); CCPR 'Concluding Observations: Viet Nam' (2019) UN Doc CCPR/C/VNM/CO/3, para. 55 (articles 2, 26-27 of the ICCPR); CCPR 'Concluding Observations: USA' (2014) UN Doc CCPR/C/USA/CO/4, para. 25 (article 27 of the ICCPR).

224 CCPR 'Guidelines for the treaty-specific document to be submitted by States parties under article 40 of the International Covenant on Civil and Political Rights' (22 November 2010) UN Doc CCPR/C/2009/1, para. 104.

Also see concluding observations, e.g., CCPR 'Concluding Observations: Colombia' (2016) UN Doc CCPR/C/COL/CO/7, para. 43; CCPR 'Concluding Observations: Costa Rica' (2016) UN Doc CCPR/C/CRI/CO/6, para. 42.

225 See CCPR 'Concluding Observations: Costa Rica' (2016) UN Doc CCPR/C/CRI/CO/6, para. 42 (c); CCPR 'Concluding Observations: Russian Federation' (2015) UN Doc CCPR/C/RUS/CO/7, para. 24; CCPR 'Concluding Observations: Colombia' (2010) UN Doc CCPR/C/COL/CO/6; CCPR 'Concluding Observations: Japan' (2014) UN Doc CCPR/C/JPN/CO/6, para. 26.

The CESCR is more explicit in this regard and addresses these obligations under Article 15 (1) (a) of the ICESCR. It specifies that

> "states must respect the right of indigenous peoples to their culture and heritage and to maintain and strengthen their spiritual relationship with their ancestral lands and other natural resources traditionally owned, occupied or used by them, and indispensable to their cultural life."[226]

Accordingly, states need to "take measures to recognize and protect the rights of indigenous peoples to own, develop, control and use their communal lands, territories and resources".[227] Moreover, the CESCR recommends "fostering an enabling environment for indigenous peoples to preserve, develop, express and share their identity, history, culture, languages, traditions and customs."[228] Concrete examples of what favourable conditions are and how they can be best achieved and maintained are not provided.

The CESCR's General Comment No. 17 on "the right of everyone to benefit from the protection of the moral and material interests resulting from any scientific, literary or artistic production of which he or she is the author" (Art. 15 (1) (c)) further highlights the necessity to adopt measures that ensure "the effective protection of the interests of indigenous peoples relating to their productions, which are often expressions of their cultural heritage and traditional knowledge".[229] The concluding observations recommend to develop and protect through the law, the right of Indigenous peoples "to their ancestral lands, natural resources and cultural heritage, including protection of their intellectual property rights to their works which are an expression of their traditional culture and knowledge."[230] The committee emphasises the need for legal protection of traditional knowledge of Indigenous peoples or communities in the case of Jamaica

---

[226] CESCR 'General Comment No. 21' (21 December 2009) UN Doc E/C.12/GC/21, para. 49 (d).

[227] *ibid* para. 36. Also see CESCR 'Concluding Observations: Costa Rica' (2016) UN Doc E/C.12/CRI/CO/5, para. 9 (c); CESCR 'Concluding Observations: Mexico' (2018) UN Doc E/C.12/MEX/CO/5-6, para. 68.

[228] CESCR 'Concluding Observations: Mexico' (2018) UN Doc E/C.12/MEX/CO/5-6, para. 68. Also see, CESCR 'Concluding Observations: Colombia' (2017) UN Doc E/C.12/COL/CO/6, para. 66.

[229] CESCR 'General Comment No. 17' (12 January 2006) UN Doc E/C.12/GC/17, para. 32.

[230] CESCR 'Concluding Observations: Russian Federation' (2011) UN Doc E/C.12/RUS/CO/5, p. 10, para. 34.

and New Zealand.[231] In the latter, the concluding observations expound that the obligation to protect the cultural rights of the Māori includes inter alia their "right to conserve, promote and develop their own culture, language and cultural heritage, traditional knowledge and traditional cultural expressions".[232]

4.4.5 Ensuring Effective Participation in Decision Making Concerning Heritage, Including Free, Prior and Informed Consent

A positive obligation under Article 27 of the ICCPR is to ensure the "effective participation of members of minority communities in decisions which affect them".[233] The concluding observations show that such decisions can concern "their rights, way of life and culture"[234] and relate to infrastructure, (development) projects and extractive industries.[235] In addition, the reporting under Article 1 of the ICCPR and ICESCR sets out duly consultations with Indigenous

[231] CESCR 'Concluding Observations: Jamaica' (2013) UN Doc E/C.12/JAM/CO/3-4, para. 32.

[232] CESCR 'Concluding Observations: New Zealand' (2016) UN Doc E/C.12/NZL/CO/3, para. 26.

[233] CCPR 'General Comment No. 23: Article 27 (Rights of Minorities)' (8 April 1994) UN Doc CCPR/C/21/Rev.1/Add.5, 7. Also see CCPR 'Guidelines for the treaty-specific document to be submitted by States parties under article 40 of the International Covenant on Civil and Political Rights' (22 November 2010) UN Doc CCPR/C/2009/1, para. 104.

[234] CCPR 'Concluding Observations: USA' (2023) UN Doc CCPR/C/USA/CO/5, para. 66.

[235] Also see: CCPR 'Concluding Observations: Ethiopia' (2022) UN Doc CCPR/C/ETH/CO/2, para. 47; CCPR 'Concluding Observations: Tanzania' (2009) UN Doc CCPR/C/TZA/CO/4, para. 26; CCPR 'Concluding Observations: Angola' (2019) UN Doc CCPR/C/AGO/CO/2, para. 49; CCPR 'Concluding Observations: Namibia (2016) UN Doc CCPR/C/NAM/CO/2, para. 44; CCPR 'Concluding Observations: Costa Rica' (2016) UN Doc CCPR/C/CRI/CO/6, para. 42; CCPR 'Concluding Observations: Mexico' (2019) UN Doc CCPR/C/MEX/CO/6, para. 44; CCPR 'Concluding Observations: Canada' (2015) UN Doc CCPR/C/CAN/CO/6, para. 16; CCPR 'Concluding Observations: Viet Nam' (2019) UN Doc CCPR/C/VNM/CO/3, para. 56; CCPR 'Concluding Observations: Russian Federation' (2022) UN Doc CCPR/C/RUS/CO/8, para. 40; CCPR 'Concluding Observations: Finland' (2021) UN Doc CCPR/C/FIN/CO/7, para. 42.

and local communities as well as the extent to which "their prior informed consent is sought in any decision-making processes affecting their rights and interests under the Covenant".[236]

The CESCR's General Comment No. 21 declares obtaining the free, prior, and informed consent of minority or Indigenous communities—especially when the preservation of their cultural resources is concerned—a core obligation of the state parties under Article 15 (1) (a) of the ICESCR.[237] The Committee further asserts that the obligation to respect requires states to ensure to participate freely, actively and in an informed manner, free from discrimination "in any important decision-making process that may have an impact on his or her way of life and [...] rights under article 15, paragraph 1 (a)".[238] In the concluding observations the legal basis of such recommendations is not always indicated but often Article 1 (2) of the ICESCR, the right to freely dispose of natural wealth and resources, is cited.[239] This suggests that any framing of a right to free, prior and informed consent is in practice applied to Indigenous communities, not all minority groups. This is supported by the comments by both Committees in this context, which are directed at Indigenous peoples.[240]

But what does effective participation entail according to the treaty bodies? The concluding observations of both Committees either refer to effectively consult communities,[241] their effective or meaningful participation,[242] seeking or

---

[236] CCPR 'Guidelines for the treaty-specific document to be submitted by States parties under article 40 of the International Covenant on Civil and Political Rights' (22 November 2010) UN Doc CCPR/C/2009/1, para. 8.

[237] CESCR 'General Comment No. 21' (21 December 2009) UN Doc E/C.12/GC/21, para. 55 (e). Also see paras. 36-37.

[238] CESCR 'General Comment No. 21' (21 December 2009) UN Doc E/C.12/GC/21, para. 49 (e). Also see paras. 55 (e), 49 (e).

[239] See, for example, CESCR 'Concluding Observations: New Zealand' (2018) UN Doc E/C.12/NZL/CO/4, para. 8; CESCR 'Concluding Observations: Costa Rica' (2016) UN Doc E/C.12/CRI/CO/5, para. 8.

[240] Also addressing (traditional) communities more generally: CCPR 'Concluding Observations: Viet Nam' (2019) UN Doc CCPR/C/VNM/CO/R.3, para. 55; CESCR 'Concluding Observations: Colombia' (2017) UN Doc E/C.12/COL/CO/6, para. 18 (b); CESCR 'Concluding Observations: Brazil' (2023) UN Doc E/C.12/BRA/CO/3, para. 20 (a); CCPR 'Concluding Observations: Australia' (2017) UN Doc CCPR/C/AUS/CO/6, para. 50.

[241] CCPR 'Concluding Observations: Costa Rica' (2016) UN Doc CCPR/C/CRI/CO/6, paras. 41-42; CCPR 'Concluding Observations: Mexico' (2010) UN Doc CCPR/C/MEX/CO/5, para. 22; CCPR 'Concluding Observations: USA' (2014) UN Doc CCPR/C/USA/CO/4, para. 25.

[242] CCPR 'Concluding Observations: New Zealand' (2016) UN Doc CCPR/C/NZL/CO/6, para. 46 (b) and (c); CCPR 'List of Issues: Australia' (2012) UN Doc CCPR/C/AUS/Q/6,

attempting to obtain their free and informed consent,[243] a "requirement of free, prior and informed consent"[244], "legal and procedural guarantees for obtaining the free, prior and informed consent"[245], and, in a few cases, recommend ensuring their *right* to free, prior, and informed consent[246].

The views of the HRC underline the importance

> "that measures that compromise or interfere with the culturally significant economic activities of an indigenous community are taken with the free, prior and informed consent of the members of the community and respect the principle of proportionality so as not to endanger the very survival of the community."[247]

Furthermore, the views of the HRC offer insight into these positive measures: Effective consultations entail strengthening decision-making powers of representatives of Indigenous communities in the relevant institutions.[248] Participation in decision-making plays an important factor in the HRC's assessment of the acceptability of the measures[249] as well as in the admissibility of those measures that

---

para. 23; CESCR 'Concluding Observations: New Zealand' (2018) UN Doc E/C.12/NZL/CO/4, para. 8.

[243] See CCPR 'Concluding Observations: Namibia' (2016) UN Doc CCPR/C/NAM/CO/2, para. 44; CCPR 'Concluding Observations: Colombia' (2016) UN Doc CCPR/C/COL/CO/7, para. 43; CCPR 'Concluding Observations: USA' (2014) UN Doc CCPR/C/USA/CO/4, para. 25, CCPR 'Concluding Observations: Russian Federation' (2015) UN Doc CCPR/C/RUS/CO/7, para. 24. Also see, CCPR 'Concluding Observations: Dominican Republic' (2020) UN Doc CCPR/C/DMA/COAR/1, para. 48 (c); CCPR 'Concluding Observations: Mexico' (2019) UN Doc CCPR/C/MEX/CO/6, para. 45.

[244] CESCR 'Concluding Observations: Russian Federation' (2016) UN Doc E/C.12/RUS/CO/6, paras. 14-15; CESCR 'Concluding Observations: New Zealand' (2018) UN Doc E/C.12/NZL/CO/4, para. 9 (e).

[245] CESCR 'Concluding observations: Finland' (2021) UN Doc E/C.12/FIN/CO/7, para. 51. Also see CESCR 'Concluding Observations: Australia' (2017) UN Doc E/C.12/AUS/CO/5, para. 16 (e).

[246] CCPR 'Concluding Observations: Canada' (2015) UN Doc CCPR/C/CAN/CO/6, para. 16; CCPR 'Concluding Observations: Viet Nam' (2019) UN Doc CCPR/C/VNM/CO/3, para. 56 (b); CESCR 'Concluding Observations: Canada' (2016) UN Doc E/C.12/CAN/CO/6, para. 14.

[247] CCPR, Communication No. 2552/2015, *Benito Oliveira Pereira et al. v. Paraguay* (12 October 2021) UN Doc CCPR/C/132/D/2552/2015, para. 8.7.

[248] CCPR 'Concluding Observations: Finland' (2013) UN Doc CCPR/C/FIN/CO/6, para. 16.

[249] CCPR, Communication No 547/1993, *Apirana Mahuika and Ors v New Zealand* (27 October 2000) UN Doc CCPR/C/70/D/547/1993, para. 9.5. Also see, CCPR, Communication No. 511/1992, *Länsman v. Finland* (14 October 1993) UN Doc CCPR/C/49/D/511/1992, paras. 9.6, 9.8.

compromise or interfere "with the culturally significant economic activities of a minority or Indigenous community"[250]. In practice, the HRC criticises individual aspects, such as the lack of consultations with the community, the absence of a study by a competent and independent body to determine the impact of the construction of the traditional economic activity, and/or the lack of steps taken to minimise the negative consequences or even to repair them.[251] Moreover, measures need to be proportionate to ensure the survival of the members of the community.[252] The Committee further assesses whether the measure will allow the minority community to continue benefitting from their traditional economy.[253]

In *Poma Poma v. Peru*, the HRC found—amid the lack of effective participation—a violation of Article 27 of the ICCPR.[254] Furthermore, the committee argued that whilst states may promote their economic development, the promotion may not undermine the rights set out in Article 27 of the ICCPR.[255] The committee shed light on what effective participation entails by requiring "not mere consultation but the free, prior, and informed consent of the members of the community".[256] The communication *Mahuika et al. v. New Zealand*, where the HRC did *not* find a violation of any of the Covenant rights, shows that the "complicated" consultation process undertaken by the state party was sufficient.[257] The Committee inter alia noted the state's consideration of the input of the Māori communities, that the state party proceeded to legislate only upon the confirmation of Māori representatives that substantial Māori support existed,

---

[250] CCPR, Communication No. 1457/2006, *Ángela Poma Poma v. Peru* (24 April 2009) UN Doc CCPR/C/95/D/1457/2006, para. 7.6.

[251] *ibid* para. 7.7.

[252] *ibid* para. 7.6. Also see, CCPR, Communication No. 2102/2011, *Paadar et al. v. Finland* (10 April 2014) UN Doc CCPR/C/110/D/2102/2011, p. 16.

[253] CCPR, Communication No 547/1993, *Apirana Mahuika and Ors v New Zealand* (27 October 2000) UN Doc CCPR/C/70/D/547/1993, p. 15, para. 9.5. Also see: CCPR, Communication No. 511/1992, *Länsman v. Finland* (14 October 1993) UN Doc CCPR/C/49/D/511/1992, paras. 9.6 and 9.8 and CCPR, Communication No. 1457/2006, *Ángela Poma Poma v. Peru* (24 April 2009) UN Doc CCPR/C/95/D/1457/2006, para. 7.6.

[254] CCPR, Communication No. 1457/2006, *Ángela Poma Poma v. Peru* (24 April 2009) UN Doc CCPR/C/95/D/1457/2006, para. 7.6.

[255] *ibid* para. 7.4.

[256] *ibid* para. 7.6.

[257] CCPR, Communication No 547/1993, *Apirana Mahuika and Ors v New Zealand* (27 October 2000) UN Doc CCPR/C/70/D/547/1993, para. 9.6.

and upon considering the compatibility of the Fisheries Settlement with Māori fishing rights under Article 27 of the ICCPR.[258]

The concluding observations of the HRC criticise interpretations of the principle of free, prior and informed consent that are restrictive[259] and regularly remind states to "guarantee" and "seek" the free, prior, and informed consent in situations where the traditional way of life, land rights, and culture are impacted[260] and where the principle is "routinely violated"[261]. Furthermore, they express concern over vague criteria for impact assessments, including in development projects concerning the culture and traditional livelihoods of the Sami, which have resulted in the "failure to engage in meaningful consultations to obtain their free, prior and informed consent".[262] The concluding observations on Ethiopia's report express concern that the continuation of development projects that may affect the rights of Indigenous peoples and that lack "prior full and meaningful consultations with affected communities" and that fail to publish independent impact assessments or implement necessary safeguards.[263] Whilst the concluding observations on the report of the United States of America recommend holding consultations "with a view to obtaining [...] free, prior and informed consent",[264] in its views, the HRC is clear that mere consultations do not suffice, stating unequivocally that effective participation *requires* "the free, prior and informed consent of the members of the community".[265] Most recently, the committee can also be seen encouraging

[258] *ibid* paras. 9.6, 9.8.

[259] CCPR 'Concluding Observations: USA' (2023) UN Doc CCPR/C/USA/CO/5, para. 66.

[260] CCPR 'Concluding Observations: Brazil' (2023) UN Doc CCPR/C/BRA/CO/3, para 68 (a); CCPR 'Concluding Observations: Finland' (2013) UN Doc CCPR/C/FIN/CO/6, para. 16; CCPR 'Concluding Observations: Namibia' (2016) UN Doc CCPR/C/NAM/CO/2, para. 44; CCPR 'Concluding Observations: Mexico (2010) UN Doc CCPR/C/MEX/CO/5, para. 22, CCPR 'Concluding Observations: Colombia' (2010) UN Doc CCPR/C/COL/CO/6, para. 25.

[261] CCPR 'Concluding Observations: Brazil' (2023) UN Doc CCPR/C/BRA/CO/3, para. 67.

[262] CCPR 'Concluding Observations: Finland' (2021) UN Doc CCPR/C/FIN/CO/7, para. 42.

[263] CCPR 'Concluding Observations: Ethiopia' (2022) UN Doc CCPR/C/ETH/CO/2, para. 47.

[264] CCPR 'Concluding Observations: USA' (2014) UN Doc CCPR/C/USA/CO/4, para. 25.

[265] "In the Committee's view, the admissibility of measures which substantially compromise or interfere with the culturally significant economic activities of a minority or indigenous community depends on whether the members of the community in question have had the opportunity to participate in the decision-making process in relation to these measures and whether they will continue to benefit from their traditional economy. The Committee considers that participation in the decision-making process must be effective, which requires not mere consultation but the free, prior and informed consent of the members of the community.

the adoption of "a methodology that guarantees Indigenous Peoples' right to free, prior and informed consultation, in accordance with international standards".[266]

In the concluding observations of the CESCR, the committee specifies that the state ought to conduct and publish independent studies that address the potential impact of large projects on the lives of Indigenous peoples and traditional communities, taking into account the social, environmental and human rights impact.[267] The CESCR also addresses consultations in the context of Indigenous knowledge and scientific development.

The CESCR's General Comment No. 25 on science and economic, social and cultural rights (article 15 (1) (b), (2), (3) and (4)' ICESCR) emphasises taking

> "all measures to respect and protect the rights of indigenous peoples, particularly their land, their identity and the protection of the moral and material interests resulting from their knowledge, of which they are authors, individually or collectively."[268]

The committee further underscores the necessity for state and non-state actors to engage in "genuine consultation in order to obtain free, prior and informed consent" whenever decisions are taken and policies are created that relate to science and impact Indigenous peoples or when their knowledge is used.[269]

4.4.6 Interim Conclusions

The analysis identifies five obligations: (1) promoting and protecting cultural diversity; (2) respecting and protecting cultural heritage at all times; (3) guaranteeing access to cultural, linguistic and religious heritage; (4) respecting, protecting, and recognising the heritage of persons belonging to minorities and Indigenous communities and creating favourable conditions; and (5) ensuring effective participation in decision making concerning heritage.

In addition, the measures must respect the principle of proportionality so as not to endanger the very survival of the community and its members." CCPR, Communication No. 1457/2006, *Ángela Poma Poma v. Peru* (24 April 2009) UN Doc CCPR/C/95/D/1457/2006, para. 7.6.

266 CCPR 'Concluding Observations: Colombia' (2021) UN Doc CCPR/C/COL/CO/8, para. 39.

267 CESCR 'Concluding Observations: Brazil' (2023) UN Doc E/C.12/BRA/CO/3, para. 20 (a).

268 CESCR 'General Comment No. 25' (30 April 2020) UN Doc E/C.12/GC/25, para. 40.

269 *ibid*.

Promoting and protecting cultural diversity, on the one hand, and respecting and protecting cultural heritage at all times, on the other, can be framed as underlying obligations that serve all other rights and obligations. They are both rooted in the CESCR's interpretation of Article 15 (1) (a) of the ICESCR.[270]

Analogous to the rights that the analysis has previously identified, the obligations focus on guaranteeing access to cultural heritage, places of worship and sacred sites, and linguistic heritage. Access to linguistic heritage is an important component of ensuring access to heritage, as reflected in the breadth of pronouncements by the committees on this subject. The HRC's interpretation of Article 27 of the ICCPR highlights the recognition of linguistic minorities and withdrawing reservations to the provisions. The views of the committee, however, also clearly illustrate limits to the scope of the right. The CESCR views the use of language as an aspect of the right to take part in cultural life both in public and in private. It underlines the legal recognition, promotion, and preservation of minority languages under the right to education and the right to take part in cultural life.

The obligation to respect, protect, and recognise the heritage of persons belonging to minorities and Indigenous communities and to create favourable conditions can be structured into obligations towards persons belonging to minorities and more specific obligations that apply to Indigenous peoples. The first category includes a plethora of obligations that seek to ensure the promotion of the heritage of persons belonging to minorities and the creation of favourable conditions to enable the preservation, development, expression and dissemination of the "identity, history, culture, language, traditions and customs" of persons belonging to minorities.[271]. The second category entails obligations tailored to the specific needs of Indigenous peoples and includes the legal recognition and effective legal protection of their rights to their lands, territories, and sacred areas that they have traditionally occupied or used as sources of livelihood and heritage.

The HRC's interpretation of Article 27 of the ICCPR underscores the legal recognition of minorities and Indigenous peoples. This includes emphasising the need for measures that ensure that Indigenous communities can lead their particular way of life. The link to cultural heritage is not strongly articulated in the obligations set out by the HRC. States' obligations of recognising, respecting and protecting minority culture as an important component of their states'

[270] See CESCR 'General Comment No. 21' (21 December 2009) UN Doc E/C.12/GC/21, para. 52 (a) and 50 (a).

[271] CESCR 'Guidelines on Treaty-Specific Documents to be Submitted by State Parties under Articles 16 and 17 of the International Covenant on Economic, Social and Cultural Rights' (24 March 2009) UN Doc E/C.12/2008/2, 68.

identity stems from the CESCR's interpretation of Article 15 (1) (a) in General Comment No. 21. Moreover, the CESCR's general comment and concluding observations serve to accentuate Article 15 (1) (c) of the ICESCR as a legal basis. They do so by underlining the adoption of measures that ensure that Indigenous peoples' interests in relation to their productions—often linked to expressions of their cultural heritage and traditional knowledge—are effectively protected.

Finally, the committees set out obligations for states to ensure the effective participation of persons belonging to minorities in decisions that affect them, including those that impact their way of life. Consultations with Indigenous and local communities are set out under Articles 1 common to the ICESCR and ICCPR, Article 15 (1) (a) of the ICESCR and Article 27 of the ICCPR. The pronouncements of the committees make clear that consultations are not enough and that states ought to obtain the free, prior and informed consent of the affected community.

4.5 Conclusions: Cultural Heritage through the Lens of the CESCR and HRC

This chapter has analysed the interpretation of two treaty bodies with a specific view to their pronouncements on meaning, rights, and obligations vis-à-vis cultural heritage. The pronouncements of the CESCR and the HRC show both a focus on threats to cultural heritage that impact rights, especially the rights of persons belonging to minorities and Indigenous communities, and threats that directly impact cultural heritage. The concluding observations address, for example, the deterioration and loss of cultural heritage that lacks protection from illegal excavations, vandalization, looting, targeting and deliberate destruction; the preservation, protection, and promotion of cultural diversity, languages, traditions, and traditional ways of life of communities as well as forcible evictions, resettlements, and displacement from inter alia ancestral or traditional lands for the purposes of large scale economic activities that have a negative cultural impact on the communities.[272]

[272] CESCR 'Concluding Observations: Afghanistan' (2010) UN Doc E/C.12/AFG/Co/2-4, para. 44; CESCR 'Concluding Observations: Mali' (2018) UN Doc E/C.12/MLI/CO/1, para. 53; CESCR 'Concluding Observations: Egypt' (2013) UN Doc E/C.12/EGY/CO/2-4, para. 23; CESCR 'Concluding Observations: Tanzania' (2012) UN Doc E/C.12/TZA/CO/1-3, para. 29; CESCR 'Concluding Observations: Turkey' (2011) UN Doc E/C.12/TUR/CO/1, para. 26.

The analysis has provided an overview of how the CESCR and the HRC understand culture and cultural heritage and how they articulate rights and obligations relating to cultural heritage. The CESCR has adopted a broad understanding of cultural life that includes cultural heritage in its various forms. Considering its evolution, the notion was initially conceptualised by reference to material manifestations, especially high culture, as evident in the *travaux préparatoires* of the ICESCR and reflected in, for example, the CESCR's reporting guidelines.

In the 1991 reporting guidelines the CESCR requested states to provide information on the "preservation and presentation of mankind's cultural heritage".[273] With the adoption of the current reporting guidelines of 2009 the wording has evolved to refer to "broad participation in, and access to, cultural goods, institutions and activities" and to enhance "access to the cultural heritage of mankind".[274] This shift can be interpreted to reflect a broader and more holistic understanding of cultural heritage protection and a deliberate inclusion of rights regarding meaningful participation and access by everyone. Indeed, the committee's recommendation that a state party goes "beyond the creation of museums [...] as a way of preserving and promoting culture" and shift their focus to consulting communities that may be negatively impacted by development projects in the concluding observations can be read as evidence of this too.[275] The evolving understanding can be considered parallel to the evolving use in terminology, specifically the shift from cultural property to cultural heritage in international cultural heritage law.

The CESCR applies Article 15 (1) (a) of the ICESCR in two ways: first, the committee promotes broad and popular cultural participation, including the need to report on measures to protect cultural diversity.[276] This is often framed as an

[273] CESCR 'Revised General Guidelines Regarding the Form and Contents of Reports to be Submitted by States Parties under Articles 16 and 17 of the International Covenant on Economic, Social and Cultural Rights' (17 June 1991) UN Doc E/C.12/1991/1, Art. 15, 1(f). Also see Human Rights Instruments, Compilation of Guidelines on The Form and Content of Reports to Be Submitted by States Parties to The International Human Rights Treaties, HRI/GEN/2/Rev.1, 9 May 2001, 66 (b) (d) (f).

[274] CESCR 'Guidelines on Treaty-Specific Documents to be Submitted by State Parties under Articles 16 and 17 of the International Covenant on Economic, Social and Cultural Rights' (24 March 2009) UN Doc E/C.12/2008/2, 67(b).

[275] CESCR 'Concluding Observations: India' (2008) UN Doc E/C.12/IND/CO/5, para. 84.

[276] See CESCR 'Guidelines on Treaty-Specific Documents to be Submitted by State Parties under Articles 16 and 17 of the International Covenant on Economic, Social and Cultural Rights' (24 March 2009) UN Doc E/C.12/2008/2, p. 15, para. 68. 'Cultural diversity' describes "the factual situation of the existence of cultural differences between and within

individual right and is in line with the wording and the focus of the article, stating that *everyone* has the right to take part in cultural life. There is however a clear collective dimension[277] to this individual right. The committee recognises that a person may exercise cultural rights in different ways and has explained that "everyone" refers to the individual or to the collective.[278] Here, it has distinguished between the exercise "(a) as an individual, (b) in association with others, or (c) within a community or group, as such."[279] Second, the CESCR applies this provision to address the rights of persons belonging to minorities and Indigenous communities to enjoy their own culture, specifically referring to their cultural identity, cultural heritage, and way of life. The Committee thereby subsumes and emphasizes cultural rights of members of minority groups and Indigenous communities under Article 15 (1) (a), offering a further legal basis.

The right to take part in cultural life, Article 15 (1) (a), thereby becomes the CESCR's legal basis for virtually all the rights and obligations identified in this analysis. Most prominently, Article 15 (1) (a) becomes the basis for access to cultural heritage.[280] This does not only refer to access to cultural heritage but also encompasses access to linguistic heritage.[281] Furthermore, Article 15 (1) (a) includes the obligations to protect and promote cultural diversity and to respect and protect cultural heritage at all times.[282]

The CESCR's focus on minority rights in Article 15 (1) (a) also covers that minorities and persons belonging to minorities have the right to "conserve, promote and develop their own culture."[283] This includes the right to follow a particular way of life.[284] Correspondingly, the committee asserts the obligations

---

states". The notion of 'cultural pluralism' describes "the way such cultural diversity is valued and translated in laws and policies." Donders (2014), p. 205.

[277] Lenzerini (2016b), p. 137.

[278] CESCR 'General Comment No. 17' (12 January 2006) UN Doc E/C.12/GC/17, para. 7; CESCR 'General Comment No. 21' (21 December 2009) UN Doc E/C.12/GC/21, para. 9.

[279] CESCR 'General Comment No. 21' (21 December 2009) UN Doc E/C.12/GC/21, para. 9.

[280] *ibid* para. 15.

[281] *ibid* para. 13.

[282] See *ibid* para. 52 (a). Also see CESCR 'Guidelines on Treaty-Specific Documents to be Submitted by State Parties under Articles 16 and 17 of the International Covenant on Economic, Social and Cultural Rights' (24 March 2009) UN Doc E/C.12/2008/2, 68. *ibid* para. 50 (a).

[283] CESCR 'General Comment No. 21' (21 December 2009) UN Doc E/C.12/GC/21, para 32.

[284] *ibid* para. 15.

to "recognize, respect and protect minority cultures as an essential component of the identity of the States themselves"[285] and to "create favourable conditions for [persons belonging to minorities] [...] to preserve, develop, express and disseminate their identity, history, culture, language, traditions and customs".[286]

In the case of Indigenous peoples, this further implies their right to "maintain, control, protect and develop their cultural heritage, traditional knowledge and traditional cultural expressions, as well as the manifestations of their sciences, technologies and cultures".[287] This means that states need to "take measures to recognize and protect the rights of indigenous peoples to own, develop, control and use their communal lands, territories and resources".[288] Article 15 (1) (c) of the ICESCR offers another legal basis for ensuring the enjoyment and benefits from Indigenous- and local communities' knowledge, innovations, and practices.[289] Articles 15 (1) (a) and (c) also form the basis for effective consultations, participation in decision-making and the principle of free, prior and informed consent "in all matters covered by their specific rights".[290]

The CESCR's reporting guidelines have evolved to highlighting Articles 13–15 ICESCR as important legal bases for linguistic heritage. The CESCR recognises the right to education, Articles 13–14 of the ICESCR, as an important means for the education on one's own culture and that of others.[291] The right to education is, furthermore, highlighted as key to transmitting elements of heritage, chiefly language and linguistic heritage, customs, and cultural identity.[292]

---

[285] *ibid* para 32.

[286] CESCR 'Guidelines on Treaty-Specific Documents to be Submitted by State Parties under Articles 16 and 17 of the International Covenant on Economic, Social and Cultural Rights' (24 March 2009) UN Doc E/C.12/2008/2, 68.

The CESCR recommends promoting awareness of the cultural heritage of minorities. See CESCR 'Concluding Observations: Mali' (2018) UN Doc E/C.12/MLI/CO/1, para. 54; CESCR 'Concluding Observations: UK' (2009) UN Doc E/C.12/GBR/CO/5, para. 37.

[287] CESCR 'General Comment No. 21' (21 December 2009) UN Doc E/C.12/GC/21, para. 37.

[288] *ibid* para. 36. Also see CESCR 'Concluding Observations: Costa Rica' (2016) UN Doc E/C.12/CRI/CO/5, para. 9 (c); CESCR 'Concluding Observations: Mexico' (2018) UN Doc E/C.12/MEX/CO/5-6, para. 68.

[289] CESCR 'General Comment No. 17' (12 January 2006) UN Doc E/C.12/GC/17, para. 9.

[290] CESCR 'General Comment No. 21' (21 December 2009) UN Doc E/C.12/GC/21, para. 37. *ibid* para. 32.

[291] CESCR 'General Comment No. 21' (21 December 2009) UN Doc E/C.12/GC/21, para. 49 (d).

[292] *ibid* para. 2.

Whilst General Comment No. 21 underscores access to linguistic heritage within the scope of Article 15 (1) (a),[293] the committee's concluding observations emphasise the official recognition of minority languages, their promotion, and their preservation.[294] The CESCR underscores access to linguistic heritage and "language, oral and written literature, music and song, [and] non-verbal communication" under the notion of "participation" in Article 15 (1) (a) of the ICESCR as well as in some cases more generally under Articles 1 and 15 of the ICESCR.[295] Recommendations concerning Indigenous and minority languages and mother-tongue instruction in the concluding observations are mainly addressed without explicit references to the transmission of cultural heritage.[296]

Article 27 of the ICCPR serves as the primary provision to protect elements of cultural heritage in the pronouncements of the HRC. Article 27 of the ICCPR protects the rights of persons belonging to ethnic, religious or linguistic minorities "to enjoy their own culture, to profess and practise their own religion, or to use their own language". The HRC's interpretation of the concept of culture in Article 27 encompasses the right of persons belonging to minorities to follow a particular or traditional way of life. This is a particularly important concept for Indigenous peoples applied in the committee's views, as it recognises the connection between lands and territories and the rights of Indigenous peoples, which go beyond cultural rights but can also intersect with their survival.[297]

The HRC uses effective consultations and participation in decision-making processes as an important yardstick in deciding on the acceptability of measures

---

[293] Emphasis added. *ibid* para. 49 (d).

[294] See e.g., CESCR 'Concluding Observations: France' (2008) UN Doc E/C.12/FRA/CO/3, para. 30. CESCR 'Concluding Observations: Costa Rica' (2008) UN Doc E/C.12/CRI/CO/4, para. 7; CESCR 'Concluding Observations: Russian Federation' (2017) UN Doc E/C.12/RUS/CO/6, para. 59; CESCR 'Concluding Observations: UK (2016) UN Doc E/C.12/GBR/CO/6, paras. 67-68; CESCR 'Concluding Observations: Canada' (2016) UN Doc E/C.12/CAN/CO/6, paras. 59-60; CESCR 'Concluding Observations: Angola' (2016) UN Doc E/C.12/AGO/CO/4-5, paras. 55-56; CESCR 'Concluding Observations: Uzbekistan' (2014) UN Doc E/C.12/UZB/CO/2, para. 26; CESCR 'Concluding Observations: Japan' (2013) UN Doc E/C.12/JPN/CO/3, para. 30.

[295] International Covenant on Economic, Social and Cultural Rights (adopted 16 Dec 1966, entered into force 3 Jan 1976) 993 UNTS 3 (*ICESCR*), para. 15 (a). Also see CESCR 'General Comment No. 17' (12 January 2006) UN Doc E/C.12/GC/17, para. 49 (b). Also see CESCR 'Concluding Observations: France' (2016) UN Doc E/C.12/FRA/CO/4, para. 57.

[296] CESCR 'Concluding Observations: Australia' (2009) UN Doc E/C.12/AUS/CO/4, para. 33 (a).

[297] See, e.g., CCPR, Communication No. 1457/2006, *Ángela Poma Poma v. Peru* (24 April 2009) UN Doc CCPR/C/95/D/1457/2006, para. 7.2.

under communications alleging violations of Article 27. In its views, the committee makes unequivocally clear that effective participation requires free, prior and informed consent of the affected community.[298]

In considering the protection of linguistic minorities and linguistic heritage under Article 27 of the ICCPR, the committee underlines the importance of the recognition of minorities, and specifically the legal recognition of linguistic minorities[299] as well as the protection of indigenous languages.[300] However, the HRC has also shown the limits of positive obligations towards linguistic minorities in its views.[301] The Committee's decision to find further cases concerning the Breton language inadmissible has been interpreted "as a negative result for the safeguarding of intangible heritage through human rights mechanisms because the Committee overlooked cultural significance of the language in order to favour an easier, less cumbersome, and economical solution for the State (...)".[302]

The HRC has expounded on the notions of privacy, family, and home within the meaning of Article 17 of the ICCPR. Amid threats and damage caused by developments, toxins, pollution, and climate change to the privacy, family, and home, the HRC has found that elements of the authors' particular or traditional way of life fall under Article 17 of the ICCPR. Reliance on privacy and family life does not appear to be common practice in the reporting mechanism, although it is interesting to note the occasional reference to the link between cultural heritage and family life in the reporting mechanism.[303]

---

298 *ibid* para. 7.6.

299 CCPR 'Concluding Observations: France' (2015) UN Doc CCPR/C/FRA/CO/5, para. 6; CCPR 'Concluding Observations: Ethiopia' (2011) UN Doc CCPR/C/ETH/CO/1, para. 26.

300 E.g., CCPR 'Concluding Observations: Canada' (2015) UN Doc CCPR/C/CAN/CO/6, para. 19.

301 See CCPR, Communication Nos. 221/1987 and 323/1988, *Yves Cadoret and Hervé Le Bihan v. France* (11 April 1991) UN Doc CCPR/C/41/D/221/1987.

302 Lixinski (2013a), p. 163.

303 See, "...an increasing trend in the number of persons underscoring the moral role of the family. That reflects the promotion of cultural continuity through the transmission of the cultural heritage to younger generations, the stabilization of the social structure and the social scrutiny of behaviour." In: CCPR 'Periodic report submitted by Uzbekistan' (2012) UN Doc CCPR/C/UZB/4, para. 787.

Also see, "...strengthening families in ethnic groups by supporting processes and activities that promote the development of families in indigenous communities, in those of African descent, and in Raizal and Roma families to enable them to reaffirm their cultural identities, uses and customs." In: CCPR 'Periodic report submitted by Colombia' (2015) UN Doc CCPR/C/COL/7, para. 217.

Whereas the reporting guidelines provide an evident legal basis for access to places of worship with Article 18 of the ICCPR[304], the provision is not commonly used in the context of religious heritage or sacred places, and often the committee instead uses Article 27 to address restrictions placed on sites of sacred or religious value and cultural significance.[305]

Reference List

Donders, Yvonne (2020): Cultural Heritage and Human Rights. In: Francesco Francioni and Ana Filipa Vrdoljak (eds.): The Oxford Handbook of International Cultural Heritage Law. Oxford (Oxford University Press), S. 379–406.

Donders, Yvonne (2016): Foundations of Collective Cultural Rights in International Human Rights Law. In: Andrzej Jakubowski (ed.): Cultural Rights as Collective Rights. Leiden (Brill | Nijhoff), S. 85–112.

Donders, Yvonne (2014): Cultural Pluralism in International Human Rights Law: The Role of Reservations. In: Ana Filipa Vrdoljak (ed.): The Cultural Dimension of Human Rights. Oxford (Oxford University Press), S. 205–239.

Lenzerini, Federico (2016b): The Safeguarding of Collective Cultural Rights through the Evolutionary Interpretation of Human Rights Treaties and Their Translation into Principles of Customary International Law. In: Andrzej Jakubowski (ed.): Cultural Rights as Collective Rights. Leiden (Brill | Nijhoff), S. 133–153.

Lixinski, Lucas (2013a): Intangible Cultural Heritage in International Law. Oxford (Oxford University Press).

Mosby (2020): Here is why we are taking the Australian government to the UN over its inaction on climate change, https://www.theguardian.com/commentisfree/2020/aug/14/here-is-why-we-are-taking-the-australian-government-to-the-un-over-its-inaction-on-climate-change, accessed 31 January 2025.

Murphy (2019): Torres Strait Islanders take climate change complaint to the United Nations, https://www.theguardian.com/australia-news/2019/may/13/torres-strait-islanders-take-climate-change-complaint-to-the-united-nations, accessed 31 January 2025.

OHCHR (2025a): Days of general discussion, https://www.ohchr.org/en/treaty-bodies/cescr/days-general-discussion, accessed 31 January 2025.

OHCHR (2025b): International standards: Special Rapporteur in the field of cultural rights, https://www.ohchr.org/en/special-procedures/sr-cultural-rights/international-standards, accessed 31 January 2025.

[304] CCPR 'Guidelines for the treaty-specific document to be submitted by States parties under article 40 of the International Covenant on Civil and Political Rights' (22 November 2010) UN Doc CCPR/C/2009/1, p. 21, para. 104.

[305] CCPR 'Concluding Observations: USA' (2014) UN Doc CCPR/C/USA/CO/4, para. 25. CCPR 'Concluding Observations: New Zealand' (2010) UN Doc CCPR/C/NZL/CO/5, para. 19.

Saul, B., Kinley, D., & Mowbray, J. (2014). *The International Covenant on Economic, Social and Cultural Rights: Commentary*. Oxford (Oxford University Press).

Taylor, Paul M. (2020): A Commentary on the International Covenant on Civil and Political Rights: The UN Human Rights Committee's Monitoring of ICCPR Rights. Cambridge (Cambridge University Press).

Vrdoljak, Ana Filipa (2014): Human Rights and Cultural Heritage in International Law. In: Federico Lenzerini and Ana Filipa Vrdoljak (eds.): International Law for Common Goods: Normative Perspectives on Human Rights, Culture and Nature. Oxford (Hart Publishing), Available at SSRN: https://ssrn.com/abstract=2200186, S. 1–39.

De Witte, Bruno (2020): Language as Cultural Heritage. In: Francesco Francioni and Ana Filipa Vrdoljak (eds.): The Oxford Handbook of International Cultural Heritage Law. Oxford (Oxford University Press), S. 371–378.

5 Cultural Heritage through the Lens of the Human Rights Council and Its Mechanisms

5.1 Introduction

In 2007, the Human Rights Council (UNHRC) adopted resolution 6/11, which included the request to the UN High Commissioner for Human Rights "to encourage all relevant human rights bodies and mechanisms to pay due attention to the issue of […] protection of cultural heritage as an important component of the promotion and protection of all human rights".[1] This chapter reviews the work of the UNHRC and selected mechanisms. This includes primarily UNHRC resolutions, the work of the Special Procedures, specifically the mandate of the Special Rapporteur in the field of cultural rights, and the Universal Periodic Review (UPR). Where appropriate, the chapter also considers resolutions of the UN Security Council[2] and the UN General Assembly that address cultural heritage.

The analysis is structured into three parts: "meaning", "rights", and "obligations". The first section sets out to understand the meaning of cultural heritage in the human rights framework. Do the charter-based bodies assert their own understanding of cultural heritage? How is cultural heritage understood? The subsequent sections attend to the rights and obligations that the charter-based bodies identify in their work. The analysis highlights three rights: (1) the ability to access and enjoy cultural heritage; (2) participation and free, prior and

[1] UNHRC Res 6/11 (28 September 2007) UN Doc A/HRC/RES/6/11, preamble, para. 8.

[2] The main responsibility of the Security Council is maintaining international peace and security. As upholding peace and human rights go hand in hand, the Security Council "is also enmeshed in the institutional framework for the protection of human rights", see Tomuschat (2019), para. 5.

L. Hofmann, *Cultural Heritage through the Lens of Human Rights Bodies*, Juridicum – Bochumer Schriften zur Friedenssicherung und zum Humanitären Völkerrecht | Bochum Studies on the International Law of Peace and Armed Conflict 73, https://doi.org/10.1007/978-3-658-51203-3_5

informed consent in decision-making; and (3) the rights of Indigenous peoples to maintain, control, protect and develop their cultural heritage. Subsequently, the analysis identifies five obligations: (1) the underlying obligation to recognise the diversity of cultural heritage; (2) to respect and protect cultural heritage and prevent its destruction; (3) to ensure the ability to access and enjoy cultural heritage; (4) to ensure effective participation, including the free, prior and informed consent, and (5) to legally recognise and protect the rights of Indigenous peoples to their cultural heritage and traditional cultural expressions and set up effective mechanism to enable these rights. The chapter identifies the legal bases of these rights and obligations and examines their scope.

5.2 Meaning: Cultural Heritage as the Resources Enabling Cultural Identification

Whilst resolutions by the UNHRC and the Security Council have addressed cultural heritage and have condemned its destruction, they have not put forward a definition of cultural heritage in the human rights framework. Instead, they refer to it in factual terms.[3] Nevertheless, Security Council resolution 2617 (2021) is notable for not only condemning the destruction of cultural heritage by third parties, i.e. terrorist groups, but by setting the destruction, looting and smuggling of cultural heritage in a bigger context. The resolution emphasises that these acts are accompanied by attempts "to deny historical roots and cultural diversity" and eventually result in "undermining the security, stability, governance, social, economic and cultural development of affected States"[4].

What is more, UNHRC resolutions and the expert mechanism have taken a critical look at the understanding of cultural heritage. UNHRC resolution 76/214 reflects a multifaceted understanding of the protection of cultural heritage

[3] See UNSC Re 2071 (12 October 2012) UN Doc S/RES/2071, 14th recital; UNSC Res 2085 (20 December 2012) UN Doc S/Res/2085, 6th recital; UNSC Res 1483 (22 May 2003) UN Doc S/RES/1483, 12th recital.

A few resolutions specify that these include artefacts, museums, libraries, and archives. See UNSC Res 2199 (12 February 2015) UN Doc S/Res/2199, paras. 15–16; UNSC Res 2347 (24 March 2017) UN Doc S/Res/2347, paras. 1, 3, 5; UNSC Res 2071 (12 October 2012) UN Doc S/Res/2071, preamble, 14th recital; UNSC Res 2085 (20 December 2012) UN Doc S/Res/2085, preamble, 6th recital; UNSC Res 2170 (15 August 2014) UN Doc S/Res/2170, preamble, 5th recital, para. 2; UNSC, Res 249 (20 November 2015) UN Doc S/Res/2249, preamble, 5th recital, para. 3.

[4] UNSC Res 2617 (30 December 2021) UN Doc S/RES/2617, preamble, 29th recital.

and the contribution of culture and cultural heritage to social development, environmental sustainability, and sustainable development.[5] UNGA resolution 79/133 emphasises the return and restitution of cultural heritage, underlining that ensuring access to these objects is key to the preservation of cultural identity in the present and the future.[6] Along the same lines, the UNHRC's expert mechanism on human rights and Indigenous peoples[7] debated the suitability of the concept of cultural heritage in the context of Indigenous heritage. They cautioned that the threefold distinction of heritage in international law, referring to tangible, intangible, and natural heritage, is unsuitable for Indigenous heritage.[8] Instead, the expert mechanism recommended that Indigenous heritage should be framed in a holistic and inter-generational manner.[9]

The Special Rapporteur in the field of cultural rights[10] has offered a definition of cultural heritage in the context of human rights law. Accordingly, cultural heritage should be understood as:

> "the resources enabling the cultural identification and development processes of individuals and groups, which they, implicitly or explicitly, wish to transmit to future generations."[11]

The definition bears resemblance to the definition of cultural heritage in the Council of Europe's Faro Convention, which understands cultural heritage as

> "a group of resources inherited from the past which people identify, independently of ownership, as a reflection and expression of their constantly evolving values, beliefs,

---

[5] See UNHRC Res 76/214 (7 January 2022) UN Doc A/RES/76/214, paras. 5, 10, 17, 18.

[6] UNGA Res 79/133 (12 December 2024) UN Doc A/RES/79/133, para. 6.

[7] UNHRC 'Expert Mechanism on Rights of Indigenous Peoples' (14 December 2007) UN Doc A/HRC/RES/6/36.

[8] UNHRC 'Promotion and Protection of the Right of Indigenous Peoples with Respect to their Cultural Heritage, Study by the Expert Mechanism on the Rights of Indigenous Peoples' (19 August 2015) UN Doc A/HRC/30/53, para. 8, Annex A.1–3., A.6.

[9] *ibid.* Also see discussion on terminology and use of "heritage" in ECOSOC 'Human Rights of Indigenous Peoples' (19 June 2000) UN Doc E/CN.4/Sub.2/2000/26.

[10] Formerly Independent Expert. UNHRC Res 10/23 (26 March 2009) UN Doc A/HRC/RES/10/23. Resolution 28/9 of 10 April 2015 and 37/12 of 22 March 2018 further extended the mandate of the Independent Expert, now the Special Rapporteur in the field of cultural rights.
See UNHRC Res 28/9 (10 April 2015) UN Doc A/HRC/RES/28/9; UNHRC Res 37/12 (22 March 2018) UN Doc A/HRC/RES/37/12.

[11] UNHRC, Report of the independent expert in the field of cultural rights, Farida Shaheed (21 March 2011) UN Doc A/HRC/17/38, para. 6.

> knowledge and traditions. It includes all aspects of the environment resulting from the interaction between people and places through time."[12]

Considering cultural heritage as a resource contingent on cultural identification and development processes presents a shift in thinking about and defining cultural heritage. The Special Rapporteur's definition does not negate other definitions of cultural heritage, but tailors the definition to the human rights context. This understanding is intended to bring to the forefront the multiple interpretations and manifestations of heritage.[13] This is also reflected in the phrasing of the Faro Convention, which speaks of "a reflection and expression of their constantly evolving values, beliefs, knowledge and traditions".[14]

Central to a human rights-centred understanding of cultural heritage is the focus on the rights of individuals and communities in relation to cultural heritage.[15] The Special Rapporteur's human rights approach to cultural heritage

> "obliges [...] [considering] the rights of individuals and communities in relation to such object or manifestation, and, in particular, to connect cultural heritage with its source of production."[16]

It positions individuals and communities at the centre of its definition and places cultural heritage in relation to them. This placement constitutes a shift in the typical focus on (physical) form, the notion of authenticity and outstanding universal value associated with the UNESCO conventions. Apart from the Intangible Cultural Heritage Convention, which gives "communities, groups and, in some cases, individuals" a role in the recognition of cultural heritage,[17] this is evidently a change from the approach of UNESCO conventions.

---

[12] Council of Europe Framework Convention on the Value of Cultural Heritage for Society (CETS No. 199).

[13] UNHRC, Report of the independent expert in the field of cultural rights, Farida Shaheed (21 March 2011) UN Doc A/HRC/17/38, para. 6.

[14] Council of Europe Framework Convention on the Value of Cultural Heritage for Society (CETS No. 199), Art. 2 (a).

[15] UNGA 'Report of the Special Rapporteur in the field of cultural rights' (9 August 2016) UN Doc A/71/317, para. 48.

[16] UNHRC, Report of the Special Rapporteur in the field of cultural rights (3 February 2016) UN Doc A/HRC/31/59, para. 70.

[17] Convention for the Safeguarding of the Intangible Cultural Heritage (adopted 17 October 2003, entered into force 20 April 2006) 2368 UNTS 3 (*Intangible Cultural Heritage Convention*), Art. 2.

The Special Rapporteur's understanding of cultural heritage forms part of the mandate holders' human rights approach to cultural heritage, which focuses on the relationship between people and heritage[18] and has addressed the intentional destruction of heritage as a human rights issue.[19] The approach considers the prevention of heritage destruction, the education about all forms of cultural heritage, the support for cultural heritage defenders [20] as well as the interconnectedness of tangible and intangible heritage, especially when under threat.[21] The approach underlines the consultation of people with a view to ensuring that cultural heritage "is considered as living heritage, with proper plans for the future use, destination and maintenance of sites".[22]

Accordingly, this definition further distinguishes itself through a holistic understanding of cultural heritage, which is inclusive of all perspectives, contributions, and practices.[23] This rejects monolithic views of heritage[24] and instead

[18] UNHRC 'Report of the Special Rapporteur in the field of cultural rights on her mission to Cyprus' (3 March 2017) UN Doc A/HRC/34/56/Add.1, para. 58.

[19] UNHRC, Report of the Special Rapporteur in the field of cultural rights (3 February 2016) UN Doc A/HRC/31/59, para. 69.

[20] UNHRC 'Report of the Special Rapporteur in the field of cultural rights on her mission to Cyprus' (3 March 2017) UN Doc A/HRC/34/56/Add.1, para. 58.

[21] UNHRC, Report of the Special Rapporteur in the field of cultural rights (3 February 2016) UN Doc A/HRC/31/59, para. 77.

[22] UNHRC 'Report of the Special Rapporteur in the field of cultural rights on her mission to Cyprus' (3 March 2017) UN Doc A/HRC/34/56/Add.1, para. 109 (a).

[23] UNHRC 'Report of the Special Rapporteur in the field of cultural rights on her mission to Serbia and Kosovo' (19 January 2018) UN Doc A/HRC/37/55/Add.1, para. 60.

The Special Rapporteur has emphasised the rights of women and their role and involvement in "identifying and interpreting cultural heritage" and deciding "which cultural traditions, values or practices are to be kept, modified or discarded". UNGA 'Report by the Special Rapporteur in the field of cultural rights' (10 August 2012) UN Doc A/67/287, p. 2. The Special Rapporteur has further underlined the inclusion of people with mental or physical disabilities and has focused on the cultural rights of migrants. See UNHRC, Report of the independent expert in the field of cultural rights, Farida Shaheed (21 March 2011) UN Doc A/HRC/17/38, paras. 80 (k); HRC, Cultural rights and migration

UNHRC 'Report of the Special Rapporteur in the field of cultural rights' (2 February 2023) UN Doc A/HRC/52/35.

[24] UNHRC 'Report of the Special Rapporteur in the field of cultural rights on her mission to Serbia and Kosovo' (19 January 2018) UN Doc A/HRC/37/55/Add.1, para. 60; UNGA 'Report of the Special Rapporteur in the field of cultural rights, Karima Bennoune' (19 July 2021) UN Doc A/76/178.

emphasises heritage as "a living practice"[25], its "plural and dynamic character" and "syncretic nature"[26]. The understanding also underlines the importance of the inclusion of minority cultures in a state's understanding of cultural heritage, by embedding communities' history, outstanding figures and linguistic heritage.[27] By addressing the need for a multifaceted and dynamic understanding, the work of the Special Rapporteur counters the idea that heritage is static and has one singular meaning to one specific group.[28]

To conclude, the Special Procedures have positioned and contextualised cultural heritage in the human rights framework by bringing forward their own understanding. Whilst the mandate holders of the Special Rapporteur in the field of cultural rights have highlighted the repercussions of the deliberate destruction of cultural heritage, they have gone beyond the threat of terrorism, situations of violence, and armed conflicts. The Special Rapporteur specifically addresses the management of heritage by cautioning that the notion of authenticity in heritage preservation "may have a stultifying and negative impact on living cultures" and by emphasising the importance of recognizing heritage diversity.[29] Here, the Special Rapporteur appears to draw on the ongoing scholarly discourse on this subject. Academic scholarship has pointed to the potentially harmful repercussions of failing to understand heritage as living heritage, specifically on the development of local communities.[30] Their cultural manifestations might run the risk of being "fossilized" if labelled as "heritage" or "traditional knowledge" and not being allowed to be further developed.[31]

---

[25] UNHRC 'Report of the Special Rapporteur in the field of cultural rights, Farida Shaheed, Addendum, Visit to Viet Nam' (29 January 2015) UN Doc A/HRC/28/57/Add.1, para. 100.

[26] UNGA 'Report of the Special Rapporteur in the field of cultural rights, Karima Bennoune' (19 July 2021) UN Doc A/76/178, paras. 31, 75.

[27] UNHRC 'Report of the Independent Expert in the field of cultural rights, Farida Shaheed, Addendum, Mission to Morocco' (2 May 2012) UN Doc A/HRC/20/26/Add.2, para. 40.

[28] The diversity of cultural heritage is similarly underlined in communication AL TUR 12/2020, which concerns the transformation of the Hagia Sofia from a museum into a mosque. Mandates of the Special Rapporteur in the field of cultural rights; and the Special Rapporteur on freedom of religion or belief, AL TUR 12/2020 (28 July 2020), p. 3.

[29] UNGA 'Report of the Special Rapporteur in the field of cultural rights, Karima Bennoune' (19 July 2021) UN Doc A/76/178, paras. 32–33.

[30] See, e.g., Lixinski (2013a).

[31] Lixinski (2013a), p. 24.

5.3 Rights

The analysis highlights three rights: (1) the ability to access and enjoy cultural heritage; (2) participation and free, prior and informed consent in decision-making; and (3) to maintain, control, protect and develop cultural heritage, traditional knowledge and traditional cultural expressions.

5.3.1 Ability to Access and Enjoy Cultural Heritage

Access and enjoyment of cultural heritage are first reflected in the Special Rapporteur's right to access and enjoy cultural heritage, which draws on several legally binding and non-binding bases. Second, rights vis-à-vis access are also found in the United Nations Declaration on the Rights of Indigenous Peoples (UNDRIP). Third, the ability to access and enjoy cultural heritage is recognised in resolutions of the UNHRC.

5.3.1.1 The Special Rapporteur's Right of Access to and Enjoyment of Cultural Heritage

Building on the work of the CESCR's General Comment No. 21, the 2011 report[32] of the first mandate holder of the Special Rapporteur in the field of cultural rights concludes that a right of access to and enjoyment of cultural heritage forms part of human rights law and is anchored in

> "the right to take part in cultural life, the right of members of minorities to enjoy their own culture, and the right of indigenous peoples to self-determination and to maintain, control, protect and develop cultural heritage."[33]

The Special Rapporteur thereby frames a right of access to and enjoyment of cultural heritage[34], which finds a legal basis in Article 15 (1) (a) of the ICESCR,

---

[32] UNHRC, Report of the independent expert in the field of cultural rights, Farida Shaheed (21 March 2011) UN Doc A/HRC/17/38.

[33] *ibid* 78.

[34] See "The Special Rapporteur stresses the importance of the right to access and enjoy cultural heritage as a human right (…)" UNHRC 'Report of the Special Rapporteur in the field of cultural rights on her mission to Serbia and Kosovo' (19 January 2018) UN Doc A/HRC/37/55/Add.1, para. 10. "In order to (…) ensure the right to access and enjoy cultural heritage" in UNHRC 'Visit to Maldives: Report of the Special Rapporteur in the field of cultural rights' (21 February 2020) UN Doc A/HRC/43/50/Add.2, para. 95. Also see, UNHRC 'Visit

Article 27 of the ICCPR and rights recognised in the UNDRIP[35] and the International Labour Organisation (ILO) Convention No. 169 concerning Indigenous and Tribal Peoples in Independent Countries.[36] The right accordingly draws on codified human rights and labour rights standards and on a declaration that sets out minimum standards for Indigenous peoples.[37]

Access and enjoyment are interpreted as interdependent concepts, meaning one implies the other.[38] Access to and enjoyment of cultural heritage is framed as both an individual and a collective human right.[39] The scope of the right is broad as it covers

to Tuvalu: Report of the Special Rapporteur in the field of cultural rights, Karima Bennoune' (24 December 2020) UN Doc A/HRC/46/34/Add.1, para. 12.

[35] The Declaration was adopted with 143 votes in favour, 4 against, 11 abstentions, and 34 non-voting. See UN Digital Library (2025): United Nations Declaration on the Rights of Indigenous Peoples : resolution / adopted by the General Assembly, https://digitallibrary.un.org/record/609197?ln=en, accessed 31 January 2025. In the meantime, the four countries voting against the adoption of the UNDRIP have reversed their decision. See, UN Department of Economic and Social Affairs (2015): United Nations Declaration On The Rights Of Indigenous Peoples, https://social.desa.un.org/issues/indigenous-peoples/united-nations-declaration-on-the-rights-of-indigenous-peoples, accessed 31 January 2025.

[36] UNHRC, Report of the independent expert in the field of cultural rights, Farida Shaheed (21 March 2011) UN Doc A/HRC/17/38, para. 58 f.

The importance of access to cultural heritage is a prevalent issue in the thematic—and country reports by the Special Rapporteur. An issue being repeated in virtually all reports. "Cultural rights also protect access to and enjoyment of cultural heritage, (…)" in UNHRC 'Report of the Special Rapporteur in the field of cultural rights on her mission to Cyprus' (3 March 2017) UN Doc A/HRC/34/56/Add.1, para. 43. As well as UNHRC 'Report of the Special Rapporteur in the field of cultural rights, Farida Shaheed, Addendum, Mission to Bosnia and Herzegovina' (3 March 2014) UN Doc A/HRC/25/49/Add.1, paras 99, 101; UNHRC 'Report of the Special Rapporteur in the field of cultural rights, Farida Shaheed, Memorialization processes' (23 January 2014) UN Doc A/HRC/25/49, para. 48; UNGA 'Report of the Special Rapporteur in the field of cultural rights' (8 August 2014) UN Doc A/69/286, para. 1.

[37] Although not enjoying the same number of ratifications as the ICCPR or ICESCR, the ILO Convention No. 169 is ratified by 24 states. See ILO (2025): Ratifications of C169—Indigenous and Tribal Peoples Convention, 1989 (No. 169), https://normlex.ilo.org/dyn/nrmlx_en/f?p=1000:11300:0::NO:11300:P11300_INSTRUMENT_ID:312314, accessed 31 January 2025.

[38] UNHRC, Report of the independent expert in the field of cultural rights, Farida Shaheed (21 March 2011) UN Doc A/HRC/17/38, para. 58.

[39] *ibid* para. 61.

> "the right of individuals and communities to, inter alia, know, understand, enter, visit, make use of, maintain, exchange and develop cultural heritage, as well as to benefit from the cultural heritage and the creation of others."[40]

It further covers participation rights, as it sets out for individuals and communities to contribute to "the identification, interpretation and development of cultural heritage [...] to the design and implementation of preservation/safeguard policies and programmes" and effectively participate "in decision-making processes relating to cultural heritage"[41].[42]

Whilst the right put forward by the Special Rapporteur entitles *everyone* to access and enjoy cultural heritage, its legal bases indicate that certain components are solely addressed to persons belonging to minorities and Indigenous communities (i.e. Article 27 of the ICCPR, ILO Convention No. 169, and the UNDRIP). It is for this reason, the Special Rapporteur notes that "varying degrees of access and enjoyment may be recognized, taking into consideration the diverse interests of individuals and communities depending on their relationship to specific cultural heritages."[43] These varying degrees distinguish between different right holders, namely:

> "(a) originators or "source communities", [...] which consider themselves as the custodians/owners of a specific cultural heritage, people who are keeping cultural heritage alive and/or have taken responsibility for it; (b) individuals and communities, including local communities, who consider the cultural heritage in question an integral part of the life of the community, but may not be actively involved in its maintenance; (c) scientists and artists; and (d) the general public accessing the cultural heritage of others."[44]

[40] *ibid* para. 79.

[41] UNHRC, Report of the independent expert in the field of cultural rights, Farida Shaheed (21 March 2011) UN Doc A/HRC/17/38, paras. 58. Also see UNHRC 'Report of the Special Rapporteur in the field of cultural rights, Farida Shaheed, Addendum, Visit to Viet Nam' (29 January 2015) UN Doc A/HRC/28/57/Add.1, para. 63.

[42] In the framework of the UPR, Ethiopia reports on the constitutional right held by "the Nation, Nationalities and Peoples of Ethiopia (...) to speak, write and develop their own language, culture and preserve their history." In: UNHRC 'National Report submitted by Ethiopia' (2019) UN Doc A/HRC/WG.6/33/ETH/1, para. 68.

[43] UNHRC, Report of the independent expert in the field of cultural rights, Farida Shaheed (21 March 2011) UN Doc A/HRC/17/38, para. 79.

[44] *ibid* 62. The varying degrees are repeated by the Special Rapporteur in several communications. See, Mandates of the Working Group of Experts on People of African Descent; the Special Rapporteur in the field of cultural rights; the Special Rapporteur on extrajudicial,

The Special Rapporteur finds that states can legitimately limit public use and access to cultural heritage on the grounds of providing for its safety, indicating conservation purposes, or to safeguard a community's right of accessing and enjoying its own cultural heritage[45] or to exercising "traditional cultural uses and practices"[46]. Communications in the Special Procedures address that restrictions on access as well as the destruction of religious sites and places of worship impact the right to take part in cultural life and, specifically, the right to enjoy and access cultural heritage.[47] Moreover, the Special Rapporteur underlines the importance that access does not have detrimental repercussions on the object or the source community.[48]

This is particularly important considering the cultural privacy of local or Indigenous communities. For example, in the country report on Viet Nam, the Special Rapporteur addresses the Cong Chieng gongs, which are played by communities in the Central Highlands and also are inscribed on the Representative List of Intangible Cultural Heritage.[49] In light of the commercialization of playing the instrument for touristic and commercial purposes, the Special Rapporteur observes that the way in which the Cong Chieng is played is being influenced by people outside the communities.[50] Whilst acknowledging the inevitable "evolution of cultural practices", she also underlines the importance of seeking the

---

summary or arbitrary executions; the Special Rapporteur on adequate housing as a component of the right to an adequate standard of living, and on the right to non-discrimination in this context; and the Independent Expert on the enjoyment of all human rights by older persons, AL BRA 2/2018 (20 April 2018), p. 6.

[45] UNHRC, Report of the independent expert in the field of cultural rights, Farida Shaheed (21 March 2011) UN Doc A/HRC/17/38, para. 75.

[46] UNGA 'Report of the Special Rapporteur in the field of cultural rights' (30 July 2019) UN Doc A/74/255, para. 81.

[47] Mandates of the Special Rapporteur in the field of cultural rights and the Special Rapporteur on freedom of religion or belief, AL, SAU 7/2015 (14 October 2015), p. 4; Mandates of the Special Rapporteur in the field of cultural rights; the Special Rapporteur on freedom of religion or belief; and the Special Rapporteur on the situation of human rights in the Islamic Republic of Iran, AL, IRN 14/2016 (31 May 2016). With regard to access to Indigenous lands, see Mandates of the Special Rapporteur in the field of cultural rights; the Special Rapporteur on the independence of judges and lawyers and the Special Rapporteur on the rights of Indigenous Peoples, AL NZL 1/2024 (1 July 2024).

[48] UNHRC, Report of the independent expert in the field of cultural rights, Farida Shaheed (21 March 2011) UN Doc A/HRC/17/38, para. 63.

[49] UNHRC 'Report of the Special Rapporteur in the field of cultural rights, Farida Shaheed, Addendum, Visit to Viet Nam' (29 January 2015) UN Doc A/HRC/28/57/Add.1, paras. 97–98.

[50] *ibid* paras. 97–98.

consent of communities "as to whether, how, when and where to perform and share aspects of their particular cultural heritage."[51]

5.3.1.2 Access in the UNDRIP

The UNDRIP was adopted by the UN General Assembly in 2007 and covers Indigenous peoples' human rights. The UNDRIP sets out rights concerning access to heritage, and principles of the UNDRIP are integrated into the Operational Guidelines for the Implementation of the World Heritage Convention[52].

Articles 12 and 13 of the UNDRIP address access to heritage with Article 12 focusing on manifesting, practising, developing, and teaching spiritual and religious traditions, customs, and ceremonies. The article further speaks of inter alia maintaining, protecting, and accessing religious and cultural sites and using and controlling ceremonial objects. Article 13 inter alia covers the right to "revitalize, use, develop and transmit to future generations their histories, languages, oral traditions, philosophies, writing systems and literatures". Articles 14 and 15 are also relevant, as they set out the rights for Indigenous peoples to establish and control their educational system and institutions, and to provide education in their own languages.[53]

5.3.1.3 The Ability to Access Cultural Heritage in UNHRC Resolutions

The 2011 UNHRC resolution 17/15 takes note of the Special Rapporteur's report on the right of access to and enjoyment of cultural heritage,[54] and subsequent resolutions of the HRC reference access and enjoyment of cultural heritage. In UNHRC resolution 25/15 of 15 April 2014, states recognise "the public right of access to culture".[55] UNHRC resolution 33/20 of 6 October 2016 refers to access to cultural heritage as part of the right to take part in cultural life.[56] The

---

[51] *ibid* para. 98.

[52] See, e.g., UNESCO 'Operational Guidelines for the Implementation of the World Heritage Convention' (31 July 2024) WHC.24/01 (Operational Guidelines), para. 40.

[53] Also see principle 7 "To protect and preserve their heritage, indigenous peoples must control their own forms of cultural transmission and education." in: ECOSOC 'Human Rights of Indigenous Peoples' (19 June 2000) UN Doc E/CN.4/Sub.2/2000/26.

[54] UNHRC Res 17/15 (14 July 2011) UN Doc A/HRC/RES/17/15, para. 9.

[55] UNHRC Res 25/15 (15 April 2014) UN Doc A/HRC/RES/25/15, para. 6 (n). The resolution was adopted by a vote of 30 to 14, with 3 abstentions.

[56] UNHRC Res 33/20 (6 October 2016) UN Doc A/HRC/RES/33/20, 5th recital, 8th recital, paras. 1, 6.

most important reference for the purposes of this analysis is likely the resolution calling

> "upon all States to respect, promote and protect the right of everyone to take part in cultural life, including the ability to access and enjoy cultural heritage."[57]

In 2018, the UNHRC unanimously adopted resolution 37/17 on "cultural rights and the protection of cultural heritage", which employs the same wording.[58] Resolution 37/17 notes in its preamble that damage to or destruction of cultural heritage

> "may have a detrimental and irreversible impact on the enjoyment of cultural rights, in particular the right of everyone to take part in cultural life, including the ability to access and enjoy cultural heritage."[59]

The wording of the resolutions can be interpreted to mark access and enjoyment of cultural heritage as an important component of the right to take part in cultural life. This is also reflected in the resolution calling upon states "to respect, promote and protect the right of everyone to take part in cultural life, including the ability to access and enjoy cultural heritage".[60]

5.3.2 Participation and the Free, Prior and Informed Consent in Decision-Making

The charter-based mechanisms reveal different bases for participation rights: the UNDRIP, the 1992 Declaration on the Rights of Persons Belonging to National or

---

[57] *ibid* para. 1. Also see UNHRC Res 49/7 (31 March 2022) UN Doc A/HRC/RES/49/7, para. 1 (also see para. 6., and preamble, 9th and 12th recital) (Adopted without a vote).

[58] UNHRC Res 37/17 (9 April 2018) UN Doc A/HRC/RES/37/17, para. 1. Also see UNHRC Res 54/1 (11 October 2023) UN Doc A/HRC/RES/54/1, para, 10 (Adopted without a vote) ("*Also calls for* respect for and the promotion and protection of the right of everyone to take part in cultural life, including the ability to have access to and to enjoy cultural heritage, and for the protection of tangible and intangible cultural heritage consistent with article 15 of the International Covenant on Economic, Social and Cultural Rights (...)").

[59] UNHRC Res 37/17 (9 April 2018) UN Doc A/HRC/RES/37/17, preamble. Also see recital "Recognizing that the violation or abuse of the right of everyone to take part in cultural life, *including the ability* to access and enjoy cultural heritage, (...)" (emphasis added).

[60] UNHRC Res 37/17 (9 April 2018) UN Doc A/HRC/RES/37/17, para. 1.

Ethnic, Religious and Linguistic Minorities, and the Special Procedures reliance on Article 1 of the ICESCR and ICCPR and on Article 15 (1) (a) of the ICESCR.

The free, prior and informed consent of Indigenous peoples is an integral part of the UNDRIP.[61] Although the UNDRIP sets out participation rights that include the principle of free, prior and informed consent, these rights are not directly linked to heritage and couched in more general terms. In accordance with Article 19 of the UNDRIP, "States shall consult and cooperate in good faith" with Indigenous peoples before adopting or implementing changes of a legislative or administrative nature that may affect Indigenous peoples and obtain their free, prior and informed consent. Article 32 sets out the same duty for states with regard to any projects that may affect Indigenous lands, territories or resources.[62] Nevertheless the wording of the articles reflects their applicability to legislative and administrative changes as well as to any sort of projects that may affect Indigenous peoples' way of life, sacred sites, and more generally their cultural heritage.

The topic of free, prior and informed consent is also part of the UPR, specifically in relation to projects that potentially impact the land and ancestral ways of life of Indigenous peoples.[63] The state reports elaborate on free, prior and

[61] UNGA Res 61/295 'United Nations Declaration on the Rights of Indigenous Peoples' (13 September 2007) UN Doc A/Res/61/295. The resolution was adopted by a majority of 144 states in favour. Australia, Canada, New Zealand, and the United States of America voted against, and 11 states abstained (Azerbaijan, Bangladesh, Bhutan, Burundi, Colombia, Georgia, Kenya, Nigeria, Russian Federation, Samoa, and Ukraine).

Also see ECOSOC 'Human Rights of Indigenous Peoples' (19 June 2000) UN Doc E/CN.4/Sub.2/2000/26, principle 9.

[62] Also see UNGA Res 61/295 'United Nations Declaration on the Rights of Indigenous Peoples' (13 September 2007) UN Doc A/Res/61/295 (*UNDRIP*), Art. 10 UNDRIP with regard to the prohibition of forcible removal and the requirements for relocation and Art. 29 with regard to the storage or disposal of hazardous materials on indigenous lands or territories.

[63] UNHRC 'Report of the Working Group on the UPR: Australia' (2021) UN Doc A/HRC/47/8, para. 146.288; UNHRC 'Report of the Working Group on the UPR: United Republic of Tanzania' (2011) UN Doc A/HRC/19/4, para. 86.49; UNHRC 'Report of the Working Group on the UPR: Brazil' (2022) UN Doc A/HRC/52/14, recommendations 149.260, 149.261, 149.269. Also see 149.200; UNHRC 'Report of the Working Group on the UPR: Colombia (2023) UN Doc A/HRC/55/7, recommendations 116.205, 116.209, 116.210; UNHRC 'Report of the Working Group on the UPR: Russian Federation' (2023) UN Doc A/HRC/55/14, recommendation 35.339.

informed consent in the context of ILO Convention No. 169[64] and consultations with Indigenous peoples in the implementation of UNDRIP[65].

Participation rights are also set out in the 1992 Declaration on the Rights of Persons Belonging to National or Ethnic, Religious and Linguistic Minorities adopted by the UN General Assembly. These, similarly, contain more general references to the development of culture and traditions.[66]

In addition, the Special Rapporteur in the field of cultural rights has underlined that the free, prior and informed consent of Indigenous peoples is grounded in the right to self-determination set out in Articles 1 of the ICCPR and ICESCR.[67] Outside the scope of the right to self-determination, the Special Rapporteur continuously emphasises the relevance and applicability of participation rights for cultural heritage in the scope of protection of Article 15 (1) (a) of the ICESCR.[68] Cognisant that the active participation and consultation of groups is not always in place,[69] these consultation and participation rights have a key function in "protecting the collective rights of indigenous peoples and other groups"[70]. Specifically, this concerns identifying, interpreting, elaborating, and developing heritage as well as designing and implementing preservation and safeguard programmes and in the reformulation of cultural identities.[71]

---

[64] UNHRC 'National report submitted by Costa Rica' (2024) UN Doc A/HRC/WG.6/47/CRI/1, para. 36; UNHRC 'Report of the Working Group on the UPR: Finland' (2024) UN Doc A/HRC/52/9, recommendation 138.14.

[65] UNHRC 'National report submitted by Canada' (2023) UN Doc A/HRC/WG.6/44/CAN/1, para. 19 f.

[66] UNGA Res 47/135 'Declaration on the Rights of Persons Belonging to National or Ethnic, Religious and Linguistic Minorities' (18 December 1992) UN Doc A/RES/47/13, Arts. 2, 4.

[67] UNGA 'Development and cultural rights: the principles, Report of the Special Rapporteur in the field of cultural rights, Alexandra Xanthaki' (15 August 2022) UN Doc A/77/290, para. 81.

[68] UNHRC 'Cultural rights: an empowering agenda, Report of the Special Rapporteur in the field of cultural rights, Alexandra Xanthaki' (2022) UN Doc A/HRC/49/54, para. 79.

[69] *ibid*, para. 19.

[70] UNGA 'Development and cultural rights: the principles, Report of the Special Rapporteur in the field of cultural rights, Alexandra Xanthaki' (15 August 2022) UN Doc A/77/290, paras. 82, 84, 98 (b).

[71] For example, UNHRC, Cultural Rights: Tenth Anniversary Report (17 January 2019) UN Doc A/HRC/40/53, para. 18. UNHRC, Report of the Special Rapporteur in the field of cultural rights (3 February 2016) UN Doc A/HRC/31/59, para. 9.Also see the emphasis on consultation in: UNHRC, COVID-19, Culture and Cultural Rights (17 February 2021) UN Doc A/HRC/46/34, para. 77. Also see, UNHRC 'Report of the Independent Expert in the field of cultural rights, Farida Shaheed, Addendum, Mission to Morocco' (2 May 2012) UN

5.3.3 Maintain, Control, Protect and Develop Cultural Heritage, Traditional Knowledge and Traditional Cultural Expressions

Among the fundamental rights set forth in the UNDRIP is the right of Indigenous peoples and individuals to belong to an Indigenous community or nation.[72] The UNDRIP affirms the right of Indigenous peoples to self-determination and "[b]y virtue of that right" the free pursuit of their cultural development.[73] More specifically, charter-based bodies speak of a right to maintain, control, protect and develop their cultural heritage. This is most prominently visible, in the UNDRIP and UNHRC resolutions, on the one hand, and, on the other, in the Special Procedures.

5.3.3.1 The Right to Maintain, Control, Protect and Develop Cultural Heritage in the UNDRIP and UNHRC Resolutions

Article 11 (1) of the UNDRIP sets out the "right to practice and revitalize [...] cultural traditions and customs", including maintaining, protecting, and developing "past, present and future manifestations of their cultures". This is exemplified by "archaeological and historical sites, artefacts, designs, ceremonies, technologies and visual and performing arts and literature."[74] Article 26 (2) of the UNDRIP affirms the right of Indigenous peoples to "own, use, develop and control the lands, territories and resources that they possess by reason of traditional ownership or other traditional occupation or use". Furthermore, Article 31 of the UNDRIP declares that:

> "[i]ndigenous peoples have the right to maintain, control, protect and develop their cultural heritage, traditional knowledge and traditional cultural expressions, as well as the manifestations of their sciences, technologies and cultures, including human and genetic resources, seeds, medicines, knowledge of the properties of fauna and flora, oral traditions, literatures, designs, sports and traditional games and visual and performing arts. They also have the right to maintain, control, protect and develop their

Doc A/HRC/20/26/Add.2, para. 90; UNHRC 'Report of the Special Rapporteur in the field of cultural rights, Farida Shaheed, Addendum, Visit to Viet Nam' (29 January 2015) UN Doc A/HRC/28/57/Add.1, para. 111.

72 UNDRIP, Art. 9.

73 *ibid* Art. 3.

74 *ibid* Art. 11 (1).

intellectual property over such cultural heritage, traditional knowledge, and traditional cultural expressions."[75]

The *travaux préparatoires* of the Declaration reveal that the original text inserted the recognition that Indigenous peoples are entitled to not only control and protect their cultural (and intellectual) property, but also to full ownership.[76]

UNHRC resolution 49/7 on "cultural rights and the protection of cultural heritage" acknowledges in its preamble inter alia "the key role of indigenous peoples and/or persons belonging to national or ethnic, religious and linguistic minorities in maintaining, controlling, protecting, developing and preserving tangible and intangible cultural heritage".[77] Whereas the focus of the UNDRIP evidently lies on Indigenous peoples, the preamble of the UNHRC resolution also includes persons belonging to national or ethnic, religious and linguistic minorities.

The UNDRIP is far-reaching, and further rights are relevant to the realisation of the rights of Indigenous peoples vis-à-vis their heritage. This similarly encompasses maintaining and strengthening their cultural institutions (Art. 5), to establish and control their educational systems and institutions and offer education in their own languages (Arts. 14–15), and in accordance with human rights standards "to promote, develop and maintain their institutional structures and their distinctive customs, spirituality, traditions, procedures, practices, [...] judicial systems or customs" (Art. 34).

5.3.3.2 The Right to Maintain, Control, Protect and Develop Cultural Heritage in the Special Procedures

The Special Rapporteur in the field of cultural rights has underlined the importance for states to ensure that their national laws are in conformity with "the right of indigenous peoples to maintain, control, protect and develop their cultural heritage, traditional knowledge and traditional cultural expressions".[78]

---

[75] It is important to note that the right enunciated in the first sentence and the second sentence differ. The second sentence pertains to a right to intellectual property. Also see, UNGA 'Report of the Special Rapporteur in the field of cultural rights' (4 August 2014) UN Doc A/70/279, para. 36.

[76] ECOSOC 'Human Rights and Indigenous Issues' (22 March 2006) UN Doc E/CN.4/2006/79, p. 59.

[77] UNHRC Res 49/7 (31 March 2022) UN Doc A/HRC/RES/49/7, preamble, 14th recital. Adopted without a vote.

[78] UNGA 'Report of the Special Rapporteur in the field of cultural rights' (8 August 2014) UN Doc A/69/286, para. 55. Also see, "States should adopt measures to ensure the right of indigenous peoples to maintain, control, protect and develop their intellectual property over

The right to maintain, control, protect and develop cultural heritage is reiterated in the UNHRC's communication UA NPL 3/2017 by drawing on Article 11 of the UNDRIP.[79] The communication concerns allegations of the "forced eviction, internal displacement, destruction of cultural property, and lack of consultation and free, prior informed consent of the Newar peoples" in the context of the Kathmandu Valley Road Improvement Project.[80]

Furthermore, the Special Rapporteur draws on Article 15 (1) (c) of the ICESCR, the right "to benefit from the protection of the moral and material interests resulting from (…) (one's) scientific, literary or artistic production". The Special Rapporteur states that the

> "right (…) safeguards the personal link between authors and their creations and between peoples, communities or other groups and their collective cultural heritage, as well as their basic material interests, which are necessary to enable authors to enjoy an adequate standard of living."[81]

The thematic report further clarifies that these are inalienable human rights, and not a human right to the protection of intellectual property rights.[82]

their cultural heritage, traditional knowledge, and traditional cultural expressions." UNHRC 'Report of the Special Rapporteur in the field of cultural rights, Farida Shaheed, Copyright policy and the right to science and culture' (21 December 2024) UN Doc A/HRC/28/57, para. 117.

[79] Mandates of the Special Rapporteur in the field of cultural rights; the Special Rapporteur on the rights of indigenous peoples and the Special Rapporteur on the human rights of internally displaced persons, UA NPL 3/2017 (14 August 2017) p. 6.

[80] Mandates of the Working Group on the issue of human rights and transnational corporations and other business enterprises; the Special Rapporteur in the field of cultural rights; the Special Rapporteur on the promotion and protection of the right to freedom of opinion and expression; the Special Rapporteur on the situation of human rights defenders; the Special Rapporteur on the independence of judges and lawyers and the Special Rapporteur on the rights of Indigenous Peoples, AL NPL 2/2023 (20 July 2023) follows this communication. It concerns the construction of a business complex on the indigenous land that destroyed cultural and religious customs and lacking prior and informed participation in the decision-making process affecting "historically significant land" of the Indigenous Pradhan Newar. p. 1. Several communications, all dated 25 June 2024, also attend to this.

[81] UNGA 'Report of the Special Rapporteur in the field of cultural rights' (4 August 2014) UN Doc A/70/279, para. 32.

[82] *ibid*.

5.3.4 Interim Conclusions

UNHRC resolutions and the Special Procedures highlight the ability to access and enjoy cultural heritage. Whereas the UNHRC resolutions infer access and enjoyment of cultural heritage from the right to take part in cultural life (Article 15 (1) (a) of the ICESCR), the Special Rapporteur derives a right to access and enjoy cultural heritage from several legal bases, namely Article 15 (1) (a) of the ICESCR, Article 27 of the ICCPR, and rights recognised in the UNDRIP and the ILO Convention No. 169. Whilst this point will be further explored in the following chapter, it is of interest to note that the resolutions repeat access and enjoyment of cultural heritage but speak of the *ability*, not of a right. In the case of the specific rights of Indigenous peoples, the UNDRIP offers several articles as bases.

Article 1 common to both the ICESCR and ICCPR and Article 15 (1) (a) of the ICESCR offer legal bases for participation rights in decision-making concerning cultural heritage in the Special Procedures. Furthermore, the UNDRIP offers an important instrument for participation rights of Indigenous peoples. Although articles in the UNDRIP do not specifically refer to the cultural heritage or way of life of Indigenous peoples in the context of participation, the wording can be interpreted to cover a multitude of situations in which state-led changes or (development) projects affect the cultural heritage, sacred sites or way of life of Indigenous peoples. This is reflected in recommendations in the UPR, which cite the rights set out in the UNDRIP and the principle of free, prior and informed consent.

The right of Indigenous peoples to maintain, control, protect and develop their cultural heritage, traditional knowledge and traditional cultural expressions is common to the UNDRIP, UNHRC resolutions, and the Special Procedures. The UNDRIP's Article 31 expounds on this right and underlines, in addition to cultural heritage, Indigenous peoples' traditional knowledge, traditional cultural expressions, and other manifestations of their sciences, technologies and cultures as well as their intellectual property over such heritage, knowledge and expressions. The preamble of UNHRC resolution 49/7 acknowledges the role of Indigenous peoples in exercising these rights in relation to tangible and intangible cultural heritage. In the framework of the Special Procedures, the Special Rapporteur mainly draws on Article 11 of the UNDRIP to assert Indigenous peoples' right "to maintain, control, protect and develop their cultural heritage, traditional knowledge and traditional cultural expressions".[83] By drawing on Article 15 (1) (c) of the ICESCR, the Special Rapporteur, furthermore, emphasises the legal basis for safeguarding "the personal link [...] between peoples, communities, or other groups and their collective cultural heritage".[84]

5.4 Obligations

The analysis identifies five obligations in the framework of the charter-based human rights bodies: (1) the underlying duty to recognise the diversity of cultural heritage, which can be understood as a prerequisite for all rights and obligations; (2) to respect and protect cultural heritage and prevent its destruction; (3) ensuring the ability to access and enjoy cultural heritage; (4) ensuring effective participation, including the free, prior and informed consent in any decisions concerning heritage; and (5) legally recognise and protect the rights of Indigenous peoples to their cultural heritage, traditional knowledge, and (traditional) cultural expressions and set up effective mechanisms to enable these rights.

5.4.1 Recognition of the Diversity of Cultural Heritage

A state's obligation to recognise the diversity of cultural heritage can be regarded as a precondition serving the protection of cultural heritage and as a corroboration of a state's general obligation to protect cultural heritage.

Several UNHRC resolutions recognise the benefits of cultural pluralism and diversity in "contributing to a wider exchange of knowledge and understanding of cultural heritage and cultural background".[85] UNHRC resolution 49/7 concerning "cultural rights and the protection of cultural heritage", for example, underlines in its preamble

[83] UNGA 'Report of the Special Rapporteur in the field of cultural rights' (8 August 2014) UN Doc A/69/286, para. 55.

[84] UNGA 'Report of the Special Rapporteur in the field of cultural rights' (4 August 2014) UN Doc A/70/279, para. 32.

[85] E.g., UNHRC Res 14/9 (23 June 2010) UN Doc A/HRC/RES/14/9, para. 6. Adopted without a vote. UNHRC Res 37/12 (22 March 2018) UN Doc A/HRC/RES/37/12, para. 6 Adopted without a vote; UNHRC Res 52/18 (3 April 2023) UN Doc A/HRC/RES/52/18, para. 6. Adopted without a vote.

Also see, UNHRC 'Mandate of the Special Rapporteur in the field of cultural rights' (26 March 2015) UN Doc A/70/53.

> "the necessity to protect the cultural heritage of persons belonging to minorities from intentional destruction aimed at erasing evidence of their presence as a key factor for the preservation of their identity."[86]

State reports and recommendations in the UPR, similarly, underline the preservation and recognition of minority cultures, cultural heritage, ways of life, and cultural identity.[87] This encompasses giving special importance to preserving the cultures of ethnic minorities,[88] the cultural and linguistic identity of Indigenous peoples,[89] recognising the heritage sites of minorities as national cultural heritage,[90] and promoting and raising awareness of the diversity of cultural heritage[91]. Australia, for example, reports on the legislative protection for the cultural heritage of Aboriginal and Torres Strait Islander peoples, including legal recognition of their intangible heritage, referring specifically to "language, traditional knowledge and arts".[92] Furthermore, Australia affirms its commitment to ensuring "the right protections" for the cultural heritage of Aboriginal and Torres Strait Islander peoples.[93]

---

86 UNHRC Res 49/7 (31 March 2022) UN Doc A/HRC/RES/49/7, preamble, 15th recital. Also see, e.g., UNHRC Res 49/9 (31 March 2022) UN Doc A/HRC/RES/49/9, preamble, 32nd recital (Adopted without a vote).

87 See UNHRC 'Report of the Working Group on the UPR: Colombia' (2009) UN Doc A/HRC/10/82, para. 13; UNHRC 'Report of the Human Rights Council on its fortieth session' (3 June 2019) UN Doc A/HRC/40/2, para. 499; UNHRC 'National report submitted by Russian Federation' (2013) UN Doc A/HRC/WG.6/16/RUS/1, p. 20, para. 189; UNHRC 'National report submitted by Viet Nam' (2009) UN Doc A/HRC/WG.6/5/VNM/1, pp. 2–3, para. 5; UNHRC 'National report submitted by Brazil' (2017) UN Doc A/HRC/WG.6/27/BRA/1, pp. 9–10, paras. 56, 61; UNHRC 'National report submitted by Mexico' (2013) UN Doc A/HRC/WG.6/17/MEX/1, p. 29, fn 118.

88 UNHRC 'National report submitted by Viet Nam' (2018) UN Doc A/HRC/WG.6/32/VNM/1, pp. 15–16, para. 83.

89 UNHRC 'Report of the Working Group on the UPR: Australia' (2021) UN Doc A/HRC/47/8, para. 146 ('Accepted' by Australia).

90 UNHRC 'National report submitted by Viet Nam' (2018) UN Doc A/HRC/WG.6/32/VNM/1, pp. 15–16, para. 83; UNHRC 'National report submitted by Viet Nam' (2013) UN Doc A/HRC/WG.6/18/VNM/1, p. 16, para. 73.

91 UNHRC 'Report of the Working Group on the Universal Periodic Review: Brazil' (2022) UN Doc A/HRC/52/14, para. 149.189. Also see UNHRC 'Report of the Working Group on the UPR: Colombia (2023) UN Doc A/HRC/55/7, para. 116.203.

92 UNHRC 'National report submitted by Australia' (2020) UN Doc A/HRC/WG.6/37/AUS/1, para. 83.

93 UNHRC 'Report of the Working Group on the UPR: Australia' (2021) UN Doc A/HRC/47/8, para. 18.

The Special Rapporteur in the field of cultural rights also emphasises the recognition of the diversity of cultural heritage. She recommends, in her inaugural report, to "recognize and value" this diversity in the states' respective territories and to "acknowledge, respect and protect the possible diverging interpretations that may arise over cultural heritage.[94] States should also respect and protect the decisions of individuals and communities to either feel associated with elements of cultural heritage or not.[95]

The recommendation to recognise the diversity of cultural heritage is also reflected in thematic reports[96] and several communications, which concern threats and damage to cultural heritage that considerably impact minority communities[97]. Communication AL TUR 13/2018, for example, addresses the construction of the Ilisu dam project that affects mainly members of the Kurdish minority and "cultural heritage of great importance to local residents".[98] Another example is communication AL BRA 2/2018, which concerns residents living in the historic neighbourhood of *Horto Florestal* in Rio de Janeiro and facing eviction due to

[94] UNHRC, Report of the independent expert in the field of cultural rights, Farida Shaheed (21 March 2011) UN Doc A/HRC/17/38, para. 80 (a). Reiterated in e.g. Mandates of the Special Rapporteur in the field of cultural rights; the Special Rapporteur on extreme poverty and human rights; and the Special Rapporteur on freedom of religion or belief, AL, BHR 6/2015 (30 October 2015) p. 11; Also see, Mandate of the Special Rapporteur in the field of cultural rights, AL, SAU 3/2016 (31 May 2016) p. 8; Mandates of the Special Rapporteur in the field of cultural rights; the Special Rapporteur on the issue of human rights obligations relating to the enjoyment of a safe, clean, healthy and sustainable environment; the Special Rapporteur on adequate housing as a component of the right to an adequate standard of living, and on the right to non-discrimination in this context; the Special Rapporteur on the human rights of internally displaced persons; and the Special Rapporteur on minority issues, AL TUR 13/2018 (23 November 2018) p. 11.

[95] UNHRC, Report of the independent expert in the field of cultural rights, Farida Shaheed (21 March 2011) UN Doc A/HRC/17/38, para. 80 (a).

[96] See e.g., the Special Rapporteur's 2023 report on 'Cultural Rights and Migration' which inter alia sheds light on migration turning individuals belonging to the majority culture to being part of a minority. UNHRC 'Report of the Special Rapporteur in the field of cultural rights' (2 February 2023) UN Doc A/HRC/52/35.

[97] Noting a "disproportionate impact on people from African descent" in: AL BRA 2/2018 (20 April 2018) p. 3.

[98] AL TUR 13/2018 (23 November 2018) p. 6.

urban planning.[99] Yet, these communications merely cite the recommendation and do not further expound on it.

5.4.2 Respect and Protect Cultural Heritage and Prevent its Destruction

Damage and destruction can concern all forms of heritage, including mixed and natural sites, religious and sacred sites, and intangible practices. The charter-based bodies primarily emphasise the prevention of destruction, respect and protection of cultural heritage, and the duty not to destroy, damage or alter cultural heritage. In addition to cultural rights, freedom of religion, the prohibition of inciting national, racial or religious hatred, freedom of expression, and the right to education are found to be important legal bases.

5.4.2.1 Legal Bases for the Protection of Cultural Heritage from Destruction or Damage

In resolution 6/1, the UNHRC highlights that protecting cultural property in the event of an armed conflict "can contribute to the full enjoyment of the right […] to take part in cultural life".[100] The resolution further finds that damage to or destruction of "cultural property may impair the enjoyment of cultural rights", specifically invoking Article 15 of the ICESCR.[101] It similarly affirms cultural heritage as "an important component of the cultural identity of communities, groups and individuals",[102] recognising "that its intentional destruction may have

[99] AL BRA 2/2018 (20 April 2018). Also see Mandates of the Working Group on Arbitrary Detention; the Special Rapporteur in the field of cultural rights; the Special Rapporteur on the promotion and protection of the right to freedom of opinion and expression; the Special Rapporteur on the rights to freedom of peaceful assembly and of association; the Special Rapporteur on adequate housing as a component of the right to an adequate standard of living, and on the right to non-discrimination in this context; the Special Rapporteur on the situation of human rights defenders and the Special Rapporteur on the human rights of internally displaced persons, AL OTH 63/2023 (21 June 2023) concerning allegations of "forced evictions and demolition of many areas of Benghazi, including historic neighbourhoods, protected heritage sites and many residential units" (p. 1). Also see AL OTH 70/2023 (21 June 2023).

[100] UNHRC Res 6/1 (27 September 2007) UN Doc A/HRC/RES/6/1, para. 4.

[101] *Ibid* 8th recital.

[102] UNHRC Res 6/11 (28 September 2007) UN Doc A/HRC/RES/6/11, preamble, 7th recital.

adverse consequences on human dignity and human rights".[103] It specifically highlights Article 20 of the ICCPR,[104] acknowledging

> "[t]hat intentional destruction of cultural heritage may constitute advocacy and incitement to national, racial or religious hatred and thereby violates fundamental principles of international human rights law, inter alia, the provisions as stipulated in article 20 of the International Covenant on Civil and Political Rights [...]."[105]

The Special Rapporteur on freedom of religion or belief has asserted an intrinsic link between Article 18 of the ICCPR and Article 15 (1) (a) of the ICESCR.[106] The Special Rapporteur has clearly stated that the destruction of religious monuments and unique relics prevents certain individuals "from enjoying their cultural and religious rights".[107] The UNHRC has, in a number of resolutions, expressed its concern for the destruction of cultural and religious sites.[108] In resolution 6/37 on the "elimination of all forms of intolerance and of discrimination based on religion or belief", the Council links religious intolerance and discrimination as well as the violation of the freedom of thought, conscience, religion, or belief to the protection of "religious places, sites, shrines and symbols", addressing their vulnerability to desecration or destruction.[109]

Communications in the Special Procedures also address the protection of religious (heritage) sites and places of worship from destruction under the right to religious freedom, the right to freedom of expression, and the right to take part in cultural life.[110] Communication MLI 1/2012 of the Special Procedures

---

[103] *ibid.*

[104] *Ibid* preamble, 4th recital; para. 3.

[105] *ibid* para. 3.

[106] UNHRC 'Report of the Special Rapporteur on freedom of religion or belief, Asma Jahangir' (6 January 2009) UN Doc A/HRC/10/8, para. 52.

[107] *ibid.*

[108] See UNHRC Res 21/25 (19 October 2012) UN Doc A/HRC/RES/21/25, para. 1.

[109] UNHRC Res 6/37 (14 December 2007) UN Doc A/HRC/RES/6/37, para. 9 (e). Also see UNHRC Res 14/11 (23 June 2010) UN Doc A/HRC/RES/14/11, paras. 5, 6 (c). Of interest is also UNGA Res 55/254 on the protection of religious sites, adopted in the Afghan context. UNGA Res 55/254 (31 May 2001) UN Doc A/RES/55/254.

[110] E.g., concerning the protection of cultural, historic and religious sites in Gaza: Mandates of the Working Group on the issue of human rights and transnational corporations and other business enterprises; the Special Rapporteur in the field of cultural rights; the Special Rapporteur on adequate housing as a component of the right to an adequate standard of living, and on the right to non-discrimination in this context and the Special Rapporteur on the rights of Indigenous Peoples, AL NPL 1/2024 (25 June 2024). Also, Rapporteur characterizes the

focuses on the destruction of religious sites and places of worship in the cities of Timbuktu, Gao, and Kidal. The communication demonstrates the impact of such destruction on the right to take part in cultural life, the right to freedom of religion or belief, and the right to freedom of expression.[111] In the context of the intentional destruction of religious sites and places of worship in Timbuktu, declared world cultural heritage sites, former mandate holder Shaheed underlines that the demolition of the tombs signify a denial of inter alia the beliefs of the local population, referring to the need to respect and protect the cultural rights and the freedom of religion and belief of the local population.[112] The Special Rapporteur on freedom of religion or belief described the attack on the religious sites as violating "the rights of not only a single believer, but also the group of individuals forming the community attached to the place in question".[113] In the UPR, recommendations concern Mali's protection of religious sites and ending grave human rights violations, including the destruction of places of worship and depriving persons of their right to freedom of religion.[114] In its latest report in the UPR, Mali addresses the restoration of the destroyed mausoleums under the right of freedom to religion.[115]

In communication AL BGR 3/2015, which concerns unauthorised excavations and the conversion of a mosque into a museum, the Special Rapporteurs underline that the right to freedom of religion or belief "obliges States to ensure that religious places, sites, shrines and symbols are fully respected and protected" and,

alleged systematic destruction of inter alia religious sites as violations of the right to freedom of religion and belief and the right to participate in cultural life. AL, SAU 7/2015 (14 October 2015).

[111] Mandates of the Special Rapporteur in the field of cultural rights and the Special Rapporteur on freedom of religion or belief, UA Cultural rights (2009) G/SO 214 (56–23), MLI 1/2012 (9 July 2012).

[112] OHCHR (2012) A very dark future for the local populations in Northern Mali', warn UN experts, https://www.ohchr.org/en/press-releases/2012/07/very-dark-future-local-populations-northern-mali-warn-un-experts, accessed 31 January 2025. Also see UNHRC 'Compilation prepared by the Office of the High Commissioner for Human Rights: Mali' (2012) UN Doc A/HRC/WG.6/15/MLI/2, paras. 35, 57.

[113] *ibid.*

[114] UNHRC 'Report of the Working Group on the UPR: Mali' (2013) UN Doc A/HRC/23/6, p. 18, recommendation 111.65; UNHRC 'Report of the Working Group on the UPR: Mali, Addendum' (2013) UN Doc A/HRC/23/6/Add.1, p. 2, recommendation 112.13 (Mali supports both recommendations).

[115] UNHRC 'National report submitted by Mali' (2017) UN Doc A/HRC/WG.6/29/MLI/1, para. 41.

where appropriate, "take additional measures in cases where they are vulnerable to desecration or destruction".[116]

Communications further address acts of vandalism of cultural heritage[117] and rely on the right to education when addressing the destruction of cultural heritage. In communication AL BHR 6/2015, the Special Rapporteurs recall that cultural and religious sites are important resources for learning, thereby ensuring "the right to education and training without any discrimination".[118]

5.4.2.2 Respect and Protect Cultural Heritage and Prevent its Destruction

The UNHRC and the Special Rapporteur in the field of cultural rights both emphasise the need to prevent abuses and violations of human rights, specifically, of cultural rights caused by the destruction, damage, looting, theft, and illicit trafficking of cultural heritage.[119] In communication SAU 3/2016 of 31 May 2016, the Special Rapporteur cites the CESCR's findings that the obligation to respect and protect cultural heritage at all times entails the positive obligations of its preservation, development, enrichment and transmission to future generations as well as its "care, preservation and restoration".[120]

UNHRC resolution 37/17 titled "cultural rights and the protection of cultural heritage" lists a number of strategies aimed at preventing the destruction of cultural heritage including

> "ensuring accountability, documenting the cultural heritage within their jurisdiction, including through digital means, implementing educational programmes on the importance of cultural heritage and cultural rights and training military forces and

[116] Mandate of the Special Rapporteur in the field of cultural rights; the Special Rapporteur on minority issues; and the Special Rapporteur on freedom of religion or belief, AL BGR 3/2015 (15 September 2015) p. 2.

[117] Mandatos del Relator Especial sobre la promoción de la verdad, la justicia, la reparación y las garantías de no repetición; de la Relatora Especial en la esfera de los derechos culturales; del Grupo de Trabajo sobre las Desapariciones Forzadas o Involuntarias; del Relator Especial sobre ejecuciones extrajudiciales, sumarias o arbitrarias yde la Relatora Especial sobre la promoción y protección del derecho a la libertad de opinión y de expresión, AL PER 3/2024, 29 April 2024.

[118] AL, BHR 6/2015 (30 October 2015) p. 11.

[119] See UNGA 'Report of the Special Rapporteur in the field of cultural rights' (9 August 2016) UN Doc A/71/317, paras.55, 77; UNHRC Res 33/20 (6 October 2016) UN Doc A/HRC/RES/33/20, para. 7, Also see 8th recital,

[120] AL, SAU 3/2016 (31 May 2016) p. 8.

> humanitarian actors in all relevant rules concerning the protection of cultural heritage, both during and in the aftermath of armed conflict."[121]

In addition, resolution 49/7 addresses further strategies that seek to prevent the destruction of cultural heritage. These include the timely development of preparatory measures and risk preparedness plans.[122] Moreover, states are encouraged

> "to address limitations of cultural rights, to take the measures necessary to prevent the destruction of historic monuments, works of art or places of worship that constitute the cultural or spiritual heritage of peoples, both in conflict and non-conflict situations, and to promote respect for cultural diversity."[123]

UNHRC resolutions also address and condemn the destruction of cultural heritage and urge governments to help prevent "all attacks on and destruction of cultural and religious sites in violation of international law", highlighting in particular world cultural and natural heritage sites.[124]

Several UN Security Council resolutions affirm the primary responsibility of states to protect their cultural heritage in the context of armed conflicts.[125] In 1999, Security Council resolution 1267 underlined "respect for Afghanistan's cultural and historical heritage".[126] Resolution 2357 (2017) unequivocally "condemns the unlawful destruction of cultural heritage" in armed conflicts, including

---

[121] UNHRC Res 37/17 (9 April 2018) UN Doc A/HRC/RES/37/17, para. 11. (Adopted without a vote).

[122] UNHRC Res 49/7 (31 March 2022) UN Doc A/HRC/RES/49/7, para. 13.

[123] *ibid* para. 14.

[124] UNHRC Res 31/27 (20 April 2016) UN Doc A/HRC/RES/31/27, para. 27 (g). Resolution adopted without a vote. Also see, for example, UNHRC Res 32/25 (15 July 2016) UN Doc A/HRC/RES/32/25, para. 26; UNHRC Res 34/26 (5 April 2017) UN Doc A/HRC/RES/34/26, para. 35 (Resolution adopted by a vote of 27 to 7, with 13 abstentions); UNHRC Res 37/29 (9 April 2018) UN Doc A/HRC/RES/37/29, preamble, para. 22 (Resolution adopted by a vote of 27 to 4, with 16 abstentions). Also see: UNHRC Res 52/32 (4 April 2023) UN Doc A/HRC/RES/52/32, preamble (Adopted by a recorded vote of 28 to 2, with 17 abstentions); UNHRC Res 49/7 (31 March 2022) UN Doc A/HRC/RES/49/7, preamble, 16th recital.

[125] See "Stresses that Member States have the primary responsibility in protecting their cultural heritage and that efforts to protect cultural heritage in the context of armed conflicts should be in conformity with the Charter, including its purposes and principles, and international law, and should respect the sovereignty of all States". UNSC Res 2347 (24 March 2017) UN Doc S/Res/2347, para. 5.

[126] UNSC Res 1267 (15 October 1999) UN Doc S/RES/1267 (1999).

the "destruction of religious sites and artefacts, as well as the looting and smuggling of cultural property [...] notably by terrorist groups".[127]

Importantly, UNHRC resolution 37/17 (2018) titled "cultural rights and the protection of cultural heritage" reaffirms the need to protect cultural heritage "in both conflict and non-conflict situations" by state and non-state actors.[128] In addition to addressing the destruction, the UNHRC also addresses looting, theft, smuggling and illicit trafficking in cultural property and encourages states to strengthen capacities to protect and preserve cultural heritage.[129]

By advocating for respecting and protecting cultural heritage, and preventing attacks, theft and destruction of cultural heritage, the 2011 report of the then independent expert puts forward the recommendation that it is the duty of states "not to destroy, damage or alter cultural heritage".[130] This is not an absolute duty, as the report clarifies that "the free, prior and informed consent of concerned communities" needs to be sought.[131] The duty not to destroy, damage or alter cultural heritage applies in both peacetime and during armed conflict. The recommendation does not cite a legal basis, but the report inter alia relies on Article 15 (1) (a) of the ICESCR and the CESCR's General Comment No. 21 throughout. The report has underlined that the duty extends to protecting cultural heritage from destruction or damage by third parties.[132] This duty is referenced in the communication to Mali of 9 July 2012, which concerns inter alia the destruction of cultural heritage in Timbuktu, Gao, and Kidal at the hands of the non-state armed group *Ansar Dine*.[133] Against the backdrop of the negative duty from

---

[127] UNSC Res 2347 (24 March 2017) UN Doc S/Res/2347, para. 1. Also see, for example, UNSC Res 2697 (15 September 2023) UN Doc S/RES/2697, preamble, 4th recital ("Condemning the commission of acts by ISIL/Da'esh involving (...) its destruction of cultural heritage, including archaeological sites, and trafficking of cultural property"); UNSC Res 2610 (17 December 2021) UN Doc S/RES/2610, preamble, 42nd recital. Also see UNSC Res 2617 (30 December 2021) UN Doc S/RES/2617, preamble, 29th recital.

[128] UNHRC Res 37/17 (9 April 2018) UN Doc A/HRC/RES/37/17, preamble. (Adopted without a vote)

[129] UNHRC Res 49/7 (31 March 2022) UN Doc A/HRC/RES/49/7, para. 5.

[130] UNHRC, Report of the independent expert in the field of cultural rights, Farida Shaheed (21 March 2011) UN Doc A/HRC/17/38, para. 80 (b).

[131] *ibid.*

[132] *ibid.*

[133] UA Cultural rights (2009) G/SO 214 (56–23), MLI 1/2012, 9 July 2012.

refraining to destroy, damage or alter cultural heritage, the Special Rapporteur asserts the duty to "respect the free development of cultural heritage".[134]

The duty not to destroy, damage or alter cultural heritage does not correspond to a general right *to* cultural heritage. It is, however, important to recall that the UNDRIP sets out a right for Indigenous peoples and individuals "not to be subjected to [the] [...] destruction of their culture".[135] The duty can be seen as an underlying obligation to ensure all rights vis-à-vis cultural heritage, including safeguarding the access and enjoyment of cultural heritage[136] and freedom of religion[137].

The duty not to destroy, damage or alter cultural heritage is repeated in almost all the reviewed communications on the topic.[138] In more recent communications, references to the 2003 UNESCO Declaration, UNHRC resolutions 33/20, 37/17 and 49/7 as well as the obligations set out in General Comment No. 21 towards the protection of cultural heritage prevail.[139] Communication AL NAM 1/2024, which addresses the impact of a planned extension of the Lüderitz seaport in Namibia on a heritage site that memorialises the 1904–1908 genocide, highlights the obligations to respect, protect and transmit cultural heritage but also does not repeat the duty.[140] Although references to the duty not to destroy, damage or alter

---

[134] UNHRC, Report of the independent expert in the field of cultural rights, Farida Shaheed (21 March 2011) UN Doc A/HRC/17/38, para. 80 (b).

[135] UNDRIP, Art. 8 (1).

[136] See "She (the Special Rapporteur) deplores all these destructions of cultural heritage, which constitute violations of the right to access and enjoy cultural heritage." UNHRC 'Report of the Special Rapporteur in the field of cultural rights on her mission to Serbia and Kosovo' (19 January 2018) UN Doc A/HRC/37/55/Add.1, para. 72.

[137] See UNGA 'Interim report of the Special Rapporteur of the Commission on Human Rights on the elimination of all forms of intolerance and of discrimination based on religion or belief' (31 July 2001) UN Doc A/56/253, para. 27.

[138] See e.g., AL OTH 63/2023 (21 June 2023), p.11; Mandates of the Special Rapporteur in the field of cultural rights; the Special Rapporteur on the right to education and the Special Rapporteur on freedom of religion or belief, AL RUS 18/2022 (20 December 2022) p. 10; Mandate of the Special Rapporteur in the field of cultural rights, AL cultural rights (2009), SGP 2/2012, (29 May 2012), p. 2; AL, BHR 6/2015 (30 October 2015) p. 10; AL, IRN 14/2016 (31 May 2016) p. 9; AL TUR 13/2018 (23 November 2018) p.11.

[139] Although this does not mean that references to the duty are completely absent. Mandate of the Special Rapporteur in the field of cultural rights, AL AZE 2/2023 (22 September 2023) pp. 9–10.

[140] Mandates of the Working Group on the issue of human rights and transnational corporations and other business enterprises; the Special Rapporteur in the field of cultural rights and the Special Rapporteur on the promotion of truth, justice, reparation and guarantees of non-recurrence, AL NAM 1/2024 (12 September 2024) p. 7.

cultural heritage are not completely absent in more recent communications[141], a change in the mandate holder might contribute to the use of different wording.

Whilst the reviewed responses to the communications usually seek to clarify, dismiss or negate aspects of allegations, they do not explicitly comment on obligations vis-à-vis cultural heritage under human rights law.

Recommendations citing a duty not to destroy, damage, or alter cultural heritage are virtually absent from the UPR. Furthermore, recommendations address both the heritage of the majority culture and heritage of persons belonging to minorities and Indigenous communities. Cuba explains that implementing the right to culture and its cultural policy inter alia entails ensuring the preservation of heritage.[142] Egypt recommends Cuba to "continue efforts to defend […] cultural heritage preservation"[143] and to protect its heritage and promote cultural rights[144]. Togo recommends Mali to "ensure better protection of victims of attacks against cultural heritage",[145] while Serbia recommends Russia to take further measures to protect the intangible cultural heritage[146]. Namibia recommends Brazil to continue "preventing violence and racial discrimination against Afro-Brazilians and at protecting their cultural heritage sites and places of worship".[147]

---

[141] AL RUS 18/2022 (20 December 2022) p. 10.

[142] UNHRC 'National report submitted by Cuba' (2018) UN Doc A/HRC/WG.6/30/CUB/1, para. 105. Also see, "The cultural policy has focused on protecting identity, preserving cultural heritage and promoting creativity, artistic and literary production and art appreciation." UNHRC 'National report submitted by Cuba' (2013) UN Doc A/HRC/WG.6/16/CUB/1, p. 13, para. 108.

[143] UNHRC 'Report of the Working Group on the UPR: Cuba' (2013) UN Doc A/HRC/24/16, p. 27, recommendation 170.261.

[144] UNHRC 'Report of the Working Group on the UPR: Cuba' (2023) UN Doc A/HRC/55/17 (2023), para. 44.252.

[145] UNHRC 'Report of the Working Group on the UPR: Mali' (2023) UN Doc A/HRC/54/8, p. 18. Recommendation supported by Mali.

[146] UNHRC 'Report of the Working Group on the UPR: Russian Federation' (2023) UN Doc A/HRC/55/14, para. 35.284. Accepted by Russia.

[147] UNHRC 'Report of the Working Group on the UPR: Brazil' (2017) UN Doc A/HRC/36/11, p. 14, recommendation 136.60.

5.4.3 Ensuring the Ability to Access and Enjoy Cultural Heritage

The charter-based bodies show that ensuring access to and enjoyment of cultural heritage has different facets. The recommendations within the framework of the UPR illustrate this well: Kyrgyzstan and Nicaragua recommend Cuba to guarantee access for all Cubans "to cultural values in all its manifestations",[148] emphasising broad cultural participation. Egypt responds to recommendations on the promotion of cultural rights by reporting on programmes encouraging involvement in cultural heritage as well as in its protection and accessibility.[149] In Turkey's review, Greece comments on limitations to the enjoyment of cultural and religious heritage,[150] highlighting the cultural participation of persons belonging to minorities. The United States of America expresses commitment to support the "tribes' efforts to recover their human remains, sacred and ceremonial objects and cultural property that has been stolen, looted or trafficked",[151] which sheds light on prevalent concerns of Indigenous communities. Benin's comment towards France highlights the long-term effects of colonialisation on cultural heritage, as it recommends continuing its "commitment to the restitution of African heritage, in order to contribute to the access by African populations to their culture".[152] State reports, within the UPR mechanism, cover efforts to guarantee

[148] UNHRC 'Report of the Working Group on the UPR: Cuba' (2013) UN Doc A/HRC/24/16p. 27, recommendations 170.259 and 170.262. In the second cycle, Cuba supports both recommendations by Kyrgyzstan and Nicaragua. See, UNHRC 'Report of the Working Group on the UPR: Cuba, Addendum' (2013) UN Doc A/HRC/24/16/Add.1, para. 6; UNHRC 'Report of the Working Group on the UPR: Cuba' (2009) UN Doc A/HRC/11/22, para. 130, recommendation 56. In the first cycle, this recommendations by the Russian Federation and Bahrain, are also supported by Cuba.

[149] UNHRC 'National Report submitted by Egypt' (2019) UN Doc A/HRC/WG.6/34/EGY/1, para. 43.

[150] UNHRC 'Report of the Working Group on the UPR: Turkey' (2015) UN Doc A/HRC/29/15, p. 22, recommendation 149.15.

[151] UNHRC 'Report of the Working Group on the UPR: USA'(2015) UN Doc A/HRC/30/12, 20 July 2015, para. 173.

[152] UNHRC 'Report of the Working Group on the UPR: France' (2023) UN Doc A/HRC/54/5, p. 9. Recommendation supported by France.

access to culture and cultural heritage[153] as well as linguistic heritage[154]. Communications in the Special Procedures also address the cultural dimension of languages and access to linguistic heritage.[155]

The ability to access and enjoy cultural heritage is explicitly addressed by UNHRC resolutions 33/20 (2016) and 37/17 (2018), which state that

> "the destruction of or damage to cultural heritage may have a detrimental and irreversible impact on the enjoyment of cultural rights, in particular the right of everyone to take part in cultural life, including the ability to access and enjoy cultural heritage."[156]

Ensuring the ability to access and enjoy cultural heritage primarily entails respecting, promoting and protecting "the right of everyone to take part in cultural life, including the ability to access and enjoy cultural heritage".[157] In addition, the UNHRC adds a positive duty of developing partnerships in pursuit of this right.[158]

The Special Rapporteur states that the ICESCR requires state parties to "respect free access by people to their own culture, heritage and other forms of expression".[159] This entails enabling the free exercise of cultural identity and

---

[153] See, e.g., "improvement of legislation in the cultural sphere, aimed (…) at (…) access to and protection of cultural assets" in: UNHRC 'National Report submitted by Russian Federation' (2023) UN Doc A/HRC/WG.6/44/RUS/1, para. 199; "access to culture" in: UNHRC 'National Report submitted by France' (2023) UN Doc A/HRC/WG.6/43/FRA/1, para. 85; the constitutional right to "enjoy and access cultural values, participate in cultural life, and use cultural facilities" in: UNHRC 'National report submitted by Viet Nam' (2018) UN Doc A/HRC/WG.6/32/VNM/1, para. 9.

[154] See, "enabling federal institutions to provide access to services in Indigenous languages where capacity and demand exist" in: UNHRC 'National report submitted by Canada' (2023) UN Doc A/HRC/WG.6/44/CAN/1, para. 17.

[155] Mandates of the Special Rapporteur in the field of cultural rights; the Special Rapporteur on the right to education and the Special Rapporteur on minority issues, AL ITA 4/2023 (18 September 2023). Also see observations concerning discrimination, minority languages, and the promotion of the Greek language: Mandate of the Special Rapporteur in the field of cultural rights, OL GRC 1/2024, 18 March 2024.

[156] UNHRC Res 33/20 (6 October 2016) UN Doc A/HRC/RES/33/20, preamble, 4th recital; UNHRC Res 37/17 (9 April 2018) UN Doc A/HRC/RES/37/17, para. 1. Also see para. 6 and preamble, 7th recital.

[157] UNHRC Res 33/20 (6 October 2016) UN Doc A/HRC/RES/33/20, 1, 6.

[158] *ibid.*

[159] AL, IRN 14/2016 (31 May 2016) p. 9.

cultural practices.[160] It also entails ensuring accessibility for persons with disabilities.[161] As the Special Rapporteur infers a quasi-independent right of access to and enjoyment of cultural heritage, she underlines the need to guarantee the rights linked and associated with this right, too, including to protect the freedom of individuals and groups to recreate and transmit cultural heritage to future generations.[162]

The Special Procedures' communications and country visits and reports provide further insight into any "undue restrictions" on access and enjoyment of cultural heritage.[163] For example, the explorative approach taken in the thematic reports of the Special Rapporteur in the field of cultural rights highlight tourism's impact on access to heritage.[164] The use of and access to cultural heritage for all people is addressed in communication AL TUR 12/2020 concerning the Hagia Sofia's transformation from a museum to a mosque.[165] The Special Procedures emphasise that restrictions on access to religious or sacred sites and places of worship affect the principle of non-discrimination, the right to take part in cultural life, freedom of religion or belief, the right to education, and the liberty of movement.[166] Communication AL IRN 14/2016 alleges that reported restrictions on access to religious sites and places of worship as well as attacks against these sites form part of a wider pattern of discrimination that seeks to marginalise the Bahá'í community "in the public space and the historical narratives" of Iran.[167] Communication AL BHR 6/2015 brings similar allegations against Bahrain with

---

[160] *ibid.*

[161] For example, UNHRC 'Visit to Poland, Report of the Special Rapporteur in the field of cultural rights' (2020) UN Doc A/HRC/43/50/Add.1, para. 72.

[162] AL, IRN 14/2016 (31 May 2016) p. 9.

[163] For example, UNHRC 'Visit to Malaysia, Report of the Special Rapporteur in the field of cultural rights' (2019) UN Doc A/HRC/40/53/Add.1, para. 64.

[164] UNGA 'Report of the Special Rapporteur in the field of cultural rights' (30 July 2019) UN Doc A/74/255, para. 80.

[165] AL TUR 12/2020 (28 July 2020) p. 3.

[166] UNHRC 'Report of the Special Rapporteur on freedom of religion or belief, Asma Jahangir' (6 January 2009) UN Doc A/HRC/10/8, para. 53. Mandate of the Special Rapporteur in the field of cultural rights, AL ARM 2/2021 (29 October 2021) p. 2. Also see Mandate of the Special Rapporteur in the field of cultural rights, AL AZE 3/2021 (29 October 2021) p. 2. Both communications addresses "the alleged destruction of and damage to many cultural heritage sites and objects of religious, historical and cultural importance in the Nagorno-Karabakh region".

[167] Mandates of the Special Rapporteur in the field of cultural rights; the Special Rapporteur on freedom of religion or belief; and the Special Rapporteur on the situation of human rights in the Islamic Republic of Iran, AL, IRN 14/2016 (31 May 2016) pp. 5–6.

regard to reported discrimination against Shias, which inter alia takes the form of destroying places of worship.[168]

The Special Rapporteur on freedom of religion or belief has described the desecration and destruction of religious sites and places of worship as "symbolic violence" by the perpetrators with the aim of excluding members of minorities from the community.[169] In accordance with this finding, states are advised to list religious sites of minorities as national and international cultural heritage to promote their preservation.[170] This underlines the importance of the formal recognition of heritage and shows how the mechanism of cultural heritage listing can be utilised to improve the protection of the religious freedom of members of minority groups.

5.4.4 Ensuring Effective Participation, Including the Free, Prior and Informed Consent

During the intersessional seminar on cultural rights and the protection of cultural heritage, organised by the Office of the High Commissioner for Human Rights and mandated by UNHRC resolution 33/20, participation was underlined as an important topic in the context of cultural heritage protection.[171] The ensuing report recommends states to adopt a human rights-based approach to cultural heritage protection that involves consulting with minority groups, local populations, and marginalised groups "in all aspects of decision-making".[172] Furthermore, the report recommended the United Nations to "facilitate the inclusion of local populations, heritage defenders and indigenous peoples in United Nations work and debates concerning the promotion and protection of cultural heritage".[173]

The principle of "free, prior and informed consent" can be characterised as a central component of any international instrument on the rights of Indigenous

[168] AL, BHR 6/2015 (30 October 2015).

[169] UNHRC 'Report of the Special Rapporteur on freedom of religion or belief, Heiner Bielefeldt' (2012) UN Doc A/HRC/22/51, para. 48.

[170] *Ibid* para. 74.

[171] "No decisions about us without us". See UNHRC 'Intersessional seminar on cultural rights and the protection of cultural heritage' (27 December 2017) UN Doc A/HRC/37/29, para. 95. Also see, UNHRC Res 33/20 (6 October 2016) UN Doc A/HRC/RES/33/20, para. 13 (a).

[172] *Ibid* para. 100 (b).

[173] *Ibid* para.107 (c).

peoples.[174] It implies that states need to consult and cooperate with Indigenous peoples in good faith in order to obtain their free, prior and informed consent before implementing any changes, decisions or projects that may affect them. The concept seeks to ensure the absence of any coercion, intimidation, or manipulation ("free"); timely consultations ("prior"); and that Indigenous communities are fully informed of all aspects of the project ("informed consent").[175]

Several provisions of the UNDRIP address this principle and require states to consult and cooperate in good faith to obtain the free, prior and informed consent of Indigenous communities. Article 11 of UNDRIP references the free, prior and informed consent, albeit in the context of setting up redress mechanism for "cultural, intellectual, religious and spiritual property" that was taken without the "free, prior and informed consent or in violation of [...] laws, traditions and customs" of Indigenous peoples.[176] The positive obligation to obtain the free, prior and informed consent is not explicitly linked to the protection of cultural heritage but Articles 10, 19, 29, and 32 are phrased in a broad manner that would allow heritage to be covered. Article 10 mandates the free, prior and informed consent for relocation from Indigenous lands and territories. Article 19 refers to "adopting and implementing legislative or administrative measures that may affect them" and Article 32 covers "any project affecting their lands or territories and other resources, particularly in connection with the development, utilization or exploitation of mineral, water or other resources". Article 29 addresses the protection of the environment of Indigenous peoples and requires free, prior and informed consent with regard to the storage or disposal of hazardous materials.

In the UPR, references to the "free, prior and informed consent" are common.[177] Recommendations in the framework of the review do not always

---

[174] See Convention (No. 169) concerning indigenous and tribal peoples in independent countries (adopted 27 June 1989, entered into force 5 December 1991) 1650 UNTS (ILO Convention No. 169); UNDRIP.

[175] All aspects of the project refers to the "nature, size, pace, reversibility and scope (...) the purpose of the project as well as its duration; locality and areas affected; a preliminary assessment of the likely economic, social, cultural and environmental impact, including potential risks; personnel likely to be involved in the execution of the project; and procedures the project may entail." OHCHR (2025): Consultation and free, prior and informed consent (FPIC), https://www.ohchr.org/en/indigenous-peoples/consultation-and-free-prior-and-informed-consent-fpic, accessed 31 January 2025.

[176] UNDRIP, Art. 11.

[177] E.g., UNHRC 'National report submitted by Mexico' (2013) UN Doc A/HRC/WG.6/17/MEX/1, p. 29, footnote 116; UNHRC 'Report of the Working Group on the Universal Periodic Review: Brazil' (2022) UN Doc A/HRC/52/14, p. 24; UNHRC 'Report of the Working Group on the UPR: Colombia (2023) UN Doc A/HRC/55/7, p. 22.

explicitly reference heritage, but often refer to (ancestral) land.[178] Examples of recommendations that address elements of the cultural heritage of Indigenous peoples include Denmark, which recommends Tanzania to "[a]dopt measures to protect and preserve the cultural heritage and traditional way of life of indigenous peoples and undertake effective consultations with indigenous peoples based on free, prior and informed consent".[179] Paraguay recommends the United States of America to "ensure free, prior and informed consultation with indigenous communities in relation to projects with a potential impact of their territories and traditional ways of life".[180] Namibia reports on enacting legislation that ensures the consultation and participation of Indigenous peoples when it comes to "decision-making processes relating to traditional sites and objects."[181] Finland addresses the safeguarding of "indigenous culture, including traditional livelihoods, especially for indigenous peoples, and adopt positive measures to protect them" to Tanzania.[182]

Addressing consultation and consultation participation rights,[183] the Special Rapporteur has emphasised the need to recognise "that indigenous peoples must give their free, prior and informed consent before any project that affects them is implemented".[184] They further clarify that

---

[178] E.g., UNHRC 'Report of the Working Group on the UPR: Canada' (2023) UN Doc A/HRC/55/12, paras. 37.299 and 37.295; UNHRC 'Report of the Working Group on the UPR: Colombia (2023) UN Doc A/HRC/55/7, para. 116.202; UNHRC 'Report of the Working Group on the UPR: Russian Federation' (2023) UN Doc A/HRC/55/14, para. 35.339.

[179] UNHRC 'Report of the Working Group on the UPR: United Republic of Tanzania' (2011) UN Doc A/HRC/19/4, p. 22, recommendation 86.49.

[180] UNHRC 'Report of the Working Group on the UPR: USA' (2020) UN Doc A/HRC/46/15, 15 December 2020, para. 26. 'Supported' by the United States of America. See UNHRC 'Report of the Working Group on the UPR: USA Addendum,' (2021) UN Doc A/HRC/46/15/Add.1, para. 6.

[181] UNHRC 'Report of the Working Group on the UPR: Namibia' (2016) UN Doc A/HRC/32/4, para. 109.

[182] UNHRC 'Report of the Working Group on the UPR: United Republic of Tanzania' (2021) UN Doc A/HRC/49/13, para. 147. 'Supported' by Tanzania. See, UNHRC 'Report of the Working Group on the Universal Periodic Review: United Republic of Tanzania, Addendum' (2021) UN Doc A/HRC/49/13/Add.1.

[183] AL TUR 13/2018 (23 November 2018) pp. 6, 11; Also see, UNGA 'Report of the Special Rapporteur in the field of cultural rights' (9 August 2016) UN Doc A/71/317, para. 58.

[184] UNGA 'Development and cultural rights: the principles, Report of the Special Rapporteur in the field of cultural rights, Alexandra Xanthaki' (15 August 2022) UN Doc A/77/290, para. 98 (e).

> "any impact assessment failing to address living heritage or the cultural significance of affected natural resources, or conducted without the free, prior and informed consent, consultation and active participation of the persons and communities affected directly or indirectly, should be rejected as insufficient and incomplete."[185]

The wealth of communications in the Special Procedures illustrate the applicability of the principle to cultural heritage.[186] Communication UA USA 7/2016 concerns the construction of an oil pipeline that affects inter alia a place of cultural and religious significance to the Tribe of the Standing Rock Sioux Reservation. The communication reprehends the lack of consultation with the Indigenous peoples and the failure to seek free, prior and informed consent.[187] Communication AL CAN 2/2022 concerns the impacts of oil and gas projects in British Columbia on the rights of Indigenous Peoples as well as the absence of the communities' consent.[188] The communication inter alia alleges that the construction has not only resulted in destruction to archaeological and cultural heritage to the Wet'suwet'en people but also in inaccessibility of their territories and impaired traditional hunting practices.[189] Communication AL OTH 136/2024 emphasizes free, prior and informed consent in the context of forced evictions and resettlement of Indigenous peoples' at World Heritage site Angkor.[190]

---

185 *ibid* para. 98 (d). Also see UNHRC, Cultural Rights: Tenth Anniversary Report (17 January 2019) UN Doc A/HRC/40/53, para. 18; UNHRC 'Report of the Independent Expert in the field of cultural rights, Farida Shaheed, Addendum, Mission to Morocco' (2 May 2012) UN Doc A/HRC/20/26/Add.2, para. 90; UNHRC 'Report of the Special Rapporteur in the field of cultural rights, Farida Shaheed, Addendum, Visit to Viet Nam' (29 January 2015) UN Doc A/HRC/28/57/Add.1, para. 111.

186 See, Mandatos del Grupo de Trabajo sobre la cuestión de los derechos humanos y las empresas transnacionales y otras empresas; de la Relatora Especial en la esfera de los derechos culturales y del Relator Especial sobre los derechos de los Pueblos Indígenas, AL ECU 4/2024 (11 June 2024).

187 See Mandate of the Working Group on the issue of human rights and transnational corporations and other business enterprises et al., UA USA 7/2016 (21 September 2016) p. 6.

188 Mandates of the Working Group on the issue of human rights and transnational corporations and other business enterprises et al., AL CAN 2/2022 (13 January 2023).

189 *ibid* p. 2.

190 See, for example, Mandates of the Special Rapporteur on adequate housing as a component of the right to an adequate standard of living, and on the right to non-discrimination in this context; the Special Rapporteur on the situation of human rights in Cambodia; the Special Rapporteur in the field of cultural rights; the Special Rapporteur on the right to food; the Special Rapporteur on the rights of Indigenous Peoples and the Special Rapporteur on the human rights of internally displaced persons, AL OTH 136/2024 (17 October 2024).

Ensuring participation in decision-making processes indeed has not been limited to destruction or alterations of heritage but is also addressed in the context of inscribing heritage nationally or internationally. In these cases, the Special Rapporteur has recommended that any such decisions "should be the result of inclusive processes and to be taken with the free and informed consent of all relevant stakeholders".[191]

5.4.5 Legally Recognise and Protect the Rights of Indigenous Peoples to their Cultural Heritage, Traditional Knowledge, and (Traditional) Cultural Expressions and Set Up Effective Mechanism to Enable these Rights

The rights set out in the UNDRIP are accompanied by obligations, including positive measures to be taken by the states.[192] Two central obligations in the UNDRIP are, on the one hand, the legal recognition of the rights of Indigenous peoples and the protection of rights, and on the other, setting up effective mechanisms to enable these rights. Article 31 (2) requires states to take effective measures for the recognition and protection of Indigenous peoples' exercise of their right to maintain, control, protect, and develop their cultural heritage, traditional knowledge, and other (traditional) cultural expressions, set out in Article 31 (1) of the UNDRIP.

In addition, the UNDRIP underlines the legal recognition and protection of Indigenous lands, territories, and resources[193] to guarantee their "right to own, use, develop and control the lands, territories and resources that they possess by reason of traditional ownership or other traditional occupation or use"[194]. In accordance with Article 27 of the UNDRIP this also implies the recognition and adjudication of these rights of Indigenous peoples by establishing "a fair, independent, impartial, open and transparent process" with the respective Indigenous communities. The participation rights of Indigenous peoples are tied to this.[195]

---

[191] UNGA 'Report of the Special Rapporteur in the field of cultural rights' (30 July 2019) UN Doc A/74/255, para. 91.

[192] Also see Jakubowski (2024), p. 505.

[193] UNDRIP, Art. 26 (3).

[194] UNDRIP, Art. 26 (2).

[195] See UNDRIP, Art. 27.

This links to the second form of obligation, namely setting up effective mechanisms, developed together with Indigenous communities. States are to set up these mechanisms to address Indigenous peoples' "cultural, intellectual, religious and spiritual property taken without their free, prior and informed consent or in violation of their laws, traditions and customs"[196] as well as to "seek to enable the access and/or repatriation of ceremonial objects and human remains in their possession".[197] These obligations aim to ensure the exercise of Indigenous peoples' right to maintain, protect and develop manifestations of their culture in accordance with Article 11 (1) and the rights set out in Article 12 (1), which inter alia include "the right to maintain, protect, and have access in privacy to their religious and cultural sites".

This is also reflected in Article 20 (2) of the UNDRIP, which sets out just and fair redress for Indigenous peoples who are deprived of their means of subsistence and development. Article 8 sets out obligation for states to ensure prevention and redress for inter alia actions that deprive or seek to deprive Indigenous peoples of their cultural values, ethnic identities, dispossession of their lands, territories, or resources. Prevention and redress are to be achieved through setting up effective mechanisms.

5.4.6 Interim Conclusions

Recognising the diversity of cultural heritage is not necessarily rooted in one legal basis but has been emphasised UNHRC resolutions as well as in the UPR in the context of cultural heritage. The Special Rapporteur ties the recognition of the diversity of cultural heritage to recommend to states to allow for the diversity of interpretation of cultural heritage in their own territories.

Respecting and protecting cultural heritage and preventing its destruction is an important topic in resolutions by the UN Security Council and the UNHRC as well as in the Special Procedures. The UNHRC resolutions primarily highlight impacts on the right to take part in cultural life (Article 15 (1) (a) of the ICESCR), the prohibition of national, racial or religious hatred (Article 20 of the ICCPR), and the right to freedom of thought, conscience and religion (Article 18 of the ICCPR). Relying on the pronouncements of the CESCR on the right to take part in cultural life, the Special Rapporteur asserts a duty not to destroy, damage, or alter cultural heritage. Although the wording of the duty is not repeated in any

---

[196] UNDRIP, Art. 11 (2)

[197] UNDRIP, Art. 12 (2).

UNHRC resolutions, resolutions highlight different ways aimed at preventing the destruction of cultural heritage.

The UNHRC resolutions address the ability to access and enjoy cultural heritage as a component of ensuring the right to take part in cultural life. The Special Procedures emphasises obligations that ensure the free access and exercise of cultural identity and practices, the official recognition of religious sites, as well as the realisation of concomitant rights.

Effective consultations with a view to obtaining the free, prior and informed consent of Indigenous communities is a central obligation in the UNDRIP. The UNDRIP does not link free, prior and informed consent to the cultural heritage of Indigenous peoples, however, the provisions' mentioning of impacts on lands, environment, and impacts resulting from development projects can be interpreted to cover any impacts on the cultural heritage and Indigenous way of life.

Finally, the UNDRIP identifies two main obligations that ensure the protection of the rights of Indigenous peoples to their cultural heritage, traditional knowledge, and (traditional) cultural expressions. The UNDRIP emphasises the legal recognition of Indigenous peoples' rights, lands, territories, and resources, on the one hand, and setting up effective mechanisms developed together with the respective Indigenous communities.

5.5 Conclusions

This analysis has applied a cultural heritage lens to the work of the charter-based bodies. The appeal by UNHRC resolution 6/11 in 2007 to all human rights bodies and mechanisms to take up cultural heritage protection and address it as part of the promotion and protection of human rights is certainly reflected in the work of the charter-based bodies. The chapter has shown that the protection of cultural heritage, and rights and obligations vis-à-vis cultural heritage are addressed in UN Security Council resolutions, UN General Assembly resolutions, UNHRC resolutions, the UPR, and in the work of the UNHRC, including its Special Procedures.

The chapter has shown that the Special Rapporteur in the field of cultural rights, a mandate established in 2009, has embedded cultural heritage in the human rights framework. Set against the assertion that "the destruction of cultural heritage is generally still not addressed by the international community as a question of human rights", the Special Rapporteur has developed a human rights approach.[198] This is most prominently reflected in the mandate holder's understanding of cultural heritage as containing the resources that enable "the cultural identification and development processes of individuals and groups, which they, implicitly or explicitly, wish to transmit to future generations".[199] The definition can be understood as part of a conceptual shift from the way cultural heritage is thought about in international cultural heritage law,[200] focusing less on the qualities and values of cultural heritage itself and more on the resources cultural heritage provides for the cultural identity of individuals and groups. In treaty law, the understanding of cultural heritage as a resource can be previously observed in the Faro Convention.[201] The definition forms part of the Special Rapporteur's larger human rights approach to cultural heritage, characterised by a holistic view of heritage as a living practice and countering narratives of a monolithic view of heritage.[202]

While state-led mechanisms of the UNHRC, such as UNHRC resolutions or the UPR, do not appear to follow the definition of the Special Rapporteur and refer to cultural heritage in factual terms, the work of the UNHRC reflects a thoughtful consideration of the concept. This particularly includes the UNHRC's expert mechanism on human rights and Indigenous peoples, which understood that the threefold definition of cultural heritage falls short of grasping the cultural heritage of Indigenous peoples and recommended framing it in a holistic and inter-generational manner.

The analysis has identified three rights in relation to cultural heritage and five corresponding state obligations, asserted by the different charter-based bodies and

---

198 UNGA 'Report of the Special Rapporteur in the field of cultural rights' (9 August 2016) UN Doc A/71/317, para. 52.

199 UNHRC, Report of the independent expert in the field of cultural rights, Farida Shaheed (21 March 2011) UN Doc A/HRC/17/38, para. 6.

200 Concern for heritage that may be excluded by this definition taps into the debate concerning whether cultural heritage has an intrinsic value. See, e.g., Frulli (2011), pp. 203–217.

201 "Cultural heritage is a group of resources inherited from the past which people identify, independently of ownership, as a reflection and expression of their constantly evolving values, beliefs, knowledge and traditions. It includes all aspects of the environment resulting from the interaction between people and places through time" Framework Convention on the Value of Cultural Heritage for Society, Signed in Faro on 27 October 2005. Council of Europe, ETS No. 199. Entry into force: 1 June 2011 (Faro Convention), Art. 2 (a). Subsequent use of defining and understanding cultural heritage as a resource also includes: ILA (2022): Participation in Global Cultural Heritage Governance, Lisbon Conference (2022), https://www.ila-hq.org/en_GB/documents/01-ila-participation-final-report-lisbonfinal, accessed 31 January 2025, para. 18.

202 See Chap. 5, Section 5.2.

mechanisms. The chapter has highlighted two underlying obligations, namely (1) recognising the diversity of cultural heritage and (2) respecting and protecting cultural heritage and preventing its destruction.

Owing to the Special Rapporteur's report, access and enjoyment of cultural heritage is a prominent topic in the Special Procedures. It is however of interest to note that the legal bases and the expression of this right differs in the mechanisms of the UNHRC. The Special Rapporteur's right to access to and enjoyment of cultural heritage is rooted in Article 15 (1) (a) of the ICESCR; Article 27 of the ICCPR, and in rights recognised in the ILO Convention No. 169 and the UNDRIP. The Special Rapporteur's conceptualization of the right of access to and enjoyment of cultural heritage is broad and also includes participation rights in identifying and developing cultural heritage as well as in contributing to the implementation of preservation policies. Although UNHRC resolution 37/17 addresses access to and enjoyment of cultural heritage, it speaks of the "ability" rather than of a right. This can be interpreted as support of the understanding that "the ability to access and enjoy cultural heritage" is a component of the right to take part in cultural life under Article 15 (1) (a) of the ICESCR.

Overall, rights and obligations relating to cultural heritage exhibit a focus on the rights of Indigenous peoples, which can in part be ascribed to the adoption of the UNDRIP in 2007. The UNDRIP presents itself as an important instrument for rights and obligations relating to Indigenous peoples' cultural heritage, traditional knowledge, and (traditional) cultural expressions. These prominently include participation rights in decision-making and the free, prior and informed consent. Whilst the UNDRIP does not specifically set out this right in relation to Indigenous peoples' cultural heritage, traditional knowledge, and (traditional) cultural expressions, the articles appear to be sufficiently broadly phrased.

The chapter has also shown that the UNDRIP serves as the basis for Indigenous peoples' "right to maintain, control, protect and develop their cultural heritage, traditional knowledge and traditional cultural expressions, as well as the manifestations of their sciences, technologies and cultures".[203] The role of Indigenous peoples and persons belonging to minorities in "maintaining, controlling, protecting, developing and preserving tangible and intangible cultural heritage", moreover, finds acknowledgement in the preamble of UNHRC resolution 49/7.[204] Corresponding obligations can be found among the articles of the UNDRIP, which overall emphasise the legal recognition of the rights of Indigenous peoples

---

[203] UNDRIP, Art. 31.

[204] UNHRC Res 49/7 (31 March 2022) UN Doc A/HRC/RES/49/7, preamble, 14th recital. (Adopted without a vote).

and their lands, territories and resources as well as setting up effective mechanisms to enable and protect these rights. Considering the status of the UNDRIP as a declaration, the reliance on rights and obligations contained in the instrument would necessitate the national implementation of the declaration.

Reference List

Frulli, Micaela (2011): The Criminalization of Offences against Cultural Heritage in Times of Armed Conflict: The Quest for Consistency. In: European Journal of International Law, Volume 22, Issue 1, pp. 203–217.

ILA (2022): Participation in Global Cultural Heritage Governance, Lisbon Conference (2022), https://www.ila-hq.org/en_GB/documents/01-ila-participation-final-report-lisbonfinal, accessed 31 January 2025.

Jakubowski, Andrzej (2024): Heritage and Human Rights. In: Lucas Lixinski and Lucie K. Morisset (eds.): The Routledge Handbook of Heritage and the Law. London (Routledge), pp. 500–516.

Lixinski, Lucas (2013a): Intangible Cultural Heritage in International Law. Oxford (Oxford University Press).

OHCHR (2025): Consultation and free, prior and informed consent (FPIC), https://www.ohchr.org/en/indigenous-peoples/consultation-and-free-prior-and-informed-consent-fpic, accessed 31 January 2025.

OHCHR (2012) A very dark future for the local populations in Northern Mali', warn UN experts, https://www.ohchr.org/en/press-releases/2012/07/very-dark-future-local-populations-northern-mali-warn-un-experts, accessed 31 January 2025.

Tomuschat, Christian (2019). Universal Periodic Review Procedure: Human Rights Council. In: Max Planck Encyclopedias of International Law.

6 Meaning, Rights, and Obligations in the Human Rights Bodies

6.1 Introduction

This thesis set out to understand if and how human rights law can provide a legal basis for the protection of cultural heritage. The analysis of human rights bodies has revealed that international human rights law offers several legal bases to rely on in matters concerning heritage for individuals and for persons belonging to minority groups and Indigenous communities, thereby answering the research question in the affirmative.

The preceding analyses have examined separately how the treaty bodies and charter-based bodies interpret the meaning of cultural heritage and rights and obligations vis-à-vis cultural heritage. Building on these analyses, this present chapter pursues two goals: First, it seeks to attain a more holistic understanding of the human rights bodies' interpretation of cultural heritage's meaning, rights, and obligations concerning cultural heritage. For these purposes, it synthesises the findings of the treaty bodies and charter-based bodies. It does so by collating and grouping the findings of both chapters. In addition, it supplements the analysis of treaty bodies by consulting the corresponding state reports, attempting to attain a better understanding of the states' views, reactions, and ultimately understandings of the scope and interpretation of the respective provisions.

The analysis is guided by the question to what extent these rights and obligations have emerged in the international human rights law framework. This does not refer to an attempt at a definitive determination if a right has emerged as a so-called "new human right". Cognisant of the difficulties in establishing the

L. Hofmann, *Cultural Heritage through the Lens of Human Rights Bodies*, Juridicum – Bochumer Schriften zur Friedenssicherung und zum Humanitären Völkerrecht | Bochum Studies on the International Law of Peace and Armed Conflict 73, https://doi.org/10.1007/978-3-658-51203-3_6

exact legal effect and significance of the pronouncements, resolutions, and declarations, as discussed in Chap. 2, the present chapter identifies and focuses on tendencies in the interpretation of the human rights bodies.

Second, the chapter appraises the more practical protection that can be offered in the international human rights framework by reviewing and appraising the use of human rights tools and mechanisms. It argues that the mechanisms of human rights bodies can strengthen rights and obligations relating to cultural heritage by providing access to additional monitoring- and complaints mechanisms.

The chapter starts with the meaning, rights and obligations vis-à-vis cultural heritage (Sect. 6.2). It first reviews the meaning of cultural heritage in the international human rights framework (Sect. 6.2.1). This includes the different legal bases the analysis has identified as well as the human rights bodies' terminology. Subsequently, the chapter addresses the rights and obligations vis-à-vis cultural heritage in the international human rights framework (Sect. 6.2.2). Finally, the chapter attends to the use of human rights mechanisms and its application to cultural heritage matters (Sect. 6.3).

6.2 Meaning, Rights and Obligations vis-à-vis Cultural Heritage in Human Rights Bodies

6.2.1 The Meaning and Scope of Cultural Heritage

The analyses of human rights bodies have identified different understandings and different legal bases for cultural heritage. Building on these findings, this section addresses the terminology employed by the human rights bodies to speak of and about cultural heritage. It highlights the two examples of culture as a way of life and the relationship between culture, cultural heritage, and language. Thereafter, the section reviews and discusses the different legal bases for cultural heritage.

6.2.1.1 The Terminology of the Human Rights Bodies

In contrast to the human rights-based definition of cultural heritage by the mandate holders of the Special Rapporteur in the field of cultural rights, most human rights bodies do not put forward their own definition of cultural heritage. The Special Rapporteur in the field of cultural rights embeds cultural heritage in the human rights framework and understands cultural heritage as containing the resources that enable "the cultural identification and development processes of individuals and groups, which they, implicitly or explicitly, wish to transmit to future generations".[1] This understanding is informed by a holistic view of heritage as a living practice and counters narratives of a monolithic view of heritage.[2] References to cultural heritage in the pronouncements of the treaty bodies as well as in the UPR and in resolutions by the UNHRC, UN General Assembly, and UN Security Council do not put forward their own understanding and are factual.

However, the analysis has shown that several human rights bodies have adopted a dynamic understanding characterised by a careful consideration of cultural heritage in the human rights context. This could be interpreted to demonstrate awareness and caution of the repercussions of adopting a rigid approach to the development of culture, ways of life, and cultural heritage.[3] This engagement is exemplified in the HRC's confirmation in *Länsman v. Finland* that the adaptation of traditional activities by means of modern technology,[4] resulting over years of practice is still covered by Article 27 of the ICCPR. It is also reflected in the CESCR's broad definition of cultural life[5] and in the reports of the Special Rapporteur in the field of cultural rights that expound on the dangers of "freezing" cultural heritage and not recognising it as a living culture and living heritage[6]. Moreover, UN Security Council resolution 2617 (2021) contributed to the contextualisation of the destruction of cultural heritage in the accompanying denial of cultural diversity and historical roots[7] and UNHRC resolution 79/133 emphasised the link between restitution and access to cultural heritage and cultural identity.[8]

This dynamic and careful consideration of cultural heritage is important vis-à-vis the notion of authenticity, attached to the requirement of "outstanding universal value", of the world heritage status.[9] The academic discourse has highlighted that the notion of authenticity is often not consistent with how Indigenous

---

[1] UNHRC, Report of the independent expert in the field of cultural rights, Farida Shaheed (21 March 2011) UN Doc A/HRC/17/38, para. 6.

[2] See Chap. 5, Sect. 5.2.

[3] Also see Gilbert (2018), p. 122.

[4] CCPR, Communication No. 511/1992, *Länsman v. Finland* (14 October 1993) UN Doc CCPR/C/49/D/511/1992, para. 9.3

[5] CESCR 'General Comment No. 21.' (21 December 2009) UN Doc E/C.12/GC/21, p. 3, fn 12.

[6] UNGA 'Report of the Special Rapporteur in the field of cultural rights, Karima Bennoune' (19 July 2021) UN Doc A/76/178, paras. 32–33.

[7] UNSC Res 2617 (30 December 2021) UN Doc S/RES/2617, preamble, 29th recital.

[8] UNGA Res 79/133 (12 December 2024) UN Doc A/RES/79/133, para. 6.

[9] UNESCO 'Operational Guidelines for the Implementation of the World Heritage Convention' (31 July 2024) WHC.24/01 (*Operational Guidelines*), para.78.

peoples view their heritage.[10] The discourse has emphasised that Indigenous "lands and territories are not relics but living cultural landscapes" that also hold an economic, spiritual and social dimension.[11]

At the level of regional human rights bodies, this was also reflected in the case of the *Mayagna (Sumo) Awas Tingni Community v. Nicaragua* at the Inter-American Court of Human Rights (IACtHR) and the *Ogiek* case of the African Commission on Human and Peoples' Rights (AfCHPR). The IACtHR emphasised the close ties to land and understood land "as the fundamental basis of their cultures, their spiritual life, their integrity, and their economic survival".[12] The *Ogiek* case emphasised that change is a naturally occurring part of the culture of Indigenous peoples by arguing

> "that stagnation or the existence of a static way of life is not a defining element of culture or cultural distinctiveness. It is natural that some aspects of indigenous populations' culture [...] could change over time."[13]

This is an important response and application of international human rights law by the AfCHPR to the claim that a group and their way of life can evolve to the point where the group loses their distinctive cultural identity.[14] As two parts of the same coin, the HRC's assertion in *Länsman v. Finland* and the reasoning of AfCHPR in the *Ogiek* case reflect a considerate engagement with the culture and way of life of Indigenous communities.

6.2.1.1.1 Culture as a Way of Life

Against the backdrop of the HRC's use of way of life in its pronouncements, the infrequent use of the term cultural heritage by the committee does not appear to be a shortcoming necessarily. The committee prominently addresses culture as particular, traditional, or Indigenous way of life in its pronouncements. As previously shown, the notion of culture as a way of life stems from the HRC's finding that in

[10] Disko (2017), p. 75; Nafziger, Kirkwood Paterson and Dundes Renteln (2010), p. 77; Lixinski (2018b), p. 477.

[11] Disko (2017), p. 76. Also see Lixinski's conservation paradigm. Lixinski (2019), p. 18.

[12] *Case of the Mayagna (Sumo) Awas Tingni Community v. Nicaragua*, Merits, reparations and costs, IACHR Series C No 79 [2001], para. 149.

[13] *African Commission on Human and Peoples' Rights v Kenya* (Application No. 006/2012) [2017] AfCHPR 2 (26 May 2017), para. 185.

[14] *ibid* para. 185. Also see Gilbert (2018), p. 122.

> "the exercise of the cultural rights protected under article 27 [...] culture manifests itself in many forms, including a particular way of life associated with the use of land resources, especially in the case of indigenous peoples [...] [including] such traditional activities as fishing or hunting."[15]

Gilbert explains that

> "in the context of a way of life of specific groups or communities, cultural rights refer to the anthropological meaning of culture, which includes any social activity or expression specific to a given population, including any customs or practices connected to natural resources."[16]

The use of ways of life by the HRC, but also by other human rights bodies, can be interpreted to approach the complex relationship of Indigenous peoples and their cultural heritage, characterised in a holistic and inter-generational way, that cannot be grasped by the legal definition of cultural heritage in the UNESCO conventions[17]. Indeed, the dichotomy between cultural and natural heritage does not reflect the reality for many Indigenous peoples, when precisely the interaction between both elements have shaped their cultural heritage.[18]

Indeed, the absence of a definition of cultural heritage in the human rights bodies—with the exception of the Special Procedures—is neither *strictly* advantageous nor disadvantageous. On the one hand, the use of the term can contribute to articulating and emphasising a cultural heritage dimension in human rights law. By not explicitly articulating this dimension, when the protection of cultural heritage is at stake, the paucity could result in its under-articulation or even omission of the dimension in the human rights framework. It has been argued that since international human rights law does not provide for a right to heritage, the reliance on human rights provisions "falls short of recognising cultural heritage as being a full member of the human rights legal framework".[19] It follow that

---

[15] CCPR 'General Comment No. 23: Article 27 (Rights of Minorities)' (8 April 1994) UN Doc CCPR/C/21/Rev.1/Add.5, para. 7.

[16] Gilbert (2018), p. 119.

[17] UNHRC 'Promotion and Protection of the Right of Indigenous Peoples with Respect to their Cultural Heritage, Study by the Expert Mechanism on the Rights of Indigenous Peoples' (19 August 2015) UN Doc A/HRC/30/53, para. 8, Annex A.1–3., A.6. Also see, Dorough and Wiessner (2020), p. 414.

[18] Gilbert (2018), p. 130.

[19] Francioni and Lixinski (2017), p. 18. Also see Novic (2016), p. 135.

cultural heritage considerations remain subordinate to the primary goals of the specific human rights provision.[20]

6.2.1.1.2 Cultural Heritage and Language

The overall broad and dynamic understanding of culture by the human rights bodies is advantageous because it can protect a broad range of cultural forms and expressions. The Intangible Cultural Heritage Convention, for example, recognises language only as a vehicle to transmit oral traditions and expressions.[21] This makes clear that the convention only protects language to the extent that it allows for the transmission of oral traditions and expressions.[22] Moreover, the lists show that language plays an important role for other inscriptions in the Representative List of the Intangible Cultural Heritage of Humanity, such as writing systems or music and song.[23]

The rather "restrictive"[24] approach of the Convention has been ascribed to states' resistance "to any external interference in national cultural policies and the codification of protection for intangible heritage, especially languages, at the international level."[25] The text of the Convention accordingly is a compromise between states that were keen to explicitly include languages and those that deemed languages to be sufficiently covered by oral traditions.[26] While any reluctance and resistance by states towards minority languages will inevitably affect the branch of international human rights law, too, language is also protected in several human rights provisions.

The protection of language in international human rights law by means of, for example, Article 15 (1) (a) of the ICESCR does not appear to be conditional

[20] *ibid.*

[21] Convention for the Safeguarding of the Intangible Cultural Heritage (adopted 17 October 2003, entered into force 20 April 2006) 2368 UNTS 3 (*Intangible Cultural Heritage Convention*), Art. 2 (a).

[22] Currently, three languages are inscribed in the Representative List of the Intangible Cultural Heritage of Humanity and the List of Intangible Cultural Heritage in Need of Urgent Safeguarding. Türkiye inscribed whistled language; Spain inscribed the Silbo Gomero, a whistled language of La Gomera Island in the Canaries; and Belize, Guatemala, Honduras, and Nicaragua inscribed the language, dance, and music of the Garifuna. UNESCO (2025): Browse the Lists of Intangible Cultural Heritage and the Register of good safeguarding practices: Language, https://ich.unesco.org/en/lists?text=language&multinational=3#tabs, accessed 31 January 2025.

[23] *ibid.*

[24] De Witte (2020), p. 372.

[25] Vrdoljak (2014), p. 10.

[26] Ubertazzi (2020), p. 62.

but covers language as an element of cultural life.[27] This can contribute to the protection of minority and Indigenous languages. Yet, the protection offered by human rights law—while broader—is also more unspecific. This means that while language is protected as e.g., an element of cultural life in Article 15 (1) (a) of the ICESCR, as mother tongue instruction in Articles 13–14 of the ICESCR or as a minority language in Article 27 of the ICCPR, monitoring by treaty bodies cannot address human rights issues in an all-encompassing and exhaustive manner. Furthermore, specific cultural elements of language that form part of the intangible cultural heritage are likely to be under-articulated in the international human rights framework.[28]

6.2.1.2 Legal Bases for Elements of Cultural Heritage

In addition, to expounding on the scope of "cultural life" in Article 15 (1) (a) of the ICESCR and "culture" in Article 27 of the ICCPR, as discussed in the preceding analyses, the treaty bodies rely on, and the Special Procedures highlights further legal bases in their pronouncements and work. In other words, elements and aspects of cultural heritage are not confined to these concepts of "cultural life" and "culture" in the above-mentioned human rights provisions.

In framing cultural heritage as "a fundamental resource for other human rights", the Special Rapporteur in the field of cultural rights has highlighted

> "the rights to freedom of opinion and expression, freedom of thought, conscience and religion, as well as the economic rights of the many people who earn a living through tourism related to such heritage, the right to education and the right to development."[29]

Moreover, communications of HRC and the Special Procedures have offered further and more specific insights into additional legal bases as well as into the interdependency of human rights and the realisation of concomitant rights. More specifically, they show how this interdependence challenges the notion of a strict separation between civil and political rights, on the one hand, and economic,

---

[27] Notwithstanding of course that General Comment No. 21 argues that certain circumstances and negative practices may render limitations necessary, "including those attributed to customs and traditions, that infringe upon other human rights". CESCR 'General Comment No. 21.' (21 December 2009) UN Doc E/C.12/GC/21, para. 19.

[28] Following the reasoning of *Lixinski* with regard to food in: Lixinski (2018b), p. 486.

[29] UNHRC 'Report of the Special Rapporteur in the field of cultural rights' (3 February 2016) UN Doc A/HRC/31/59, para. 51.

social, and cultural rights, on the other.[30] Five legal bases can be highlighted: Article 6 of the ICCPR (right to life and survival); Article 17 of the ICCPR (protection from the unlawful interference with privacy, family, and home); Article 18 of the ICCPR (freedom of thought, conscience and religion) Article 11 of the ICESCR (the right to an adequate standard of living); and Articles 13–14 of the ICESCR (the right to education). In the following, the section shows how these rights have been linked to cultural heritage by the human rights bodies.

6.2.1.2.1 Article 6 of the ICCPR and the Environmental Impact on Living a Life with Dignity

In *Benito Oliveira Pereira et al. v Paraguay,*[31] the communication that considered the contamination of the *Campo Agua'ẽ* indigenous community and the state's failure to provide protection from illegal fumigation, the joint concurring opinion of two committee members noted with regret that neither the parties nor the committee had raised (*proprio motu* in the case of the committee) Article 6, the right to life, as the primary consequence of the environmental pollution.[32] The opinion highlighted its previous finding, namely that the right to enjoy a life with dignity forms part of the right to life.[33]

The pronouncements of the HRC have also underlined that states have a duty to "take appropriate measures to address the general conditions in society that may give rise to direct threats to life or prevent individuals from enjoying their right to life with dignity".[34] The degradation of the environment and the deprivation of Indigenous peoples' land, territories, and resources, all form part of these general conditions.[35] As highlighted previously, Indigenous peoples' land, territories and resources are intimately connected to their way of life, cultural heritage, traditional knowledge, and (traditional) cultural expressions.

The notion of the right of life with dignity has long been reflected in the concept of "*vida digna*" of the IACtHR, which describes and has found the right to

---

[30] Chenwi (2009), p. 25.

[31] CCPR, Communication No. 2552/2015, *Benito Oliveira Pereira et al. v. Paraguay* (12 October 2021) UN Doc CCPR/C/132/D/2552/2015, para. 8.4.

[32] *ibid* Annex I, para. 2.

[33] CCPR 'General Comment No. 36: Article 6: right to life' (3 September 2019) UN Doc CCPR/C/GC/36, para. 3.

[34] CCPR, Communication No. 2552/2015, *Benito Oliveira Pereira et al. v. Paraguay* (12 October 2021) UN Doc CCPR/C/132/D/2552/2015, para. 2; Referring to and citing *ibid* para. 26; CCPR, Communication No. 2751/2016, *Norma Portillo Cáceres v. Paraguay et al.* (20 September 2019) UN Doc CCPR/C/126/D/2751/2016, para 7.3.

[35] *ibid.*

life to include the right to have "access to the conditions that guarantee a dignified existence".[36] The concurring opinion of members of the HRC further highlighted that in *Yakye Axa Indigenous Community v. Paraguay* and the subsequent Advisory opinion on the environment and human rights, the IACtHR found and established a close relationship between the right to life and economic, social, and cultural rights.[37]

Whilst the right to enjoy a life with dignity is not explicitly linked to the protection of cultural heritage, the communication of *Benito Oliveira Pereira et al. v Paraguay* has shown the environmental impact on an Indigenous peoples' enjoyment of their culture. Furthermore, the communication has emphasised that the right of Indigenous peoples to enjoy their culture is "directed towards the *survival* and continued development of the cultural identity".[38]

6.2.1.2.2 Article 17 of the ICCPR and the Notions of Privacy, Family, Home, and a Particular, Traditional, or Indigenous Way of Life

The HRC has interpreted Article 17 of the ICCPR in at least two significant ways for the purposes of this analysis. It has expanded the understanding of family by affirming to consider cultural traditions[39] and a person's relationship to their ancestors as an important element of their identity and of their family life[40]. Whereas the expansion of the concept of family was critique by the dissenting opinion,[41] more recent communications have applied the protective scope of Article 17 to protect elements of heritage, albeit in a slightly different way.

In *Norma Portillo Cáceres et al. v. Paraguay, Daniel Billy et al. v. Australia*, and *Benito Oliveira Pereira et al. v. Paraguay* the HRC found the scope of protection of Article 17 to include components of the (traditional or Indigenous) way of life

---

[36] Case of the 'Street Children' (Villagrán-Morales et al.) v. Guatemala, Merits, Judgment. Series C No. 63 [1999], para. 144.

[37] CCPR, Communication No. 2552/2015, *Benito Oliveira Pereira et al. v. Paraguay* (12 October 2021) UN Doc CCPR/C/132/D/2552/2015, para. 3 referring to and citing *Case of the Yakye Axa Indigenous Community v. Paraguay,* Merits, Reparations, and Costs, Judgment. Series C No. 125 [2005], para. 163.

[38] Emphasis added by author. CCPR, Communication No. 2552/2015, *Benito Oliveira Pereira et al. v. Paraguay* (12 October 2021) UN Doc CCPR/C/132/D/2552/2015, para. 7.2.

[39] *ibid.*

[40] *ibid.*

[41] *ibid* para. 4.

of the authors.[42] *Portillo Cáceres et al. v. Paraguay* concerned *campesino* authors, sharing "a special attachment to and dependency on the land".[43] *Daniel Billy et al. v. Australia* highlighted the impacts of climate change on the traditional ways of life of Torres Strait Islander peoples. In *Benito Oliveira Pereira et al. v. Paraguay*, the authors belonged to an Indigenous community, whose subsistence depended on "crops, livestock, fruit trees, hunting, foraging, fishing and water resources" which the committee all found to be elements of their territory.[44] The committee found these elements to form part of authors' way of life and their special relationship with their territory, and fall under the scope of protection of Article 17 of the ICCPR.[45]

The notion of home is also found in communication AL OTH 136/2024 of the Special Procedures, which focuses on the multifaceted impact of forced evictions at Angkor on Indigenous peoples. It argues, inter alia, that in the case of Angkor, the displacement from Indigenous lands results in a severing of "cultural, spiritual and physical connection to their lands, territories and resources and to what they consider to be 'home'".[46] The communication also refers to the HRC's findings on the meaning of home for Indigenous peoples in *Benito Oliveira Pereira et al. v. Paraguay*.[47]

The interpretation of the HRC implies that a particular, traditional, or Indigenous way of life is not only protected under Article 27 of the ICCPR but also finds protection under Article 17 of the ICCPR. Article 27 link to the enjoyment

---

[42] CCPR, Communication No. 2751/2016, *Norma Portillo Cáceres v. Paraguay et al.* (20 September 2019) UN Doc CCPR/C/126/D/2751/2016, para. 7.8; CCPR, Communication No. 3624/2019, *Daniel Billy et al. v. Australia* (22 September 2022) UN Doc CCPR/C/135/D/3624/2019, para. 8.10; CCPR, Communication No. 2552/2015, *Benito Oliveira Pereira et al. v. Paraguay* (12 October 2021) UN Doc CCPR/C/132/D/2552/2015, para. 8.3.

[43] CCPR, Communication No. 2751/2016, *Norma Portillo Cáceres v. Paraguay et al.* (20 September 2019) UN Doc CCPR/C/126/D/2751/2016, para. 7.8.

[44] CCPR, Communication No. 2552/2015, *Benito Oliveira Pereira et al. v. Paraguay* (12 October 2021) UN Doc CCPR/C/132/D/2552/2015, para. 8.3.

[45] *ibid.*

[46] Mandates of the Special Rapporteur on adequate housing as a component of the right to an adequate standard of living, and on the right to non-discrimination in this context; the Special Rapporteur on the situation of human rights in Cambodia; the Special Rapporteur in the field of cultural rights; the Special Rapporteur on the right to food; the Special Rapporteur on the rights of Indigenous Peoples and the Special Rapporteur on the human rights of internally displaced persons, AL OTH 136/2024 (17 October 2024), p. 6.

[47] "... the notion of "home" must be understood in the context of the special relationship that they have with their territories and their ways of life, including their subsistence activities such as livestock-raising." *ibid* p. 9.

of culture, practice of religion, and use of language. Article 17, while protecting the "same" way of life, adds an additional layer of family, home and privacy. This can contribute to a more holistic understanding of the notion of ways of life that links it to family life and by underlining cultural privacy. It also adds a group dimension to the right.[48]

At the regional level, the European Court of Human Rights (ECtHR) found in the case of *Sargsyan v. Azerbaijan* the applicant's rights to his "private family life" and home to be affected by his inability to return to his former place of residence. The ECtHR took the applicant's "cultural and religious attachment to his late relatives' graves" into account and argued that this attachment may fall within the notion of "private and family life".[49]

6.2.1.2.3 Article 18 of the ICCPR and the Protection of Religious Heritage and Sacred Sites

Article 18 of the ICCPR protects the right to freedom of thought, conscience and religion. The analyses have shown that the reliance on Article 18 for the protection of religious heritage and sacred sites is not frequent, especially in the treaty bodies. Whilst the HRC addresses restrictions on places of worship under Article 18 of the ICCPR, the link to cultural heritage protection is not articulated in the pronouncements of the HRC. Instead, the committee appears to rely more on Article 27 of the ICCPR to address any restrictions or interferences with religious, spiritual, and sacred places,[50] which specifically protects the rights of religious minorities to practise their religion.

In contrast, the work of the Special Rapporteur on freedom of religion or belief has highlighted the intrinsic link between cultural rights and the right to freedom of thought, conscience and religion. This includes finding that the destruction of religious monuments and relics can prevent the enjoyment of cultural and religious rights.[51]

In addition to the UNHRC resolutions addressing the destruction of places of worship, it is also worth noting that resolution 2001/13 by the then-Commission on Human Rights described the intentional destruction of Afghan cultural heritage

---

[48] Lixinski (2013a), p. 164.

[49] *Sargsyan v. Azerbaijan,* Judgment, App. No. 40167/06 (Eur. Ct. H.R., 2015), para. 257.

[50] E.g., CCPRR 'Concluding Observations: USA' (2014) UN Doc CCPR/C/USA/CO/4, para. 25; CCPR 'Concluding Observations: New Zealand' (2010) UN Doc CCPR/C/NZL/CO/5, para. 19.

[51] UNHRC 'Report of the Special Rapporteur on freedom of religion or belief, Asma Jahangir' (6 January 2009) UN Doc A/HRC/10/8, para. 52.

as a "grave violation of, *inter alia,* article 15, paragraph 2 of the" ICESCR.[52] This is a reference to the obligation of the state party "to achieve the full realization of" the right to take part in cultural life, including "those necessary for the conservation, the development and the diffusion of [...] culture".[53]

6.2.1.2.4 Article 11 of the ICESCR and the Expression of Cultural Identity

The right to an adequate standard of living has a cultural component. This is, for example, reflected in the CESCR's interpretation of the cultural component of water,[54] the cultural acceptability of food,[55] and in the cultural adequacy of the right to adequate housing[56].

Communications of the Special Procedures highlight the expression of cultural identity as an important component of Article 11 of the ICESCR and illustrate its link to cultural heritage. Communications AL TUR 13/2018, AL BRA 2/2018, and UA PAK 1/2016 in the Special Procedures concern inter alia the right to take part in cultural life and the right to an adequate standard of living and housing.[57] Communication AL TUR 13/2018 addresses the construction of the Ilisu dam, reported to result in, among other impacts, resettlements of the population—mainly members of the Kurdish minority—and in the destruction of cultural heritage in the city of Hasankeyf and the surrounding.[58] Communication AL BRA 2/2018 focuses on the importance of the residents of *Horto Florestal*, a

---

[52] Commission on Human Rights Res 2001/13 (18 April 2001), 15th recital. International Covenant on Economic, Social and Cultural Rights (adopted 16 Dec 1966, entered into force 3 Jan 1976) 993 UNTS 3 (*ICESCR*), Art. 15 (2) states "[t]he steps to be taken by the States Parties to the present Covenant to achieve the full realization of this right shall include those necessary for the conservation, the development and the diffusion of science and culture".

[53] ICESCR, Art. 15 (2).

[54] CESCR 'General Comment No. 15 (2002)' (20 January 2003), UN Doc E/C.12/2002/11, paras 6, 11.

[55] CESCR 'General Comment No. 12' (12 May 1999), UN Doc E/C.12/1999/5, paras. 8, 11.

[56] CESCR 'General Comment No. 4' (13 December 1991) UN Doc E/1992/23, para. 8(g).

[57] Also see, Mandates of the Special Rapporteur in the field of cultural rights; the Special Rapporteur on the rights of indigenous peoples and the Special Rapporteur on the human rights of internally displaced persons, UA NPL 3/2017 (14 August 2017).

[58] Mandates of the Special Rapporteur in the field of cultural rights; the Special Rapporteur on the issue of human rights obligations relating to the enjoyment of a safe, clean, healthy and sustainable environment; the Special Rapporteur on adequate housing as a component of the right to an adequate standard of living, and on the right to non-discrimination in this context; the Special Rapporteur on the human rights of internally displaced persons; and the Special Rapporteur on minority issues, AL TUR 13/2018 (23 November 2018).

historical neighbourhood in Rio de Janeiro, in creating, developing, and protecting cultural heritage, which is at the risk of destruction due to forced evictions and urban planning.[59] Communication PAK 1/2016 concerns the construction of a metro line in Lahore and reports that the construction will result in the destruction of historic buildings and neighbourhoods, alleging a violation of the right to take part in cultural life and the right to housing.[60]

Although the communications do not interpret or expound further on the right, all three communications underline the inadequacy of implementing the right to an adequate standard of living and housing if the state's implementation fails to respect and consider "the expression of cultural identity".[61] This was previously emphasised in the CESCR'S General Comment No. 4, which states that any developments or modernisation of housing "should ensure that the cultural dimensions of housing are not sacrificed".[62] General Comment No. 4 and the Special Procedures thereby underline the link between the right to an adequate standard of living and cultural identity. The communications by the Special Procedures go beyond this by highlighting the impacts of development projects and urban planning on communities that live in and around cultural heritage.

6.2.1.2.5 Articles 13–14 of the ICESCR and the Transmission of Heritage

The pronouncements of the human rights bodies show that the right to education is instrumental to individuals and persons belonging to minorities and Indigenous peoples to access and transmit language, customs and traditions, and other cultural values.[63] Education is an integral part of access to cultural heritage, within

---

[59] Mandates of the Working Group of Experts on People of African Descent; the Special Rapporteur in the field of cultural rights; the Special Rapporteur on extrajudicial, summary or arbitrary executions; the Special Rapporteur on adequate housing as a component of the right to an adequate standard of living, and on the right to non-discrimination in this context; and the Independent Expert on the enjoyment of all human rights by older persons, AL BRA 2/2018, 20 April 2018. Communication remained unanswered.

[60] Mandates of the Special Rapporteur on adequate housing as a component of the right to an adequate standard of living, and on the right to non-discrimination in this context and the Special Rapporteur in the field of cultural rights, UA, PAK 1/2016 (12 January 2016). Communication remained unanswered.

[61] *ibid* pp. 4–5.

[62] CESCR 'General Comment No. 4' (13 December 1991) UN Doc E/1992/23, para. 8 (g).

[63] UNHRC 'Report of the independent expert in the field of cultural rights, Ms. Farida Shaheed, submitted pursuant to resolution 10/23 of the HRC' (22 March 2010) UN Doc A/HRC/14/36, para. 9; CESCR 'General Comment No. 21.' (21 December 2009) UN Doc E/C.12/GC/21, para. 15. Also see, CESCR 'Excerpt from the Report on the Eighteenth and

the meaning of General Comment No. 21, too.[64] The Special Rapporteur in the field of cultural rights underlines the importance of the right to education for access to cultural heritage in country- and thematic reports,[65] including on the topic of the teaching of history.[66] Moreover, the Special Rapporteurs emphasise in communications UA NPL 3/2017 and AL BGR 3/2015 cultural heritage's role as a "critical resource for safeguarding, questioning and transmitting historical knowledge and narratives of the past", making clear that it thereby ensures "the right to education and training without any discrimination".[67]

In addition to the right to education in the ICESCR, the UNDRIP offers a further basis for Indigenous peoples' rights to maintain and strengthen their cultural institutions[68] and to establish and control their educational systems and institutions and offer education in their own languages.[69]

6.2.1.3 Interim Conclusions: A Refined and Holistic Understanding of Cultural Heritage's and its Link to Human Rights

The analysis has shown that the human rights bodies' approach to cultural heritage contributes to a refined and holistic understanding of cultural heritage. It argues that the human rights bodies do this in at least two ways. First, the Special Rapporteur in the field of cultural rights follows the lead of the Faro Convention

---

Nineteenth Sessions' (30 November 1998) UN Doc E/1999/22 -E/C.12/1998/26, para. 482. CESCR 'Summary Record of the 49th Meeting: General Discussion: "Right to education (Articles 13 and 14 of the Covenant)' (2 December 1998) UN Doc E/C.12/1998/SR.49, para. 54; UNHRC 'Report of the Special Rapporteur in the field of cultural rights, Farida Shaheed, Addendum, Mission to Bosnia and Herzegovina' (3 March 2014) UN Doc A/HRC/25/49/ Add.1, para. 43.

64 CESCR 'General Comment No. 21.' (21 December 2009) UN Doc E/C.12/GC/21, para. 15.

65 UNHRC 'Report of the Special Rapporteur in the field of cultural rights, Farida Shaheed, Addendum, Mission to Bosnia and Herzegovina' (3 March 2014) UN Doc A/HRC/25/49/ Add.1, para. 104 (b); UNGA 'Report of the Special Rapporteur in the field of cultural rights' (9 August 2013) UN Doc A/68/296, para. 7.

66 UNGA 'Report of the Special Rapporteur in the field of cultural rights' (9 August 2013) UN Doc A/68/296, p. 2.

67 UA NPL 3/2017, 14 August 2017, pp. 6–7; Mandate of the Special Rapporteur in the field of cultural rights; the Special Rapporteur on minority issues; and the Special Rapporteur on freedom of religion or belief, AL BGR 3/2015 (15 September 2015), p. 5.

68 UNGA Res 61/295 'United Nations Declaration on the Rights of Indigenous Peoples' (13 September 2007) UN Doc A/Res/61/295, Art. 5.

69 *ibid* Arts. 14–15.

and posits the conceptual understanding of cultural heritage as containing the resources that enable "the cultural identification and development processes of individuals and groups, which they, implicitly or explicitly, wish to transmit to future generations".[70]

Second, the CESCR, HRC, Special Procedures and individual resolutions of the UNHRC, UN General Assembly, and even the UN Security Council approach cultural heritage in a holistic manner and often consider its dynamic and evolving nature. In addition to the most common legal bases employed by the human rights bodies, i.e. Article 15 (1) (a) of the ICESCR and Article 27 of the ICCPR, the work and pronouncements of human rights bodies have also relied on (in the case of the pronouncements) or highlighted (in the case of the Special Procedures' communications) further (legal) bases. These contribute to a better understanding of cultural heritage in international human rights law, as they illustrate inter alia the proximity of cultural heritage to family life, privacy and home as well as to the right to life in the case of Indigenous peoples.

6.2.2 Rights and Obligations vis-à-vis Cultural Heritage

The analyses of the preceding two chapters have identified several rights and obligations, articulated by human rights bodies. This has left us with individual puzzle pieces of a right or obligation relating to cultural heritage. In most cases, these pieces stem from two legal bases: Article 15 (1) (a) of the ICESCR and Article 27 of the ICCPR. In addition, the previous section has shown that further legal bases are relevant for rights and obligations relating to cultural heritage.

The analyses have also shown that these pieces, i.e. the rights and obligations, are fragmented and, especially in the case of the rights of Indigenous peoples, "scattered across the human rights legal framework".[71] Collating the rights and obligations previously identified, this chapter focuses on four rights and obligations in international human rights law framework: (1) respect and protect cultural heritage at all times and prevent destruction and damage, (2) access to cultural heritage, (3) respect, protect, and (formally) recognise the rights and cultural heritage of persons belonging to minorities and Indigenous peoples and to create favourable conditions for the enjoyment and development of culture and cultural

[70] UNHRC, Report of the independent expert in the field of cultural rights, Farida Shaheed (21 March 2011) UN Doc A/HRC/17/38, para. 6.

[71] Gilbert (2017), p. 24. For a detailed breakdown of the rights of Indigenous peoples in the human rights framework, see Gilbert (2017), p. 23 f.

heritage, and the right of Indigenous peoples to (4) effective consultations, participation and free, prior and informed consent in decision-making affecting cultural heritage and their way of life.

The analysis recounts how human rights bodies set these rights and obligations out and which legal bases they stem from. The analysis is guided by the question to what extent these rights and obligations have emerged in the international human rights framework. For these purposes it further contextualises the findings of the treaty bodies by drawing on periodic state reports under the treaty bodies' monitoring mechanism.

6.2.2.1 Respect and Protect Cultural Heritage at all Times and Prevent Destruction and Damage

The human rights bodies show that damage and destruction can concern all forms of heritage, including mixed and natural sites, religious and sacred sites, or intangible practices. The CESCR articulates the fundamental obligation to respect and protect cultural heritage at all times, whether in times of war, peace or natural disasters, as a specific legal obligation under the right to take part in cultural life.[72] This is repeated by the Special Rapporteur and built on, by asserting the duty not to destroy, damage or alter[73] cultural heritage. Beyond the specific scope of Article 15 (1) (a) of the ICESCR, i.e. participation in cultural life, this thesis argues that it can be regarded as a general obligation that serves the implementation of any right and obligations relating to cultural heritage and articulated by the human rights bodies.

Outside the context of the prohibition of the destruction of cultural property in international humanitarian law,[74] international human rights law does not set out a specific right to be protected from the destruction of cultural heritage. General Comment No. 21 emphasises as part of the obligation to respect to "not […] be subjected to any form of discrimination based on cultural identity, exclusion or forced assimilation".[75] In the context of the rights of Indigenous peoples,

---

[72] CESCR 'General Comment No. 21' (21 December 2009) UN Doc E/C.12/GC/21, para. 50 (a).

[73] As outlined in Chap. 5, the Special Rapporteur specifies that any destruction, damage or alterations should be preceded by "the free, prior and informed consent of concerned communities". UNHRC, Report of the independent expert in the field of cultural rights, Farida Shaheed (21 March 2011) UN Doc A/HRC/17/38, para. 80 (b).

[74] On the criminalisation of these acts and the civilian use vs cultural value approach in international criminal law, see Frulli (2020), pp. 100–120.

[75] CESCR 'General Comment No. 21' (21 December 2009) UN Doc E/C.12/GC/21, para. 49 (a). More generally on this, see Blake (2015), p. 301.

however, the UNDRIP addresses the right not to be subjected to the destruction of one's culture,[76] which can be interpreted as a specific articulation of the flip side of the obligation.

What do respect and protection of cultural heritage imply? The CESCR sets out that the obligation to respect and protect cultural heritage at all times requires preserving, developing, enriching, and transmitting cultural heritage to future generations, including caring, preserving, and restoring historical sites, monuments, works of art and literary works.[77] Whilst these positive actions, of course, serve the right to participate in cultural life, the CESCR's interpretation of Article 15 (1) (a) also sets out obligations directly towards heritage. In other words, the obligation to respect and protect cultural heritage at all times is not directed immediately at individuals but is for their benefit. *Saul, Kinley* and *Mowbray* argue that the CESCR has made clear "that Article 15 (1) (a) imposes obligations on states to preserve and promote cultural heritage".[78] The state reports under the ICESCR appear to confirm this, as they consistently address the protection of cultural heritage.[79] One example is the state report of the Dominican Republic, which speaks of its efforts "to preserve and promote their tangible and intangible heritage from a history full of interactions between highly diverse cultural groups: indigenous peoples, Africans, Europeans, Asians and Americans."[80]

Despite the unequivocal nature in which the obligation is phrased in General Comment No. 21, the analysis has found that the concluding observations of the CESCR are couched in much softer terms, thereby showing a gap with how much vigour the obligation is expressed.[81] In the reporting under the ICESCR, states

---

[76] UNGA Res 61/295 'United Nations Declaration on the Rights of Indigenous Peoples' (13 September 2007) UN Doc A/Res/61/295, Art. 8 (1).

[77] CESCR 'General Comment No. 21.' (21 December 2009) UN Doc E/C.12/GC/21, para. 50 (a). Repeated by the Special Rapporteur in e.g., UNHRC, Report of the independent expert in the field of cultural rights, Farida Shaheed (21 March 2011) UN Doc A/HRC/17/38, para. 80 (b); Mandate of the Special Rapporteur in the field of cultural rights, AL, SAU 3/2016 (31 May 2016), p. 8.

[78] Saul, Kinley and Mowbray (2014), p. 1190.

[79] CESCR 'Periodic report submitted by Australia' (2023) UN Doc E/C.12/AUS/6, para. 352 f; CESCR 'Periodic report submitted by Mexico' (2023) UN Doc E/C.12/MEX/7, para. 324 f; CESCR 'Periodic Reports submitted by Afghanistan' (2009) UN Doc E/C.12/AFG/2–4, para. 206 f; CESCR 'Periodic report submitted by Japan,' (2011) UN Doc E/C.12/JPN/3, para. 382; CESCR 'Periodic report submitted by Turkey' (2009) UN Doc E/C.12/TUR/1, para. 639 f, especially para. 648 f.

[80] CESCR 'Periodic report submitted by Dominican Republic' (2015) UN Doc E/C.12/DOM/4, para. 223.

[81] See Chap. 4, Sect. 4.4.2.

address implementation measures for the protection of cultural heritage, the ratification of or accession to cultural heritage instruments,[82] and affirm the aim to protect cultural heritage[83]. However, there is a paucity in the state reports of verbatim repetitions of this obligation. The same holds true for recommendations in the framework of the UPR.[84] The absence in both mechanisms could be ascribed to states seeking to protect their sovereignty over decisions affecting cultural heritage that is situated in their respective territories,[85] such as how to respond to the effects of climate change or decisions over (development) projects that might affect heritage sites or practices.

UNHRC resolution 49/7 *encourages* states

> "to take the measures necessary to prevent the destruction of historic monuments, works of art or places of worship that constitute the cultural or spiritual heritage of peoples, *both in conflict and non-conflict situations*, and to promote respect for cultural diversity."[86]

The preamble of HRC resolution 37/17 reaffirms the need to protect all forms of cultural heritage "in both conflict and non-conflict situations".[87] The resolution

[82] See, e.g., CESCR 'Periodic report submitted by New Zealand' (2011) UN Doc E/C.12/NZL/3, p. 109, para. 643; CESCR 'Periodic report submitted by Brazil' (2020) UN Doc E/C.12/BRA/3, p. 36, para. 329; CESCR 'Periodic report submitted by Vietnam' (2013) UN Doc E/C.12/VNM/2–4, para. 545; CESCR 'Periodic report submitted by Uzbekistan' (2019) UN Doc E/C.12/UZB/3, para. 166 f.; CESCR 'Periodic report submitted by Turkey' (2009) UN Doc E/C.12/TUR/1, para. 646 f.; CESCR 'Periodic report submitted by Japan' (2011) UN Doc E/C.12/JPN/3, para. 416.

This also happens in the framework of the HRC. See, CCPR 'Periodic report submitted by France' (2013) UN Doc CCPR/C/FRA/5, para. 76; CCPR 'Periodic report submitted by Brazil' (2021) UN Doc CCPR/C/BRA/3, para. 261; CCPR 'Periodic report submitted by Egypt' (2020) UN Doc CCPR/C/EGY/5, para. 149.

[83] CESCR 'Periodic report submitted by Turkey' (2009) UN Doc E/C.12/TUR/1, para. 647; CESCR 'Periodic report submitted by Japan' (2011) UN Doc E/C.12/JPN/3, para. 414; CESCR 'Periodic report submitted by Angola' (2014) UN Doc E/C.12/AGO/4–5, para. 272; CESCR 'Periodic report submitted by Uzbekistan' (2012) UN Doc E/C.12/UZB/2, para. 939; CESCR 'Periodic report submitted by Ethiopia' (2011) UN Doc, E/C.12/ETH/1–3, para. 372.

[84] See Chap. 5, Sect. 5.4.2.2.

[85] *Francioni* has elaborated on "the difficult balance between the principle of sovereignty of each State Party over the World Heritage Sites located in its territory" in the context of the implementation of the World Heritage Convention. Francioni (2020c), p. 262.

[86] Emphasis added. UNHRC Res 49/7 (31 March 2022) UN Doc A/HRC/RES/49/7, para. 14.

[87] UNHRC Res 37/17 (9 April 2018) UN Doc A/HRC/RES/37/17, preamble. (Adopted without a vote)

focuses on preparatory measures and risk preparedness, documenting cultural heritage, education programmes and military training on the importance of and rules concerning cultural heritage and cultural rights.[88] Other UNHRC resolutions highlight the positive link between protecting cultural heritage in armed conflict and the right to take part in cultural life.[89]

However, rather than asserting a more general obligation, the resolutions tend to encourage taking measures that prevent the destruction of cultural heritage both in times of peace and armed conflict or repeat established rules of international humanitarian law and customary international humanitarian law,[90] by urging the "parties to armed conflicts to refrain from any unlawful military use or targeting of cultural property".[91]

The general obligation to protect cultural heritage in times of armed conflict, peace, and natural disasters is set out in the UNESCO instruments on the protection of cultural heritage.[92] The obligation to respect and protect cultural heritage and prevent its destruction as a corollary of the right to take part in cultural life, does not add a "new" legal obligation per se. However, it links the impacts of such destruction to cultural rights. To the extent that such destruction interferes with the enjoyment of the right to take part in cultural life, the international human rights framework might be able to offer a legal basis for a (soft) enforcement tool of the protection of the cultural rights and by extension of the protection of heritage.

---

88 *ibid* para. 11. (Adopted without a vote); UNHRC Res 49/7 (31 March 2022) UN Doc A/HRC/RES/49/7, paras. 13–14.

89 UNHRC Res 6/1 (27 September 2007) UN Doc A/HRC/RES/6/1, para. 4.

90 Convention for the Protection of Cultural Property in the Event of Armed Conflict (adopted 14 October 1954, entered into force 7 August 1956) 249 UNTS 215 (*1954 Hague Convention*), Art. 4; Second Protocol to The Hague Convention of 1954 for the Protection of Cultural Property in the Event of Armed Conflict (adopted 26 March 1999, entered into force 9 March 2004) 2253 UNTS 172 (*Second Protocol*); Protocol Additional to the Geneva Conventions of 12 August 1949, and Relating to the Protection of Victims of International Armed Conflicts (adopted 8 June 1977, entered into force 7 Dec 1978) 1125 UNTS 3 (*AP I*), Art. 53; Protocol Additional to the Geneva Conventions of 12 August 1949, and Relating to the Protection of Victims of Non-International Armed Conflicts (adopted 8 June 1977, entered into force 7 Dec 1978) 1125 UNTS 609 (*AP II*), Art. 16; Also see ICRC (2025): Rule 40: Respect for Cultural Property, https://ihl-databases.icrc.org/en/customary-ihl/v1/rule40, accessed 31 January 2025.

91 UNHRC Res 33/20 (6 October 2016) UN Doc A/HRC/RES/33/20, para. 2.

92 See, e.g., Convention concerning the Protection of the World Cultural and Natural Heritage (adopted 16 November 1972, entered into force 17 December 1975) 1037 UNTS 151(*World Heritage Convention*), Art. 4.

6.2.2.2 Access and Enjoy Cultural Heritage

Chief among the rights and obligations vis-à-vis cultural heritage, articulated by the human rights bodies, is access to cultural heritage. Access to cultural heritage emanates from the CESCR's interpretation of Article 15 (1) (a) of the ICESCR. Access is a broad concept that includes (1) the right to know, understand, learn *and* being taught about one's culture and forms of expression through quality education and training that considers one's cultural identity. It encompasses (2) the right "to follow a way of life associated with the use of cultural goods and resources" and (3) the right to benefit from cultural heritage.[93] The committee's interpretation of the right to take part in cultural life covers both broad cultural participation and the cultural rights of persons belonging to minorities or Indigenous peoples, including culture as a way of life.

The approach by the Special Rapporteur in the field of cultural rights, characterised by the articulation of a right of access to *and enjoyment o*f cultural heritage, derives the right from Article 15 (1) (a) of the ICECR; Article 27 of the ICCPR, and rights recognised in the UNDRIP and set out in ILO Convention No. 169 concerning Indigenous and Tribal Peoples in Independent Countries.[94] UNHRC resolution 17/15 took note of the Special Rapporteur's report on the right of access to and enjoyment of cultural heritage[95] and subsequent resolutions of the UNHRC reference access and enjoyment of cultural heritage, albeit not as a right.[96]

The analysis has differentiated between different legal bases to provide access.[97] Rights ensuring access to cultural heritage for Indigenous peoples are indeed an important component of the UNDRIP, especially in the Declaration's Articles 12 and 13. Moreover, the analysis has pointed out that access to cultural heritage is not articulated in the same way in the pronouncements of the HRC. In a few instances, however, the HRC addresses restrictions and limitations on access to cultural and sacred sites under Article 27 of the ICCPR (and to a lesser extent under Article 18 of the ICCPR). Article 27 also serves as the legal basis

---

[93] CESCR 'General Comment No. 21' (21 December 2009) UN Doc E/C.12/GC/21, para. 15.

[94] UNHRC, Report of the independent expert in the field of cultural rights, Farida Shaheed (21 March 2011) UN Doc A/HRC/17/38, para. 58 f. Also see "The Special Rapporteur stresses the importance of the right to access and enjoy cultural heritage as a human right" in: UNHRC 'Report of the Special Rapporteur in the field of cultural rights on her mission to Cyprus' (3 March 2017) UN Doc A/HRC/34/56/Add.1, para. 6.

[95] UNHRC Res 17/15 (14 July 2011) UN Doc A/HRC/RES/17/15, para. 9.

[96] See Chap. 5, Sect. 5.4.3.

[97] See Chap. 4, sects. 4.3.1 and 4.4.3 and Chap. 5, Sect. 5.3.1 and 5.4.3.

for access to linguistic heritage, as it sets out the right of persons belonging to minorities to use their own language. The right to education (Articles 13–14 of the ICESCR) and the right to take part in cultural life (Article 15 (1) (a) of the ICESCR) provide further legal bases for access to languages and linguistic heritage, including the teaching of languages and the transmission of linguistic heritage.[98]

Although few recommendations in the concluding observations by the CESCR attend to access to cultural heritage, the review of the state reports suggests that access has become an important focus of the reporting system under the CESCR. The few recommendations can indeed be ascribed to an emphasis on access or on measures ensuring and improving access to cultural heritage and cultural goods in the state reports.[99] Reporting includes ensuring physical access to cultural heritage for persons with disabilities,[100] including through legal protection[101] as well as digital access[102]. Furthermore, the state report of the Russian Federation asserts that "the right to have access to cultural properties" is enshrined in its constitution.[103] Brazil describes "the obligation to provide the means of access

98 See Chap. 4, sects. 4.3.1 and 4.4.3.

99 See CESCR 'Periodic report submitted by Brazil' (2008) UN Doc E/C.12/BRA/2, para. 524; CESCR 'Periodic report submitted by Canada' (2013) UN Doc E/C.12/CAN/6, para. 264; CESCR 'Periodic report submitted by United Kingdom of Great Britain and Northern Ireland' (2008) UN Doc E/C.12/GBR/5, para. 362 f.; CESCR 'Periodic report submitted by United Kingdom of Great Britain and Northern Ireland' (2014) UN Doc E/C.12/GBR/6, para. 221; CESCR 'Periodic report submitted by Italy' (2021) UN Doc E/C.12/ITA/6, para. 213; CESCR 'Periodic report submitted by Dominican Republic' (2009) UN Doc E/C.12/DOM/3, para. 347 (b); CESCR 'Periodic report submitted by Colombia' (2022) UN Doc E/C.12/COL/7, para. 259 f.

In the context of the implementation of the ICCPR and in the context of minorities, see CCPR 'Periodic report submitted by Colombia' (2009) UN Doc CCPR/C/COL/6, para. 666.

100 CESCR 'Periodic report submitted by United Kingdom of Great Britain and Northern Ireland' (2008) UN Doc E/C.12/GBR/5, para. 370; CESCR 'Periodic report submitted by Russian Federation' (2016) UN Doc E/C.12/RUS/6, para. 362.

101 CESCR 'Periodic report submitted by Russian Federation' (2016) UN Doc E/C.12/RUS/6, paras. 362–63.

102 CESCR 'Periodic report submitted by Mexico' (2016) UN Doc E/C.12/MEX/5–6, para. 223.

103 CESCR 'Periodic report submitted by Russian Federation' (2016) UN Doc E/C.12/RUS/6, para. 352.

to culture, education, and science" as incumbent under the implementation of Article 15 (1) (b) of the ICESCR[104].

States also report on the promotion and legal recognition of Indigenous languages and minority languages.[105] Australia ties arts and languages to "the wellbeing, culture and identity" of Indigenous communities and explains that the rationale for its continuous support rests in their preservation as part of Australia's cultural heritage and their safeguarding and transmission to future generations.[106] Canada, under the implementation of the right to education, commits to preserving and promoting the culture, language, and heritage of each official linguistic community, especially those of First Nations.[107] Colombia reports on national legislation that "declares the native languages of Colombia to be an integral part of the non-material cultural heritage of the peoples that speak them".[108]

The pervasiveness of access to cultural heritage in the pronouncements of CESCR, the work of the UNHRC's Special Procedures, and in the UNDRIP leaves little doubt that ensuring access to cultural heritage, in a broad sense, forms part of *several* (legal)[109] bases. Cultural participation in Article 27 of the UDHR reinforces these legal bases.[110]

Considering the prominence of access and enjoyment of cultural heritage in the pronouncements warrants further investigation into the nature and emergence of this right at the international level.[111] "New human rights" can be considered as those that in their initial conception "are not expressly recognised in any human

---

[104] CESCR 'Periodic report submitted by Brazil' (2008) UN Doc E/C.12/BRA/2, para. 538. Emphasising the importance of access to cultural goods, see CESCR 'Periodic report submitted by Canada' (2013) UN Doc E/C.12/CAN/6, para. 264.

[105] E.g., CESCR 'Periodic report submitted by United Kingdom of Great Britain and Northern Ireland' (2014) UN Doc E/C.12/GBR/6, para. 328; CESCR 'Periodic report submitted by the United Kingdom of Great Britain and Northern Ireland' (2022) UN Doc E/C.12/GBR/7, para. 179; CESCR 'Periodic report submitted by France' (2014) UN Doc E/C.12/FRA/4, para. 502.

[106] CESCR 'Periodic report submitted by Australia' (2014) UN Doc E/C.12/AUS/5, para. 287.

[107] CESCR 'Periodic report submitted by Canada' (2013) UN Doc E/C.12/CAN/6, para. 486, paras. 458–460.

[108] CESCR 'Periodic report submitted by Colombia' (2016) UN Doc E/C.12/COL/6, para. 225.

[109] Cognisant that the UNDRIP is, in contrast to the ICESCR, ICCPR or ILO Convention No. 169, not a legally binding instrument.

[110] UNGA Res 217A(III) (10 December 1948) UN Doc. A/810, 71.

[111] For the seminal criticism of the creation of new human rights, see: Alston (1984), pp. 607–621.

rights treaty and are not in any other way recognised as rights in a legal sense".[112] Legal recognition, accordingly, implies that such rights need to be grounded "in a formal source of public international law".[113]

The chapter argues that while it appears to be widely accepted that access and enjoyment of cultural heritage form part of the implementation of the right to take part in cultural life, determining the emergence of "new human right" in this respect requires caution on several grounds. First, there appears to be a general acceptance that access to cultural heritage forms part of the implementation of inter alia the right to take part in cultural life. This can be inferred from both the above-outlined pronouncements,[114] state reports as well as recommendations in the UPR[115], which all address access to cultural heritage. With the exception of the Special Procedures, however, independent statements concerning this right remain rare.[116] It is also possible to consider that national understandings of the scope of the right may differ from each other, i.e. the question whether this goes beyond physical access, and may not be synonymous to the meaning of access in the human rights bodies.

Second, although HRC resolutions have taken note of the Special Rapporteur's report on the right of access to and enjoyment of cultural heritage[117], the resolutions explicitly refer to the "*ability* to access and enjoy cultural heritage" as part of the right to take part in cultural life.[118]

For "access to and enjoyment of cultural heritage" to set on the path of emerging as a "new" human right, similar to the cases of the right to water, "conceptualised through the technique of legal derivation"[119] or, more recently,

---

[112] von der Decken and Koch (2020), p. 8.

[113] *ibid.*

[114] Chief among those: CESCR 'General Comment No. 21.' (21 December 2009) UN Doc E/C.12/GC/21, paras. 6, 15 (b).

[115] See Chap. 5, Sect. 5.4.3.

[116] Armenia "reiterates its full adherence to the relevant international norms and principles on the respect, preservation, *right of access to* and enjoyment of all forms of cultural heritage." Permanent Mission of the Republic of Armenia, Information provided by the Government of the Republic of Armenia in response to the questions contained in letter AL ARM 2/2021, dated October 29, 2021, of the UN Special Rapporteur in the field of cultural rights, 2203/1969/2021, 28 December 2021, pp. 1–2; CESCR 'Periodic report submitted by Russian Federation' (2016) UN Doc E/C.12/RUS/6, para. 352; CESCR 'Periodic report submitted by Brazil' (2008) UN Doc E/C.12/BRA/2, para. 538.

[117] UNHRC Res 17/15 (14 July 2011) UN Doc A/HRC/RES/17/15, para. 9.

[118] Emphasis added. UNHRC Res 33/20 (6 October 2016) UN Doc A/HRC/RES/33/20, para. 1; UNHRC Res 37/17 (9 April 2018) UN Doc A/HRC/RES/37/17.

[119] Thielbörger (2020), p. 73. Also see, Thielbörger (2014).

the right to a healthy environment, the right would need to undergo different phases of recognition.[120] Whilst both examples are still regarded to be in emergence[121], particularly pertinent milestones for the right to a healthy environment were the adoption of specific UNHRC resolutions on this right and a formal recognition of the right by the UN General Assembly, such as through resolution 76/300 on the human right to a clean, healthy and sustainable environment[122].[123]

Further international recognition of a right to access and enjoy cultural heritage through, for example UNHRC and UN General Assembly resolutions would have the power to elevate[124] the heritage dimension of human rights and might encourage individual communications under Article 15 (1) (a) of the ICESCR. In other words, such resolutions would serve as "a catalyst for action".[125] This point was reflected in the report of the UNHRC's intersessional seminar on cultural rights and the protection of cultural heritage, which recommended to civil society organisations to submit more

> "individual complaints related to article 15 of the International Covenant on Economic, Social and Cultural Rights and under the Optional Protocol thereto, to help expand the Committee's jurisprudence regarding the right to take part in cultural life and the right to access and enjoy cultural heritage."[126]

Third, the CESCR's remarks in General Comment No. 21 do not frame access to cultural heritage as a right, but make it implicit in the concepts of "taking part" or "participating" in cultural life.[127] Subsequent assertions of a right to access to

---

[120] Phases of recognition can be structured into 1) the idea, 2) the emergence, and 3) full recognition. See von der Decken and Koch (2020), p. 8 f.

[121] *ibid* p. 15.

[122] UNGA Res 76/300 (28 July 2022) UN Doc A/RES/76/300. Preceded by: UNHRC Res 48/13 (18 October 2021) UN Doc A/HRC/RES/48/13.

[123] For reference to the right to water, see specifically UNHRC Res 7/22 (28 March 2008) UN Doc A/HRC/RES/7/22; UNGA Res 64/292 (3 August 2010) UN Doc A/RES/64/292; UNHRC Res 15/9 (6 October 2010) UN Doc A/HRC/RES/15/9. Although it is important to note that efforts for the recognition of a right to water had a significantly longer lead time.

[124] As argued with regard to the right to a healthy environment, see Kotzé (2018), p. 144.

[125] Argued in an analogous manner for the right to a healthy environment: UN News (2022): Why the UN General Assembly must back the right to a healthy environment, https://news.un.org/en/story/2022/07/1123142, accessed 31 January 2025.

[126] UNHRC 'Intersessional seminar on cultural rights and the protection of cultural heritage' (27 December 2017) UN Doc A/HRC/37/29, para. 110 (a).

[127] CESCR 'General Comment No. 21.' (21 December 2009) UN Doc E/C.12/GC/21, para. 15 (b).

cultural goods and services[128] and of a right to "have access to their own cultural and linguistic heritage and to that of others"[129] are inferred from the obligation to respect and the adoption of measures that are aim at respecting these rights.[130]

To conclude, the chapter argues that the pronouncements, resolutions, and work of international human rights bodies suggest that rights and obligations vis-à-vis access to cultural heritage form part of the states' proper implementation of Article 15 (1) (a) of the ICESCR. General Comment No 21 and the adoption of UNHRC resolutions 33/20 and 37/17 are of importance, as they set out rights and obligations associated with the concept of access and enjoyment and refer to the ability to access and enjoy cultural heritage as forming part of the right to take part in cultural life, respectively. The recognition in the UNHRC resolutions is important but also silent on the scope of access and enjoyment.

6.2.2.3 Respect, Promote, Protect, and Formally Recognise the Cultural Diversity, Cultural Heritage and Rights of Persons Belonging to Minorities and Indigenous Peoples and Create Favourable Conditions

This section groups rights and obligations relating to the cultural heritage of persons belonging to minorities and Indigenous peoples under broad heading of rights and obligations that relate to the respect, the protection and the formal recognition of the rights and cultural heritage of persons belonging to minorities and Indigenous peoples, and the creation of favourable conditions for the enjoyment and development of culture and cultural heritage.

The analyses have highlighted the fragmentation of these rights, particularly for Indigenous peoples, in the international human rights framework. In a general manner, these rights relate to the pursuit of cultural development. As part of the right to self-determination, the pursuit of cultural development is laid down in Articles 1 of the ICCPR and ICESCR. Outside of the context of the right to self-determination, the pursuit of cultural development is prominently set out in the Universal Declaration of Human Rights (UDHR).[131] Statements and recommendations concerning the recognition, promotion and protection of cultural diversity, cultural pluralism and the diversity of cultural heritage play a central

---

[128] *ibid* para. 49(b).

[129] *ibid* para. 49(d).

[130] *ibid* para. 49.

[131] UNGA Res 217A(III) (10 December 1948) UN Doc. A/810, 71, Art. 22. Art. 22 speaks of the "realization [...] of the economic, social and cultural rights indispensable for his [or her, their] dignity and the free development of his personality".

role in all human rights bodies. This includes the pronouncements of treaty bodies, UNHRC resolutions, the UPR, and in the Special Procedures in context of the preservation of heritage.[132]

The CESCR anchors the obligation to protect and promote cultural diversity in the obligation to fulfil under Article 15 (1) (a) of the ICESCR.[133] The state reports submitted under the periodic reporting of the CESCR reflect this[134] and in a few cases tie this obligation to the protection and promotion of cultural heritage. Brazil, for example, reports on measures adopted to "safeguard the constitutional right to culture" and to protect and promote cultural heritage and diversity.[135] Reporting on the implementation of the ICCPR, Colombia expounds on its policy concerning the recognition of cultural diversity and the preservation of the cultural heritage of Indigenous peoples. This entails seeking "to protect, promote and guarantee access to cultural content of all ethnic minorities"[136].

The legal recognition and protection of the rights of persons belonging to minorities "to enjoy their own culture, to profess and practise their own religion, or to use their own language" is set out in Article 27 of the ICCPR. Whereas the rights of Indigenous peoples are distinct from the rights of minorities, the pronouncements of the HRC have shown that the scope of protection of Article 27 also applies to Indigenous peoples.[137] Protecting the right to lead a particular, traditional, or Indigenous way of life is a central part of the committee's understanding of culture. Whilst emanating from the HRC's interpretation of Article 27, the analysis has highlighted that the protection of ways of life also form part of

---

[132] See Chap. 5, Sect. 5.4.1.

[133] See CESCR 'General Comment No. 21.' (21 December 2009) UN Doc E/C.12/GC/21, para. 52 (a). This is also reflected in the reporting guidelines of the CESCR. See CESCR 'Guidelines on Treaty-Specific Documents to be Submitted by State Parties under Articles 16 and 17 of the International Covenant on Economic, Social and Cultural Rights' (24 March 2009) UN Doc E/C.12/2008/2, para. 68.

[134] CESCR 'Periodic report submitted by United Kingdom of Great Britain and Northern Ireland' (2008) UN Doc E/C.12/GBR/5, para. 369.

[135] CESCR 'Periodic report submitted by Brazil' (2020) UN Doc E/C.12/BRA/3, para. 316. Reporting on the constitutional right to culture, also see CESCR 'Periodic report submitted by Namibia' (2023) UN Doc E/C.12/NAM/2, para. 203.

[136] CCPR 'Periodic report submitted by Colombia' (2009) UN Doc CCPR/C/COL/6, para. 666. Also see, CCPR 'Periodic report submitted by Colombia' (2015) UN Doc CCPR/C/COL/7, para. 9.

[137] See CCPR 'Guidelines for the treaty-specific document to be submitted by States parties under article 40 of the International Covenant on Civil and Political Rights' (22 November 2010) UN Doc CCPR/C/2009/1, para. 104; CCPR 'General Comment No. 23: Article 27 (Rights of Minorities)' (8 April 1994) UN Doc CCPR/C/21/Rev.1/Add.5.

the HRC's interpretation of Article 17 of the ICCPR as well as the CESCR's scope of protection under Article 15 (1) (a).[138]

The pronouncements of the HRC and the CESCR recognise and reflect that persons belonging to minorities and Indigenous communities have the right to develop and enjoy their culture, cultural heritage, diversity, and identity, and that states have the obligations to "recognize, respect and protect minority cultures as an essential component of the identity of the States themselves."[139] This entails both positive and negative state obligations, including non-interference, on the one hand, and formal recognition, the promotion of awareness and creation of favourable conditions for minorities and Indigenous peoples "to preserve, develop, express and disseminate their identity, history, culture, language, traditions and customs"[140].

The promotion of cultural diversity, cultural heritage and linguistic heritage of minorities and Indigenous peoples[141] as well as the creation of favourable conditions[142] is addressed in state reports under both the ICESCR and ICCPR.

---

[138] CCPR 'General Comment No. 23: Article 27 (Rights of Minorities)' (8 April 1994) UN Doc CCPR/C/21/Rev.1/Add.5, paras. 3.2, 7; CESCR 'General Comment No. 21.' (21 December 2009) UN Doc E/C.12/GC/21, para. 15.

[139] CESCR 'General Comment No. 21.' (21 December 2009) UN Doc E/C.12/GC/21, para. 32.

Also reflected in CESCR 'Periodic report submitted by Russian Federation' (2016) UN Doc E/C.12/RUS/6, para. 387.

See, CESCR 'Concluding Observations: Finland' (2021) UN Doc E/C.12/FIN/CO/7, paras. 50–1; CESCR 'Concluding Observations: Namibia' (2016) UN Doc E/C.12/NAM/CO/1, para. 15; CESCR 'Concluding Observations: Viet Nam' (2014) UN Doc E/C.12/VNM/CO/2–4, para. 33 and CCPR 'Concluding Observations: Tanzania' (2009) UN Doc CCPR/C/TZA/CO/4, para. 26; CCPR 'Concluding Observations: Italy' (2017) UN Doc CCPR/C/ITA/CO/6, para. 15 (e); CCPR 'Concluding Observations: Japan' (2015) UN Doc CCPR/C/JPN/CO/7, para. 42; CCPR 'Concluding Observations: Russian Federation' (2015) UN Doc CCPR/C/RUS/CO/7, para. 24; CCPR 'Concluding Observations: Ethiopia' (2022) UN Doc CCPR/C/ETH/CO/2, para. 47.

[140] CESCR 'Guidelines on Treaty-Specific Documents to be Submitted by State Parties under Articles 16 and 17 of the International Covenant on Economic, Social and Cultural Rights' (24 March 2009) UN Doc E/C.12/2008/2, para. 68.

[141] See e.g., CESCR 'Periodic report submitted by Brazil' (2020) UN Doc E/C.12/BRA/3, para. 316; CCPR 'Periodic report submitted by Canada' (2013) UN Doc CCPR/C/CAN/6, para. 56; CESCR 'Periodic report submitted by Canada' (2024) UN Doc E/C.12/CAN/7, para. 147; CESCR 'Periodic Report submitted by Australia' (2023) UN Doc E/C.12/AUS/6, para. 344.

[142] See, CESCR 'Periodic Report submitted by Canada' (2004) UN Doc E/C.12/4/Add.15, para. 513–514; CESCR 'Periodic report submitted by Vietnam' (2013) UN Doc E/C.12/VNM/2–4, para. 27.

Examples include reporting on the constitutional safeguarding of the right to culture.[143] Ethiopia refers to the right to "express, to develop and to promote their culture and preserve their historical heritages" as provided for constitutionally.[144] Viet Nam's report affirms the constitutional right of ethnic minorities "to use their voice, written language, to preserve their national identity, to promote their fine practices, custom, traditions and culture"[145] and reports on the preservation and promotion of "typical values of cultural heritage of ethnic minorities in Viet Nam."[146].

The legal recognition of heritage sites by minority groups or Indigenous communities is also reported on.[147] Similarly, with regard to recognizing, respecting, and protecting minority cultures as an essential component of a state's identity, Viet Nam explicitly states that it recognises the cultural heritage of ethnic minorities as national cultural heritage and that several sites have been listed on the World Heritage List.[148] Australia's report identifies Indigenous peoples as "the primary source of information on the value of their heritage" and underlines the essence of their "active participation [...] to the effective management of Indigenous heritage values".[149] Uzbekistan reports that "particular emphasis is placed on guarding the cultural heritage of the country's ethnic group"[150] and

---

[143] See e.g., Colombia referring to the right to culture in their constitution. CCPR 'Periodic report submitted by Colombia' (2002) UN Doc CCPR/C/COL/2002/5, para. 959.

[144] CESCR 'Periodic report submitted by Ethiopia' (2011) UN Doc E/C.12/ETH/1–3, para. 372.

This is reinforced in its reporting under Article 1, stating that "Nations, Nationalities and Peoples have the right to speak, write and develop their own language, to express, develop and preserve their culture and their history". See para. 5.

[145] CCPR 'Periodic report submitted by Vietnam' (2018) UN Doc CCPR/C/VNM/3, para. 238. CESCR 'Periodic report submitted by Vietnam' (2013) UN Doc E/C.12/VNM/2–4, para. 27.

[146] CCPR 'Periodic report submitted by Viet Nam' (2024) UN Doc CCPR/C/VNM/4, para. 130. Also see: CESCR 'Periodic report submitted by Vietnam' (2013) UN Doc E/C.12/VNM/2–4, para. 552.

[147] See in UPR: UNHRC 'National report submitted by Viet Nam' (2018) UN Doc A/HRC/WG.6/32/VNM/1, pp. 15–16, para. 83; UNHRC 'National report submitted by Viet Nam' (2013) UN Doc A/HRC/WG.6/18/VNM/1, p. 16, para. 73.

[148] CCPR 'Periodic report submitted by Vietnam' (2018) UN Doc CCPR/C/VNM/3, para. 242.

[149] CCPR 'Periodic report submitted by Australia' (2016) UN Doc CCPR/C/AUS/6, para. 201.

[150] CCPR 'Periodic report submitted by Uzbekistan' (2012) UN Doc CCPR/C/UZB/4, para. 933.

reports on the adoption of measures preserving their cultural identity.[151] Mexico reports on measures to ensure respect for the culture and heritage of indigenous and Afro-descendant populations.[152] Although there is a strong emphasis on the legal recognition of minorities by the committees, it is important to note that in individual cases, states continue reporting on the non-recognition of minorities.[153]

The formal recognition, promotion and protection of cultural diversity and the diversity of cultural heritage is of particular importance to support, secure, and strengthen the cultural rights of persons belonging to minorities, and the corresponding state obligations. This also applies to the specific (cultural) right of Indigenous peoples and prominently includes the rights to "belong to an indigenous community or nation"[154] and "not to be subjected to forced assimilation or destruction of their culture"[155], as stated in the UNDRIP.[156] The more generic obligations to recognise, promote and protect cultural diversity, pluralism and heritage serve the objectives of these rights.[157]

The UNDRIP and the pronouncements of the treaty bodies[158] also include the right of Indigenous peoples to maintain, control, protect, and develop their cultural heritage, traditional knowledge, and other (traditional) cultural expressions.[159] The central role of Indigenous peoples "in maintaining, controlling,

---

[151] *ibid* para. 924. It is interesting to note that the report adds that "members of the public have a duty to conserve historical, spiritual and cultural artefacts", whilst the state protects cultural monuments (para. 933).

[152] CESCR 'Periodic report submitted by Mexico' (2016) UN Doc E/C.12/MEX/5–6, para. 222.

[153] See, e.g., Egypt, as emphasised in its most recent report under the HRC. CCPR 'Periodic report submitted by Egypt' (2020) UN Doc CCPR/C/EGY/5, para. 149.

[154] UNGA Res 61/295 'United Nations Declaration on the Rights of Indigenous Peoples' (13 September 2007) UN Doc A/Res/61/295, Art. 9.

[155] *ibid* Art. 8 (1).

[156] Also see Asia Pacific Forum of National Human Rights Institutions and the Office of the United Nations High Commissioner for Human Rights (2013) The United Nations Declaration on the Rights of Indigenous Peoples: A Manual for National Human Rights Institutions, p. 13; Gilbert (2017), pp. 23–24.

The Special Procedures also directly refers to the UNDRIP in its reports and communications. See Chap. 5, Sect. 5.3.3.2.

[157] See Chap. 5, Sect. 5.4.1.

[158] By relying on UNGA Res 61/295 'United Nations Declaration on the Rights of Indigenous Peoples' (13 September 2007) UN Doc A/Res/61/295, Art. 31. See CESCR 'General Comment No. 21.' (21 December 2009) UN Doc E/C.12/GC/21, para. 37.

[159] UNGA Res 61/295 'United Nations Declaration on the Rights of Indigenous Peoples' (13 September 2007) UN Doc A/Res/61/295, Art. 31.

protecting, developing and preserving tangible and intangible cultural heritage" is acknowledged in the preamble of UNHRC resolution 49/7.[160]

In the state reports under the treaty bodies, France reports on creating the Amazonian Park of French Guiana, which allows "indigenous people to maintain and further develop the tangible and intangible aspects of their cultures."[161] In its state report under the ICESCR, the Russian Federation reports on the "right of indigenous peoples [...] to preserve and develop their native language, traditions and culture".[162] It further elaborates on the legal safeguards for the protection of inter alia the cultural development of Indigenous groups, including their traditional way of life.[163] The report also links this to "the preservation of the cultural properties of these peoples, including their traditional knowledge".[164] The state report by New Zealand reflects this by reporting on government measures that assist "Iwi/Māori rights to conserve, promote and develop their own culture, language and cultural heritage".[165] This can be seen as contributing to the creation of favourable conditions. In the state reports submitted under the CESCR, New Zealand calls it a duty that its Department of Conservation manages sacred places of the Māori without compromising their values.[166] The relevant policy, mentioned in the state report, states that the Māori community "will be invited to participate in the identification, preservation and management of heritage of significance to them".[167]

Other state reports do not necessarily establish a link to cultural heritage but affirm the right of Indigenous peoples to enjoy their culture and to maintain and

---

[160] UNHRC Res 49/7 (31 March 2022) UN Doc A/HRC/RES/49/7, preamble, 14th recital.

[161] CESCR 'Periodic report submitted by France' (2014) UN Doc E/C.12/FRA/4, para. 34. A further generic affirmation without a link to heritage can be found here: CESCR 'Periodic report submitted by Finland' (2013) UN Doc E/C.12/FIN/6, para. 5.

[162] CESCR 'Periodic report submitted by Russian Federation' (2010) UN Doc E/C.12/RUS/5, para. 405.

[163] CESCR 'Periodic report submitted by Russian Federation' (2016) UN Doc E/C.12/RUS/6, para. 386. Also see the report under the HRC referring to rights relating to "the preservation of their traditional activities and way of life" under Art. 1 ICCPR. CCPR 'Periodic report submitted by Russian Federation' (2013) UN Doc CCPR/C/RUS/7, para. 8.

[164] As opposed to their protection as elements of intellectual property. CESCR 'Periodic report submitted by Russian Federation' (2016) UN Doc E/C.12/RUS/6, para. 387.

[165] CESCR 'Periodic report submitted by New Zealand' (2017) UN Doc E/C.12/NZL/4, p. 17, para. 99.

[166] CESCR 'Periodic report submitted by New Zealand' (2011) UN Doc E/C.12/NZL/3, p. 111, para. 656.

[167] *ibid* p. 111, para. 657.

develop their own language and culture.[168] This ties in with rights set out in Articles 14–15 of the UNDRIP, namely for Indigenous peoples "to establish and control their education systems and institutions" and to provide "education in their languages in a manner appropriate to their cultural methods of teaching and learning".

Article 31 of the UNDRIP also covers "the right to maintain, control, protect and develop their intellectual property over such cultural heritage, traditional knowledge, and traditional cultural expressions." The CESCR's interpretation of Article 15 (1) (c) of the ICESCR is also assertive in protecting the interests of Indigenous communities in relation their (cultural) productions.[169] This is reflected in a few state reports with regard to the knowledge and culture of Indigenous peoples.[170] Mexico's most recent state report, which refers to the adoption of the Federal Act for the Protection of the Cultural Heritage of Indigenous and Afro-Mexican Peoples and Communities, and outlines the recognition of property rights and collective intellectual property rights over their cultural heritage, traditional knowledge, and cultural expressions.[171]

In conclusion, the pronouncements of the treaty bodies show that both rights and obligations emanate from Article 15 (1) (a) of the ICESCR and Article 27 of the ICCPR that are of particular relevance for the cultural heritage of persons belonging to minorities or Indigenous communities. The adoption of the UNDRIP, often referred to and relied on by the treaty bodies, has compiled rights of Indigenous peoples, and stated their "right to maintain, control, protect and develop their cultural heritage",[172] thereby going further than for instance ILO Convention No. 169.[173]

---

[168] See, CCPR 'Periodic report submitted by Finland' (2011) UN Doc CCPR/C/FIN/6, paras. 54, 56, 242.

[169] CESCR 'General Comment No. 17' (12 January 2006) UN Doc E/C.12/GC/17, para. 32.

[170] See, CESCR 'Periodic report submitted by Australia' (2014) UN Doc E/C.12/AUS/5, para. 294–95; CESCR 'Periodic report submitted by New Zealand' (2017) UN Doc E/C.12/NZL/4, para. 100. To a lesser extent: CESCR 'Periodic report submitted by Russian Federation' (2016) UN Doc E/C.12/RUS/6, para. 385.

[171] CESCR 'Periodic report submitted by Mexico' (2023) UN Doc E/C.12/MEX/7, para. 340.

[172] UNGA Res 61/295 'United Nations Declaration on the Rights of Indigenous Peoples' (13 September 2007) UN Doc A/Res/61/295, Art. 31 (1). Reinforced in CESCR 'General Comment No. 21.' (21 December 2009) UN Doc E/C.12/GC/21, para. 37 and in UNHRC Res 49/7 (31 March 2022) UN Doc A/HRC/RES/49/7, preamble, 14th recital.

[173] Jakubowski (2024), p. 505.

Whilst the rights contained in the UNDRIP "may be on a path to recognition" as legally binding human rights, they are likely to "fall short of full recognition as free-standing customary rights".[174] Similarly, while the states report on legal safeguards for the rights of persons belonging to minorities and Indigenous communities vis-à-vis their culture and cultural heritage, state reports do not always offer insights into how the actual enjoyment of these rights or on obstacles in the implementation.[175] The same reasoning can be applied to recommendations in the UPR.

6.2.2.4 Effective Consultations, Participation and Free, Prior and Informed Consent in Decision-Making Affecting Cultural Heritage and one's Way of Life

The inclusion of free, prior and informed consent in the nomination process of world heritage sites in the Operational Guidelines for Indigenous peoples was an important step in enhancing participation rights in the World Heritage Convention.[176] As the recommendation is not obligatory, human rights law provisions may lend further support to enforcing participatory rights. The participation and consultation of Indigenous peoples and local communities in decision-making processes concerning heritage has become very topical and is reflected in efforts towards inclusive, participatory governance at the regional and international level.[177]

Cultural rights serve an important function in protecting participation rights for the interpretation, elaboration and development of cultural heritage and cultural identities.[178] Indeed, the "mutually reinforcing" nature of the branches of international cultural heritage law and international human rights law with regard to access and participation rights in decision-making concerning cultural heritage is subject of the academic discourse.[179]

---

174 von der Decken and Koch (2020), p. 15.

175 Rodley (2013), pp. 627–628.

176 Disko (2017), p. 69.

177 See Jakubowski et al. (2022).

178 UNHRC, Report of the Special Rapporteur in the field of cultural rights (3 February 2016) UN Doc A/HRC/31/59, para. 9.

179 E.g., Lixinski (2016): ESIL – International Human Rights Law Symposium: International Human Rights Law and International Cultural Heritage Law: Cooperation, Conflict or Cooption?, https://www.ejiltalk.org/esil-international-human-rights-law-symposium-international-human-rights-law-and-international-cultural-heritage-law-cooperation-conflict-or-cooption/, accessed 31 January 2025. Also see Vrdoljak (2018), p. 246.

The Special Rapporteur has underlined participation rights of individuals and persons belonging to groups[180] and has shown their applicability vis-à-vis the identification, interpretation, development, and safeguarding of heritage[181] as well as in the context of inscriptions on the World Heritage List.[182]

The HRC's views in communications concerning Article 27 of the ICCPR have expounded on the positive measures a state needs to adopt to ensure effective participation. Lack of consultations with the community, the absence of a study by a competent and independent body to determine the impact of the construction of the traditional economic activity, the lack of steps taken to minimise the negative consequences or even to repair them, disproportionate measures, and whether a community can continue to benefit from their traditional economy all form part of the Committee's assessment.[183] In the reporting under the HRC, Australia's report addresses the "integral" role of Indigenous peoples in the management of heritage and expounds on "the right to negotiate", applicable when "acts such as mining and mining related exploration are to occur on land that is subject to native title".[184] This requires the parties to negotiate in good faith over a period of at least six months with the aim of reaching agreement on the proposal.[185]

---

[180] UNHRC 'Cultural Rights: Tenth Anniversary Report' (17 January 2019) UN Doc A/HRC/40/53, para. 18.

[181] For example, UNHRC 'Cultural Rights: Tenth Anniversary Report' (17 January 2019) UN Doc A/HRC/40/53, para. 18. Also see the emphasis on consultation in: UNHRC 'COVID-19, Culture and Cultural Rights' (17 February 2021) UN Doc A/HRC/46/34, para. 77. Also see, UNHRC 'Report of the Independent Expert in the field of cultural rights, Farida Shaheed, Addendum, Mission to Morocco' (2 May 2012) UN Doc A/HRC/20/26/Add.2, para. 90; UNHRC 'Report of the Special Rapporteur in the field of cultural rights, Farida Shaheed, Addendum, Visit to Viet Nam' (29 January 2015) UN Doc A/HRC/28/57/Add.1, para. 111.

[182] UNHRC 'Report of the Special Rapporteur in the field of cultural rights on her visit to Botswana' (12 January 2016) UN Doc A/HRC/31/59/Add.1, para. 91.

[183] CCPR, Communication No. 1457/2006, *Ángela Poma Poma v. Peru* (24 April 2009) UN Doc CCPR/C/95/D/1457/2006, paras. 7.6–7.7; Also see, CCPR, Communication No. 2102/2011, *Paadar et al. v. Finland*, (10 April 2014) CCPR/C/110/D/2102/2011, p. 16. CCPR, Communication No 547/1993, *Apirana Mahuika and Ors v New Zealand* (27 October 2000) UN Doc CCPR/C/70/D/547/1993, p. 15, para. 9.5. CCPR, Communication No. 511/1992, *Länsman v. Finland* (14 October 1993) UN Doc CCPR/C/49/D/511/1992, paras. 9.6 and 9.8.

[184] CCPR 'Periodic report submitted by Australia' (2016) UN Doc CCPR/C/AUS/6, paras. 201, 203.

[185] "If agreement cannot be reached during the period, an application may be made for a determination about whether the mining or exploration may proceed which may be subject to conditions. If negotiations have not been conducted in good faith, the determination must not be made and parties must continue to negotiate in good faith." *ibid* para. 203.

Finland's report elaborates on the "obligation to negotiate" in the context of "all far-reaching and important measures, which may directly and in a specific way affect the status of the Sámi as an indigenous peoples", including in matters of their language and culture.[186] This obligation further requires that the Sámi parliament has "an opportunity to be heard and to engage in discussions on the matter."[187]

In 2009, the HRC asserted in its views on *Poma Poma v. Peru* that effective participation entails "not mere consultation but the free, prior, and informed consent of the members of the community".[188] The principle of free, prior and informed consent of Indigenous peoples is an integral part of the UNDRIP.[189] Whereas the positive obligation to obtain the free, prior and informed consent is not explicitly mentioned with regard to cultural heritage[190], it is suggested that the provisions of the UNDRIP covering the free, prior and informed consent are phrased in a sufficiently broad manner that cover the heritage of Indigenous peoples.[191]

The HRC highlights the principle of free, prior, and informed consent in its views on Article 27 of the ICCPR as a requirement of effective participation.[192] The committee's concluding observations either recommend effectively consulting communities,[193] their effective participation,[194] seeking or attempting

[186] CCPR 'Periodic report submitted by Finland' (2020) UN Doc CCPR/C/FIN/7, para. 264.

[187] *ibid.*

[188] CCPR, Communication No. 1457/2006, *Ángela Poma Poma v. Peru* (24 April 2009) UN Doc CCPR/C/95/D/1457/2006, para. 7.6.

[189] See UNGA Res 61/295 'United Nations Declaration on the Rights of Indigenous Peoples' (13 September 2007) UN Doc A/Res/61/295, Art. 10 in relation to ancestral lands or territories, Art. 19 in relation to the adoption and implementation of legislative or administrative measures, Art. 29 with regard to the storage or disposal of hazardous material, and Art. 32 with regard to (development) projects project affecting Indigenous lands, territories and resources.

[190] Expect for *ibid* Art. 11 (2), which establishes redress mechanisms for heritage taken without the free, prior and informed consent of Indigenous peoples.

[191] See *ibid* Arts. 10, 19, 29 (2), 32.

[192] CCPR, Communication No. 1457/2006, *Ángela Poma Poma v. Peru* (24 April 2009) UN Doc CCPR/C/95/D/1457/2006, para. 7.6.

[193] CCPR 'Concluding Observations: Costa Rica' (2016) UN Doc CCPR/C/CRI/CO/6, p. 7, paras. 41–42; CCPR 'Concluding Observations: Mexico' (2010) CCPR/C/MEX/CO/5, p. 7, para 22; CCPR 'Concluding Observations: Colombia' (2004) UN Doc CCPR/CO/80/COL, p. 5, para. 20 (with respect to Indigenous communities, land and property); CCPR 'Concluding Observations: United States of America' (2014) UN Doc CCPR/C/USA/CO/4, p. 11, para. 25.

[194] CCPR 'Concluding Observations: New Zealand' (2016) UN Doc CCPR/C/NZL/CO/6, p. 9, para. 46 (b) and (c) (with respect to Indigenous communities); CCPR 'List of Issues:

to obtain their free and informed consent,[195] and only in a few cases recommend ensuring their *right* to free, prior, and informed consent[196]. A right to the free, prior, and informed consent is not frequently mentioned in the concluding observations in matters concerning the heritage or cultural practices of Indigenous peoples under Article 27 of the ICCPR. One example are the concluding observations on Australia's state report, where the HRC explicitly mentions the "effective protection and management of indigenous heritage sites" in this context, albeit without reference to free, prior and informed consent.[197]

Several state reports under the HRC's reporting mechanism elaborate on the legal recognition of consultations with Indigenous peoples by means of constitutional recognitions,[198] Supreme Court decisions,[199] Executive Orders,[200] acts

Australia' (2012) UN Doc CCPR/C/AUS/Q/6, p. 4, para. 23 (with regard to Indigenous communities).

[195] See CCPR 'Concluding Observations: Namibia' (2016) UN Doc CCPR/C/NAM/CO/2, p. 9, para. 44; CCPR 'Concluding Observations: Colombia' (2016) UN Doc CCPR/C/COL/CO/7, p. 9, para. 43; CCPR 'Concluding Observations: USA' (2014) UN Doc CCPR/C/USA/CO/4, p. 11, para. 25, CCPR 'Concluding Observations: Russian Federation' (2015) UN Doc CCPR/C/RUS/CO/7, p. 12, para. 24.

[196] CCPR 'Concluding Observations: Canada' (2015) UN Doc CCPR/C/CAN/CO/6, p. 6, para. 16; CCPR 'Concluding Observations: Viet Nam' (2019) UN Doc CCPR/C/VNM/CO/3, p. 10, para. 56 (b).

[197] CCPR 'Concluding Observations: Australia' (2017) UN Doc CCPR/C/AUS/CO/6, pp. 9–10, para. 50.

[198] Under Art. 1 of the ICCPR and with reference to development initiatives. CCPR 'Periodic report submitted by Ethiopia' (2020) UN Doc CCPR/C/ETH/2, para. 7; with reference to a constitutional duty to consult when conduct "might adversely impact potential or established Aboriginal or treaty rights". CCPR 'List of issues: Canada' (2015) UN Doc CCPR/C/CAN/Q/6/Add.1, para. 122.

[199] Without reference to heritage, but with references to cases that arguably have a heritage dimension. CCPR 'Periodic report submitted by Mexico' (2015) UN Doc CCPR/C/MEX/6, para. 227.

[200] Cognisant that an Executive Order is not legislation, and with reference to Executive Order 13175. CCPR 'Periodic report submitted by USA' (2012) UN Doc CCPR/C/USA/4, para. 691. Also see, Executive Order 13175 of November 6, 2000, Consultation and Coordination With Indian Tribal, The White House; Memorandum on Uniform Standards for Tribal Consultation, November 30, 2022 (available at: https://www.whitehouse.gov/briefing-room/presidential-actions/2022/11/30/memorandum-on-uniform-standards-for-tribal-consultation/). Further efforts described in most recent state report: CCPR 'Periodic report submitted by USA' (2021) CCPR/C/USA/5, para. 121 f.

by Parliament,[201] with reference to ratification of ILO Convention No. 169,[202] and, in one instance, with direct reference to the right to free, prior and informed consent[203].

The CESCR's General comment No. 21 refers to obtaining the free, prior and informed consent as a core obligation under Article 15 (1) (a) of the ICESCR[204] and the committee's General Comment No. 17 cites respecting the principles of free, prior and informed consent as a specific legal obligation under Article 15 (1) (c) of the ICESCR. [205] In the context of the right to take part in cultural life, this particularly concerns cases when laws and policies might put the preservation of the cultural resources of persons belonging to minority groups, indigenous peoples or other communities at risk, "especially those associated with their way of life and cultural expression".[206] The wording of the complete core obligation is, however, more vague, as it refers to allowing and encouraging participation of "persons belonging to minority groups, indigenous peoples or other communities".[207] Along the same lines, in the concluding observations obtaining the right to free, prior and informed consent is often couched in softer terms and not always expressed with the vigour of an obligation. Frequently, it is framed as a recommendation, omitting the word "right" or referring to the full participation of local communities[208]. For example, the CESCR recommends "meaningful participation in debates related to nature conservation, commercial hunting, tourism and other uses of the land, based on free, prior and informed consent."[209]

---

[201] With explicit reference to language and culture. CCPR 'Periodic report submitted by Finland' (2020) UN Doc CCPR/C/FIN/7, para. 264.

[202] CCPR 'Periodic report submitted by Brazil' (2021) UN Doc CCPR/C/BRA/3, para. 28; CCPR 'Periodic report submitted by Colombia' (2021) UN Doc CCPR/C/COL/8, para. 6.

[203] "There are many mechanisms for effective Māori participation in decision-making processes on issues that affect Māori and their right to free, prior and informed consent." CCPR 'Periodic report submitted by New Zealand' (2015) UN Doc CCPR/C/NZL/6, para. 251.

[204] CESCR 'General Comment No. 21' (21 December 2009) UN Doc E/C.12/GC/21, para. 55(e).

[205] CESCR 'General Comment No. 17' (12 January 2006) UN Doc E/C.12/GC/17, para. 32.

[206] CESCR 'General Comment No. 21' (21 December 2009) UN Doc E/C.12/GC/21, para. 55 (e). Also see paras. 36–37. Also see CESCR 'General Comment No. 25' (30 April 2020) UN Doc E/C.12/GC/25.

[207] CESCR 'General Comment No. 21' (21 December 2009) UN Doc E/C.12/GC/21, para. 55(e). Also see, Pineschi (2012), p. 45.

[208] See e.g., CESCR 'Concluding Observations: Uzbekistan' (2022) UN Doc E/C.12/UZB/CO/3, para. 57.

[209] CESCR 'Concluding Observations: Tanzania' (2012) UN Doc E/C.12/TZA/CO/1–3, para. 29.

In addition to referring to consultation with Indigenous peoples,[210] state reports under the ICESCR name free, prior and informed consent more frequently[211]. Whilst there appears to be a paucity in establishing a link between free, prior and informed consent and the impacts on the heritage of Indigenous peoples, more recent state reports do refer to the implementation of ILO Convention No. 169 and the UNDRIP[212]. Under both the Convention and the Declaration the safeguarding of Indigenous cultural and traditional values and heritage is explicitly addressed.[213]

The human rights standard-setting in the UNDRIP can be read as a recognition by the international community "that the existing human rights regime was not fully responsive to the diverse cultural contexts of Indigenous peoples and their distinct status and rights."[214] The UNDRIP does not only set out essential human rights for Indigenous peoples, it also make clear that the heritage of Indigenous peoples cannot be treated separately "from questions concerning land rights, control over natural resources, and self-determination".[215]

Notwithstanding the positive affirmations of consultations and free, prior and informed consent in the reporting under Article 27, invoking free, prior and informed consent under Article 27 was meet with refusal in Australia's response in *Alisa Roy v. Australia*, where the state argues that "article 27 did not provide for Indigenous Peoples' right to free, prior and informed consent".[216]

The terminology in the state reports and the committee's pronouncement is indeed heterogenous on the matter of consultations and free, prior and informed consent. State reports might not refer to free, prior and informed consent but rather refer to effective consultations. While it is positive that they report on undertaking consultations, there is a significant distinction between participation being consultative or refers to obtaining consent. Similarly, the paucity in the concluding observations on establishing a link between the free, prior and informed

---

[210] CESCR 'Periodic report submitted by Mexico' (2023) UN Doc E/C.12/MEX/7, para. 6; CESCR 'Periodic report submitted by Australia' (2014) UN Doc E/C.12/AUS/5, para. 283.

[211] CESCR 'Periodic report submitted by Mexico' (2023) UN Doc E/C.12/MEX/7, para. 20; CESCR 'Periodic report submitted by Colombia' (2022) UN Doc E/C.12/COL/7, para. 7; CESCR 'Periodic report submitted by Angola' (2023) UN Doc E/C.12/AGO/6, para. 57.

[212] CESCR 'Periodic report submitted by Namibia' (2023) UN Doc E/C.12/NAM/2, para. 18; CESCR 'Periodic report submitted by Canada' (2024) UN Doc E/C.12/CAN/7, para. 9.

[213] See, ILO Convention No. 169, Arts. 13, 14–15, 23; UNDRIP, Art. 31.

[214] Dorough and Wiessner (2020), p. 408.

[215] Blake (2015), p. 281.

[216] CCPR, Communication No. 3585/2019, *Alisa Roy v. Australia* (13 March 2024) UN Doc CCPR/C/137/D/3585/2019, paras. 5.3 and 6.3.

consent and explicitly addressing the impacts on the heritage of Indigenous peoples, suggests that the application to safeguard elements of heritage is not widely applied.

6.2.3 Interim Conclusions

In the absence of one holistic provision that encapsulates the notion of cultural heritage, the analysis has shown the diversity of rights and obligations surrounding cultural heritage in the international human rights framework. Rather than making the case for the emergence or the ambition of "new" human rights, the chapter argues that the interpretation of human rights bodies, specifically the treaty bodies and UNHRC resolutions, shows that rights and obligations vis-à-vis cultural heritage form part of several (legal) bases.

Similarly, the international human rights framework does not necessarily offer a distinct right but endows individuals and persons belonging to a group with a tool to strengthen their claims and offer soft enforcement tools for the respective rights. While the international human rights law framework does not offer an international human rights law court, especially the pronouncements of the human rights bodies can contribute to standard-setting at the international level and cross-fertilisation with the regional human rights commissions and courts.

The focus of the analysis is informed and at the same time limited by the sample, which is why the chapter does not make claims of exhaustiveness. Consulting the state reports under the periodic reporting systems of the CESCR and the HRC has sought to contribute to better insights into how states view the rights and obligations articulated by the treaty bodies. As pointed out previously, information provided in the state reports can often be self-congratulatory and not respond to the Covenants encouragement to also share difficulties in the implementation and fulfilment of their obligations.[217] This, in turn, fail to offer insights into the actual enjoyment of rights.[218] This is reflected in, for instance, simple statements of how rights are enshrined and protected in the constitution or through the adoption of a law.[219] The state report are of course insightful in terms of understanding, which rights and obligations states formally accept and protect, however the practical enjoyment of the rights falls short. This can be countered by consulting additional information from civil society organisations and national

[217] Rodley (2013), p. 627. ICCPR, Art. 40(2); ICESCR Art. 17 (2).

[218] Rodley (2013), pp. 627–628.

[219] See *ibid* p. 628.

human rights institutions in the framework of the periodic monitoring as well as by drawing on the compilation of UN information and stakeholders' information in the UPR.

Especially for Indigenous peoples, this diversity and fragmentation implies the necessity to rely on several human rights provisions "scattered across the human rights legal framework".[220] As this can fail to recgonise the holistic nature of the cultural heritage of Indigenous peoples that also encompasses spiritual, economic, and social bonds to lands and territories, the academic discourse has put forward a right to cultural integrity for Indigenous peoples.[221] The concept of a right to cultural integrity understands "that cultural heritage includes indigenous peoples' rights to culture, religion, health, development and natural resources".[222] Finding support and mentioning in pronouncements of the international and regional human rights bodies[223], it will be of interest to see how this concept is further developed and applied.

6.3 Applying Human Rights Tools and Mechanisms to Cultural Heritage

The wording of the research question does not only ask *if* human rights law can provide a legal basis for cultural heritage matters but *how*. Whilst the preceding analysis discussed how human rights bodies interpret human rights provisions and set out rights and obligations concerning cultural heritage, this part of the analysis explores how the mechanisms inherent in the international human rights system are being used and could be used in cultural heritage matters.

At the state level, the international human rights framework offers an independent monitoring mechanisms through the treaty bodies' concluding observations on states' reports and a peer review mechanism through the UPR. In the design of human rights to protect the enjoyment of human rights and freedoms from state

---

[220] Gilbert (2017), p. 24.

[221] *ibid* pp. 21, 25–26. Also see, Disko (2017), p. 75; Blake (2015), p. 293; Dorough and Wiessner (2020), p. 416.

[222] Gilbert (2017), p. 21.

[223] Gilbert's analysis mentions allusions to this right by the CESCR, the Inter-American Commission on Human Rights, and the African Commission on Human and Peoples' Rights. *ibid* p. 25 f. See, *Centre for Minority Rights Development and Minority Rights Group International and Centre on Housing Rights and Evictions (intervening) (on behalf of Endorois Welfare Council) v Kenya* (Communication 276/2003), (ACHPR 2009), November 2009.

power,[224] the international human rights framework equips individuals and persons belonging to groups with additional tools of varying legal effect to enforce their rights by lodging individual complaints primarily through communications in the treaty body system, and also in the Special Procedures. This tool is reinforced at regional levels through the availability of regional human rights commissions and, more importantly, courts.

Against the backdrop of the horizontal character of international cultural heritage law and scholarships' critical views towards the lack of enforcement mechanisms,[225] this section argues that the tools and mechanisms of the human rights framework can contribute to strengthening (1) the accountability of states, (2) rights claims in matters concerning cultural heritage, and (3) the enforcement system of cultural heritage protection at the international level. This includes the use of the two main monitoring mechanisms (see Sect. 6.3.1), namely the treaty bodies' periodic reporting and the UPR, and (see Sect. 6.3.2) the individual complaints mechanism under the treaty bodies and, on a smaller scale, the use of communications in the Special Procedures.

6.3.1 Monitoring Mechanisms

Considering that threats to cultural heritage are multifaceted, ranging from state-supported development projects to the intentional targeting of cultural heritage in the form of a terrorist attack, periodic reporting and monitoring can contribute to identifying early warning signs. While the World Heritage Convention and the Intangible Cultural Heritage Convention contain a periodic reporting system,[226] these systems only concern listed heritage sites, making the mechanism more limited and simultaneously more specific. The monitoring mechanisms in the human rights framework provides a more general account of the human rights situation in a specific state but will only address cultural heritage to the extent that protection is linked to specific rights. Furthermore, the analysis has found that states also report on the general preservation and promotion of cultural heritage under Article 15 (1) (a) of the ICESCR.[227] While this means that periodic reporting under the treaty bodies can address all forms and elements of cultural heritage,

[224] See Jakubowski (2024), p. 501.

[225] See Chap. 3, section 5.

[226] As well as a reactive monitoring system in the case of the World Heritage Convention.

[227] Also see Saul, Kinley and Mowbray (2014), p. 1190.

and are not limited by the world heritage status, in most cases addressing cultural heritage will hinge on its link to restrictions or obstacles to specific rights.

6.3.1.1 The Treaty Bodies' Concluding Observations

States ought to report regularly on the implementation of human rights treaties, and the analysis has shown that treaty bodies use the list of issues and the concluding observations to address cultural heritage matters. The approach of the CESCR and HRC, however, differs in approaching the topic.

The CESCR addresses the preservation and promotion of the culture, cultural heritage,[228] customs, history, identity, and language of minority groups and of the traditional way of life of Indigenous peoples.[229] It also covers the need for further legislative measures to protect, preserve, and promote cultural heritage and the traditional ways of life of minority groups and Indigenous communities.[230] In this context, strengthening the legal framework is similarly addressed.[231] Recommendations regarding the revitalisation of languages or linguistic rights and minorities, directly connected to the preservation of cultural diversity and heritage, occur more frequently in the European context than in other regions.[232] Issues and recommendations are (almost) exclusively addressed under the heading of cultural rights or by reference to Article 15 of the ICESCR.

In contrast, the term "cultural heritage" finds almost no mentioning in the concluding observations of the HRC.[233] Instead, the committee approaches forms and elements of and associated with cultural heritage almost exclusively through the lens of the rights of Indigenous peoples. The committee addresses the legal recognition of minorities and Indigenous peoples, measures affecting the traditional

---

[228] Including world heritage sites. See, for example, CESCR 'Concluding Observations: Uzbekistan' (2022) UN Doc E/C.12/UZB/CO/3, paras. 56–57; CESCR 'List of issues: Afghanistan' (2009) UN Doc E/C.12/AFG/Q/2–4, para. 42.

[229] CESCR 'Concluding Observations: United Republic of Tanzania' (2012) UN Doc E/C.12/TZA/CO/1–3, para. 29; CESCR 'List of Issues: Colombia' (2009) UN Doc E/C.12/COL/Q/5, para. 42; CESCR 'Concluding Observations: New Zealand' (2012) UN Doc E/C,12/NZL/CO/3, para. 26.

[230] CESCR 'List of issues: United Republic of Tanzania' (2011) UN Doc E/C.12/TZA/Q/1–3, para. 28.

[231] CESCR 'Concluding Observations: Mali' (2018) UN Doc E/C.12/MLI/CO/1, para. 54.

[232] CESCR 'Concluding Observations: Finland' (2014) UN Doc E/C.12/FIN/CO/6, para. 29; CESCR 'Concluding Observations: UK' (2009) UN Doc E/C.12/GBR/CO/5, para. 37.

[233] Exceptions: CCPR 'Concluding Observations: Tanzania' (2009) UN Doc CCPR/C/TZA/CO/4, para. 26; CCPR 'Concluding Observations: Australia' (2017) UN Doc CCPR/C/AUS/CO/6, para. 50.

ways of life of minority groups and Indigenous communities, ensuring participation in decision-making, including the free, prior and informed consent of Indigenous peoples, the safeguarding of minority and Indigenous languages, and access to Indigenous sites of cultural or sacred significance.[234] Adverse impacts on traditional ways of life or cultural rights are not always,[235] but very frequently mentioned in this context. The same applies to connecting, for example, the risk of disappearance of Indigenous languages to culture, heritage, and identity.[236] The HRC does not solely rely on Article 27 of the ICCPR in its concluding observations, but similarly refers to the right to participate in public affairs, electoral rights and access to public service (Article 25 of the ICCPR), equality before the law (Article 26 of the ICCPR) and non-discrimination (Article 2 (1) of the ICCPR).[237]

6.3.1.2 The Universal Periodic Review's Peer Review Mechanism

In addition to the monitoring mechanism of the treaty bodies, the UNHRC makes use of a peer review mechanisms. In the UPR, each UN member state is subject to the peer review mechanism every four and a half years. Instead of using independent experts, the states review each other and provide recommendations based on each other's human rights record. While the strict timeframe has been pointed out to hamper the dialogue and often "resemble [...] a juxtaposition of two monologues,"[238] the UPR offers valuable insights into how states understand or broach the topic of cultural heritage.

States are taking up "cultural heritage" as a multifaceted topic in their reporting under the UPR. During the 2017 UNHRC's intersessional seminar on cultural

---

[234] See, for example, CCPR 'Concluding Observations: Angola' (2019) UN Doc CCPR/C/AGO/CO/2, para. 49; CCPR 'Concluding Observations: USA' (2023) UN Doc CCPR/C/USA/CO/5, para. 66 f.; CCPR 'Concluding Observations: Costa Rica' (2016) UN Doc CCPR/C/CRI/CO/6, para. 41 f.; CCPR 'Concluding Observations: Viet Nam' (2019) UN Doc CCPR/C/VNM/CO/3, para. 55 f.; CCPR' Concluding Observations: Finland' (2021) UN Doc CCPR/C/FIN/CO/7, para. 42; CCPR 'Concluding Observations: New Zealand' (2016) UN Doc CCPR/C/NZL/CO/6, para. 45 f.

[235] See, for example, CCPR 'List of issues: Angola' (2018) UN Doc CCPR/C/AGO/Q/2, para. 30; CCPR 'Concluding Observations: Mexico' (2010) CCPR/C/MEX/CO/5, para. 22.

[236] See CCPR 'Concluding Observations: Canada' (2015) UN Doc CCPR/C/CAN/CO/6, para. 19.

[237] See, for example, CCPR 'Concluding Observations: Angola' (2019) UN Doc CCPR/C/AGO/CO/2, para. 49; CCPR 'Concluding Observations: Ethiopia' (2022) UN Doc CCPR/C/ETH/CO/2, para. 47.

[238] Tomuschat (2019), para. 44.

rights and the protection of cultural heritage, the proposal to address cultural heritage concerns in the framework of the UPR found the support of the delegates.[239] The outcome document of the seminar accordingly recommended states to use the UPR to encourage ratification of cultural heritage treaties.[240] The recommendations also addressed mainstreaming the human rights-centred approach to the protection of cultural heritage in relevant human rights mechanisms.[241]

Reporting and recommendations on cultural heritage range from challenges, such as the impacts of climate change, the safeguarding of linguistic heritage, the restitution of African heritage, limitations to the enjoyment of cultural and religious heritage, destruction and looting of cultural heritage and places of worship impacting the implementation of human rights, restoration efforts, and measures to protect the heritage of minority groups and Indigenous peoples[242].

There are at least three ways in which states address heritage in the UPR: (1) general measures taken to protect cultural heritage, (2) linking heritage protection to minority rights, and (3) reporting on the preservation of Indigenous heritage.

First, states report on general measures concerning heritage, making use of a less rights-focused rhetoric. One such example is that in their national reports, states detail the general measures implemented to protect cultural heritage, museums, libraries, or other cultural institutions.[243] States are similarly reporting on passing national cultural heritage laws and their ratification of international cultural heritage law instruments[244] as well as on listing cultural and natural sites,

---

[239] UNHRC 'Intersessional seminar on cultural rights and the protection of cultural heritage' (27 December 2017) UN Doc A/HRC/37/29, para. 84.

[240] *ibid* para. 99 (c).

[241] *ibid* para. 107 (a).

[242] E.g. UNHRC 'National report submitted by Vanuatu' (2018) UN Doc A/HRC/WG.6/32/VUT/1, para. 12; UNHRC 'National report submitted by Federated States of Micronesia' (2010) UN Doc A/HRC/WG.6/9/FSM/1, para. 97; UNHRC 'National report submitted by France' (2012) UN Doc A/HRC/WG.6/15/FRA/1, paras. 52–53; UNHRC 'Report of the Working Group on the UPR: France, A/HRC/54/5, 17 July 2023, para. 45.77; UNHRC 'Report of the Working Group on the UPR: Turkey' (2015) UN Doc A/HRC/29/15, para. 149.15; UNHRC 'National report submitted by Afghanistan' (2009) UN Doc A/HRC/WG.6/5/AFG/, para. 61; UNHRC 'Report of the Working Group on the UPR: Colombia (2023) UN Doc A/HRC/55/7, para. 116.203; UNHRC 'National report submitted by Mali' (2012) UN Doc A/HRC/WG.6/15/MLI/1, para. 76; UNHRC 'National report submitted by Mali' (2023) UN Doc A/HRC/WG.6/43/MLI/1, para. 24.

[243] UNHRC 'National report submitted by Cuba' (2018) UN Doc A/HRC/WG.6/30/CUB/1, paras. 108–109; UNHRC 'National report submitted by Uzbekistan (2008) UN Doc A/HRC/WG.6/3/UZB/1, p. 9, paras. 38–40.

[244] See, UNHRC 'National report submitted by Afghanistan' (2018) UN Doc A/HRC/WG.6/32/AFG/1, para. 9; UNHRC 'National report submitted by Uzbekistan' (2013) UN Doc A/

either as national heritage or on the UNESCO lists.[245] This suggests that states have continued adhering to the recommendations of the outcomes of the intersessional seminar. It is, however, interesting to note that the current guidance document for states does not specifically include a specific subitem to report on cultural rights.[246]

Second, recommendations in the UPR link to the protection of minority rights. Observations and recommendations concerning minority groups and cultural heritage usually relate to the protection of culture, language, and identity[247] but also to equal access to the enjoyment of cultural and religious heritage[248]. Recommendations further cover awareness raising of the cultural heritage of minority groups as part of combatting discrimination.[249] States detail efforts to foster traditions and preserve other forms of intangible cultural heritage of minority groups.[250] Furthermore, state reports include examples of the protection and promotion of

---

HRC/WG.6/16/UZB/1, paras. 37, 40, UNHRC 'National report submitted by Costa Rica' (2014) UN Doc A/HRC/WG.6/19/CRI/1, para. 15; UNHRC 'National report submitted by Viet Nam' (2024) UN Doc A/HRC/WG.6/46/VNM/1, paras. 87, 90.

This can also be seen in reflections of states within the framework of the working groups: UNHRC 'Report of the Working Group on the UPR: UK' (2023) A/HRC/52/10, para. 41.

[245] E.g. in all state reports 2018, 2013, and 2009 submitted by Viet Nam. The most recent is UNHRC 'National report submitted by Viet Nam' (2024) UN Doc A/HRC/WG.6/46/VNM/1, paras. 52, 90. UNHRC 'National report submitted by Dominican Republic' (2009) UN Doc A/HRC/WG.6/6/DOM/1, para. 95; UNHRC 'National report submitted by Mali' (2012) UN Doc A/HRC/WG.6/15/MLI/1, para. 36.

[246] See 4th Cycle Universal Periodic Review National Report – Guidance Note, https://www.ohchr.org/sites/default/files/documents/hrbodies/upr/infonotes/4thCycle-Guidance-Note-National-Report-EN.pdf.

[247] E.g. UNHRC 'Report of the Working Group on the UPR: Italy' (2010) UN Doc A/HRC/14/4, para. 25; UNHRC 'Report of the Working Group of the UPR. Japan' (2008) UN Doc A/HRC/8/44, para. 46; UNHRC 'Report of the Working Group on the UPR: Russian Federation' (2018) UN Doc A/HRC/39/13, p. 26, 147.302 (accepted by the Russian Federation); UNHRC 'Report of the Working Group on the UPR: Viet Nam' (2024) UN Doc A/HRC/57/7, para. 42.312–42.314; UNHRC 'Report of the Working Group on the UPR: Turkey', (2010) UN Doc A/HRC/15/13, para. 101.3.

[248] UNHRC 'Report of the Working Group on the UPR: Turkey' (2015) UN Doc A/HRC/29/15, para. 149.15.

[249] UNHRC 'Report of the Working Group of the UPR: Cuba' (2018) UN Doc A/HRC/39/16, para. 24.85; UNHRC 'National report submitted by Colombia' (2018) UN Doc A/HRC/WG.6/30/COL/1, para. 96.

[250] E.g., UNHRC 'National report submitted by Viet Nam' (2024) UN Doc A/HRC/WG.6/46/VNM/1 paras. 115 f.

the languages of minority communities through national cultural heritage legislation.[251] Criticism and recommendations are received rather positively by states, who in response affirm, for example, their commitment to respecting minority rights, including the preservation minority culture and identity[252] or to "organize campaigns and educational programmes" with the aim of raising "awareness of the importance of cultural heritage in all its diversity"[253]. Comments or recommendations on the protection of (elements of) majority culture or cultural heritage are sporadic.[254]

Third, states report and issue recommendations on the preservation and protection of Indigenous heritage. Aspects include the fostering, the preservation and protection of cultural diversity and heritage, including cultural heritage rights, land rights, language and linguistic diversity, cultural archaeology, values, spiritual and religious and traditional practices and traditional knowledge.[255] Nicaragua, for example, recommends the Russian Federation to increase its promotion and preservation of Indigenous languages "through the education system and via the adoption of relevant measures to safeguard intangible cultural heritage".[256] Australia shows itself committed to supporting the preservation and protection of arts, languages and cultures of the Aboriginal peoples and Torres

---

[251] *ibid* paras. 117–118.

[252] See UNHRC 'Report of the Working Group on the UPR: Italy' (2010) UN Doc A/HRC/14/4, para. 55.

[253] UNHRC 'Report of the Working Group on the UPR: Brazil' (2022) UN Doc A/HRC/52/14, para. 149.189; UNHRC 'Report of the Working Group on the UPR; Egypt' (2019) UN Doc A/HRC/43/16, para. 31.297.

[254] UNHRC 'Report of the Working Group on the UPR: Tonga' (2008) UN Doc A/HRC/8/48, para. 30. Addressing more general and specific campaigns to promote cultural heritage: UNHRC 'National report submitted by Egypt' (2024) UN Doc A/HRC/WG.6/48/EGY/1, para. 61.

[255] UNHRC 'National report submitted by Australia' (2020) UN Doc A/HRC/WG.6/37/AUS/1, paras. 54, 83; UNHRC 'National report submitted by Russian Federation' (2013) UN Doc A/HRC/WG.6/16/RUS/1, para. 189; UNHRC 'National report submitted by Brazil' (2017) UN Doc A/HRC/WG.6/27/BRA/1, para. 61.

UNHRC 'Report of the Working Group on the UPR: Australia, Addendum' (2021) UN Doc A/HRC/47/8/Add.1, para. 18. UNHRC 'Report of the Working Group on the UPR: Vanuatu' (2014) UN Doc A/HRC/57/10, paras. 85, 110.115–110.117; UNHRC 'Report of the Working Group on the UPR: New Zealand' (2024) UN Doc A/HRC/57/4, paras. 33, 45. Also see 2014 report by New Zealand. UNHRC 'Report of the Working Group in the UPR: Brazil' (2017) UN Doc A/HRC/36/11, para. 36.

[256] UNHRC 'Report of the Working Group on the UPR: Russian Federation' (2018) UN Doc A/HRC/39/13, p. 25, 147.293 (accepted).

Strait Islander peoples.[257] Recommendation similarly link the preservation of cultural heritage to education strategies.[258] Moreover, they address free, prior and informed consent in the context of projects with a potential impact on the land and ancestral ways of life of Indigenous peoples.[259]

6.3.2 Communications and Views

Human rights bodies offer two types of communications. Communications under the individual complaint procedure of the treaty bodies and communications under the UNHRC's Special Procedures.

6.3.2.1 The Use of the Communications Mechanism under the Treaty Bodies

The CESCR has not yet had the chance to address any issues relating to cultural heritage directly in its views.[260] In fact, most communications brought before the committee appear to concern evictions and the right to housing in Spain.[261] One communication, *A.M.B v Ecuador*, brought under the right to take part in cultural life concerns discrimination in respect to participation in football tournaments but was declared inadmissible due to the failure to exhaust domestic

[257] UNHRC 'Report of the Working Group on the UPR: Australia. Addendum' (2016) UN Doc A/HRC/31/14/Add.1, para. 27.

[258] UNHRC 'Report of the Working Group on the UPR: Australia' (2021) UN Doc A/HRC/47/8, para. 146.284; UNHRC 'Report of the Working Group on the UPR: Russian Federation' (2018) UN Doc A/HRC/39/13, para. 167.293.

[259] UNHRC 'Report of the Working Group on the UPR: Australia' (2021) UN Doc A/HRC/47/8, para. 146.288; UNHRC 'Report of the Working Group on the UPR: United Republic of Tanzania' (2011) UN Doc A/HRC/19/4, 8 December 2011, para. 86.49; UNHRC 'Report of the Working Group on the UPR: Brazil' (2022) UN Doc A/HRC/52/14, recommendations 149.260, 149.261, 149.269. Also see 149.200; UNHRC 'Report of the Working Group on the UPR: Colombia (2023) UN Doc A/HRC/55/7, recommendations 116.205, 116.209, 116.210; UNHRC 'Report of the Working Group on the UPR: Russian Federation' (2023) UN Doc A/HRC/55/14, recommendation 35.339.

[260] In fact, most communications brought before the Committee appear to concern evictions and the right to housing in *inter alia* Spain. UN Human Rights Treaty Body Database (2025): UN Treaty Body Database, https://tbinternet.ohchr.org/_layouts/15/TreatyBodyExternal/Home.aspx, 31 January 2025.

[261] See jurisprudence database at: OHCHR (2025): OHCHR Juris Database, https://juris.ohchr.org, 31 January 2025.

remedies.[262] Cultural rights have played a role in *López Albán v. Spain*, too, in which the committee expounded on what the right to adequate housing implies. The CESCR concluded that cultural adequacy needs to be taken into account when assessing the adequacy of housing to ensure "that expressions of cultural identity and diversity may be respected".[263] The case concerned a violation of Article 11 (1) of the ICESCR stemming from a forced eviction without providing alternative accommodation and the state's failure to comply with the committee's order of interim measures.[264]

The absence of any communication regarding rights and obligations vis-à-vis cultural heritage is likely attributable to the relatively young age of the Optional Protocol to the ICESCR. The Optional Protocol only entered into force in 2013 and is at the time of writing only ratified by 31 state parties.[265] A communication that concerns cultural heritage matters under Article 15 of the ICESCR could further expound on and clarify states' obligations on a case-to-case basis and strengthen cultural rights relating to cultural heritage. A communication would also be of symbolic value and reinforce the link between cultural rights and cultural heritage in the international human rights framework.

A further reason for the paucity of communications under the CESCR's complaints mechanism might be the long-established complaints procedure under the HRC, specifically under Article 27 of the ICCPR. Article 27 has served as a legal basis for communications concerning the rights of persons belonging to minorities and Indigenous peoples that exhibit a cultural heritage dimension. General comment No. 23 explains that Article 27 of the ICCPR establishes and recognises a distinct and additional right for persons belonging to minorities, which is conferred upon individuals.[266]

Although it is an individual right, persons belonging to minorities can enjoy the right alone or in community with other members.[267] The HRC has underlined the communal nature by stating that the exercise of the rights "depend[s] [...] on

[262] CESCR, Communication No. 3/2014, *A.M.B v Ecuador* (8 August 2016) UN Doc E/C.12/58/D/3/2014.

[263] CESCR, Communication No. 37/2018, *Maribel Viviana López Albán v. Spain* (29 November 2019) UN Doc E/C.12/66/D/37/2018, para. 9.3.

[264] *ibid* para. 15.

[265] UN Human Rights Treaty Body Database (2025): UN Treaty Body Database, https://tbinternet.ohchr.org/_layouts/15/TreatyBodyExternal/Home.aspx., accessed 31 January 2025).

[266] CCPR 'General Comment No. 23' (8 April 1994) UN Doc CCPR/C/21/Rev.1/Add.5, paras. 1; 3.1.

[267] International Covenant on Civil and Political Rights (adopted 16 Dec 1966, entered into force 23 Mar 1976) 999 UNTS 171 (ICCPR), Art. 27.

the ability of the minority group to maintain its culture, language or religion."[268] This may render it necessary "to protect the identity of a minority and the rights of its members to enjoy and develop their culture and language [...] in community with the other members of the group."[269]

The committee established in its 1994 General Comment No 23 that Article 27 covers many manifestations of culture, including a traditional or particular way of life. The Committee's views, such as the 1988 communication of *Kitok v Sweden* or the 1993 communication of *Länsman et al. v Finland*, addressed the protection of traditional activities and means of livelihoods. They recognise that economic activities can form an essential element in the culture of a community and can be adapted with the help of modern technology.[270]

In the past three decades, the HRC has dealt with communications that both address intangible and tangible forms of cultural heritage.[271] These communications are not limited to the scope of Article 27, but also rely on inter alia Articles 17 (1), 23 (1), and 18 of the ICCPR to name the most prominent examples.[272] In this context, the views of the HRC have significantly contributed to our understanding that components of someone's traditional and/or Indigenous ways of life[273] and attachment and dependency on the land[274] can also fall under the sphere of protection of Article 17 of the ICCPR. Similarly, it has shown

[268] CCPR 'General Comment No. 23' (8 April 1994) UN Doc CCPR/C/21/Rev.1/Add.5, para. 6.2.

[269] *ibid.*

[270] See Chap. 4, Sect. 4.4.2.

[271] See Chap. 4, Sect. 4.2 f.

[272] See CCPR, Communication No. 822/1998, *Mathieu Vakoumé et al v. France* (2000) UN Doc CCPR/C/70/D/822/1998; CCPR, Communication No. 549/1993, *Hopu and Bessert v. France* (29 December 1997) UN Doc CCPR/C/60/D/549/1993/Rev.1; CCPR, Communication No. 3624/2019, *Daniel Billy et al. v. Australia* (22 September 2022) UN Doc CCPR/C/135/D/3624/2019; CCPR, Communication No. 2552/2015, *Benito Oliveira Pereira et al. v. Paraguay* (12 October 2021) UN Doc CCPR/C/132/D/2552/2015; CCPR, Communication No. 2751/2016, *Norma Portillo Cáceres et al. v. Paraguay* (20 September 2019) UN Doc CCPR/C/126/D/2751/2016; CCPR, Communication No. 3585/2019, *Alisa Roy v. Australia* (13 March 2024) UN Doc CCPR/C/137/D/3585/2019.

[273] CCPR, Communication No. 3624/2019, *Daniel Billy et al. v. Australia* (22 September 2022) UN Doc CCPR/C/135/D/3624/2019, para. 8.10; CCPR, *Benito Oliveira Pereira et al. v. Paraguay* (12 October 2021) UN Doc CCPR/C/132/D/2552/2015, para. 8.3.

[274] See CCPR, Communication No. 3585/2019, *Norma Portillo Cáceres et al. v. Paraguay* (20 September 2019) UN Doc CCPR/C/126/D/2751/2016; para. 7.8.

how development projects can affect rights relating to cultural heritage in these communications.[275]

Considering the legal basis, it is not surprising that most communications with a cultural heritage dimension have a group dimension. However, the group dimension also reveals a hurdle in using the complaints mechanism, as the international human rights system is structured around individual rights.[276] This means that the recognition of the standing of an individual as a rights holder is generally unproblematic.[277] This is clearly reflected in the Optional Protocol to the ICCPR, providing "a procedure under which individuals can claim that their individual rights have been violated".[278] The views of the HRC have recognised that a group of individuals, claiming to be similarly affected, can submit a communication.[279] Yet, for these purposes explicit authorisation by the affected individuals is required or arguments showing that such authorisation is not needed.[280]

The HRC established early on that the rights included in Article 1 of the ICCPR are excluded from the scope of these individual rights.[281] The HRC has

> "repeatedly declared the non-justiciability of rights protected under Article 1, and at most has transformed these claims into claims under other, individual rights protected by the Covenant."[282]

The HRC has held that it may interpret Article 1 "in determining whether rights protected in parts II and III of the Covenant have been violated".[283] This means

---

[275] See Sect. 6.2.1.2 above and Chap. 4, Sect. 4.3.4.

[276] See, "… the traditional human rights regime, which in the eyes of many is constructed around the interests and concerns of individual human beings." Wiessner (2011), p. 124.

[277] Vrdoljak (2016), p. 3.

[278] CCPR, Communication No 547/1993, *Apirana Mahuika and Ors v New Zealand* (27 October 2000) UN Doc CCPR/C/70/D/547/1993, p. 16, para. 9.2.

[279] CCPR, Communication No 167/1984, *Chief Bernard Ominayak and the Lubicon Lake Band v Canada* (26 March 1990) UN Doc CCPR/C/38/D/167/1984, p. 27, para. 32.1.

[280] CCPR Communication No. 879/1999, *Howard v. Canada* (4 August 2005) UN Doc CCPR/C/84/D/879/1999, p. 9, para. 8.3.

[281] CCPR, Communication No 167/1984, *Chief Bernard Ominayak and the Lubicon Lake Band v Canada* (26 March 1990) UN Doc CCPR/C/38/D/167/1984, p. 27, para. 32.1; CCPR, Communication No 547/1993, *Apirana Mahuika and Ors v New Zealand* (27 October 2000) UN Doc CCPR/C/70/D/547/1993, para. 9.2.

[282] Lixinski (2013a), p. 148.

[283] CCPR, Communication No. 3585/2019, *Alisa Roy v. Australia (13 March 2024)* UN Doc CCPR/C/137/D/3585/2019, para. 7.3 with reference to "precedents" in CCPR, Communication No. 932/2000, *Ms. Marie-Hélène Gillot et al. v. France* (21 July 2002) UN Doc CCPR/C/

that Article 1 may be taken into account when interpreting alleged violations of Covenant rights.

Groups and communities "cannot be easily accommodated in individualistic rights' narratives".[284] The reluctance to advance the rights paradigm towards group or peoples' rights can be partially attributed to the notion that groups can pose challenges to state sovereignty.[285] Furthermore, the position of communities vis-à-vis states "is usually assumed to be one of opposition, rather than cooperation" making it difficult in "entering into conversation about community rights in international law".[286]

The HRC has considered communications by individual members of a community or group, and found "in principle, no objection to a group of individuals, who claim to be similarly affected, collectively to submit a communication" in *Howard v. Canada*.[287] Whilst this allows individual members to be heard, the claims of communities are "filtered through individuals"[288].

6.3.2.2 The Use of the Communications Mechanism under the Special Procedures

Amid the absence of any communications concerning cultural heritage in the complaints procedure of the ICESCR, the communications emanating from the Special Procedures suggest that this absence cannot be ascribed to lack of potential complaints, although it is important to emphasise that the threshold for communications and their nature in both mechanisms significantly differs.

The complaints procedure under the ICCPR and the UNHRC's Special Procedures both shed light on the impacts of development projects, which often fail to consider cultural rights, the impact on particular or traditional ways of life, and the heritage of persons belonging to minorities and Indigenous peoples.[289] Communications have also addressed the destruction of cultural heritage and states'

75/D/932/2000, para. 13.4; CCPR, Communication No. 2668/2015, *Sanila-Aikio v. Finland* (16 October 2018) UN Doc CCPR/C/119/D/2668/2015, para. 8.6.

[284] Jakubowski et al. (2022), para. 72.

[285] *ibid.*

[286] *ibid.*

[287] CCPR Communication No. 879/1999, *Howard v. Canada* (4 August 2005) UN Doc CCPR/C/84/D/879/1999, para. 8.3. Also see Vrdoljak, 2014, p. 26.

[288] Lixinski (2018b), p. 486.

[289] See for example UA NPL 3/2017 (14 August 2017) p. 3; Mandates of the Special Rapporteur in the field of cultural rights; and the Special Rapporteur on adequate housing as a component of the right to an adequate standard of living, and on the right to non-discrimination in this context, UA UZB 3/2019 (9 August 2019) p. 6.

duties not to destroy, damage or alter cultural heritage on their own territory outside of the context of an armed conflict.[290]

The Special Procedures' communications are not restricted to individual cases, but can address a wide array of situations, including general patterns of human rights violations as well as draft or adopted legislation, policy and practice. This flexibility is advantageous and allows for a variety of communications to be considered, including those prompted by, for example, the inscription of traditions on the national list of intangible cultural heritage[291] or those addressing the Director of the World Heritage Centre to ensure the rights, interests, and concerns of Indigenous communities are considered in the nomination process of world heritage sites[292]. Communication UA UZB 3/2019, for example, dealt with an urban development project resulting in the destruction of historic neighbourhoods in Samarkand, Uzbekistan, as well as the ongoing demolitions of houses in inter alia Uzbekistan's capital Tashkent. The communication addressed both "the rights to adequate housing and the right to access and enjoy cultural heritage".[293]

Communications under the Special Procedures do not emulate legal proceedings or constitute a quasi-judicial mechanism.[294] Communications outline the alleged facts and list the applicable international legal norms or pronouncements[295]. One or more Special Rapporteur(s) invite(s) the addressee to respond to allegations on interferences with human rights guarantees by inter alia commenting on the accuracy of the alleged facts, by providing details on the situation, and by reporting on measures taken to protect the concerned rights. In this sense, they are rather an (open) letter that is subsequently published online and expose governments to public scrutiny,[296] thereby offering a political tool.

---

[290] E.g., Mandate of the Special Rapporteur in the field of cultural rights, AL cultural rights (2009), SGP 2/2012, 29 May 2012, p. 2. See Chap. 5, Sect. 5.4.2.

[291] Mandates of the Working Group of experts on people of African descent; the Special Rapporteur in the field of cultural rights; the Special Rapporteur on issues; and the Special Rapporteur on contemporary forms of racism, racial discrimination, xenophobia and related intolerance, OL, NLD 1 /2015 (28 January 2015).

[292] Mandate of the Special Rapporteur on the rights of indigenous peoples, OL Indigenous (2001–8), OTH 10/2013 (18 November 2013).

[293] UA UZB 3/2019 (9 August 2019), p. 6.

[294] Also see, OHCHR (2025a): What are communications, https://www.ohchr.org/en/special-procedures-human-rights-council/what-are-communications, accessed 31 January 2025.

[295] See, e.g., AL OTH 136/2024 (17 October 2024) annex.

[296] Bielefeldt (2011), p. 231.

Communications vary in either identifying the identities of alleged victims by referring to a specific community or highlight an affected family[297] or refraining from doing so. Communication SGP 2/2012, for example, concerns the construction of a highway through Bukit Brown Cemetery, a mixed cultural and natural heritage site in Singapore, and does not clearly identify individuals or communities impacted by the project.[298]

Communications under the Special Procedures come with clear advantages of flexibility, speediness, and fewer procedural hurdles. Furthermore, they can be directed to states, intergovernmental organisations, and business too.[299] Whilst the communications of the Special Procedures are not a quasi-judicial individual complaints procedure, for the purposes of cultural heritage protection in the international human rights framework, the contribution of the communications lie in continuously emphasising and demonstrating the close link between cultural heritage and the enjoyment of human rights. Against the background of the under-articulation or, at times, omission of the cultural heritage dimension, this is an important contribution.

The available documentation of the communications indicates that states not always respond to communications,[300] confirming the finding that communication rather have a "limited impact on influencing state behaviour".[301] Whilst some states seek to clarify the alleged facts and do not dismiss them altogether, they do not tend to comment on any obligations towards cultural heritage.[302]

---

[297] AL BRA 2/2018 (20 April 2018), p. 2. See "concerned communities" in OL, NLD 1 / 2015 (28 January 2015), p. 2.

[298] AL cultural rights (2009), SGP 2/2012 (29 May 2012).

[299] OHCHR (2025a): What are communications, https://www.ohchr.org/en/special-procedures-human-rights-council/what-are-communications, accessed 31 January 2025.

[300] Communications and responses can be found here, OHCHR (2025): Communication report and search, https://spcommreports.ohchr.org/TmSearch/Mandates?m=19, accessed 31 January 2025.

[301] Subedi (2014), p. 260. Also see Piccone (2010), p. ix.

[302] One of the exceptions is Libya. It is interesting to note that in responding to the allegations, Libya separates between its obligations towards ensuring the protection of human rights, on the one hand, and towards cultural heritage, on the other. Mission permanente de la Libye auprès de l'Office des Nations Unies à Genève et des autres organisations internationales en Suisse, No. 284.12, 7 September 2012. (In response to UA Cultural rights (2009) G/SO 214 (67–17) Assembly & Association (2010–1) G/SO 214 (56–23) G/SO 214 (107–9) Minorities (2005–4) G/SO 214 (53–24) LBY 2/2012, 5 September 2012)

6.3.3 Interim Conclusions

By considering how the international human rights law framework can provide a legal basis for rights and obligations relating to cultural heritage, this section has explored the use of the monitoring and on the complaints mechanism. Human rights mechanisms can contribute to strengthening cultural heritage in at least three ways. Both, periodic reporting and the UPR, monitor the implementation of human rights and thereby contribute to holding states accountable. The analysis has shown that both treaty bodies and states address recommendations vis-à-vis cultural heritage to (other) states. The complaints procedure also contributes to the accountability of states by addressing alleged violations in an international forum.

Second, it strengthens rights claims in matter concerning cultural heritage for individuals and individuals as part of a group. This primarily only offer these persons access to a (soft) enforcement tool at the international level. It also provides treaty bodies with the opportunity to expound on the scope of Covenant rights. Treaty bodies concretise the meaning of individual rights and state obligations through inter alia their three principal activities.[303] Treaty bodies, however, do not act completely independently and the interaction between states and treaty bodies results in the pronouncements, specifically the general comments, not being a pure reflection of a committee's position.[304] As cultural heritage is "generally associated with collective, group or community experience, identity, and values",[305] the individualistic nature of the international human rights system provides a limited approach[306]. Against this backdrop, the analysis has shown how the views of the HRC can protect the rights of individuals belonging to a minority group or Indigenous community.

While these tools and mechanisms apply a human rights lens, their availability can support the enforcement system of international cultural heritage law at the international level. The UPR and the Special Procedures are universal mechanisms, although the country missions of the Special Rapporteurs will depend on

[303] OHCHR (2015): Statement by Mr. Zeid Ra'ad Al Hussein, United Nations High Commissioner for Human Rights, at the International Law Commission, https://www.ohchr.org/EN/NewsEvents/Pages/DisplayNews.aspx?NewsID=16254&LangID=E, accessed 31 January 2025.

[304] See CCPR 'Rules of Procedure of the Human Rights Committee' (9 January 2019) UN Doc CCPR/C/3/Rev.11., Rule 76(4).

[305] Jakubowski (2024), p. 500.

[306] The reference to "limited approach" is borrowed from *Solntsev*, who uses it in context of the environment and human rights in Solntsev (2024), p.365.

the invitation by the respective country.[307] The work of the treaty bodies hinges, on the one hand, on the ratification of the Covenants, and, on the other hand, on the ratification of the the First Optional Protocols to the ICCPR and the ICESCR.

6.4 Conclusions

This chapter set out to gauge if and how the international human rights framework can offer a legal basis for the protection of cultural heritage. The analysis has examined one possible answer: through the work and pronouncements of international human rights bodies. Analysing the protection of cultural heritage through the lens of international human rights law and the associated framework, the analysis has focused on the meaning of cultural heritage and on rights and obligations relating to cultural heritage.

The chapter has argued that the Special Rapporteur's conceptual understanding, on the one hand, and the pronouncements and overall work of human rights bodies, on the other, haves contributed to a more refined and holistic understanding of the link between cultural heritage and human rights. It has suggested that the reliance on different human rights provisions in cultural heritage matters has resulted illustrated inter alia the proximity of cultural heritage to family life, privacy and home as well as to the right to life in the case of Indigenous peoples. Yet, as international human rights law does not provide for right to cultural heritage, the reliance on different human rights provisions can be attributed to the constraints of their mandates and cultural heritage considerations are generally subordinated to the respective rights.[308]

Considering the protection of cultural heritage in the international human rights framework might at first sight appear to complement the protection by offering exactly what international cultural heritage law is missing: rights and (soft) enforcement tools that are built in the framework.

The human rights bodies indeed articulate several rights and obligations that form part of primarily the implementation of Article 15 (1) (a) of the ICESCR and Article 27 of the ICCPR. Moreover, the analysis has highlighted other important rights of the ICCPR and ICESCR as well as of the UNDRIP, giving the

---

[307] OHCHR (2025b): Country Visits, https://www.ohchr.org/en/special-procedures/sr-freedom-of-assembly-and-association/country-visits, accessed 31 January 2025.

[308] Francioni and Lixinski (2017), p. 18.

impression of a fragmentation of such rights, especially in the case of Indigenous peoples[309].

The analysis has sought to arrange these puzzle pieces of rights and obligations and has highlighted (1) respect and protect cultural heritage at all times and prevent destruction and damage, (2) access to cultural heritage, (3) respect, protect, and (formally) recognise the rights and cultural heritage of persons belonging to minorities and Indigenous peoples and to create favourable conditions for the enjoyment and development of culture and cultural heritage, and the right of Indigenous peoples to (4) effective consultations, participation and free, prior and informed consent in decision-making affecting cultural heritage and their way of life.

In the absence of a human right to cultural heritage, international human rights law offers several legal bases for cultural heritage matters. The human rights bodies apply human rights norms to cultural heritage matters, making clear that rights and obligations relating to cultural heritage form part of human rights law. As a consequence, when speaking of rights and obligations in the context of cultural heritage at the international level, we do not speak of self-standing human rights that have already emerged as "new" human rights. Rights and obligations vis-à-vis cultural heritage exist but—in most cases—form part of the proper implementation of human rights. The pronouncements by the treaty bodies suggest that certain rights and obligations are seen as forming part of the state parties in the implementation of human rights provisions. The specific rights of Indigenous peoples do not solely emanate from human rights norms in the ICESCR and ICCPR but are most prominently codified in ILO Convention No. 169 and have been further developed through the UNDRIP. Reporting has pointed out the implementation of the UNDRIP at the national level.[310]

The chapter puts forward that although the human rights framework does not necessarily offer specific rights *to* cultural heritage, it equips individuals and persons belonging to a group with legal bases they can rely on to strengthen their claims and to potentially enforce rights relating to cultural heritage. At the international level, the human rights framework offers periodic reporting and a peer review system that have both attended to a variety cultural heritage matters in their concluding observations and recommendations. It equips individuals and individuals of a group with a (soft) enforcement tool through the individual complaints procedure. While the Special Procedures' communications have been

---

[309] Gilbert (2017), p. 24.

[310] CCPR 'Periodic report submitted by Brazil' (2021) UN Doc CCPR/C/BRA/3, para. 28; CCPR 'Periodic report submitted by Colombia' (2021) UN Doc CCPR/C/COL/8, para. 6.

found to be of "limited impact",[311] the contribution of the Special Procedures lies inter alia in continuously tabling the topic of heritage protection in reports and communications, and showing how cultural heritage matters and human rights concerns intersect.

While the international human rights framework provides a limited approach and a limited avenue, it has been described as likely " the most effective avenue" for individuals and persons belonging to minority groups and Indigenous peoples "to protect and safeguard their cultural heritage".[312] In addition, the international human rights framework can strengthen the general conditions that ensure the safeguarding and protection of cultural diversity and heritage, its contribution primarily lies in improving and strengthening rights and obligations relating to cultural heritage. This is not to say that the international human rights framework *should* remain the only avenue. Turning back to the human dimension of international cultural heritage law, it is important to highlight efforts of advancing the concept of "participatory governance" in scholarship and cultural policy frameworks with the objective of stakeholder engagement and empowering individuals, groups, and communities with a link to cultural heritage.[313]

Reference List

Alston, Philip (1984): Conjuring Up New Human Rights: A Proposal for Quality Control. In: American Journal of International Law, Vol. 78 No. 3, S. 607–621.

Bielefeldt, Heiner (2011): Protecting and Implementing the Right to Freedom of Religion or Belief: Interview with Heiner Bielefeldt. In: Journal of Human Rights Practice, Vol. 3, Issue 3, S. 229–244.

Blake, Janet (2015): International Cultural Heritage Law. Oxford (Oxford University Press).

Chenwi, Lilian (2009): Correcting the Historical Asymmetry between Rights: The Optional Protocol to the International Covenant on Economic, Social and Cultural Rights. In: African Human Rights Law Journal, Vol. 9, No.1, S. 23–51.

von der Decken, Kerstin and Koch, Nikolaus (2020): Recognition of New Human Rights: Phases, Techniques and the Approach of 'Differentiated Traditionalism.' In: Andreas von Arnauld, Kerstin von der Decke and Mart Susi (eds.): The Cambridge Handbook of New Human Rights. Cambridge (Cambridge University Press 2020), S. 7–20.

---

[311] Subedi (2014), p. 260.

[312] See Blake (2015), p. 311.

[313] Jakubowski (2024), p. 511. With reference to the regional EU level, see Council of European Union, Council conclusions on participatory governance of cultural heritage (2014/C 463/01).

Also see, Jakubowski et al. (2022).

Disko, Stefan (2017): Indigenous Cultural Heritage in the Implementation of UNESCO's World Heritage Convention: Opportunities, Obstacles and Challenges. In: Indigenous Peoples' Cultural Heritage: Rights, Debates and Challenges. Leiden (Brill I Nijhoff), S. 39–77.

Dorough, Dalee Sambo and Wiessner, Siegfried (2020): Indigenous Peoples and Cultural Heritage. In: Francesco Francioni and Ana Filipa Vrdoljak (eds.): The Oxford Handbook of International Cultural Heritage Law. Oxford (Oxford University Press), S. 407–430.

Francioni, Francesco (2020c): World Cultural Heritage. In: Francesco Francioni and Ana Filipa Vrdoljak (eds.): The Oxford Handbook of International Cultural Heritage Law. Oxford (Oxford University Press), S. 250–271.

Francioni, Francesco and Lixinski, Lucas (2017): Opening the Toolbox of International Human Rights Law in the Safeguarding of Cultural Heritage. In: Andrea Durbach and Lucas Lixinski (eds.): Heritage, Culture and Rights: Challenging Legal Discourses. Oxford (Hart Publishing.

Frulli, Micaela (2020): International Criminal Law and the Protection of Cultural Heritage. In: Francesco Francioni and Ana Filipa Vrdoljak (eds.) The Oxford Handbook of International Cultural Heritage Law. Oxford (Oxford University Press), S. 100–120.

Gilbert, Jérémie (2018): Natural Resources and Human Rights: An Appraisal. Oxford (Oxford University Press).

Gilbert, Jérémie (2017): Indigenous Peoples, Human Rights, and Cultural Heritage: Towards a Right to Cultural Integrity. In: Alexandra Xanthaki, Sanna Valkonen, Leena Heinämäki and Piia Nuorgam (eds.): Indigenous Peoples' Cultural Heritage: Rights, Debates and Challenges. Leiden (Brill), S. 20–38.

ICRC (2025): Rule 40: Respect for Cultural Property, https://ihl-databases.icrc.org/en/customary-ihl/v1/rule40, accessed 31 January 2025.

Jakubowski, Andrzej (2024): Heritage and Human Rights. In: Lucas Lixinski and Lucie K. Morisset (eds.): The Routledge Handbook of Heritage and the Law. London (Routledge), S. 500–516.

Jakubowski et al. (2022): International Law Association Committee on Participation in Global Cultural Heritage Governance - Final Report (2022).

Kotzé, Louis J. (2018): In Search of a Right to a Healthy Environment in International Law: Jus Cogens Norms. In: John H. Knox and Ramin Pejan (eds.): The Human Right to a Healthy Environment. Cambridge (Cambridge University Press), S. 136–54.

Lixinski, Lucas (2019): International Heritage Law for Communities: Exclusion and Re-Imagination. Oxford (Oxford University Press).

Lixinski, Lucas (2018b): Food as Heritage and Multi-Level International Legal Governance. In: International Journal of Cultural Property, Vol. 25, No. 4, S. 469–490.

Lixinski (2016): ESIL – International Human Rights Law Symposium: International Human Rights Law and International Cultural Heritage Law: Cooperation, Conflict or Cooption?, https://www.ejiltalk.org/esil-international-human-rights-law-symposium-international-human-rights-law-and-international-cultural-heritage-law-cooperation-conflict-or-cooption/, accessed 31 January 2025.

Lixinski, Lucas (2013a): Intangible Cultural Heritage in International Law. Oxford (Oxford University Press).

Nafziger, James A. R., Robert Kirkwood Paterson and Alison Dundes Renteln (2010): Cultural Law: International, Comparative, and Indigenous. Cambridge (Cambridge University Press), S. 1–99.

Novic, Elisa (2016): The Concept of Cultural Genocide. Oxford (Oxford University Press).

OHCHR (2025): OHCHR Juris Database, https://juris.ohchr.org, 31 January 2025.

OHCHR (2025a): What are communications, https://www.ohchr.org/en/special-procedures-human-rights-council/what-are-communications, accessed 31 January 2025.

OHCHR (2025b): Country Visits, https://www.ohchr.org/en/special-procedures/sr-freedom-of-assembly-and-association/country-visits, accessed 31 January 2025.

OHCHR (2015): Statement by Mr. Zeid Ra'ad Al Hussein, United Nations High Commissioner for Human Rights, at the International Law Commission, https://www.ohchr.org/EN/NewsEvents/Pages/DisplayNews.aspx?NewsID=16254&LangID=E, accessed 31 January 2025.

O'Flaherty, Michael (2006): Concluding Observations of United Nations Human Rights Treaty Bodies. In: Human Rights Law Review, Vol. 6, No. 1, S. 27–52.

Piccone, Ted (2010): Catalysts for Rights: The Unique Contribution of the UN's Independent Experts on Human Rights. Final Report of the Brookings Research Project on Strengthening U.N. Special Procedures. The Brookings Institution, S. 1–90.

Pineschi, L. (2012): Cultural Diversity as a Human Right? General Comment No. 21 of the Committee on Economic, Social and Cultural Rights. Silvia Borelli and Federico Lenzerini (eds.): In Cultural Heritage, Cultural Rights, Cultural Diversity. Leiden (Brill | Nijhoff) S. 27–53.

Rodley, Sir Nigel (2013): The Role and Impact of Treaty Bodies. In: Shelton, Dinah (ed.): The Oxford Handbook of International Human Rights Law. Oxford (Oxford University Press), S. 621–648.

Saul, B., Kinley, D., & Mowbray, J. (2014). *The International Covenant on Economic, Social and Cultural Rights: Commentary*. Oxford (Oxford University Press).

Solntsev, Alexander (2024): The Growing Importance of Human Rights Treaty Bodies in Environmental Dispute Resolution. In: Christina Voigt and Caroline Foster (eds.): International Courts versus Non-Compliance Mechanisms: Comparative Advantages in Strengthening Treaty Implementation, Cambridge (Cambridge University Press), S. 361–382.

Subedi, Surya P. (2014): Human Rights Experts in the United Nations: A Review of the Role of United Nations Special Procedures. In: Monika Ambrus, Karin Arts, Ellen Hey and Helena Raulus (eds.): The Role of 'Experts' in International and European Decision-Making Processes: Advisors, Decision Makers or Irrelevant Actors?. Cambridge (Cambridge University Press), S. 241–262.

Thielbörger, Pierre (2020): Something Old, Something New, Something Borrowed and Something Blue: Lessons to Be Learned from the Oldest of the 'New' Rights – the Human Right to Water. In: Andreas von Arnauld, Kerstin von der Decke and Mart Susi (eds.): The Cambridge Handbook of New Human Rights. Cambridge (Cambridge University Press 2020), S. 70–78.

Thielbörger, Pierre (2014): The Right(s) to Water: The Multi-Level Governance of a Unique Human Right. Berlin (Springer).

Tomuschat, Christian (2019). Universal Periodic Review Procedure: Human Rights Council. In: Max Planck Encyclopedias of International Law.

Ubertazzi, B. (2020). Art. 2(2) Manifesting Intangible Cultural Heritage. In J. Blake & L. Lixinski (Eds.), *The 2003 UNESCO Intangible Heritage Convention: A Commentary* (pp. 58–80). Oxford University Press.

UN News (2022): Why the UN General Assembly must back the right to a healthy environment, https://news.un.org/en/story/2022/07/1123142, accessed 31 January 2025.

Vrdoljak, Ana Filipa (2018): Indigenous Peoples, World Heritage, and Human Rights. In: International Journal of Cultural Property, Vol. 25, No. 3, S. 245–281.

Vrdoljak, Ana Filipa (2016): Standing and Collective Cultural Rights, in: Andrej Jakubowski (ed.) Cultural Rights as Collective Rights: An International Law Perspective. Leiden (Brill Publishers), Available at SSRN: https://ssrn.com/abstract=2518054, S. 1–16.

Vrdoljak, Ana Filipa (2014): Human Rights and Cultural Heritage in International Law. In: Federico Lenzerini and Ana Filipa Vrdoljak (eds.): International Law for Common Goods: Normative Perspectives on Human Rights, Culture and Nature. Oxford (Hart Publishing), Available at SSRN: https://ssrn.com/abstract=2200186, S. 1–39.

Wiessner, Siegfried (2011): The Cultural Rights of Indigenous Peoples: Achievements and Continuing Challenges. In: European Journal of International Law, Vol. 22, Issue 1, S. 121–140.

De Witte, Bruno (2020): Language as Cultural Heritage. In: Francesco Francioni and Ana Filipa Vrdoljak (eds.): The Oxford Handbook of International Cultural Heritage Law. Oxford (Oxford University Press), S. 371–378.

Conclusions—Cultural Heritage Protection through the Lens of Human Rights Bodies

7

This thesis has shown that international human rights law offers several legal bases for rights and obligations vis-à-vis cultural heritage. It argues that in spite of shortcomings in the application to cultural heritage, the mechanism and tools of the international human rights framework can contribute to strengthening the holistic protection of cultural heritage.

The thesis is set against the backdrop of two primary observations, which the first chapter outlines. First, a shift in thinking in and about international cultural heritage law that increasingly considers the relationship between people and cultural heritage and their participation in international cultural heritage law. Second, the encouragement by the UN High Commissioner for Human Rights in 2007 to consider "cultural heritage *as an important component of the promotion and protection of all human rights*"[1], and a subsequent mainstreaming of cultural heritage in the human rights bodies.

The first chapter sheds light on the evolving challenges of protecting cultural heritage. It has highlighted that in addition to threats to cultural heritage owing to armed conflicts or its intentional targeting, threats that typically occur in peacetime can have detrimental impacts on cultural heritage. These include the effects of development projects, tourism, natural disasters, climate change, and state-driven readings and understandings of culture to the exclusion of minority groups and Indigenous peoples.

In light of the absence of an explicit right to cultural heritage in international human rights law, this thesis examines the work and pronouncements of human

[1] UNHRC Res 6/11 (28 September 2007) UN Doc A/HRC/RES/6/11, para. 8. (Emphasis added).

L. Hofmann, *Cultural Heritage through the Lens of Human Rights Bodies*, Juridicum – Bochumer Schriften zur Friedenssicherung und zum Humanitären Völkerrecht | Bochum Studies on the International Law of Peace and Armed Conflict 73, https://doi.org/10.1007/978-3-658-51203-3_7

rights bodies as one possible answer to the research question. Specifically, it focuses on the CESCR and HRC, as examples of two treaty bodies, and broadly speaking on the work of the UNHRC and its mechanisms, as an example of the charter-based bodies. Cognisant of its reliance on formally non-binding sources of different human rights bodies, Chap. 2 has considered the role and function of human rights bodies in international law and examined their interpretive authority and how their pronouncements and work fit within the sources of international law. This analysis has considered the concept of the most highly qualified publicists, the contribution to the crystallisation of customary international law, and subsequent practice in the application of a treaty. Against the backdrop of international law's binary understanding of legally binding or non-binding sources, this thesis has built on the distinction of the legally binding nature, on one hand, and legal effect, on the other, and submits that the formally non-binding sources emanating from the work and pronouncements of human rights bodies should be considered on a continuum of legal effect. Consequently, sources are understood as being able to exert a legal effect that can—but does not have to—contribute to or even result in a rule, right or obligation becoming legally binding.

Considering cultural heritage protection in the international human rights framework can contribute to a so-called "rhetorical paradigm shift"[2] that demonstrates the need to protect *rights* vis-à-vis cultural heritage. Chapter 3 has traced this shift in different aspects of international cultural heritage law. The purpose of this was not only to demonstrate the impetus for this thesis' focus on cultural heritage protection in the international human rights framework. The aim was also to demonstrate the link between cultural heritage and the enjoyment of human rights, and how international cultural heritage law has incrementally considered the role and rights of individuals and groups. The thesis has exemplified this shift firstly, in the evolving meaning of cultural heritage; secondly, in the evolving rationale to protect cultural heritage and the role of communities, groups, and individuals in protection mechanisms; and thirdly, in the enforcement of the protection of cultural heritage in international criminal law.

By reference to the definitions of cultural property and cultural heritage in the UNESCO Conventions and the academic discourse, the chapter has traced the incremental expansion of the concept of cultural heritage. It has shown how it evolved from the notion of cultural heritage as high culture to a more open concept of culture as a living culture and as emphasising cultural diversity in both material and dematerialised forms.

---

[2] See von Arnauld and Theilen (2020), p. 41.

The evolving rationales to protect cultural heritage have reflected a shift in the impetus to protect cultural heritage. With preambles of early international cultural heritage instruments underlining the international community, humanity and scientific and educational advancements, the 2003 Intangible Cultural Heritage Convention mentions these aspects, too, but also underlines protecting cultural heritage for its importance to peoples, groups, communities, and individuals. The shift to the increasing role and rights of persons is reinforced by the efforts of the Intangible Cultural Heritage Convention to integrate participation of communities and groups, albeit couched in soft terms.

Finally, Chap 3 underlines the contribution of international criminal law in revealing people or human communities as the victims of cultural heritage destruction. By reference to the jurisprudence of the ICTY and ICC, the chapter shows how the enforcement of cultural heritage protection has contributed to a better understanding of the human impact of cultural heritage destruction, and the close link between such destruction and evidence of an intent of the crime of genocide and crimes against humanity.

Given that the horizontal character of UNESCO treaties does not create rights for individuals, and that translating the language of rights, increasingly employed in international cultural heritage law, remains at the discretion of each state, Chaps 4 and 5 analyse the human rights bodies' articulation of meaning, rights, and obligations vis-à-vis cultural heritage in their pronouncements and work.

The treaty bodies, subject of Chap. 4, approach cultural heritage in different ways. The CESCR promotes broad and popular participation in cultural life while at the same time focusing on the cultural rights of persons belonging to minorities and Indigenous peoples under Article 15 (1) (a) of the ICESCR. This is also reflected in the CESCR's focus on culture as cultural products *and* as a way of life.

The HRC's interpretation of the concept of "culture" within the meaning of Article 27 of the ICCPR encompasses the right of persons belonging to minorities and Indigenous communities to follow a particular, traditional, or Indigenous way of life. This is a particularly important concept for Indigenous peoples, as it recognises the connection between lands and territories and inter alia the cultural heritage and related rights of Indigenous peoples. Lands and territories are typically not only characterised by natural or cultural components, but an amalgamation of cultural, economic, natural spiritual and social values. Similarly, the rights of Indigenous peoples in this context go beyond cultural rights and intersect with, among others, a life with dignity and the survival of the community (Article 6 of the ICCPR) and the notions of family life, home, and privacy (Article 17 of the ICCPR).

The analysis of the treaty bodies has identified the underlying obligations to promote and protect cultural diversity and to respect and protect cultural heritage at all times, both emanating primarily from the CESCR's pronouncements on Article 15 (1) (a). Furthermore, the analysis has presented three rights and corresponding obligations. The analysis has highlighted the right of access to cultural heritage and the obligation to guarantee access to cultural, linguistic and religious or sacred heritage. This right is an important component of Article 15 (1) (a) of the ICESCR and applies to and is owed to everyone.

Second, the findings have included the rights of persons belonging to minorities and Indigenous communities to conserve, promote, and develop their culture, including their cultural heritage and for the state to respect, protect, and recognise the heritage of persons belonging to minorities and indigenous communities and create favourable conditions. This further encompasses the right to lead a way of life associated with the use of land resources and traditional activities for both persons belonging to minorities and Indigenous communities, as articulated by the HRC in Articles 17 and 27. Directed more specifically at Indigenous peoples, this implies their right to maintain, control, protect, and develop cultural heritage, traditional knowledge, and expressions, as articulated by the CESCR under Article 15.

Third, the analysis has presented the right of Indigenous and local communities to effective consultations, the principle of free, prior and informed consent and participation in decision-making. For both right and obligations, the treaty bodies draw on Article 1 common to the ICESCR and ICCPR, Article 15 (1) (a) of the ICESCR and Article 27 of the ICCPR. The corresponding obligation is for states to ensure effective participation in decision making concerning cultural heritage matters, including obtaining their free, prior and informed consent.

The chapter has revealed that these rights can be grounded in several legal bases that highlight different nuances of the human rights provision. Access to cultural heritage, for example, can stem from cultural participation (Article 15 (1) (a) of the ICESCR), from professing and practicing one's religion as a person belonging to a minority (Article 27 of the ICCPR), from freedom of worship (Article 18 of the ICCPR), and from linguistic rights and using one's language as a person belonging to a minority (Article 27 of the ICCPR) and the right to education (Articles 13–14 of the ICESCR). The protection of a particular, traditional or Indigenous way of life constitutes another example. The pronouncement of the HRC reveal that this right it not only protected under Article 27 of the ICCPR but also under Article 17 of the ICCPR. While Article 17 protects the "same" way of life it does so through the prism of family, home and privacy, contributing to a holistic understanding of the notion.

As the counterpart to the analysis on treaty bodies, Chap. 5 has focused on the UNHRC and its mechanisms. It has similarly identified two underlying obligations: First, recognising the diversity of cultural heritage, as referred to in UNHRC resolutions and underlined in recommendations in the UPR and Special Procedures, and second, respecting and protecting cultural heritage and preventing its destruction. In the context of the latter, UNHRC resolutions do not only refer to the right to take part in cultural life (Article 15 of the ICESCR) but also to the prohibition of incitement to national, racial or religious hatred (Article 20 of the ICCPR). The Special Procedures also highlight the link between cultural and religious rights.

The analysis has emphasised three rights and obligations. These include, first, the ability to access and enjoy cultural heritage and the corresponding state obligation to ensure this right. By reference to the Special Procedures, UNHRC resolutions, and the rights contained in the UNDRIP, the analysis shows different approaches and (legal) bases of the right.

Second, participation rights and the free, prior and informed consent have been highlighted, including for the state to ensure these rights. The chapter has implied that participation rights and free, prior and informed consent are an integral part of the UNDRIP, albeit without a direct link to cultural heritage. In contrast, this link has been underlined in intersessional seminars of the UNHRC and the Special Procedures and occasionally addressed in the UPR.

Third, the analysis has identified the right of Indigenous peoples to maintain, control, protect and develop their cultural heritage, traditional knowledge, and (traditional) cultural expressions and for states to legally recognise and protect these rights and set up effective mechanism to enable these rights. The rights are grounded in the UNDRIP and reflect the Declaration's focus on positive measures for states to protect, by inter alia legally recognising rights, and by setting up effective mechanisms (together with Indigenous peoples) that enable these rights.

The analyses of Chaps 4 and 5 leave us with a puzzle of rights and obligations relating to cultural heritage in the international human rights law framework. In setting out to arrange these pieces in Chap. 6 and answering the research question, the thesis argues that although international human rights law does not provide for an explicit right to cultural heritage, rights and obligations vis-à-vis cultural heritage already form part of international human right law. These rights and obligations are important components of inter alia the right to take part in cultural life (Article 15 (1) (a) of the ICESCR), the right of persons belonging to minorities to enjoy their own culture, profess and practise their religion or use their language (Article 27 of the ICCPR), the protection from the unlawful interference with privacy, family, and home (Article 17 of the ICCPR) or the right to education (Articles 13–14 of the ICESCR). The pronouncements of the treaty bodies and especially the monitoring of human rights bodies suggest that these rights and obligations form part of the implementation of these rights.

Particularly in the case of Indigenous peoples, rights and obligations are especially fragmented, and the UNDRIP is arguably the most important instrument to supplement the rights and obligations in the international human rights framework.

Whilst human rights bodies refer to and rely on these bases to address rights and obligations vis-à-vis cultural heritage, they do not always make use of the language or the vocabulary of cultural heritage. In other words, cultural heritage is not always clearly articulated and often remains implicit in relevance, such as in the case of access to religious heritage and sacred sites or situations concerning participation rights and free, prior and informed consent. This paucity can result in an under-articulation or omission of the cultural heritage dimension and might be reflective that human rights bodies have not yet fully developed an approach to cultural heritage.[3] Whilst the HRC long established its approach to culture as a way of life and has continuously shown that forms of cultural heritage can fall under this notion, other ways of addressing cultural heritage in the framework of human rights bodies are still evolving.

Considering the research question, if and how human rights law can provide a legal basis for cultural heritage, this thesis argues that the international human rights framework can be considered as a complementary tool for the international protection of cultural heritage in at least two ways: First, it offers legal bases for rights and obligations relating to cultural heritage, as demonstrated above. Second, in the absence of enforcement tools in the World Heritage Convention and the Intangible Cultural Heritage Convention, the international human rights framework offers (soft) tools to strengthen rights and obligations vis-à-vis cultural heritage.

---

[3] See Jakubowski (2024), p. 508. *Jakubowski* argues this with regard to human rights law instruments and the jurisprudence of international courts and tribunals.

The international human rights framework indeed does not offer a distinct right to cultural heritage but equips individuals and persons belonging to a group with legal bases and tools to strengthen their claims and offer soft enforcement tools for the respective rights at the international level. Amid absence of an international human rights law court, especially the pronouncements of the human rights bodies can contribute to standard-setting at the international level and cross-fertilisation with the regional human rights commissions and courts, which can thereby exert a legal effect.

Inherent in this argument is the contribution of human rights bodies in articulating these rights and obligations through their pronouncements and instruments. This sheds light on the involvement of individual and independent experts working alone or together in the international human rights framework, and in contributing to the interpretation—and by extension—the development of international human rights law.

In spite of arguing that the international human rights framework and human rights bodies can contribute to strengthening rights and obligations vis-à-vis cultural heritage, the analysis has been cognisant of shortcomings in applying the international human rights framework to cultural heritage matters. This has inter alia included the individualistic nature of international human rights law and the absence of a primary focus on cultural heritage. More generally, the application of human rights law can be objected to on the grounds that it would exercise "competence over areas it was not originally meant to interfere with". [4] Although this argument can be met with the human rights systems' use of the "'evolutionary interpretation' of their instruments" to widen their general scope.[5]

The international nature of the thesis' investigation seeks to provide a broad overview but also presents limitations. As noted throughout the thesis, the analysis makes no claims of exhaustiveness owing to the selected sample and human rights bodies. This concluding chapter addresses two such limitations with the objective of illustrating the impetus for future research.

First, this thesis has focused on the international level. This has come with the advantage of providing a broad overview of states and by shedding light on the pronouncements and work of arguably the most important human rights bodies internationally. This focus has resulted in passing references to the regional human rights frameworks and human rights bodies. The regional human rights bodies, however, are particularly relevant for rights and obligations relating to

[4] Lixinski (2013a), p. 147.

[5] *ibid.*

cultural heritage.[6] This has been touched upon in the thesis in, for example, references to the recognition "that rights relating to cultural heritage are inherent in the right to participate in cultural life, as defined in the Universal Declaration of Human Rights" in the Faro Convention.[7] The European Court of Human Rights indeed has been found to recognise through dynamic interpretation rights that relate to culture and cultural heritage.[8] The analysis similarly highlighted the *Ogiek* case of the African Commission of Human and Peoples' Rights, which interpreted Article 17 of the Banjul Charter to go "beyond the duty, not to destroy or deliberately weaken minority groups, but requires respect for, and protection of, their cultural heritage essential to the group's identity".[9] Finally, it included references to the jurisprudence of the Inter-American Court of Human Rights, which can be highlighted for addressing different aspects of heritage protection in its judgements, particularly with regard to the rights of Indigenous peoples.[10] The examples are cursory insights and are worth exploring in much more depth. This argument can be reinforced by reference to the availability of human rights courts at the regional level that can strengthen the enforcement in a significant way.

The second limitation pertains to the selection of human rights bodies. Considering recent developments, it will be of utmost interest to further engage with the International Convention on the Elimination of All Forms of Racial Discrimination (ICERD) and the Committee on the Elimination of Racial Discrimination (CERD). The thesis has mentioned the HRC's reliance on the CERD's pronouncements in its views. In *Benito Oliveira Pereira et al. v Paraguay*, the HRC referred to the CERD's assertion that "the close ties of indigenous peoples to the land must be recognized and understood as the fundamental basis of their cultures,

[6] See e.g., Jakubowski (2024), p. 507 f.

[7] Framework Convention on the Value of Cultural Heritage for Society, Signed in Faro on 27 October 2005. Council of Europe, ETS No. 199. Entry into force: 1 June 2011, Art. 1 (a). The Faro Convention is ratified by 25 states of the Council of Europe: https://www.coe.int/en/web/conventions/full-list?module=signatures-by-treaty&treatynum=199. Although the Faro Convention contains a right to heritage, the Convention also clarifies that it does not create any enforceable rights. Art. 6 (c).

[8] See Jakubowski (2024), p. 507 f.

[9] African Commission on Human and Peoples' Rights v Kenya (Application No. 006/2012) [2017] AfCHPR 2 (26 May 2017), para. 179.

[10] *Case of the Yakye Axa Indigenous Community v. Paraguay*, Merits, Reparations, and Costs, Judgment. Series C No. 125 [2005], paras. 168, 176 (with regard to the right to life); *Case of the Mayagna (Sumo) Awas Tingni Community v. Nicaragua*, Merits, reparations and costs, IACHR Series C No 79 [2001], para. 148 (right to property and safeguarding Indigenous heritage)

spiritual life, integrity and economic survival."[11] Furthermore, in *Alisa Roy v Australia* the HRC noted "that the close ties of Indigenous Peoples to the land must be recognized and understood as the fundamental basis of their cultures" citing the CERD's finding that relied on regional jurisprudence.[12] The CERD has also focused on the issue of free, prior and informed consent in its pronouncements.[13]

In addition, access to cultural heritage has found mentioning in the ICJ's *Request for the Indication of Provisional Measures* in *Armenia v. Azerbaijan.*[14] In its request for provisional measures, Armenia asserts the right to access and enjoy historic, cultural and religious heritage, grounding them "in a possible interpretation" of the Convention on the Elimination of All Forms of Racial Discrimination.[15] The order of the ICJ does not repeat the right asserted by Armenia under the Convention but considered the plausible argument by the state party and requested Azerbaijan to "prevent and punish acts of vandalism and desecration affecting Armenian cultural heritage".[16] These development strongly indicate further research into as well as a critical engagement[17] with the pronouncements and application of the ICERD and, more generally, understanding cultural heritage through the lens of human rights bodies.

[11] See CCPR, Communication No. 2552/2015, *Benito Oliveira Pereira et al. v. Paraguay* (12 October 2021) UN Doc CCPR/C/132/D/2552/2015, para. 8.6, citing *Ågren et al. v. Sweden* (CERD/C/102/D/54/2013), para. 6.6, citing the *Case of the Mayagna (Sumo) Awas Tingni Community v. Nicaragua*, para. 149, and *Case of the Saramaka People v. Suriname*, judgment of 28 November 2007, para. 121.

[12] CCPR, *Communication No. 3585/2019, Alisa Roy v. Australia* (13 March 2024) UN Doc CCPR/C/137/D/3585/2019, para. 8.3.

[13] See Lenzerini (2014), p. 164.

[14] *Application of the International Convention on the Elimination of All Forms of Racial Discrimination (Armenia v. Azerbaijan), Order on provisional measures, [2021].*

[15] Specifically, International Convention on the Elimination of All Forms of Racial Discrimination (adopted 7 March 1966, entered into force 4 January 1969) 660 UNTS 195 (*ICERD*), Arts. 2 and 5. See, *Application of the International Convention on the Elimination of All Forms of Racial Discrimination (Armenia v. Azerbaijan), Order on provisional measures, [2021],* paras. 5, 46, 50.

[16] *Application of the International Convention on the Elimination of All Forms of Racial Discrimination (Armenia v. Azerbaijan), Order on provisional measures, [2021]*, para. 98 (1) (c). For an analysis of earlier jurisprudence at the ICJ with regard to the CERD and cultural heritage, see Zaręba (2024), p. 267.

[17] Very noteworthy in this respect is: *Application of the CERD (Armenia v. Azerbaijan),* Dissenting Opinion Judge Yusuf.

Reference List

von Arnauld, Andreas and Theilen, Jens T. (2020): Rhetoric of Rights: A Topical Perspective on the Functions of Claiming a 'Human Right to ...". In: Andreas von Arnauld, Kerstin von der Decke and Mart Susi (eds.): The Cambridge Handbook of New Human Rights. Cambridge (Cambridge University Press 2020), pp. 34–50.

Jakubowski, Andrzej (2024): Heritage and Human Rights. In: Lucas Lixinski and Lucie K. Morisset (eds.): The Routledge Handbook of Heritage and the Law. London (Routledge), pp. 500–516.

Lenzerini, Federico (2014): The Culturalization of Human Rights Law. Oxford (Oxford University Press).

Lixinski, Lucas (2013a): Intangible Cultural Heritage in International Law. Oxford (Oxford University Press).

Zaręba, Szymon (2024): The ICJ, Racial Discrimination, and the Protection of Cultural Heritage. In: Santander Art and Culture Law Review, Vol. 2 No. 10, pp. 263–282.

Bibliography

All webpages were last accessed on 31 January 2025.

Affolder, Natasha (2007): Democratising or Demonising the World Heritage Convention?. In: Victoria University of Wellington Law Review, Vol. 38, No. 2, pp. 341–362.

van Alebeek, Rosanne and Nollenkaemper, André (2012): The Legal Status of Decisions by Human Rights Treaty Bodies In National Law. In: Helen Keller and Geir Ulfstein (eds.): UN Human Rights Treaty Bodies. Law and Legitimacy. Cambridge (Cambridge University Press), pp. 356–413.

Alston, Philip (1984): Conjuring Up New Human Rights: A Proposal for Quality Control. In: American Journal of International Law, Vol. 78 No. 3, pp. 607–621.

von Arnauld, Andreas and Theilen, Jens T. (2020): Rhetoric of Rights: A Topical Perspective on the Functions of Claiming a 'Human Right to ...". In: Andreas von Arnauld, Kerstin von der Decke and Mart Susi (eds.): The Cambridge Handbook of New Human Rights. Cambridge (Cambridge University Press 2020), pp. 34–50.

Asamoah, Obed (1964): The Legal Effect of Resolutions of the General Assembly. In: Columbia Journal of Transnational Law, Vol. 3, No. 2, pp. 210–230.

Azaria, Danae (2020): The Legal Significance of Expert Treaty Bodies Pronouncements for the Purpose of the Interpretation of Treaties. In: International Community Law Review, Vol. 22, No.1, pp. 33–60.

Bassiouni, M. C. (1983). Reflections on Criminal Jurisdiction in International Protection of Cultural Property. *Syracuse Journal of International Law and Commerce, 10*(2), 281–322.

Battini, Stefano (2011): The Procedural Side of Legal Globalization: The Case of the World Heritage Convention. In: International Journal of Constitutional Law, Vol. 9, Issue 2, pp. 340–368.

Bertacchini, E., Liuzza, C., Meskell, L., & Saccone, D. (2016). The Politicization of UNESCO World Heritage Decision Making. *Public Choice, 167*, 95–129.

Bianchi, Andrea (2016): International Law Theories: An Inquiry into Different Ways of Thinking. Oxford (Oxford University Press).

Bielefeldt, Heiner (2011): Protecting and Implementing the Right to Freedom of Religion or Belief: Interview with Heiner Bielefeldt. In: Journal of Human Rights Practice, Vol. 3, Issue 3, pp. 229–244.

L. Hofmann, *Cultural Heritage through the Lens of Human Rights Bodies*, Juridicum – Bochumer Schriften zur Friedenssicherung und zum Humanitären Völkerrecht | Bochum Studies on the International Law of Peace and Armed Conflict 73, https://doi.org/10.1007/978-3-658-51203-3

Blake, Janet (2020b): The Preamble. In: Janet Blake and Lucas Lixinski (eds.): The 2003 UNESCO Intangible Heritage Convention: A Commentary. Oxford (Oxford University Press), pp. 19–35.

Blake, Janet (2020a): Safeguarding Intangible Cultural Heritage. In: Francesco Francioni and Ana Filipa Vrdoljak (eds.): The Oxford Handbook of International Cultural Heritage Law. Place of Publication (Oxford University Press), pp. 347–370.

Blake, Janet (2015): International Cultural Heritage Law. Oxford (Oxford University Press).

Blake, Janet (2011): Taking a Human Rights Approach to Cultural Heritage Protection. In: Heritage & Society, Vol. 4 No. 2, pp. 199–238.

Blake, Janet (2006): Commentary on the UNESCO 2003 Convention on the Safeguarding of the Intangible Cultural Heritage. Leicester (Institute of Art and Law).

Bódig, Mátyás (2016): Soft Law, Doctrinal Development, and the General Comments of the UN Committee on Economic, Social and Cultural Rights. In: Stéphanie Lagoutte, Thomas Gammeltoft-Hansen and John Cerone (eds.): Tracing the Roles of Soft Law in Human Rights. Oxford (Oxford University Press), pp. 69–88.

Borlini, Leonardo and Crema, Luigi (2020): The Legal Status of Decisions by Human Rights Treaty Bodies: Authoritative Interpretation or *mission éducatrice*?. In: Giuliana Ziccardi Capaldo (ed.): Global Community Yearbook of International Law and Jurisprudence. Oxford (Oxford University Press).

Brown, Nicholas E., Liuzza, Claudia and Meskell, Lynn (2019): The Politics of Peril: UNESCO's List of World Heritage in Danger. In: Journal of Field Archaeology, Vol. 44, No. 5, pp. 287–303.

Buzzini, Gionata P. and Condorelli, Luigi (2008): Art.11: List of World Heritage in Danger and Deletion of a Property from the World Heritage List. In Francesco Francioni and Federico Lenzerini (eds.) The 1972 World Heritage Convention: A Commentary. 2nd edition, Oxford (Oxford University Press).

Casini, Lorenzo (2018): Cultural Sites Between Nationhood and Mankind. In: Eyal Benvenisti and Georg Nolte (eds.): Community Interests Across International Law. Oxford (Oxford University Press), pp. 176–188.

Chenwi, Lilian (2009): Correcting the Historical Asymmetry between Rights: The Optional Protocol to the International Covenant on Economic, Social and Cultural Rights. In: African Human Rights Law Journal, Vol. 9, No.1, pp. 23–51.

Chinkin, C. (2022). Sources. In D. Moeckli, S. Shah, S. Sivakumaran, & D. Harris (Eds.), *International Human Rights Law* (4th ed., pp. 65–88). Oxford University Press.

Clapperton, Matthew, Jones, David Martin and Smith, M. L. R. (2017): Iconoclasm and Strategic Thought: Islamic State and Cultural Heritage in Iraq and Syria. In: International Affairs, Vol. 93, No. 5, pp. 1205–1231.

Coomans, Fons (2018): The UN Committee on Economic, Social, and Cultural Rights. In: Oberleitner, Gerd (ed): International Human Rights Institutions, Tribunals, and Courts: International Human Rights. Singapore (Springer).

von der Decken, Kerstin and Koch, Nikolaus (2020): Recognition of New Human Rights: Phases, Techniques and the Approach of 'Differentiated Traditionalism.' In: Andreas von Arnauld, Kerstin von der Decke and Mart Susi (eds.): The Cambridge Handbook of New Human Rights. Cambridge (Cambridge University Press 2020), pp. 7–20.

Dijkstal, Haydee J (2019): Destruction of Cultural Heritage before the ICC: The Influence of Human Rights on Reparations Proceedings for Victims and the Accused. In: Journal of International Criminal Justice, Vol. 17, Issue 2, pp. 391–412.

Disko, Stefan (2017): Indigenous Cultural Heritage in the Implementation of UNESCO's World Heritage Convention: Opportunities, Obstacles and Challenges. In: Indigenous Peoples' Cultural Heritage: Rights, Debates and Challenges. Leiden (Brill | Nijhoff), pp. 39–77.

Dörr, Oliver and Schmalenbach, Kirsten (2018): Vienna Convention on the Law of Treaties. A Commentary. 2nd edition, Berlin/Heidelberg (Springer).

Domínguez-Redondo, Elvira (2020): In Defense of Politicization of Human Rights: The UN Special Procedures. Oxford (Oxford University Press).

Dorough, Dalee Sambo and Wiessner, Siegfried (2020): Indigenous Peoples and Cultural Heritage. In: Francesco Francioni and Ana Filipa Vrdoljak (eds.): The Oxford Handbook of International Cultural Heritage Law. Oxford (Oxford University Press), pp. 407–430.

Donders, Yvonne (2020): Cultural Heritage and Human Rights. In: Francesco Francioni and Ana Filipa Vrdoljak (eds.): The Oxford Handbook of International Cultural Heritage Law. Oxford (Oxford University Press), pp. 379–406.

Donders, Yvonne (2018): Protection and Promotion of Cultural Heritage and Human Rights through International Treaties: Two Worlds of Difference. In: Charlotte Waelde, Catherine Cummings, Mathilde Pavis and Helena Enright (eds.): Research Handbook on Contemporary Intangible Cultural Heritage. Cheltenham/Northampton (Edward Elgar), pp. 54–77.

Donders, Yvonne (2017): Cultural Diversity and Cultural Identity in Human Rights. In: Andreas J. Wiesand, Kalliopi Chainoglou und Anna Sledzinska-Simon (eds.): Culture and Human Rights: The Wroclaw Commentaries. Berlin, Boston (De Gruyter), pp. 23–32.

Donders, Yvonne (2016): Foundations of Collective Cultural Rights in International Human Rights Law. In: Andrzej Jakubowski (ed.): Cultural Rights as Collective Rights. Leiden (Brill | Nijhoff), pp. 85–112.

Donders, Yvonne (2014): Cultural Pluralism in International Human Rights Law: The Role of Reservations. In: Ana Filipa Vrdoljak (ed.): The Cultural Dimension of Human Rights. Oxford (Oxford University Press), pp. 205–239.

Francioni, Francesco and Lenzerini, Federico (2023): The Future of the World Heritage Convention: Problems and Prospects. In: Francesco Francioni (ed.): The 1972 World Heritage Convention: A Commentary. 2nd edition, Oxford (Oxford University Press), pp. 421–428.

Francioni, Francesco (2023): The Preamble. In: Francesco Francioni (ed.): The 1972 World Heritage Convention—A Commentary. 2nd edition, Oxford (Oxford University Press), pp. 21–29.

Francioni, Francesco (2020a): Custom and General Principles of International Cultural Heritage Law. In: Francesco Francioni and Ana Filipa Vrdoljak (eds.): The Oxford Handbook of International Cultural Heritage Law. Oxford (Oxford University Press), pp. 531–550.

Francioni, Francesco (2020b): Art.2(1) Defining Intangible Cultural Heritage. In: Janet Blake and Lucas Lixinski (eds.): The 2003 UNESCO Intangible Heritage Convention: A Commentary. Oxford (Oxford University Press), pp. 48–57.

Francioni, Francesco (2020c): World Cultural Heritage. In: Francesco Francioni and Ana Filipa Vrdoljak (eds.): The Oxford Handbook of International Cultural Heritage Law. Oxford (Oxford University Press), pp. 250–271.

Francioni, Francesco and Lixinski, Lucas (2017): Opening the Toolbox of International Human Rights Law in the Safeguarding of Cultural Heritage. In: Andrea Durbach and Lucas Lixinski (eds.): Heritage, Culture and Rights: Challenging Legal Discourses. Oxford (Hart Publishing.

Francioni, Francesco (2013b): Plurality and Interaction of Legal Orders in the Enforcement of Cultural Heritage Law. In: Francesco Francioni and James Gordley (eds.): Enforcing International Cultural Heritage Law. Oxford (Oxford University Press), pp. 9–21.

Francioni, Francesco (2013a): Cultural Heritage. In: Max Planck Encyclopaedia of International Law.

Francioni, Francesco (2012): The Evolving Framework for the Protection of Cultural Heritage in International Law. In: Silvia Borelli and Federico Lenzerini (eds.): In Cultural Heritage, Cultural Rights, Cultural Diversity. Leiden (Brill | Nijhoff).

Francioni, Francesco (2011): The Human Dimension of International Cultural Heritage Law: An Introduction. In: European Journal of International Law, Vol. 22, Issue 1, pp. 9–16.

Francioni, Francesco (2008): Culture, Heritage and Human Rights: An Introduction. In: Francesco Francioni and Martin Scheinin (eds.): Cultural Human Rights. Leiden (Brill / Nijhoff, pp. 1–15.

Francioni, Francesco (2004): Beyond State Sovereignty: The Protection of Cultural Heritage as a Shared Interest in Humanity. In: Michigan Journal of International Law Vol. 25, No. 1, pp. 1209–1228.

Francioni, Francesco and Lenzerini, Federico (2003): The Destruction of the Buddhas of Bamiyan in International Law. In: European Journal of International Law, Vol. 14, No. 4, pp. 619–651.

Francioni, Francesco (1993): World Cultural Heritage List and National Sovereignty. In: Humanitäres Völkerrecht—Informationsschriften, Vol. 4, Issue 4, pp. 195–198.

Forman, Lisa (2016): Can Minimum Core Obligations Survive a Reasonableness Standard of Review under the Optional Protocol to the International Covenant on Economic, Social and Cultural Rights. In: Ottawa L. Rev., Vol. 47, No. 2, pp. 561–573.

Forrest, Craig (2010): International Law and the Protection of Cultural Heritage. Abingdon/ New York, NY (Routledge).

Frulli, Micaela (2020): International Criminal Law and the Protection of Cultural Heritage. In: Francesco Francioni and Ana Filipa Vrdoljak (eds.) The Oxford Handbook of International Cultural Heritage Law. Oxford (Oxford University Press), pp. 100–120.

Frulli, Micaela (2011): The Criminalization of Offences against Cultural Heritage in Times of Armed Conflict: The Quest for Consistency. In: European Journal of International Law, Volume 22, Issue 1, pp. 203–217.

Gerstenblith, P. (2008). Protecting Cultural Heritage in Armed Conflict: Looking Back, Looking Forward. *Cardozo Public Law & Policy Ethics Journal, 7*, 677–708.

Gilbert, Jérémie (2018): Natural Resources and Human Rights: An Appraisal. Oxford (Oxford University Press).

Gilbert, Jérémie (2017): Indigenous Peoples, Human Rights, and Cultural Heritage: Towards a Right to Cultural Integrity. In: Alexandra Xanthaki, Sanna Valkonen, Leena Heinämäki and Piia Nuorgam (eds.): Indigenous Peoples' Cultural Heritage: Rights, Debates and Challenges. Leiden (Brill), pp. 20–38.

Helmersen, Sondre Torp (2019): Finding 'the Most Highly Qualified Publicists': Lessons from the International Court of Justice. In: European Journal of International Law, Vol. 30, Issue 2, pp. 509–535.

Hølleland, Herdis, Hamman, Evan and Phelps, Jessica (2019): Naming, Shaming and Fire Alarms: The Compilation, Development and Use of the List of World Heritage in Danger. In: Transnational Environmental Law, Vol. 8, No. 1, pp. 35–57.

Hulme, Max H. (2016): Preambles in Treaty Interpretation. In: University of Pennsylvania Law Review, Vol. 164, Issue 5, pp. 1281–1343.

ILA. (2004). Final Report on the Impact of Findings of the United Nations Human Rights Treaty Bodies. *Int'l l. Ass'n Rep. Conf., 71*, 627.

ILA (2022): Participation in Global Cultural Heritage Governance, Lisbon Conference (2022), https://www.ila-hq.org/en_GB/documents/01-ila-participation-final-report-lisbonfinal

Jakubowski, Andrzej (2024): Heritage and Human Rights. In: Lucas Lixinski and Lucie K. Morisset (eds.): The Routledge Handbook of Heritage and the Law. London (Routledge), pp. 500–516.

Jakubowski et al. (2022): International Law Association Committee on Participation in Global Cultural Heritage Governance—Final Report (2022).

Jakubowski, Andrzej (2016): Cultural Heritage and the Collective Dimension of Cultural Rights in the Jurisprudence of the European Court of Human Rights. In: Jakubowski, Andrzej (ed.): Cultural Rights as Collective Rights. Leiden (Brill | Nijhoff), pp. 155–179.

Jokilehto, Jukka (2012): Human Rights and Cultural Heritage. Observations on the Recognition of Human Rights in the International Doctrine. In: International Journal of Heritage Studies, Vol. 18, No. 3, pp. 1–5.

Joseph, Sarah (2018): Committees: Human Rights Bodies. In: Max Planck Encyclopedias of International Law.

Kälin, Walter and Künzli, Jörg (2019). The Law of International Human Rights Protection. 2nd edition, Oxford (Oxford University Press).

Kalshoven, Frits (2005): The Protection of Cultural Property in the Event of Armed Conflict within the Framework of International Humanitarian Law. In: Museum International, Vol. 57, No. 4, pp. 61–70.

Keane, David (2004): The Failure to Protect Cultural Property in Wartime. In: DePaul Journal of Art, Technology & Intellectual Property Law, Vol. 14, Issue 1, pp. 1–38.

Keller, H., & Ulfstein, G. (2012). *UN Human Rights Treaty Bodies.* Cambridge (Cambridge University Press).

Keller, Helen and Grover, Leena (2012): General Comments of the Human Rights Committee and their Legitimacy. In: Helen Keller and Geir Ulfstein (eds.): UN Human Rights Treaty Bodies: Law and Legitimacy. Cambridge (Cambridge University Press), pp. 116–198.

Kothari, Miloon (2013): From Commission to the Council: Evolution of UN Charter Bodies. In: Dinah Shelton (ed.): The Oxford Handbook of International Human Rights Law. Oxford (Oxford University Press), pp. 587–620.

Kotzé, Louis J. (2018): In Search of a Right to a Healthy Environment in International Law: Jus Cogens Norms. In: John H. Knox and Ramin Pejan (eds.): The Human Right to a Healthy Environment. Cambridge (Cambridge University Press), pp. 136–54.

Kraak, Anne-Laura and Aykan, Bahar (2018): Editorial: The Possibilities and Limitations of Rights-Based Approaches to Heritage Practice. In: International Journal of Cultural Property, Vol. 25, No. 1, pp. 1–10.

Lafrenz Samuels, Kathryn and Platts, Ellen J. (2022): Global Climate Change and the UNESCO World Heritage Committee. In: International Journal of Cultural Property, Vol. 29, pp. 409–432.

Lemkin, Raphael (2008): Axis Rule in Occupied Europe: Laws of Occupation, Analysis of Government, Proposals for Redress. 2nd edition (first published 1944), Clark, NJ (Lawbook Exchange).

Lenzerini, Federico (2020): Intentional Destruction of Cultural Heritage. In: Francesco Francioni and Ana Filipa Vrdoljak (eds.) The Oxford Handbook of International Cultural Heritage Law. Oxford (Oxford University Press), pp. 75–99.

Lenzerini, Federico (2016b): The Safeguarding of Collective Cultural Rights through the Evolutionary Interpretation of Human Rights Treaties and Their Translation into Principles of Customary International Law. In: Andrzej Jakubowski (ed.): Cultural Rights as Collective Rights. Leiden (Brill I Nijhoff), pp. 133–153.

Lenzerini, Federico (2016a): Terrorism, Conflicts and the Responsibility to Protect Cultural Heritage. In: The International Spectator: Special Issue on the Responsibility to Protect, Vol. 51, No. 2, pp. 70–85.

Lenzerini, Federico (2015): Protecting the Tangible, Safeguarding the Intangible: A Same Conventional Model for Different Needs. In: Sabine von Schorlemer and Sylvia Maus (eds.): Climate Change as a Threat to Peace: Impacts on Cultural Heritage and Cultural Diversity. Frankfurt am Main (Peter Lang Group AG), pp. 141–160.

Lenzerini, Federico (2014): The Culturalization of Human Rights Law. Oxford (Oxford University Press).

Lenzerini, Federico (2013): Suppressing and Remedying Offences against Culture. In: Ana Filipa Vrdoljak (ed.): The Cultural Dimension of Human Rights. Oxford (Oxford University Press), pp. 240–272.

Lenzerini, Federico (2003): The UNESCO Declaration Concerning the Intentional Destruction of Cultural Heritage: One Step Forward and Two Steps Back. In: The Italian Yearbook of International Law, Vol. 3, Issue 1, pp. 131–145.

Litton, Sam (2011): The World Heritage "In Danger" Listing as a Taking. In: New York University Journal of International Law & Politics, Vol. 44, No. 1, pp. 219–265.

Lixinski, Lucas (2019): International Heritage Law for Communities: Exclusion and Re-Imagination. Oxford (Oxford University Press).

Lixinski, Lucas (2018b): Food as Heritage and Multi-Level International Legal Governance. In: International Journal of Cultural Property, Vol. 25, No. 4, pp. 469–490.

Lixinski, Lucas (2018a): Moving Cultures: Engaging Refugee and Migrant Cultural Rights in International Heritage Law. In: Indonesian Journal of International Law, Vol 16, No. 1, pp. 1–27.

Lixinski, Lucas and Schreiber, Hanna (2017): Guest Editorial, Intangible Cultural Heritage: Successes, Disappointments, and Challenges. In: Santander Art and Culture Law Review, Vol. 2, No. 4, pp. 17–20.

Lixinski, (2017): Heritage Listing as Self-determination. In: Andrea Durbach and Lucas Lixinski (eds.) Heritage, Culture and Rights: Challenging Legal Discourses. Oxford (Hart Publishing), pp. 227–249.

Lixinski, Lucas (2016): Intangible Cultural Heritage. In: Anna Sledzinska-Simon, Kalliopi Chainoglou and Andreas J Wiesand (eds.): Culture and Human Rights: The Wroclaw Commentaries. Berlin/Boston (De Gruyter), pp. 189–190.

Lixinski, Lucas (2015): Heritage Listing as a Tool for Advocacy: The Possibilities for Dissent, Contestation, and Emancipation in International Law Through International Cultural Heritage Law. In: Asian Journal of International Law, Vol. 5, No. 2, pp. 387–409.

Lixinski, Lucas (2013a): Intangible Cultural Heritage in International Law. Oxford (Oxford University Press).

Lixinski, Lucas (2013b): Luci E Ombre: The Bright and Dark Sides of International Heritage Law. In: The Italian Yearbook of International Law Online, Vol. 22, Issue 1, pp. 133–153.

Linxinski, Lucas (2011): Selecting Heritage: The Interplay of Art, Politics and Identity. In: European Journal of International Law, Vol. 22, No. 1, pp. 81–100.

Mechelm, Kerstin (2009): Treaty Bodies and the Interpretation of Human Rights. In: Vanderbilt Journal of Transnational Law, Vol. 42, Issue 3, pp. 905–947.

Medina Quiroga, Cecilia (2013): The Role of International Tribunals: Law-Making or Creative Interpretation?. In: Dinah Shelton (ed.): The Oxford Handbook of International Human Rights Law. Oxford (Oxford University Press), pp. 649–669.

Merryman, John Henry (1986): Two Ways of Thinking About Cultural Property. In: The American Journal of International Law, Vol. 80, Issue 4, pp. 831–853.

Merryman, John Henry (2005): Cultural Property Internationalism. In: International Journal of Cultural Property, Vol. 12, Issue 1, pp. 11–39.

Nafziger, James A. R. (2021): Reading: The World Court, the World Heritage, and the World of Non-state Actors. In: James A. R. Nafziger: Frontiers of Cultural Heritage Law. Leiden (Brill | Nijhoff), pp. 378–400.

Nafziger, James A. R., Robert Kirkwood Paterson and Alison Dundes Renteln (2010): Cultural Law: International, Comparative, and Indigenous. Cambridge (Cambridge University Press), pp. 1–99.

Neuman, Gerald L. (2016): Giving Meaning and Effect to Human Rights: The Contributions of Human Rights Committee Members. In: Human Rights Program Research Working Paper Series.

Novic, Elisa (2020): Remedies. In: Francesco Francioni and Ana Filipa Vrdoljak (eds.): The Oxford Handbook of International Cultural Heritage Law. Oxford (Oxford University Press), pp. 642–662.

Novic, Elisa (2016): The Concept of Cultural Genocide. Oxford (Oxford University Press).

O'Flaherty, Michael (2006): Concluding Observations of United Nations Human Rights Treaty Bodies. In: Human Rights Law Review, Vol. 6, No. 1, pp. 27–52.

O'Keefe, Roger (2007): The Protection of Cultural Property in Armed Conflict. Cambridge (Cambridge University Press).

O'Keefe, R. (2004). World Cultural Heritage: Obligations to The International Community as A Whole? *International and Comparative Law Quarterly, 53*, 189–209.

O'Keefe, Roger (1999): The Meaning of 'Cultural Property' Under the 1954 Hague Convention. In: Netherlands International Law Review, Vol. 46, No. 1, pp. 26–56.

Piccone, Ted (2010): Catalysts for Rights: The Unique Contribution of the UN's Independent Experts on Human Rights. Final Report of the Brookings Research Project on Strengthening U.N. Special Procedures. The Brookings Institution, pp. 1–90.

Pineschi, L. (2012): Cultural Diversity as a Human Right? General Comment No. 21 of the Committee on Economic, Social and Cultural Rights. Silvia Borelli and Federico Lenzerini (eds.): In Cultural Heritage, Cultural Rights, Cultural Diversity. Leiden (Brill | Nijhoff) pp. 27–53.

Pinto, Mónica (2018): Inquiry Procedures: United Nations Human Rights Bodies. In: Max Planck Encyclopedias of International Law.

Pinto, Mónica (2018): Special Procedures: Human Rights Council. Max Planck Encyclopedias of International Law.

Provost, René (2015): Interpretation in International Law as a Transcultural Project. In: Andrea Bianchi, Daniel Peat and Matthew Windsor (eds.): Interpretation in International Law. Cambridge (Cambridge University Press), pp. 290–308.

Rodley, Sir Nigel (2013): The Role and Impact of Treaty Bodies. In: Shelton, Dinah (ed.): The Oxford Handbook of International Human Rights Law. Oxford (Oxford University Press), pp. 621–648.

Rodley, Sir Nigel (2009): The United Nations Human Rights Council, Its Special Procedures, and Its Relationship with the Treaty Bodies: Complementarity or Competition?. In: Kevin Boyle (ed.): New Institutions for Human Rights Protection, Collected Courses of the Academy of European Law. Oxford (Oxford University Press), pp. 49–74.

Saul, B., Kinley, D., & Mowbray, J. (2014). *The International Covenant on Economic, Social and Cultural Rights: Commentary*. Oxford (Oxford University Press).

Schachter, Oscar (1991): International Law in Theory and Practice. Dordrecht (Martinus Nijhoff).

Scheinin, Martin (2017): The Art and Science of Interpretation in Human Rights Law. In: Bård A. Andreassen, Claire Methven O'Brien and Hans-Otto Sano (eds.): Research Methods in Human Rights: A Handbook. Cheltenham/Northampton (Edward Elgar), pp. 17–37.

Scheinin, Martin (2007): Access to Justice before International Human Rights Bodies: Reflections on the Practice of the UN Human Rights Committee and the European Court of Human Rights. In: Francesco Francioni (ed.): Access to Justice as a Human Right. Oxford (Oxford University Press), pp. 135–152.

Scheinin, Martin (1997): International Mechanisms and Procedures for Implementation. In: Raija Hanski Markku Suks (ed.): An Introduction to the International Protection of Human Rights: A Textbook. Turku (Institute for Human Rights).

Scovazzi, Tullio (2020): International Cultural Heritage Law: The Institutional Aspects. In: Francesco Francioni and Ana Filipa Vrdoljak (eds.): The Oxford Handbook of International Cultural Heritage Law, Oxford (Oxford University Press), pp. 737–759.

Seibert-Fohr, Anja and Weniger, Christine (2019): Compliance Monitoring under the International Covenant on Civil and Political Rights: To Comply, or not to Comply, that is the Question. https://papers.ssrn.com/sol3/papers.cfm?abstract_id=3337435.

Shelton, Dinah (2012): The Legal Status of Normative Pronouncements of Human Rights Treaty Bodies. In: Hestermeyer, Holger Paul, König, Doris, Matz-Lück, Nele, Röben, Volker, Seibert-Fohr, Anja, Stoll, Peter-Tobias, Vöneky, Silja (eds.): Coexistence, Cooperation and Solidarity—Liber Amicorum Rüdiger Wolfrum (vol. I). Martinus Nijhoff Publishers, pp. 553–575.

Silverman, Helaine and Fairchild Ruggles, D. (2007): Cultural Heritage and Human Rights. In. Helaine Silverman and D. Fairchild Ruggles (eds.): Cultural Heritage and Human Rights, New York, NY (Springer), pp. 3–29.

Sivakumaran, S. (2017). The Influence of Teachings of Publicists on the Development of International Law. *International and Comparative Law Quarterly, 66*, 1–37.

Smith, Laurajane (2006): Uses of Heritage. (Routledge).

Solntsev, Alexander (2024): The Growing Importance of Human Rights Treaty Bodies in Environmental Dispute Resolution. In: Christina Voigt and Caroline Foster (eds.): International Courts versus Non-Compliance Mechanisms: Comparative Advantages in Strengthening Treaty Implementation, Cambridge (Cambridge University Press), pp. 361–382.

Steiner, Lasse and Frey, Bruno S. (2011): Imbalance of World Heritage List: Did the UNESCO Strategy Work? University of Zurich, Economics Working Paper No. 14, https://ssrn.com/abstract=1807889 or https://doi.org/10.2139/ssrn.1807889.

Subedi, Surya P. (2014): Human Rights Experts in the United Nations: A Review of the Role of United Nations Special Procedures. In: Monika Ambrus, Karin Arts, Ellen Hey and Helena Raulus (eds.): The Role of 'Experts' in International and European Decision-Making Processes: Advisors, Decision Makers or Irrelevant Actors?. Cambridge (Cambridge University Press), pp. 241–262.

Taylor, Paul M. (2020): A Commentary on the International Covenant on Civil and Political Rights: The UN Human Rights Committee's Monitoring of ICCPR Rights. Cambridge (Cambridge University Press)

Thielbörger, Pierre (2020): Something Old, Something New, Something Borrowed and Something Blue: Lessons to Be Learned from the Oldest of the 'New' Rights—the Human Right to Water. In: Andreas von Arnauld, Kerstin von der Decke and Mart Susi (eds.): The Cambridge Handbook of New Human Rights. Cambridge (Cambridge University Press 2020), pp. 70–78.

Thielbörger, Pierre (2014): The Right(s) to Water: The Multi-Level Governance of a Unique Human Right. Berlin (Springer).

Tistounet, Eric (2020): The UN Human Rights Council—A Practical Anatomy. Cheltenham/ Northampton (Edward Elgar).

Tobin, J. (2010). Seeking to Persuade: A Constructive Approach to Human Rights Treaty Interpretation. *Harvard Human Rights Journal, 23*, 201–250.

Tomuschat, Christian (2019). Universal Periodic Review Procedure: Human Rights Council. In: Max Planck Encyclopedias of International Law.

Tomuschat, Christian (2003): Human Rights: Between Idealism and Realism. Oxford (Oxford University Press).

Tünsmeyer, Vanessa (2020): Bridging the Gap Between International Human Rights and International Cultural Heritage Law Instruments: A Functions Approach. In: Anne-Marie Carstens and Elizabeth Varner (eds.): Intersections in International Cultural Heritage Law. Oxford (Oxford University Press), pp. 319–342.

Ubertazzi, B. (2020). Art. 2(2) Manifesting Intangible Cultural Heritage. In J. Blake & L. Lixinski (Eds.), *The 2003 UNESCO Intangible Heritage Convention: A Commentary* (pp. 58–80). Oxford University Press.

Uerpmann-Wittzack, Robert (2018): Introduction: Cultural Heritage Law and the Quest for Human Identities. In: Evelyne Lagrange, Stefan Oeter and Robert Uerpmann-Wittzack (eds.): Cultural Heritage and International Law. Cham (Springer), pp. 1–11.

Vrdoljak, Ana Filipa and Francioni, Francesco (2020): Introduction. In: Ana Filipa Vrdoljak and Francesco Francioni (eds.): The Oxford Handbook of International Cultural Heritage Law. Oxford (Oxford University Press), pp. 1–10.

Vrdoljak, Ana Filipa and Meskell, Lynn (2020): Intellectual Cooperation Organisation, UNESCO, and the Culture Conventions. In: Francesco Francioni and Ana Filipa Vrdoljak (eds.): The Oxford Handbook of International Cultural Heritage Law, Oxford (Oxford University Press), pp. 13–40.

Vrdoljak, Ana Filipa (2018): Indigenous Peoples, World Heritage, and Human Rights. In: International Journal of Cultural Property, Vol. 25, No. 3, pp. 245–281.

Vrdoljak, Ana Filipa (2016): Standing and Collective Cultural Rights, in: Andrej Jakubowski (ed.) Cultural Rights as Collective Rights: An International Law Perspective. Leiden (Brill Publishers), Available at SSRN: https://ssrn.com/abstract=2518054, pp. 1–16.

Vrdoljak, Ana Filipa (2014): Human Rights and Cultural Heritage in International Law. In: Federico Lenzerini and Ana Filipa Vrdoljak (eds.): International Law for Common Goods: Normative Perspectives on Human Rights, Culture and Nature. Oxford (Hart Publishing), Available at SSRN: https://ssrn.com/abstract=2200186, pp. 1–39.

Vrdoljak, Ana Filipa (2013a): Introduction. In: Ana Filipa Vrdoljak (ed.): The Cultural Dimension of Human Rights. Oxford (Oxford University Press), pp. 1–10.

Vrdoljak, Ana Filipa (2013b): Liberty, Equality, Diversity: States, Cultures, and International Law. In: Ana Filipa Vrdoljak (ed.) The Cultural Dimension of Human Rights. Oxford (Oxford University Press), pp. 26–70.

Vrdoljak, Ana Filipa (2007): Intentional Destruction of Cultural Heritage and International Law. In: Kalliopi Koufa (ed.): Multiculturalism and International Law, XXXV Thesaurus Acroasium. Thessaloniki (Sakkoulas Publications), pp. 377–396.

Wiessner, Siegfried (2011): The Cultural Rights of Indigenous Peoples: Achievements and Continuing Challenges. In: European Journal of International Law, Vol. 22, Issue 1, pp. 121–140.

De Witte, Bruno (2020): Language as Cultural Heritage. In: Francesco Francioni and Ana Filipa Vrdoljak (eds.): The Oxford Handbook of International Cultural Heritage Law. Oxford (Oxford University Press), pp. 371–378.

Xanthaki, Alexandra (2017): International Instruments on Cultural Heritage: Tales of Fragmentation. In: Alexandra Xanthaki, Sanna Valkonen, Leena Heinämäki and Piia Kristiina Nuorgam (eds.): Indigenous Peoples' Cultural Heritage: Rights, Debates, Challenges. Leiden (Brill | Nijhoff.), pp. 1–19.

Zaręba, Szymon (2024): The ICJ, Racial Discrimination, and the Protection of Cultural Heritage. In: Santander Art and Culture Law Review, Vol. 2 No. 10, pp. 263–282.

Websites and Online Sources

BBC (2018): Taj Mahal colour change worries India Supreme Court, https://www.bbc.com/news/world-asia-india-43967685.

Cambridge University Press & Assessment (2025): Dictionaries, https://dictionary.cambridge.org/dictionary/english/protection.

Casla, Koldo (2018): Supreme Court of Spain: UN Treaty Body Individual Decisions Are Legally Binding, https://www.ejiltalk.org/supreme-court-of-spain-un-treaty-body-individual-decisions-are-legally-binding/.

Citroni, Gabriella (2015): The Human Rights Committee and its Role in Interpreting the International Covenant on Civil and Political Rights vis-à-vis States Parties, https://www.ejiltalk.org/the-human-rights-committee-and-its-role-in-interpreting-the-international-covenant-on-civil-and-political-rights-vis-a-vis-states-parties/.

The Chicago Manual of Style Online (2017): Capitalization (https://www.chicagomanualofstyle.org/qanda/data/faq/topics/Capitalization/faq0106.html)

Daley, Jason (2018): Indian Supreme Court Orders Government to Restore the Taj Mahal—or Demolish It, https://www.smithsonianmag.com/smart-news/indian-supreme-court-restore-taj-mahal-or-demolish-it-180969668.

Fletcher, Pascal (2012): Timbuktu tomb destroyers pulverise Islam's history, https://www.reuters.com/article/business/timbuktu-tomb-destroyers-pulverise-islams-history-idUSLNE86202H/.

Government of British Columbia (2024): Capitalization and formatting of Indigenous terms (https://www2.gov.bc.ca/gov/content/governments/services-for-government/service-experience-digital-delivery/web-content-development-guides/web-style-guide/writing-guide-for-indigenous-content/capitalization-and-formatting-of-indigenous-terms).

Government of the United States of America (2008): Comments of the United States on the Human Rights Committee's 'Draft General Comment 33: The Obligations of States Parties under the Optional Protocol to the International Covenant on Civil and Political Rights, https://2009-2017.state.gov/documents/organization/138851.pdf.

The Guardian (2021): 'The lobbying push that killed off a fight to save the Great Barrier Reef', https://www.theguardian.com/news/audio/2021/jul/29/the-lobbying-push-that-killed-off-a-fight-to-save-the-great-barrier-reef-podcast.

Harrington, Joanna (2015): The Human Rights Committee, Treaty Interpretation, and the Last Word, https://www.ejiltalk.org/the-human-rights-committee-treaty-interpretation-and-the-last-word/.

Hobbs, Harry (2016): The Uluru decision highlights what's so wrong with Indigenous policy, https://www.theguardian.com/commentisfree/2016/apr/21/the-uluru-decision-highlights-whats-so-wrong-with-indigenous-policy.

Hofmann, Laura (2024): The Great Barrier Reef and the List of World Heritage in Danger, https://voelkerrechtsblog.org/the-great-barrier-reef-and-the-list-of-world-heritage-in-danger/.

ICRC (2025): Rule 40: Respect for Cultural Property, https://ihl-databases.icrc.org/en/customary-ihl/v1/rule40.

ILO (2025): Ratifications of C169—Indigenous and Tribal Peoples Convention, 1989 (No. 169), https://normlex.ilo.org/dyn/nrmlx_en/f?p=1000:11300:0::NO:11300:P11300_INSTRUMENT_ID:312314.

Lixinski, Lucas (2016): ESIL—International Human Rights Law Symposium: International Human Rights Law and International Cultural Heritage Law: Cooperation, Conflict or Cooption?, https://www.ejiltalk.org/esil-international-human-rights-law-symposium-international-human-rights-law-and-international-cultural-heritage-law-cooperation-conflict-or-cooption/.

Maurel, Chloé (2017): Whose World Heritage? The problem with UNESCO's famous list, https://www.equaltimes.org/whose-world-heritage-the-problem?lang=en#.WZmqN62B33B.

Mosby (2020): Here is why we are taking the Australian government to the UN over its inaction on climate change, https://www.theguardian.com/commentisfree/2020/aug/14/here-is-why-we-are-taking-the-australian-government-to-the-un-over-its-inaction-on-climate-change.

Murphy (2019): Torres Strait Islanders take climate change complaint to the United Nations, https://www.theguardian.com/australia-news/2019/may/13/torres-strait-islanders-take-climate-change-complaint-to-the-united-nations.

OHCHR (2025): Communication report and search, https://spcommreports.ohchr.org/TmSearch/Mandates?m=19.

OHCHR (2025): Consultation and free, prior and informed consent (FPIC), https://www.ohchr.org/en/indigenous-peoples/consultation-and-free-prior-and-informed-consent-fpic.

OHCHR (2025): Country Visits, https://www.ohchr.org/en/special-procedures/sr-freedom-of-assembly-and-association/country-visits.

OHCHR (2025): Days of general discussion, https://www.ohchr.org/en/treaty-bodies/cescr/days-general-discussion.

OHCHR (2025): International standards: Special Rapporteur in the field of cultural rights, https://www.ohchr.org/en/special-procedures/sr-cultural-rights/international-standards.

OHCHR (2025): OHCHR Juris Database, https://juris.ohchr.org.

OHCHR (2025): What are communications, https://www.ohchr.org/en/special-procedures-human-rights-council/what-are-communications.

OHCHR (2015): Statement by Mr. Zeid Ra'ad Al Hussein, United Nations High Commissioner for Human Rights, at the International Law Commission, https://www.ohchr.org/EN/NewsEvents/Pages/DisplayNews.aspx?NewsID=16254&LangID=E.

OHCHR (2012) A very dark future for the local populations in Northern Mali', warn UN experts, https://www.ohchr.org/en/press-releases/2012/07/very-dark-future-local-populations-northern-mali-warn-un-experts.

Parks Australia (Commonwealth of Australia) (2019): Uluṟu climb closure, https://parksaustralia.gov.au/uluru/discover/culture/uluru-climb/.

Readfearn, Graham (2023): Unesco recommends against Great Barrier Reef 'in danger' listing but Australia warned more action needed, https://www.theguardian.com/environment/2023/aug/01/climate-crisis-great-barrier-reef-unesco-australia.

Readfearn, Graham (2021): World Heritage Committee agrees not to place Great Barrier Reef on 'in danger' list, https://www.theguardian.com/environment/2021/jul/23/world-heritage-committee-agrees-not-to-place-great-barrier-reef-on-in-danger-list.

Readfearn, Graham (2021): 'Low point' in world heritage committee history as politics 'tramples' human rights of the Karen people, https://www.theguardian.com/environment/2021/aug/06/low-point-in-world-heritage-committee-history-as-politics-tramples-human-rights-of-the-karen-people.

Somers Cocks, Anna (2023): For the second time, Venice escapes listing as at risk, but now Unesco is the enemy, https://www.theartnewspaper.com/2023/09/15/for-the-second-time-venice-escapes-listing-as-at-risk-but-now-unesco-is-the-enemy.

Somers Cocks, Anna (2021): Seven sites, including Venice, may be added to UNESCO in Danger list, https://www.theartnewspaper.com/news/seven-sites-may-be-added-to-unesco-in-danger-list.

Somers Cocks, Anna (2021): Italy bans cruise ships from the Venice lagoon from 1 August, https://www.theartnewspaper.com/2021/07/14/italy-bans-cruise-ships-from-the-venice-lagoon-from-1-august.

Somers Cocks, Anna (2021): Venice is not in danger, according to UNESCO, https://www.theartnewspaper.com/2021/07/22/venice-is-not-in-danger-according-to-unesco.

Somers Cocks, Anna (2021): The IPCC's predictions of sea-level rise are why Italy should beg UNESCO to put Venice on its at risk list, https://www.theartnewspaper.com/comment/the-ipcc-s-predictions-of-sea-level-rise-are-why-italy-should-beg-unesco-to-put-venice-on-its-at-risk-list.

UN Department of Economic and Social Affairs (2015): United Nations Declaration On The Rights Of Indigenous Peoples, https://social.desa.un.org/issues/indigenous-peoples/united-nations-declaration-on-the-rights-of-indigenous-peoples.

UN Digital Library (2025): United Nations Declaration on the Rights of Indigenous Peoples: resolution / adopted by the General Assembly, https://digitallibrary.un.org/record/609197?ln=en.

UNESCO (2025): World Heritage in Danger. https://whc.unesco.org/en/158.

UNESCO (2025): World Heritage List: Delisted, https://whc.unesco.org/en/list/?delisted=1.

UNESCO (2025): World Heritage List: Mixed Properties, https://whc.unesco.org/en/list/?search=&type=mixed&components=0

UN Statistics Division (2025): Methodology: Standard country or area codes for statistical use, https://unstats.un.org/unsd/methodology/m49/.

UNESCO (2025): Uluru-Kata Tjuta National Park, https://whc.unesco.org/en/list/447/.

UNESCO (2025): Uluru-Kata Tjuta National Park: Documents, whc.unesco.org/en/list/447/documents/.

UNESCO (2025): Browse the Lists of Intangible Cultural Heritage and the Register of good safeguarding practices: Language, https://ich.unesco.org/en/lists?text=language&multinational=3#tabs.

UNHRC (2025): Documentation by Country, https://www.ohchr.org/en/hr-bodies/upr/documentation.

UNHRC (2025): Human Rights Council Complaint Procedure, https://www.ohchr.org/en/hr-bodies/hrc/complaint-procedure/hrc-complaint-procedure-index.

UN Human Rights Treaty Body Database (20205): UN Treaty Body Database, https://tbinternet.ohchr.org/_layouts/15/TreatyBodyExternal/Home.aspx.

U.S. Department of State (2014): Threats to Cultural Heritage in Iraq and Syria: Fact Sheet, https://2009-2017.state.gov/r/pa/prs/ps/2014/09/232028.htm.

UN News (2022): Why the UN General Assembly must back the right to a healthy environment, https://news.un.org/en/story/2022/07/1123142.

List of Treaties and Documents

Multilateral Treaties and Instruments

Charter of the United Nations (adopted 26 June 1945, entered into force 24 October 1945) 1 UNTS XVI (*UN Charter*).

Convention (No. 169) Concerning Indigenous and Tribal Peoples in Independent Countries (adopted 27 June 1989, entered into force 5 December 1991) 1650 UNTS (*ILO Convention No. 169*)

Convention Concerning the Protection of the World Cultural and Natural Heritage (adopted 16 November 1972, entered into force 17 December 1975) 1037 UNTS 151(*World Heritage Convention*)

Convention (II) with Respect to the Laws and Customs of War on Land and its annex: Regulations concerning the Laws and Customs of War on Land. The Hague, 29 July 1899.

Convention (IV) Respecting the Laws and Customs of War on Land (adopted 18 October 1907, entered into force 26 January 1910) 205 CTS 277 (*Hague Convention IV*).

Convention on the Means of Prohibiting and Preventing the Illicit Import, Export and Transfer of Ownership of Cultural Property (adopted 14 November 1970, entered into force 24 April 1972) 823 UNTS 231 (*1970 Convention*)

Convention for the Protection of Cultural Property in the Event of Armed Conflict (adopted 14 October 1954, entered into force 7 August 1956) 249 UNTS 215 (*1954 Hague Convention*)

Convention for the Safeguarding of the Intangible Cultural Heritage (adopted 17 October 2003, entered into force 20 April 2006) 2368 UNTS 3 (*Intangible Cultural Heritage Convention*)

Convention on the Protection of the Underwater Cultural Heritage (adopted 2 November 2001, entered into force 2 January 2009) 2562UNTS (*Underwater Cultural Heritage Convention*)

Framework Convention on the Value of Cultural Heritage for Society, Signed in Faro on 27 October 2005. Council of Europe, ETS No. 199. Entry into force: 1 June 2011 (*Faro Convention*).

International Covenant on Civil and Political Rights (adopted 16 Dec 1966, entered into force 23 Mar 1976) 999 UNTS 171 (*ICCPR*).

International Covenant on Economic, Social and Cultural Rights (adopted 16 Dec 1966, entered into force 3 Jan 1976) 993 UNTS 3 (*ICESCR*).

International Convention on the Elimination of All Forms of Racial Discrimination (adopted 7 March 1966, entered into force 4 January 1969) 660 UNTS 195 (*ICERD*)

Optional Protocol to the International Covenant on Economic, Social and Cultural Rights (adopted 10 December 2008, entered into force 5 May 2013) 2922 UNTS 29 (*Optional Protocol to the ICESCR*).

Optional Protocol to the International Covenant on Civil and Political Rights (adopted 16 December 1966, entered into force 23 March 1976) 999 UNTS 171 (Optional Protocol to the ICCPR

Protocol Additional to the Geneva Conventions of 12 August 1949, and Relating to the Protection of Victims of International Armed Conflicts (adopted 8 June 1977, entered into force 7 Dec 1978) 1125 UNTS 3 (*AP I*).

Protocol Additional to the Geneva Conventions of 12 August 1949, and Relating to the Protection of Victims of Non-International Armed Conflicts (adopted 8 June 1977, entered into force 7 Dec 1978) 1125 UNTS 609 (*AP II*).
Regulations Respecting the Laws and Customs of War on Land, Annex to Convention (IV) Respecting the Laws and Customs of War on Land (adopted 18 Oct 1907, entered into force 26 Jan 1910) 205 CTS 277 (*Hague Regulations*).
Rome Statute of the International Criminal Court (adopted 17 July 1998, entered into force 1 July 2002) 2187 UNTS 3 (*ICC Statute*).
Second Protocol to The Hague Convention of 1954 for the Protection of Cultural Property in the Event of Armed Conflict (adopted 26 March 1999, entered into force 9 March 2004) 2253 UNTS 172 (*Second Protocol*)
Statute of the International Court of Justice (18th April 1946) 33 UNTS 993
Treaty on the Protection of Artistic and Scientific Institutions and Historic Monuments (Roerich Pact) (adopted 15 April 1935, entered into force 26 August 1935)
Vienna Convention on the Law of Treaties (adopted 23 May 1969, entered into force 23 Jan 1980) 1155 UNTS 331 (*VCLT*).

Domestic Legislation and Domestic Documents

Executive Order 13175 of November 6, 2000, Consultation and Coordination With Indian Tribal, The White House; Memorandum on Uniform Standards for Tribal Consultation, November 30, 2022, https://www.whitehouse.gov/briefing-room/presidential-actions/2022/11/30/memorandum-on-uniform-standards-for-tribal-consultation/.
General Orders No 100: Instructions for the Government of Armies of the United States in the Field (prepared by Francis Lieber) (signed 24 Apr 1863) (*Lieber Code*).
Mission permanente de la Libye auprès de l'Office des Nations Unies à Genève et des autres organisations internationales en Suisse, No. 284.12, 7 September 2012. (In response to UA Cultural rights (2009) G/SO 214 (67–17) Assembly & Association (2010–1) G/SO 214 (56–23) G/SO 214 (107–9) Minorities (2005–4) G/SO 214 (53–24) LBY 2/2012, 5 September 2012).
Permanent Mission of the Republic of Armenia, Information provided by the Government of the Republic of Armenia in response to the questions contained in letter AL ARM 2/2021, dated October 29, 2021, of the UN Special Rapporteur in the field of cultural rights, 2203/1969/2021, 28 December 2021.
Tribunal Supremo, Sala de lo Contencioso-Administrativo Sección Cuarta, Sentencia núm. 1263/2018, Madrid, a 17 de julio de 2018.

List of Cases

International Court of Justice

Ahmadou Sadio Diallo (Guinea v DRC) (Merits, Judgment) [2010] ICJ Rep 639.

Application of the Convention on the Prevention and Punishment of the Crime of Genocide (Bosnia and Herzegovina v Serbia and Montenegro) [2007] ICJ Rep 43.

Application of the Convention on the Prevention and Punishment of the Crime of Genocide (Croatia v Serbia) [2015] ICJ Rep 3.

Application of the International Convention on the Elimination of All Forms of Racial Discrimination (Armenia v. Azerbaijan), Order on provisional measures, [2021].

Application of the International Convention on the Elimination of All Forms of Racial Discrimination (Armenia v. Azerbaijan), Order on provisional measures, [2021] (*Dissenting Opinion Judge Yusuf*).

Legal Consequences of the Construction of a Wall in the Occupied Palestinian Territory (Advisory Opinion) [2004] ICJ Rep 136.

Legality of the Threat or Use of Nuclear Weapons (Advisory Opinion) [1996] ICJ Rep 226.

Request for Interpretation of the Judgement of 15 June 1962 in the Case Concerning the Temple of Preah Vihear (Cambodia v. Thailand) (judgement) [2013] ICJ Rep 281.

Request for Interpretation of the Judgment of 15 June 1962 in the Case Concerning the Temple of Preah Vihear (Cambodia v. Thailand) (provisional measures) [2011] ICJ Rep 537 (*separate opinion of Judge Cançado Trindade*).

South-West Africa—Voting Procedure (Advisory Opinion) [1955] ICJ Rep 67 (Separate Opinion of Judge Lauterpacht).

ICTY

Prosecutor v. Blaškić (Trial Judgement) IT-95-14 (3 March 2000)

Prosecutor v Dusko Tadić aka 'Dule' (Jurisdiction) ICTY-94-1 (2 Oct 1995)

Prosecutor v. Kordic & Cerkez (Judgement) IT-95-14/2 (26 February 2011)

Prosecutor v. Krstić (Trial Judgement) IT-98-33 (2 August 2001)

Prosecutor v. Kupreškić et al. (Trial Judgement) IT- -95-16 (14 January 2000)

Prosecutor v. Miodrag Jokić (Sentencing Judgement) IT-01-42/1-S (18 March 2004)

Prosecutor v. Pavle Strugar (Trial Judgment) IT-01-42 (31 Jan 2005)

ICC

Prosecutor v. Ahmad Al Faqi Al Mahdi (Reparations Order) ICC-01/12-01/15 (17 August 2017)

Prosecutor v. Ahmad Al Faqi Al Mahdi (Judgment and Sentence) ICC-01/12-01/15 (27 September 2016)

Regional Human Rights Courts and Commissions

Sargsyan v. Azerbaijan, Judgment, App. No. 40167/06 (Eur. Ct. H.R., 2015)
The Environment and Human Rights, IACHR, Advisory Opinion OC-23/17, 15 November 2017.
Case of the Yakye Axa Indigenous Community v. Paraguay, Merits, Reparations, and Costs, Judgment. Series C No. 125 [2005]
Case of the Mayagna (Sumo) Awas Tingni Community v. Nicaragua, Merits, reparations and costs, IACHR Series C No 79 [2001]
Case of the 'Street Children' (Villagrán-Morales et al.) v. Guatemala, Merits, Judgment. Series C No. 63 [1999]
African Commission on Human and Peoples' Rights v Kenya (Application No. 006/2012) [2017] AfCHPR 2 (26 May 2017)
Centre for Minority Rights Development and Minority Rights Group International and Centre on Housing Rights and Evictions (intervening) (on behalf of Endorois Welfare Council) v Kenya, (Communication 276/2003), (ACHPR 2009), November 2009

United Nations Documents

United Nations General Assembly

UNGA Res 79/133 (12 December 2024) UN Doc A/RES/79/133.
UNGA Res 76/300 (28 July 2022) UN Doc A/RES/76/300.
UNGA 'Development and cultural rights: the principles, Report of the Special Rapporteur in the field of cultural rights, Alexandra Xanthaki' (15 August 2022) UN Doc A/77/290.
UNGA 'Report of the Special Rapporteur in the field of cultural rights, Karima Bennoune' (19 July 2021) UN Doc A/76/178.
UNGA 'Report of the Special Rapporteur in the Field of Cultural Rights, Karima Bennoune' (10 August 2020) UN Doc A/75/298.
UNGA 'Report of the Special Rapporteur in the field of cultural rights' (30 July 2019) UN Doc A/74/255.
UNGA 'Report of the Special Rapporteur in the field of cultural rights' (9 August 2016) UN Doc A/71/317.
UNGA 'Report of the Special Rapporteur in the field of cultural rights' (8 August 2014) UN Doc A/69/286.
UNGA 'Report of the Special Rapporteur in the field of cultural rights' (4 August 2014) UN Doc A/70/279.
UNGA 'Report of the Special Rapporteur in the field of cultural rights' (9 August 2013) UN Doc A/68/296, para. 7.
UNGA 'Report by the Special Rapporteur in the field of cultural rights' (10 August 2012) UN Doc A/67/287.
UNGA Res 64/292 (3 August 2010) UN Doc A/RES/64/292.
UNGA Res 61/295 'United Nations Declaration on the Rights of Indigenous Peoples' (13 September 2007) UN Doc A/Res/61/295.

UNGA Res 60/251 (3 April 2006) UN Doc A/RES/60/251.
UNGA Res 55/254 (31 May 2001) UN Doc A/RES/55/254.
UNGA 'Interim report of the Special Rapporteur of the Commission on Human Rights on the elimination of all forms of intolerance and of discrimination based on religion or belief' (31 July 2001) UN Doc A/56/253.
UNGA 'Draft International Covenants on Human Rights' (1 July 1995) UN Doc A/2929.
UNGA Res 47/135 'Declaration on the Rights of Persons Belonging to National or Ethnic, Religious and Linguistic Minorities' (18 December 1992) UN Doc A/RES/47/13.
UNGA General Assembly 'Third Committee' (31 October 1957) UN Doc, A/C.3/SR.797.
UNGA Res 217A(III) (10 December 1948) UN Doc. A/810, 71.
UNGA Sixth Committee 'Genocide: Draft Convention and Report of the Economic and Social Council' (3 December 1948) UN Doc A/760.

United Nations Security Council

UNSC Res 2697 (15 September 2023) UN Doc S/RES/2697.
UNSC Res 2617 (30 December 2021) UN Doc S/RES/2617.
UNSC Res 2610 (17 December 2021) UN Doc S/RES/2610.
UNSC Res 2347 (24 March 2017) UN Doc S/Res/2347.
UNSC Res 249 (20 November 2015) UN Doc S/Res/2249.
UNSC Res 2199 (12 February 2015) UN Doc S/Res/2199.
UNSC Res 2170 (15 August 2014) UN Doc S/Res/2170.
UNSC Res 2085 (20 December 2012) UN Doc S/Res/2085.
UNSC Re 2071 (12 October 2012) UN Doc S/RES/2071.
UNSC Res 1483 (22 May 2003) UN Doc S/RES/1483.
UNSC Res 1267 (15 October 1999) UN Doc S/RES/1267 (1999).
UNSC Res 827 (25 May 1993) UN Doc S/RES/827.
Statute of the International Tribunal for the Former Yugoslavia, (as amended on 17 May 2002), UN Security Council, 25 May 1993.

Human Rights Council

UNHRC Res 54/1 (11 October 2023) UN Doc A/HRC/RES/54/1.
UNHRC Res 52/18 (3 April 2023) UN Doc A/HRC/RES/52/18.
UNHRC Res 49/7 (31 March 2022) UN Doc A/HRC/RES/49/7.
UNHRC Res 49/9 (31 March 2022) UN Doc A/HRC/RES/49/9.
UNHRC Res 76/214 (7 January 2022) UN Doc A/RES/76/214.
UNHRC Res 48/13 (18 October 2021) UN Doc A/HRC/RES/48/13.
UNHRC Res 37/17 (9 April 2018) UN Doc A/HRC/RES/37/17.
UNHRC Res 37/29 (9 April 2018) UN Doc A/HRC/RES/37/29.
UNHRC Res 37/12 (22 March 2018) UN Doc A/HRC/RES/37/12.
UNHRC Res 34/26 (5 April 2017) UN Doc A/HRC/RES/34/26.
UNHRC Res 33/20 (6 October 2016) UN Doc A/HRC/RES/33/20.

UNHRC Res 32/25 (15 July 2016) UN Doc A/HRC/RES/32/25.
UNHRC Res 31/27 (20 April 2016) UN Doc A/HRC/RES/31/27.
UNHRC Res 28/9 (10 April 2015) UN Doc A/HRC/RES/28/9.
UNHRC Res 25/15 (15 April 2014) UN Doc A/HRC/RES/25/15.
UNHRC Res 21/25 (19 October 2012) UN Doc A/HRC/RES/21/25.
UNHRC Res 17/15 (14 July 2011) UN Doc A/HRC/RES/17/15.
UNHRC Res 15/9 (6 October 2010) UN Doc A/HRC/RES/15/9.
UNHRC Res 14/11 (23 June 2010) UN Doc A/HRC/RES/14/11.
UNHRC Res 14/9 (23 June 2010) UN Doc A/HRC/RES/14/9.
UNHRC Res 10/23 (26 March 2009) UN Doc A/HRC/RES/10/23.
UNHRC Res 7/22 (28 March 2008) UN Doc A/HRC/RES/7/22.
UNHRC Res 6/37 (14 December 2007) UN Doc A/HRC/RES/6/37.
UNHRC Res 6/11 (28 September 2007) UN Doc A/HRC/RES/6/11.
UNHRC Res 6/1 (27 September 2007) UN Doc A/HRC/RES/6/1.
UNHRC Res 5/2 (18 June 2007) UN Doc A/HRC/RES/5/2.
UNHRC, Res 5/1 (18 June 2007) UN Doc A/HRC/RES/5/1.
UNHRC Res 2001/13 (18 April 2001).
UNHRC 'Report of the Special Rapporteur in the field of cultural rights, Farida Shaheed, Copyright policy and the right to science and culture' (21 December 2024) UN Doc A/HRC/28/57.
UNHRC 'Report of the Special Rapporteur in the field of cultural rights' (2 February 2023) UN Doc A/HRC/52/35.
UNHRC 'Cultural rights: an empowering agenda, Report of the Special Rapporteur in the field of cultural rights, Alexandra Xanthaki' (2022) UN Doc A/HRC/49/54.
UNHRC, COVID-19, Culture and Cultural Rights (17 February 2021) UN Doc A/HRC/46/34.
UNHRC 'Visit to Tuvalu: Report of the Special Rapporteur in the field of cultural rights, Karima Bennoune' (24 December 2020) UN Doc A/HRC/46/34/Add.1.
UNHRC 'Visit to Poland, Report of the Special Rapporteur in the field of cultural rights' (2020) UN Doc A/HRC/43/50/Add.1.
UNHRC 'Report of the Human Rights Council on its fortieth session' (3 June 2019) UN Doc A/HRC/40/2.
UNHRC 'Visit to Malaysia, Report of the Special Rapporteur in the field of cultural rights' (2019) UN Doc A/HRC/40/53/Add.1.
UNHRC, Cultural Rights: Tenth Anniversary Report (17 January 2019) UN Doc A/HRC/40/53.
UNHRC 'Report of the Special Rapporteur in the field of cultural rights on her mission to Serbia and Kosovo' (19 January 2018) UN Doc A/HRC/37/55/Add.1.
UNHRC 'Intersessional seminar on cultural rights and the protection of cultural heritage' (27 December 2017) UN Doc A/HRC/37/29.
UNHRC 'Report of the Special Rapporteur in the field of cultural rights on her mission to Cyprus' (3 March 2017) UN Doc A/HRC/34/56/Add.1.
UNHRC, Report of the Special Rapporteur in the field of cultural rights (3 February 2016) UN Doc A/HRC/31/59.
UNHRC 'Report of the Special Rapporteur in the field of cultural rights on her visit to Botswana' (12 January 2016) UN Doc A/HRC/31/59/Add.1.

UNHRC 'Promotion and Protection of the Right of Indigenous Peoples with Respect to their Cultural Heritage, Study by the Expert Mechanism on the Rights of Indigenous Peoples' (19 August 2015) UN Doc A/HRC/30/53.

UNHRC 'Mandate of the Special Rapporteur in the field of cultural rights' (26 March 2015) UN Doc A/70/53.

UNHRC 'Report of the Special Rapporteur in the field of cultural rights, Farida Shaheed, Addendum, Visit to Viet Nam' (29 January 2015) UN Doc A/HRC/28/57/Add.1.

UNHRC 'Report of the Special Rapporteur in the field of cultural rights, Farida Shaheed, Addendum, Mission to Bosnia and Herzegovina' (3 March 2014) UN Doc A/HRC/25/49/Add.1.

UNHRC 'Report of the Special Rapporteur in the field of cultural rights, Farida Shaheed, Memorialization processes' (23 January 2014) UN Doc A/HRC/25/49.

UNHRC 'Report of the Independent Expert in the field of cultural rights, Farida Shaheed, Addendum, Mission to Morocco' (2 May 2012) UN Doc A/HRC/20/26/Add.2.

UNHRC 'Report of the Special Rapporteur on freedom of religion or belief, Heiner Bielefeldt' (2012) UN Doc A/HRC/22/51.

UNHRC, Report of the independent expert in the field of cultural rights, Farida Shaheed (21 March 2011) UN Doc A/HRC/17/38.

UNHRC 'Report of the independent expert in the field of cultural rights, Ms. Farida Shaheed, submitted pursuant to resolution 10/23 of the HRC' (22 March 2010) UN Doc A/HRC/14/36.

UNHRC 'Report of the Special Rapporteur on freedom of religion or belief, Asma Jahangir' (6 January 2009) UN Doc A/HRC/10/8.

UNHRC 'Expert Mechanism on Rights of Indigenous Peoples' (14 December 2007) UN Doc A/HRC/RES/6/36.

UNCHR 'Report of the Sub-Commission on Prevention of Discrimination and Protection of Minorities on its Forty-Sixth Session' (28 October 1994) UN Doc E/CN.4/Sub.2/1994/56.

UPR

UNHRC 'National report submitted by Costa Rica' (2024) UN Doc A/HRC/WG.6/47/CRI/1.

UNHRC 'National report submitted by Egypt' (2024) UN Doc A/HRC/WG.6/48/EGY/1.

UNHRC 'National report submitted by Viet Nam' (2024) UN Doc A/HRC/WG.6/46/VNM/1.

UNHRC 'National report submitted by Canada' (2023) UN Doc A/HRC/WG.6/44/CAN/1.

UNHRC 'National Report submitted by Russian Federation' (2023) UN Doc A/HRC/WG.6/44/RUS/1.

UNHRC 'National report submitted by Mali' (2023) UN Doc A/HRC/WG.6/43/MLI/1.

UNHRC 'National Report submitted by France' (2023) UN Doc A/HRC/WG.6/43/FRA/1.

UNHRC 'National report submitted by Australia' (2020) UN Doc A/HRC/WG.6/37/AUS/1.

UNHRC 'National Report submitted by Ethiopia' (2019) UN Doc A/HRC/WG.6/33/ETH/1.

UNHRC 'National Report submitted by Egypt' (2019) UN Doc A/HRC/WG.6/34/EGY/1.

UNHRC 'National report submitted by Viet Nam' (2018) UN Doc A/HRC/WG.6/32/VNM/1.
UNHRC 'National report submitted by Vanuatu' (2018) UN Doc A/HRC/WG.6/32/VUT/1.
UNHRC 'National report submitted by Cuba' (2018) UN Doc A/HRC/WG.6/30/CUB/1.
UNHRC 'National report submitted by Colombia' (2018) UN Doc A/HRC/WG.6/30/COL/1.
UNHRC 'National report submitted by Afghanistan' (2018) UN Doc A/HRC/WG.6/32/AFG/1.
UNHRC 'National report submitted by Brazil' (2017) UN Doc A/HRC/WG.6/27/BRA/1.
UNHRC 'National report submitted by Mali' (2017) UN Doc A/HRC/WG.6/29/MLI/1.
UNHRC 'National report submitted by Costa Rica' (2014) UN Doc A/HRC/WG.6/19/CRI/1.
UNHRC 'National report submitted by Uzbekistan' (2013) UN Doc A/HRC/WG.6/16/UZB/1.
UNHRC 'National report submitted by Viet Nam' (2013) UN Doc A/HRC/WG.6/18/VNM/1.
UNHRC 'National report submitted by Mexico' (2013) UN Doc A/HRC/WG.6/17/MEX/1.
UNHRC 'National report submitted by Cuba' (2013) UN Doc A/HRC/WG.6/16/CUB/1.
UNHRC 'National report submitted by Russian Federation' (2013) UN Doc A/HRC/WG.6/16/RUS/1.
UNHRC 'National report submitted by Mali' (2012) UN Doc A/HRC/WG.6/15/MLI/1.
UNHRC 'National report submitted by France' (2012) UN Doc A/HRC/WG.6/15/FRA/1.
UNHRC 'National report submitted by Federated States of Micronesia' (2010) UN Doc A/HRC/WG.6/9/FSM/1.
UNHRC 'National report submitted by Viet Nam' (2009) UN Doc A/HRC/WG.6/5/VNM/1.
UNHRC 'National report submitted by Afghanistan' (2009) UN Doc A/HRC/WG.6/5/AFG.
UNHRC 'National report submitted by Dominican Republic' (2009) UN Doc A/HRC/WG.6/6/DOM/1.
UNHRC 'National report submitted by Uzbekistan (2008) UN Doc A/HRC/WG.6/3/UZB/1.
UNHRC 'Report of the Working Group on the UPR: New Zealand' (2024) UN Doc A/HRC/57/4.
UNHRC 'Report of the Working Group on the UPR: Viet Nam' (2024) UN Doc A/HRC/57/7.
UNHRC 'Report of the Working Group on the UPR: Finland' (2024) UN Doc A/HRC/52/9.
UNHRC 'Report of the Working Group on the UPR: Colombia (2023) UN Doc A/HRC/55/7.
UNHRC 'Report of the Working Group on the UPR: UK' (2023) A/HRC/52/10.
UNHRC 'Report of the Working Group on the UPR: Russian Federation' (2023) UN Doc A/HRC/55/14.
UNHRC 'Report of the Working Group on the UPR: Mali' (2023) UN Doc A/HRC/54/8.
UNHRC 'Report of the Working Group on the UPR: Cuba' (2023) UN Doc A/HRC/55/17 (2023).
UNHRC 'Report of the Working Group on the UPR: France' (2023) UN Doc A/HRC/54/5.
UNHRC 'Report of the Working Group on the UPR: Canada' (2023) UN Doc A/HRC/55/12.

UNHRC 'Report of the Working Group on the UPR: Brazil' (2022) UN Doc A/HRC/52/14.
UNHRC 'Report of the Working Group on the UPR: USA Addendum,' (2021) UN Doc A/HRC/46/15/Add.1.
UNHRC 'Report of the Working Group on the UPR: United Republic of Tanzania' (2021) UN Doc A/HRC/49/13.
UNHRC 'Report of the Working Group on the UPR: Australia' (2021) UN Doc A/HRC/47/8.
UNHRC 'Report of the Working Group on the UPR: Australia, Addendum' (2021) UN Doc A/HRC/47/8/Add.1.
UNHRC 'Report of the Working Group on the UPR: USA' (2020) UN Doc A/HRC/46/15.
UNHRC 'Report of the Working Group on the UPR; Egypt' (2019) UN Doc A/HRC/43/16.
UNHRC 'Report of the Working Group of the UPR: Cuba' (2018) UN Doc A/HRC/39/16.
UNHRC 'Report of the Working Group on the UPR: Russian Federation' (2018) UN Doc A/HRC/39/13.
UNHRC 'Report of the Working Group on the UPR: Brazil' (2017) UN Doc A/HRC/36/11.
UNHRC 'Report of the Working Group on the UPR: Namibia' (2016) UN Doc A/HRC/32/4.
UNHRC 'Report of the Working Group on the UPR: Australia' (2016) UN Doc A/HRC/31/14.
UNHRC 'Report of the Working Group on the UPR: Turkey' (2015) UN Doc A/HRC/29/15
UNHRC 'Report of the Working Group on the UPR: USA' (2015) UN Doc A/HRC/30/12
UNHRC 'Report of the Working Group on the UPR: Vanuatu' (2014) UN Doc A/HRC/57/10
UNHRC 'Report of the Working Group on the UPR: Mali' (2013) UN Doc A/HRC/23/6.
UNHRC 'Report of the Working Group on the UPR: Cuba' (2013) UN Doc A/HRC/24/16.
UNHRC 'Report of the Working Group on the UPR: Cuba, Addendum' (2013) UN Doc A/HRC/24/16/Add.1
UNHRC 'Report of the Working Group on the UPR: Mali, Addendum' (2013) UN Doc A/HRC/23/6/Add.1
UNHRC 'Compilation prepared by the Office of the High Commissioner for Human Rights: Mali' (2012) UN Doc A/HRC/WG.6/15/MLI/2.
UNHRC 'Report of the Working Group on the UPR: United Republic of Tanzania' (2011) UN Doc A/HRC/19/4.
UNHRC 'Report of the Working Group on the UPR: Italy' (2010) UN Doc A/HRC/14/4.
UNHRC 'Report of the Working Group on the UPR: Turkey', (2010) UN Doc A/HRC/15/13.
UNHRC 'Report of the Working Group on the UPR: Colombia' (2009) UN Doc A/HRC/10/82.
UNHRC 'Report of the Working Group on the UPR: Cuba' (2009) UN Doc A/HRC/11/22.
UNHRC 'Report of the Working Group of the UPR. Japan' (2008) UN Doc A/HRC/8/44.
UNHRC 'Report of the Working Group on the UPR: Tonga' (2008) UN Doc A/HRC/8/48.

Special Procedures: Communications

Mandates of the Special Rapporteur on adequate housing as a component of the right to an adequate standard of living, and on the right to non-discrimination in this context; the Special Rapporteur on the situation of human rights in Cambodia; the Special Rapporteur in the field of cultural rights; the Special Rapporteur on the right to food; the Special Rapporteur on the rights of Indigenous Peoples and the Special Rapporteur on the human rights of internally displaced persons, AL OTH 136/2024 (17 October 2024)

Mandates of the Working Group on the issue of human rights and transnational corporations and other business enterprises; the Special Rapporteur in the field of cultural rights and the Special Rapporteur on the promotion of truth, justice, reparation and guarantees of non-recurrence, AL NAM 1/2024 (12 September 2024).

Mandates of the Special Rapporteur in the field of cultural rights; the Special Rapporteur on the independence of judges and lawyers and the Special Rapporteur on the rights of Indigenous Peoples, AL NZL 1/2024 (1 July 2024).

Mandates of the Working Group on the issue of human rights and transnational corporations and other business enterprises; the Special Rapporteur in the field of cultural rights; the Special Rapporteur on adequate housing as a component of the right to an adequate standard of living, and on the right to non-discrimination in this context and the Special Rapporteur on the rights of Indigenous Peoples, AL NPL 1/2024 (25 June 2024).

Mandatos del Grupo de Trabajo sobre la cuestión de los derechos humanos y las empresas transnacionales y otras empresas; de la Relatora Especial en la esfera de los derechos culturales y del Relator Especial sobre los derechos de los Pueblos Indígenas, AL ECU 4/2024 (11 June 2024).

Mandatos del Relator Especial sobre la promoción de la verdad, la justicia, la reparación y las garantías de no repetición; de la Relatora Especial en la esfera de los derechos culturales; del Grupo de Trabajo sobre las Desapariciones Forzadas o Involuntarias; del Relator Especial sobre ejecuciones extrajudiciales, sumarias o arbitrarias yde la Relatora Especial sobre la promoción y protección del derecho a la libertad de opinión y de expresión, AL PER 3/2024 (29 April 2024).

Mandate of the Special Rapporteur in the field of cultural rights, OL GRC 1/2024 (18 March 2024).

Mandate of the Special Rapporteur in the field of cultural rights, AL AZE 2/2023 (22 September 2023).

Mandates of the Special Rapporteur in the field of cultural rights; the Special Rapporteur on the right to education and the Special Rapporteur on minority issues, AL ITA 4/2023 (18 September 2023).

Mandates of the Working Group on the issue of human rights and transnational corporations and other business enterprises; the Special Rapporteur in the field of cultural rights; the Special Rapporteur on the promotion and protection of the right to freedom of opinion and expression; the Special Rapporteur on the situation of human rights defenders; the Special Rapporteur on the independence of judges and lawyers and the Special Rapporteur on the rights of Indigenous Peoples, AL NPL 2/2023 (20 July 2023).

Mandates of the Working Group on Arbitrary Detention; the Special Rapporteur in the field of cultural rights; the Special Rapporteur on the promotion and protection of the right to freedom of opinion and expression; the Special Rapporteur on the rights to freedom of peaceful assembly and of association; the Special Rapporteur on adequate housing as a component of the right to an adequate standard of living, and on the right to non-discrimination in this context; the Special Rapporteur on the situation of human rights defenders and the Special Rapporteur on the human rights of internally displaced persons, AL OTH 63/2023 (21 June 2023).

Mandates of the Working Group on the issue of human rights and transnational corporations and other business enterprises; the Special Rapporteur in the field of cultural rights; the Special Rapporteur on the issue of human rights obligations relating to the enjoyment of a safe, clean, healthy and sustainable environment; the Special Rapporteur on the promotion and protection of the right to freedom of opinion and expression; the Special Rapporteur on the rights to freedom of peaceful assembly and of association; the Special Rapporteur on the situation of human rights defenders; the Special Rapporteur on the rights of Indigenous Peoples; the Independent Expert on protection against violence and discrimination based on sexual orientation and gender identity and the Special Rapporteur on violence against women and girls, its causes and consequences, AL CAN 2/2022 (13 January 2023).

Mandates of the Special Rapporteur in the field of cultural rights; the Special Rapporteur on the right to education and the Special Rapporteur on freedom of religion or belief, AL RUS 18/2022 (20 December 2022).

Mandate of the Special Rapporteur in the field of cultural rights, AL ARM 2/2021 (29 October 2021).

Mandate of the Special Rapporteur in the field of cultural rights, AL AZE 3/2021 (29 October 2021).

Mandates of the Special Rapporteur in the field of cultural rights; and the Special Rapporteur on freedom of religion or belief, AL TUR 12/2020 (28 July 2020).

Mandates of the Special Rapporteur in the field of cultural rights; and the Special Rapporteur on adequate housing as a component of the right to an adequate standard of living, and on the right to non-discrimination in this context, UA UZB 3/2019 (9 August 2019).

Mandates of the Special Rapporteur in the field of cultural rights; the Special Rapporteur on the issue of human rights obligations relating to the enjoyment of a safe, clean, healthy and sustainable environment; the Special Rapporteur on adequate housing as a component of the right to an adequate standard of living, and on the right to non-discrimination in this context; the Special Rapporteur on the human rights of internally displaced persons; and the Special Rapporteur on minority issues, AL TUR 13/2018 (23 November 2018).

Mandates of the Working Group of Experts on People of African Descent; the Special Rapporteur in the field of cultural rights; the Special Rapporteur on extrajudicial, summary or arbitrary executions; the Special Rapporteur on adequate housing as a component of the right to an adequate standard of living, and on the right to non-discrimination in this context; and the Independent Expert on the enjoyment of all human rights by older persons, AL BRA 2/2018 (20 April 2018).

Mandates of the Special Rapporteur in the field of cultural rights; the Special Rapporteur on the rights of indigenous peoples and the Special Rapporteur on the human rights of internally displaced persons, UA NPL 3/2017 (14 August 2017).

Mandate of the Working Group on the issue of human rights and transnational corporations and other business enterprises; the Special Rapporteur in the field of cultural rights; the

Special Rapporteur on the issue of human rights obligations relating to the enjoyment of a safe, clean, healthy and sustainable environment; the Special Rapporteur on the rights to freedom of peaceful assembly and of association; the Special Rapporteur on the situation of human rights defenders; the Special Rapporteur on the rights of indigenous peoples; and the Special Rapporteur on the human right to safe drinking water and sanitation, UA USA 7/2016 (21 September 2016).
Mandates of the Special Rapporteur in the field of cultural rights; the Special Rapporteur on freedom of religion or belief; and the Special Rapporteur on the situation of human rights in the Islamic Republic of Iran, AL, IRN 14/2016 (31 May 2016).
Mandate of the Special Rapporteur in the field of cultural rights, AL, SAU 3/2016 (31 May 2016).
Mandates of the Special Rapporteur on adequate housing as a component of the right to an adequate standard of living, and on the right to non-discrimination in this context and the Special Rapporteur in the field of cultural rights, UA, PAK 1/2016 (12 January 2016).
Mandates of the Special Rapporteur in the field of cultural rights; the Special Rapporteur on extreme poverty and human rights; and the Special Rapporteur on freedom of religion or belief, AL, BHR 6/2015 (30 October 2015).
Mandates of the Special Rapporteur in the field of cultural rights and the Special Rapporteur on freedom of religion or belief, AL, SAU 7/2015 (14 October 2015).
Mandate of the Special Rapporteur in the field of cultural rights; the Special Rapporteur on minority issues; and the Special Rapporteur on freedom of religion or belief, AL BGR 3/ 2015 (15 September 2015).
Mandates of the Working Group of experts on people of African descent; the Special Rapporteur in the field of cultural rights; the Special Rapporteur on issues; and the Special Rapporteur on contemporary forms of racism, racial discrimination, xenophobia and related intolerance, OL, NLD 1 /2015 (28 January 2015).
Mandate of the Special Rapporteur on the rights of indigenous peoples, OL Indigenous (2001–8), OTH 10/2013 (18 November 2013).
Mandates of the Special Rapporteur in the field of cultural rights and the Special Rapporteur on freedom of religion or belief, UA Cultural rights (2009) G/SO 214 (56–23), MLI 1/ 2012 (9 July 2012).
Mandate of the Special Rapporteur in the field of cultural rights, AL cultural rights (2009), SGP 2/2012 (29 May 2012).

Committee on Economic, Social and Cultural Rights

CESCR 'Rules of procedure under the Optional Protocol to the International Covenant on Economic, Social and Cultural Rights' (3 May 2022) UN Doc E/C.12/5.
CESCR 'General Comment No. 25' (30 April 2020) UN Doc E/C.12/GC/25.
CESCR 'General discussion on article 15, paragraph 1 (a), of the Covenant, on the right to take part in cultural life (continued)' (16 May 2018) UN Doc E/C.12/2008/SR.18.
CESCR 'Provisional Rules of Procedure under the Optional Protocol to the International Covenant on Economic, Social and Cultural Rights, Adopted by the Committee at its Forty-Ninth Session (12–30 November 2012)' (15 January 2013) UN Doc E/C.12/49/3.

CESCR 'Report on the Forty-Fourth and Forty-Fifth Sessions' (3–21 May 2010, 1–19 November 2010) UN Doc E/2011/22—E/C.12/2010/3.
CESCR 'General Comment No. 21' (21 December 2009) UN Doc E/C.12/GC/21.
CESCR 'Guidelines on Treaty-Specific Documents to be Submitted by State Parties under Articles 16 and 17 of the International Covenant on Economic, Social and Cultural Rights' (24 March 2009) UN Doc E/C.12/2008/2.
CESCR 'Summary Record of the Seventeenth Meeting, Substantive Issues Arising on the Implementation of the International Covenant on Economic, Social and Cultural Rights, Day of General Discussion on Paragraph 1 (a) of Article 15 of the Covenant on the Right to Take Part in Cultural Life' (19 May 2008) UN Doc E/C.12/2008/SR.17.
CESCR 'General Comment No. 17' (12 January 2006) UN Doc E/C.12/GC/17.
CESCR 'General Comment No. 15 (2002): The right to water' (20 January 2003) UN Doc E/C.12/2002/11.
CESCR 'General comment No. 13' (8 December 1999) UN Doc E/C.12/1999/10.
CESCR 'General Comment No. 12: The right to adequate food' (12 May 1999) UN Doc E/C.12/1999/5.
CESCR 'Excerpt from the Report on the Eighteenth and Nineteenth Sessions' (30 November 1998) UN Doc E/1999/22 -E/C.12/1998/26.
CESCR 'Summary Record of the 49th Meeting: General Discussion: "Right to education (Articles 13 and 14 of the Covenant)' (2 December 1998) UN Doc E/C.12/1998/SR.49.
CESCR 'Chapter VII, General Discussion on the Right to Take Part in Cultural Life as Recognized in Article 15 of the International Covenant on Economic, Social and Cultural Rights' (1993) UN Doc E/1993/22.
CESCR 'Excerpt from the Report on the Seventeenth Session, General Discussion on the Right to Take Part in Cultural Life as Recognized in Article 15 of the International Covenant on Economic, Social and Cultural Rights' (1993) UN Doc E/1993/22—E/C.12/1992/2.
CESCR 'Summary Record of the Seventeenth Meeting, General Discussions on the Right to Take Part in Cultural Life as Recognized in Article 15 of the Covenant' (12 December 1992) UN Doc E/C.12/1992/SR.17.
CESCR 'General Comment No. 4, The Right to Adequate Housing' (13 December 1991) UN Doc E/1992/23.
CESCR 'Revised General Guidelines Regarding the Form and Contents of Reports to be Submitted by States Parties under Articles 16 and 17 of the International Covenant on Economic, Social and Cultural Rights' (17 June 1991) UN Doc E/C.12/1991/1.
CESCR 'Rules of Procedure of The Committee: Provisional rules of procedure adopted by the Committee at its third session (1989)' (1 September 1993) UN Doc E/C.12/1990/4/Rev.1.
CESCR 'General Comment No. 3: The nature of States parties' obligations (art. 2, para. 1, of the Covenant' (14 December 1990) UN Doc E/1991/23.
CESCR 'General Comment No. 1: Reporting by States Parties' (27 July 1981) UN Doc E/1989/22.

Communications

CESCR, Communication No. 37/2018, *Maribel Viviana López Albán v. Spain* (29 November 2019) UN Doc E/C.12/66/D/37/2018.

CESCR, Communication No. 3/2014, *A.M.B v Ecuador* (8 August 2016) UN Doc E/C.12/58/D/3/2014.

Periodic Monitoring

CESCR 'Periodic report submitted by Canada' (2024) UN Doc E/C.12/CAN/7.

CESCR 'Periodic report submitted by Australia' (2023) UN Doc E/C.12/AUS/6.

CESCR 'Periodic report submitted by Mexico' (2023) UN Doc E/C.12/MEX/7.

CESCR 'Periodic report submitted by Namibia' (2023) UN Doc E/C.12/NAM/2.

CESCR 'Periodic report submitted by Angola' (2023) UN Doc E/C.12/AGO/6.

CESCR 'Periodic report submitted by Colombia' (2022) UN Doc E/C.12/COL/7.

CESCR 'Periodic report submitted by the UK' (2022) UN Doc E/C.12/GBR/7.

CESCR 'Periodic report submitted by Italy' (2021) UN Doc E/C.12/ITA/6.

CESCR 'Periodic report submitted by France' (2021) UN Doc E/C.12/FRA/5.

CESCR 'Periodic report submitted by Brazil' (2020) UN Doc E/C.12/BRA/3.

CESCR 'Periodic report submitted by Uzbekistan' (2019) UN Doc E/C.12/UZB/3.

CESCR 'Periodic report submitted by New Zealand' (2017) UN Doc E/C.12/NZL/4.

CESCR 'Periodic report submitted by Russian Federation' (2016) UN Doc E/C.12/RUS/6.

CESCR 'Periodic report submitted by Mexico' (2016) UN Doc E/C.12/MEX/5-6.

CESCR 'Periodic report submitted by Dominican Republic' (2015) UN Doc E/C.12/DOM/4.

CESCR 'Periodic report submitted by Angola' (2014) UN Doc E/C.12/AGO/4-5.

CESCR 'Periodic report submitted by UK' (2014) UN Doc E/C.12/GBR/6.

CESCR 'Periodic report submitted by France' (2014) UN Doc E/C.12/FRA/4.

CESCR 'Periodic report submitted by Australia' (2014) UN Doc E/C.12/AUS/5.

CESCR 'Periodic report submitted by Vietnam' (2013) UN Doc E/C.12/VNM/2-4.

CESCR 'Periodic report submitted by Canada' (2013) UN Doc E/C.12/CAN/6.

CESCR 'Periodic report submitted by Finland' (2013) UN Doc E/C.12/FIN/6.

CESCR 'Periodic report submitted by Uzbekistan' (2012) UN Doc E/C.12/UZB/2.

CESCR 'Periodic report submitted by Ethiopia' (2011) UN Doc, E/C.12/ETH/1-3.

CESCR 'Periodic report submitted by Japan,' (2011) UN Doc E/C.12/JPN/3.

CESCR 'Periodic report submitted by New Zealand' (2011) UN Doc E/C.12/NZL/3.

CESCR 'Periodic report submitted by Russian Federation' (2010) UN Doc E/C.12/RUS/5.

CESCR 'Periodic report submitted by Dominican Republic' (2009) UN Doc E/C.12/DOM/3.

CESCR 'Periodic report submitted by Afghanistan' (2009) UN Doc E/C.12/AFG/2-4.

CESCR 'Periodic report submitted by Turkey' (2009) UN Doc E/C.12/TUR/1.

CESCR 'Periodic report submitted by Brazil' (2008) UN Doc E/C.12/BRA/2.

CESCR 'Periodic report submitted by UK' (2008) UN Doc E/C.12/GBR/5.

CESCR 'Periodic Report submitted by Canada' (2004) UN Doc E/C.12/4/Add.15.

CESCR 'Concluding Observations: Brazil' (2023) UN Doc E/C.12/BRA/CO/3.
CESCR 'Concluding Observations: France' (2023) UN Doc E/C.12/FRA/CO/5.
CESCR 'Concluding Observations: Uzbekistan' (2022) UN Doc E/C.12/UZB/CO/3.
CESCR 'Concluding Observations: Finland' (2021) UN Doc E/C.12/FIN/CO/7.
CESCR 'Concluding Observations: New Zealand' (2018) UN Doc E/C.12/NZL/CO/4.
CESCR 'Concluding Observations: Mexico' (2018) UN Doc E/C.12/MEX/CO/5-6.
CESCR 'Concluding Observations: Mali' (2018) UN Doc E/C.12/MLI/CO/1.
CESCR 'Concluding Observations: Russian Federation' (2017) UN Doc E/C.12/RUS/CO/6.
CESCR 'Concluding Observations: Australia' (2017) UN Doc E/C.12/AUS/CO/5.
CESCR 'Concluding Observations: Colombia' (2017) UN Doc E/C.12/COL/CO/6.
CESCR 'Concluding Observations: New Zealand' (2016) UN Doc E/C.12/NZL/CO/3.
CESCR 'Concluding Observations: France' (2016) UN Doc E/C.12/FRA/CO/4.
CESCR 'Concluding Observations: Costa Rica' (2016) UN Doc E/C.12/CRI/CO/5.
CESCR 'Concluding Observations: Canada' (2016) UN Doc E/C.12/CAN/CO/6.
CESCR 'Concluding Observations: Angola' (2016) UN Doc E/C.12/AGO/CO-4-5.
CESCR 'Concluding Observations: UK' (2016) UN Doc E/C.12/GBR/CO/6.
CESCR 'Concluding Observations: Namibia' (2016) UN Doc E/C.12/NAM/CO/1.
CESCR 'List of Issues: Costa Rica' (2016) UN Doc E/C.12/CRI/Q/5.
CESCR 'List of Issues: Australia' (2016) UN Doc E/C.12/AUS/Q/5.
CESCR 'Concluding Observations: Italy' (2015) UN Doc E/C.12/ITA/CO/5.
CESCR 'Concluding Observations: Uzbekistan' (2014) UN Doc E/C.12/UZB/CO/2.
CESCR 'Concluding Observations: Viet Nam' (2014) UN Doc E/C.12/VNM/CO/2-4.
CESCR 'Concluding Observations: Finland' (2014) UN Doc E/C.12/FIN/CO/6.
CESCR 'Concluding Observations: Jamaica' (2013) UN Doc E/C.12/JAM/CO/3-4.
CESCR 'Concluding Observations: Japan' (2013) UN Doc E/C.12/JPN/CO/3.
CESCR 'Concluding Observations: Egypt' (2013) UN Doc E/C.12/EGY/CO/2-4.
CESCR 'Concluding Observations: Tanzania' (2012) UN Doc E/C.12/TZA/CO/1-3.
CESCR 'Concluding Observations: New Zealand' (2012) UN Doc E/C,12/NZL/CO/3.
CCPR 'List of Issues: Australia' (2012) UN Doc CCPR/C/AUS/Q/6.
CESCR 'List of Issues: United Republic of Tanzania' (2011) UN Doc E/C.12/TZA/Q/1-3.
CESCR 'Concluding Observations: Russian Federation' (2011) UN Doc E/C.12/RUS/CO/5.
CESCR 'Concluding Observations: Colombia' (2010) UN Doc E/C.12/COL/CO/5.
CESCR 'Concluding Observations: Afghanistan' (2010) UN Doc E/C.12/AFG/CO/2-4.
CESCR 'Concluding Observations: Australia' (2009) UN Doc E/C.12/AUS/CO/4.
CESCR 'Concluding Observations: UK' (2009) UN Doc E/C.12/GBR/CO/5.
CESCR 'List of Issues: Colombia' (2009) UN Doc E/C.12/COL/Q/5.
CESCR 'List of issues: Afghanistan' (2009) UN Doc E/C.12/AFG/Q/2-4.
CESCR 'Concluding Observations: Angola' (2008) UN Doc E/C.12/AGO/CO/3.
CESCR 'Concluding Observations: Finland' (2008) UN Doc E/C.12/FIN/CO/5.
CESCR 'Concluding Observations: France' (2008) UN Doc E/C.12/CO/FRA/CO/3.

Human Rights Committee

CCPR 'Rules of Procedure of the Human Rights Committee' (9 January 2019) UN Doc CCPR/C/3/Rev.11.

CCPR 'General Comment No. 36: Article 6: right to life' (3 September 2019) UN Doc CCPR/C/GC/36.

CCPR 'Guidelines on measures of reparation under the Optional Protocol to the International Covenant on Civil and Political Rights' (30 November 2016) UN Doc CCPR/C/158.

CCPR 'Rules of Procedure of the Human Rights Committee' (11 January 2012) UN Doc CCPR/C/3/Rev.10.

CCPR 'Guidelines for the treaty-specific document to be submitted by States parties under article 40 of the International Covenant on Civil and Political Rights' (22 November 2010) UN Doc CCPR/C/2009/1.

CCPR 'General Comment No. 33 The Obligations of State Parties under the Optional Protocol to the International Covenant on Civil and Political Rights' (5 November 2008) UN Doc CCPR/C/GC/33.

CCPR 'Draft General Comment No. 33' (25 August 2008) UN Doc CCPR/C/GC/33/CRP.3.

CCPR 'General Comment No. 23: Article 27 (Rights of Minorities)' (8 April 1994) UN Doc CCPR/C/21/Rev.1/Add.5.

Communications

CCPR, Communication No. 3585/2019, *Alisa Roy v. Australia* (13 March 2024) UN Doc CCPR/C/137/D/3585/2019.

CCPR, Communication No. 3624/2019, *Daniel Billy et al. v. Australia* (22 September 2022) UN Doc CCPR/C/135/D/3624/2019.

CRC, Communication No. 104/2019, *Sacchi et al v. Argentina et al.* (8 October 2021) UN Doc CRC/C/88/D/104/2019.

CCPR, Communication No. 2552/2015, *Benito Oliveira Pereira et al. v. Paraguay* (12 October 2021) UN Doc CCPR/C/132/D/2552/2015.

CCPR, Communication No. 2751/2016, *Norma Portillo Cáceres et al v. Paraguay* (20 September 2019) UN Doc CCPR/C/126/D/2751/2016.

CCPR, Communication No. CCPR/ C/124/D/2668/2015, *Tiina Sanila-Aikio v. Finland* (20 March 2019) UN Doc CCPR/ C/124/D/2668/2015.

CCPR, Communication No. 2668/2015, *Sanila-Aikio v. Finland* (16 October 2018) UN Doc CCPR/C/119/D/2668/2015.

CCPR, Communication No. 2102/2011, *Paadar et al. v. Finland*, (10 April 2014) CCPR/C/110/D/2102/2011.

CCPR, Communication No. 1621/2007, *Raihman v. Latvia* (30 November 2010) UN Doc CCPR/C/100/D/1621/2007.

CCPR, Communication No. 1457/2006, *Ángela Poma Poma v. Peru* (24 April 2009) UN Doc CCPR/C/95/D/1457/2006.

CCPR Communication No. 879/1999, *Howard v. Canada* (4 August 2005) UN Doc CCPR/C/84/D/879/1999.

CCPR, Communication No. 932/2000, *Ms. Marie-Hélène Gillot et al. v. France* (21 July 2002) UN Doc CCPR/C/75/D/932/2000.
CCPR, Communication No. 779/1997, *Äärelä and Näkkäläjärvi v Finland* (24 October 2001) UN Doc CCPR/C/73/D/779/1997.
CCPR, Communication No 547/1993, *Apirana Mahuika and Ors v New Zealand* (27 October 2000) UN Doc CCPR/C/70/D/547/1993.
CCPR, Communication No. 822/1998, *Mathieu Vakoumé et al v. France* (2000) UN Doc CCPR/C/70/D/822/1998.
CCPR, Communication No. 549/1993, *Hopu and Bessert v. France* (29 December 1997) UN Doc CCPR/C/60/D/549/1993/Rev.1.
CCPR, Communication No. 511/1992, *Länsman v. Finland* (14 October 1993) UN Doc CCPR/C/49/D/511/1992.
CCPR, Communication Nos. 221/1987 and 323/1988, *Yves Cadoret and Hervé Le Bihan v. France*, (11 April 1991) UN Doc CCPR/C/41/D/221/1987.
CCPR, Communication No. 219/1986, *Dominique Guesdon v. France*, (25 July 1990) UN Doc CCPR/C/39/D/219/1986.
CCPR, Communication No 167/1984, *Chief Bernard Ominayak and the Lubicon Lake Band v Canada* (26 March 1990) UN Doc CCPR/C/38/D/167/1984.
CCPR, Communication No. 220/1987, *T. K. v. France* (8 November 1989) UN Doc CCPR/C/37/D/220/1987.
CCPR, Communication No. 222/1987, *H. K. v. France* (8 November 1989) UN Doc CCPR/C/37/D/222/1987.
CCPR, Communication No. 197/1985, *Kitok v. Sweden* (27 July 1988) UN Doc CCPR/C/33/D/197/1985.
CCPR, Communication No. 24/1977, *Sandra Lovelace v. Canada* UN Doc CCPR/C/OP/1 at 10 (1984).

Periodic Monitoring

CCPR 'Concluding Observations: Egypt' (2023) UN Doc CCPR/C/EGY/CO/5.
CCPR 'Concluding Observations: USA' (2023) UN Doc CCPR/C/USA/CO/5.
CCPR 'Concluding Observations: Brazil' (2023) UN Doc CCPR/C/BRA/CO/3.
CCPR 'List of issues: Namibia' (2023) UN Doc CCPR/C/NAM/Q/3.
CCPR 'Concluding Observations: Russian Federation' (2022) UN Doc CCPR/C/RUS/CO/8.
CCPR 'Concluding Observations: Japan' (2022) UN Doc CCPR/C/JPN/CO/7.
CCPR 'Concluding Observations: Ethiopia' (2022) UN Doc CCPR/C/ETH/CO/2.
CCPR 'Concluding Observations: Dominican Republic' (2020) UN Doc CCPR/C/DMA/COAR/1.
CCPR 'Concluding Observations: Angola' (2019) UN Doc CCPR/C/AGO/CO/2.
CCPR 'Concluding Observations. Mexico' (2019) UN Doc CCPR/C/MEX/CO/6.
CCPR 'Concluding Observations: Viet Nam' (2019) UN Doc CCPR/C/VNM/CO/3.
´CCPR 'List of issues: Angola' (2018) UN Doc CCPR/C/AGO/Q/2.
CCPR 'Concluding Observations: Australia' (2017) UN Doc CCPR/C/AUS/CO/6.
CCPR 'Concluding Observations: Italy' (2017) UN Doc CCPR/C/ITA/CO/6.

CCPR 'Concluding Observations: Bosnia and Herzegovina' (2017) UN Doc CCPR/C/BIH/CO/3.
CCPR 'Concluding Observations: Colombia' (2016) UN Doc CCPR/C/COL/CO/7.
CCPR 'Concluding Observations: Costa Rica' (2016) UN Doc CCPR/C/CRI/CO/6.
CCPR 'Concluding Observations: Namibia' (2016) UN Doc CCPR/C/NAM/CO/2.
CCPR 'Concluding Observations: New Zealand' (2016) UN Doc CCPR/C/NZL/CO/6.
CCPR 'Concluding Observations: Russian Federation' (2015) UN Doc CCPR/C/RUS/CO/7.
CCPR 'Concluding Observations: Japan' (2015) UN Doc CCPR/C/JPN/CO/7.
CCPR 'Concluding Observations: France' (2015) UN Doc CCPR/C/FRA/CO/5.
CCPR 'Concluding Observations: Uzbekistan' (2015) UN Doc CCPR/C/UZB/CO/4.
CCPR 'Concluding Observations: Canada' (2015) UN Doc CCPR/C/CAN/CO/6.
CCPR 'List of issues: Canada' (2015) UN Doc CCPR/C/CAN/Q/6/Add.1.
CCPR 'Concluding Observations: USA' (2014) UN Doc CCPR/C/USA/CO/4.
CCPR 'Concluding Observations: Finland' (2013) UN Doc CCPR/C/FIN/CO/6.
CCPR 'Concluding Observations: Turkey' (2012) UN Doc CCPR/C/TUR/CO/1.
CCPR 'Concluding Observations: Bosnia and Herzegovina' (2012) UN Doc CCPR/C/BIH/CO/2.
CCPR 'List of Issues: Australia' (2012) UN Doc CCPR/C/AUS/Q/6.
CCPR 'Concluding Observations: Mexico' (2010) CCPR/C/MEX/CO/5.
CCPR 'Concluding Observations: New Zealand' (2010) UN Doc CCPR/C/NZL/CO/5.
CCPR 'Concluding Observations: Tanzania' (2009) UN Doc CCPR/C/TZA/CO/4.
CCPR 'Concluding Observations: Russian Federation' (2009) UN Doc CCPR/C/RUS/CO/6.
CCPR 'Concluding Observations: Japan' (2008) UN Doc CCPR/C/JPN/CO/5.
CCPR 'Concluding Observations: Bosnia and Herzegovina' (2006) UN Doc CCPR/C/BIH/Q/1.
CCPR 'Concluding Observations: Colombia' (2004) UN Doc CCPR/CO/80/COL.
CCPR 'Periodic report submitted by Viet Nam' (2024) UN Doc CCPR/C/VNM/4.
CCPR 'Periodic report submitted by Brazil' (2021) UN Doc CCPR/C/BRA/3.
CCPR 'Periodic report submitted by USA' (2021) CCPR/C/USA/5.
CCPR 'Periodic report submitted by Colombia' (2021) UN Doc CCPR/C/COL/8.
CCPR 'Periodic report submitted by Egypt' (2020) UN Doc CCPR/C/EGY/5.
CCPR 'Periodic report submitted by Finland' (2020) UN Doc CCPR/C/FIN/7.
CCPR 'Periodic report submitted by Ethiopia' (2020) UN Doc CCPR/C/ETH/2.
CCPR 'Periodic report submitted by Vietnam' (2018) UN Doc CCPR/C/VNM/3.
CCPR 'Periodic report submitted by Australia' (2016) UN Doc CCPR/C/AUS/6.
CCPR 'Periodic report submitted by Mexico' (2015) UN Doc CCPR/C/MEX/6.
CCPR 'Periodic report submitted by New Zealand' (2015) UN Doc CCPR/C/NZL/6.
CCPR 'Periodic report submitted by Colombia' (2015) UN Doc CCPR/C/COL/7.
CCPR 'Periodic report submitted by Canada' (2013) UN Doc CCPR/C/CAN/6.
CCPR 'Periodic report submitted by France' (2013) UN Doc CCPR/C/FRA/5.
CCPR 'Periodic report submitted by Uzbekistan' (2012) UN Doc CCPR/C/UZB/4.
CCPR 'Periodic report submitted by Finland' (2011) UN Doc CCPR/C/FIN/6.
CCPR 'Periodic report submitted by Colombia' (2009) UN Doc CCPR/C/COL/6.
CCPR 'Periodic report submitted by Colombia' (2002) UN Doc CCPR/C/COL/2002/5.

International Law Commission

ILC 'Draft Conclusions on Subsequent Agreements and Subsequent Practice in Relation to The Interpretation of Treaties, With Commentaries' (2018) UN Doc A/73/10.
ILC 'Draft Conclusions on Identification of Customary International Law, with Commentaries (2018)' UN Doc A/73/10.
ILC 'Fourth Report on Subsequent Agreements and Subsequent Practice in Relation to The Interpretation of Treaties by Georg Nolte, Special Rapporteur' (7 March 2016) UN Doc A/CN.4/694.
ILC, 'Third Report on Identification of Customary International Law, by Michael Wood, Special Rapporteur' (27 March 2015) UN Doc A/CN.4/682.

Other Documents

Asia Pacific Forum of National Human Rights Institutions and the Office of the United Nations High Commissioner for Human Rights (2013) The United Nations Declaration on the Rights of Indigenous Peoples: A Manual for National Human Rights Institutions.
Council of Europe 'Explanatory Report to the Council of Europe Framework Convention on the Value of Cultural Heritage for Society' (27 October 2005) ETS No. 199.
ECOSOC 'Human Rights and Indigenous Issues' (22 March 2006) UN Doc E/CN.4/2006/79.
ECOSOC Res 1985/17 (28 May 1985) UN Doc E/RES/1985/17.
ECOSOC 'Human Rights of Indigenous Peoples' (19 June 2000) UN Doc E/CN.4/Sub.2/2000/26.
Human Rights Instruments, Compilation of Guidelines on The Form and Content of Reports to Be Submitted by States Parties to The International Human Rights Treaties, HRI/GEN/2/Rev.1, 9 May 2001.
ICC (2017): 'Brief by Ms. Karima Bennoune: Expert Appointed by the ICC in the Case of The Prosecutor v. Ahmad Al Faqi Al Mahdi, ICC-01/12-01/15, Reparations Phase'.
Institute of International Law, The Laws of War on Land, Manual published by the Institute of International Law (Oxford Manual), 9 September 1880.
See 4th Cycle Universal Periodic Review National Report—Guidance Note https://www.ohchr.org/sites/default/files/documents/hrbodies/upr/infonotes/4thCycle-Guidance-Note-National-Report-EN.pdf.
Project of an International Declaration concerning the Laws and Customs of War. Brussels, 27 August 1874
UNESCO 'Operational Guidelines for the Implementation of the World Heritage Convention' (31 July 2024) WHC.24/01 (*Operational Guidelines*).
UNESCO 'Operational Directives for the implementation of the Convention for the Safeguarding of the Intangible Heritage' (7 July 2022) (*Operational Directives*).
UNESCO and Asia/Pacific Cultural Centre for UNESCO 'Expert Meeting on Community Involvement in Safeguarding Intangible Cultural Heritage: Towards the Implementation of the 2003 Convention' (13–15 March 2006) CLT/CH/ITH/DOCEM0306 REV.1.

UNESCO General Conference 'Declaration concerning the Intentional Destruction of Cultural Heritage' (17 October 2003).
UNESCO 'Draft declaration concerning the intentional Destruction of Cultural Heritage' (17 July 2003) UNESCO Doc. 32 C/25.
WHC 'Sixth extraordinary session' (3 December 2002) WHC-03/6 EXT.COM/INF.4A.
Decision 7 COM VIII Inscription Taj Mahal (India) in WHC 'Seventh Ordinary Session' SC83/CONF.009/2 ICOMOS (17 November 1983).

MIX
Papier aus verantwortungsvollen Quellen
Paper from responsible sources
FSC® C105338

If you have any concerns about our products,
you can contact us on
ProductSafety@springernature.com

In case Publisher is established outside the EU,
the EU authorized representative is:
Springer Nature Customer Service Center GmbH
Europaplatz 3, 69115 Heidelberg, Germany

Printed by Libri Plureos GmbH
in Hamburg, Germany